Lecture Notes in Computer Science 12601

More information about this subseries at http://www.springer.com/series/7407

Apurva Mudgal · C. R. Subramanian (Eds.)

Algorithms and Discrete Applied Mathematics

7th International Conference, CALDAM 2021 Rupnagar, India, February 11–13, 2021 Proceedings

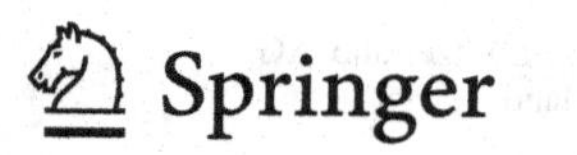

Editors
Apurva Mudgal
Indian Institute of Technology Ropar
Rupnagar, India

C. R. Subramanian
The Institute of Mathematical Sciences
Chennai, India

ISSN 0302-9743 ISSN 1611-3349 (electronic)
Lecture Notes in Computer Science
ISBN 978-3-030-67898-2 ISBN 978-3-030-67899-9 (eBook)
https://doi.org/10.1007/978-3-030-67899-9

LNCS Sublibrary: SL1 – Theoretical Computer Science and General Issues

This Springer imprint is published by the registered company Springer Nature Switzerland AG
The registered company address is: Gewerbestrasse 11, 6330 Cham, Switzerland

Preface

This volume contains the papers presented at CALDAM 2021 (the 7th International Conference on Algorithms and Discrete Applied Mathematics) held during February 11–13, 2021 at IIT Ropar, Rupnagar, Punjab, India. CALDAM 2021 was organised by the Department of Computer Science and Engineering, Indian Institute of Technology Ropar, and the Association for Computer Science and Discrete Mathematics (ACSDM), India. The program committee consisted of 31 highly experienced and active researchers from various countries.

The conference had papers in the areas of algorithms, graph theory, combinatorics, computational geometry, discrete geometry, and computational complexity. We received 82 submissions with authors from all over the world. Each paper was extensively reviewed by program committee members and other expert reviewers. The committee decided to accept 39 papers for presentation. The program included two Google invited talks by Professors Martin Fürer (of Pennsylvania State University) and Anil Maheshwari (of Carleton University).

As volume editors, we would like to thank the authors of all submissions for considering CALDAM 2021 for potential presentation of their works. We are very much indebted to the program committee members and the external reviewers for providing serious reviews within a very short period of time. We thank Springer for publishing the proceedings in the Lecture Notes in Computer Science series. Our sincerest thanks are due to the invited speakers Martin Fürer and Anil Maheshwari for accepting our invitation to give a talk. We thank the organizing committee chaired by Nitin Auluck and Arti Pandey of Indian Institute of Technology Ropar for the smooth conduct of CALDAM 2021 and Indian Institute of Technology Ropar for providing the necessary facilities. We are very grateful to the chair of the steering committee, Subir Ghosh, for his active help, support, and guidance throughout. We thank our sponsors Google Inc. for their financial support. We also thank Springer for its support for the best paper presentation awards. We thank the EasyChair and Springer OCS conference management systems, which were very effective in handling the entire process.

February 2021

Apurva Mudgal
C. R. Subramanian

Organization

Steering Committee

Subir Kumar Ghosh (Chair)	Ramakrishna Mission Vivekananda Educational and Research Institute, India
Gyula O. H. Katona	Alfréd Rényi Institute of Mathematics, Hungarian Academy of Sciences, Hungary
János Pach	École Polytechnique Fédérale de Lausanne (EPFL), Lausanne, Switzerland
Nicola Santoro	School of Computer Science, Carleton University, Canada
Swami Sarvattomananda	Ramakrishna Mission Vivekananda Educational and Research Institute, India
Chee Yap	Courant Institute of Mathematical Sciences, New York University, USA

Program Committee

Amitabha Bagchi	Indian Institute of Technology Delhi, India
Aritra Banik	National Institute of Science Education and Research, Bhubaneswar, India
Niranjan Balachandran	Indian Institute of Technology Bombay, India
Boštjan Brešar	University of Maribor, Slovenia
Manoj Changat	University of Kerala, India
Sandip Das	ISI Kolkata, India
Josep Diaz	Polytechnic University of Catalonia, Spain
Martin Fürer	The Pennsylvania State University, USA
Sumit Ganguly	Indian Institute of Technology Kanpur, India
Daya Gaur	University of Lethbridge, Canada
Sathish Govindarajan	Indian Institute of Science Bangalore, India
Pavol Hell	Simon Fraser University, Canada
R. Inkulu	Indian Institute of Technology Guwahati, India
Christos Kaklamanis	University of Patras, Greece
S. Kalyanasundaram	Indian Institute of Technology Hyderabad, India
Van Bang Le	Universität Rostock, Germany
Andrzej Lingas	Lund University, Sweden
Anil Maheshwari	Carleton University, Canada
Bodo Manthey	University of Twente, Netherlands
Bojan Mohar	Simon Fraser University, Canada
Apurva Mudgal (Co-chair)	Indian Institute of Technology Ropar, India
Rahul Muthu	Dhirubhai Ambani Institute of Information and Communication Technology, India

N. S. Narayanaswamy	Indian Institute of Technology Madras, India
B. S. Panda	Indian Institute of Technology Delhi, India
Iztok Peterin	University of Maribor, Slovenia
S. Francis Raj	Pondicherry University, India
Abhiram Ranade	Indian Institute of Technology Bombay, India
Sagnik Sen	Indian Institute of Technology Dharwad, India
Michiel Smid	Carleton University, Canada
Éric Sopena	University of Bordeaux, France
C. R. Subramanian (Co-chair)	The Institute of Mathematical Sciences, Chennai, India

Organizing Committee

Nitin Auluck (Co-chair)	Indian Institute of Technology Ropar, India
Swami Dhyanagamyananda	Ramakrishna Mission Vivekananda Educational and Research Institute, India
Shweta Jain	Indian Institute of Technology Ropar, India
Pritee Khanna	Indian Institute of Information Technology, Design and Manufacturing Jabalpur, India
Kaushik Mondal	Indian Institute of Technology Ropar, India
Arti Pandey (Co-chair)	Indian Institute of Technology Ropar, India
M. Prabhakar	Indian Institute of Technology Ropar, India
Somitra K. Sanadhya	Indian Institute of Technology Jodhpur, India
Tarkeshwar Singh	BITS Pilani Goa, India
Rishi Ranjan Singh	Indian Institute of Technology Bhilai, India

Additional Reviewers

Hossein Abdollahzadeh Ahangar
Bijo S. Anand
N. R. Aravind
Pradeesha Ashok
Jasine Babu
Kannan Balakrishnan
Susobhan Bandopadhyay
Sayan Bandyapadhyay
Julien Bensmail
Benjamin Bergougnoux
Srimanta Bhattacharya
Sujoy Bhore
Dragana Božović
Christoph Brause
Sergio Cabello
Franco Chiaraluce
Pavan P. D.
Arun Kumar Das
Hiranya Dey
Andrzej Dudek
Brice Effantin
David Eppstein
Rudolf Fleischer
Florent Foucaud
Maria Francis
Iqra Altaf Gillani
Daniel Gonçalves
Sushmita Gupta
Gregory Gutin
Gowramma B. H.
Shenwei Huang
Jesper Jansson
Anjeneya Swami Kare
Aleksander Kelenc

Linda Kleist
Mirosław Kowaluk
Christian Laforest
Juho Lauri
Christos Levcopoulos
Carlos Vinícius G. C. Lima
Daniel Lokshtanov
Tomas Madaras
Rogers Mathew
Neeldhara Misra
Kaushik Mondal
William K. Moses, Jr.
Soumen Nandi
Narayanan Narayanan
Francis P.
Sajith Padinhatteeri
Anantha Padmanabha
Sagartanu Pal
Sudebkumar Prasant Pal
Fahad Panolan
Subhabrata Paul
Bernard Ries
Leonardo Sampaio Rocha
Aniket Basu Roy
Taruni S.
Vladimir Samodivkin
Brahadeesh Sankaranarayanan
Pradeep Sarvepalli
Saket Saurabh
Ingo Schiermeyer
Jin Sima
Vaishnavi Sundararajan
Kavaskar T.
Aleksandra Tepeh
Rakesh Venkat
S. Venkitesh
Koichi Wada
Ismael González Yero
Paweł Żyliński

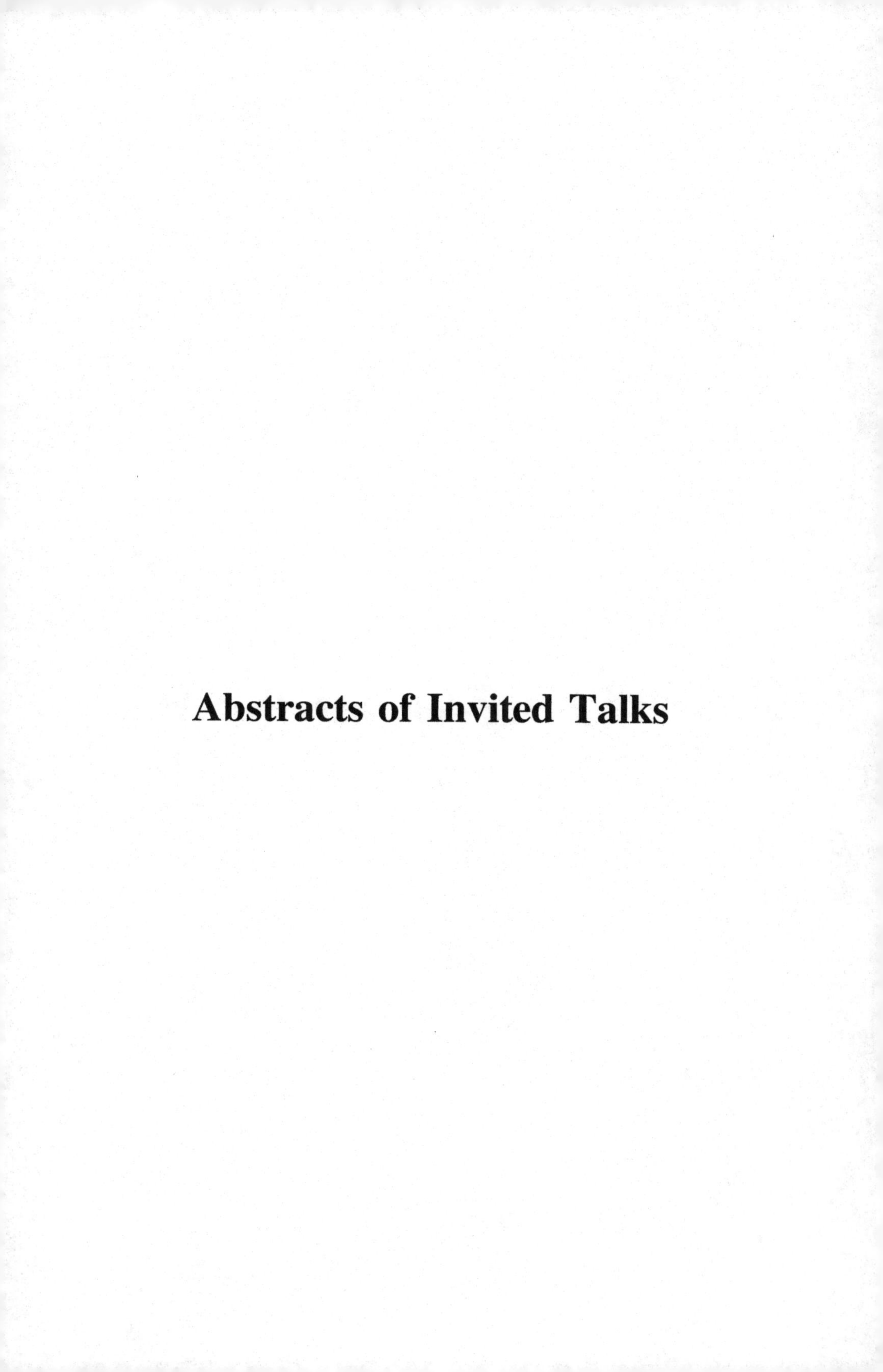

Abstracts of Invited Talks

Width Parameters for Hard and Easy Problems

Martin Fürer

Department of Computer Science and Engineering,
Pennsylvania State University, University Park, PA 16802, USA
fhs@psu.edu

Abstract. The most obvious success of width parameters is the abundance of algorithms that make NP-hard problems FPT (fixed parameter tractable). These FPT algorithms are often very efficient for small parameter values. However, the origin of the notion of treewidth is also tied to solving systems of sparse linear equations. For large sparse system, a cubic algorithm is not good enough. Traditionally, such systems have been approached by heuristics trying to minimize the fill-in by appropriate pivot strategies for Gaussian elimination.

In the most prevalent case of systems of linear equations, the matrix is symmetric and positive definite, always allowing a diagonal pivot strategy. This results in an $O(k^2 n)$ algorithm for an $n \times n$ matrix with treewidth k. If the matrix is symmetric, but not positive definite, then off-diagonal pivots are sometimes required. Nevertheless, the matrix can be kept symmetric throughout the algorithm. However, for a long time, it seemed impossible to control the fill-in for treewidth k matrices. Recently, this has been achieved for cliquewidth k and for treewidth k, by a delaying method. This results in an $O(k^2 n)$ algorithm for determining the number of eigenvalues of a graph in a given interval, an important task in spectral graph theory.

A major obstacle for employing treewidth as a tool for efficient algorithms is the construction of tree decompositions of small width and the computation of the treewidth itself. There has been significant progress in this respect, with many challenges still ahead.

A simple modification of the definition of cliquewidth results in the notion of multi-cliquewidth. For many graphs, the clique-width is exponentially larger than the multi-cliquewidth. Nevertheless for some fundamental problems, like Maximum Independent Set and Chromatic Number, the running times of the standard dynamic programming algorithms are the same functions of the multi-cliquewidth as of the cliquewidth. Thus an exponential speed-up is achieved for these graphs by using multi-cliquewidth instead of cliquewidth, assuming the corresponding tree decompositions are known.

Matching and Spanning Trees in Geometric Graphs

Anil Maheshwari

School of Computer Science, Carleton University, Ottawa ON, Canada
anil@scs.carleton.ca

Abstract. In this talk, we survey some recent work on matching and spanning trees in geometric graphs.

The *matching problem* is to find the largest set of independent edges in a graph. We are especially interested in graphs whose edge set is defined with respect to geometrical shapes. For a given shape S, and a point set P in the plane, the graph $G_S = (P, E)$ has an edge between two points $p, q \in P$ if there exists a shape S that has p and q on its boundary, and it does not contain any point of P in its interior. The Delaunay triangulation, L_∞-Delaunay, Θ_6 graph, and Gabriel graphs are obtained by considering S to be a circle, a square, an equilateral triangle, and a diametral disk, respectively. We will outline results on matchings in G_S where S is a circle, a square, an equilateral triangle, a diametral-disk, etc. We will consider variants of geometric matching problems such as the *bottle-neck matching* - find a perfect matching that minimizes the length of the longest edge; the *plane matching* - find a maximum matching so that the edges in the matching are pairwise non-crossing; the *strong matching* - find a maximum matching so that the shapes representing the edges of the matchings are pairwise disjoint; *local-to-global* - matching M is said to be k- *local optimal* if for any subset $M' \subset M$ of k edges, the optimal matching of the endpoints of M' is M'. Do k-local matchings approximate global matchings?

We will highlight some recent algorithmic results on the computation of spanning trees in bipartite and complete geometric graphs for a point set in the plane. We wish to compute spanning trees that optimize the total weight and are plane, or have bounded degree, or minimize the bottleneck length and are of bounded degree, or have all incident edges within a cone of a specific angle.

Research supported by the Natural Sciences and Engineering Research Council of Canada.

Contents

Combinatorics and Algorithms

Graph Algorithms

Computational Complexity

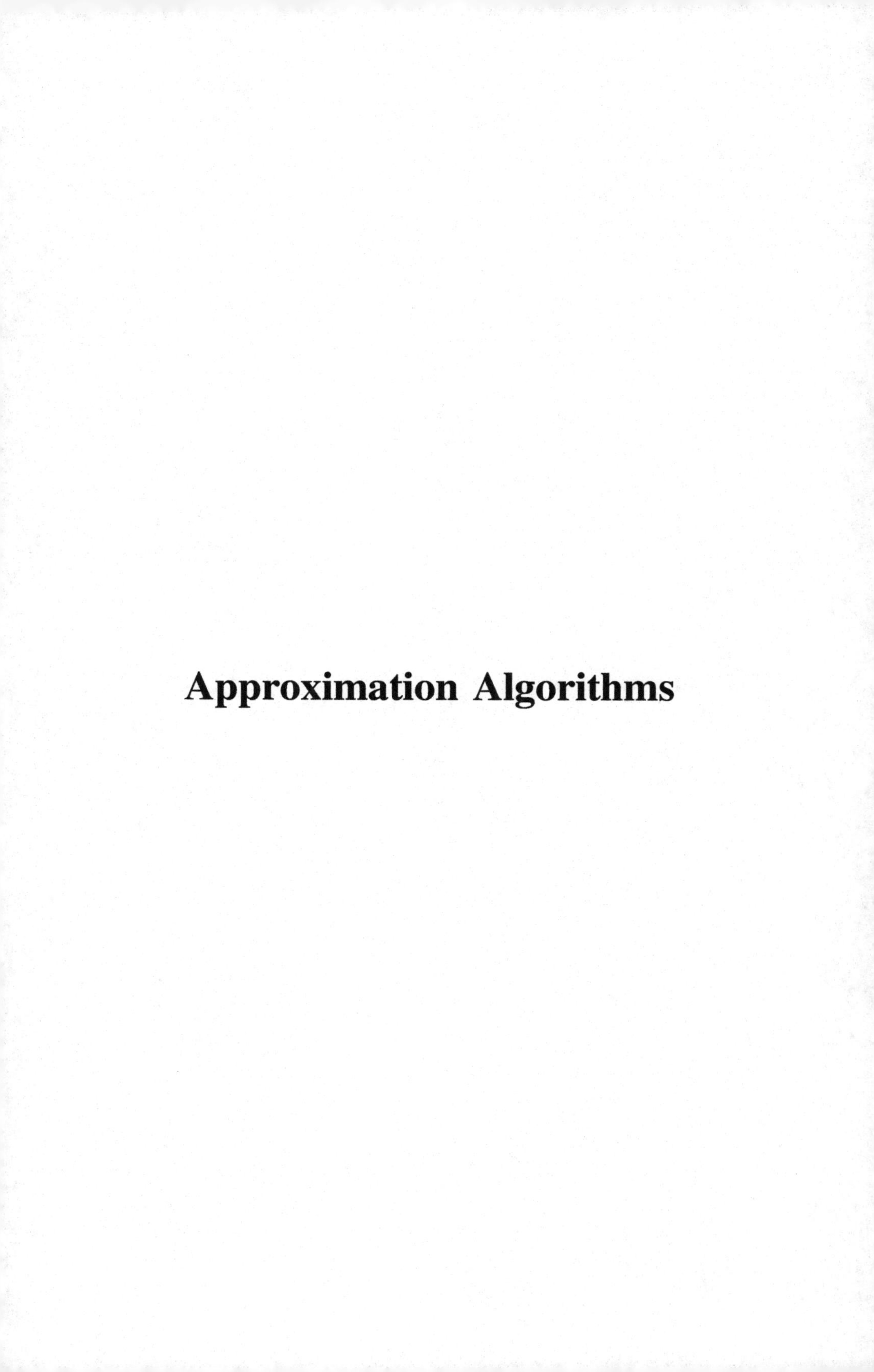

Approximation Algorithms

Online Bin Packing with Overload Cost

Kelin Luo([✉]) and Frits C. R. Spieksma

Department of Mathematics and Computer Science,
Eindhoven University of Technology, Eindhoven, The Netherlands
{k.luo,f.c.r.spieksma}@tue.nl

Abstract. In the classical online bin packing problem, items arriving one by one with a given size not greater than 1 must be packed into unit-capacity bins such that the total size of items packed in a bin does not exceed its capacity; the objective is to minimize the total number of used bins. In this paper, we allow the total size of items packed in a bin to exceed the capacity, and there is a cost for each bin that depends on the total size of items assigned to it; in particular, *overloading* a bin, i.e., exceeding the capacity of a bin, comes at a prescribed cost. The corresponding goal is to minimize the total cost corresponding to the used bins. We pay 1 to open a bin with capacity 1, and we additionally pay c for each unit with which the bin is overloaded, i.e, the overload cost is linear in the size of the overload.

For each c, we present lower bounds on the competitive ratio achievable by deterministic algorithms. Further, we give an algorithm, called First-Fit Algorithm with Fixed Overload (FFO) that achieves the best possible competitive ratio for $c \leq 3/2$. Furthermore, we sketch how the lower bounds apply to more general convex cost functions.

Keywords: Online algorithms · Bin packing · Competitive analysis

1 Introduction

The online bin packing problem (BP) is a classical online optimization problem in which a sequence of *items* of size between 0 and 1 are presented one by one. Each item must be placed in a unit-capacity *bin* before the next item arrives. The goal is to pack the items into the minimum number of bins such that the total weight of items packed in each bin is at most 1. In this paper, we investigate a variant of this classic problem, called the online bin packing problem with *overload cost*; in this problem, we allow the total weight of items packed in a bin to exceed 1.

Problem Definition. We receive a sequence of n items; notice we do not know n in advance. Each item has rational size $p_i \in (0, 1]$ for each $1 \leq i \leq n$. Each

This project has received funding from the European Union's Horizon 2020 research and innovation programme under the Marie Skłodowska-Curie grant agreement number 754462 and funding from the NWO Gravitation Project NETWORKS, Grant Number 024.002.003.

A. Mudgal and C. R. Subramanian (Eds.): CALDAM 2021, LNCS 12601, pp. 3–15, 2021.
https://doi.org/10.1007/978-3-030-67899-9_1

item must be packed irrevocably into a bin before the next item arrives. We have an infinite number of identical bins. Each bin is characterized by a fixed capacity of 1, and a fixed, unit cost for opening a bin; in addition, there is a given, rational, *overload cost* $c \geq 0$ that applies to each unit of the overload. A bin i is *used* (or *opened*) when it contains at least one item; its cost equals $cost_i = 1 + c \cdot \max\{w_i - 1, 0\}$, where w_i is the total weight of the items in bin i. The problem is to pack each item into a bin so that we minimize the sum of the costs of all used bins. We refer to this problem as the *online bin packing problem with linear overload cost* (BPOC). In this paper we prove lower bounds and upper bounds on the competitive ratio of online algorithms for BPOC, for any fixed overload cost c.

Related Work. The bin packing (BP) problem has been extensively investigated. Two famous online algorithms for bin packing are First Fit (FF; FF packs the next item into the first bin where it can be packed), and Best Fit (BF; BF packs the next item into the fullest bin that can accommodate the item). The analysis of these algorithms goes back to Ullman [17]; we refer to Sgall [16] for an overview on online bin packing.

The performance of an online algorithm can be measured by the *asymptotic competitive ratio* as well as by the *absolute competitive ratio*. The asymptotic competitive ratio is defined as: $c_A := \limsup_{n\to\infty} \sup_\sigma\{\frac{ALG(\sigma)}{OPT(\sigma)} | OPT(\sigma) = n\}$, where $OPT(\sigma)$ denotes the number of bins used by an optimal solution and $ALG(\sigma)$ denotes the number of bins used by an algorithm ALG for any input σ. Johnson et al. [12] prove that both FF and BF have asymptotic performance ratios of 1.7. Currently, the best known lower bound on the asymptotic performance ratio of any online algorithm for bin packing is 1.54278 (see Balogh et al. [2]), and the best-known performance ratio is 1.578, due to Balogh et al. [1].

The absolute competitive ratio is defined as: $c := \sup_\sigma\{\frac{ALG(\sigma)}{OPT(\sigma)}\}$. Like the asymptotic competitive result, the absolute competitive ratio of FF and BF both are shown to be 1.7 (Dósa and Sgall [7,8]). Recently, Balogh et al. [3] designed an online bin packing algorithm with an absolute competitive ratio of 5/3, which is best possible. The main idea of their algorithm, called Five-Thirds algorithm (FT), is to use FF and reserve a number of bins specifically for the large items of size more than 1/2.

Many variants of online bin packing have been considered. One variant is the problem where bins have arbitrary capacities, known as the *online variable-sized bin packing* problem. Kinnerly and Langston [13] propose a modified FF algorithm, FF with a user-specified fill factor (FFf), and prove that it is $1.5 + \frac{f}{2}$-competitive when $f \geq \frac{1}{2}$. Csirik [5] proposes the Variable Harmonic (VH) algorithm and shows that it is 1.4-competitive, see also Seiden [15] for a precise analysis.

Another variant is the *open-end packing* problem that allows to violate the capacity in a specific way. Yang and Leung [18] consider the online *ordered open-end bin packing* problem (OOBP). In OOBP it is allowed to violate the capacity in a way that the weight of items in each bin is less than 1 after removing the heaviest item in it. Epstein and Levin [9] further consider two other variants of the open-end bin packing problem: the *strong open-end bin packing* problem (SOBP), in which the weight of the items in each bin must be less than 1 after removing the lightest item; and the *lazy bin covering* problem (LBC) which has the additional constraint that the total weight of items in each bin (except at most one) is not less than 1.

A problem related to our problem is the *extensible bin packing problem*. It arises when $c = 1$ and the number of bins is given. This problem is studied, both in the offline as in the online version, for arbitrary (instead of unit) bin capacities by Dell'Olmo and Speranza [6], Coffman and Lueker [4], and Ye and Zhang [19].

Finally, there is a quite some literature that focuses on the cost of the used bins. Li and Chen [14], Epstein and Levin [10,11], Cambazard et al. [7] all consider different cost structures of the items packed in a bin. In particular, the work by Epstein and Levin [11] is relevant for the offline version of our problem. They deal with a general setting where a set of bin types with different sizes and costs, and a set of items is given and the goal is to minimize the total cost of the used bins. Epstein and Levin [11] design an AFPTAS for this problem. Here, we focus exclusively on the performance of online algorithms.

Motivation. Consider the following possible application. In a multi-processor system, tasks requiring capacity enter the system one by one, and each task needs to be assigned to one of the processors. Each processor has a given service capacity, and charges a fixed turning on/off cost. However, an overload cost will be charged if the sum of capacities required by the tasks assigned to it exceeds the service capacity. We aim to minimize the total cost of serving all tasks. This is the motivation for our online bin packing problem with overload cost: each bin represents a processor with a given capacity, and there is an overload cost if the total size of items packed in this bin exceeds its capacity.

Our Results. In this paper, we consider the absolute competitive ratio for BPOC. We present lower and upper bounds on the competitive ratio of any deterministic algorithm for BPOC. Summarizing, we prove the following theorem. Let $g(c)$ be defined as follows:

$$g(c) = \begin{cases} \max(1, c) & \text{if } 0 \leq c < \frac{3}{2} \\ \frac{3}{2} = 1.5 & \text{if } \frac{3}{2} \leq c < 1 + 2\sqrt{3} \\ 1 + \frac{\sqrt{3}}{3} \approx 1.577 & \text{if } 1 + 2\sqrt{3} < c < 17 \\ \frac{5}{3} \approx 1.667 & \text{if } 17 \leq c. \end{cases}$$

Theorem 1. *For any $c \geq 0$: no deterministic online algorithm for BPOC can achieve a competitive ratio smaller than $g(c)$.*

Further, we propose an online algorithm called *First-Fit Algorithm with Fixed Overload* (in short, FFO). Similar to the First-Fit algorithm (FF) for the classical online bin packing problem, FFO packs each item into the first opened bin where it fits, or opens a new bin if the item does not fit into any currently opened bin. The difference is that, in FFO, the total size of the items assigned to any bin may exceed its capacity 1. Let $h(c)$ be defined as follows:

$$h(c) = \begin{cases} \max(1, c) & \text{if } 0 \leq c \leq \frac{3}{2} \\ \frac{3+c}{3} & \text{if } \frac{3}{2} < c \leq \frac{9}{5} \\ \frac{8}{5} = 1.6 & \text{if } \frac{9}{5} < c \leq \frac{8}{3} \\ \frac{6c}{3c+2} & \text{if } \frac{8}{3} < c \leq \frac{14}{3} \\ \frac{7}{4} = 1.75 & \text{if } \frac{14}{3} < c. \end{cases}$$

Theorem 2. *For any $c \geq 0$: FFO is a $h(c)$-competitive algorithm for BPOC.*

A pictorial overview of these results is shown in Fig. 1; notice that for some values of c, lower and upper bounds coincide. In such a situation we call FFO *best possible*, as no deterministic algorithm with a better competitive ratio exists.

Corollary 1. *FFO is best possible for each c with $0 \leq c \leq 3/2$.*

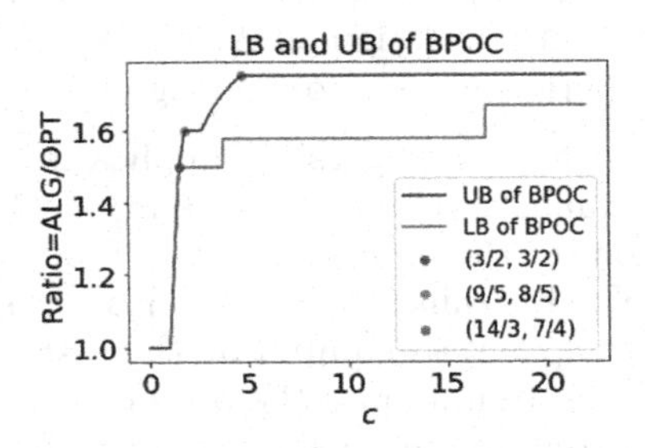

Fig. 1. Illustration of LB and UB as c changes.

2 Lower Bounds

In Sect. 2.1, we prove Theorem 1 by analyzing each of the four segments of $g(c)$. In Sect. 2.2 we sketch how the lower bounds from Theorem 1 apply to cost functions that are more general than the linear overload costs.

2.1 Proving Theorem 1

We denote the cost of packing the items in σ using algorithm A by $cost_A$, i.e., $cost_A = \sum_i cost_i$. We use ALG to denote an arbitrary on-line algorithm and OPT to denote an optimal algorithm. Since, if $c \leq 1$, BPOC becomes trivial as opening a single bin is optimal, we assume henceforth that $c > 1$.

Lemma 1. *If $1 < c \leq \frac{3}{2}$, no deterministic online algorithm for BPOC can achieve a competitive ratio smaller than c.*

Proof. As c is rational, we write $c = \frac{a}{b}$, $a, b \in \mathbb{N}$, and we choose $\varepsilon = \frac{1}{a}$. Further, we use $N_1 = a$, $N_2 = a + b$; notice that $N_1 \cdot \varepsilon = 1$ and $N_2 \cdot \varepsilon = 1 + 1/c$.

The adversary releases items of size ε until either $N_1 \cdot N_2$ items have been released, or when ALG opens a second bin.

In the former situation, ALG packs all $N_1 \cdot N_2$ items into a single bin. Hence, we have $cost_{ALG} = 1 + c(N_1 \cdot N_2 \cdot \varepsilon - 1) = c \cdot N_2 - c + 1$. Another feasible solution is one that uses N_2 bins, each bin containing N_1 items. It follows that $cost_{OPT} \leq N_2$. We have

$$\frac{cost_{ALG}}{cost_{OPT}} \geq \frac{c \cdot N_2 - c + 1}{N_2} = c - \frac{(c-1)\varepsilon}{1 + 1/c} \geq c - \varepsilon.$$

Thus, we can assume that ALG opens a second bin; let us denote the number of items released by the adversary i, $i \leq N_1 \cdot N_2$. Note that ALG packs $i - 1$ items into the first bin and packs one item into the second bin. We distinguish two cases.

Case 1: $1 \leq i \leq N_1$. Obviously, $cost_{OPT} = 1$, $cost_{ALG} = 2$, and it follows immediately that $\frac{cost_{ALG}}{cost_{OPT}} \geq 2 > c$.

Case 2: $N_1 + 1 \leq i \leq N_1 \cdot N_2$. We use that $i = \lfloor \frac{i}{N_1} \rfloor \cdot N_1 + i \bmod N_1$. Since ALG opens two bins and the second bin contains exactly one item, we have $cost_{ALG} = 2 + c(i\varepsilon - \varepsilon - 1) = 2 + c((\lfloor \frac{i}{N_1} \rfloor + i \bmod N_1) \cdot \varepsilon - \varepsilon - 1)$. We distinguish two subcases based on $i \bmod N_1$.

2.1: $(i \bmod N_1) \leq \frac{1/c}{\varepsilon}$.
A feasible solution is one that uses $\lfloor \frac{i}{N_1} \rfloor$ bins: $\lfloor \frac{i}{N_1} \rfloor - 1$ bins each containing N_1 items and the last bin containing $N_1 + (i \bmod N_1)$ items, which causes an overload $(i \bmod N_1) \cdot \varepsilon$. Thus $cost_{OPT} \leq \lfloor \frac{i}{N_1} \rfloor + c(i \bmod N_1) \cdot \varepsilon$.

$$\begin{aligned}
\frac{cost_{ALG}}{cost_{OPT}} &\geq \frac{2 + c((\lfloor \frac{i}{N_1} \rfloor + i \bmod N_1) \cdot \varepsilon - \varepsilon - 1)}{\lfloor \frac{i}{N_1} \rfloor + c(i \bmod N_1) \cdot \varepsilon} \\
&\geq \frac{2 + c((\lfloor \frac{i}{N_1} \rfloor + \frac{1/c}{\varepsilon}) \cdot \varepsilon - \varepsilon - 1)}{\lfloor \frac{i}{N_1} \rfloor + c \cdot \frac{1/c}{\varepsilon} \cdot \varepsilon} \\
&= c + \frac{3 - c - c \cdot \varepsilon - c}{\lfloor \frac{i}{N_1} \rfloor + 1} \geq c - \frac{c \cdot \varepsilon}{\lfloor \frac{i}{N_1} \rfloor + 1} > c - \varepsilon.
\end{aligned}$$

The second inequality follows from the fact that $\frac{2 + c((\lfloor \frac{i}{N_1} \rfloor + i \bmod N_1) \cdot \varepsilon - \varepsilon - 1)}{\lfloor \frac{i}{N_1} \rfloor + c(i \bmod N_1) \cdot \varepsilon}$ decreases as $(i \bmod N_1)$ increases. The third inequality follows from $c \leq \frac{3}{2}$. The last inequality follows from $\frac{c}{\lfloor \frac{i}{N_1} \rfloor + 1} < 1$ since $c < 2 \leq \lfloor \frac{i}{N_1} \rfloor + 1$.

2.2: $(i \bmod N_1) > \frac{1/c}{\varepsilon}$.
A feasible solution is one that uses $\lfloor \frac{i}{N_1} \rfloor + 1$ bins: $\lfloor \frac{i}{N_1} \rfloor$ bins each containing N_1 items and the last bin containing $(i \bmod N_1)$ items. Thus $cost_{OPT} \leq \lfloor \frac{i}{N_1} \rfloor + 1$.

$$\frac{cost_{ALG}}{cost_{OPT}} \geq \frac{2 + c((\lfloor \frac{i}{N_1} \rfloor + i \bmod N_1) \cdot \varepsilon - \varepsilon - 1)}{\lfloor \frac{i}{N_1} \rfloor + 1}$$
$$\geq \frac{2 + c((\lfloor \frac{i}{N_1} \rfloor + \frac{1/c}{\varepsilon} + 1) \cdot \varepsilon - \varepsilon - 1)}{\lfloor \frac{i}{N_1} \rfloor + 1} = c + \frac{3 - c - c}{\lfloor \frac{i}{N_1} \rfloor + 1} > c.$$

The second inequality follows from the fact that $\frac{2+c((\lfloor \frac{i}{N_1} \rfloor + i \bmod N_1)\cdot\varepsilon-\varepsilon-1)}{\lfloor \frac{i}{N_1} \rfloor+1}$ increases as ($i \bmod N_1$) increases. The last inequality follows from $c \leq \frac{3}{2}$. □

Lemma 2. *If $c \geq \frac{3}{2}$, no deterministic online algorithm for BPOC can achieve competitive ratio smaller than $\frac{3}{2}$.*

The proof goes along the lines of Lemma 1: the adversary continues to release items of very small size until ALG opens a second bin. If ALG packs all items into a single bin, then the ratio is close to c; Otherwise, ALG packs all items except the last one into one single bin and packs the last item into the second bin, then the ratio is $\frac{3}{2}$.

Lemma 3. *If $c \geq 1 + 2\sqrt{3}$, no deterministic online algorithm for BPOC can achieve competitive ratio smaller than $\frac{3+\sqrt{3}}{3}$.*

The proof follows the following idea: the adversary releases infinite many *small* items of total size $1+(\sqrt{3}-1)/c-\epsilon$ until ALG opens a second bin. If ALG opens the second bin when the total size of released items is smaller than 1, then the ratio is 2; If ALG opens the second bin when the total size of the released items is greater 1, then the ratio is $\frac{3+\sqrt{3}}{3}$ as OPT opens a single bin; Otherwise, ALG packs all items into one single bin. In the last case, the adversary releases another three *big* items with size $1/2 + y$ such that $(1 + (\sqrt{3} - 1)/c - \epsilon)/3 + 1/2 + y = 1$. OPT packs a big item and some small items of size $1/2 - y$ into one bin, which gives a total cost of 3. No matter how ALG packs the three big items, its total cost is greater than $3 + \sqrt{3}$.

Lemma 4. *If $c \geq 17$, no deterministic online algorithm for BPOC can achieve a competitive ratio smaller than $\frac{5}{3}$.*

Proof. Initially, the adversary releases six items, each with size $\frac{2}{17}$. In case ALG opens two bins to pack these six items, the adversary stops. $cost_{OPT} = 1$ since $W = 6 \cdot \frac{2}{17} < 1$ and we conclude that $\frac{cost_{ALG}}{cost_{OPT}} = 2$.

Thus, we can assume ALG has opened so far a single bin. Then the adversary releases six items, each with weight $\frac{1}{3} + x$, where $x = \frac{1}{3c}$ (referred to as *medium* items). Since $2 \cdot (\frac{2}{17} + \frac{1}{3} + x) < 1$, it follows that as long as the adversary does not release any more items, we have $cost_{OPT} = 3$; consequently, if we manage to show that $cost_{ALG} \geq 5$, the lemma has been proved. We enumerate four possible cases: ALG opens 0, 1, 2, 3 bins. Note that obviously $cost_{ALG} \geq 5$ if ALG opens at least four additional bins.

Case 1: ALG does not open an additional bin. Then the total overload is $6 \cdot (\frac{1}{3}+x)-(1-\frac{12}{17}) = 1+6x+\frac{12}{17}$, and hence $cost_{ALG} = 1+c(1+6x+\frac{12}{17}) > 5$ (since $c \geq 17$).

Case 2: ALG opens one additional bin. The overload is at least $6 \cdot \frac{2}{17} + 6 \cdot (\frac{1}{3} + x) - 2 = \frac{12}{17} + 6x$, then $cost_{ALG} \geq 2 + c(\frac{12}{17} + 6x) > 5$ because $c \geq 17$.

Case 3: ALG opens two additional bins. We distinguish three subcases.

3.1. If ALG does not pack any medium item into the first bin, then ALG distributes the 6 medium items in one of three ways over the two additional bins, namely $\{(5,1),(4,2),(3,3)\}$. Hence, $cost_{ALG} \geq 3+c\cdot\min\{\frac{2}{3}+5x, \frac{1}{3}+4x, 6x\} \geq 5$ because $c \geq 17$ and $x = \frac{1}{3c}$.

3.2. If ALG packs one medium item into the first bin, then ALG distributes 5 medium items in one of two ways over the two additional bins, namely $\{(4,1),(3,2)\}$. Hence, $cost_{ALG} \geq 3 + c(\frac{1}{3} + x - (1 - \frac{12}{17})) + c \cdot \min\{\frac{1}{3} + 4x, 3x\} \geq 5$ because $c \geq 17$ and $x = \frac{1}{3c}$.

3.3. If ALG packs two or more medium items into the first bin, then $cost_{ALG} \geq 3 + c(\frac{2}{3} + 2x - (1 - \frac{12}{17})) > 5$ since $c \geq 17$.

Case 4: ALG opens three additional bins.

4.1. If ALG packs at least one medium item into the first bin, then $cost_{ALG} \geq 4 + c(\frac{1}{3} + x - (1 - \frac{12}{17})) \geq 5$ because $c \geq 17$ and $x = \frac{1}{3c}$.

4.2. If ALG does not pack any medium item into the first bin and ALG packs at least three medium items in an additional bin, then $cost_{ALG} \geq 4 + c \cdot 3x = 5$ because $x = \frac{1}{3c}$.

4.3. If ALG does not pack any medium item into the first bin (excluding Case 4.1), and packs every two medium items into an additional bin (excluding Case 4.2), then the adversary releases six more items, each with weight $\frac{1}{2} + y$, where $y = \frac{1}{2c}$, referred to as *big* items. Note that $\frac{2}{17} + \frac{1}{3} + x + \frac{1}{2} + y \leq 1$ since $c \geq 17$. Therefore the entire input can be packed into 6 bins (as shown in Fig. 2) without any overload, thus we have $cost_{OPT} = 6$, necessitating us to show that $cost_{ALG} \geq 10$.

4.3.1. If ALG opens six additional bins and packs each big item into a bin (as shown in Fig. 2), then $cost_{ALG} = 10$.

4.3.2. If ALG opens i additional bins ($0 \leq i \leq 5$), we claim that the total overload is not smaller than $\frac{6-i}{c}$. Since each of these i additionally opened bins contains at least one big item, it follows that the weight in each bin is at least $\frac{1}{2} + y$. As we still have $6 - i$ big items to pack, we claim that, no matter how these are distributed over the bins, each of these $6 - i$ big items will cause an overload of $\frac{1}{c}$. Indeed, $\frac{1}{2}+y+\frac{1}{2}+y = 1+\frac{1}{c}$, leading to an overload of $\frac{1}{c}$ for each of the $6-i$ remaining big items.
Thus, the claim that the total overload is at least $\frac{6-i}{c}$ holds. We have $cost_{ALG} \geq 4 + i + c \cdot \frac{6-i}{c} = 10$.

Note that the equation $\frac{cost_{ALG}}{cost_{OPT}} = 5/3$ displayed in Case 3.1, Case 3.2, Case 4.1, Case 4.2, Case 4.3. □

Theorem 1 now follows from Lemma's 1, 2, 3, and 4.

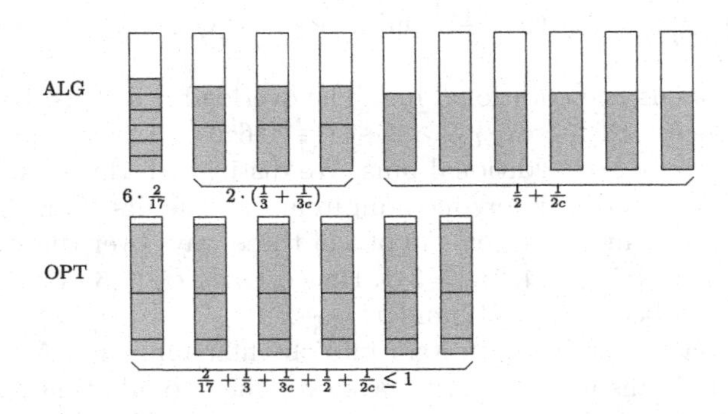

Fig. 2. Illustration of the packings found by ALG and OPT.

2.2 Lower Bounds for Convex Cost Functions

We consider a generalization of our problem BPOC, where the cost of each bin is given by a convex function $f(w)$ (where $w > 0$ represents the total size of the items packed), with $f(w) = 1$ when $0 < w \leq 1$ (as shown in Fig. 3(1)). We refer to such functions as *lin-1 convex* cost functions.

We claim that the proofs of Lemma's 1, 2, 3, and 4, can be generalized to arrive at the following conclusions that we state here without proof. Notice that the quantity $f^{-1}(2)$ plays a crucial role - this quantity refers to an overload cost that is equal to the cost of opening a second bin.

- For any lin-1 convex cost function f satisfying $\frac{5}{3} \leq f^{-1}(2) < 2$, no deterministic algorithm for BPOC can achieve a competitive ratio smaller than $\frac{1}{f^{-1}(2)-1}$.
- For any lin-1 convex cost function f satisfying $f^{-1}(2) > \frac{5}{3}$, no deterministic algorithm for BPOC can achieve competitive ratio smaller than $\frac{3}{2}$.
- For any lin-1 convex cost function f satisfying $f^{-1}(2) > \frac{10+2\sqrt{3}}{11}$, no deterministic algorithm for BPOC can achieve competitive ratio smaller than $\frac{3+\sqrt{3}}{3}$.
- For any lin-1 convex cost function f satisfying $f^{-1}(2) > \frac{18}{17}$, no deterministic algorithm for BPOC can achieve competitive ratio smaller than $\frac{5}{3}$. See the proof of Lemma 4 using $x = \frac{f^{-1}(2)-1}{3}$ and $y = \frac{f^{-1}(2)-1}{2}$.

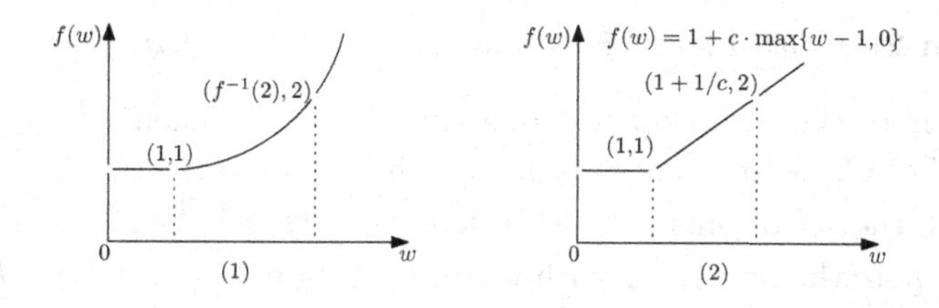

Fig. 3. (1) lin-1 convex cost function; (2) linear cost function.

3 Upper Bounds: Proving Theorem 2

We propose the *First-Fit Algorithm with Fixed Overload* (in short, FFO). First-Fit algorithm packs each item into the first opened bin where it fits, or opens a new bin if the item does not fit into any currently opened bin. The main idea of our algorithm, FFO, is to use First-Fit, but permits a fixed overload, $O(c)$, for each bin.

$$O(c) = \begin{cases} \infty, & \text{if } 0 < c \leq \frac{3}{2} \\ \frac{1}{c}, & \text{if } \frac{3}{2} < c \leq \frac{9}{5} \\ \frac{2}{3c}, & \text{if } \frac{9}{5} < c \leq \frac{14}{3} \\ \frac{1}{3c}, & \textit{otherwise} \end{cases} \tag{1}$$

Let k denote the number of bins and let w_j denote the total weight of items in bin j. The *capacity* of bin j in FFO is defined as $s_j = 1 + O(c)$, which is equal to fixed space 1 plus the overload.

Algorithm 1. First-Fit Algorithm with Fixed Overload (FFO)

1: *Input*: overload cost c.
2: *Initialization*: $O(c)$, $h = 1$, $w_1 \leftarrow 0$, $s_1 \leftarrow 1 + O(c)$.
3: For each item i do
4: For $j = 1, 2, ...h$ do
5: $d_j = s_j - (w_j + p_i)$
6: end for
7: If $\exists j \in \{1, 2, ..., h\}$, $d_j \geq 0$ do
8: pack i into bin $j = \min\{j | d_j \geq 0\}$, $w_j = w_j + p_i$
9: Otherwise do
10: $h = h + 1$, $s_h = 1 + O(c)$, pack i into bin h, and $w_h = p_i$
11: end if
12: end for
13: *Output*: $k = h$ bins and w_j for $j = 1, 2, ..., k$.

Proposition 1. *The total weight of any two bins opened by FFO is greater than* $1 + O(c)$.

Proposition 2. *At least $k-1$ bins have weight greater than $\frac{1+O(c)}{2}$.*

Let C denote the set of opened bins and $|C| = k$. When $c > 3/2$, we have $\frac{1+O(c)}{2} < 1$. Let C_1 denote the set of bins with weight no less than 1 ($|C_1| = k_1$), let C_2 denote the set of bins with weight in the interval $[\frac{1+O(c)}{2}, 1)$ ($|C_2| = k_2$), and let C_3 denote the set of bins with weight less than $\frac{1+O(c)}{2}$ ($|C_3| = k_3$). Notice that $\{C_1, C_2, C_3\}$ is a partition of C. According to Proposition 2, we know that $k_3 \leq 1$. If $k_3 = 1$, suppose the weight of that bin is δ; otherwise $\delta = 0$.

In the proof of the forthcoming lemmas, we denote the total weight of the items by W and the total overload of FFO by O. Let o be the average overload, $o = O/k_1$. Note that $0 \leq o \leq O(c)$. Note that $cost_{OPT} \geq W$ for $c \geq 1$.

Lemma 5. *FFO is 1-competitive for BPOC when $c \leq 1$.*

Since $c \leq 1$, according to (1) and Line 7–8 in Algorithm 1, we know that FFO will pack all items into one bin, i.e., $k = 1$, and hence $cost_{FFO} = \max\{1, 1+c(W-1)\}$. Obviously, $cost_{OPT} = 1 + c \cdot \max\{0, (W-1)\} = cost_{FFO}$.

Lemma 6. *FFO is c-competitive for BPOC when $1 < c \leq \frac{3}{2}$.*

Proof. Since $c > 1$, according to (1) and Line 7–8 in Algorithm 1, we know that FFO will pack all items into one bin, i.e., $k = 1$, and hence $cost_{FFO} = \max\{1, 1 + c(W-1)\}$. Suppose OPT uses h ($h \geq 1$) bins, we have $cost_{OPT} \geq \max\{h, h+c(W-h)\}$. W.l.o.g, suppose $W \geq 1$, otherwise, $cost_{FFO} = cost_{OPT} = 1$. If $W \leq h$, then $\frac{cost_{FFO}}{cost_{OPT}} \leq \frac{1+c(W-1)}{h} \leq \frac{1+c(h-1)}{1+(h-1)} \leq c$; If $W > h$, then $\frac{cost_{FFO}}{cost_{OPT}} \leq \frac{1+c(W-1)}{h+c(W-h)} = \frac{1+c(h-1)+c(W-h)}{1+(h-1)+c(W-h)} \leq c$. Thus $\frac{cost_{FFO}}{cost_{OPT}} \leq c$. □

Lemma 7. *FFO is $\frac{3+c}{3}$-competitive for BPOC when $\frac{3}{2} < c \leq \frac{9}{5}$.*

Proof. Notice that $cost_{FFO} = k + cO = k + c \cdot k_1 \cdot o$. If $k_2 \geq 1$, then $W > k_1 + k_1 \cdot o + \frac{1+1/c}{2}(k - k_1)$ (based on Proposition 1); Otherwise $W = k_1 + k_1 \cdot o + \delta$. We distinguish two cases based on k_2.

Case 1: $k_2 \geq 1$. Note that $cost_{OPT} \geq W \geq k_1 + k_1 \cdot o + \frac{1+1/c}{2}(k - k_1)$. We have

$$\begin{aligned}\frac{cost_{FFO}}{cost_{OPT}} &\leq \frac{k + c \cdot k_1 \cdot o}{k_1 + k_1 \cdot o + (1/2 + 1/(2c))(k - k_1)} \\ &\leq \frac{k + k_1}{(1/2 + 1/(2c))k + (1/2 + 1/(2c))k_1} = \frac{2c}{c+1} < \frac{3+c}{3},\end{aligned}$$

where the second inequality follows from $0 \leq o \leq 1/c$ and the fact that $\frac{k+c \cdot k_1 \cdot o}{k_1 + k_1 \cdot o + (1/2+1/(2c))(k-k_1)}$ increases as o increases; the last inequality follows from $\frac{2c}{c+1} - \frac{3+c}{3} = \frac{-(c-1)^2-2}{3(c+1)}$.

Case 2: $k_2 = 0$. In the case of $k_1 = 0$, we have $cost_{FFO} = cost_{OPT} = 1$ since $W = \delta < 1$. We distinguish three cases with respect to $k_1 \geq 1$.

2.1: $k_1 = 1$. If $k_3 = 0$, we have $cost_{FFO} = cost_{OPT} = 1 + c \cdot \max\{0, W - 1\}$; Otherwise $k_3 = 1$, we have $cost_{OPT} \geq 2$ because $W \geq 1 + \frac{1}{c}$, and hence $\frac{cost_{FFO}}{cost_{OPT}} \leq \frac{1+1/c \cdot c+1}{2} = \frac{3}{2}$.

2.2: $k_1 = 2$. Note that $cost_{FFO} \leq 2 + c \cdot O + 1 = 3 + c \cdot O$. If $O \leq \frac{1}{c}$, we have $cost_{OPT} \geq 2 + c \cdot O$ since $W \geq 2 + O$, and hence $\frac{cost_{FFO}}{cost_{OPT}} \leq \frac{3+c \cdot O}{2+c \cdot O} \leq \frac{3}{2}$. If $\frac{1}{c} < O \leq 1$, we have $cost_{OPT} \geq 2 + 1 = 3$ since $W > 2 + \frac{1}{c}$ and $cost_{FFO} \leq 2 + c + 1 = 3 + c$ since $O \leq 1$, and hence $\frac{cost_{FFO}}{cost_{OPT}} \leq \frac{3+c}{3}$. Finally, If $1 < O \leq \frac{2}{c} \leq 1 + \frac{1}{c}$, we have $cost_{OPT} \geq 3 + c(O - 1)$ since $W > 3 + (O - 1)$. Thus

$$\frac{cost_{FFO}}{cost_{OPT}} \leq \frac{3 + cO}{3 + c(O - 1)} < \frac{3 + c}{3},$$

where it follows from the fact that $\frac{3+cO}{3+c(O-1)}$ decreases as O increases.

2.3: $k_1 \geq 3$. Note that $cost_{FFO} \leq k_1 + c \cdot k_1 \cdot o + 1$ and $cost_{OPT} \geq W \geq k_1 + k_1 \cdot o$. We have

$$\frac{cost_{FFO}}{cost_{OPT}} \leq \frac{k_1 + c \cdot k_1 \cdot o + 1}{k_1 + k_1 \cdot o} \leq \frac{2k_1 + 1}{k_1 + k_1/c} \leq \frac{2 \times 3 + 1}{3 + 3/c} < \frac{3 + c}{3},$$

where the second inequality follows from $0 \leq o \leq 1/c$ and the fact that $\frac{k_1+c \cdot k_1 \cdot o+1}{k_1+k_1 \cdot o}$ increases as o increases; the third inequality follows from $k_1 \geq 3$ and the fact that $\frac{2k_1+1}{k_1+k_1/c}$ decreases as k_1 increases; the last inequality follows from $\frac{2 \times 3+1}{3+3/c} - \frac{3+c}{3} = \frac{3c-c^2-3}{3c+3} = \frac{-(c-3/2)^2-3/4}{3c+3} < 0$.

Concluding, the competitive ratio is $\max\{\frac{3}{2}, \frac{3+c}{3}\} = \frac{3+c}{3}$ (since $c \geq \frac{3}{2}$). □

Lemma 8. *FFO is* $\max\{\frac{8}{5}, \frac{6c}{6c+2}\}$*-competitive for BPOC when* $\frac{9}{5} < c \leq \frac{14}{3}$. *Specifically, the competitive ratio is* $\frac{8}{5}$ *when* $\frac{9}{5} < c \leq \frac{8}{3}$*; and the competitive ratio is* $\frac{6c}{3c+2}$ *when* $\frac{8}{3} < c \leq \frac{14}{3}$.

Proof. Notice that $cost_{FFO} = k + c \cdot k_1 \cdot o$. Recall that $k_3 \leq 1$. Similar to the analysis in Case 1 of Lemma 7, we claim that $W \geq k_1 + k_1 \cdot o + \frac{1+2/(3c)}{2}(k - k_1)$ if $k_2 \geq 1$. We distinguish two cases based on k_2.

Case 1: $k_2 \geq 1$. Note that $cost_{OPT} \geq W \geq k_1 + k_1 \cdot o + \frac{1+2/(3c)}{2}(k - k_1)$. We have

$$\frac{cost_{FFO}}{cost_{OPT}} \leq \frac{k + c \cdot k_1 \cdot o}{k_1 + k_1 \cdot o + (1/2 + 1/(3c))(k - k_1)}$$

$$\leq \frac{k + 2k_1/3}{(1/2 + 1/(3c))k + (1/2 + 1/(3c))k_1} \leq \frac{6c}{3c + 2},$$

where the second inequality follows from $0 \leq o \leq 2/(3c)$ and $f'(o) > 0$ in which $f = \frac{k+c \cdot k_1 \cdot o}{k_1+k_1 \cdot o+(1/2+1/(3c))(k-k_1)}$.

Case 2: $k_2 = 0$. Since $k_3 \leq 1$, $cost_{FFO} \leq k_1 + c \cdot k_1 \cdot o + 1$. Obviously, in the case of $k_1 = 0$, we have $cost_{FFO} = cost_{OPT} = 1$. We distinguish three cases with respect to $k_1 \geq 1$.

2.1: $k_1 = 1$.
If $k_3 = 0$, we have $cost_{FFO} = cost_{OPT} = 1 + c \cdot \max\{0, W - 1\}$. If $k_3 = 1$, we have $cost_{OPT} \geq 1 + c \cdot \frac{2}{3c} = 5/3$ because $W \geq 1 + O + \delta \geq 1 + \frac{2}{3c}$. Since $cost_{FFO} \leq 1 + c \cdot \frac{2}{3c} + 1 = 8/3$. Thus $\frac{cost_{FFO}}{cost_{OPT}} \leq \frac{8}{5}$.

2.2: $k_1 = 2$.

If $O \leq \frac{1}{c}$, we have $cost_{OPT} \geq 2 + c \cdot O$ since $W \geq 2 + O$ and $cost_{FFO} \leq 2 + c \cdot O + 1 = 3 + c \cdot O$, and hence $\frac{cost_{FFO}}{cost_{OPT}} \leq \frac{3+cO}{2+cO} \leq \frac{3}{2}$. If $\frac{1}{c} \leq O \leq \frac{4}{3c}$ (Note that $O \leq \frac{4}{3c}$ because $k_1 = 2$), we have $cost_{OPT} \geq 2+1 = 3$ because $W \geq 2 + \frac{1}{c}$ and $cost_{FFO} \leq 2 + 4/3 + 1 = 13/3$ because $O \leq \frac{4}{3c}$, and hence $\frac{cost_{FFO}}{cost_{OPT}} \leq \frac{13}{9} \leq \frac{8}{5}$.

2.3: $k_1 \geq 3$. Since $cost_{OPT} \geq W \geq k_1 + k_1 \cdot o$, we have

$$\begin{aligned} \frac{cost_{FFO}}{cost_{OPT}} &\leq \frac{k_1 + c \cdot k_1 \cdot o + 1}{k_1 + k_1 \cdot o} \\ &\leq \frac{5k_1/3 + 1}{k_1 + 2k_1/(3c)} \\ &\leq \frac{6}{3 + 2/c} = \frac{6c}{3c + 2}, \end{aligned}$$

where the second inequality follows from $0 \leq o \leq 2/(3c)$ and $f'(o) = \frac{k_1(k_1(c-1)-1)}{(k_1+k_1 \cdot o)^2} > 0$ in which $f = \frac{k_1 + c \cdot k_1 \cdot o + 1}{k_1 + k_1 \cdot o}$; the third inequality follows from $k_1 \geq 3$ and $g'(k_1) < 0$ in which $g = \frac{5k_1/3+1}{k_1+2k_1/(3c)}$.

In sum, the competitive ratio is $\max\{\frac{8}{5}, \frac{6c}{3c+2}\}$. Specifically, the competitive ratio is $\frac{8}{5}$ when $\frac{9}{5} < c \leq \frac{8}{3}$; and the competitive ratio is $\frac{6c}{3c+2}$ when $\frac{8}{3} < c \leq \frac{14}{3}$. □

Next we analyze the competitive ratio of FFO when $c > 14/3$. Notice that $\frac{3(1+O(c))}{4} < 1$ because $O(c) = \frac{1}{3c}$ and $c > 14/3$. We partition C_2 as follows: let C_{21} denote the set of bins with weight in the interval $[\frac{3(1+O(c))}{4}, 1)$ ($|C_{21}| = k_{21}$), let C_{22} denote the set of bins with weight in the interval $[\frac{2(1+O(c))}{3}, \frac{3(1+O(c))}{4})$ ($|C_{22}| = k_{22}$), and let C_{23} denote the set of bins with weight in the interval $[\frac{1+O(c)}{2}, \frac{2(1+O(c))}{3})$ ($|C_{23}| = k_{23}$). Notice that $\{C_{21}, C_{22}, C_{23}\}$ is a partition of C_2. The following proposition is useful.

Proposition 3. *Each bin in C_{22} (except the earliest opened bin of C_{22}) contains at most two items and each of them is greater than $\frac{1+O(c)}{4}$; and each bin in C_{23} (except the earliest opened bin of C_{23}) contains exactly one item.*

Based on Proposition 3, we can formulate Lemma 9 that we state here without proof.

Lemma 9. *FFO is $\frac{7}{4}$-competitive for BPOC when $c > 14/3$.*

Theorem 2 now follows from Lemma's 5, 6, 7, 8, and 9.

References

1. Balogh, J., Békési, J., Dósa, G., Epstein, L., Levin, A.: A new and improved algorithm for online bin packing. In: Azar, Y., Bast, H., Herman, G. (eds.) 26th Annual European Symposium on Algorithms, ESA 2018, 20–22 August 2018, Helsinki, Finland. LIPIcs, vol. 112, pp. 5:1–5:14. Schloss Dagstuhl - Leibniz-Zentrum für Informatik (2018)
2. Balogh, J., Békési, J., Dósa, G., Epstein, L., Levin, A.: A new lower bound for classic online bin packing. In: Bampis, E., Megow, N. (eds.) WAOA 2019. LNCS, vol. 11926, pp. 18–28. Springer, Cham (2020). https://doi.org/10.1007/978-3-030-39479-0_2
3. Balogh, J., Békési, J., Dósa, G., Sgall, J., van Stee, R.: The optimal absolute ratio for online bin packing. J. Comput. Syst. Sci. **102**, 1–17 (2019)
4. Coffman, E., Lueker, G.S.: Approximation algorithms for extensible bin packing. J. Sched. **9**(1), 63–69 (2006)
5. Csirik, J.: An on-line algorithm for variable-sized bin packing. Acta Inf. **26**(8), 697–709 (1989)
6. Dell'Olmo, P., Speranza, M.G.: Approximation algorithms for partitioning small items in unequal bins to minimize the total size. Discrete Appl. Math. **94**(1–3), 181–191 (1999)
7. Dósa, G., Sgall, J.: First fit bin packing: a tight analysis. In: 30th International Symposium on Theoretical Aspects of Computer Science (STACS 2013). Schloss Dagstuhl-Leibniz-Zentrum fuer Informatik (2013)
8. Dósa, G., Sgall, J.: Optimal analysis of best fit bin packing. In: Esparza, J., Fraigniaud, P., Husfeldt, T., Koutsoupias, E. (eds.) ICALP 2014. LNCS, vol. 8572, pp. 429–441. Springer, Heidelberg (2014). https://doi.org/10.1007/978-3-662-43948-7_36
9. Epstein, L., Levin, A.: Asymptotic fully polynomial approximation schemes for variants of open-end bin packing. Inf. Process. Lett. **109**(1), 32–37 (2008)
10. Epstein, L., Levin, A.: Bin packing with general cost structures. Math. Program. **132**(1–2), 355–391 (2012)
11. Epstein, L., Levin, A.: An AFPTAS for variable sized bin packing with general activation costs. J. Comput. Syst. Sci. **84**, 79–96 (2017)
12. Johnson, D.S., Demers, A.J., Ullman, J.D., Garey, M.R., Graham, R.L.: Worst-case performance bounds for simple one-dimensional packing algorithms. SIAM J. Comput. **3**(4), 299–325 (1974)
13. Kinnersley, N.G., Langston, M.A.: Online variable-sized bin packing. Discrete Appl. Math. **22**(2), 143–148 (1989)
14. Li, C.L., Chen, Z.L.: Bin-packing problem with concave costs of bin utilization. Naval Res. Logist. (NRL) **53**(4), 298–308 (2006)
15. Seiden, S.S.: An optimal online algorithm for bounded space variable-sized bin packing. SIAM J. Discrete Math. **14**(4), 458–470 (2001)
16. Sgall, J.: Online bin packing: old algorithms and new results. In: Beckmann, A., Csuhaj-Varjú, E., Meer, K. (eds.) CiE 2014. LNCS, vol. 8493, pp. 362–372. Springer, Cham (2014). https://doi.org/10.1007/978-3-319-08019-2_38
17. Ullman, J.D.: The performance of a memory allocation algorithm. Technical report 100, Princeton University, Prinston, NJ (1971)
18. Yang, J., Leung, J.Y.: The ordered open-end bin-packing problem. Oper. Res. **51**(5), 759–770 (2003)
19. Ye, D., Zhang, G.: On-line extensible bin packing with unequal bin sizes. Discret. Math. Theor. Comput. Sci. **11**(1), 141–152 (2009)

Scheduling Trains with Small Stretch on a Unidirectional Line

Apoorv Garg[1(✉)] and Abhiram Ranade[2(✉)]

[1] Coupa Software India Pvt. Ltd., Pune 411016, Maharashtra, India
apoorv.garg@gmail.com
[2] Indian Institute of Technology Bombay, Mumbai 400076, Maharashtra, India
ranade@cse.iitb.ac.in

Abstract. We investigate the problem of scheduling trains to minimize *max-stretch*, where the stretch suffered by a train is the ratio of its actual finishing time to its minimum possible finishing time. This metric is presumably more appropriate for train scheduling because it is fairer. Our target network, introduced in [11], is an *in-comb*: a unidirectional railway line with equidistant stations, each initially having at most one train; in addition, there can be at most one train poised to enter each station. A train takes unit time to enter a station or to move from one to the next. Trains must move to their destinations such that at any time there can be at most one train at any station and on the track connecting it to the next. We prove that minimizing max-stretch is NP-hard even on this simple network. We also give an O(1)-approximation algorithm. Our problem can also be interpreted as packet scheduling on in-comb, a special case of in-trees. Packet scheduling on general graphs and some special topologies has been studied earlier with different objective functions, e.g., makespan, flowtime, and max-delay, but there has been little work on max-stretch.

Keywords: Approximation algorithms · Combinatorial optimization · In-comb network · Max-stretch minimization · NP-hard · Packet scheduling · Train scheduling · Unidirectional line

1 Introduction

In the train scheduling problem [2,8,19,21], we are interested in moving a set of trains to their destinations, respecting track capacity and minimizing an appropriate cost metric. A natural expectation is: if trains are to be late, they should be late in proportion to their planned travel times. This can be achieved by minimizing *max-stretch*, where the stretch of a train is the ratio of its actual finishing time to its minimum possible finishing time. Such fairness is not guaranteed by minimizing other metrics such as *max-delay* or *makespan*.

The problem is modeled using a graph: nodes represent stations, links represent tracks. Initially, each station may hold one or more trains—represented as point objects—to be moved to specified stations using specified paths. On each

A. Mudgal and C. R. Subramanian (Eds.): CALDAM 2021, LNCS 12601, pp. 16–31, 2021.
https://doi.org/10.1007/978-3-030-67899-9_2

link there can be at most one train at a time and it takes a specified time for a train to traverse a link. There is also a buffering constraint: each station can hold at most a specified number of trains. Our goal is to schedule the trains such as to minimize their max-stretch, where in any schedule the stretch of a train is defined as the ratio (f/f_m) between its actual finishing time f in that schedule and its minimum finishing time f_m across all possible schedules which, in our simple problem, equals the train's path-length.

This problem is also studied in packet scheduling literature, using the terms *routers*, *channels*, and *packets* in place of stations, tracks, and trains. We do not know of any work on max-stretch; for other metrics, the problem is known to be NP-hard in various versions. With bounded buffering at nodes, minimizing makespan or max-delay—even to a constant factor—is NP-hard even on *leveled directed graphs* in which packets move from the lowest numbered level to the highest numbered level [6]. Since path-lengths of all packets are the same, the hardness result also applies to max-stretch but this is rather degenerate. Assuming unbounded buffering at nodes, minimizing makespan is NP-hard on trees as well as on general graphs [20]. On trees, a 2-approximation can be obtained [20]. However, on unidirectional rings, in-trees, and out-trees, the optimal makespan can be achieved [16]. For a variant of the problem where buffers are available in links rather than at nodes, $O(1)$-approximations, using only $O(1)$ buffers in each link are known on general graphs [14,15,23,26].

There is also a large body of experimental work on train scheduling using various approaches, such as simulation, heuristics, mixed integer linear programming, multi-agent systems, genetic algorithms, reinforcement-learning, etc. [3–5,7,9,12,13,17,18,24,25,27,29], but we do not know of papers which consider stretch. In any case our goal in this paper is to establish provable bounds.

Given the practical importance of the problem, it is worth asking whether good scheduling is possible for simpler networks. We have not found any such results for max-stretch. In this paper we begin such a study by considering an *in-comb* network, a special case of in-trees. The in-comb, defined formally later, is a directed path with an extra, *branch* edge entering every node on the path. There can be a train in each node and one poised to enter it from the branch. Trains may exit the network in any node.

Our motivation for studying this network is two fold. First, for ease of management, a large railway network is often broken into sub-networks, each consisting of a *trunk route* with trains entering and exiting from and to branch lines. Each of the two directions of the route is like an in-comb. Second, the in-comb is perhaps the simplest interesting network, and it would be good to understand the computational complexity of scheduling on it with minimum max-stretch. We further simplify it, assuming identical traversal times for all trains on all links. Minimizing max-delay is known to be NP-hard on this simple network [11].

Main Results

1. Minimizing max-stretch is NP-hard for train scheduling on in-comb (Sect. 4).
2. A polytime algorithm to schedule trains on the in-comb with a max-stretch $O(1)$ times the optimal (Sect. 3).

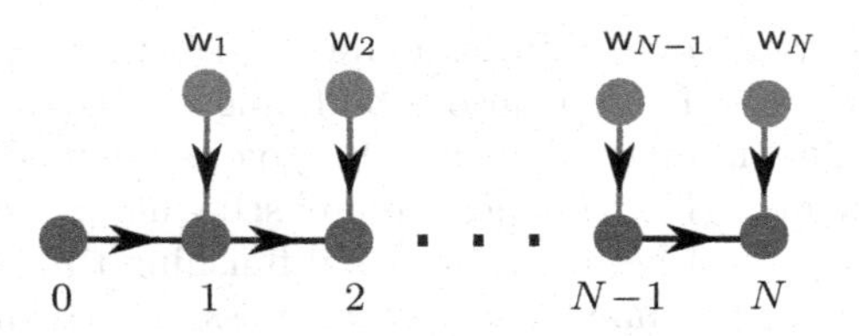

Fig. 1. The in-comb network

2 Preliminaries

2.1 Network Definition and Problem Statement

Our target network is an in-comb (Fig. 1). It consists of:

1. *Line*: sequence of stations, labeled $0, 1, \ldots, N$, and links $(s-1, s)$ $\forall s \in [N]$
2. *Branches*: for each station $s > 0$, an *outer* w_s and a link (w_s, s)

For every node (station or outer), at most one train is given to be there at time 0, along with its destination. No more trains are introduced into the network later. Trains starting at stations are called *internal trains*; those starting at outers *external trains*. Each node can hold at most one train at a time. Any train takes one *step*[1] to enter a station from its outer or to move from a station to the next, and vanishes (exits the network) on reaching its destination. An external train entering the line is called an *entry*, to distinguish it from a *movement* which means a train moving from one station to the next. The required output is a schedule for the trains, minimizing the max-stretch. (Note: A train of path-length ℓ finishing its journey at time f is said to have suffered a stretch f/ℓ).

2.2 The Chain-Hole View

Our arguments to prove the claimed results are based on a *chain-hole view* of schedules [11], summarized next.

Hole refers to a vacancy at a station s. The hole might have been at s since the very beginning (time 0), or *created* later by the exit of a train at s, or it might have come to s from upstream. Some clarifications are needed regarding holes and their progress on the line. First, for convenience we assume an infinite number of suitably numbered artificial stations[2] to the left and right of the line, with no external or internal trains, i.e., each artificial station having a hole. Second, suppose a station u has a hole h at time t. For any $v > u$, if trains at stations $u+1, \ldots, v-1$ remain stationary but the trains at stations $u-1$ and v move, the hole in station u will vanish and a hole will appear in station v.

[1] $\forall t \in \mathbb{Z}^+$, *step* t is the unit time duration $(t-1, t]$ that ends at time t.

[2] Stations $-1, \ldots, -\infty$ upstream of station 0, and $N+1, \ldots, \infty$ downstream of N.

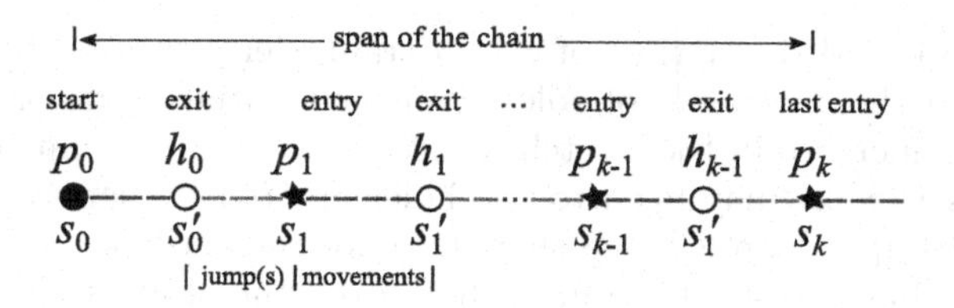

Fig. 2. Spatial view of a chain $\langle p_0, h_0, p_1, h_1, \ldots, p_{k-1}, h_{k-1}, p_k \rangle$. Station s_0 is the origin ($\bullet$) of the internal train p_0 with which the chain begins. Stations $s_1, \ldots, s_k$ are the entry points ($\star$) of external trains $p_1, \ldots, p_k$. $s_0', s_1', \ldots, s_{k-1}'$ are the destinations ($\circ$) of non-terminal trains $p_0, p_1, \ldots, p_{k-1}$, where the holes $h_0, h_1, \ldots, h_{k-1}$ get created when those trains exit. Links (–) crossed by train-movements of the chain are shown in red while those crossed by hole-jumps are in green. (Color figure online)

We define this as the hole h *jumping* from station u to station v. In this view both holes and trains can move forward, but only a hole can jump across several stations in a step. It is also useful to consider that holes and trains contend for links in order to progress down the line: across any link in any step, either a train can move or a hole can jump but not both.

An external train can enter a station s only by *filling* a hole that might have already been at s, or might jump to s from upstream. When a train p_1 exits the line at a station s_1', it leaves behind a hole which can be used for the entry of another train p_2 at a station $s_2 > s_1'$; p_2 would exit at some station $s_2' \geq s_2$, re-creating the hole; and so on. Such a sequence of trains is called a *chain*. Thus, a chain consists of a preexisting hole or an internal train followed by a sequence of some k external trains $p_1 \ldots p_k$; p_1 must fill a hole h_0 which is a preexisting hole or a hole created by the exit of an internal train p_0, and for $j > 1$, p_j must fill the hole h_{j-1} created by the exit of p_{j-1} (See Fig. 2). The chain is said to *begin* with the preexisting hole h_0 (or the train p_0). Clearly, in any schedule, every external train will be placed in some chain.

In order to build a schedule, we must somehow form such chains of trains. After we form chains, we can worry about how to move the trains so that the entries happen as per the chains. This is the structure of our algorithms.

3 An O(1)-Approximation of the Optimal Max-Stretch

The first ingredient of our algorithm is a strategy for scheduling any single class of external trains in an optimal manner, where trains with path-lengths between 2^{i-1} and $2^i - 1$ constitute class i. This is discussed in Sect. 3.1.

The second ingredient is: schedule classes in increasing class order. Since the path-lengths of trains in classes 1 through $i-1$ roughly add up to the path-length of a class i train, the delay caused to the class i trains by lower class trains itself does not substantially affect the stretch of class i trains. However, the movement of lower class trains causes the holes to move. So this makes it harder to apply the lower bound on delivery time of class i trains derived in Sect. 3.1. We show in Sect. 3.2 that the movement of holes from their initial positions only causes a constant factor increase in the stretch. A second problem is that in the optimal schedule, class i trains may need to use holes created by departure of internal trains of classes up to some j. We show that we can estimate j through a precomputation. Thus, before scheduling the movement of class i external trains, we deliver all class j internal trains. Again, this may not create the holes we need in the same positions as in the optimal. In Sect. 3.2, we also show that the drift of these holes also does not matter too much.

3.1 Schedule for a Single Class Using only Preexisting Holes

We consider how to minimize the makespan of a class i using only preexisting holes and holes created by the exits of class i external trains. For ease of exposition, rather than say "we do not use the holes created by the exit of any internal trains", we modify all internal trains to have the last station N as their destination. Positions of the preexisting holes remain unchanged. Note that since the path-lengths in a class differ at most by a factor of 2, minimizing makespan instead of max-stretch may worsen the latter at most by a factor of 2.

We show below that good schedules are possible if and only if the initial holes are well distributed among the external trains, and to the extent they are well distributed. We begin with the lower bound: good schedules are not possible if some region with many external trains has very few holes.

Lemma 1. *Suppose all internal trains go to the last station N. Suppose a contiguous sequence S of stations has w external trains of class i and h holes. Then, class i trains have a makespan at least* $\max\left\{2^{i-1}, \min\left\{\frac{w \cdot 2^{i-1}}{2h}, \sqrt{w \cdot 2^{i-1}}\right\}\right\}$.

Proof. Suppose all w external trains of class i enter by time T. Their makespan F is at least $T + d - 1$, where $d = 2^{i-1}$, and one of the following must be true:

1. At least $\frac{w}{2}$ trains enter in chains beginning with the holes within S: at most h can enter in step 1 (and exit in step d, re-creating those h holes), h more in step $d+1$, and so on. By time T, at most $\left(1 + \frac{T-1}{d}\right) h$ trains can enter. But all do enter by time T, i.e., $\left(1 + \frac{T-1}{d}\right) h \geq \frac{w}{2} \implies T + d - 1 \geq \frac{wd}{2h} \implies F \geq \frac{wd}{2h}$.
2. At least $\frac{w}{2}$ trains enter in chains beginning with the holes upstream of S. Since in each step no more than one hole may jump into S from upstream, in each of the first d steps at most one train can enter. Each can exit d steps later to re-create a hole, so in the 2nd set of d steps, trains can fill these holes and another d holes from upstream, i.e., two entries per step. In n^{th} set, at most n entries per step. Thus, at most $q(q+1)\frac{d}{2} + (q+1)r$ entries by time T, where $q = \lfloor \frac{T}{d} \rfloor$ and $r = T \bmod d$, i.e., $q(q+1)\frac{d}{2} + (q+1)r \geq \frac{w}{2}$ and:
   (a) If $r = 0$ then $T = qd$ and $(q + \frac{1}{2})^2 d^2 > wd \Rightarrow T + \frac{d}{2} > \sqrt{wd} \Rightarrow F \geq \sqrt{wd}$
   (b) If $r \geq 1$ then $(qd+d)(qd+2r) \geq wd \Rightarrow (qd+r+d-1)^2 \geq wd \Rightarrow F \geq \sqrt{wd}$

Thus, $F \geq \min\left\{\frac{wd}{2h}, \sqrt{wd}\right\}$. But $F \geq d$, the minimum class i path-length.

We next prove that if every region, or *segment* as defined below, has a large number of holes as compared to the number of external trains within it then all trains can be scheduled with small makespan.

Definition 1 (Class i segment of size w). *Any contiguous sequence of stations initially having w external trains of class i (at their outers), such that:*

1. *The first station of the sequence has a class i external train.*
2. *Either the downstream neighbor of its last station has a class i external train, or the sequence includes the downstream artificial stations $N+1, ...\infty$.*

Lemma 2. *Suppose all internal trains go to the last station N, and i, w, and h are given such that $h \geq \min\{w, 2^i\}$ and every class i segment of size w has at least h holes. Then in polytime we can schedule class i trains to finish by time $2h + \frac{w \cdot 2^i}{h} + 2^i$.*

Proof Sketch. Here we only give the main idea (the details are in Appendix A): We partition the network into a sequence of class i segments of size w and in each we form h chains, every chain having at most $w/h + 1$ trains. In the first h steps, the chain-heads enter in parallel in all segments—each filling one of the h or more initial holes of the w-sized segment upstream of it. Afterwards, the chains progress in parallel; conflicts are resolved by prioritizing external trains over internal, and downstream chains over upstream. We can show that all chains finish before time $2h + w \cdot 2^i/h + 2^i$.

Next we define *grain-size* of an instance, which tells how to apply the lemmas.

Definition 2 (Grain-size for class i). *The smallest $w \in \{1, ..., W_i\}$ for which every class i segment of size w initially has at least $\sqrt{w \cdot 2^i}$ holes, where W_i is the number of class i external trains.*

Note: Since the segment of size W_i has infinite holes, grain-size is well defined.

Theorem 1. *Suppose all internal trains go to the last station N. In polytime, we can schedule class i trains to finish by time $O(F^*)$, where F^* is their optimal makespan. Moreover, $F^* \geq \widetilde{F} = \max\left\{2^{i-1},\ \frac{1}{4}\sqrt{w \cdot 2^{i-1}}\right\}$ where w is the grain-size for class i.*

Proof. Let $d = 2^{i-1}$, the minimum class i path-length, and w be the grain-size for class i, i.e., each class i segment of size w has at least $h = \lceil\sqrt{2wd}\rceil$ holes. Then $\sqrt{2wd} \leq h < \sqrt{w \cdot 2d} + 1$ and Lemma 2 gives a schedule where class i trains finish by time $F < 3\sqrt{w \cdot 2d} + 2d + 2$. For $w \geq 2$, we know that some segment of size $\frac{w}{2}$ must have less than $\sqrt{wd}$ holes, so applying Lemma 1 we have: $F^* \geq \max\left\{d, \min\left\{\frac{1}{4}\sqrt{wd}, \frac{1}{\sqrt{2}}\sqrt{wd}\right\}\right\} = \max\left\{d, \frac{\sqrt{wd}}{4}\right\}$. For $w = 1$, $F^* \geq d \geq \max\left\{d, \frac{\sqrt{wd}}{4}\right\}$. Thus, $F^* \geq \widetilde{F} = \max\left\{d, \frac{\sqrt{wd}}{4}\right\} \geq \frac{1}{8}\sqrt{wd} + \frac{1}{2}d$.

Algorithm 1: Preprocessing

Input: Π

1 **for** $i = 1, ..., \lfloor \log N \rfloor + 1$ **do**
2 $\Pi_0 = \Pi$;
3 For $j > 0$: $\Pi_j = \Pi$ with all internal trains of classes $0, ..., j$ replaced by holes, and destinations of all internal trains of classes $> j$ set to N;
4 **for** $j = 0, ..., \log N$ **do**
5 $w(j) \longleftarrow$ the grain-size for class i in Π_j as per Definition 2;
6 $M(j) \longleftarrow \max\left\{2^{j-1},\ 2^{i-1},\ \frac{1}{4}\sqrt{w(j)\cdot 2^{i-1}}\right\}$;
7 **end**
8 $J_i \longleftarrow \text{argmin}_j\ M(j);\ \ w_i \longleftarrow w(J_i);\ \ h_i \longleftarrow \lceil\sqrt{w_i \cdot 2^i}\rceil$;
9 **end**

Output: (J_i, w_i, h_i) for each class $i \in \{1, ..., \lfloor \log N \rfloor + 1\}$

3.2 The Overall Scheduling Algorithm

We now consider the scheduling of all trains, using holes created by the exits of internal trains as well as the preexisting holes. As mentioned earlier, the classes are scheduled one after another in ascending order. Scheduling of each class i is as in the previous section but with the following two crucial differences.

First, for the entry of external trains, now we can also use holes created by the exits of internal trains (in addition to the preexisting holes). Which of them to use for class i has to be carefully decided, and the makespan lower-bound accordingly adjusted. We do that in a preprocessing module.

Second, delivering the previous classes $1, ..., i-1$ delays class i and also alters the distribution of holes (preexisting as well as created) relative to its external trains. However, all those holes do become available for class i as they are re-created at the exits of the trains of previous classes, although they appear shifted somewhat downstream of their initial positions. We show in a scheduling module that the delay and the shifts are small enough—relative to the class i path-lengths—for us to still schedule class i with a max-stretch which is a weighted sum of the max-stretch lower-bounds of classes $i, i-1, ..., 1$ with the corresponding weights in a decreasing geometric progression. Since the optimal max-stretch for all trains can be no smaller than the maximum of the class-wise lower-bounds, the overall max-stretch we achieve is only a constant times the optimal (Theorem 2).

Preprocessing. This module (Algorithm 1) answers the following question: for entering class i external trains, which holes should we use? In principle, we could use the holes left behind by internal trains of any class j as well as the preexisting holes. So we create an instance Π_j by removing internal trains of classes $1, \ldots, j$ and find the grain-size $w(j)$ for class i trains in Π_j, and then use Theorem 1 to determine a lower bound $M(j)$ on the class i makespan in Π_j. Clearly, $\min_j M(j)$ is a lower bound on the class i makespan in Π. The

Algorithm 2: Scheduling

Input: $\Pi, (J_i, w_i, h_i)$ for each class $i \in \{1, \ldots, \lfloor \log N \rfloor + 1\}$
1 **for** $i = 1, ..., \lfloor \log N \rfloor + 1$ **do**
2 $\quad F_{i-1} \longleftarrow$ the number of steps already executed for classes $1, ..., i-1$;
3 $\quad$ Execute 2^{J_i} movement steps, with no entries;
4 $\quad$ Schedule class i using Lemma 2 with $w = \widehat{w} = r_i w_i$ and $h = \hat{h} = r_i h_i - (F_{i-1} + 2^{J_i})$, where $r_i = \lceil \frac{F_{i-1}+2^{J_i}}{h_i} + \max\{\frac{F_{i-1}+2^{J_i}}{4h_i}, \frac{2^i}{h_i}\} \rceil$;
5 **end**

corresponding j is returned as J_i. The corresponding grain-size $w(j)$ is returned as w_i, and the promised number of holes per grain as h_i. We summarize this as follows.

Lemma 3. *Let F_i^* denote the optimal makespan if only class i trains are to be delivered. Then $F_i^* \geq \max\{2^{J_i - 1}, 2^{i-1}, \frac{1}{4}\sqrt{w_i \cdot 2^{i-1}}\}$, where J_i, w_i are as per Algorithm 1.*

Proof. The last two terms are as per Theorem 1. The first term arises as $2^{J_i - 1}$ steps have to elapse in order to use the holes created by exit of class J_i trains.

Scheduling. We schedule the classes one after another in ascending order. Class i trains use the holes left behind by internal trains of classes $1, \ldots, J_i$ where J_i is as determined during preprocessing. To ensure that these trains have exited, we run 2^{J_i} movement steps. Note that these holes will not be present at the same positions as in Π_j. To account for this and also to account for all the movements that occurred while delivering class $1, \ldots, i-1$ trains, we use a somewhat larger grain-size than w_i. (See Algorithm 2.)

Lemma 4. *Let F_i denote the makespan for class i trains as per our algorithm, F_i^* the optimal class i makespan, and $F_0 = 0$. Then $F_i = \frac{3}{2}F_{i-1} + O(F_i^*)$.*

Proof. In Π_{J_i}, every class i segment of size w_i has at least $h_i \geq \sqrt{w_i \cdot 2^i}$ holes. Thus, in Π, every class i segment of size $\widehat{w} = r_i w_i$ has $r_i h_i$ or more *potential holes*, i.e., actual holes and internal trains of classes $1, ..., J_i$. During the first F_{i-1} steps, some of them turn into holes and get used for entries of previous i–1 classes but then also get re-created. By the end of the following 2^{J_i} movement steps, all of them would be available as holes but possibly downstream from their initial positions. At most one train or hole may move out of any segment in a step, hence at time $F_{i-1} + 2^{J_i}$, every segment of size $\widehat{w}$ must have at least $\hat{h} = r_i h_i - (F_{i-1} + 2^{J_i})$ holes. Simplifying, we get: $\max\left\{\frac{1}{4}(F_{i-1} + 2^{J_i}), 2^i\right\} \leq \hat{h} < \frac{1}{4}(F_{i-1} + 2^{J_i}) + 2^i + h_i$, which implies: $\hat{h} \geq 2^i \geq \min\{\widehat{w}, 2^i\}$. Thus, Lemma 2 can indeed be used with segment-size $\widehat{w}$, and $\hat{h}$ holes per segment, to let class i trains finish by time:

$$F_i = (F_{i-1} + 2^{J_i}) + 2\hat{h} + \widehat{w} \cdot 2^i/\hat{h} + 2^i$$

$$< (F_{i-1} + 2^{J_i}) + \frac{1}{2}(F_{i-1} + 2^{J_i}) + 2^{i+1} + 2h_i + \frac{(\hat{h} + F_{i-1} + 2^{J_i}) \cdot w_i \cdot 2^i}{h_i \cdot \hat{h}} + 2^i$$

$$\because \hat{h} < \tfrac{1}{4}(F_{i-1} + 2^{J_i}) + 2^i + h_J \, , \, \widehat{w} = rw_i \, , \, r = \tfrac{\hat{h}+F_{i-1}+2^{J_i}}{h_i}$$

$$< \frac{3}{2}F_{i-1} + \frac{3}{2}2^{J_i} + 2h_i + 5w_i \cdot 2^i/h_i + 3 \cdot 2^i \qquad \because \tfrac{1}{4}(F_{i-1} + 2^{J_i}) < \hat{h}$$

$$< \frac{3}{2}F_{i-1} + \frac{3}{2}2^{J_i} + 2\sqrt{w_i \cdot 2^i} + 2 + 5\sqrt{w_i \cdot 2^i} + 3 \cdot 2^i$$

$$\because \sqrt{w_i \cdot 2^i} \le h_i < \sqrt{w_i \cdot 2^i} + 1$$

$$= \frac{3}{2}F_{i-1} + O(F_i^*)$$

The last line follows from that $F_i^* \ge \max\{2^{J_i-1}, 2^{i-1}, \frac{1}{4}\sqrt{w_i \cdot 2^{i-1}}\}$ by Lemma 3.

Theorem 2. *Our schedule has a max-stretch $O(1)$ times the optimal.*

Proof. Let X_i^* be the optimal class i max-stretch. Clearly, the overall max-stretch $X^* \ge X_i^*$ for all i. Let X_i be the class i max-stretch in our schedule.

We know that $F_i = \frac{3}{2}F_{i-1} + O(F_i^*)$. Thus $F_i = O(1)\sum_{k=1}^{i}\left(\frac{3}{2}\right)^{i-k}F_k^*$. Since trains of class k have path-lengths $< 2^k$, we have $X_k^* > \frac{F_k^*}{2^k} \Rightarrow F_k^* < 2^k X_k^*$. Since trains of class i have path-lengths at least 2^{i-1}, we get $X_i \le \frac{F_i}{2^{i-1}} = 2^{-i+1}F_i$. Substituting we have: $X_i = 2^{-i+1} \cdot O(1)\sum_{k=1}^{i}\left(\frac{3}{2}\right)^{i-k}2^k X_k^* = O(1)\sum_{k=1}^{i}\left(\frac{3}{4}\right)^{i-k}X_k^* = O(1)X^*\sum_{k=1}^{i}\left(\frac{3}{4}\right)^{i-k} = O(X^*)$, because $X_k^* \le X^*$.

4 NP-Hardness

We reduce from the strongly NP-hard 3-Partition problem [10], defined below.

Definition 3 (Problem 3P). *Let U be a set of positive integers, and $S(U) = \sum_{u \in U} u$. Let $B = \frac{|U|}{3}$ and $C = \frac{S(U)}{B}$ be integers and $\frac{C}{4} < u < \frac{C}{2}$ $\forall u \in U$. Can U be partitioned into B triples such that each triple adds up to the same value C?*

The core of the reduction is a *solver* widget. This contains a train for each integer in the 3P instance. If and only if the 3P instance has a solution, the three trains corresponding to each triple in the partition get linked into a single chain. For making sure that only B chains get formed, we use a *hole-blocker* widget to prevent too many holes reaching the solver. The widgets, defined next, have size polynomial in $S(U)$, the size of the 3P instance in unary.

Lemma 5. *For any 3P instance U, there exists a widget* Solver(U) *with N_U stations and integers T and L such that N_U, T, L are polynomial in $S(U)$, $N_U > 4L$, and:*

1. *If U has a solution then the trains of the widget can be scheduled with a max-stretch at most $X_U = 1 + \frac{T}{L}$, and in each step, a train can enter the solver from upstream and subsequently move forward non-stop.*
2. *If the trains can be scheduled with max-stretch $\leq X_U$ and no hole enters the widget from upstream during the first $2T$ steps, then U must have a solution.*

Proof. Let $U = \{u_1, \ldots, u_n\}$ be the 3P instance, where $n = |U|$. Suppose $\alpha = 4B$ and $N_U = 2B + \alpha BC + \alpha\frac{C}{4}$. Define *Solver($U$)* as stations $s_1, ..., s_{N_U}$ such that:

1. The first B stations $s_1, ..., s_B$ have holes, labelled respectively as $h_1, ..., h_B$.
2. For each u_i, we have an external train Q_i with path-length αu_i. These trains wait at outers downstream of s_B such that paths of all Q_is are node-disjoint. Let s_D be the destination of the most downstream of Q_is, i.e., $D = B + \alpha BC$.
3. Outers of $s_{D+1} ... s_{D+B}$ have trains $R_1 ... R_B$, each with path-length $L = \alpha\frac{C}{4}$.
4. $s_{B+1} ... s_{D+B}$ have trains going to the last station N, while $s_{D+B+1} ... s_{N_U}$ have holes.

Clearly, $N_U > 4L$ and L, N_U are polynomial in $S(U)$.

Now suppose U has a solution $\{U_1, \ldots, U_B\}$. Then schedule the trains as follows. For each $k \in \{1, \ldots, B\}$, construct a chain c'_k consisting of:

1. the hole h_k at station s_k,
2. the three external trains (Q_is) for the three integers in U_k, and
3. the external train R_k.

In each step $k \in [B]$, let the first train of c'_k enter using h_k. Then let all chains progress in parallel—prioritize entries over movements and arbitrarily resolve the conflicts among entries. Entries occur in at most $n + B$ steps. In other steps, for each c'_k, a non-terminal train moves on the line unless R_k has already entered; there can be at most $\sum_{u \in U_k}(\alpha u - 1) = \alpha C - 3$ such steps.[3] Then at most $T = (\alpha C - 3) + (n + B) = \alpha(C + 1) - 3$ steps occur by the time R_k has entered, i.e., all Q_is and R_ks have entered by time T. (Clearly, T is polynomial in $S(U)$.) So no train needs to halt after time T, i.e., max-delay$\leq T$. Since every train has a path-length $\geq L$, max-stretch $\leq 1 + \frac{T}{L} = X_U$. Moreover, since the entries do not use any holes from upstream of the widget, in each step a train from upstream can enter the widget and then also move ahead non-stop.

Finally, suppose the trains can be scheduled with max-stretch $\leq X_U$ such that no upstream hole enters the widget in the first $2T$ steps. Consider the set of chains induced by the schedule. Every Q_i has a path-length $\alpha u_i < \alpha\frac{C}{2} = 2L$, and R_k has path-length L. So each suffers a delay $< 2L(X_U - 1) = 2T$, i.e., enters by time $2T$, hence it can not fill a hole from upstream of the widget. All internal trains go to the last station N. Therefore, entries can use the B holes $h_1, ..., h_B$ or exit holes of other external trains, i.e., the chain-set has at most B chains, say $c'_1, ..., c'_B$. Clearly, no two R_ks can be in same chain, so each must be the terminal train of a chain. Then Q_is must be the non-terminal trains. $\forall k \in [B]$, let U_k be the set of integers corresponding to the non-terminal trains of c'_k. Then

[3] An external train with path-length l moves only $l - 1$ steps on the line.

$\{U_1, ..., U_B\}$ is a partition of U. Since R_k has a path-length L and a stretch at most $1+\frac{T}{L}$, it must enter by time $T+1$, i.e., the path-lengths of the non-terminal trains of c'_k add up to at most $T = \alpha(C+1) - 3$, i.e., U_k adds up to at most $\frac{T}{\alpha} = (C+1) - \frac{3}{\alpha} < C+1$, i.e., at most C. Then, since $\frac{C}{4} < u < \frac{C}{2}$ $\forall u \in U$, the partition $\{U_1, ..., U_B\}$ must be a valid solution to the 3P instance U.

Lemma 6. *Given integers $\ell \geq 1$ and $\tau \geq 2$, $\exists$ a widget* HoleBlocker(ℓ, τ) *of size polynomial in ℓ and τ such that:*

1. *Suppose there are at least 2ℓ stations downstream of the widget, and the widget's trains can move downstream from the widget and then continue non-stop. Then they can be scheduled with a max-stretch at most $X_{HB} = 1 + \frac{\tau-1}{\ell}$.*
2. *Suppose the widget's trains can be scheduled with max-stretch at most X_{HB}. Then no holes go downstream from the widget during the first τ steps.*

Proof. Let *HoleBlocker*(ℓ, τ) consist of blocks $F_0, \ldots, F_{q-1}, E, F_q$ from upstream to downstream, where $q = \lfloor \tau/\ell \rfloor$. For each $b \in \{0, ..., q\}$, F_b has $\tau - b\ell$ stations, each with an external train going a distance ℓ and an internal train going to the last station N. The external trains are labeled $P_{b,\tau-bl}$ to $P_{b,1}$ from upstream to downstream. E has ℓ stations with only internal trains, each going to the last station N. Overall, the widget has $w = (q+1)(q\ell/2 + r)$ external trains (where $r = \tau \bmod \ell$) and $w + \ell$ internal trains. Clearly, its size is polynomial in ℓ and τ.

To prove part 1, we make τ chains $c_1, ..., c_\tau$. Each c_j consists of the external trains $P_{b,j}$ $\forall b \in \{0, ..., q\}$. In each step $j \in \{1 ... \tau\}$, we let the first train $P_{0,j}$ of c_j enter using a hole from the artificial stations $-1, -2, ...$. After the entry of its first train, each chain *progresses non-stop*. It is easy to see that all chains can do so in parallel, and that the number of external trains that will enter by time $\tau = q\ell + r$ is $\ell q(q+1)/2 + r(q+1) = (q+1)(q\ell/2 + r) = w$, i.e., all of the $P_{b,j}$s will enter by time τ. Then all trains move non-stop to their destinations. So any external train suffers a delay at most $\tau - 1$ and hence a stretch at most $1 + \frac{\tau-1}{\ell} = X_{HB}$; any internal train suffers a delay at most τ and hence a stretch at most $1 + \tau/2\ell \leq X_{HB}$.

To prove part 2, we note that since there were no holes in the widget and all internal trains go to the last station N, entries may fill the *external* holes (from artificial stations $-1, -2, ...$) or the exit holes of other external trains. Moreover, in any step at most one external hole can enter the widget. Then, it is easy to see (similar to part 2 of the proof of Lemma 1) that by time $\tau = q\ell + r$, at most $(q\ell/2 + r)(q+1)$ trains can enter, but that is exactly the count w of the external trains. That means if in any of the first τ steps a hole goes downstream from the widget, i.e., we miss using that hole to enter one of the widget's external trains, then not all of those trains can enter by time τ, which implies at least one of them will suffer a delay of τ or more and hence a stretch at least $1 + \frac{\tau}{\ell} > X_{HB}$.

With the two widgets defined above, we can now prove the hardness result.

Theorem 3. *Minimizing max-stretch on in-comb is NP-hard.*

Proof. The proof is by reduction from the 3-Partition problem. For an instance U of the 3-Partition problem, our Train Scheduling instance is as follows.

The in-comb network consists of a HoleBlocker(ℓ, τ) followed by a Solver(U). We choose $\ell = 2L$ and $\tau = 2T + 1$, where T, L are as promised by Lemma 5. Note that the size of our solver widget is $N_U > 4L$, so the path-length of every internal train of the hole-blocker is at least $4L = 2\ell$, as required by Lemma 6. Then it can be seen from Lemmas 5 and 6 that the size of the train scheduling instance is polynomial in $S(U)$. We fix the target max-stretch X to $1 + \frac{T}{L}$.

Now, suppose the 3-Partition instance U has a solution. From Lemma 5 part 1, we can schedule trains of the solver widget with max-stretch at most X such that any trains coming into the solver from the hole-blocker can move ahead non-stop. Then, from Lemma 6 part 1, trains of the hole-blocker can also be scheduled with max-stretch at most $1 + \frac{\tau-1}{\ell} = X$.

Conversely, suppose all trains can be scheduled with stretch at most X. From Lemma 6 part 2, we know that no hole goes downstream from the hole-blocker in the first $2T$ steps. Then, by Lemma 5 part 2, the 3-partition instance U must have a solution.

Finally, note that the above construction is broadly similar to the one used in [11] for the max-delay minimization problem, but now the crucial hole-blocker widget has to be designed more carefully in order to prove the hardness of minimizing max-stretch.

5 Conclusion

The train scheduling problem on in-comb was introduced in [11]. That work also gave the chain-hole view as an important insight into the problem and used it fruitfully to prove the hardness of max-delay minimization as well as design an $O(\log N)$ approximation algorithm for it. We have used the same chain-hole view but with entirely new ideas for lower and upper bounds to give an $O(1)$ approximation algorithm and the hardness proof for a lesser explored but presumably more relevant metric—the max-stretch.

Further work can focus on tighter bounds as well as on possible generalizations of the problem, e.g., multiple tracks between stations, multiple trains at every station, variable train speeds, etc. Parametric formulations of the problem—e.g., with a given maximum number of external trains—should also be studied.

A Proof of Lemma 2

Without loss of generality, extend the path-length of every external train (of class i) to $\ell = 2d - 1$, where $d = 2^{i-1}$. Starting from upstream, partition the network into groups of stations, each group being a class i segment of size w, i.e., containing w external trains.

The schedule is trivial for the case when $h \geq w$: external trains of one set of alternate groups can enter in the first $w \leq h$ steps and those of the other set in the next w steps, trains of each group filling the holes of its upstream neighbour; afterwards all trains can move to their destinations during the next $\ell - 1$ steps, thus giving a makespan of $2w + \ell - 1 < 2h + 2^i$. Next, let us consider the non-trivial case: $w > h > 2d - 1$.

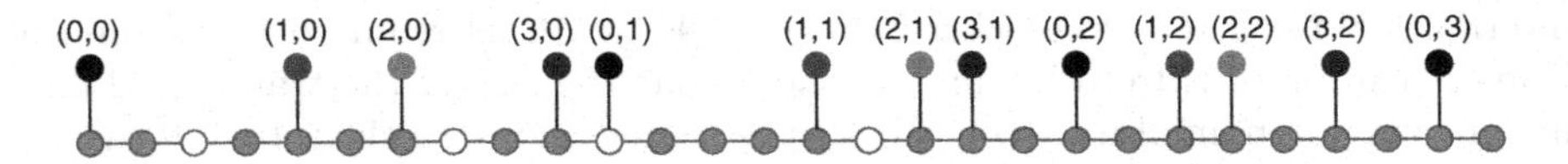

Fig. 3. Chains in a group ($i = 2, w = 13, h = 4$). Circles denote holes and grey disks denote internal trains. The other, labeled disks denote external trains, disks of a color representing trains of a chain.

Within each group, number the external trains from 0 to $w - 1$ starting upstream, and then form h chains, allocating the trains to the chains in a round-robin fashion: train p becomes the jth train of the kth chain, where $j = \lfloor p/h \rfloor$ and $k = p \bmod h$. We may refer to train p also as train (k, j). Thus, each chain will have some J trains where $J \leq \lceil w/h \rceil$. (See Fig. 3 for an example of chain formation in a group.)

For $k \in \{0, ..., h - 1\}$, consider train $(k, 0)$ in any group G. Let q denote the first unassigned hole downstream of train $(k, 0)$ in the previous group $G - 1$. Assign this is as the initial hole of chain k in group G. Because each segment of size w starting at any external train is guaranteed to have h holes, it can be easily seen that this assignment succeeds in finding a distinct initial hole for every chain.

Each external train has to perform one entry and $2d - 2$ movements. Let us also consider the jump of the hole that it fills as a movement of the entering train, i.e., the train has $2d-1$ movements. In each chain, number the movements from 1 to $(2d - 1) \times J$, starting upstream and numbering consecutively for all trains of the chain. We will use $\langle k, m\rangle$ to denote m^{th} movement in chain k. Note that among the movements of different chains, we have an *overlap condition*: if $k' > k$ then the movement $\langle k, m\rangle$ may overlap with $\langle k', m'\rangle$ only for $m' \leq m$.

The train movements are scheduled in three phases, as follows:

1. This has h steps: $1, ..., h$. In step i, the initial hole of chain $h - i$ in every group jumps to the first train $(h - i, 0)$ of the chain and gets filled by the train. It is easily seen that paths of the jumps in each step are disjoint.
2. In this phase, every train—except the last—in a chain completes its journey and exits the network; the re-created hole then jumps to the next train of the chain. We discuss this in more detail below.
3. In this last phase, the last trains of all chains move to their destinations. It is easily seen that there are no conflicts and this phase takes $2d-1$ steps.

In Phase 2, the paths of trains in one group do not overlap with those in other groups. So, we can consider each group separately. To resolve the conflict between overlapping movements due to take place in the same step, we use a very simple *scheduling rule*: higher numbered chains have higher priority. Now we can use a simple *delay sequence argument*—a proof technique used earlier in [1,22,28]—to prove the time bound as follows.

Suppose $\langle k, m\rangle$ occurs in step t. Then, for $t-1$, one of the following is true:

1. Movement $\langle k, m-1\rangle$ occurred in step $t-1$.
2. Movement $\langle k', m'\rangle$ occurred in step $t-1$, delaying the (lower priority) movement $\langle k, m\rangle$. Here $k' > k$ because of the scheduling rule and $m' \leq m$ because of the overlap condition.

Thus, if the last movement of Phase 2 occurs in step $h+T$ then we can find a sequence of movements $\langle k, m\rangle$, one for each of the steps $h+T$, $h+T-1, ..., h+1$, such that:

1. k never decreases,
2. m never increases, and
3. at least one of the two does change.

But this can happen only $h+(2d-1)(J-1) < h+2d\cdot w/h$ times. So, the overall time for all the three phases is only less than $2h + 2d \cdot w/h + 2d$.

Figure 4, on page 16, illustrates the schedule for a small example as a space-time diagram. Note that while the space-time trajectories of trains can not cross one-another, trajectories of holes may cross those of trains because a hole can jump over stationary trains.

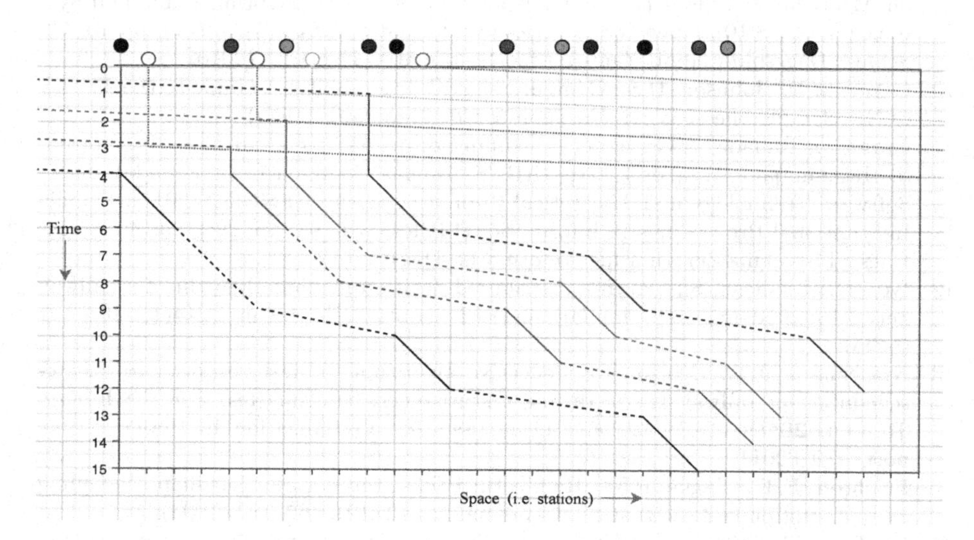

Fig. 4. Space-time diagram of a group, say G, for $i = 2, w = 12, h = 4$. Colored disks and circles at the top respectively represent external trains (of G) and holes (matched to the entries of the next group $G+1$) at their initial positions. External trains of a color belong to same chain. Thus, black lines mark the space-time trajectory of the 0th chain, red lines the trajectory of the 1st chain, and so on. Solid lines represent train-movements. Dashed lines represent hole-jumps for the entries of G, while dotted lines represent the hole-jumps for the entries of $G+1$. (Color figure online)

References

1. Aleliunas, R.: Randomized parallel communication (preliminary version). In: Proceedings of the First ACM SIGACT-SIGOPS Symposium on Principles of Distributed Computing, PODC 1982, pp. 60–72. ACM, New York (1982)
2. Cacchiani, V., et al.: An overview of recovery models and algorithms for real-time railway rescheduling. Transp. Res. Part B: Methodol. **63**, 15–37 (2014)
3. Cai, X., Goh, C., Mees, A.I.: Greedy heuristics for rapid scheduling of trains on a single track. IIE Trans. **30**(5), 481–493 (1998)
4. Caimi, G., Chudak, F., Fuchsberger, M., Laumanns, M., Zenklusen, R.: A new resource-constrained multicommodity flow model for conflict-free train routing and scheduling. Transp. Sci. **45**(2), 212–227 (2011)
5. Chiang, T., Hau, H., Chiang, H.M., Kob, S.Y., Hsieh, C.H.: Knowledge-based system for railway scheduling. Data Knowl. Eng. **27**(3), 289–312 (1998)
6. Clementi, A., Ianni, M.D.: Optimum schedule problems in store and forward networks. In: 13th Proceedings of IEEE Networking for Global Communications, INFOCOM 1994, vol. 3, pp. 1336–1343, June 1994
7. D'Ariano, A.: Improving real-time train dispatching: models, algorithms and applications. Doctoral thesis, TRAIL Research School, Deft, The Netherlands (2008)
8. Fang, W., Yang, S., Yao, X.: A survey on problem models and solution approaches to rescheduling in railway networks. IEEE Trans. Intell. Transp. Syst. **16**(6), 2997–3016 (2015)
9. Flier, H., Mihalák, M., Schöbel, A., Widmayer, P., Zych, A.: Vertex disjoint paths for dispatching in railways. In: Erlebach, T., Lübbecke, M. (eds.) 10th Workshop on Algorithmic Approaches for Transportation Modelling, Optimization, and Systems (ATMOS'10). OpenAccess Series in Informatics (OASIcs), vol. 14, pp. 61–73. Schloss Dagstuhl-Leibniz-Zentrum fuer Informatik, Dagstuhl (2010)
10. Garey, M.R., Johnson, D.S.: Computers and Intractability: A Guide to the Theory of NP-Completeness. W. H. Freeman & Co., New York (1979)
11. Garg, A., Ranade, A.G.: Train scheduling on a unidirectional path. In: Lokam, S., Ramanujam, R. (eds.) 37th IARCS Annual Conference on Foundations of Software Technology and Theoretical Computer Science (FSTTCS 2017). Leibniz International Proceedings in Informatics (LIPIcs), vol. 93, pp. 29:1–29:14. Schloss Dagstuhl-Leibniz-Zentrum fuer Informatik, Dagstuhl (2018)
12. Iyer, R.V., Ghosh, S.: Daryn-a distributed decision-making algorithm for railway networks: modeling and simulation. IEEE Trans. Veh. Technol. **44**(1), 180–191 (1995)
13. Krasemann, J.T.: Design of an effective algorithm for fast response to the rescheduling of railway traffic during disturbances. Transp. Res. Part C: Emerg. Technol. **20**(1), 62–78 (2012). Special issue on Optimization in Public Transport+ISTT2011
14. Leighton, F.T., Maggs, B.M., Rao, S.B.: Packet routing and job-shop scheduling in o(congestion+dilation) steps. Combinatorica **14**(2), 167–186 (1994)
15. Leighton, T., Maggs, B., Richa, A.W.: Fast algorithms for finding o(congestion + dilation) packet routing schedules. Combinatorica **19**(3), 375–401 (1999)
16. Leung, J.Y.T., Tam, T.W., Young, G.H.: On-line routing of real-time messages. J. Parallel Distrib. Comput. **34**(2), 211–217 (1996)
17. Mannino, C., Mascis, A.: Optimal real-time traffic control in metro stations. Oper. Res. **57**(4), 1026–1039 (2009)

18. Mascis, A., Pacciarelli, D.: Job-shop scheduling with blocking and no-wait constraints. Eur. J. Oper. Res. **143**(3), 498–517 (2002)
19. Narayanaswami, S., Rangaraj, N.: Scheduling and rescheduling of railway operations: a review and expository analysis. Technol. Oper. Manage. **2**(2), 102–122 (2011)
20. Peis, B., Skutella, M., Wiese, A.: Packet routing: complexity and algorithms. In: Bampis, E., Jansen, K. (eds.) WAOA 2009. LNCS, vol. 5893, pp. 217–228. Springer, Heidelberg (2010). https://doi.org/10.1007/978-3-642-12450-1_20
21. Pellegrini, P., Rodriguez, J.: Single European sky and single european railway area: a system level analysis of air and rail transportation. Transp. Res. Part A: Policy Pract. **57**, 64–86 (2013)
22. Ranade, A.G.: Fluent parallel computation. Ph.D. thesis, Department of Computer Science, Yale University, New Haven, CT, USA (1989). aAI9010675
23. Rothvoß, T.: A simpler proof for o(congestion + dilation) packet routing. CoRR abs/1206.3718 (2012)
24. Sahin, I.: Railway traffic control and train scheduling based oninter-train conflict management. Transp. Res. Part B: Methodol. **33**(7), 511–534 (1999)
25. Salim, V., Cai, X.: A genetic algorithm for railway scheduling with environmental considerations. Environ. Model Softw. **12**(4), 301–309 (1997)
26. Scheideler, C.: Universal routing strategies for interconnection networks. In: Goos, G., Hartmanis, J., van Leeuwen, J. (eds.) Lecture Notes in Computer Science, vol. 1390, pp. 57–67. W. H. Freeman & Co., New York (1998)
27. Tormos, P., Lova, A., Barber, F., Ingolotti, L., Abril, M., Salido, M.A.: A genetic algorithm for railway scheduling problems. In: Xhafa, F., Abraham, A. (eds.) Metaheuristics for Scheduling in Industrial and Manufacturing Applications. Studies in Computational Intelligence, vol. 128, pp. 255–276. Springer, Heidelberg (2008). https://doi.org/10.1007/978-3-540-78985-7_10
28. Upfal, E.: Efficient schemes for parallel communication. J. ACM **31**(3), 507–517 (1984)
29. Šemrov, D., Marsetič, R., Žura, M., Todorovski, L., Srdic, A.: Reinforcement learning approach for train rescheduling on a single-track railway. Transp. Res. Part B: Methodol. **86**, 250–267 (2016)

Algorithmic Aspects of Total Roman and Total Double Roman Domination in Graphs

Chakradhar Padamutham(✉) and Venkata Subba Reddy Palagiri

Department of Computer Science and Engineering, National Institute of Technology, Warangal, Warangal 506 004, Telangana, India
corneliusp7@gmail.com, pvsr@nitw.ac.in

Abstract. For a simple, undirected and connected graph $G = (V, E)$, a total Roman dominating function (TRDF) $f : V \to \{0, 1, 2\}$ has the property that, every vertex u with $f(u) = 0$ is adjacent to at least one vertex v for which $f(v) = 2$ and the subgraph induced by the set of vertices labeled one or two has no isolated vertices. A total double Roman dominating function (TDRDF) on G is a function $f : V \to \{0, 1, 2, 3\}$ such that for every vertex $v \in V$ if $f(v) = 0$, then v has at least two neighbors x, y with $f(x) = f(y) = 2$ or one neighbor w with $f(w) = 3$, and if $f(v) = 1$, then v must have at least one neighbor w with $f(w) \geq 2$ and the subgraph induced by the set $\{u_i : f(u_i) \geq 1\}$ has no isolated vertices. The weight of a T(D)RDF f is the sum $f(V) = \sum_{v \in V} f(v)$. The minimum total (double) Roman domination problem (MT(D)RDP) is to find a T(D)RDF of minimum weight of the input graph. In this article, we show that MTRDP and MTDRDP are polynomial time solvable for bounded treewidth graphs, chain graphs and threshold graphs. We design a $2(\ln(\Delta - 0.5) + 1.5)$-approximation algorithm (APX-AL) for the MTRDP and $3(\ln(\Delta - 0.5) + 1.5)$-APX-AL for the MTDRDP, where Δ is the maximum degree of G, and show that the same cannot have $(1 - \delta) \ln |V|$ ratio APX-AL for any $\delta > 0$ unless $P = NP$. Finally, we show that MT(D)RDP is APX-hard for graphs with $\Delta = 5$.

Keywords: Total Roman domination · Total double Roman domination · APX-complete

1 Introduction

Let $G(V, E)$ be a simple, undirected and connected graph. For a vertex u of G, the *(open) neighborhood* denoted $N_G(u)$ is the set $\{v : (v, u) \in E\}$ and its *degree* is $|N_G(u)|$. The *closed neighborhood* of u is $N_G[u] = \{u\} \cup N_G(u)$. *Maximum degree* of G denoted Δ (or clearly $\Delta(G)$) is $max_{u \in V} |N_G(u)|$. A vertex v is called *isolated vertex* if $|N_G(v)| = 0$. A vertex v of G is called *universal vertex* if $N_G[v] = V(G)$. A graph formed with the vertex set $S \subseteq V$ of graph $H(V, E)$ and the edge set $\{(u, v) \in E : u, v \in S\}$ is called an *induced subgraph* of H denoted $\langle S \rangle$. For undefined terminology and notations we refer to [35].

A. Mudgal and C. R. Subramanian (Eds.): CALDAM 2021, LNCS 12601, pp. 32–42, 2021.
https://doi.org/10.1007/978-3-030-67899-9_3

A *dominating set* (DS) of a graph G is a set D such that $D \subseteq V$ and $\cup_{w \in D} N_G[w] = V$ and further D is called a *total dominating set* (TDS) of G if every vertex in V is adjacent to at least one vertex in D. The (*total*) *domination number* of G denoted by $(\gamma_t(G))$ $\gamma(G)$ is $min\{|Q| : Q$ is a (T)DS of $G\}$. The problem of finding a (T)DS of smallest cardinality in a graph is called the minimum (total) dominating set (M(T)DS) problem. Literature on the concept of, domination has been surveyed in [16], total domination has been surveyed in [17].

In 2004, Cockayne et al. in [11] introduced the concept of Roman domination (RDOM). A function $f : V \rightarrow \{0, 1, 2\}$ is a *Roman Dominating Function* (RDF) on G if every vertex with label zero is adjacent to at least one vertex with label two. We refer to [3,11,13,14,18–21,25,28,31,32] for the literature on RDOM in graphs.

The notion of total Roman domination (TRDOM) was introduced in 2013 by Liu et al. in [22]. A total Roman dominating function (TRDF) is a Roman dominating function with the additional property that the subgraph of G induced by the set $\{k \in V : f(k) \geq 1\}$ is without isolated vertices. The concept of TRDOM has been studied in [1,7,9,27].

Double Roman domination was introduced in 2016 by Beeler et al. in [30]. A Double Roman Dominating Function (DRDF) on G is a function $g : V \rightarrow \{0, 1, 2, 3\}$ such that for every vertex $k \in V$ if $g(k) = 0$, then k has at least two neighbors $x, y \in N_G(k)$ with $g(x) = g(y) = 2$ or one neighbor w with $g(w) = 3$, and if $g(k) = 1$, then k must have at least one neighbor w with $g(w) \geq 2$. The double Roman domination has been studied in [2,4,5,8,24].

Total double Roman domination (TDRDOM) was introduced in 2019 by Shao et al. in [33], which is a variant of double Roman domination. A total double Roman dominating function (TDRDF) is a double Roman dominating function with the additional property that the subgraph of G induced by the set $\{k \in V : g(k) \geq 1\}$ is without isolated vertices. The concept TDRDOM has been studied in [15,33].

The weight of a RDF (TRDF, DRDF, TDRDF) g is the value $g(V) = \sum_{v \in V} g(v)$. The *Roman domination number*, *total Roman domination number*, *double Roman domination number*, *total double Roman domination number*, respectively, equals the minimum weight of a RDF, TRDF, DRDF and TDRDF, respectively, denoted by $\gamma_R(G)$, $\gamma_{tR}(G)$, $\gamma_{dR}(G)$ and $\gamma_{tdR}(G)$. The minimum total (double) Roman domination problem (MT(D)RDP) is to find a T(D)RDF of minimum weight in the input graph.

2 Bounded Tree-Width Graphs

A *tree decomposition* of a graph H is a tree T_1 with the vertex set $V(T_1) = \{Z_1, Z_2, \ldots, \}$, where each Z_i is a subset of $V(H)$ with the following requirements.

i) $V(H) = \bigcup_{Z_k \in V(T_1)} Z_k$
ii) $\forall (u, v) \in E(H)$, there exists a vertex $Z_t \in V(T_1)$ such that $u, v \in Z_t$ and

iii) $\forall v \in V(H)$, the induced subgraph $\{Z_t : v \in Z_t \text{ and } Z_t \in V(T_1)\}$ is a subtree of T_1.
Then the tree decomposition T_1 of H is said to have *width* equals to $max\{|Z_t|-1 : Z_t \in V(T_1)\}$ [29]. The *treewidth* is the smallest width of a tree decomposition of a graph.

Theorem 1. *Given a graph G and a positive integer k, TRDP can be expressed in CMSOL.*

Proof. Let $f : V \to \{0,1,2\}$ be a function on a graph G, where $V_i = \{v|f(v) = i\}$ for $i \in \{0,1,2\}$. The CMSOL formula for the RDF problem is expressed as follows.

$Rom\_Dom(V) = \exists V_0, V_1, V_2, \forall p(p \in V_1 \vee p \in V_2 \vee (p \in V_0 \wedge \exists q \in V_2 \wedge adj(p,q)))$,
where $adj(p,q)$ is the binary adjacency relation which holds if and only if, p, q are two adjacent vertices of G.

Next, we give a CMSOL formula for the $Total\_Rom(V)$, which says that every vertex $p \in V_1 \cup V_2$ is adjacent to some vertex q in $V_1 \cup V_2$, as follows.

$Total\_Rom(V) = \exists V_0, V_1, V_2, \forall p, \exists q(p \in (V_1 \cup V_2) \wedge q \in (V_1 \cup V_2) \wedge adj(p,q))$.

Let k be a positive integer, then the CMSOL formula for the TRDP is expressed as follows.

$Total\_Rom\_Dom(V) = (f(V) \leq k) \wedge Rom\_Dom(V) \wedge Total\_Rom(V)$.

Now, from Theorem 1 and Courcelle's result in [12], the theorem below follows.

Theorem 2. *MTRDP for graphs with treewidth at most a constant is solvable in linear time.*

Theorem 3. *Given a graph G and a positive integer k, TDRDP can be expressed in CMSOL.*

Proof. Let $g : V \to \{0,1,2,3\}$ be a function on a graph G, where $V_i = \{v|g(v) = i\}$ for $i \in \{0,1,2,3\}$. The CMSOL formula for the DRDF problem is expressed as follows.

$Double\_Rom\_Dom(V) = \exists V_0, V_1, V_2, V_3, \forall p((p \in V_0 \wedge ((\exists q, r \in V_2 \wedge adj(p,q) \wedge adj(p,r)) \vee (\exists s \in V_3 \wedge adj(p,s))) \vee (p \in V_1 \wedge (\exists t \in V_2 \wedge adj(p,t) \vee (\exists u \in V_3 \wedge adj(p,u))))) \vee (p \in V_2) \vee (p \in V_3))$,
where $adj(p,q)$ is the binary adjacency relation which holds if and only if, p, q are two adjacent vertices of G.

Next, we give a CMSOL formula for the $Total\_Double\_Rom(V)$, which says that every vertex $p \in V_1 \cup V_2 \cup V_3$ is adjacent to some vertex q in $V_1 \cup V_2 \cup V_3$, as follows.

$Total\_Double\_Rom(V) = \exists V_0, V_1, V_2, V_3, \forall p, \exists q(p \in (V_1 \cup V_2 \cup V_3) \wedge q \in (V_1 \cup V_2 \cup V_3) \wedge adj(p,q))$.
Let k be a positive integer, then the CMSOL formula for the TDRDP is expressed as follows.

$Total\_Double\_Rom\_Dom(V) = (g(V) \leq k) \wedge Double\_Rom\_Dom(V) \wedge Total\_Double\_Rom(V)$.

Now, from Theorem 3 and Courcelle's result in [12], the theorem below follows.

Theorem 4. *MTDRDP for graphs with treewidth at most a constant is solvable in linear time.*

3 Threshold Graphs

Here, we solve MTRDP and MTDRDP for connected threshold graphs in linear time. A graph G is *threshold* iff the following conditions hold, see [23]

i) Vertex set of G is partitioned into two disjoint sets, a clique Q and an independent set R
ii) There exists a permutation $(q_1, q_2, \dots, q_p)$ of vertices of Q such that $N_G[q_1] \subseteq N_G[q_2] \subseteq \dots \subseteq N_G[q_p]$ and
iii) There exists a permutation $(r_1, r_2, \dots, r_i)$ of vertices of R such that $N_G(r_1) \supseteq N_G(r_2) \supseteq \dots \supseteq N_G(r_i)$.

Theorem 5. *Let G be a connected threshold graph. Then,*

$$\gamma_{tR}(G) = \begin{cases} 2, & \text{if } G \cong K_2 \\ 3, & \text{otherwise} \end{cases} \tag{1}$$

and

$$\gamma_{tdR}(G) = \begin{cases} 3, & \text{if } G \cong K_2 \\ 4, & \text{otherwise} \end{cases} \tag{2}$$

Proof. Let G be a connected threshold graph with p clique vertices and i independent vertices as described above. Since, q_p is a universal vertex of G, clearly, this implies that $\gamma_{tR}(G) = 3$ and $\gamma_{tdR}(G) = 4$, except when $G \cong K_2$ where $\gamma_{tR}(G) = 2$ and $\gamma_{tdR}(G) = 3$.

Now, the following result is immediate from Theorem 5 and the fact that the ordering of clique vertices of threshold graph can be found in linear time [23].

Theorem 6. *MTRDP and MTDRDP for connected threshold graphs are linear time solvable.*

If threshold graph G is disconnected i.e., G contains isolated vertices, then TRDF and TDRDF can not be defined on G.

4 Chain Graphs

Here, we solve MTRDP and MTDRDP for connected chain graphs in linear time. An ordering $\alpha = (y_1, y_2, \dots, y_p, z_1, z_2, \dots, z_q)$ of vertex set of a bipartite graph $G(Y, Z, E)$ is a *chain ordering* if $N_G(y_1) \subseteq N_G(y_2) \subseteq \dots \subseteq N_G(y_p)$ and $N_G(z_1) \supseteq N_G(z_2) \supseteq \dots \supseteq N_G(z_q)$. A bipartite graph is a *chain graph* iff it has a chain ordering [36].

Theorem 7. *Let $G(Y,Z,E)$ be a connected chain graph. Then,*

$$\gamma_{tR}(G) = \begin{cases} 2, & \textit{if } G \textit{ is } K_2 \\ 3, & \textit{if } G \textit{ is } K_{1,s}, \textit{ where } s \geq 2 \\ 4, & \textit{otherwise} \end{cases} \tag{3}$$

and

$$\gamma_{tdR}(G) = \begin{cases} 3, & \textit{if } G \textit{ is } K_2 \\ 4, & \textit{if } G \textit{ is } K_{1,s}, \textit{ where } s \geq 2 \\ 6, & \textit{otherwise} \end{cases} \tag{4}$$

Proof. Let $G(Y,Z,E)$ be a connected chain graph with $|Y| = p$ and $|Z| = q$ where $p, q \geq 1$. If $G \cong K_2$ or $G \cong K_{1,s}$, where $s \geq 2$, then $\gamma_{tR}(G)$ and $\gamma_{tdR}(G)$ can be determined directly from Theorem 5. Otherwise, define functions $f : V \rightarrow \{0,1,2\}$ and $g : V \rightarrow \{0,1,2,3\}$ as follows.

$$f(v) = \begin{cases} 2, & \text{if } v \in \{y_p, z_1\} \\ 0, & \text{otherwise} \end{cases} \tag{5}$$

$$g(v) = \begin{cases} 3, & \text{if } v \in \{y_p, z_1\} \\ 0, & \text{otherwise} \end{cases} \tag{6}$$

Clearly, f (g) is a TRDF (TDRDF) and $\gamma_{tR}(G) \leq 4$ ($\gamma_{tdR}(G) \leq 6$). By contradiction, it can be easily verified that $\gamma_{tR}(G) \geq 4$ ($\gamma_{tdR}(G) \geq 6$). Therefore $\gamma_{tR}(G) = 4$ ($\gamma_{tdR}(G) = 6$).

Now, from Theorem 7 and the fact that chain ordering can be computed in linear time [34], the theorem below follows.

Theorem 8. *MTRDP and MTDRDP for connected chain graphs are solvable in linear time.*

If chain graph G is disconnected i.e., G contains isolated vertices, then TRDF and TDRDF can not be defined on G.

5 Approximation Algorithm and Complexity

Here, results related to obtaining approximate solutions to MTRDP and MTDRDP are presented.

5.1 Approximation Bounds

An existing result obtained on lower bound of approximation ratio of MDS is given below.

Theorem 9 ([10]). *For a graph $G = (V,E)$, unless $P = NP$, the MDS problem cannot have a solution with approximation ratio $(1-\delta)\ln|V|$ for any $\delta > 0$.*

Theorem below provides a lower bound on approximation ratio of MTRDP.

Theorem 10. *For a graph H, unless $P = NP$, the MTRDP cannot have a solution with approximation ratio $(1-\delta)\ln|V|$ for any $\delta > 0$.*

Proof. We propose a reduction which preserves the approximation. Let $H(V, E)$, where $V = \{v_1, v_2, \ldots, v_n\}$ be an instance of the MDS problem. From H, an instance H' of MTRDP is constructed as follows.

Create n copies of P_3 with b_i as the central vertex and a_i, c_i as terminal vertices. Add the edges $\{(v_i, a_i), (v_i, c_i) : 1 \leq i \leq n\}$. An example construction of H' from H is shown in Fig. 1. Next, we prove a claim.

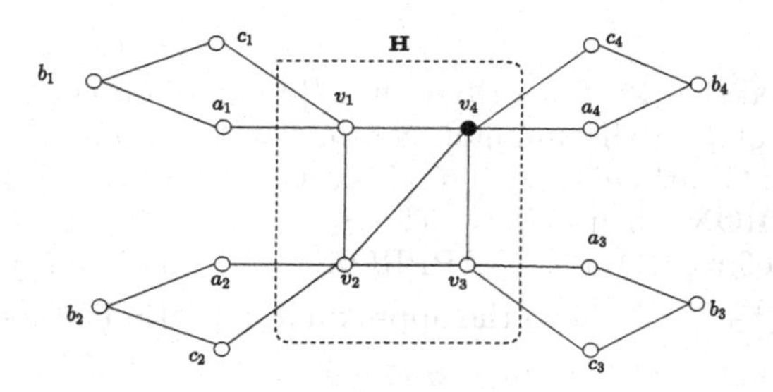

Fig. 1. Construction of H' from H

Claim. $\gamma_{tR}(H') = 3n + \gamma(H)$.

Proof. Let $H(V, E)$, where $V = \{v_1, v_2, \ldots, v_n\}$ be a graph and $H' = (V', E')$ is a graph constructed from H.

Let M^* be a MDS of H i.e., $|M^*| = \gamma(H)$ and f be a function on H', defined as

$$f(v) = \begin{cases} 2, & \text{if } v \in \{v_i, a_i : v_i \in M^*\} \text{ or } v \in \{b_i : v_i \notin M^*\} \\ 1, & \text{if } v \in \{a_i : v_i \notin M^*\} \\ 0, & \text{otherwise} \end{cases} \tag{7}$$

Clearly, f is a TRDF and $\gamma_{tR}(H') \leq 3n + |M^*|$.

Next, we show that $\gamma_{tR}(H') \geq 3n + |M^*|$. Let g be a TRDF on graph H'. Clearly if $g(v_i) = 0$, then $g(a_i) + g(b_i) + g(c_i) \geq 3$ and if $g(v_i) \geq 1$, then $g(v_i)+g(a_i)+g(b_i)+g(c_i) \geq 4$. Therefore $\gamma_{tR}(H') \geq 3n+|M^*|$. Hence $\gamma_{tR}(H') = 3n + \gamma(H)$.

Suppose that the MTRDP has an approximation algorithm (APX-AL) A which runs in polynomial time with approximation ratio β, where $\beta = (1-\delta)\ln|V|$ for some fixed $\delta > 0$. Let l be a fixed positive integer. Next, we design an APX-AL, say DOM-SET-APPROX which runs in polynomial time to find a DS of a given graph H.

Algorithm 1. DOM-SET-APPROX(G)

Require: A simple and undirected graph H.
Ensure: A DS M of H.
1: **if** there exists a DS M' of size at most l, **then**
2: $M \leftarrow M'$
3: **else**
4: Build the graph H'
5: Calculate a TRDF f on H' by using algorithm A
6: Find a DS M of H from TRDF f (as illustrated in the proof of Claim in
7: Sect. 5.1)
8: **end if**
9: return M.

It can be noted that if M is a DS with $|M| \leq l$, then it is optimal. Otherwise, let M^* be a DS of H with minimum cardinality and g be a TRDF of H' with $g(V') = \gamma_{tR}(H')$. Clearly $g(V) \geq l$. If M is a DS of H obtained by the algorithm DOM-SET-APPROX, then $|M| \leq f(V) \leq \beta(g(V)) \leq \beta(3n + |M^*|) = \beta(1 + \frac{3n}{|M^*|})|M^*|$. Therefore, DOM-SET-APPROX approximates a MDS within a ratio $\beta(1 + \frac{3n}{|M^*|})$. If $\frac{1}{|M^*|} < \delta/2$, then the approximation ratio becomes $\beta(1 + \frac{3n}{|M^*|}) < (1-\delta)(1 + \frac{3n\delta}{2}) \ln n = (1-\delta') \ln n$, where $\delta' = \frac{3n\delta^2}{2} - \frac{3n\delta}{2} + \delta$.

By Theorem 9, if there exists an APX-AL for MDS problem with approximation ratio $(1-\delta)\ln|V|$, then $P = NP$. Similarly, if there exists an APX-AL for MTRDP with approximation ratio $(1-\delta)\ln|V|$, then $P = NP$. For large values of n, $\ln n \approx \ln(4n)$. Hence, in a graph $H'(V', E')$, where $|V'| = 4|V|$, the MTRDP cannot have an approximation algorithm with a ratio of $(1-\delta)\ln|V'|$ unless $P = NP$.

Theorem 11. *For a graph H, unless $P = NP$, the MTDRDP cannot have a solution with approximation ratio $(1-\delta)\ln|V|$ for any $\delta > 0$.*

Proof. The proof is obtained with similar arguments as in Theorem 10, in which replace the assigned value, for the vertices, 2 with 3.

5.2 Approximation Algorithm

Here, an APX-AL for MT(D)RDP is designed based on the approximation result known for MTDS problem below.

Theorem 12 ([37]). *The MTDS problem can be approximated with an approximation ratio of* $\ln(\Delta - 0.5) + 1.5$.

Let APP-TD-SET be an APX-AL that produces a TDS D of a graph G such that $|D| \leq (\ln(\Delta - 0.5) + 1.5)\gamma_t(G)$.

Next, we designe APP-TRDF algorithm to determine an approximate solution of MTRDP. In our algorithm, first we determine a TDS D of G using the APX-AL APP-TD-SET. Next, we build a total Roman dominating triple

Algorithm 2. APP-TRDF(G)

Input: A simple, undirected graph G.
Output: A TRDT T_r of G.
1: $D \leftarrow$ APP-TD-SET(G)
2: $T_r \leftarrow (V \setminus D, \emptyset, D)$
3: return T_r.

(TRDT) T_r such that weight 2 is assigned for all vertices in D and weight 0 is assigned for the remaining vertices.

Now, let $T_r = (D', \emptyset, D)$ be the TRDT obtained from the APP-TRDF algorithm. Clearly, every vertex in G is assigned with weight either 2 or 0, T_r gives a TRDF of G and APP-TRDF computes a TRDT T_r of G in polynomial time. Hence, the result follows.

Theorem 13. *The MTRDP in a graph can be approximated with an approximation ratio of* $2(\ln(\Delta - 0.5) + 1.5)$.

Proof. Let D be the TDS from APP-TD-SET algorithm, T_r be the TRDT produced by the APP-TRDF algorithm and W_r be the weight of T_r. Clearly, $W_r = 2|D|$. It is known that $|D| \leq (\ln(\Delta - 0.5) + 1.5)\gamma_t(G)$. Therefore, $W_r \leq 2(\ln(\Delta - 0.5) + 1.5)\gamma_t(G)$. Since $\gamma_t(G) \leq \gamma_{tR}(G)$ [1], it follows that $W_r \leq 2(\ln(\Delta - 0.5) + 1.5)\gamma_{tR}(G)$.

The corollary below follows from Theorem 13.

Corollary 1. *MTRDP* $\in$ *APX for graphs with* $\Delta = O(1)$.

Similar to the Algorithm 2, we propose an APX-AL APP-TDRDF which produces a total double Roman dominating quadruple (TDRDQ).

Algorithm 3. APP-TDRDF(G)

Input: A simple, undirected graph G.
Output: A TDRDQ Q_r of G.
1: $D \leftarrow$ APP-TD-SET(G)
2: $Q_r \leftarrow (V \setminus D, \emptyset, \emptyset, D)$
3: return Q_r.

We also note that the algorithm APP-TDRDF computes a TDRDQ Q_r of a given graph G in polynomial time and the following theorem holds.

Theorem 14. *The MTDRDP in a graph can be approximated with an approximation ratio of* $3(\ln(\Delta - 0.5) + 1.5)$.

Proof. The proof is obtained with similar arguments as in Theorem 13.

The corollary below follows from Theorem 14.

Corollary 2. *MTDRDP* $\in$ *APX for graphs with* $\Delta = O(1)$.

5.3 Approximation Completeness

Here, we prove that the MTRDP and MTDRDP are APX-complete (APXC) for graphs with $\Delta = 5$ using the L-reduction [26]. An optimization problem X is said to be APXC if X belongs to APX and APX-hard classes. By providing an L-reduction from MDS problem with $\Delta = 3$ i.e., DOM-3 which is known to be APXC [6], we show that the MTRDP and MTDRDP belongs to APX-hard for graphs with $\Delta = 5$.

Theorem 15. *MTRDP* $\in$ *APXC for graphs with* $\Delta = 5$.

Proof. From Corollary 1, it is clear that MTRDP is in APX. Given an instance $G = (V, E)$ of DOM-3, where $V = \{v_1, v_2, \ldots, v_n\}$, we construct an instance $G' = (V', E')$ of MTRDP same as in Sect. 5.1. Note that G' is a graph with $\Delta = 5$. First we prove the following claim.

Claim. $\gamma_{tR}(G') = 3n + \gamma(G)$, where $n = |V|$.

Proof. The proof is same as in the Claim in Sect. 5.1.

Let D^* be a MDS of G and $f : V' \rightarrow \{0, 1, 2\}$ be a minimum TRDF of G'. It is known that for any graph $G = (V, E)$ with maximum degree Δ, $\gamma(G) \geq \frac{n}{\Delta+1}$, where $n = |V|$. Thus, $|D^*| \geq \frac{n}{4}$. From the above claim it is evident that $f(V') = |D^*| + 3n \leq |D^*| + 12|D^*| = 13|D^*|$.

Now consider a TRDF $g : V' \rightarrow \{0, 1, 2\}$ of G'. Clearly, the set $D = \{v_i : g(v_i) \geq 1 \text{ or } g(a_i) \geq 1 \text{ or } g(c_i) \geq 1\}$ is a DS of G. Therefore, $|D| \leq g(V') - 3n$. Hence, $|D| - |D^*| \leq g(V') - 3n - |D^*| \leq g(V') - f(V')$. This implies that there exists an L-reduction with $\alpha = 13$ and $\beta = 1$.

Theorem 16. *MTDRDP* $\in$ *APX-complete for graphs with* $\Delta = 5$.

Proof. The proof is obtained with similar arguments as in Theorem 15, in which replace the assigned value 2 with 3. We get an L-reduction with $\alpha = 18$ and $\beta = 1$.

References

1. Abdollahzadeh Ahangar, H., Henning, M.A., Samodivkin, V., Yero, I.G.: Total Roman domination in graphs. Appl. Anal. Discrete Math. **10**, 501–517 (2016). https://doi.org/10.2298/AADM160802017A
2. Abdollahzadeh Ahangar, H., Chellali, M., Sheikholeslami, S.M.: On the double Roman domination in graphs. Discrete Appl. Math. **232**, 1–7 (2017). https://doi.org/10.1016/j.dam.2017.06.014
3. Abdollahzadeh Ahangar, H., Álvarez, M.P., Chellali, M., Sheikholeslami, S.M., Valenzuela-Tripodoro, J.C.: Triple Roman domination in graphs. Appl. Math. Comput. **391**, 125444 (2021). https://doi.org/10.1016/j.amc.2020.125444
4. Abdollahzadeh Ahangar, H., Chellali, M., Sheikholeslami, S.M.: Outer independent double Roman domination. Appl. Math. Comput. **364**, 124617 (2020). https://doi.org/10.1016/j.amc.2019.124617

5. Abdollahzadeh Ahangar, H., Chellali, M., Sheikholeslami, S.M.: Signed double Roman domination in graphs. Discrete Appl. Math. **257**, 1–1 (2019). https://doi.org/10.1016/j.dam.2018.09.009
6. Alimonti, P., Kann, V.: Some APX-completeness results for cubic graphs. Theor. Comput. Sci. **237**, 123–134 (2000). https://doi.org/10.1016/S0304-3975(98)00158-3
7. Amjadi, J., Nazari-Moghaddam, S., Sheikholeslami, S.M.: Total Roman domination number of trees. Australas. J. Combin. **69**, 271–285 (2017)
8. Anu, V., Aparna Lakshmanan, S.: Double Roman domination number. Discrete Appl. Math. **244**, 198–204 (2018). https://doi.org/10.1016/j.dam.2018.03.026
9. Campanelli, N., Kuziak, D.: Total Roman domination in the lexicographic product of graphs. Discrete Appl. Math. **263**, 88–95 (2019). https://doi.org/10.1016/j.dam.2018.06.008
10. Chlebík, M., Chlebíková, J.: Approximation hardness of dominating set problems in bounded degree graphs. Inf. Comput. **206**, 1264–1275 (2008). https://doi.org/10.1016/j.ic.2008.07.003
11. Cockayne, E.J., Dreyer, P.A., Hedetniemi, S.M., Hedetniemi, S.T.: Roman domination in graphs. Discrete Math. **278**, 11–22 (2004). https://doi.org/10.1016/j.disc.2003.06.004
12. Courcelle, B.: The monadic second-order logic of graphs. I. Recognizable sets of finite graphs. Inf. Comput. **85**, 12–75 (1990). https://doi.org/10.1016/0890-5401(90)90043-H
13. Dreyer, P.A.: Applications and variations of domination in graphs. Ph.D. thesis, Rutgers University, The State University of New Jersey, New Brunswick, New Jersey (2000)
14. Favaron, O., Karami, H., Khoeilar, R., Sheikholeslami, S.M.: On the Roman domination number of a graph. Discrete Math. **309**, 3447–3451 (2009). https://doi.org/10.1016/j.disc.2008.09.043
15. Hao, G., Volkmann, L., Mojdeh, D.A.: Total double Roman domination in graphs. Commun. Comb. Optim. **5**, 27–39 (2020). https://doi.org/10.22049/CCO.2019.26484.1118
16. Haynes, T.W., Hedetniemi, S., Slater, P.: Fundamentals of Domination in Graphs. CRC Press, Boca Raton (1998)
17. Henning, M.A., Yeo, A.: Total Domination in Graphs. Springer, Heidelberg (2013). https://doi.org/10.1007/978-1-4614-6525-6
18. Henning, M.: Defending the Roman empire from multiple attacks. Discrete Math. **271**, 101–115 (2003). https://doi.org/10.1016/S0012-365X(03)00040-2
19. Henning, M.A., Hedetniemi, S.T.: Defending the Roman empire–a new strategy. Discrete Math. **266**, 239–251 (2003). https://doi.org/10.1016/S0012-365X(02)00811-7
20. Henning, M.: A characterization of Roman trees. Discuss. Math. Graph Theory **22**, 325–334 (2002). https://doi.org/10.7151/dmgt.1178
21. Liedloff, M., Kloks, T., Liu, J., Peng, S.H.: Roman domination in some special classes of graphs. Report TR-MA-04-01 (2004)
22. Liu, C.H., Chang, G.J.: Roman domination on strongly chordal graphs. J. Comb. Optim. **26**, 608–619 (2013). https://doi.org/10.1007/s10878-012-9482-y
23. Mahadev, N., Peled, U.: Threshold Graphs and Related Topics. Elsevier, North Holland (1995)
24. Padamutham, C., Palagiri, V.S.R.: Complexity of Roman {2}-domination and the double Roman domination in graphs. AKCE Int. J. Graphs Comb. 1–6 (2020). https://doi.org/10.1016/j.akcej.2020.01.005

25. Padamutham, C., Palagiri, V.S.R.: Algorithmic aspects of Roman domination in graphs. J. Appl. Math. Comput. 89–102 (2020). https://doi.org/10.1007/s12190-020-01345-4
26. Papadimitriou, C.H., Yannakakis, M.: Optimization, approximation, and complexity classes. J. Comput. Syst. Sci. **43**, 425–440 (1991). https://doi.org/10.1016/0022-0000(91)90023-X
27. Poureidi, A., Rad, N.J.: Algorithmic and complexity aspects of problems related to total Roman domination for graphs. J. Combin. Optim. **39**(3), 747–763 (2019). https://doi.org/10.1007/s10878-019-00514-x
28. Rad, N.J., Volkmann, L.: Roman domination perfect graphs. An. Stiint. Univ. Ovidius Constanta Ser. Mat. **19**, 167–174 (2019)
29. Röhrig, H.: Tree decomposition: a feasibility study, Masters's thesis, Max-Planck-Institut Für Informatik, Citeseer (1998)
30. Robert, A., Haynes, T.W., Hedetniemi, S.T.: Double Roman domination. Discrete Appl. Math. **211**, 23–29 (2016). https://doi.org/10.1016/j.dam.2016.03.017
31. ReVelle, C.S., Rosing, K.E.: Defendens imperium romanum: a classical problem in military strategy. Am. Math. Mon. **107**, 585–594 (2000). https://doi.org/10.2307/2589113
32. Stewart, I.: Defend the Roman empire!. Sci. Am. **281**, 136–138 (1999). https://doi.org/10.1038/scientificamerican1299-136
33. Shao, Z., Amjadi, J., Sheikholeslami, S., Valinavaz, M.: On the total double Roman domination. IEEE Access **7**, 52035–52041 (2019). https://doi.org/10.1109/ACCESS.2019.2911659
34. Uehara, R., Uno, Y.: Efficient algorithms for the longest path problem. In: Fleischer, R., Trippen, G. (eds.) ISAAC 2004. LNCS, vol. 3341, pp. 871–883. Springer, Heidelberg (2004). https://doi.org/10.1007/978-3-540-30551-4_74
35. West, D.B.: Introduction to Graph Theory, vol. 2. Prentice Hall, Upper Saddle River (2001)
36. Yannakakis, M.: Node-and edge-deletion NP-complete problems. In: Proceedings of the Tenth Annual ACM Symposium on Theory of Computing, pp. 253–264 (1978). https://doi.org/10.1145/800133.804355
37. Zhu, J.: Approximation for minimum total dominating set. In: Proceedings of ICIS 09, Seoul, Korea, 24–26 November 2009, pp. 119–124 (2009). https://doi.org/10.1145/1655925.1655948

Approximation Algorithms for Orthogonal Line Centers

Arun Kumar Das[1(✉)], Sandip Das[1], and Joydeep Mukherjee[2]

[1] Indian Statistical Institute, Kolkata, India
arund426@gmail.com, sandipdas@isical.ac.in
[2] Ramakrishna Mission Vivekananda Educational and Research Institute, Howrah, India
joydeep.m1981@gmail.com

Abstract. k orthogonal line center problem computes a set of k axis-parallel lines for a given set of points in $2D$ such that the maximum among the distance between each point to its nearest line is minimized. A 2-factor approximation algorithm and a $(\frac{7}{4}, \frac{3}{2})$ bi-criteria approximation algorithm is presented for the problem. Both of them are deterministic approximation algorithms, having sub-quadratic running time and not based on linear programming.

1 Introduction

A classical problem in computer science is *data clustering*. One has to group a given set of data points such that every point in the same group is similar with respect to some optimizing criteria. This problem finds application in learning theory, data-mining, spatial range searching, etc. [10]. *k-line center* problem for a given set of points is a type of clustering problem. A set of points and a positive integer k are given as input. A set of k lines needs to be computed such that the maximum among the distances between each point to its nearest line is minimized. These lines are called the *line centers* for the given set of points. Many variants of this problem are well studied due to their enormous applications in the domain of facility location [18, 19], and machine learning [10], etc. In this paper, we study one such variant, where axis-parallel line centers are computed for a given point set in 2-dimension. We call this problem as *k-ORTHOGONAL-LINE-CENTER (kOLC)* problem. Some real-life applications of our problem are designing transport networks, where the tracks are orthogonal to each other, or to design circuit boards where the wires need to be embedded in orthogonal orientations. The $kOLC$ is stated as follows. Here the *distance* between a line and a point is the perpendicular distance between them.

A. Mudgal and C. R. Subramanian (Eds.): CALDAM 2021, LNCS 12601, pp. 43–54, 2021.
https://doi.org/10.1007/978-3-030-67899-9_4

Problem 1. k-ORTHOGONAL-LINE-CENTER (kOLC)
Input: A set $\mathcal{S}$ of n points in 2-dimensional plane and a positive integer k.
Output: A set of k axis-parallel lines such that the maximum among the distances from every point to its nearest line is minimized.

In order to solve *kOLC* we state another problem called *k-ORTHOGONAL-LINE-CENTER-WITH-RADIUS (kOLCR)* problem. In fact Lemma 1 shows that *kOLC* can be solved in polynomial time whenever *kOLCR* can be solved in polynomial time.

Problem 2. k-ORTHOGONAL-LINE-CENTER-WITH-RADIUS (kOLCR)
Input: A set $\mathcal{S}$ of n points in 2-dimensional plane and a positive integer k and a positive real number r.
Output: A set of k axis-parallel lines such that the maximum among the distances from every point to its nearest line is at most r.

We state another problem to establish the hardness of *kOLCR*.

Problem 3. STABBING-AXIS-PARALLEL-SQUARES-OF-SAME-SIZE (SASS)
Input: A set $\mathcal{S}$ of n axis-parallel squares of side length $2r$ and a positive integer k.
Output: A set of k axis-parallel lines such that each square is intersected with at least one line.

Consider the center of the squares (intersection point of two diagonals) and half of the side length of the squares in a given instance of *SASS* as the input point set and input radius to *kOLCR* respectively. Then *SASS* reduces to *kOLCR*. Observe that *kOLCR* has a solution if and only if *SASS* has a solution. *SASS* is known to be **W[1]**-hard [9]. This problem is a special case of a well known **NP**-hard problem of *rectangle stabbing* [12]. In the problem of rectangle stabbing, one has to find a set of lines for a given set of rectangles such that each rectangle is intersected by at least one line. This problem is well studied due to its applications in data analysis [8], sensor networks [14], radiotherapy [17] etc. The best known constant factor approximation algorithm for this problem is designed by Gaur et al [12], which uses linear programming (LP) to solve the problem. Dom et al. [9] presented a $(4k+1)^k n^{O(1)}$ time fixed-parameter tractable (FPT) algorithm for stabbing axis-parallel disjoint squares of the same size in the same paper in which they proved the **W**[1]-hardness of *SASS*.

Agarwal and Procopiuc [1] studied the problem of covering a point set in $\mathbb{R}^d$ with k cylinders. For $d = 2$, they designed an $O(nk^2 \log^4 n)$ expected time randomized algorithm, when $k^2 \log k < n$ to compute $O(k \log k)$ strips of width at most w^* that cover the given set of points. Here w^* is the optimal radius of the cylinders for the problem. But the expected running time is $O(n^{\frac{2}{3}}k^{\frac{8}{3}}\log^4 n)$ for higher values of k. The expected time is $O(n^{\frac{3}{2}}k^{\frac{9}{4}}polylog(n))$, for $d=3$. They also presented an $O(dnk^3\log^4 n)$ expected time randomized algorithm to compute a set of $O(dk\log k)$ d-cylinders of diameter at most $8w^*$ for points in $\mathbb{R}^d$. Aggarwal et al. [3] designed an $O(n\log n)$ expected time randomized algorithm to find k cylinders of radius $(1+\varepsilon)w^*$ that cover a set of n given points in $\mathbb{R}^d$. The constant of proportionality depends on k, d and ε. Practical implementation and heuristics for these problems are also well-studied [4–6,16]. Jaromczyk and Kowaluk [13] presented a $O(n^2\log^2 n)$ time algorithm for a special case like computing 2 line center. An $O(n(\log n+\varepsilon^{-2}\log\frac{1}{\varepsilon})+\varepsilon^{-\frac{7}{2}}\log\frac{1}{\varepsilon})$ time $(1+\varepsilon)$ approximation scheme for computing 2 line center is designed by Agarwal et al. [2]. Feldman et al. [10] presented a randomized linear time bi-criteria approximation for generalized k-center, mean and median problems.

One can devise a 2-factor approximation algorithm for solving *kOLC* by using the LP-based approach of Gaur et al. [12]. In this paper, we present a 2-factor approximation algorithm for *kOLC*, which is not LP-based. Next, we present a deterministic bi-criteria approximation algorithm for *kOLC*, based on our 2-factor approximation algorithm. To the best of our knowledge, the only deterministic bi-criteria approximation algorithm known for *kOLC* is a $(\frac{3}{2}, 16)$ bi-criteria approximation by Chakraborty et al. [7], which is based on local search technique.

The paper is organized in the following manner. Section 2 consists of some definitions and the lemma showing the relation between the running times of *kOLC* and *kOLCR*. We devise a 2-factor approximation algorithm for *kOLC* in Sect. 3. A $(\frac{7}{4}, \frac{3}{2})$ bi-criteria approximation for the same is described in Sect. 4.

2 Preliminaries

A set of axis-parallel lines is said to be a *set of line centers* for a set of points with *radius* r, whenever the maximum among the distances between each point to its nearest line is at most r. Let $\mathcal{C}$ be a set of line centers for $\mathcal{S}$ with radius r. For a point p of $\mathcal{S}$, $c(p)$ denotes the nearest member of $\mathcal{C}$ from p. We say that this line center is *assigned* to the point p and p is *served* by this line center. Thus *Client set of a line center*, l, where $l \in \mathcal{C}$, denoted as $s(l)$, is defined as $\{p|p \in \mathcal{S} \text{ and } c(p)=l\}$.

A set of line centers is called an (α, β) bi-criteria approximate solution for *kOLC* if the cardinality of the set is at most αk and the radius is at most β times the optimal radius.

We show that the existence of a polynomial-time algorithm for *kOLCR* implies the existence of a polynomial-time solution for *kOLC*.

Lemma 1. *kOLC can be solved in* $O((T+n)\log n)$ *time, if kOLCR can be solved in* $O(T)$ *time.*

Proof. Let r^* be the maximum among all the distances between the points and the line centers in the optimal solution of *kOLC*. Consider a line center l^* in the optimal solution of $kOLC$, such that the distance between the farthest point from l in $s(l)$ is r^*. Then there must be two such points in $s(l)$ such that they are at a distance r^* from l^*. Otherwise, we can translate l^* towards the farthest one to reduce the distance. Thus the optimal radius is determined by two points in the given set $\mathcal{S}$. Since $|\mathcal{S}| = n$, we have $O(n^2)$ possible candidates for the optimal radius. We can perform a binary search on them to get the optimal solution for $kOLC$ by using the algorithm for $kOLCR$. We use the technique by Frederickson and Johnson [11] to determine the median element in the binary search which requires $O(m \log(\frac{2n}{m}))$ time for a $n \times m$ sorted matrix. Since our matrix is of order $n \times n$ the running time is $O(n)$ in our case. Thus the lemma holds. □

An (i, j, t)-*grid* is a set of i vertical and j horizontal lines where the distance between two consecutive lines of same orientation is t. We call it simply a grid and denote as W. Let l be a vertical line in a grid W and the x-coordinate of every point on l be x_l. A *right-shift* operation on l by an amount ξ produces a new vertical line $R(l)$ such that the x-coordinate of every point on $R(l)$ is $l_x + \xi$. Similarly a right-shift operation on an (i, j, t)-grid W produces another (i, j, t)-grid $R(W)$ where every vertical line of $R(W)$ is produced by performing right-shift operation on the vertical lines of W. Similarly we define a *left-shift* operation by an amount ξ on a vertical line l (or a grid W) to produce a new line $L(l)$ (or a grid $L(W)$). We also define *up-shift* and *down-shift* operations on a horizontal line l as well as on a grid W by translating the horizontal lines vertically upward and downward respectively. The newly produced lines are denoted as $U(l)$ and $D(l)$ respectively and the newly produced grids are denoted as $U(W)$ and $D(W)$ respectively. Furthermore we define the composition of two shift operations. They are performed one after another on a grid. So an *up-right-shift* on a grid W produces a new grid $UR(W)$, where $UR(W)$ is obtained by performing an up-shift operation on $R(W)$. Similarly we define the other combinations. Clearly these combinations are commutative and $DU(W) = UD(W) = W$ and $RL(W) = LR(W) = W$.

In Sect. 3 we design an algorithm that produces a 2-factor approximation for the $kOLC$ problem in terms of the number of centers. The algorithm chooses a subset of the grid lines which serve as line centers for the given set of points. Let the difference between the x-coordinates of the leftmost and rightmost points of $\mathcal{S}$ be w and the difference between the y-coordinates of the topmost and bottommost points of $\mathcal{S}$ be h.

3 A 2-factor Approximation Algorithm

We begin with an instance of $kOLCR$, where the radius r is also given as input along with all other input of $kOLC$. We construct a $(\lceil \frac{w}{2r} \rceil, \lceil \frac{h}{2r} \rceil, 2r)$-grid W. The leftmost vertical line of the grid passes through the leftmost member of $\mathcal{S}$ and the topmost horizontal line passes through the topmost member of $\mathcal{S}$. Then every point in $\mathcal{S}$ lies between two horizontal and two vertical lines which are $2r$ distance apart from each other. This fact follows from the construction of the

$(\lceil \frac{w}{2r} \rceil, \lceil \frac{h}{2r} \rceil, 2r)$-grid W. Thus each of the points is within a distance r from at least one line of both orientations. Only the points which are equidistant from both lines of the same orientation are within a distance r from two lines of the same orientation. But a point can not be within a distance r from more than 2 lines of the same orientation as the lines are $2r$ distance apart from each other. We state this fact as an observation.

Observation 1. *Every point in $\mathcal{S}$ is within a distance r from at least one vertical line and at least one horizontal line of the grid W. A point in $\mathcal{S}$ can be exactly at a distance r from at most two horizontal lines and at most two vertical lines.*

Our algorithm finds a subset of the grid W (ie. a set of lines from the grid W) and returns that as a set of axis-parallel line centers for $\mathcal{S}$ with radius r. So for a line l, where $l \in W$, we define the client set of l in W as the set of points in $\mathcal{S}$ such that they are within a distance r from l. We denote it by $s_W(l)$. It follows from Lemma 1 that each point is assigned to at least one line of one orientation. But in our algorithm we do not assign a point to more than one line of the same orientation. In other words we want $s_W(l_i) \cap s_W(l_j) = \phi$, for every l_i and l_j of the same orientation, i.e. both of them are either vertical or horizontal in W. It follows from the definition of $s_W(l)$ that the points which are equidistant from two lines of same orientation in W, belong to the client sets of both the lines. We break the tie in the following manner.

Let Q be the set of points in $\mathcal{S}$ such that every member in Q is at a distance r from two vertical lines of the grid W. We allot every points in Q in the client set of the line which is serving more points other than the points in Q. In case of tie we choose arbitrarily but we allot every member of Q to the same line. Similarly we allot points of $\mathcal{S}$ which are equidistant from two horizontal lines. Thus a point only belongs to client sets of one vertical and one horizontal line, after this assignment.

Now we construct a graph with $\lceil \frac{w}{2r} \rceil + \lceil \frac{h}{2r} \rceil$ vertices such that each vertex represents a unique line of W. Let l^v denote the corresponding line of W for a vertex v of the graph. Two vertices u and v have an edge between them if and only if they satisfy the following conditions:

1. l^u and l^v are of different orientations in W. In other words one of them is vertical and the other one is horizontal.
2. $s_W(l^u) \cap s_W(l^v) \neq \phi$, where $s_W(l^u)$ and $s_W(l^v)$ are the client set of the lines of W, corresponding to u and v respectively.

We call this graph as the graph of W and denote it by $G(W)$.

Lemma 2. *The graph $G(W)$, defined above, is a bipartite graph.*

Proof. The edges of $G(W)$ are given between two vertices u and v if they correspond to two lines of W in different orientations. We can divide the set of vertices of $G(W)$ into two partitions, namely, the vertices corresponding to the vertical lines of W and the vertices corresponding to the horizontal lines of W, with no edge between two vertices in the same partition. Thus the graph $G(W)$ is a bipartite graph. □

Now we show a relation between the vertex cover of $G(W)$ and a set of line centers, chosen from W, for $\mathcal{S}$. The following lemma states the relation, which helps us to choose a set of line centers for $\mathcal{S}$ with radius r from the lines present in W.

Lemma 3. *A subset C of W is a set of line centers for $\mathcal{S}$ with radius r, if and only if the vertices in $G(W)$, corresponding to the lines in C, form a vertex cover of $G(W)$.*

Proof. Every point of $\mathcal{S}$ is assigned to a unique set $s_W(l)$, where $l \in W$, in one orientation. So exactly one horizontal and exactly one vertical line of W can serve a point p, where $p \in \mathcal{S}$, as a center. This fact together with the structure of the graph $G(W)$ imply that each point can be assigned to a unique edge in $G(W)$. Let p^e denote the edge corresponding to a point p, where $p \in \mathcal{S}$.

Let C' a vertex cover of $G(W)$. Then p^e must have an end vertex, v(say), in C'. Then from the structure of $G(W)$ it follows that $p \in s_W(l^v)$. This fact holds for all the points $p \in \mathcal{S}$. Hence the lines in W corresponding to the vertices in C' is a set of line centers with radius r.

Conversely let C be a set of line centers for $\mathcal{S}$ with radius r where $C \subseteq W$. Consider an edge e of $G(W)$. There exists a point p in $\mathcal{S}$ such that $e = p^e$. Now C being a set of line centers with radius r, a line of W must be present in C such that $p \in s_W(l)$. In other words, one end vertex of e is present in the set of vertices corresponding to the lines in C. Since this fact holds for every edge of $G(W)$, the vertices in $G(W)$ corresponding to the lines of C in W form a vertex cover of $G(W)$. Thus the lemma holds. □

Now we wish to choose a set of lines from the grid W to serve as centers for $\mathcal{S}$ with radius r. We establish a relation between the lines in an optimal solution to a given instance of problem $kOLCR$ and the line centers that can be chosen from the grid W for this instance. Let $\mathcal{C}^*$ be an optimal solution consisting k orthogonal line centers for $\mathcal{S}$. The following lemma states the relation between the lines in $\mathcal{C}^*$ and a set of lines that can be chosen from W.

Lemma 4. *Let l^* be any line belonging to $\mathcal{C}^*$. At most 2 lines of W are sufficient to serve all the points in $s(l^*)$ as centers with radius r, where these lines of W have the same orientation as l^*. The vertices of $G(W)$, corresponding to these lines also form a vertex cover.*

Proof. Without loss of generality let l^* has vertical orientation. If it coincides with a vertical line of W then that line is sufficient to serve all members of $s(l^*)$ as centers with radius r. Now we assume that l^* lies between two consecutive vertical lines of the grid W, say, l_i and l_{i+1}. Let p be a point of $\mathcal{S}$, such that $p \in s(l^*)$. Since p lies within a distance r from l^*, p must belong to either $s_W(l_i)$ or $s_W(l_{i+1})$. This imply $s(l^*) \subseteq s_W(l_i) \cup s_W(l_{i+1})$. The property of forming a vertex cover follows from Lemma 3. Thus the lemma holds. □

Now we are in a position to devise an approximation algorithm to compute k axis-parallel line centers with radius r for $\mathcal{S}$. We compute the grid W as mentioned above and compute the minimum vertex cover of the graph $G(W)$.

It follows from Lemma 3 that a minimum set of lines of W, which can serve $\mathcal{S}$ with radius r, form a minimum vertex cover of $G(W)$. Furthermore this minimum vertex cover is of size at most $2k$ if there exists a solution for k*OLCR*, by Lemma 4. We check the cardinality of the minimum vertex cover of $G(W)$. For the cardinality more than $2k$, there does not exist a set of orthogonal line centers with radius r for $\mathcal{S}$, which has cardinality at most k. In that case, the algorithm returns "NOT POSSIBLE". Otherwise, it returns the lines of W which corresponds to the vertices in the minimum vertex cover of $G(W)$ as our output. Algorithm 1 describes the steps to compute a solution for problem k*OLCR*. The following lemma states the correctness of the algorithm as a 2-factor approximation for k*OLCR*.

```
Input: S, k, r
Output: A set of axis parallel lines C
Construct the (⌈w/2r⌉, ⌈h/2r⌉, 2r)-grid W and the graph G(W);
Compute the minimum vertex cover C' of the graph G(W), using the
algorithm described in [15] ;
if |C'| ≤ 2k then
  | C ← The lines of W which correspond to the vertices in C' ;
  | return C
else
  | return "NOT POSSIBLE";
end
```

Algorithm 1: 2-FACTOR-kOLCR()

Lemma 5. *Algorithm 1 returns a 2-factor approximation of the problem* k*OLCR.*

Proof. The proof follows from combining Lemma 3 and Lemma 4. □

Now we analyze the running time of Algorithm 1. Although there are $\lceil \frac{w+h}{2r} \rceil$ grid lines, we compute only those grid lines which serve at least one point of $\mathcal{S}$. Thus we only deal with $O(n)$ lines of W. Also note that the number of edges in the graph is also $O(n)$, since each member of $\mathcal{S}$ corresponds to at most one edge. We can determine the graph by spending a constant amount of time with each member of $\mathcal{S}$. This takes $O(n)$ time as well as $O(n)$ space. Then using the algorithm described in [15] we can compute the minimum vertex cover in $O(n^{\frac{3}{2}})$ time and $O(n)$ space. The rest steps of the algorithm can be done in constant time. So the overall running time of the algorithm is $O(n^{\frac{3}{2}})$. Hence we conclude the following theorem.

Theorem 1. *A 2-factor approximate solution for* k*OLC, described in Problem 1 for a given set with* n *points, can be computed in* $O(n^{\frac{3}{2}} \log n)$ *time.*

Proof. The proof follows from the above discussion and Lemma 1. □

In the following section, we design a bi-criteria approximation algorithm using the shifting of the grid W, as defined in Sect. 2.

4 A $(\frac{7}{4}, \frac{3}{2})$ bi-criteria approximation algorithm

We design a bi-criteria approximation algorithm in this section. We allow the radius of the centers to be within a constant factor of the given radius to achieve a tighter approximation factor in terms of the number of centers. We allow the radius of this new set of line centers to be $(r + \frac{r}{2})$, ie. $\frac{3r}{2}$ instead of r.

We begin with an instance of $kOLCR$ as before. We initially construct a $(\lceil\frac{2w}{3r}\rceil, \lceil\frac{2h}{3r}\rceil, 3r)$-grid, similar to the grid constructed in Sect. 3. The distance between two lines of the same orientation is $3r$ instead of $2r$ like the previous one. We denote this grid by W from now for notational simplicity. Clearly, these grid lines can serve the members of $\mathcal{S}$ as centers with radius $\frac{3r}{2}$. Then we construct 8 more $(\lceil\frac{2w}{3r}\rceil, \lceil\frac{2h}{3r}\rceil, 3r)$-grids by performing the shift operations with amount r on W, defined in Sect. 2. We construct the corresponding bipartite graphs and check the cardinality of the minimum vertex covers. We return the lines corresponding to the vertex cover, which has the minimum cardinality among them, as output. We establish that this solution is a $(\frac{7}{4}, \frac{3}{2})$ bi-criteria approximation to $kOLCR$.

We use similar arguments as in Lemma 4 and conclude that at most two lines of W (as well as the other grids constructed by shift operations on W) are sufficient to serve the client set $s(l^*)$ of a line center l^*, present in the optimal solution $\mathcal{C}^*$ of the given instance of $kOLCR$. Now we wish to reduce this number by locating a set of lines $\mathcal{L}$, where $\mathcal{L} \subseteq \mathcal{C}^*$, such that only one line from a grid is sufficient to serve $s(l^*)$, for every $l^* \in \mathcal{L}$. We show that $|\mathcal{L}| \geq \frac{|\mathcal{C}^*|}{4}$ for at least one grid among all the nine grids we constructed. Note that the line centers in the optimal solution to the problem serve the points of $\mathcal{S}$ with radius r and the lines of the grids serve them with radius $\frac{3r}{2}$.

Now we analyze the changes in the client sets of the vertical lines when we perform a right-shift operation on W to produce $R(W)$. Note that the client sets of the horizontal lines remain the same for W and $R(W)$. Let $\mathcal{C}_v^*$ and $\mathcal{C}_h^*$ denote the set of vertical and horizontal line centers present in $\mathcal{C}^*$ respectively. The following cases can happen in the shifted grid $R(W)$.

1. $s(l^*)$, which was served by a single vertical line in W, is still being served by a single vertical line in $R(W)$. We denote the set of these lines, present in $\mathcal{C}^*$, as $\mathcal{C}_1$.
2. $s(l^*)$, which was served by a single vertical line in W, needs two vertical lines in $R(W)$ to be served. We denote the set of these lines, present in $\mathcal{C}^*$, as $\mathcal{C}_2$.
3. $s(l^*)$, which was served by two vertical lines in W, can be served with by a single vertical line in $R(W)$. We denote the set of these lines, present in $\mathcal{C}^*$, as $\mathcal{C}_3$.
4. $s(l^*)$, which was served by two vertical lines in W, still requires two vertical lines in $R(W)$ to be served. We denote the set of these lines, present in $\mathcal{C}^*$, as $\mathcal{C}_4$.

Since $\mathcal{C}_1 \cup \mathcal{C}_2 \cup \mathcal{C}_3 \cup \mathcal{C}_4 = \mathcal{C}_v^*$, we can conclude that the cardinality of at least one of these four sets has at least one-fourth of the cardinality of $\mathcal{C}_v^*$. Utilizing this fact we show in Lemma 6 that the client set of the vertical line centers of the optimal solution can be served with at most $\frac{7}{4}|\mathcal{C}_v^*|$ vertical lines from one of the grids among W, $R(W)$ and $L(W)$.

Lemma 6. *There is at least one grid among* W*,* $R(W)$ *and* $L(W)$ *such that at most* $\frac{7}{4}|\mathcal{C}_v^*|$ *vertical lines from that grid is sufficient to serve all the points in the client set* $\cup_{l^* \in \mathcal{C}_v^*} s(l^*)$*.*

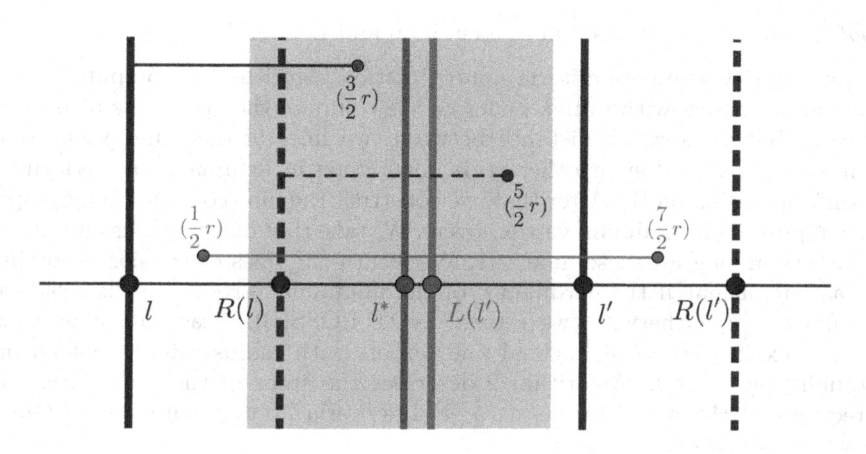

Fig. 1. Proof of Lemma 6

Proof. We prove this lemma by considering the cardinalities of the sets $\mathcal{C}_1$, $\mathcal{C}_2$, $\mathcal{C}_3$ and $\mathcal{C}_4$. The lemma holds for the grid W, if either $|\mathcal{C}_1| \geq \frac{|\mathcal{C}_v^*|}{4}$ or $|\mathcal{C}_2| \geq \frac{|\mathcal{C}_v^*|}{4}$. The lemma holds for the grid $R(W)$, if $|\mathcal{C}_3| \geq \frac{|\mathcal{C}_v^*|}{4}$. The only case left to consider is when $|\mathcal{C}_4| \geq \frac{|\mathcal{C}_v^*|}{4}$ and $|\mathcal{C}_1| < \frac{|\mathcal{C}_v^*|}{4}$, $|\mathcal{C}_2| < \frac{|\mathcal{C}_v^*|}{4}$, $|\mathcal{C}_3| < \frac{|\mathcal{C}_v^*|}{4}$.

Let l and l' be two vertical lines of W, which are necessary to serve $s(l^*)$, for some $l^* \in \mathcal{C}_4$, where l lies on the left of l^*. Let $R(l)$ and $R(l')$ be the lines of $R(W)$, which are necessary to serve $s(l^*)$. This implies that l^* is not lying within a distance $\frac{3}{2}r$ from l. Since the distance between l and l' is $3r$, l^* is lying within a distance $\frac{3}{2}r$ form l'. This implies that $L(l')$ can serve all members of $s(l^*)$. Then the lemma holds for the grid $L(W)$. The scenario is depicted in Fig. 1. □

We argue similarly and we can state the following lemma which states about serving all the points which are served by the members of $\mathcal{C}_h^*$.

Lemma 7. *There is at least one grid among* W*,* $U(W)$ *and* $D(W)$ *such that at most* $\frac{7}{4}|\mathcal{C}_h^*|$ *horizontal lines from that grid is sufficient to serve all the points in the client set* $\cup_{l^* \in \mathcal{C}_h^*} s(l^*)$*.*

Proof. Proof follows from similar arguments in Lemma 6. □

Now consider all the possible shifting operations on W which consist of left, right, up, and down shifts as defined in Sect. 2. The following lemma states that we can locate a grid among them where at most $\frac{7}{4}k$ lines are sufficient to serve all the points as line centers with radius $\frac{3}{2}r$.

Lemma 8. *There is at least one grid among W, $U(W)$, $D(W)$, $R(W)$, $L(W)$, $LU(W)$, $RU(W)$, $LD(W)$, $RD(W)$, such that at most $\frac{7}{4}k$ lines from that grid is sufficient to serve all the points in the client set $\cup_{l^* \in C^*} s(l^*)$.*

Proof. Proof follows by combining Lemma 6 and Lemma 7. □

Now we devise our bi-criteria approximation algorithm to compute k axis-parallel line centers with radius $\frac{3}{2}r$ for $\mathcal{S}$. We compute the grid W as mentioned in Sect. 3 but we keep $3r$ distance between two lines of the same orientation. Then we compute all eight other grids, mentioned in Lemma 8, by performing the shift operation on W. After that, we construct the nine corresponding graphs and compute their minimum vertex covers. We take that one which has minimum cardinality among all these nine sets. We return the lines of the corresponding grid as our output if the cardinality of the minimum vertex cover is less than or equal to $\frac{7}{4}k$. Otherwise, we return "NOT POSSIBLE" as output as there does not exist a set of orthogonal line centers with radius r for $\mathcal{S}$, which has cardinality at most k. Algorithm 2 describes the steps of the procedure. The correctness of the algorithm as a $(\frac{7}{4}, \frac{3}{2})$ bi-criteria approximation for $kOLCR$ follows from Lemma 8.

Input: $\mathcal{S}, k, r$
Output: A set of lines C
Construct the $(\lceil \frac{w}{3r} \rceil, \lceil \frac{h}{3r} \rceil, 3r)$-grids W, $R(W)$, $L(W)$, $U(W)$, $D(W)$, $RU(W)$, $RD(W)$, $LU(W)$, $LD(W)$;
Construct the graphs $G(W)$, $G(R(W))$, $G(L(W))$, $G(U(W))$, $G(D(W))$, $G(RU(W))$, $G(RD(W))$, $G(LU(W))$, $G(LD(W))$ and store them in an array $\mathcal{G}$;
$\mathcal{H}$ is an array to store the vertex covers ;
for *each member* $\mathcal{G}[i]$ *of* $\mathcal{G}$ **do**
| $\mathcal{H}[i] \leftarrow$ Minimum vertex cover of $\mathcal{G}[i]$
end
$C \leftarrow$ the lines corresponding to the member of $\mathcal{H}$ with the minimum cardinality ;
if $|C| \leq \frac{7}{4}k$ **then**
| **return** C ;
else
| **return** "NOT POSSIBLE" ;
end

Algorithm 2: BI-CRITERIA-kOLCR()

The running time analysis of Algorithm 2 is similar to Algorithm 1. We compute the graphs and their vertex covers 9 times in the same way. So the running time is still $O(n^{\frac{3}{2}})$. Thus we conclude the following theorem.

Theorem 2. *A* $(\frac{7}{4}, \frac{3}{2})$ *bi-criteria approximation algorithm for* $kOLC$ *problem can be computed in* $O(n^{\frac{3}{2}} \log n)$ *time.*

Proof. Proof follows from the above discussion combining with Lemma 1. □

References

1. Agarwal, P.K., Procopiuc, C.M.: Approximation algorithms for projective clustering. J. Algorithms **46**(2), 115–139 (2003)
2. Agarwal, P.K., Procopiuc, C.M., Varadarajan, K.R.: A (1+ ε)-approximation algorithm for 2-line-center. Comput. Geomet. **26**(2), 119–128 (2003)
3. Agarwal, P.K., Procopiuc, C.M., Varadarajan, K.R.: Approximation algorithms for a k-line center. Algorithmica **42**(3), 221–230 (2005)
4. Aggarwal, C.C., Wolf, J.L., Yu, P.S., Procopiuc, C., Park, J.S.: Fast algorithms for projected clustering, p. 61–72. Association for Computing Machinery, New York, June 1999
5. Aggarwal, C.C., Yu, P.S.: Finding generalized projected clusters in high dimensional spaces. In: Proceedings of the 2000 ACM SIGMOD International Conference on Management of Data, SIGMOD 2000, pp. 70–81. Association for Computing Machinery, New York (2000)
6. Agrawal, R., Gehrke, J., Gunopulos, D., Raghavan, P.: Automatic subspace clustering of high dimensional data for data mining applications. ACM SIGMOD Rec. **27**(2), 94–105 (1998)
7. Chakraborty, B., Das, A.K., Das, S., Mukherjee, J.: Approximating k-orthogonal line center. In: Wu, W., Zhang, Z. (eds.) COCOA 2020. LNCS, vol. 12577, pp. 47–60. Springer, Cham (2020). https://doi.org/10.1007/978-3-030-64843-5_4
8. Călinescu, G., Dumitrescu, A., Karloff, H., Wan, P.J.: Separating points by axis-parallel lines. Int. J. Comput. Geomet. Appl. **15**(06), 575–590 (2005)
9. Dom, M., Fellows, M.R., Rosamond, F.A.: Parameterized complexity of stabbing rectangles and squares in the plane. In: Das, S., Uehara, R. (eds.) WALCOM 2009. LNCS, vol. 5431, pp. 298–309. Springer, Heidelberg (2009). https://doi.org/10.1007/978-3-642-00202-1_26
10. Feldman, D., Fiat, A., Sharir, M., Segev, D.: Bi-criteria linear-time approximations for generalized k-mean/median/center. In: Proceedings of the Twenty-Third Annual Symposium on Computational Geometry, SCG 2007, pp. 19–26. Association for Computing Machinery, New York (2007)
11. Frederickson, G.N., Johnson, D.B.: Generalized selection and ranking: sorted matrices. SIAM J. Comput. **13**(1), 14–30 (1984)
12. Gaur, D.R., Ibaraki, T., Krishnamurti, R.: Constant ratio approximation algorithms for the rectangle stabbing problem and the rectilinear partitioning problem. J. Algorithms **43**(1), 138–152 (2002)
13. Jaromczyk, J.W., Kowaluk, M.: The two-line center problem from a polar view: a new algorithm and data structure. In: Akl, S.G., Dehne, F., Sack, J.-R., Santoro, N. (eds.) WADS 1995. LNCS, vol. 955, pp. 13–25. Springer, Heidelberg (1995). https://doi.org/10.1007/3-540-60220-8_47
14. Koushanfar, F., Slijepcevic, S., Potkonjak, M., Sangiovanni-Vincentelli, A.: Error-tolerant multimodal sensor fusion. In: IEEE CAS Workshop on Wireless Communication and Networking, pp. 5–6 (2002)

15. Lovász, L., Plummer, M.D.: Matching Theory, vol. 367. American Mathematical Soc. (2009)
16. Procopiuc, C.M., Jones, M., Agarwal, P.K., Murali, T.M.: A Monte Carlo algorithm for fast projective clustering. In: Proceedings of the 2002 ACM SIGMOD International Conference on Management of Data, SIGMOD 2002, pp. 418–427. Association for Computing Machinery, New York (2002)
17. Renner, W.D., Pugh, N.O., Ross, D.B., Berg, R.E., Hall, D.C.: An algorithm for planning stereotactic brain implants. Int. J. Radiat. Oncol. Biol. Phys. **13**(4), 631–637 (1987)
18. Tansel, B.C., Francis, R.L., Lowe, T.J.: State of the art-location on networks: a survey. Part I: the p-center and p-median problems. Manage. Sci. **29**(4), 482–497 (1983)
19. Zanjirani Farahani, R., Hekmatfar, M.: Facility Location: Concepts, Models, Algorithms and Case Studies. Springer, Heidelberg (2009). https://doi.org/10.1007/978-3-7908-2151-2

Semitotal Domination on AT-Free Graphs and Circle Graphs

Ton Kloks[1] and Arti Pandey[2(✉)]

[1] Department of Computer Science, National Tsing Hua University, Hsinchu, Taiwan
klokston@gmail.com

[2] Department of Mathematics, Indian Institute of Technology Ropar, Nangal Road, Rupnagar 140001, Punjab, India
arti@iitrpr.ac.in

Abstract. For a graph $G = (V, E)$ with no isolated vertices, a set $D \subseteq V$ is called a semitotal dominating set of G if (i) D is a dominating set of G, and (ii) every vertex in D has another vertex in D at a distance at most two. The minimum cardinality of a semitotal dominating set of G is called the semitotoal domination number of G, and is denoted by $\gamma_{t2}(G)$. The Minimum Semitotal Domination problem is to find a semitotal dominating set of G of cardinality $\gamma_{t2}(G)$. In this paper, we present some algorithmic results on Semitotal Domination. We show that the decision version of the Minimum Semitotal Domination problem is NP-complete for circle graphs. On the positive side, we show that the Minimum Semitotal Domination problem is polynomial-time solvable for AT-free graphs. We also prove that the Minimum Semitotal Domination for AT-free graphs can be approximated within approximation ratio of 3 in linear-time. Our results answer the open questions posed by Galby et al. in their recent paper.

Keywords: Domination · Semitotal domination · AT-free graphs · Circle graphs · Graph algorithms · NP-completeness · Approximation algorithm

1 Introduction

For a graph $G = (V, E)$, a *dominating set* in G is a set $D \subseteq V$ such that every vertex in $V \setminus D$ is adjacent to at least one vertex in D. The *domination number* of G, denoted by $\gamma(G)$, is the minimum cardinality of a dominating set of G. More details about the domination problem can be found in the books [1,2]. An important variation of domination is connected domination, see [3]. The connected domination has various applications in wireless networks. A dominating set D is called a *connected dominating set* of G if $G[D]$ is connected. The cardinality of a minimum connected dominating set of G is called the *connected domination number* of G, and is denoted by $\gamma_c(G)$.

A *total dominating set*, of a graph G with no isolated vertex is a set S of vertices of G such that every vertex in G is adjacent to at least one vertex

A. Mudgal and C. R. Subramanian (Eds.): CALDAM 2021, LNCS 12601, pp. 55–65, 2021.
https://doi.org/10.1007/978-3-030-67899-9_5

in S. The *total domination number* of G, denoted by $\gamma_t(G)$, is the minimum cardinality of a total dominating set of G. Total domination is well studied problem in graph theory. The literature on the subject of total domination in graphs has been surveyed and detailed in the recent book [4]. A survey of total domination in graphs can also be found in [5].

A relaxed form of total domination called semitotal domination was introduced by Goddard, Henning and McPillan [6], and studied further in [7–11] and elsewhere. A set D of vertices in a graph G with no isolated vertices is a *semitotal dominating set* of G if D is a dominating set of G and every vertex in D is within distance 2 of another vertex of D. The *semitotal domination number* of G, denoted by $\gamma_{t2}(G)$, is the minimum cardinality of a semitotal dominating set of G. Since every total dominating set is a semitotal dominating set, and every semitotal dominating set is a dominating set, for every graph G with no isolated vertex, $\gamma(G) \leq \gamma_{t2}(G) \leq \gamma_t(G)$. Therefore, the semitotal domination number is squeezed between the two most important domination parameters, namely the domination number and the total domination number. However as remarked in [9], the semitotal domination number behaves very differently to both the domination and total domination number. For example, the total domination number is generally incomparable with the matching number, while the semitotal domination is comparable with the matching number and is bounded by at most the matching number plus one (see [7]). The PhD thesis by A. Marcon is entirely on this topic of semitotal domination in graphs, and provides further motivation and importance for its study.

The MINIMUM DOMINATION problem is to find a dominating set of cardinality $\gamma(G)$. Given a graph G and an integer k, the DOMINATION DECISION problem is to determine whether G has a dominating set of cardinality at most k. The MINIMUM TOTAL DOMINATION problem is to find a total dominating set of cardinality $\gamma_t(G)$. The MINIMUM SEMITOTAL DOMINATION problem (MSDP) is to find a semitotal dominating set of minimum cardinality. The SEMITOTAL DOMINATION DECISION problem (SDDP) is the decision version of the MINIMUM SEMITOTAL DOMINATION problem.

Goddard et al. [6] initiated the algorithmic study of the MINIMUM SEMITOTAL DOMINATION problem. They proved that the SEMITOTAL DOMINATION DECISION problem is NP-complete for general graphs [6], and they proposed a linear-time algorithm to find a minimum cardinality semitotal dominating set in trees [6]. Henning et al. [11] proved that the SEMITOTAL DOMINATION DECISION problem remains NP-complete for split graphs, planar graphs and chordal bipartite graphs. In addition, they proposed an $O(n^2)$-time algorithm to compute a minimum cardinality semitotal dominating set in interval graphs. Galby et al. [12] proposed polynomial-time algorithm for graphs of bounded mim-width. They also proved that it is NP-complete to recognise the graphs such that $\gamma_{t2}(G) = \gamma_t(G)$, even if restricted to be planar graphs with maximum degree 4. Galby et al. [12] also posed the problem of determining the complexity status of the MINIMUM SEMITOTAL DOMINATION problem for AT-free graphs, circle graphs, dually chordal graphs and tolerance graphs as open problems. In this paper, we resolve the complexity status of the problem for AT-free graphs and circle graphs. We propose a polynomial-time algorithm to compute a minimum cardinality semitotal dominating set in AT-free graphs, and we prove that the SEMITOTAL DOMINATION DECISION problem is NP-complete for circle graphs.

The main contributions of the paper are summarized below. In Sect. 2, we discuss some pertinent definitions. In Sect. 3, we show that there exists a linear-time 3-approximation algorithm for the MINIMUM SEMITOTAL DOMINATION problem in AT-free graphs. In Sect. 4, we propose a polynomial-time algorithm for the MINIMUM SEMITOTAL DOMINATION problem in AT-free graphs. In Sect. 5, we show that the SEMITOTAL DOMINATION DECISION problem is NP-complete for circle graphs. Finally, Sect. 6, concludes the paper.

2 Preliminaries

Let $G = (V, E)$ be a graph. For a vertex $v \in V$, $N_G(v) = \{u \in V \mid uv \in E\}$ and $N_G[v] = N_G(v) \cup \{v\}$ denote the *open neighborhood* and the *closed neighborhood* of v, respectively. For a vertex v, *degree* of v is $|N_G(v)|$, and is denoted by $d_G(v)$. A vertex of degree one is called a *pendant vertex* and a vertex of degree zero is called an *isolated vertex*. For a set $S \subseteq V$, the sets $N_G(S) = \bigcup_{u \in S} N_G(u)$ and $N_G[S] = N_G(S) \cup S$ are called *open neighborhood* and the *closed neighborhood* of S, respectively. A sequence of vertices of G is a path $P = (v_0, v_1, \dots, v_k)$ of G if $v_i v_{i+1} \in E$ for every $i \in \{0, 1, 2, \dots, k-1\}$. The length of a path $P = (v_0, v_1, \dots, v_k)$ is the number of edges in it. For two distinct vertices $u, v \in V$, the distance $d_G(u, v)$ between u and v is the length of a shortest path between u and v. For a set $S \subseteq V(G)$, the subgraph induced by S is denoted by $G[S]$. A set $S \subseteq V$ is an *independent set* if $G[S]$ has no edge. The *square* of G denoted as G^2 is obtained from G by adding new edges between every two vertices having distance two in G.

The vertices u, v and w of a graph G form an *asteroidal triple* (AT) if $\{x, y, z\}$ is an independent set and for any two of these vertices there is a path between them that avoids the neighborhood of the third. A graph G is said to be *asteroidal triple-free* (AT-free) if it does not contain an asteroidal triple. A graph G is called a *circle graph* if there is a one to one correspondence between the vertex set $V(G) = \{v_1, v_2, \dots, v_n\}$ of the circle graph G and a set $C = \{c_1, c_2, \dots, c_n\}$ of chords on a circle such that two vertices are adjacent if and only if the corresponding chords intersect.

Let n and m denote the number of vertices and edges of G, respectively. In this paper, we only consider connected graphs with at least two vertices.

3 Linear-Time Approximation Algorithm

In this section, we show that there exists a linear-time 3-approximation algorithm to compute a minimum cardinality semitotal dominating set of an AT-free graph. We first recall the definition of *dominating pair* and *dominating shortest path* of a graph.

Definition 1. *For a graph $G = (V, E)$ and $u, v \in V$, the pair (u, v) is called a* **dominating pair** *of G if the vertex set of any path between u and v is a dominating set of G.*

Definition 2. *A path $P = (u = u_0, u_1, u_2, \ldots, u_d = v)$ is a* **dominating shortest path** *of a graph $G = (V, E)$ if $d_G(u, v) = d$ and $\{u_0, u_1, u_2, \ldots, u_d\}$ is a dominating set of G.*

The following result is already known for AT-free graphs.

Theorem 1 [13]. *A connected AT-free graph G has a dominating pair, and hence a dominating shortest path. Also, a dominating pair of G can be computed in linear-time.*

The following relation between the domination number and connected domination number is well known.

Theorem 2 [14]. *For a graph G, $\gamma_c(G) \leq 3\gamma(G) - 2$.*

Since $\gamma(G) \leq \gamma_{t2}(G)$, we have the following relation between the connected domination number and semitotal domination number of a graph.

Theorem 3. *For a graph G, $\gamma_c(G) \leq 3\gamma_{t2}(G) - 2$.*

Now, we are ready to present the main result of this section.

Theorem 4. *A minimum semitotal dominating set of an AT-free G can be approximated in linear-time within an approximation ratio of* 3.

Proof. Let D be the vertex set of a dominating shortest path between u and v in G, then diameter of G must be at least $|D| - 1$. We also know that any connected dominating set has cardinality at least $diam(G) - 1$. Hence, the cardinality of connected dominating set must be at least $|D| - 2$. That is, $\gamma_c(G) \geq |D| - 2$. Also, by Theorem 3, $\gamma_c(G) \leq 3\gamma_{t2}(G) - 2$. Hence $|D| - 2 \leq 3\gamma_{t2}(G) - 2$, which implies that $|D| \leq 3\gamma_{t2}(G)$. Note that D is the set of vertices of a path. Hence, D is also a semitotal dominating set of G, and $|D| \leq 3\gamma_{t2}(G)$. Hence D approximates the Minimum Semitotal Domination problem within an approximation ratio of 3. Also, D can be computed in linear-time for AT-free graphs. □

4 Polynomial-Time Exact Algorithm

In this section, we present a polynomial-time exact algorithm to compute a minimum cardinality semitotal dominating set of an AT-free graph G. We first describe the main idea behind our algorithm.

Galby et al. [12] have shown that given a graph G, we can construct a graph G' such that $\gamma_{t2}(G) = \gamma_t(G')$. In addition, given a minimum cardinality total dominating set of G', we can construct a minimum cardinality semitotal dominating set of G in polynomial-time. In our algorithm, we have the following three steps: (i) Given an AT-free graph G, construct G', (ii) Compute a minimum cardinality total dominating set D' of G', (iii) From D', construct a minimum cardinality semitotal dominating set D of G. Next we first describe how to construct G' from G, then we will present an algorithm to find a minimum cardinality total dominating set of G'.

4.1 Construction of G' from G

Given a graph $G = (V, E)$ where $V = \{v_1, v_2, \ldots, v_n\}$, we construct G' in the following way: for each vertex v_i in G, take two vertices v_i^1 and v_i^2 in G'. If v_i and v_j are at distance at most 2 in G, then add an edge between v_i^1 and v_j^1 in G'. Also, if $v_i \in N_G[v_j]$ in G, then add an edge between v_i^1 and v_j^2. Formally $G' = (V', E')$ where

$$V' = \{v_i^1, v_i^2 \mid 1 \leq i \leq n\}, \text{ and}$$
$$E' = \{v_i^1 v_j^1 \mid d_G(v_i, v_j) \leq 2\} \cup \{v_i^1 v_j^2 \mid v_i \in N_G[v_j]\}.$$

If we define $V^1 = \{v_i^1 \mid 1 \leq i \leq n\}$ and $V^2 = \{v_i^2 \mid 1 \leq i \leq n\}$, then $G'[V^1]$ is isomorphic to G^2 and V^2 is an independent set. Note that G' can be constructed from G in $O(|V|.|E|)$-time. Note that if G is an AT-free graph, then G' is not necessarily AT-free. For example consider the graph G shown in Fig. 1. Then G is AT-free, but G' has an asteroidal triple. The set of three vertices v_1^2, v_5^2, v_6^2 form an asteroidal triple in G'.

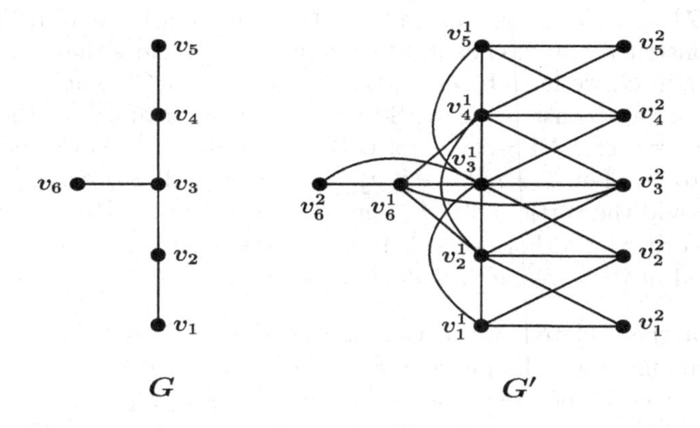

Fig. 1. An illustration to the construction of G' from G: G is AT-free, but G' has an asteroidal triple.

Theorem 5 [12]. *For a graph G, $\gamma_{t2}(G) = \gamma_t(G')$. Moreover, a minimum semi-total dominating set of G can be obtained from a minimum total dominating set of G' in linear-time.*

In the next section, we will present the algorithm to compute a minimum cardinality total dominating set in G', which is not an AT-free graph, but constructed from an AT-free graph G. To find an algorithm for minimum cardinality total dominating set in G', we have used the idea given by Kratsch [15] for finding the minimum cardinality total dominating of an AT-free graph.

4.2 Algorithm to Find a Minimum Total Dominating Set of G'

Let G be a graph and T be a minimum total dominating set of G. Kratsch has shown that if there exists an $x \in V$ such that any three consecutive BFS-levels of x contain at most k elements of T, then there is an $O(n^{k+2})$-time algorithm to compute a minimum cardinality total dominating set of G (see [15, Theorem 8]).

If G is an AT-free graph and G' is the graph constructed from G as discussed in Sect. 4.1, then we will show that there exists a minimum total dominating set D_t of G' and a vertex $x \in V(G')$ such that any j-consecutive BFS-levels of x contain at most $2j+9$ elements of D_t. This will imply that any three consecutive BFS-levels of x contain at most 15 elements of D_t, and hence we will get an $O(n^{17})$-time algorithm to compute a minimum cardinality total dominating set of G'.

Lemma 1. *If G is an AT-free graph and G' is the graph constructed from G as discussed in Sect. 4.1, then there exists a minimum total dominating set D_t of G' and a vertex $x \in V(G')$ such that any j-consecutive BFS-levels of x contain at most $2j + 9$ elements of D_t.*

Proof. Let G be an AT-free graph, and (x, y) be a dominating pair in G. Let G' be the graph constructed from G as illustrated in Sect. 4.1. Note that corresponding to a vertex x in G, we have taken vertices x^1 and x^2 in G', where $x^1 \in V^1$ and $x^2 \in V^2$. Next we construct a BFS-tree of G' rooted at x^1 in the following way: First construct a BFS-tree T of $G'[V^1]$ rooted at x^1. Note that $G'[V^1]$ is isomorphic to G^2. Let $\alpha = (x^1 = v_1^1, v_2^1, \ldots, v_n^1)$ be a BFS ordering of T. Now, in tree T we add the vertices of V^2 to get a BFS-tree of G'. Process the vertices of V^1 in the ordering α. For each v_i^1, find the vertices in $N_{G'}(v_i^1) \cap V_2$ which are not yet added in BFS-tree, and make them child of v_i^1.

Note that the updated tree T is a BFS-tree of G'. Now we show that there exists a minimum total dominating set D_t of G' such that any j consecutive levels of BFS-tree T contains at most $2j + 9$ vertices of D_t.

Let $x = x_0, x_1, x_2, \ldots, x_d = y$ be a shortest path between the vertices of dominating pair (x, y). Corresponding to this path, we get the following two shortest paths in $G'[V^1]$:

$$P_1 = x_0^1, x_2^1, x_4^1, \ldots \qquad \textit{and} \qquad P_2 = x_0^1, x_1^1, x_3^1, \ldots$$

Note that the vertex set of $P_1 \cup P_2$ is a total dominating set of G'. We refer to this total dominating set as A.

Let $L_0, L_1, L_2, \ldots L_l$ denote the levels of BFS-tree T of G' with root $x^1 \in V^1$. So, $L_0 = x^1$, $L_1 = N_{G'}(x^1)$, and $L_i = \{v \in G' \mid d(x^1, v) = i\}$ for any $i \in \{1, 2, \ldots, l\}$, where l is the depth of the tree T. By contradiction, assume that there exists j consecutive levels containing more than 2j + 9 vertices of a minimum total dominating set D_t. Now, choose i minimum and j maximum with respect to i such that $L_i \cup \ldots \cup L_{i+j-1}$ has more than $2j + 9$ vertices of a minimum total dominating set D_t. We call (i, j) as a *bad segment*.

Let $D_t^{i,j} = D_t \cap (L_i \cup \ldots \cup L_{i+j-1})$. Then $|D_t^{i,j}| \geq 2j + 10$. Let $A^{i,j} = A \cap \{L_{i-2} \cup \ldots \cup L_{i+j+1}\}$. Since $|A \cap L_k| \leq 2$ for any k, $|A^{i,j}| \leq 2(j+4) = 2j+8$. Note that $|D_t \cap L_{i-1}| \leq 1$, if not then $(i-1, j)$ will also be a bad segment, which is contradiction to the minimality of i. Similarly, $|D_t \cap L_{i+j}| \leq 1$, if not then $(i, j+1)$ will also be a bad segment, which is contradiction to the maximality of j with respect to i.

Now, we modify the set D_t to another total dominating set in the following way: Define $D_t^* = (D_t \backslash D_t^{i,j}) \cup A^{i,j}$. Note that $N_{G'}[D_t^{i,j}] \subseteq N_{G'}[A^{i,j}]$, hence D_t^* still dominates all the vertices of G'. But there may exist a vertex $v \in L_{i-1} \cap D_t$, which is an isolated vertex in $G'[D_t^*]$. This case arises if there exists a neighbor w of v such that $w \in L_i \cap D_t$ but $w \notin D_t^*$. In this case we also add w in D_t^*. Similarly, there may exist a vertex $s \in L_{i+j} \cap D_t$, which is an isolated vertex in $G'[D_t^*]$. Again, this case arises if there exists a neighbor t of s such that $t \in L_{i+j-1} \cap D_t$ but $t \notin D_t^*$. In this case, we add t in D_t^*. Now, the set D_t^* is a total dominating set of G'. Since $|A^{i,j} \cup \{w, t\}| \leq 2j + 10$, $|D_t^*| \leq |D_t|$. Hence D_t^* is also a minimum total dominating set of G'.

Note that the boundary cases $i \in \{0, 1\}$ and $j \in \{l-1, l\}$ are not possible. Because in these cases, $A^{i,j} \leq 2j + 6$, and hence the cardinality of D_t^* will be smaller than the cardinality of D_t, a contradiction arises.

Next we prove the following claim.

Claim. If there exists a bad segment (i', j') in the BFS-tree T with respect to the minimum total dominating set D_t^*, then $i' > i$.

Proof. Suppose (i', j') is a bad segment in T with respect to D_t^*. Suppose $i' \leq i$. Then $i' + j' - 1 \geq i - 2$, otherwise (i', j') should also be a bad segment in T with respect to D_t, contradicting the choice of i. By construction $|D_t^* \cap L_k| \geq 2$ for any $k \in \{i-2, i-1, i, \ldots, i+j-1, i+j, i+j+1\}$. Hence if (i', j') is a bad segment with $i' \leq i$ and $i' + j' - 1 \geq i - 2$, then there is a j^* such that (i', j^*) is a bad segment with respect to D_t^* and $i' + j^* - 1 \geq i + j + 1$. By the construction of D_t^*, this implies $|D_t^* \cap (L_{i'} \cup L_{i'+1} \cup \ldots \cup L_{i'+j^*-1})| = |D_t \cap (L_{i'} \cup L_{i'+1} \cup \ldots \cup L_{i'+j^*-1})|$, and hence (i', j^*) should be a bad segment in D_t, contradicting the choice of either i or j.

Hence $i' > i$. This completes the proof of the claim. □

Using the above claim, we can say that if a bad segment in the BFS-tree T with respect to the minimum total dominating set D_t starts at level i, then we can modify the set D_t to get another minimum total dominating set D_t^* such that if a bad segment still exists, then it will start from level i', where $i' > i$. If we continue doing the same process again and again, then after finite number of steps, we will get a minimum total dominating set D of G' such that there will be no bad segment with respect to D. Note that the number of steps will be bounded by the depth of the BFS-tree. Hence, any j-consecutive BFS-levels of x will contain at most $2j + 9$ elements of D. This completes the proof of our lemma. □

Now, we may directly state the main theorem of this section.

Theorem 6. *There exists a polynomial-time algorithm to compute a minimum cardinality semitotal dominating set of an AT-free graph.*

5 NP-Completeness Result on Circle Graphs

In this section, we show that the SEMITOTAL DOMINATION DECISION problem is NP-complete for circle graphs. To prove the NP-completeness result, we provide a polynomial reduction from the DOMINATION DECISION problem in circle graphs, which is already known to be NP-complete [16]. We first prove the following lemma.

Lemma 2. *Let G be a circle graph. Then, the graph G' obtained by adding a pendant vertex to any vertex $v \in V(G)$ is also a circle graph.*

Proof. Let G be a circle graph with vertex set $V(G)$. Note that there is a one to one correspondence between the vertex set $V(G) = \{v_1, v_2, \ldots, v_n\}$ of the circle graph G and a set $C = \{c_1, c_2, \ldots, c_n\}$ of chords on a circle such that two vertices are adjacent if and only if the corresponding chords intersect. An illustration is given in Fig. 2. If two chords have same endpoint, we can slightly shift one of the chord without changing any intersection relation and representing the same graph. So without loss of generality, we may assume that the end points of different chords are distinct. Now, to prove the result it is enough to show that if c_k is any chord on the circle then we can draw another chord on the same circle which intersects only c_k.

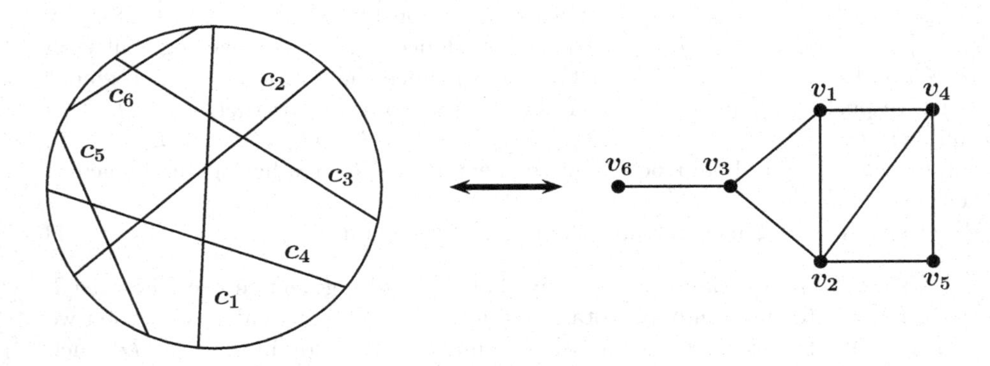

Fig. 2. An illustration of a circle graph and its corresponding circle model.

As we have n chords on the circle satisfying the property that no two chord have a common end point, we have total $2n$ points on the circle, representing the end of the chords. Starting with any end point, let $a_1, a_2, \ldots, a_{2n}$ are the consecutive and distinct end points of the chords on the circle in clockwise orientation. Let c_k be a chord having a_i as one of its end point. Now take a point a

on the arc $a_i a_{i+1}$ other than a_i, a_{i+1} and another point b on the arc $a_{i-1}a_i$ other than a_{i-1}, a_i. Now make a chord with end points a and b (see Fig. 3). We may observe that this new chord will intersect only c_k. Hence, the result is proved. □

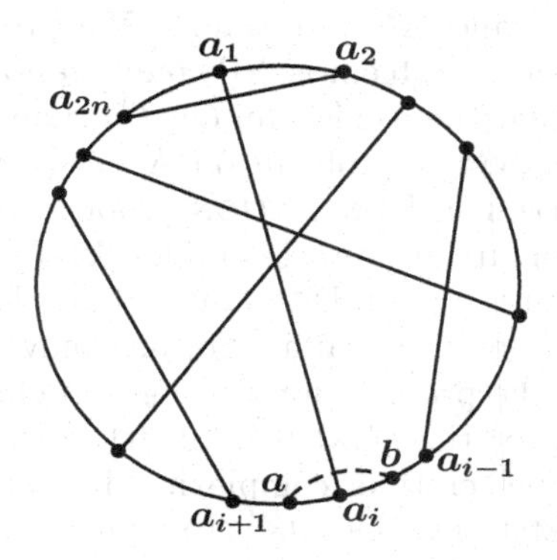

Fig. 3. Illustration of making additional chord.

Theorem 7. *The* SEMITOTAL DOMINATION DECISION *problem is NP-complete for circle graphs.*

Proof. Clearly, the SEMITOTAL DOMINATION DECISION problem is in NP for circle graphs. To prove the NP-hardness result, we give a polynomial reduction from the MINIMUM DOMINATION problem in circle graphs.

Given a circle graph $G = (V, E)$, where $V = \{v_1, v_2, \ldots, v_n\}$, we construct another circle graph $G' = (V', E')$ as follows: For each i, $1 \leq i \leq n$, we first add a pendant vertex a_i at v_i and then add a pendant vertex b_i at a_i. Formally, $V' = V \cup \{a_i, b_i \mid 1 \leq i \leq n\}$ and $E' = E \cup \{v_i a_i, a_i b_i \mid 1 \leq i \leq n\}$. It clearly follows from Lemma 1 that G' is also a circle graph. Next, we will show that G has a dominating set of size atmost k if and only if G' has a semitotal dominating set of size atmost $n + k$.

Clearly, if D is a dominating set G of cardinality atmost k, then the set $D' = D \cup \{a_1, a_2, \ldots, a_n\}$ is a semi-TD-set of G', and $|D'| \leq n + k$.

Conversely, suppose that D' is a semi-TD-set of G' of cardinality atmost $n + k$. For each i, $1 \leq i \leq n$ to dominate the vertex b_i, either a_i or b_i must belong to D'. If both a_i and b_i belong to D', then b_i can be replaced with v_i in D'. If $b_i \in D'$ but $a_i \notin D'$, then b_i can be replaced with a_i. Note that the updated set D' still remains a semi-TD-set of G'. So, without loss of generality, we may assume that $\{a_1, a_2, \ldots, a_n\} \subseteq D'$ and $\{b_1, b_2, \ldots, b_n\} \cap D' = \emptyset$. Now, for each i, $1 \leq i \leq n$, D' should also contain a vertex at distance at most 2 from a_i. Since $b_i \notin D'$, either v_i or one of the neighbor of v_i in G must be present in D'. Hence, the set $D' \setminus \{a_1, a_2, \ldots, a_n\}$ is a dominating set of G of cardinality at most k. This completes the proof of the theorem. □

6 Conclusion

Galby et al. [12] mentioned that the complexity of the Minimum Semitotal Domination problem is open for AT-free graphs, circle graphs, dually graphs and tolerance graphs. In this paper, we resolved the complexity status of the problem for circle graphs and AT-free graphs. We proved that the Semitotal Domination Decision problem is NP-complete for circle graphs. On the positive side, we proved that there exists an $O(n^{19})$-time algorithm for the Minimum Semitotal Domination problem on AT-free graphs. We also proved that the Minimum Semitotal Domination problem can be approximated in linear-time within an approximation ratio of 3 for AT-free graphs. But the complexity of the Minimum Semitotal Domination problem is still open for doubly chordal graphs and tolerance graphs. So, one may further try to find out the complexity status of the problem for these graph classes. One may also try to propose an algorithm for the Minimum Semitotal Domination problem in AT-free graphs with better time complexity. It will also be interesting to study the Minimum Semitotal Domination problem for k-polygon graphs. *k-polygon graphs* are the intersection graphs of chords in a k-sided polygon. Elmallah et al. [17] have shown that the Minimum Domination problem can be solved in $O(n^{4k^2}+3)$-time in k-polygon graphs. One may try to check whether the Minimum Semitotal Domination problem is also polynomially solvable for k-polygon graphs or not.

References

1. Haynes, T., Hedetniemi, S., Slater, P.: Fundamentals of Domination in Graphs, vol. 208. Marcel Dekker Inc., New York (1998)
2. Haynes, T., Hedetniemi, S., Slater, P.: Domination in Graphs: Advanced Topics, vol. 209. Marcel Dekker Inc., New York (1998)
3. Du, D., Wan, P.: Connected Dominating Set: Theory and Applications. Springer, New York (2013). https://doi.org/10.1007/978-1-4614-5242-3
4. Henning, M.A., Yeo, A.: Total Domination in Graphs. Springer, New York (2013). https://doi.org/10.1007/978-1-4614-6525-6
5. Henning, M.A.: A survey of selected recent results on total domination in graphs. Discrete Math. **309**, 32–63 (2009)
6. Goddard, W., Henning, M.A., McPillan, C.A.: Semitotal domination in graphs. Util. Math. **94**, 67–81 (2014)
7. Henning, M.A., Marcon, A.J.: On matching and semitotal domination in graphs. Discrete Math. **324**, 13–18 (2014)
8. Henning, M.A., Marcon, A.J.: Vertices contained in all or in no minimum semitotal dominating set of a tree. Discuss. Math. Graph Theory **36**, 71–93 (2016)
9. Henning, M.A., Marcon, A.J.: Semitotal domination in claw-free cubic graphs. Ann. Combin. **20**, 799–813 (2016)
10. Henning, M.A.: Edge weighting functions on semitotal dominating sets. Graphs Combin. **33**, 403–417 (2017)
11. Henning, M.A., Pandey, A.: Algorithmic aspects of semitotal domination graphs. Theor. Comput. Sci. **766**, 46–57 (2019)

12. Galby, E., Munaro, A., Ries, D.: Semitotal domination: new hardness results and a polynomial-time algorithm for graphs of bounded mim-width. Theor. Comput. Sci. **814**, 28–48 (2020)
13. Corneil, D.G., Olariu, S., Stewart, L.: Asteroidal triple-free graphs. SIAM J. Discrete Math. **10**, 399–430 (1997)
14. Duchet, P., Meyneil, H.: On hadwiger's number and the stability number. Discrete Math. **13**, 71–74 (1982)
15. Kratsch, D.: Domination and total domination on asteroidal-triple free graphs. Discrete Appl. Math. **99**, 111–123 (2000)
16. Keil, J.: The complexity of domination problems in circle graphs. Discrete Appl. Math. **42**, 51–63 (1993)
17. Elmallah, E., Stewart, L.: Independence and domination in polygon graphs. Discrete Math. **44**, 65–77 (1993)

Burning Grids and Intervals

Arya Tanmay Gupta[1(✉)], Swapnil A. Lokhande[2], and Kaushik Mondal[3(✉)]

[1] Michigan State University, East Lansing, MI, USA
guptaar3@msu.edu
[2] Indian Institute of Information Technology Vadodara, Gandhinagar, India
[3] Indian Institute of Technology Ropar, Rupnagar, India
kaushik.mondal@iitrpr.ac.in

Abstract. Graph burning runs on discrete time steps. The aim is to burn all the vertices in a given graph in the least number of time steps. This number is known to be the burning number of the graph. The spread of social influence, an alarm, or a social contagion can be modeled using graph burning. The less the burning number, the faster the spread.

Optimal burning of general graphs is NP-Hard. There is a 3-approximation algorithm to burn general graphs where as better approximation factors are there for many sub classes. Here we study burning of grids; provide a lower bound for burning arbitrary grids and a 2-approximation algorithm for burning square grids. On the other hand, burning path forests, spider graphs, and trees with maximum degree three is already known to be NP-Complete. In this article we show burning problem to be NP-Complete on connected interval graphs.

1 Introduction

The spread of social influence in order to analyze a social network is an important topic of study [4,17,18]. Kramer et al. [19] have highlighted that the underlying network plays an essential role in the spread of an emotional contagion; they have nullified the necessity of in-person interaction and non-verbal cues. With the aim to model such problems, *Graph Burning* was introduced in [10]. Graph burning is also inspired by other contact processes like *firefighting* [12], *graph cleaning* [2], and *graph bootstrap percolation* [3]. Burning a graph can be used to model the spread of a meme, gossip, or a social contagion, influence or emotion. It can also be used to model the viral infections: the exposure to infections and proliferation of virus.

Graph burning runs on discrete time-steps (or rounds) as follows: in each time-step t, first (a) an arbitrary vertex is *burnt* from "outside" (it is selected as a *fire source*), and then, (b) the fire spreads to the vertices that are one hop neighbors of the already burnt vertices (burnt by round $t-1$); this process continues till all the vertices of the given graph are burned. Observe that some

A.T. Gupta—This study was conducted when A T Gupta was affiliated with the Indian Institute of Information Technology Vadodara, India.

A. Mudgal and C. R. Subramanian (Eds.): CALDAM 2021, LNCS 12601, pp. 66–79, 2021.
https://doi.org/10.1007/978-3-030-67899-9_6

fire source selected at round t does not spread fire to its one hop neighbors in the same round. The sequence of fire sources, selected one in each round until a graph is completely burnt, is called a *burning sequence* of that graph. The minimum time steps (equivalently, number of fire sources) required to burn a graph G is called the *burning length* or the *burning number* of G, and is denoted by $b(G)$. The less the value of $b(G)$, the faster it is to spread the fire, and therefore burn all the vertices in G. The graph burning problem is to find an optimal burning sequence for a given graph G. At places, we use burning problem to refer the same.

The underlying decision problem for graph burning is as follows: the given input is an arbitrary graph G and an integer k, the task is to determine if G can be burned in k or less rounds. Bessy et al. [5] showed that optimal graph burning is an NP-Complete problem. They also showed that burning spider graphs, trees with maximum degree three and path forests is NP-Complete. In this article, we study the graph burning problem on *interval graphs* and *grids*. Interval graphs are formed from a set of closed intervals on the real line such that each interval corresponds to a vertex and the vertices corresponding to two such intervals are connected only if they overlap on the real line. Grids are formed by a set of equidistant horizontal and vertical lines intersecting at right angles such that each intersection point corresponds to a vertex and all the (induced) line segments joining those vertices are considered as edges. We prove NP-Completeness results for interval graphs. Our construction and proof technique are similar to [5]. We also provide matching bounds for burning grids.

Our Contribution: We provide a lower bound for the burning number of grids of arbitrary size and a 2-approximation algorithm for burning square grids (Sect. 4). We prove burning connected interval graphs to be NP-Complete(Sect. 5). We also report hardness results on some more graph classes (Sect. 6).

2 Preliminary Definitions and Symbols

We mention below some of the notations used in this article. Let G be a graph. We denote the set of vertices in G by $V(G)$. The distance between two vertices imply the number of edges contained in the shortest path between those two vertices in G. The radical center of a graph means the vertex from which the shortest distance to the furthest vertex is minimum. We define $\cup_{\backslash s}$ to be the *left sequential union*. This operation can add a single element to a sequence, or merge two sequences. As an example, let $P = (a, b)$ be a sequence, then after executing the statement $P = P \cup_{\backslash s} (c)$, P becomes (c, a, b). Similarly $\cup_{s/}$ is defined as the *right sequential union*. Let Q_1, Q_2 be two paths. By *joining* these two paths in order Q_1, Q_2, we mean adding an edge between the last vertex of Q_1 and the first vertex of the Q_2. Let A be a set of natural numbers. We denote the sum of all numbers in A as $s(A)$. The largest element in A is denoted by $\max(A)$.

Let W be a non-empty set of vertices such that $W \subset V(G)$. The set of vertices that are at most at a distance i from W in G, including W is denoted by $G.N_i[W]$. The set W may be a set containing a single element. Let $S = (x_1, x_2, ..., x_k)$ be a burning sequence of G of size k such that, x_i is chosen as the fire source in round i. The *burning cluster* (or simply *cluster*, when it is clear from the context) of a fire source x_i is the set $G.N_{k-i}[x_i]$. Precisely, it is the set of vertices, to whom the fire source x_i is able to spread fire to in the remaining $(k-i)$ rounds. Observe that, x_i would be able to burn all of it's $(k-i)$ hop neighbors. Now it is easy to see that, if S is able to burn G completely, then Eq. 1 must hold true [5].

$$G.N_{k-1}[x_1] \cup G.N_{k-2}[x_2] \cup ... \cup G.N_0[x_k] = V(G) \tag{1}$$

For the NP-Completeness proof, we reduce *distinct 3-partition problem* to our problem. The input of the distinct 3-partition problem is a set of distinct natural numbers, $X = \{a_1, a_2, ..., a_{3n}\}$, such that $\sum_{i=1}^{3n} a_i = nB$ where $\frac{B}{4} < a_i < \frac{B}{2}$. The task is to determine if X can be partitioned into n sets, each containing 3 elements such that sum of each set equals B. Note that, B can only be a natural number as it is a sum of 3 natural numbers. It is well known that the distinct 3-partition problem is NP-Complete in the strong sense (see [5,13]).

3 Related Works

The burning problem was introduced by Bonato et al. (2014) [10]. This work showed that the burning number of a path or cycle of length p is $\lceil \sqrt{p} \rceil$ along with some other properties and results. Bessy et al. [5] showed that burning a general graph is NP-Complete: they showed that burning spider graphs, trees, and path-forests are NP-Complete. A 3-approximation algorithm for burning general graphs was described in [5]. Bonato et al. [8] proposed a 2-approximation algorithm for burning trees. A 1.5-approximation algorithm for burning path-forests was described in [9]. A 2-approximation algorithm for burning graphs that are bounded by a diameter of constant length was described in [15]. There are works providing upper bounds on the burning number of some classes of graphs. Authors in [11] as well as [9] showed that burning number of spider graphs of order n is at most $\sqrt{n}$. Bessy et al. [6] provided a bound on the burning number of a connected graph of order n, and a special class of trees. Simon et al. [1] presented systems that utilize burning in the spread of an alarm through a network. Simon et al. [20] provided heuristics to minimize the time steps in *burning* a graph. Kamali et al. [15] provides upper bound on burning number for the graphs with bounded path length and also for the graphs with minimum degree δ. Along with this, authors in [15] (also [16]) discussed bounds on the burning number of interval graphs and showed almost tight bounds. Although, they have not provided any algorithm to find an optimal burning sequence. Despite the known bounds on the burning number and the fact that most of other properties of interval graphs can be computed in polynomial time, we show that burning connected interval graphs turns out to be NP-Complete. Also we study graph burning on grids which is a graph of constant minimum degree.

4 Burning Grids

In this section we study graph burning problem on grids by providing a lower bound for grids of arbitrary size and an 2-approximation algorithm for square grids. According to [15], the upper bound on burning number of graphs with constant minimum degree is $O(\sqrt{n})$. Here we provide a better upper bound and a matching lower bound for this specific class of graphs.

First we analyze at most how many nodes can be burnt by an arbitrary fire source inside the grid. We show an example in Fig. 1 (a). Let the maximum number of vertices that can be burned by a single fire source in k rounds be denoted by f_k. We compute f_k using the recurrence relations as follows.

$$f_1 = 1$$
$$f_k = 4(k-1) + f_{k-1}$$

At k^{th} time step after a fire source is placed, the number of vertices which can be burned is $1+4+8+12+...+4(k-1) = 1+4(1+2+3+...+(k-1)) = 1+4\times\frac{k(k-1)}{2} = 2k(k-1)+1$.

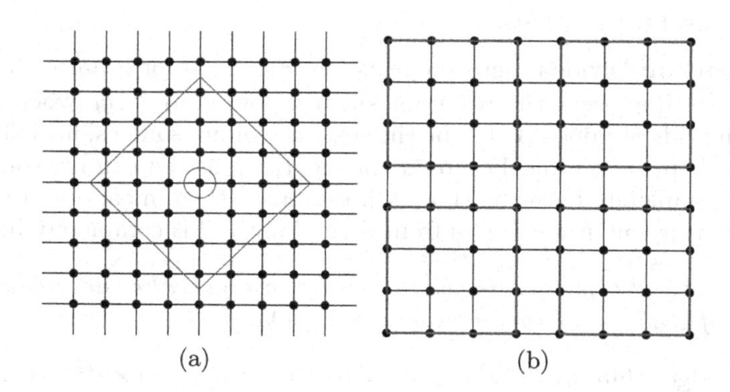

Fig. 1. (a) On a grid, a fire source (circled) is able to burn 1 vertex in 1 round, 5 vertices in 2 rounds, 13 vertices in 3 rounds, 25 vertices in 4 rounds, and so on. (b) An 8×8 grid divided into four 4×4 subgrids.

The Lower Bound: We prove the following lemma on the lower bound of burning number on any arbitrary grid of size $l\times b$.

Lemma 1. *To burn a grid of size $l\times b$, we need a burning sequence containing at least $(l\times b)^{\frac{1}{3}}$ fire sources.*

Proof. Consider a grid of size $l\times b$, where l and b are any positive integers. As discussed above, $f_k = 2k(k-1)+1$. So, if i be the burning number of the grid, then the total number of vertices that are burned by this burning sequence will be $1+5+13+25+...+2i(i-1)+1 = (2(0+2+6+12+20+...+(i^2-i))+i)$ $= 2((1^2-1)+(2^2-2)+(3^2-3)+...+(i^2-i))+i = 2((1^2+2^2+3^2+...+i^2)-(1+2+3+...+i))+i = 2\left(\frac{i(i+1)(2i+1)}{6} - \frac{i(i+1)}{2}\right)+i = \frac{i(i+1)(2i+1)}{3} - i(i+1)+i$ $= \frac{i(i+1)(2i-2)}{3} + i = \frac{i(2i^2-2+3)}{3} = \frac{i(2i^2+1)}{3}$.

A burning sequence of size i shall be able to burn at most $\frac{2i^3+i}{3}$ vertices on the grid. So to burn the grid in i rounds, we must have,

$$\frac{2i^3+i}{3} \geq l \times b \tag{2}$$

Since $\frac{2i^3+i}{3} \leq i^3 \ \forall i \geq 1$ and the burning sequence burns all the vertices, we have $i^3 \geq \frac{2i^3+i}{3} \geq l \times b$. This implies $i \geq (l \times b)^{\frac{1}{3}}$. □

We state the following corollary.

Corollary 1. *To burn a square grid of size n, the burning number needs to be at least $n^{\frac{1}{3}}$.*

Now we describe the following algorithm to burn an arbitrary grid G of size $l \times b$. We further show that this is a 2-approximation algorithm for burning any $l \times l$ square grid with $l \geq 403$.

The Algorithm: Divide G into subgrids (see Fig. 1 (b) for example) of dimensions $l^{\frac{2}{3}} \times b^{\frac{2}{3}}$. Represent the resultant subgrids by $g_1, g_2, ..., g_k$ where k is the count of subgrids obtained. Let S be the sequence of fire sources, initially empty. For $1 \leq i \leq k$, put the radical center x_i of subgrid g_i as the i-th fire source in S. If G is not completely burnt by those k fire sources, then in each step $i \geq k+1$, continue putting unburnt vertices from G in S until G is completely burnt.

Theorem 1. *Our algorithm is able to burn a grid G within an approximation factor of 2 if G is a $l \times l$ square grid with $l \geq 403$.*

Proof. The algorithm divides the grid in to at most $(l^{\frac{1}{3}}+1) \times (l^{\frac{1}{3}}+1)$ subgrids. In each round i, x_i is set to be the i-th fire source. As the fire source is placed at the radical center of a subgrid, and the radius of the subgrid is $l^{\frac{2}{3}}$, so it takes $l^{\frac{2}{3}}$ rounds to burn the corresponding subgrid. As the last fire source may take up to $l^{\frac{2}{3}}$ rounds to burn the respective subgrid, our algorithm takes a total of at most $\lceil (l^{\frac{1}{3}}+1) \times (l^{\frac{1}{3}}+1) + l^{\frac{2}{3}} \rceil$ rounds i.e., $\lceil 2l^{\frac{2}{3}} + 2l^{\frac{1}{3}} + 1 \rceil$ rounds to burn G completely. Next we see what can not be burnt using half of the rounds that our algorithm takes in worst case.

It is impossible to burn any $l \times l$ square grid with $l \geq 403$ in less than $\frac{\lceil 2l^{\frac{2}{3}}+2l^{\frac{1}{3}}+1 \rceil}{2}$ rounds as $i = \frac{\lceil 2l^{\frac{2}{3}}+2l^{\frac{1}{3}}+1 \rceil}{2}$ value does not satisfy Eq. 2 for $b = l$ where $l \geq 403$. Remember, to burn G completely in i rounds, Eq. 2 must be satisfied. Hence the proof. □

5 Burning Interval Graphs

We show that burning connected interval graphs is NP-Complete by giving a reduction from the distinct 3-partition problem. We construct interval graph from any given instance of the distinct 3-partition problem. We do so by replacing each spider structure by a "comb structure" in the construction of the NP-Completeness proof for burning trees in [5], which we elaborate later in this section. But before going to that, we have the following discussion that tries to relate burning an interval graph to burning a path.

Bonato et al. (2016) [7] proved that a path or a cycle of n vertices can be *burned* in $\lceil\sqrt{n}\rceil$ steps. Note the following observation from the above fact.

Observation 1. *The burning clusters of each of the n fire sources of any optimal burning sequence of a path of n^2 vertices are pairwise disjoint.*

We would like to recall another result from [15,16] on the bounds on burning number of interval graphs as the following observation. We provide an example of burning a path of size nine as shown in the Fig. 2.

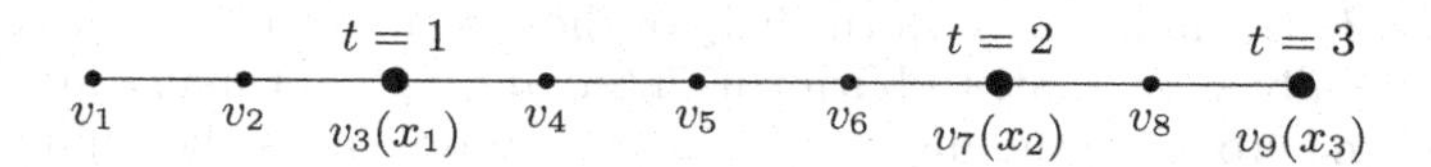

Fig. 2. Path of length nine is burnt in three steps. The vertex v_3 is chosen as fire source x_1 in time $t = 1$ and x_1 is burned at this step. In time $t = 2$, v_7 is chosen as the next fire source x_2 and it is burned in this step. Along with this, the one hop neighbors v_2, v_4 of the already burnt (by step 1) vertex v_3 also are burned in this step. In time $t = 3$, v_9 is selected as the third fire source and subsequently is burned in this step. Also the one hop neighbors v_1, v_5, v_6 and v_8 are burned in this step by the spread of fire from the already burnt vertices (by step 2).

Observation 2. *Let L be a maximum length path among the all pair shortest paths of an interval graph G. Then $b(L) \leq b(G) \leq b(L) + 1$.*

Also note that, finding such L is easy to do in polynomial time. We can simply compute all pair shortest path and choose the maximum length path among all. Then we can see from the proof of Observation 2 that burning an interval graph in $(b(L)+1)$ rounds, i.e., at most in $(b(G)+1)$ rounds is trivial. We study whether finding a burning sequence of length $b(G)$ is possible in polynomial time, especially if $b(G) = b(L)$ for an interval graph G. We show that determining whether $b(G) = b(L)$ is NP-Complete.

General Idea: First we provide a general idea behind our approach. We prove the NP-Completeness of burning interval graphs by giving a reduction from the distinct 3-partition problem. We construct interval graph from any given instance of the distinct 3-partition problem. We show that burning this interval graph is possible optimally in polynomial time if and only if one can solve the distinct 3-partition problem. While describing the idea, we refer to few notations here which are defined in Sect. 5.1.

We start with any input X of the distinct 3-partition problem. First we construct another set X' from X such that all the elements of X' are odd. The reason behind moving to X' is, we aim to use the fact that the sizes of the burning clusters of the fire sources on a path are all odds if the length of the path is a perfect square. First we construct a path P_I of length $(2m+1)^2$ (where $m = \max(X)$) by combining few subpaths of shorter lengths. Note that $b(P_I) = 2m+1$. Then we add few vertices and corresponding edges to some of the subpaths T_j of P_I in such a way that it remains an interval graph. We call it $IG(X)$ (Sect. 5.1). The optimal burning number $b(IG(X))$ takes the value $2m+1$ whenever X'(and eventually X) can be partitioned according to the distinct 3-partition problem (Lemma 3). So we keep the burning number of the path P_I and the interval graph $IG(X)$ same.

Additional vertices and edges are added to the sub paths T_j to form structures T_j^c (refer Fig. 3) in such a way that to burn $IG(X)$ optimally, one must have to burn each T_j^c only with one fire source (Lemma 2). Not only that, one must have to put that fire source on T_j^c in a particular round depending on the length of the subpath T_j (Lemma 5). With the help of these results and another couple of results, we finally show that, to burn this interval graph optimally in $b(IG(X))$ steps, one needs to solve the distinct 3-partition problem on the input X. This makes our problem an NP-Complete problem (Sect. 5.3 Theorem 2).

5.1 Interval Graph Construction

Let n be a natural number. Let $X = \{a_1, a_2, \cdots, a_{3n}\}$ be an input to a distinct 3-partition problem. So, $n = \frac{|X|}{3}$ and $B = \frac{s(X)}{n}$. Let $m = \max(X)$, and $k = m - 3n$. Let F_m be the set of first m natural numbers, $F_m = \{1, 2, 3, ..., m\}$. Also let F'_m be the set of first m odd numbers, $F'_m = \{2\ f_i - 1 : f_i \in F_m\} = \{1, 3, 5, ..., 2m-1\}$. Let $X' = \{2\ a_i - 1 : a_i \in X\}$, $B' = \frac{s(X')}{n}$. Observe that $s(X') = \sum_{i=1}^{3n} 2\ a_i - 1 = 2nB - 3n$, so $B' = 2B - 3$. It is easy to observe that any solution of X gives a solution of X' and vice versa. Let $Y = F'_m \setminus X'$.

Let there be n paths $Q_1, Q_2 ..., Q_n$, each of order B'. Consider k paths Q'_1, Q'_2, ..., Q'_k such that each Q'_j ($\forall\ 1 \leq j \leq k$) is of order of jth largest number in Y, where $k = |Y|$. Clearly the total number of vertices in $Q_1, Q_2 ..., Q_n, Q'_1, Q'_2, ..., Q'_k$ is m^2, i.e., equals $s(F'_m)$. Consider another $m+1$ paths $T_1, T_2, ..., T_{m+1}$ such that each T_j ($\forall 1 \leq j \leq m+1$) is of order of $2(2m+1-j)+1$. Total number of vertices in $T_1, T_2, ..., T_{m+1}$ is $\sum_{j=1}^{m+1}(2(2m+1-j)+1) = (3m^2+4m+1)$. We join these paths in the following order to form a larger path:

Q_1, T_1, Q_2, T_2, ..., Q_n, T_n, Q'_1, T_{n+1}, Q'_2, T_{n+2}, ..., Q'_k, T_{n+k}, T_{n+k+1}, ..., T_{m+1}. We denote this path as P_I. The total number of vertices in P_I is $m^2 + 3m^2 + 4m + 1 = (2m+1)^2$. Hence $b(P_I) = (2m+1)$.

Now we add few more vertices to P_I in such a way that it remains an interval graph and the optimal burning number of the graph remains same as $b(P_I)$. We add a distinct vertex connected to each vertex from $2nd$ to $2nd$-last vertices of T_j, $\forall 1 \leq j \leq m+1$ (Fig. 3 illustrates an example T_j along with the added vertices and edges (vertically upwards w.r.t. T_j). This forms a kind of comb structure; we call it T_j^c). Let this graph be called $IG(X)$. Now we calculate total number of vertices in $IG(X)$. Number of vertices added to each T_j is $(2(2m+1-j)+1)-2 = (4m+1-j)$. Hence total number of vertices added to P_I is $\sum_{j=1}^{m+1}(4m+1-j) = (3m^2+2m-1)$. So, total number of vertices in $IG(X)$ is $(2m+1)^2+(3m^2+2m-1) = (7m^2+6m)$. One such example of $IG(X)$ is shown in Fig. 4 corresponding to the numerical example given in Sect. 5.2.

Observe that P_I is a diameter of $IG(X)$ and there is no cycle in $IG(X)$. Also, all the vertices which are not in P_I are connected to some vertex of P_I by an edge. Hence $IG(X)$ is a valid interval graph.

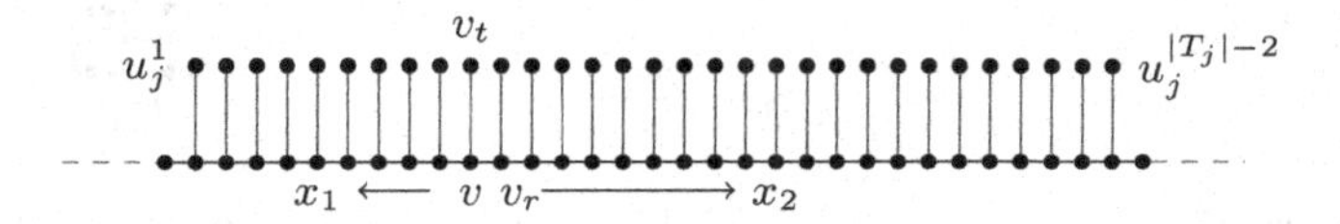

Fig. 3. Structure of a T_j with 33 vertices, along with the extra vertices connected to it. The dashed line represents the fact that other subpaths may be connected to a T_j on either or both ends.

Let u_j^1 be the vertex connected to the $2nd$ vertex of each T_j and $u_j^{|T_j|-2}$ be the vertex connected to its $2nd$-last vertex of T_j, where $|T_j|$ stands for the number of vertices in the subpath T_j. Let $A^{T_j} = \{u_j^1, u_j^2, ..., u_j^{|T_j|-2}\}$ be the set of all $|T_j|-2$ additional vertices corresponding to T_j. Now we mention an important observation regarding burning T_j^c.

Observation 3. *If T_j^c is burnt by putting $m \geq 2$ fire sources on T_j, then the burning clusters of at least two of these fire sources overlap (i.e., contain common vertices) of T_j.*

Proof. Let that some T_j^c be completely burnt by two or more fire sources and yet there is no overlap between the burning clusters of any of those fire sources. Since all the fire sources are on T_j, which is a sub path of P_I, we say two fire sources on T_j are *adjacent* if there is a path in T_j between those two fire sources such that the path does not contain any other fire sources. For any two adjacent fire sources let us assume that there is no vertex which lies in the burning clusters of both the fire sources. Let v be a vertex on the path joining those two adjacent fire sources x_1 and x_2, such that the vertices in the left side of v including it (vertices towards x_1 as shown in Fig. 3 using the left arrow) are burnt by x_1 and the vertices in the right of v (excluding v) are burnt by x_2.

Let the vertex that is just right to v is v_r. By pigeonhole principle, we have that at least one of v or v_r having a neighbor v_t in T_j^c which is not in T_j. Without the loss of generality, let that v is having such a neighbor. Since the

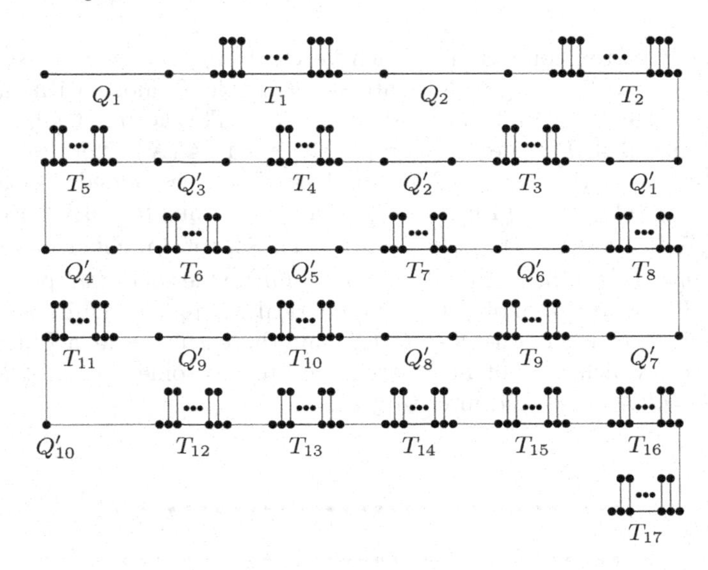

Fig. 4. Construction of an example $IG(X)$.The corresponding numerical example is in Sect. 5.2

burning cluster of x_1 extends till v and not to its one hop neighbor v_r ($\in T_j$), so it does not burn the other one hop neighbor v_t ($\notin T_j$) too. It is easy to see that the second fire source can not burn v_t. This is contradiction to our assumption that T_j^c is burnt completely without overlapping clusters. □

The following observation is immediate.

Observation 4. *If a single fire source is able to burn T_j in t rounds, then T_j^c would also be burnt by it in the same number of rounds.*

Lemma 2. *If at least one T_j is burnt using more than one fire sources, then P_I can not be burnt optimally, i.e., in $b(P_I) = 2m + 1$ steps.*

Proof. Since P_I is a simple path of length $(2m+1)^2$, according to Observation 1, each fire source in an optimal burning sequence must burn disjoint set of vertices of P_I. Let $x_1, x_2, ..., x_{2m+1}$ be an optimal burning sequence of P_I such that some T_j is burnt using more than one fire sources, then according to Observation 3, at least two fire sources burn at least one common vertex of P_I and hence $x_1, x_2, ..., x_{2m+1}$ can not be an optimal burning sequence for P_I. □

5.2 Example Construction

Here we show the construction of $IG(X)$ from a particular input set X. Let $X = \{10, 11, 12, 14, 15, 16\}$. Then $n = 2$, $m = 16$, $B = 39$, and $k = 10$. Also $F_m = \{1, 2, ..., 16\}$ and $F'_m = \{1, 3, ..., 31\}$. Further, $X' = \{19, 21, 23, 27, 29, 31\}$, $B' = 75 = 2B - 3$ and $Y = \{1, 3, 5, 7, 9, 11, 13, 15, 17, 25\}$. Observe that Q_1 and

Q_2 are paths of size 75, and each $Q'_1, Q'_2, ..., Q'_k$ are paths of order of 25, 17, 15, 13, 11, 9, 7, 5, 3, 1 respectively. $T_1, T_2, T_3, ...T_{m+1}$ are of order of $65, 63, 61..., 33$ respectively. We add a vertex connected to each vertex from *2nd* to *2nd*-last vertices of $T_j (1 \leq j \leq m+1)$. Observe that this is a valid interval graph. The constructed example $IG(X)$ is shown in Fig. 4.

5.3 NP-Completeness

In this section we proceed to the NP-Completeness proof through a series of lemmas. We start with the following.

Lemma 3. *If X' has a solution for the distinct 3-partition problem, then burning number of $IG(X)$ is $2m+1$.*

Proof. If X' has a solution for the distinct 3-partition problem, there would be n sets of three numbers each, sum of which is B'. Recall that length of each Q_i is B'. Hence, $Q_1, ..., Q_n$ can be partitioned into further subpaths $Q''_1, ..., Q''_{3n}$. Let us call the partitions of Q_i as $Q''_{3(i-1)+1}$, $Q''_{3(i-1)+2}$, and Q''_{3i}. Since X' is a set of odd numbers, length of each of these $3n$ subpaths are odd.

Let $P' = \{Q''_1, ..., Q''_{3n}, Q'_1, ..., Q'_k, T_1, ..., T_{m+1}\}$. Let r_i be the $((2m+1)-i+1)^{th} = (2m-i+2)^{th}$ vertex on the ith largest subpath in P'. Then, the burning sequence $S' = (r_1, r_2, .., r_{2m+1})$ can burn P_I and subsequently $IG(X)$. This implies that $b(IG(X)) \leq 2m+1$. Since $IG(X)$ has a subpath of length $(2m+1)^2$ in form of P_I, we have $b(IG(X)) \geq 2m+1$. Hence, $b(IG(X)) = 2m+1$. □

Lemma 4. *Each fire source y_i of any optimal burning sequence $(y_1, y_2, ..., y_{2m+1})$ of $IG(X)$ must be on P_I.*

Proof. We prove it by contradiction. If for any i, y_i is on P_I, then subgraph induced by $G.N_{2m+1-i}[y_i] \cap P_I$ has length at most $2(2m+1-i)+1$. Let we put a fire source y_i on some vertex of A^{T_j} for some j, which is not on P_I, and still burn $IG(X)$ in $2m+1$ steps. Then subgraph induced by $G.N_{2m+1-i}[y_i] \cap P_I$ is a path of length less than $2(2m+1-i)+1$. This along with Eq. 1 implies that $|\cup_{i=1}^{2m+1} G.N_{2m+1-i}[y_i] \cap P_I| < (2m+1)^2$. So, even P_I is not burnt. This is a contradiction to our assumption that $IG(X)$ is burnt in $2m+1$ steps. Therefore each y_i must be a put on some vertex in P_I. □

Let $S' = (y_1, y_2, ..., y_{2m+1})$ be any optimal *burning sequence*. Let r_i be the $(2m-i+2)^{th}$ vertex on the ith largest sub path in P' as described in the proof of Lemma 3. Observe that T_j's are the largest $m+1$ sub paths in P'.

Lemma 5. *We must have $y_i = r_i$, $\forall\ 1 \leq i \leq m+1$.*

Proof. We are going to prove this lemma using the strong induction hypothesis. We have that each $u_j^k \in A^{T_j}$ for some j must receive fire from some y_i in P_I, as all fire sources must be on P_I (Lemma 4). For $i = 1$, the only vertex connected to both u_1^1 and $u_1^{|T_1|-2}$ and within a distance $2m+1-i = 2m$, is r_1. Now we must have $y_1 = r_1$, else, if we put y_1 somewhere else, then neither y_1 nor any other fire source can burn T_1^c alone. Also, we can not use multiple fire sources to burn T_1^c as an optimal burning of $IG(X)$ does not allow that (Lemma 2). So, we must have that $y_1 = r_1$. Now to establish strong induction, let that we need to have y_k on r_k for $1 \leq k \leq m$.

Since r_k is already used to burn T_k^c, the only fire source that can burn T_{k+1}^c alone, is r_{k+1}. Recall that T_{k+1} has the largest length among the remaining subpaths after $T_1, T_2, \cdots, T_k$ are burnt. And we can not use multiple fire sources to burn T_{k+1}^c (Lemma 2). Also the only vertex connected to both u_{k+1}^1 and $u_{k+1}^{|T_{k+1}|-2}$ within distance $2m+1-(k+1)$ is r_{k+1}. So, we must have that $y_{k+1} = r_{k+1}$. This completes the proof. □

Let $P'' = IG(X) \backslash (T_1^c \cup T_2^c \cup ... \cup T_{m+1}^c)$. Now we present the following lemma on burning this remaining subgraph P''. That is P'' is a path forest consists of the subpaths $Q_1, Q_2 ..., Q_n, Q_1', Q_2', ..., Q_k'$. Now we present the following lemma on burning P''.

Lemma 6. *There is a partition of P'', induced by the fire sources y_i $(m+1 \leq i \leq 2m+1)$ of the optimal burning sequence S', into paths of orders in F_m'.*

Proof. From Lemma 5, we have that $\forall\, 1 \leq i \leq m+1$, all the vertices in T_i^c, would be burnt by y_i. Therefore, we have to burn the vertices in $Q_1, ..., Q_n, Q_1', ..., Q_k'$ by the fire sources $\{y_{m+2}, y_{m+3}, \cdots, y_{2m+1}\}$ (the remaining m sources of fire). Since P'' is a disjoint union of paths, so we have that $\forall i$ such that $m+2 \leq i \leq 2m+1$, the subgraph induced by the vertices in $G.N_{2m+1-i}[y_i]$ is a path of length at most $2(2m+1-i)+1$. Moreover, we have that the path forest P'' is of order $\sum_{i=1}^{m}(2i-1) = m^2$. This implies that for each i with $m+2 \leq i \leq 2m+1$, the subgraph induced by the vertices in $G.N_{2m+1-i}[y_i]$ is a path of order equal to $2(2m+1-i)+1$, otherwise we cannot burn all the vertices of P'' by these m fire sources which is a contradiction to the fact that S' is a optimal burning sequence. Therefore there must be a partition of P'', induced by the burning sequence $y_{m+2}, y_{m+3}, ..., y_{2m+1}$, into subpaths of length as per each element in $F_m' = \{1, 3, 5, ..., 2m-1\}$. □

Theorem 2. *Optimal burning of an interval graph is NP-Complete.*

Proof. one part is already proved in Lemma 3. Here we show the other part. Let say we have a $2m+1$ round optimal solution of the burning problem. Each T_j^c must get burned by exactly one fire source as per Lemma 5. From Lemma 6, we claim that the remaining path forest must be burned by the rest of available fire sources corresponding to the set F_m'.

Now, if each of the Q_i' is burned by a single fire source, then they must be burned by the fire sources corresponding to the integers belonging to set Y. Hence the remaining fire sources burning Q_i's are burnt by fire sources corresponding to the integers belonging to the set X'. As the size of each Q_i is B', which is always odd ($B' = 2B-3$), and also $B' > 2m-1$ (from the definition of distinct 3-partition problem, $m < B/2$), so no Q_i can be burnt by a single fire source. Again it can not be burnt by two fire sources as sum of any two numbers in X'

are even. Also no Q_i can be burnt by 4 or more fire sources as then by pigeon whole principle there would be at least one Q_i which needs to be burned by at most 2 fire sources, which is not possible. Hence each Q_i must be burnt by exactly three fire sources.

Else, if Q_i''s are not burned by single fire sources, we apply the following process subject to each subpath Q_j' for $1 \leq j \leq k$. Let that some subpath Q_j' is burned using multiple fire sources. Since the given solution is optimal so these burning clusters are non overlapping. Not only that, the sum of the cluster sizes of these fire sources is exactly same as order of Q_j'. Now some fire source with cluster size equal to order of Q_j' must be present on some other subpath. We can interchange that fire source (whose cluster size is Q_j') by these fire sources (which are presently burning Q_j'). This way we can make each subpath Q_j' to be burnt by a single fire source whose cluster size is equal to Q_j'. This takes $O(m)$ time. After this we again arrived to the case discussed above and we can see that each Q_i are burnt by exactly three fire sources corresponding to the integers in X'. Therefore we have a solution of the distinct 3-partition problem whose input set is X'. This, in turn, gives us the solution of the distinct 3-partition problem on the input set X.

Therefore, we have reduced the burning problem of $IG(X)$ from the distinct 3-partition problem in pseudo-polynomial time. Since, the distinct 3-partition problem is NP-Complete in the strong sense, burning $IG(X)$ is also NP-Complete in the strong sense. □

6 More Hardness Results

In this section we report hardness results on few more graph classes that mostly follow from our result on the interval graph. A *disc graph* is formed from an arrangement of discs on a Euclidean plane such that there is a vertex in the disc graph corresponding to each disc, and if there is an overlap between a pair of discs, then there shall be an edge between their corresponding vertices in the disc graph. Since any interval graph is valid to be a disc graph, we have the following.

Corollary 2. *Optimal burning of disc graphs is NP-Complete even if the underlying disc representation is given.*

In a *unit distance graph*, the edges can be drawn in a euclidean plane such that each edge is of unit length. In *matchstick graph*, the edges can be drawn in a euclidean plane such that each edge is of unit length and they do not intersect each other. The graph class that we have constructed is valid to be a unit distance graph and a matchstick graph. So we have Corollary 3 as follows.

Corollary 3. *Optimal burning of unit distance graphs and matchstick graphs is NP-Complete.*

A *permutation graph* is constructed from an original sequence of objects $O = (1, 2, 3, ..., k)$ which are numbers here and its permutation $P = (p_1, p_2, p_3, ..., p_k)$ such that there is an edge between two vertices corresponding to number i and j

respectively, if $i < j$ and j occurs before i in P. We can compute a permutation of an arbitrary original sequence of numbers such that the corresponding permutation graph becomes a path forest (see [14] for a simple construction formula), which leads to the following Corollary 4.

Corollary 4. *Burning of general permutation graphs is NP-Complete.*

7 Conclusion

In this article we show that the graph burning problem is NP-Complete on interval graphs which completes the study of burning interval graph given the already existing results. We also show a lower bound for the burning number of grids of arbitrary size and give a two approximation algorithm for burning square grids. It remains an open question whether burning grids is an NP-Complete problem. Another very much related direction is to try and improve the 3-approximation algorithm provided in [5] for burning general graphs.

References

1. S̃imon, M., Huraj, L., Dirgovã Luptãkovã, I., Pospíchal, J.: Heuristics for spreading alarm throughout a network. Appl. Sci. **9**(16), 3269 (2019)
2. Alon, N., PraLat, P., Wormald, N.: Cleaning regular graphs with brushes. SIAM J. Disc. Math. **23**(1), 233–250 (2009)
3. Balogh, J., Bollobás, B., Morris, R.: Graph bootstrap percolation. Random Struct. Algorithms **41**(4), 413–440 (2012)
4. Banerjee, S., Gopalan, A., Das, A., Shakkottai, S.: Epidemic spreading with external agents. IEEE Trans. Inf. Theory **60**, 06 (2012)
5. Bessy, S., Bonato, A., Janssen, J., Rautenbach, D., Roshanbin, E.: Burning a graph is hard. Disc. Appl. Math. **232**(C), 73–87 (2017)
6. Bessy, S., Bonato, A., Janssen, J., Rautenbach, D., Roshanbin, E.: Bounds on the burning number. Disc. Appl. Math. **235**, 16–22 (2018)
7. Bonato, A., Janssen, J., Roshanbin, E.: How to burn a graph. Internet Math. **12**(1–2), 85–100 (2016)
8. Bonato, A., Kamali, S.: Approximation algorithms for graph burning. In: Gopal, T.V., Watada, J. (eds.) TAMC 2019. LNCS, vol. 11436, pp. 74–92. Springer, Cham (2019). https://doi.org/10.1007/978-3-030-14812-6_6
9. Bonato, A., Lidbetter, T.: Bounds on the burning numbers of spiders and path-forests. Theor. Comput. Scie. **794**, 12–19 (2019)
10. Bonato, A., Janssen, J., Roshanbin, E.: Burning a graph as a model of social contagion. In: Bonato, A., Graham, F.C., Prałat, P. (eds.) WAW 2014. LNCS, vol. 8882, pp. 13–22. Springer, Cham (2014). https://doi.org/10.1007/978-3-319-13123-8_2
11. Das, S., Dev, S.R., Sadhukhan, A., Sahoo, U., Sen, S.: Burning spiders. In: Panda, B.S., Goswami, P.P. (eds.) CALDAM 2018. LNCS, vol. 10743, pp. 155–163. Springer, Cham (2018). https://doi.org/10.1007/978-3-319-74180-2_13
12. Finbow, S., Macgillivray, G.: The firefighter problem: a survey of results, directions and questions. Australas. J. Comb. [Electron. only] **43**, 57–78 (2009)

13. Garey, M.R., Johnson, D.S.: Computers and Intractability: A Guide to the Theory of NP-Completeness. W. H. Freeman & Co., New York (1979)
14. Gupta, A.T., Lokhande, S.A., Mondal, K.: Np-completeness results for graph burning on geometric graphs. arXiv: 2003.07746 (2020)
15. Kamali, S., Miller, A., Zhang, K.: Burning two worlds. In: SOFSEM (2020)
16. Kare, A.S., Vinod Reddy, I.: Parameterized algorithms for graph burning problem. In: Colbourn, C.J., Grossi, R., Pisanti, N. (eds.) IWOCA 2019. LNCS, vol. 11638, pp. 304–314. Springer, Cham (2019). https://doi.org/10.1007/978-3-030-25005-8_25
17. Kempe, D., Kleinberg, J., Tardos, É.: Maximizing the spread of influence through a social network. In: ACM SIGKDD, pp. 137–146. Association for Computing Machinery, New York (2003)
18. Kempe, D., Kleinberg, J., Tardos, É.: Influential nodes in a diffusion model for social networks. In: Caires, L., Italiano, G.F., Monteiro, L., Palamidessi, C., Yung, M. (eds.) ICALP 2005. LNCS, vol. 3580, pp. 1127–1138. Springer, Heidelberg (2005). https://doi.org/10.1007/11523468_91
19. Kramer, A.D.I., Guillory, J.E., Hancock, J.T.: Experimental evidence of massive-scale emotional contagion through social networks. Proc. Natl. Acad. Sci. **111**(24), 8788–8790 (2014)
20. Simon, M., Huraj, L., Dirgova, L., Pospichal, J.: How to burn a network or spread alarm. MENDEL **25**(2), 11–18 (2019)

Parameterized Algorithms

On Parameterized Complexity of Liquid Democracy

Palash Dey, Arnab Maiti, and Amatya Sharma(✉)

Indian Institute of Technology Kharagpur, Kharagpur, India
palash.dey@cse.iitkgp.ac.in, arnabmaiti@iitkgp.ac.in,
amatya65555@iitkgp.ac.in

Abstract. In *liquid democracy*, each voter either votes herself or delegates her vote to some other voter. This gives rise to what is called a delegation graph. To decide the voters who eventually votes along with the subset of voters whose votes they give, we need to *resolve* the cycles in the delegation graph. This gives rise to the Resolve Delegation to MinMaxWeight problem where we need to find an acyclic sub-graph of the delegation graph such that the number of voters whose votes they give is bounded above by some integer λ. Putting a cap on the number of voters whose votes a voter gives enable the system designer restrict the power of any individual voter. The Resolve Delegation to MinMaxWeight problem is already known to be NP-hard. In this paper we study the parameterized complexity of this problem. We show that Resolve Delegation to MinMaxWeight is para-NP-hard with respect to parameters λ, number of sink nodes and the maximum degree of the delegation graph. We also show that Resolve Delegation to MinMaxWeight is W[1]-hard even with respect to the treewidth of the delegation graph. We complement our negative results by exhibiting FPT algorithms with respect to some other parameters. We finally show that a related problem, which we call Resolve Fractional Delegation, is polynomial time solvable.

Keywords: Liquid democracy · Resolve Delegation to MinMaxWeight · Parameterized complexity

1 Introduction

In a *direct democracy*, agents vote for a candidate by themselves. In *liquid democracy*, the voters can delegate their votes to other agents who can vote on their behalf. Suppose voter 1 delegates her vote to voter 2 and voters 2 and 3 delegate their votes to voter 4. Then voter 4 has a voting power equivalent to 4 individual votes. That is delegations are transitive. This particular feature can make liquid democracy a disruptive approach to democratic voting system. This happens because such a voting system can lead to what we call a super-voter who has a lot of voting power. So now the candidates instead of trying to appease the general public can do behind the closed door dealings with the super-voters and try

A. Mudgal and C. R. Subramanian (Eds.): CALDAM 2021, LNCS 12601, pp. 83–94, 2021.
https://doi.org/10.1007/978-3-030-67899-9_7

to win the election in an unfair manner. In order to deal with this issue, a central mechanism ensures that no super-voter has a lot of voting power. Formally we do it as follows. We create a delegation graph where the set of vertices is the set of voters and we have a directed edge from vertex i to vertex j if voter i delegates her vote to voter j. We observe that delegation graph may contain cycles. Every voter is also allowed to delegate her vote to more than one other voters and let the system decide her final delegation. We use a central mechanism to find a acyclic sub-graph of the delegation graph such that no super-voter (the vertices having out-degree 0) has a lot of voting power. We call this problem Resolve Delegation to MinMaxWeight.

1.1 Related Work

An empirical investigation of the existence and influence of super-voters was done by [9]. They showed that the super-voters can be powerful although they seem to act in a responsible manner according to their results. There have been a few theoretical work in this area by [4,7] and [8]. A detailed theoretical work especially on the approximation algorithms in this setting was done by [6]. Some other important work in Liquid democracy includes [2] and [3].

1.2 Our Contribution

We study parameterized complexity of the Resolve Delegation to MinMaxWeight problem with respect to various natural parameters. In particular, we consider the number of sink vertices (t), maximum allowed weight λ of any sink in the final delegation graph, maximum degree (Δ), tree-width, number of edges deleted in optimal solution (e_{rem}), number of non-sink vertices ($|\mathcal{V} \setminus \mathcal{T}|$). The number of sink vertices corresponds to the number of influential voters which is often a small number in practice. This makes the number of sink vertices an important parameter to study. Similarly, the parameter λ corresponds to the "power" of a voter. Since the input to the problem is a graph, it is natural to study parameters, for example, tree-width (by ignoring the directions of the edges) and the number of edges that one needs to delete in an optimal solution. We summarize our results in Table 1. We finally show that Resolve Delegation to MinMaxWeight is polynomial time solvable if we allow fractional delegations [Theorem 6].

2 Preliminaries

A directed graph $\mathcal{G}$ is a tuple $(\mathcal{V}, \mathcal{E})$ where $\mathcal{E} \subseteq \{(x, y) : x, y \in \mathcal{V}, x \neq y\}$. For a graph $\mathcal{G}$, we denote its set of vertices by $\mathcal{V}[\mathcal{G}]$, its set of edges by $\mathcal{E}[\mathcal{G}]$, the number of vertices by n, and the number of edges by m. Given a graph $\mathcal{G} = (\mathcal{V}, \mathcal{E})$, a sub-graph $\mathcal{H} = (\mathcal{V}', \mathcal{E}')$ is a graph such that (i) $\mathcal{V}' \subseteq \mathcal{V}$, (ii) $\mathcal{E}' \subseteq \mathcal{E}$, and (iii) for every $(x, y) \in \mathcal{E}'$, we have $x, y \in \mathcal{V}'$. A sub-graph $\mathcal{H}$ of a graph $\mathcal{G}$ is called a *spanning sub-graph* if $\mathcal{V}[\mathcal{H}] = \mathcal{V}[\mathcal{G}]$ and *induced sub-graph* if $\mathcal{E}[\mathcal{H}] = \{(x, y) \in \mathcal{E}[\mathcal{G}] : x, y \in \mathcal{V}[\mathcal{H}]\}$. Given an induced path P of a graph, we define *end vertex* as vertex with 0 outdegree in P and *start vertex* as a vertex with 0 indegree in P.

Table 1. Summary of results.

Parameter	*Result*
t	para-NP-hard [Theorem 1]
(λ, Δ)	para-NP-hard [Theorem 2]
(λ, t)	quadratic vertex kernel [Observation 1]
tree-width	W[1]-Hard [Theorem 3]
e_{rem}	FPT by bounded search tree technique [Theorem 5]
$\lvert \mathcal{V} \setminus \mathcal{T} \rvert$	FPT by bounded search tree technique [Theorem 4]
Problem under Assumption	*Result*
fractional delegation	Reduction to LP [Theorem 6]
DAG,Bipartite Graph	W[1]-Hard w.r.t treewidth [Corollary 2]
DAG,Bipartite Graph	para-NP-hard w.r.t λ, Δ [Corollary 1]

2.1 Problem Definition

We now define our problem formally.

Definition 1 (Resolve Delegation to MinMaxWeight). *Given a directed graph $\mathcal{G} = (\mathcal{V}, \mathcal{E})$ (also known as delegation graph) with the set $\mathcal{T} \subseteq \mathcal{V}$ as its set of sink vertices and an integer λ, decide if there exists a spanning sub-graph $\mathcal{H} \subseteq \mathcal{G}$ such that*

(i) The out-degree of every vertex in $\mathcal{V} \setminus \mathcal{T}$ is exactly 1
(ii) For every sink vertex $t \in \mathcal{T}$, the number of vertices (including t) in $\mathcal{V}$ which has a path to t in the sub-graph $\mathcal{H}$ is at most λ

We denote an arbitrary instance of Resolve Delegation to MinMaxWeight *by $(\mathcal{G}, \lambda)$.*

In the spanning sub-graph $\mathcal{H} \subseteq \mathcal{G}$, if there is a path from u to v in $\mathcal{H}$ such that all the vertices on this path except v has out-degree 1, then we say that vertex u *delegates* to vertex v. In any spanning sub-graph $\mathcal{H} \subseteq \mathcal{G}$ with the out-degree of every vertex in $\mathcal{V} \setminus \mathcal{T}$ is exactly 1 (we call sub-graph $\mathcal{H}$ a feasible solution), weight of a tree rooted at the sink vertex u is the number of vertices (including u) that have a directed path to u. We study parameterized complexity of Resolve Delegation to MinMaxWeight with respect to t, λ, and the maximum degree Δ of the input graph as our parameters. In the optimization version of Resolve Delegation to MinMaxWeight, we aim to minimize λ.

3 Results: Algorithmic Hardness

Our first result shows that Resolve Delegation to MinMaxWeight is NP-complete even if we have only 3 sink vertices. For that, we exhibit reduction from the Two Vertex Disjoint Paths problem.

Definition 2 (Two Vertex Disjoint Paths). *Given a directed graph* $\mathcal{G} = (\mathcal{V}, \mathcal{E})$*, two pairs* (s_1, t_1) *and* (s_2, t_2) *of vertices which are all different from each other, compute if there exists two vertex disjoint paths* $\mathcal{P}_1$ *and* $\mathcal{P}_2$ *where* $\mathcal{P}_i$ *is a path from* s_i *to* t_i *for* $i \in [2]$*. We denote an arbitrary instance of it by* $(\mathcal{G}, s_1, t_1, s_2, t_2)$.

We know that Two Vertex Disjoint Paths is NP-complete [5]. The idea is to add paths containing large number of nodes in the instance of Resolve Delegation to MinMaxWeight which we are creating using the instance of Two Vertex Disjoint Paths. This key idea will make both the instances equivalent.

Theorem 1. *The* Resolve Delegation to MinMaxWeight *problem is* NP*-complete even if we have only* 3 *sink vertices. In particular,* Resolve Delegation to MinMaxWeight *is para-*NP*-hard with respect to the parameter* t.

Proof. The Resolve Delegation to MinMaxWeight problem clearly belongs to NP. To show its NP-hardness, we reduce from Two Vertex Disjoint Paths. Let $(\mathcal{G} = (\mathcal{V}, \mathcal{E}), s_1, t_1, s_2, t_2)$ be an arbitrary instance of Two Vertex Disjoint Paths. Let $n = |\mathcal{V}|$. We consider the following instance $(\mathcal{G}' = (\mathcal{V}', \mathcal{E}'), \lambda)$.

$$\mathcal{V}' = \{a_v : v \in \mathcal{V}\} \cup \mathcal{D}_1 \cup \mathcal{D}_1' \cup \mathcal{D}_2 \cup \mathcal{D}_2' \cup \mathcal{D}_3 \text{ where}$$
$$|\mathcal{D}_1| = |\mathcal{D}_2'| = 10n, |\mathcal{D}_1'| = |\mathcal{D}_2| = 5n, |\mathcal{D}_3| = 15n$$
$$\mathcal{E}' = \{(a_u, a_v) : (u, v) \in \mathcal{E}\} \cup \mathcal{F}$$

We now describe the edges in $\mathcal{F}$. Each $\mathcal{D}_1, \mathcal{D}_1', \mathcal{D}_2, \mathcal{D}_2'$ and $\mathcal{D}_3$ induces a path in $\mathcal{G}'$ and thus the edges in these paths are part of $\mathcal{F}$. The end vertices of the path induced on $\mathcal{D}_1$ and $\mathcal{D}_2$ be respectively d_1 and d_2. The start vertices of the path induced on $\mathcal{D}_1'$ and $\mathcal{D}_2'$ be respectively d_1' and d_2'. The end vertices of the path induced on $\mathcal{D}_1', \mathcal{D}_2'$ and $\mathcal{D}_3$ be t_1', t_2' and t_3' respectively. The set $\mathcal{F}$ also contains the edges in $\{(d_1, a_{s_1}), (d_2, a_{s_2}), (a_{t_1}, d_1'), (a_{t_2}, d_2')\}$. $\mathcal{F}$ also contains edge (a_v, t_3') $\forall v \in \mathcal{V}$. This finishes the description of $\mathcal{F}$ and thus the description of $\mathcal{G}'$. We observe that $\mathcal{G}'$ has exactly 3 sink vertices, namely t_1', t_2' and t_3'. Finally we define $\lambda = 17n$. We claim that the two instances are equivalent.

In one direction, let us assume that the Two Vertex Disjoint Paths instance is a yes instance. For all $i \in [2]$, let $\mathcal{P}_i$ be a path from s_i to t_i in $\mathcal{G}$ such that $\mathcal{P}_1$ and $\mathcal{P}_2$ are vertex disjoint. We build the solution $\mathcal{H}$ for Resolve Delegation to MinMaxWeight by first adding the set of edges $\{(u, v) | \text{outdegree of } u \text{ is } 1\}$. Then we add the paths $\mathcal{P}_1$ and $\mathcal{P}_2$. Then we add the edges $(a_{t_1}, d_1'), (a_{t_2}, d_2')$. Then for each vertex u in the set $\mathcal{V}^r = \{a_v | v \in \mathcal{V}\} \setminus \mathcal{V}[\mathcal{P}_1 \cup \mathcal{P}_2]$, add the edge (u, t_3') to $\mathcal{H}$.

We observe that the out degree of every vertex is exactly 1 in $\mathcal{H}$ except the sink vertices in $\mathcal{G}'$ (which are t_1', t_2' and t_3'). Also since $\mathcal{H}$ contains the path $\mathcal{P}_i$, every vertex in $\mathcal{D}_i$ has a path to t_i' for $i \in [2]$. Of course, every vertex in $\mathcal{D}_i'$ has a path to t_i' for $i \in [2]$ and every vertex in $\mathcal{D}_3$ delegates to t_3'. Hence $\forall i \in [3]$, the number of vertices which has a path to t_i in $\mathcal{H}'$ is at most $16n$ which is less than λ. Hence the RESOLVE DELEGATION TO MINMAXWEIGHT instance is a YES instance.

In the other direction, let us assume that the RESOLVE DELEGATION TO MINMAXWEIGHT instance is a YES instance. Let $\mathcal{H}' = (\mathcal{V}', \mathcal{E}'') \subseteq \mathcal{G}'$ be a spanning sub-graph of $\mathcal{G}'$ such that (i) the out degree of every vertex which is not a sink is exactly 1, (ii) there are at most λ $(= 17n)$ vertices (including the sink nodes) in $\mathcal{H}'$ which has a path to t'_i for $i \in [3]$. Note that a_{s_1} must have a path $\mathcal{P}'_1$ to a_{t_1} in $\mathcal{H}'$ otherwise at least $20n$ vertices have path to either t'_2 or t'_3 in $\mathcal{H}'$ which is a contradiction (since $\lambda = 17n$). Similarly a_{s_2} must have a path $\mathcal{P}'_2$ to a_{t_2} in $\mathcal{H}'$ otherwise at least $20n$ vertices have path to either t'_1 or t'_3 in $\mathcal{H}'$ which is a contradiction (since $\lambda = 17n$). Since, for $i \in [2]$, we have a path $\mathcal{P}'_i$ from a_{s_i} to a_{t_i} in $\mathcal{H}'$ and the out-degree of every vertex in $\mathcal{H}'$ except t'_1, t'_2 and t'_3 is 1, the paths $\mathcal{P}'_1$ and $\mathcal{P}'_2$ are vertex disjoint. We define path $\mathcal{P}_i = \{(u, v) : (a_u, a_v) \in \mathcal{P}'_i\}$ in $\mathcal{G}$ for $i \in [2]$. Since $\mathcal{P}'_1$ and $\mathcal{P}'_2$ are vertex disjoint, it follows that $\mathcal{P}_1$ and $\mathcal{P}_2$ are also vertex disjoint. Thus the RESOLVE DELEGATION TO MINMAXWEIGHT instance is a YES instance. □

We next show that RESOLVE DELEGATION TO MINMAXWEIGHT is NP-complete even if we have $\lambda = 3$ and $\Delta = 3$. For that we exhibit a reduction from (3, B2)-SAT which is known to be NP-complete [1].

Definition 3 ((3, *B*2)-SAT). *Given a set* $\mathcal{X} = \{x_i : i \in [n]\}$ *of* n *variables and a set* $\mathcal{C} = \{C_j : j \in [m]\}$ *of* m *3-CNF clauses on* $\mathcal{X}$ *such that, for every* $i \in [n]$, x_i *and* $\bar{x}_i$ *each appear in exactly* 2 *clauses, compute if there exists any Boolean assignment to the variables which satisfy all the* m *clauses simultaneously. We denote an arbitrary instance of* (3, *B*2)-SAT *by* $(\mathcal{X}, \mathcal{C})$.

For each literal and clause in (3, B2)-SAT we add a node in the instance of RESOLVE DELEGATION TO MINMAXWEIGHT and we add some special set of edges and nodes so that $\lambda = 3$ and both the out-degree and in-degree of every vertex is at most 3

Theorem 2. *The* RESOLVE DELEGATION TO MINMAXWEIGHT *problem is* NP-*complete even if we have* $\lambda = 3$ *and both the out-degree and in-degree of every vertex is at most* 3. *In particular,* RESOLVE DELEGATION TO MINMAXWEIGHT *is para-*NP*-hard with respect to the parameter* (λ, Δ).

Proof. The RESOLVE DELEGATION TO MINMAXWEIGHT problem clearly belongs to NP. To show its NP-hardness, we reduce from (3, B2)-SAT. Let $(\mathcal{X} = \{x_i : i \in [n]\}, \mathcal{C} = \{C_j : j \in [m]\})$ be an arbitrary instance of (3, B2)-SAT. We define a function $f : \{x_i, \bar{x}_i : i \in [n]\} \longrightarrow \{a_i, \bar{a}_i : i \in [n]\}$ as $f(x_i) = a_i$ and $f(\bar{x}_i) = \bar{a}_i$ for $i \in [n]$. We consider the following instance $(\mathcal{G} = (\mathcal{V}, \mathcal{E}), \lambda)$.

$$\mathcal{V} = \{a_i, \bar{a}_i, d_{i,1}, d_{i,2} : i \in [n]\} \cup \{y_j : j \in [m]\}$$
$$\mathcal{E} = \{(y_j, f(l^j_1)), (y_j, f(l^j_2)), (y_j, f(l^j_3)) : C_j = (l^j_1 \vee l^j_2 \vee l^j_3), j \in [m]\}$$
$$\cup \{(d_{i,2}, d_{i,1}), (d_{i,1}, a_i), (d_{i,1}, \bar{a}_i) : i \in [n]\}$$
$$\lambda = 3$$

We observe that both the in-degree and out-degree of every vertex in $\mathcal{G}$ is at most 3. Also $\Delta = 3$. We now claim that the two instances are equivalent.

Suppose the $(3, B2)$-SAT instance is a YES instance. Let $g : \{x_i : i \in [n]\} \longrightarrow \{\text{TRUE}, \text{FALSE}\}$ be a satisfying assignment of the $(3, B2)$-SAT instance. We define another function $h(g, j) = f(l), j \in [m]$, for some literal l which appears in the clause C_j and g sets it to TRUE. We consider the following sub-graph $\mathcal{H} \subseteq \mathcal{G}$

$$\begin{aligned}\mathcal{E}[\mathcal{H}] = &\{(d_{i,2}, d_{i,1}) : i \in [n]\} \\ &\cup \{(d_{i,1}, a_i) : i \in [n], g(x_i) = \text{FALSE}\} \\ &\cup \{(d_{i,1}, \bar{a}_i) : i \in [n], g(x_i) = \text{TRUE}\} \\ &\cup \{(y_j, h(g, j)) : j \in [m]\}\end{aligned}$$

We observe that $\mathcal{H}$ is a spanning sub-graph of $\mathcal{G}$ such that (i) every non-sink vertices in $\mathcal{G}$ has exactly one outgoing edge in $\mathcal{H}$ and (ii) for each sink vertex in $\mathcal{G}$, there are at most 3 vertices (including the sink itself) which has a path to it. Hence the RESOLVE DELEGATION TO MINMAXWEIGHT instance is a YES instance.

In the other direction, let the RESOLVE DELEGATION TO MINMAXWEIGHT instance is a YES instance. Let $\mathcal{H} \subseteq \mathcal{G}$ be a sub-graph of $\mathcal{G}$ such that (i) every non-sink vertices in $\mathcal{G}$ has exactly one outgoing edge in $\mathcal{H}$ and (ii) for each sink vertex in $\mathcal{G}$, there are at most 3 vertices (including the sink itself) which has a path to it. We define an assignment $g : \{x_i : i \in [n]\} \longrightarrow \{\text{TRUE}, \text{FALSE}\}$ as $g(x_i) =$ FALSE if $(d_{i,1}, a_i) \in \mathcal{E}[\mathcal{H}]$ and TRUE otherwise. We claim that g is a satisfying assignment for the $(3, B2)$-SAT instance. Suppose not, then there exists a clause $C_j = (l_1^j \vee l_2^j \vee l_3^j)$ for some $j \in [m]$ whom g does not satisfy. We define functions $f_1, f_2 : \{x_i, \bar{x}_i : i \in [n]\} \longrightarrow \{d_{i,1}, d_{i,2} : i \in [n]\}$ as $f_1(x_i) = f_1(\bar{x}_i) = d_{i,1}$ and $f_2(x_i) = f_2(\bar{x}_i) = d_{i,2}$. We observe that the sink vertex $f(l_i^j)$ is reachable from both $f_1(l_i^j)$ and $f_2(l_i^j)$ in $\mathcal{H}$ for every $i \in [3]$. Since $\lambda = 3$, we do not have a path from y_j to any of $f(l_i), i \in [3]$ which is a contradiction since the non-sink vertex y_j must have out-degree 1 in $\mathcal{H}$. Hence g is a satisfying assignment for the $(3, B2)$-SAT instance and thus the instance is a YES instance. □

Corollary 1. *Given that the input graph is both bipartite and directed acyclic graph, the* RESOLVE DELEGATION TO MINMAXWEIGHT *problem is* NP*-complete even if we have* $\lambda = 3$ *and both the out-degree and in-degree of every vertex is at most* 3 *which concludes that* RESOLVE DELEGATION TO MINMAXWEIGHT *is para-*NP*-hard with respect to the parameter* (λ, Δ).

Proof. The corollary follows as the resulting graph $\mathcal{G}$ from reduction of $(3, B2)$-SAT instance in Theorem 2 is bipartite as $\mathcal{V}$ can be partitioned into 2 independent sets $\mathcal{V}_1 = \{y_j : j \in [m]\} \cup \{d_{i,1} : i \in [n]\}$ and $\mathcal{V}_2 = \{a_i, \bar{a}_i, d_{i,2} : i \in [n]\}$. Also $\mathcal{G}$ is Directed Acyclic graph as it doesn't have directed cycles. □

Definition 4. *A (positive integral)* edge weighting *of a graph* G *is a mapping* w *that assigns to each edge of* G *a positive integer.*

Definition 5. *An* orientation *of* G *is a mapping* $\Lambda : E(G) \to V(G) \times V(G)$ *with* $\Lambda((u,v)) \in \{(u,v),(v,u)\}$.

Definition 6. *The* weighted outdegree *of a vertex* $v \in V(G)$ *w.r.t an edge weighting* w *and an orientation* Λ *is defined as* $d^+_{G,w,\Lambda}(v) = \sum_{(v,u)\in E(G) \text{ with } \Lambda((v,u))=(v,u)} w((v,u))$.

Definition 7. (MINIMUM MAXIMUM OUTDEGREE). *Given a graph* G, *an edge weighting* w *of* G *in unary and a positive integer* r, *is there an orientation* Λ *of* G *such that* $d^+_{G,w,\Lambda}(v) \leqslant r$ *for each* $v \in V(G)$?

Lemma 1. [10] MINIMUM MAXIMUM OUTDEGREE *is W[1]-hard when parameterized by the treewidth of the instance graph.*

We now show that RESOLVE DELEGATION TO MINMAXWEIGHT is W[1]-hard when parameterized by the treewidth of the instance graph. We reduce from MINIMUM MAXIMUM OUTDEGREE with instance graph G to RESOLVE DELEGATION TO MINMAXWEIGHT by first creating a replica of the G and then taking an edge (u,v) with weight w and replacing it with a path of w nodes with the end vertex having edges to u and v.

Theorem 3. RESOLVE DELEGATION TO MINMAXWEIGHT *is W[1]-hard when parameterized by the treewidth of the instance graph*

Proof. To prove W[1]-Hardness we reduce from MINIMUM MAXIMUM OUTDEGREE to RESOLVE DELEGATION TO MINMAXWEIGHT. Let a graph $G(V,E)$ with an edge weighting w in unary and a positive integer r be an arbitrary instance of MINIMUM MAXIMUM OUTDEGREE. MINIMUM MAXIMUM OUTDEGREE is considered to be a YES instance if the weighted outdegree of every vertex is upper bounded by r. Now using the instance of MINIMUM MAXIMUM OUTDEGREE we create an instance $(\mathcal{H}, r+1)$ of RESOLVE DELEGATION TO MINMAXWEIGHT. Let us construct a graph $\mathcal{H} = (\mathcal{V}, \mathcal{E})$ where $\mathcal{V} = V_1 \cup V_2$. $V_1 = \{b_u : u \in V\}$. $\forall (u,v) \in E$ add the set of vertices $\{a_{uv_1}, a_{uv_2}, \dots, a_{uv_{w(u,v)}}\}$ to V_2. $\forall (u,v) \in E$, $(a_{uv_1}, b_u) \in \mathcal{E}$, $(a_{uv_1}, b_v) \in \mathcal{E}$ and $\forall i \in [w(u,v)] \setminus \{1\}$, $(a_{uv_i}, a_{uv_{i-1}}) \in \mathcal{E}$. This completes the construction of $\mathcal{H}$ with V_1 as the sink nodes. It is trivial to observe the fact that $tw(\mathcal{H}) \leqslant tw(G) + 2$. We now prove that the MINIMUM MAXIMUM OUTDEGREE is an YES instance iff the RESOLVE DELEGATION TO MINMAXWEIGHT is an YES instance

Let MINIMUM MAXIMUM OUTDEGREE be a YES instance. Let Λ be the orientation of G which makes MINIMUM MAXIMUM OUTDEGREE an YES instance. We consider the following sub-graph $\mathcal{H}' \subseteq \mathcal{H}$

$$\begin{aligned}\mathcal{E}[\mathcal{H}'] = &\{(a_{uv_i}, a_{uv_{i-1}}) : i \in [w(u,v)] \setminus \{1\}, (u,v) \in E\}\\ &\cup \{(a_{uv_1}, b_u) : (u,v) \in E, \Lambda((u,v)) = (u,v)\}\end{aligned}$$

We observe that $\mathcal{H}'$ is a spanning sub-graph of $\mathcal{H}$ such that (i) every non-sink vertices in $\mathcal{H}$ has exactly one outgoing edge in $\mathcal{H}'$ and (ii) for each sink vertex in $\mathcal{H}$, there are at most $r+1$ vertices (including the sink itself) which has a path to it. Hence the RESOLVE DELEGATION TO MINMAXWEIGHT instance is a YES instance.

Let RESOLVE DELEGATION TO MINMAXWEIGHT be a YES instance. Let $\mathcal{H}'$ be the spanning sub-graph of $\mathcal{H}$ which make RESOLVE DELEGATION TO MINMAXWEIGHT a YES instance. Let the edges in $\mathcal{H}'$ be denoted by $\mathcal{E}'$. We consider the following orientation Λ of G

$$\Lambda((u,v)) = \begin{cases} (u,v) & \text{if} \quad (a_{uv_1}, b_u) \in \mathcal{E}' \\ (v,u) & \text{otherwise} \end{cases}$$

Clearly weighted outdegree of every vertex in G is atmost r. Therefore MINIMUM MAXIMUM OUTDEGREE is an YES instance.
This concludes the proof of this theorem. □

Corollary 2. RESOLVE DELEGATION TO MINMAXWEIGHT *is W[1]-hard when parameterized by the treewidth even when the input graph is both Bipartite and Directed Acyclic Graph.*

Proof. In the instance of RESOLVE DELEGATION TO MINMAXWEIGHT created in Theorem 3, graph $\mathcal{H}$ is Bipartite as there is no odd cycle in the underlying undirected graph. Also graph $\mathcal{H}$ is Directed Acyclic Graph (DAG) as there is no directed cycle. □

4 FPT Algorithms

We now prresent our FPT algorithms.

Observation 1. *There is a kernel for* RESOLVE DELEGATION TO MINMAXWEIGHT *consisting of at most* λt *vertices. In particular, there is an* FPT *algorithm for the* RESOLVE DELEGATION TO MINMAXWEIGHT *problem parameterized by* (λ, t).

Proof. If the number n of vertices in the input graph is more than λt, then the instance is clearly a NO instance. Hence, we have $n \leqslant \lambda t$. □

In this section we define the notion of weights for the nodes in the subgraph $\mathcal{H}$ of the delegation graph $\mathcal{G}$. We define weight of all nodes u in $\mathcal{G}$ to be 1. To get a notion of weight of a vertex u in a subgraph $\mathcal{H}$, it can be considered as a number which is one more than the number of nodes who have delegated their vote to u and then have been removed from the graph $\mathcal{G}$ during the construction of $\mathcal{H}$. If $\mathcal{H}$ is a forest such that every non-sink node has an outdegree 1, then clearly the weight of the tree rooted at a sink node say t is sum of the weights of the nodes in the tree. We now show RESOLVE DELEGATION TO MINMAXWEIGHT is FPT w.r.t number of non-sink nodes by using the technique of bounded search tree by the branching on set of vertices satisfying some key properties.

Theorem 4. *The* RESOLVE DELEGATION TO MINMAXWEIGHT *problem has a FPT with respect to the parameter* k *which is the number of non-sink nodes in* $\mathcal{G}$ *(delegation graph).*

Proof. Let us denote the problem instance by $(\mathcal{G}, \lambda, k)$. Now we present the following reduction and branching rules.

Reduction RD.1. If there is a vertex v in $\mathcal{V}$ with only one outgoing edge to a vertex u (u, v are distinct), delete v from graph and increase weight of u by the weight of v. The incoming edges which were incident on v (except the self loops if any) are now incident on u.

Safeness of Reduction RD.1. is trivial as a node v with single outgoing edge can only delegate the votes it has got (this includes v's own vote and the votes of other nodes who have delegated to v so far) to the only neighbor u it has got.

Reduction RD.2 Remove self loops if any.
Safeness of Reduction RD.2. follows from the fact that no non-sink node can delegate to itself

Reduction RD.3. If $\mathcal{G}$ contains a non-sink node v with outdegree more than $2(k-1)$ and indegree 0, delete v from $\mathcal{G}$. The new instance is $(\mathcal{G} - v, \lambda, k-1)$
Safeness of Reduction RD.3. is due to the fact that if we have a vertex v with outdegree greater than $2(k-1)$, it implies that it has an outgoing edge to at least k sink nodes. Let us denote these sink nodes by set S. So, irrespective of the delegations made by other vertices, there will exist one sink node $t' \in S$ such that none of the other $k-1$ non-sink nodes have delegated to t' and hence we can delegate v to t' and still not increase the maximum weight of the sink node.

Branching B.1. Pick a vertex v such that the outdegree is more than $2(k-1)$ and indegree is $k' > 0$. Note that $k' \leqslant k-1$. Each of k' nodes having an outgoing edge to v can either delegate to v or not delegate it. So we have $2^{k'}$ possibilities and hence we can create $2^{k'}$ subproblems. In each possibility if a node u_1 is delegating to v then we delete all the outgoing edges of u_1 expect (u_1, v) and if we have a node u_2 which doesn't delegate to v then we delete the outgoing edge from u_2 to v. In each of the $2^{k'}$ instances of graph created first apply R.D.1, then R.D.2, and then finally R.D.3. Now solve the problem recursively for each of the $2^{k'}$ instances created by considering each of them as a subproblem. If a non-sink node u has delegated to v then u gets deleted due to R.D.1 and if none of the non-sink nodes delegate to v then v gets deleted to R.D.3. So therefore, the new parameter (number of non-sink nodes) for the smaller subproblems gets reduced by at least 1.

Given a directed delegation graph $\mathcal{G}$, the algorithm works as follows. It first applies Reductions RD.1., RD.2.,RD.3. and Branching Rule B.1 exhaustively and in the same order. The parameter (number of non-sink nodes) decreases by at least 1 for each of the subproblems as explained earlier. If we can't apply the branching rule B.1 to a given subproblem it implies that there is no non-sink node such that the outdegree is more than $2(k-1)$ and indegree is greater than 0. Also due to R.D.3 we don't have any non-sink node with outdegree more than

$2(k-1)$ and indegree equal to 0. So we can do a brute force by considering every possible delegations and solve this instance in $O(k^k \cdot n^{O(1)})$ running time. Note that our algorithm will only look at the feasible solutions of RESOLVE DELEGATION TO MINMAXWEIGHT while brute forcing for a subproblem.

Also since every node of bounded search tree splits into at most 2^{k-1} subproblems and height of the tree is $O(k)$, we get $f(k)$ leaves (where $f(k)$ is a function of k only). Clearly the time taken at every node is bounded by $g(k) \cdot n^{O(1)}$ where $g(k)$ is a function of k only. Thus, the total time used by the algorithm is at-most $O(f(k) \cdot g(k) \cdot n^{O(1)})$ which gives us an FPT for RESOLVE DELEGATION TO MINMAXWEIGHT. □

We now show RESOLVE DELEGATION TO MINMAXWEIGHT is FPT w.r.t number of edges to be deleted from delegation graph by using the technique of bounded search tree by the branching on set of edges satisfying some key properties.

Theorem 5. *The* RESOLVE DELEGATION TO MINMAXWEIGHT *problem has a FPT with respect to the parameter* k *which is the number of edges to be deleted from delegation graph.*

Proof. The parameter k is the number of edges to be deleted. Given any instance $\mathcal{G}$ of problem , every feasible solution graph $\mathcal{G}_\mathcal{T}$ is a forest with trees with set of roots as set of all sink nodes $\mathcal{T}$. Clearly then $k = |\mathcal{E}| - |\mathcal{V}| + |\mathcal{T}|$. Let us denote the problem instance by $(\mathcal{G}, \lambda, k)$.

Observation 2. *If* $k > 0$ *and only the sink nodes have outdegree 0, then there is a non-sink node with outdegree atleast* 2.

Proof. Sum of outdegree of all the non-sink nodes is greater than $|\mathcal{V}| - |\mathcal{T}|$. Hence the observation follows from pigeon hole principle.

Branching B.1. Let $k > 0$. Consider the vertex with maximum outdegree. If l is the outdegree of one such vertex v, delete one of the two groups of edges $\{1, \ldots, \lfloor l/2 \rfloor\}$ and $\{\lfloor l/2 \rfloor + 1, \ldots, l\}$ outgoing from v. Then solve the problem recursively for two new subproblems with new parameter $k' \leqslant k - 1$.

Now we describe why the Branching B.1 is safe. Note that the Branching B.1 is triggered only when $k > 0$. It follows from Observation 2 that outdegree of v is at least 2. Consider the degree of v to be l and the corresponding outgoing edges from v to be $\{1, \ldots, l\}$. Since v can delegate only to exactly one of its neighbours connected by $\{1, \ldots, l\}$, other $l - 1$ edges need to be deleted from delegation graph as they can not be a part of feasible solution. If we partition the set of edges into two disjoint sets $\{1, \ldots, \lfloor l/2 \rfloor\}$ and $\{\lfloor l/2 \rfloor + 1, \ldots, l\}$, only one out of the two groups can be a part of feasible solution. This allows us to delete the other half set say $\{\lfloor l/2 \rfloor + 1, \ldots, l\}$. As we know that $|l| \geqslant 2$ which comes from the fact that outdegree of vertex v is at least 2. The problem now reduces to a smaller instance $\mathcal{G}'$ with edges $\mathcal{E}'[\mathcal{G}'] = \mathcal{E}[\mathcal{G}] \setminus \{\lfloor l/2 \rfloor + 1, \ldots, l\}$ and parameter number of edges to be deleted as $k' \leqslant k - 1$. Thus way we get a bounded search tree with only constant number of subproblems at each branch such that at each recursive step the height of search tree reduces by at least one.

Given a directed delegation graph $\mathcal{G}$, the algorithm works as follows. As long as $k > 0$, Branching Rule B.1 is applied exhaustively in the bounded search tree. Note that Branching Rule B.1 brings down the parameter k in every call by at least 1. Whenever the parameter k becomes 0, we have a feasible solution as the non-sink nodes have the outdegree of 1. Now we can easily check in polynomial time whether the feasible solution is a YES instance or a NO instance. At every recursive call we decrease the parameter by at least 1 and thus the height of the tree is at most k. Also since every node of bounded search tree splits into two, we get $O(2^k)$ leaves. Clearly the time taken at every node is bounded by $n^{O(1)}$. Thus if $f(k) = O(2^k)$ be the number of nodes in the bounded search tree, the total time used by the algorithm is at most $O(2^k n^{O(1)})$ which gives us an FPT for Resolve Delegation to MinMaxWeight. □

5 Structural Results

Theorem 6. *There exists a linear programming formulation for the optimization version of* Resolve Delegation to MinMaxWeight *where fractional delegation of votes is allowed. Thus the fractional variant is solvable in polynomial time.*

Proof. We consider the fractional variant of Liquid Democracy Delegation Problem where it is allowed to fractionally delegate votes of a source (delegator) to multiple nodes such that total number of votes being delegated is conserved at the delegator. We formally define conservation while formulating the LP for the problem.
LP formulation follows similar to the LP formulation of flow-problems (e.g. Max-FLow-MinCut etc). We assign $x_{u,v}$ as weight to every edge $(u, v) \in \mathcal{E}[\mathcal{G}]$ which corresponds to the fractional weight of votes delegated from vertex u to v (for all $u, v \in \mathcal{V}[\mathcal{G}]$. For all other $x_{u,v}$ where (u, v) pair doesn't correspond to an edge of delegation graph we assign value 0. It immediately follows that for all sink nodes $t \in \mathcal{T}[\mathcal{G}]$, total weight of fractional votes being delegated to each sink-node t (including that of the sink node t) is $\sum_{v \in \mathcal{V} \setminus \mathcal{T}} x_{v,t} + 1 \; \forall t \in \mathcal{T}$. For all other non-sink nodes $s \in \mathcal{V} \setminus \mathcal{T}$, node s obeys conservation as follows :

$$\sum_{u \in \mathcal{V} \setminus \mathcal{T}} x_{u,s} + 1 = \sum_{v \in \mathcal{V}} x_{s,v}, \forall s \in \mathcal{V} \setminus \mathcal{T}$$

Our aim is to minimize the maximum weight of votes delegated to any sink node (including that of the sink node). The corresponding LP formulation is:

$$\begin{aligned} &\text{minimize } z \\ &z \geqslant \sum_{v \in \mathcal{V} \setminus \mathcal{T}} x_{v,t} + 1, \; \forall t \in \mathcal{T} \\ \sum_{u \in \mathcal{V} \setminus \mathcal{T}} x_{u,s} + 1 = &\sum_{v \in \mathcal{V}} x_{s,v} \; , \; \forall s \in \mathcal{V} \setminus \mathcal{T} \text{ [Follows from conservation]} \\ &x_{u,v} \geqslant 0 \; , \forall (u, v) \in \mathcal{E}[\mathcal{G}] \\ &x_{u,v} = 0 \; , \forall (u, v) \notin \mathcal{E}[\mathcal{G}] \end{aligned}$$

□

6 Conclusion and Future Direction

We have studied the parameterized complexity of a fundamental problem in liquid democracy, namely Resolve Delegation to MinMaxWeight. We considered various natural parameters for the problem including the number of sink vertices, maximum allowed weight of any sink in the final delegation graph, maximum degree of any vertex, tree-width, the number of edges that one deletes in an optimal solution, number of non-sink vertices. We also show that a related problem which we call Resolve Fractional Delegation is polynomial time solvable.

An important future work is to resolve the complexity of Resolve Delegation to MinMaxWeight if the input graph is already acyclic or tree. We know that there exists a $\Omega(\log n)$ lower bound on the approximation factor of optimizing the maximum allowed weight of any sink [6]. It would be interesting to see if there exists FPT algorithms achieving a approximation factor of $o(\log n)$.

References

1. Berman, P., Karpinski, M., Scott, A.: Approximation hardness of short symmetric instances of max-3sat. Technical report (2004)
2. Brill, M., Talmon, N.: Pairwise liquid democracy. In: IJCAI, vol. 18, pp. 137–143 (2018)
3. Caragiannis, I., Micha, E.: A contribution to the critique of liquid democracy. In: IJCAI, pp. 116–122 (2019)
4. Christoff, Z., Grossi, D.: Binary voting with delegable proxy: An analysis of liquid democracy. arXiv preprint arXiv:1707.08741 (2017)
5. Fortune, S., Hopcroft, J., Wyllie, J.: The directed subgraph homeomorphism problem. Theoret. Comput. Sci. **10**(2), 111–121 (1980)
6. Gölz, P., Kahng, A., Mackenzie, S., Procaccia, A.D.: The fluid mechanics of liquid democracy. In: Christodoulou, G., Harks, T. (eds.) WINE 2018. LNCS, vol. 11316, pp. 188–202. Springer, Cham (2018). https://doi.org/10.1007/978-3-030-04612-5_13
7. Green-Armytage, J.: Direct voting and proxy voting. Constitutional Political Econ. **26**(2), 190–220 (2014). https://doi.org/10.1007/s10602-014-9176-9
8. Kahng, A., Mackenzie, S., Procaccia, A.D.: Liquid democracy: an algorithmic perspective. In: AAAI 2018 (2018)
9. Kling, C.C., Kunegis, J., Hartmann, H., Strohmaier, M., Staab, S.: Voting behaviour and power in online democracy: A study of liquidfeedback in Germany's pirate party. arXiv preprint arXiv:1503.07723 (2015)
10. Szeider, S.: Not so easy problems for tree decomposable graphs. arXiv preprint arXiv:1107.1177 (2011)

Acyclic Coloring Parameterized by Directed Clique-Width

Frank Gurski(✉), Dominique Komander, and Carolin Rehs

Institute of Computer Science, Algorithmics for Hard Problems Group, Heinrich-Heine-University Düsseldorf, 40225 Düsseldorf, Germany
frank.gurski@hhu.de

Abstract. An acyclic r-coloring of a directed graph $G = (V, E)$ is a partition of the vertex set V into r acyclic sets. The dichromatic number of a directed graph G is the smallest r such that G allows an acyclic r-coloring. For symmetric digraphs the dichromatic number equals the well-known chromatic number of the underlying undirected graph. This allows us to carry over the W[1]-hardness and lower bounds for running times of the chromatic number problem parameterized by clique-width to the dichromatic number problem parameterized by directed clique-width. We introduce the first polynomial-time algorithm for the acyclic coloring problem on digraphs of constant directed clique-width. From a parameterized point of view our algorithm shows that the Dichromatic Number problem is in XP when parameterized by directed clique-width and extends the only known structural parameterization by directed modular width for this problem. Furthermore, we apply defineability within monadic second order logic in order to show that Dichromatic Number problem is in FPT when parameterized by the directed clique-width and r. For directed co-graphs, which is a class of digraphs of directed clique-width 2, we even show a linear time solution for computing the dichromatic number.

Keywords: Acyclic coloring · Directed clique-width · Directed co-graphs · Polynomial time algorithms

1 Introduction

In this paper, we consider an approach for coloring the vertices of digraphs. An *acyclic r-coloring* of a digraph $G = (V, E)$ is a partition of the vertex set V into r sets such that all sets induce an acyclic subdigraph in G. The *dichromatic number* of G is the smallest integer r such that G has an acyclic r-coloring. Acyclic colorings of digraphs received a lot of attention in [4,28,29] and also in recent works [26,27,32]. The dichromatic number is one of two basic concepts for the class of perfect digraphs [1] and can be regarded as a natural counterpart of the well known chromatic number for undirected graphs.

In the Dichromatic Number problem (DCN) there is given a digraph G and an integer r and the question is whether G has an acyclic r-coloring. If r is constant

A. Mudgal and C. R. Subramanian (Eds.): CALDAM 2021, LNCS 12601, pp. 95–108, 2021.
https://doi.org/10.1007/978-3-030-67899-9_8

and not part of the input, the corresponding problem is denoted by DCN_r. Even DCN_2 is NP-complete [12], which motivates to consider the Dichromatic Number problem on special graph classes. Up to now, only few classes of digraphs are known, for which the dichromatic number can be found in polynomial time. The set of DAGs is obviously equal to the set of digraphs of dichromatic number 1. Further, every odd-cycle free digraph [29] and every non-even digraph [27] has dichromatic number at most 2.

The Dichromatic Number problem remains hard even for inputs of bounded directed feedback vertex set size [27]. This result implies that there are no XP-algorithms[1] for the Dichromatic Number problem parameterized by directed width parameters such as directed path-width, directed tree-width, DAG-width or Kelly-width. The first positive result concerning structural parameterizations of the Dichromatic Number problem is the existence of an FPT-algorithm[2] for the Dichromatic Number problem parameterized by directed modular width [31].

In this paper, we introduce the first polynomial-time algorithm for the Dichromatic Number problem on digraphs of constant directed clique-width. Therefore, we consider a directed clique-width expression X of the input digraph G of directed clique-width k. For each node t of the corresponding rooted expression-tree T we use label-based reachability information about the subgraph G_t of the subtree rooted at t. For every partition of the vertex set of G_t into acyclic sets $V_1, \ldots, V_s$ we compute the multi set $\langle \text{reach}(V_1), \ldots, \text{reach}(V_s) \rangle$, where $\text{reach}(V_i)$, $1 \leq i \leq s$, is the set of all label pairs (a, b) such that the subgraph of G_t induced by V_i contains a vertex labeled by b, which is reachable by a vertex labeled by a. By using bottom-up dynamic programming along expression-tree T, we obtain an algorithm for the Dichromatic Number problem of running time $n^{2^{\mathcal{O}(k^2)}}$ where n denotes the number of vertices of the input digraph. Since any algorithm with running time in $n^{2^{o(k)}}$ would disprove the Exponential Time Hypothesis (ETH), the exponential dependence on k in the degree of the polynomial cannot be avoided, unless ETH fails.

From a parameterized point of view, our algorithm shows that the Dichromatic Number problem is in XP when parameterized by directed clique-width. Further, we show that the Dichromatic Number problem is W[1]-hard on symmetric digraphs when parameterized by directed clique-width. Inferring from this, there is no FPT-algorithm for the Dichromatic Number problem parameterized by directed clique-width under reasonable assumptions. The best parameterized complexity, which can be achieved, is given by an XP-algorithm. Furthermore, we apply defineability within monadic second order logic (MSO) in order to show that Dichromatic Number problem is in FPT when parameterized by the directed clique-width and r, which implies that for every integer r it holds that DCN_r is in FPT when parameterized by directed clique-width.

[1] XP is the class of all parameterized problems which can be solved by algorithms that are polynomial if the parameter is considered as a constant [9].

[2] FPT is the class of all parameterized problems which can be solved by algorithms that are exponential only in the size of a fixed parameter while being polynomial in the size of the input size [9].

Since the directed clique-width of a digraph is at most its directed modular width [32], we reprove the existence of an XP-algorithm for DCN and an FPT-algorithm for DCN_r parameterized by directed modular width [31]. On the other hand, there exist several classes of digraphs of bounded directed clique-width and unbounded directed modular width, which implies that directed clique-width is the more powerful parameter and thus, the results of [31] does not imply any parameterized algorithm for directed clique-width.

In Table 1 we summarize the known results for DCN and DCN_r parameterized by width parameters.

Table 1. Complexity of DCN and DCN_r parameterized by width parameters. We assume that $\text{P} \neq \text{NP}$. The "///" entries indicate that by taking r out of the instance the considered parameter makes no sense.

Parameter	DCN		DCN_r	
Directed modular width	FPT	[31]	FPT	[31]
Directed clique-width	W[1]-hard	Corollary 1	FPT	Corollary 5
	XP	Corollary 3		
Directed clique-width + r	FPT	Theorem 4	///	
Directed tree-width	$\not\in$ XP	[27]	$\not\in$ XP	[27]
Directed path-width	$\not\in$ XP	[27]	$\not\in$ XP	[27]
DAG-width	$\not\in$ XP	[27]	$\not\in$ XP	[27]
Kelly-width	$\not\in$ XP	[27]	$\not\in$ XP	[27]
Clique-width of $un(G)$	$\not\in$ FPT	by Corollary 1	open	

For directed co-graphs, which is a class of digraphs of directed clique-width 2 [23], we even show a linear time solution for computing the dichromatic number and an optimal acyclic coloring.

2 Preliminaries

We use the notations of Bang-Jensen and Gutin [2] for graphs and digraphs.

2.1 Directed Graphs

A *directed graph* or *digraph* is a pair $G = (V, E)$, where V is a finite set of *vertices* and $E \subseteq \{(u, v) \mid u, v \in V,\ u \neq v\}$ is a finite set of ordered pairs of distinct vertices called *arcs* or *directed edges*. For a vertex $v \in V$, the sets $N^+(v) = \{u \in V \mid (v, u) \in E\}$ and $N^-(v) = \{u \in V \mid (u, v) \in E\}$ are called the *set of all successors* and the *set of all predecessors* of v. The *outdegree* of v, outdegree(v) for short, is the number of successors of v and the *indegree* of v, indegree(v) for short, is the number of predecessors of v.

A digraph $G' = (V', E')$ is a *subdigraph* of digraph $G = (V, E)$ if $V' \subseteq V$ and $E' \subseteq E$. If every arc of E with both end vertices in V' is in E', we say that G' is an *induced subdigraph* of digraph G and we write $G' = G[V']$.

For some given digraph $G = (V, E)$ we define its *underlying undirected graph* by ignoring the directions of the arcs, i.e. $un(G) = (V, \{\{u, v\} \mid (u, v) \in E, u, v \in V\})$. There are several ways to define a digraph $G = (V, E)$ from an undirected graph $G' = (V, E')$. If we replace every edge $\{u, v\} \in E'$ by

- both arcs (u, v) and (v, u), we refer to G as a *complete biorientation* of G'. Since in this case G is well defined by G' we also denote it by $\overleftrightarrow{G'}$. Every digraph G which can be obtained by a complete biorientation of some undirected graph G' is called a *complete bioriented graph* or *symmetric digraph.*
- one of the arcs (u, v) and (v, u), we refer to G as an *orientation* of G'. Every digraph G which can be obtained by an orientation of some undirected graph G' is called an *oriented graph.*

For a digraph $G = (V, E)$ an arc $(u, v) \in E$ is *symmetric* if $(v, u) \in E$. Thus, each bidirectional arc is symmetric. Further, an arc is *asymmetric* if it is not symmetric. We define the symmetric part of G as $\text{sym}(G)$, which is the spanning subdigraph of G that contains exactly the symmetric arcs of G. Analogously, we define the asymmetric part of G as $\text{asym}(G)$, which is the spanning subdigraph with only asymmetric arcs.

By $\overrightarrow{P_n} = (\{v_1, \dots, v_n\}, \{(v_1, v_2), \dots, (v_{n-1}, v_n)\}), n \geq 2$, we denote the directed path on n vertices, by $\overrightarrow{C_n} = (\{v_1, \dots, v_n\}, \{(v_1, v_2), \dots, (v_{n-1}, v_n), (v_n, v_1)\}), n \geq 2$, we denote the directed cycle on n vertices.

A *directed acyclic graph (DAG)* is a digraph without any $\overrightarrow{C_n}$, for $n \geq 2$, as subdigraph. A vertex v is *reachable* from a vertex u in G if G contains a $\overrightarrow{P_n}$ as a subdigraph having start vertex u and end vertex v. A digraph is *odd cycle free* if it does not contain a $\overrightarrow{C_n}$, for odd $n \geq 3$, as subdigraph. A digraph G is planar if $un(G)$ is planar.

A digraph is *even* if for every 0-1-weighting of the edges it contains a directed cycle of even total weight.

2.2 Acyclic Coloring of Directed Graphs

We consider the approach for coloring digraphs given in [29]. A set V' of vertices of a digraph G is called *acyclic* if $G[V']$ is acyclic.

Definition 1 (Acyclic graph coloring [29]**).** *An* acyclic r-coloring *of a digraph $G = (V, E)$ is a mapping $c : V \to \{1, \dots, r\}$, such that the color classes $c^{-1}(i)$ for $1 \leq i \leq r$ are acyclic. The* dichromatic number *of G, denoted by $\vec{\chi}(G)$, is the smallest r, such that G has an acyclic r-coloring.*

There are several works on acyclic graph coloring [4,28,29] including several recent works [26,27,32]. The following observations support that the dichromatic number can be regarded as a natural counterpart of the well known chromatic number $\chi(G)$ for undirected graphs G.

Observation 1. *For every symmetric directed graph G it holds that $\vec{\chi}(G) = \chi(un(G))$.*

Observation 2. *For every directed graph G it holds that $\vec{\chi}(G) \leq \chi(un(G))$.*

Observation 3. *Let G be a digraph and H be a subdigraph of G, then $\vec{\chi}(H) \leq \vec{\chi}(G)$.*

Name: Dichromatic Number (DCN)
Instance: A digraph $G = (V, E)$ and a positive integer $r \leq |V|$.
Question: Is there an acyclic r-coloring for G?

If r is a constant and not part of the input, the corresponding problem is denoted by r-Dichromatic Number (DCN_r). Even DCN_2 is NP-complete [12].

3 Acyclic Coloring of Directed Co-graphs

As recently mentioned in [31], only few classes of digraphs for which the dichromatic number can be found in polynomial time are known. The set of DAGs is obviously equal to the set of digraphs of dichromatic number 1. Every odd-cycle free digraph [29] and every non-even digraph [27] has dichromatic number at most 2. Thus, for DAGs, odd-cycle free digraphs, and non-even digraphs the dichromatic number can be computed in linear time. Furthermore, for every perfect digraph the dichromatic number can be found in polynomial time [1].

We next show how to find an optimal acyclic coloring for directed co-graphs, which are defined below, in linear time.

Definition 2 (Directed co-graphs [8]). *The class of* directed co-graphs *is recursively defined as follows.*

1. *Every digraph with a single vertex $(\{v\}, \emptyset)$, denoted by v, is a* directed co-graph.
2. *If $G_1 = (V_1, E_1)$ and $G_2 = (V_2, E_2)$ are vertex-disjoint directed co-graphs, then*
   (a) *the disjoint union $G_1 \oplus G_2$, which is defined as the digraph with vertex set $V_1 \cup V_2$ and arc set $E_1 \cup E_2$,*
   (b) *the series composition $G_1 \otimes G_2$, which is defined by their disjoint union plus all possible directed edges between V_1 and V_2, and*
   (c) *the order composition $G_1 \oslash G_2$, which is defined by their disjoint union plus all possible directed edges from V_1 to V_2, are* directed co-graphs.

Every expression X using the four operations of Definition 2 is called a *di-co-expression*. For every directed co-graph we can define a tree structure denoted as *di-co-tree*. This is an ordered rooted tree whose leaves represent the vertices of the digraph and whose inner nodes correspond to the operations applied on the subexpressions defined by the subtrees. For every directed co-graph one can construct a di-co-tree in linear time [8]. Directed co-graphs are interesting from an algorithmic point of view since several hard graph problems can be solved in polynomial time by dynamic programming along the tree structure of the input graph, see [3,18,19].

Lemma 1 ($\bigstar^3$). *Let G_1 and G_2 be two vertex-disjoint directed graphs. Then, the following equations hold:*

1. $\vec{\chi}(G_1 \oplus G_2) = \max(\vec{\chi}(G_1), \vec{\chi}(G_2))$
2. $\vec{\chi}(G_1 \oslash G_2) = \max(\vec{\chi}(G_1), \vec{\chi}(G_2))$
3. $\vec{\chi}(G_1 \otimes G_2) = \vec{\chi}(G_1) + \vec{\chi}(G_2)$

Lemma 1 can be used to obtain the following result.

Theorem 1. *Let G be a directed co-graph. Then, an optimal acyclic coloring for G and $\vec{\chi}(G)$ can be computed in linear time.*

The *clique number* $\omega_d(G)$ of a digraph G is the number of vertices in a largest complete bioriented subdigraph of G and the *clique number* $\omega(G)$ of a (-n undirected) graph G is the number of vertices in a largest complete subgraph of G. Since the results of Lemma 1 also hold for ω_d instead of $\vec{\chi}$ we obtain the following result.

Proposition 1. *Let G be a directed co-graph. Then, it holds that*

$$\vec{\chi}(G) = \chi(un(sym(G))) = \omega(un(sym(G))) = \omega_d(G)$$

and all values can be computed in linear time.

4 Parameterized Algorithms for Directed Clique-Width

For undirected graphs the clique-width [7] is one of the most important parameters. Clique-width measures how difficult it is to decompose the graph into a special tree-structure. From an algorithmic point of view, only tree-width [30] is a more studied graph parameter. Clique-width is more general than tree-width since graphs of bounded tree-width have also bounded clique-width [5]. The tree-width can only be bounded by the clique-width under certain conditions [22]. Many NP-hard graph problems admit polynomial-time solutions when restricted to graphs of bounded tree-width or graphs of bounded clique-width.

For directed graphs there are several attempts to generalize tree-width such as directed tree-width, DAG-width, or Kelly-width, which are representative for what people are working on, see the surveys [16,17]. Unfortunately, none of these attempts allows polynomial-time algorithms for a large class of problems on digraphs of bounded width [16, Table 2]. This also holds for DCN_r and DCN since even for bounded size of a directed feedback vertex set, deciding whether a

[3] The proofs of the results marked with a ★ are omitted due to space restrictions, see [20].

directed graph has dichromatic number at most 2 is NP-complete [27]. This result rules out XP-algorithms for DCN and DCN_r by directed width parameters such as directed path-width, directed tree-width, DAG-width or Kelly-width, since all of these are upper bounded by the feedback vertex set number.

Next, we discuss parameters which allow XP-algorithms or even FPT-algorithms for DCN and DCN_r. The first positive result concerning structural parameterizations of DCN was recently given in [31] using the directed modular width (dmw).

Theorem 2 ([31]). *The Dichromatic Number problem is in FPT when parameterized by directed modular width.*

By [16], directed clique-width performs much better than directed path-width, directed tree-width, DAG-width, and Kelly-width from the parameterized complexity point of view. Hence, we consider the parameterized complexity of DCN parameterized by directed clique-width.

Definition 3 (Directed clique-width [7]**).** *The* directed clique-width *of a digraph G, d-cw(G) for short, is the minimum number of labels needed to define G using the following four operations:*

1. *Creation of a new vertex v with label a (denoted by $a(v)$).*
2. *Disjoint union of two labeled digraphs G and H (denoted by $G \oplus H$).*
3. *Inserting an arc from every vertex with label a to every vertex with label b ($a \neq b$, denoted by $\alpha_{a,b}$).*
4. *Change label a into label b (denoted by $\rho_{a \to b}$).*

An expression X built with the operations defined above using k labels is called a directed clique-width k-expression. *Let digraph(X) be the digraph defined by k-expression X.*

In [23] the set of directed co-graphs is characterized by excluding two digraphs as a proper subset of the set of all graphs of directed clique-width 2, while for the undirected versions both classes are equal.

By the given definition every graph of directed clique-width at most k can be represented by a tree structure, denoted as *k-expression-tree.* The leaves of the k-expression-tree represent the vertices of the digraph and the inner nodes of the k-expression-tree correspond to the operations applied to the subexpressions defined by the subtrees. Using the k-expression-tree many hard problems have been shown to be solvable in polynomial time when restricted to graphs of bounded directed clique-width [16,23].

Directed clique-width is not comparable to the directed variants of tree-width mentioned above, which can be observed by the set of all complete biorientations of cliques and the set of all acyclic orientations of grids. The relation of directed clique-width and directed modular width [32] is as follows.

Lemma 2 ([32]). *For every digraph G it holds that d-cw(G) $\leq$ dmw(G).*

On the other hand, there exist several classes of digraphs of bounded directed clique-width and unbounded directed modular width, e.g. even the set of all directed paths $\{\overrightarrow{P_n} \mid n \geq 1\}$, the set of all directed cycles $\{\overrightarrow{C_n} \mid n \geq 1\}$, and the set of all minimal series-parallel digraphs [33]. Thus, the result of [31] does not imply any XP-algorithm or FPT-algorithm for directed clique-width.

Corollary 1. *The Dichromatic Number problem is W[1]-hard on symmetric digraphs and thus, on all digraphs when parameterized by directed clique-width.*

Proof. The Chromatic Number problem is W[1]-hard when parameterized by clique-width [13]. An instance consisting of a graph $G = (V, E)$ and a positive integer r for the Chromatic Number problem can be transformed into an instance for the Dichromatic Number problem on digraph $\overleftrightarrow{G}$ and integer r. Then, G has an r-coloring if and only if $\overleftrightarrow{G}$ has an acyclic r-coloring by Observation 1. Since for every undirected graph G its clique-width equals the directed clique-width of $\overleftrightarrow{G}$ [23], we obtain a parameterized reduction. □

Thus, under reasonable assumptions there is no FPT-algorithm for the Dichromatic Number problem parameterized by directed clique-width and an XP-algorithm is the best that can be achieved. Next, we introduce such an XP-algorithm.

Let $G = (V, E)$ be a digraph which is given by some directed clique-width k-expression X. For some vertex set $V' \subseteq V$, we define reach(V') as the set of all pairs (a, b) such that there is a vertex $u \in V'$ labeled by a and there is a vertex $v \in V'$ labeled by b and v is reachable from u in $G[V']$.

Within a construction of a digraph by directed clique-width operations only the edge insertion operation can change the reachability between the present vertices. Next, we show which acyclic sets remain acyclic when performing an edge insertion operation and how the reachability information of these sets have to be updated due to the edge insertion operation.

Lemma 3 (★). *Let $G = (V, E)$ be a vertex labeled digraph defined by some directed clique-width k-expression X, $a \neq b$, $a, b \in \{1, \ldots, k\}$, and $V' \subseteq V$ be an acyclic set in G. Then, vertex set V' remains acyclic in digraph$(\alpha_{a,b}(X))$ if and only if $(b, a) \notin$ reach(V').*

Lemma 4 (★). *Let $G = (V, E)$ be a vertex labeled digraph defined by some directed clique-width k-expression X, $a \neq b$, $a, b \in \{1, \ldots, k\}$, $V' \subseteq V$ be an acyclic set in G, and $(b, a) \notin$ reach(V'). Then, reach(V') for digraph$(\alpha_{a,b}(X))$ can be obtained from reach(V') for digraph(X) as follows:*

- *For every pair $(x, a) \in$ reach(V') and every pair $(b, y) \in$ reach(V'), we extend reach(V') by (x, y).*

For a disjoint partition of V into acyclic sets $V_1, \ldots, V_s$, let $\mathcal{M}$ be the multi set[4] $\langle$reach$(V_1), \ldots,$ reach$(V_s)\rangle$. Let $F(X)$ be the set of all mutually different

[4] We use the notion of a *multi set*, i.e., a set that may have several equal elements. For a multi set with elements $x_1, \ldots, x_n$ we write $\mathcal{M} = \langle x_1, \ldots, x_n \rangle$. The number

multi sets $\mathcal{M}$ for all disjoint partitions of vertex set V into acyclic sets. Every multi set in $F(X)$ consists of nonempty subsets of $\{1,\ldots,k\} \times \{1,\ldots,k\}$. Each subset can occur 0 times and not more than $|V|$ times. Thus, $F(X)$ has at most

$$(|V|+1)^{2^{k^2}-1} \in |V|^{2^{\mathcal{O}(k^2)}}$$

mutually different multi sets and is polynomially bounded in the size of X.

In order to give a dynamic programming solution along the recursive structure of a directed clique-width k-expression, we show how to compute $F(a(v))$, $F(X \oplus Y)$ from $F(X)$ and $F(Y)$, as well as $F(\alpha_{a,b}(X))$ and $F(\rho_{a\to b}(X))$ from $F(X)$.

Lemma 5 (★). *Let $a,b \in \{1,\ldots,k\}$, $a \neq b$.*

1. *$F(a(v)) = \{\langle\{(a,a)\}\rangle\}$.*
2. *Starting with set $D = \{\langle\rangle\} \times F(X) \times F(Y)$ extend D by all triples that can be obtained from some triple $(\mathcal{M},\mathcal{M}',\mathcal{M}'') \in D$ by removing a set L' from $\mathcal{M}'$ or a set L'' from $\mathcal{M}''$ and inserting it into $\mathcal{M}$, or by removing both sets and inserting $L' \cup L''$ into $\mathcal{M}$. Finally, we choose $F(X \oplus Y) = \{\mathcal{M} \mid (\mathcal{M},\langle\rangle,\langle\rangle) \in D\}$.*
3. *$F(\alpha_{a,b}(X))$ can be obtained from $F(X)$ as follows. First, we remove from $F(X)$ all multi sets $\langle L_1,\ldots,L_s\rangle$ such that $(b,a) \in L_t$ for some $1 \leq t \leq s$. Afterwards, we modify every remaining multi set $\langle L_1,\ldots,L_s\rangle$ in $F(X)$ as follows:*
   - *For every L_i which contains a pair (x,a) and a pair (b,y), we extend L_i by (x,y).*
4. *$F(\rho_{a\to b}(X)) = \{\langle\rho_{a\to b}(L_1),\ldots,\rho_{a\to b}(L_s)\rangle \mid \langle L_1,\ldots,L_s\rangle \in F(X)\}$, where we use $\rho_{a\to b}(L_i) = \{(\rho_{a\to b}(c),\rho_{a\to b}(d)) \mid (c,d) \in L_i\}$ and $\rho_{a\to b}(c) = b$, if $c = a$, and $\rho_{a\to b}(c) = c$, if $c \neq a$.*

Since every possible coloring of G is realized in the set $F(X)$, where X is a directed clique-width k-expression for G, it is easy to find a minimum coloring for G.

Corollary 2. *Let $G = (V,E)$ be a digraph given by a directed clique-width k-expression X. There is a partition of V into r acyclic sets if and only if there is some $\mathcal{M} \in F(X)$ consisting of r sets of label pairs.*

Theorem 3. *The Dichromatic Number problem on digraphs on n vertices given by a directed clique-width k-expression can be solved in $n^{2^{\mathcal{O}(k^2)}}$ time.*

Proof. Let $G = (V,E)$ be a digraph of directed clique-width at most k and T be a k-expression-tree for G with root w. For some vertex u of T we denote by T_u the subtree rooted at u and X_u the k-expression defined by T_u. In order to solve

how often an element x occurs in $\mathcal{M}$ is denoted by $\psi(\mathcal{M},x)$. Two multi sets $\mathcal{M}_1$ and $\mathcal{M}_2$ are *equal* if for each element $x \in \mathcal{M}_1 \cup \mathcal{M}_2$, $\psi(\mathcal{M}_1,x) = \psi(\mathcal{M}_2,x)$, otherwise they are called *different*. The empty multi set is denoted by $\langle\rangle$.

the Dichromatic Number problem for G, we traverse k-expression-tree T in a bottom-up order. For every vertex u of T we compute $F(X_u)$ following the rules given in Lemma 5. By Corollary 2 we can solve our problem by $F(X_w) = F(X)$.

Our rules given Lemma 5 show the following running times. For every $v \in V$ and $a \in \{1, \ldots, k\}$ set $F(a(v))$ can be computed in $\mathcal{O}(1)$. The set $F(X \oplus Y)$ can be computed in time $(n+1)^{3(2^{k^2}-1)} \in n^{2^{\mathcal{O}(k^2)}}$ from $F(X)$ and $F(Y)$. The sets $F(\alpha_{a,b}(X))$ and $F(\rho_{a \to b}(X))$ can be computed in time $(n+1)^{2^{k^2}-1} \in n^{2^{\mathcal{O}(k^2)}}$ from $F(X)$.

In order to bound the number and order of operations within directed clique-width expressions, we can use the normal form for clique-width expressions defined in [11]. The proof of Theorem 4.2 in [11] shows that also for directed clique-width expression X, we can assume that for every subexpression, after a disjoint union operation first there is a sequence of edge insertion operations followed by a sequence of relabeling operations, i.e. between two disjoint union operations there is no relabeling before an edge insertion. Since there are n leaves in T, we have $n-1$ disjoint union operations, at most $(n-1) \cdot (k-1)$ relabeling operations, and at most $(n-1) \cdot k(k-1)$ edge insertion operations. This leads to an overall running time of $n^{2^{\mathcal{O}(k^2)}}$. □

The running time shown in Theorem 3 leads to the following result.

Corollary 3. *The Dichromatic Number problem is in XP when parameterized by directed clique-width.*

Up to now there are only very few digraph classes for which we can compute a directed clique-width expression in polynomial time. This holds for directed co-graphs, digraphs of bounded directed modular width, and orientations of trees. For such classes we can apply the result of Theorem 3. In order to find directed clique-width expressions for general digraphs one can use results on the related parameter bi-rank-width [24]. By [2, Lemma 9.9.12] we can use approximate directed clique-width expressions obtained from rank-decomposition with the drawback of a single-exponential blow-up on the parameter.

Next, we give a lower bound for the running time of parameterized algorithms for Dichromatic Number problem parameterized by the directed clique-width.

Corollary 4. *The Dichromatic Number problem on digraphs on n vertices parameterized by the directed clique-width k cannot be solved in time $n^{2^{o(k)}}$, unless ETH fails.*

Proof. In order to show the statement we apply the following lower bound for the Chromatic Number problem parameterized by clique-width given in [14]. Any algorithm for the Chromatic Number problem parameterized by clique-width with running in $n^{2^{o(k)}}$ would disprove the Exponential Time Hypothesis. By Observation 1 and since for every undirected graph G its clique-width equals the directed clique-width of $\overleftrightarrow{G}$ [23], any algorithm for the Dichromatic Number problem parameterized by directed clique-width can be used to solve the Chromatic Number problem parameterized by clique-width. □

In order to show fixed parameter tractability for DCN_r w.r.t. the parameter directed clique-width one can use its defineability within monadic second order logic (MSO). We restrict to MSO_1-logic, which allows propositional logic, variables for vertices and vertex sets of digraphs, the predicate $\text{arc}(u,v)$ for arcs of digraphs, and quantifications over vertices and vertex sets [6]. In [16, Theorem 4.2] it has been shown that for every integer k and MSO_1 formula ψ, every ψ-LinEMSO_1 optimization problem (see [16]) is fixed-parameter tractable on digraphs of clique-width k w.r.t. the parameters k and length of the formula $|\psi|$. Next, we will apply this result to DCN.

Theorem 4. *The Dichromatic Number problem is in FPT when parameterized by directed clique-width and r.*

Proof. Let $G = (V, E)$ be a digraph. We can define DCN_r by an MSO_1 formula

$$\psi = \exists V_1, \ldots, V_r : \left(\text{Partition}(V, V_1, \ldots, V_r) \wedge \bigwedge_{1 \leq i \leq r} \text{Acyclic}(V_i) \right)$$

with

$$\begin{aligned} \text{Partition}(V, V_1, \ldots, V_r) = \forall v \in V &: (\textstyle\bigvee_{1 \leq i \leq r} v \in V_i) \wedge \\ \nexists v \in V &: (\textstyle\bigvee_{i \neq j,\ 1 \leq i,j \leq r} (v \in V_i \wedge v \in V_j)) \end{aligned}$$

and

$$\text{Acyclic}(V_i) = \forall V' \subseteq V_i, V' \neq \emptyset : \exists v \in V'(\text{outdegree}(v) = 0 \vee \text{outdegree}(v) \geq 2)$$

For the correctness we note the following. For every induced cycle V' in G it holds that for every vertex $v \in V'$ we have $\text{outdegree}(v) = 1$ in G. This does not hold for non-induced cycles. But since for every cycle V'' in G there is a subset $V' \subseteq V''$, such that $G[V']$ is a cycle, we can verify by $\text{Acyclic}(V_i)$ whether $G[V_i]$ is acyclic. Since it holds that $|\psi| \in \mathcal{O}(r)$, the statement follows by the result of [16] stated above. □

Corollary 5. *For every integer r the r-Dichromatic Number problem is in FPT when parameterized by directed clique-width.*

5 Conclusions and Outlook

The presented methods allow us to compute the dichromatic number on directed co-graphs in linear time and on graph classes of bounded directed clique-width in polynomial time.

The shown parameterized solutions of Corollary 3 and Theorem 4 also hold for any parameter which is larger or equal than directed clique-width, such as the parameter directed modular width [32] (which even allows an FPT-algorithm by [31,32]) and directed linear clique-width [21].

Further, the hardness result of Corollary 1 rules out FPT-algorithms for the Dichromatic Number problem parameterized by width parameters which can be bounded by directed clique-width. Among these are the clique-width and rank-width of the underlying undirected graph, which also have been considered in [15] on the Oriented Chromatic Number problem.

From a parameterized point of view width parameters are so-called structural parameters, which are measuring the difficulty of decomposing a graph into a special tree-structure. Beside these, the standard parameter, i.e. the threshold value given in the instance, is well studied. Unfortunately, for the Dichromatic Number problem the standard parameter is the number of necessary colors r and does even not allow an XP-algorithm, since DCN_2 is NP-complete [27]. A positive result can be obtained for parameter "number of vertices" n. Since integer linear programming is fixed-parameter tractable for the parameter "number of variables" [25] the existence of an integer program for DCN using $\mathcal{O}(n^2)$ variables implies an FPT-algorithm for parameter n, see [20].

It remains to verify whether the running time of our XP-algorithm for DCN can be improved to $n^{2^{\mathcal{O}(k)}}$, which is possible for the Chromatic Number problem by [10]. Further, it remains open whether the hardness of Corollary 1 also holds for special digraph classes and for directed linear clique-width [21]. Additionally, the existence of an FPT-algorithm for DCN_r w.r.t. parameter clique-width of the underlying undirected graph is open.

Acknowledgments. This work was funded by the Deutsche Forschungsgemeinschaft (DFG, German Research Foundation) – 388221852.

References

1. Andres, S., Hochstättler, W.: Perfect digraphs. J. Graph Theory **79**(1), 21–29 (2015)
2. Bang-Jensen, J., Gutin, G.: Classes of Directed Graphs. SMM. Springer, Cham (2018). https://doi.org/10.1007/978-3-319-71840-8
3. Bang-Jensen, J., Maddaloni, A.: Arc-disjoint paths in decomposable digraphs. J. Graph Theory **77**, 89–110 (2014)
4. Bokal, D., Fijavz, G., Juvan, M., Kayll, P., Mohar, B.: The circular chromatic number of a digraph. J. Graph Theory **46**(3), 227–240 (2004)
5. Corneil, D., Rotics, U.: On the relationship between clique-width and treewidth. SIAM J. Comput. **4**, 825–847 (2005)
6. Courcelle, B., Engelfriet, J.: Graph Structure and Monadic Second-Order Logic. A Language-Theoretic Approach. Encyclopedia of Mathematics and its Applications. Cambridge University Press, Cambridge (2012)
7. Courcelle, B., Olariu, S.: Upper bounds to the clique width of graphs. Discret. Appl. Math. **101**, 77–114 (2000)
8. Crespelle, C., Paul, C.: Fully dynamic recognition algorithm and certificate for directed cographs. Discret. Appl. Math. **154**(12), 1722–1741 (2006)
9. Downey, R.G., Fellows, M.R.: Fundamentals of Parameterized Complexity. TCS. Springer, London (2013). https://doi.org/10.1007/978-1-4471-5559-1

10. Espelage, W., Gurski, F., Wanke, E.: How to solve NP-hard graph problems on clique-width bounded graphs in polynomial time. In: Brandstädt, A., Le, V.B. (eds.) WG 2001. LNCS, vol. 2204, pp. 117–128. Springer, Heidelberg (2001). https://doi.org/10.1007/3-540-45477-2_12
11. Espelage, W., Gurski, F., Wanke, E.: Deciding clique-width for graphs of bounded tree-width. In: Dehne, F., Sack, J.-R., Tamassia, R. (eds.) WADS 2001. LNCS, vol. 2125, pp. 87–98. Springer, Heidelberg (2001). https://doi.org/10.1007/3-540-44634-6_9
12. Feder, T., Hell, P., Mohar, B.: Acyclic homomorphisms and circular colorings of digraphs. SIAM J. Discret. Math. **17**(1), 161–163 (2003)
13. Fomin, F., Golovach, P., Lokshtanov, D., Saurabh, S.: Intractability of clique-width parameterizations. SIAM J. Comput. **39**(5), 1941–1956 (2010)
14. Fomin, F., Golovach, P., Lokshtanov, D., Saurabh, S., Zehavi, M. Cliquewidth III: the odd case of graph coloring parameterized by cliquewidth. ACM Trans. Algorithms **15**(1), 9:1–9:27 (2018)
15. Ganian, R.: The parameterized complexity of oriented colouring. In: Proceedings of Doctoral Workshop on Mathematical and Engineering Methods in Computer Science, MEMICS. OASICS, vol. 13. Schloss Dagstuhl - Leibniz-Zentrum fuer Informatik, Germany (2009)
16. Ganian, R., Hlinený, P., Kneis, J., Langer, A., Obdrzálek, J., Rossmanith, P.: Digraph width measures in parameterized algorithmics. Discret. Appl. Math. **168**, 88–107 (2014)
17. Ganian, R., et al.: Are there any good digraph width measures? J. Comb. Theory Ser. B **116**, 250–286 (2016)
18. Gurski, F., Hoffmann, S., Komander, D., Rehs, C., Rethmann, J., Wanke, E.: Computing directed Steiner path covers for directed co-graphs (extended abstract). In: Chatzigeorgiou, A., et al. (eds.) SOFSEM 2020. LNCS, vol. 12011, pp. 556–565. Springer, Cham (2020). https://doi.org/10.1007/978-3-030-38919-2_45
19. Gurski, F., Komander, D., Rehs, C.: Computing digraph width measures on directed co-graphs. In: Gąsieniec, L.A., Jansson, J., Levcopoulos, C. (eds.) FCT 2019. LNCS, vol. 11651, pp. 292–305. Springer, Cham (2019). https://doi.org/10.1007/978-3-030-25027-0_20
20. Gurski, F., Komander, D., Rehs, C.: Acyclic coloring of special digraphs. ACM Computing Research Repository (CoRR), abs/2006.13911, p. 16 (2020)
21. Gurski, F., Rehs, C.: Comparing linear width parameters for directed graphs. Theory Comput. Syst. **63**(6), 1358–1387 (2019)
22. Gurski, F., Wanke, E.: The tree-width of clique-width bounded graphs without $K_{n,n}$. In: Brandes, U., Wagner, D. (eds.) WG 2000. LNCS, vol. 1928, pp. 196–205. Springer, Heidelberg (2000). https://doi.org/10.1007/3-540-40064-8_19
23. Gurski, F., Wanke, E., Yilmaz, E.: Directed NLC-width. Theor. Comput. Sci. **616**, 1–17 (2016)
24. Kanté, M., Rao, M.: The rank-width of edge-coloured graphs. Theory Comput. Syst. **52**(4), 599–644 (2013)
25. Lenstra, H.: Integer programming with a fixed number of variables. Math. Oper. Res. **8**, 538–548 (1983)
26. Li, Z., Mohar, B.: Planar digraphs of digirth four are 2-colorable. SIAM J. Discret. Math. **31**, 2201–2205 (2017)
27. Millani, M., Steiner, R., Wiederrecht, S.: Colouring non-even digraphs. ACM Computing Research Repository (CoRR), abs/1903.02872, p. 37 (2019)
28. Mohar, B.: Circular colorings of edge-weighted graphs. J. Graph Theory **43**(2), 107–116 (2003)

29. Neumann-Lara, V.: The dichromatic number of a digraph. J. Comb. Theory Ser. B **33**(2), 265–270 (1982)
30. Robertson, N., Seymour, P.: Graph minors II. Algorithmic aspects of tree width. J. Algorithms **7**, 309–322 (1986)
31. Steiner, R., Wiederrecht, S.: Parameterized algorithms for directed modular width. ACM Computing Research Repository (CoRR), abs/1905.13203, p. 37 (2019)
32. Steiner, R., Wiederrecht, S.: Parameterized algorithms for directed modular width. In: Changat, M., Das, S. (eds.) CALDAM 2020. LNCS, vol. 12016, pp. 415–426. Springer, Cham (2020). https://doi.org/10.1007/978-3-030-39219-2_33
33. Valdes, J., Tarjan, R., Lawler, E.: The recognition of series-parallel digraphs. SIAM J. Comput. **11**, 298–313 (1982)

On Structural Parameterizations of Load Coloring

I. Vinod Reddy(✉)

Department of Electrical Engineerng and Computer Science,
Indian Institute of Technology Bhilai, Raipur, India
vinod@iitbhilai.ac.in

Abstract. Given a graph G and a positive integer k, the 2-Load Coloring problem is to check whether there is a 2-coloring $f : V(G) \rightarrow \{r, b\}$ of G such that for every $i \in \{r, b\}$, there are at least k edges with both end vertices colored i. It is known that the problem is NP-complete even on special classes of graphs like regular graphs. Gutin and Jones (Inf Process Lett 114:446-449, 2014) showed that the problem is fixed-parameter tractable by giving a kernel with at most $7k$ vertices. Barbero et al. (Algorithmica 79:211-229, 2017) obtained a kernel with less than $4k$ vertices and $O(k)$ edges, improving the earlier result.

In this paper, we study the parameterized complexity of the problem with respect to structural graph parameters. We show that 2-Load Coloring cannot be solved in time $f(w)n^{o(w)}$, unless ETH fails and it can be solved in time $n^{O(w)}$, where n is the size of the input graph, w is the clique-width of the graph and f is an arbitrary function of w. Next, we consider the parameters distance to cluster graphs, distance to co-cluster graphs and distance to threshold graphs, which are weaker than the parameter clique-width and show that the problem is fixed-parameter tractable (FPT) with respect to these parameters. Finally, we show that 2-Load Coloring is NP-complete even on bipartite graphs and split graphs.

1 Introduction

Given a graph G and a positive integer c, the load distribution of a c-coloring $f : V(G) \rightarrow [c]$ is a tuple $(f_1, \ldots, f_c)$, where f_i is the number of edges with at least one end point colored with i. The c-Load Coloring problem is to find a coloring f such that the function $\ell_f(G) = \max\{f_i : i \in [c]\}$ is minimum. We denote this minimum by $\ell(G)$. Ahuja et al [1] showed that the problem is NP-hard on general graphs when $c = 2$. They also gave a polynomial time algorithm for 2-Load Coloring on trees.

In a 2-coloring $f : V(G) \rightarrow \{r, b\}$, an edge is called red (resp. blue) if both end vertices are colored with r (resp. b). We use r_f and b_f to denote the number of red and blue edges in a 2-coloring f of G. Let $\mu_f(G) = \min\{r_f, b_f\}$ and $\mu(G)$ is the maximum of $\mu_f(G)$ over all possible 2-colorings of G. Ahuja et al. [1] showed that the 2-Load Coloring problem is equivalent to maximizing $\mu(G)$ over all possible 2-colorings of G, in particular they showed that $\ell(G) = |E(G)| - \mu(G)$.

A. Mudgal and C. R. Subramanian (Eds.): CALDAM 2021, LNCS 12601, pp. 109–121, 2021.
https://doi.org/10.1007/978-3-030-67899-9_9

Input: A graph $G = (V, E)$ and an integer k
Question: Does there exists a coloring $f : V(G) \to \{r, b\}$ such that $\mu(G) \geq k$? (i.e, $r_f \geq k$ and $b_f \geq k$)

The above version of the load coloring problem has been studied from the parameterized complexity perspective. Gutin and Jones [10] proved that the problem admits a polynomial kernel (with at most $7k$ vertices) parameterized by k. They also showed that the problem is fixed-parameter tractable when parameterized by the tree-width of the input graph. More recently, Barbero et al. [2] obtained a kernel for the problem with at most $4k$ vertices improving the result of [10].

In this paper, we study the following variant of the load coloring problem.

2-LOAD COLORING
Input: A graph $G = (V, E)$ and integers k_1 and k_2
Question: Does there exists a coloring $f : V(G) \to \{r, b\}$ such that $r_f \geq k_1$ and $b_f \geq k_2$?

Our Contributions. In this paper, we study the 2-LOAD COLORING problem from the viewpoint of parameterized complexity. A parameterized problem with input size n and parameter k is called fixed-parameter tractable (FPT) if it can be solved in time $f(k)n^{O(1)}$, where f is a function only depending on the parameter k (for more details on parameterized complexity refer to the textbook [5]). There are many possible parameterizations for 2-LOAD COLORING. One such parameter is the size of the solution. The problem admits a linear kernel [10] with respect to the size of the solution. In this paper, we study the 2-LOAD COLORING problem with respect to various structural graph parameters. These parameters measure the complexity of the input rather than the problem itself. Tree-width is one of the well-known structural graph parameters. The 2-LOAD COLORING problem is FPT when parameterized by tree-width [10] of the input graph.

Even though tree-width is a widely used graph parameter for sparse graphs, it is not suitable for dense graphs, even if they have a simple structure. In Sect. 3, we consider the graph parameter clique-width introduced by Courcelle and Olariu [4], which is a generalization of the parameter tree-width. We show that 2-LOAD COLORING can be solved in time $n^{O(w)}$, and cannot be solved in $f(w)n^{o(w)}$ unless ETH fails, where w is the clique-width of the n vertex input graph and f is an arbitrary function of w. Next, we consider the parameters distance to cluster graphs, distance to co-cluster graphs, and distance to threshold graphs. These parameters are weaker than the parameter clique-width in the sense that they are a subclass of bounded clique-width graphs. Thus studying the parameterized complexity of 2-LOAD COLORING with respect to these

Table 1. Known and new parameterized results for 2-LOAD COLORING

Parameter	Results
Size of the solution	linear kernel [2,10]
Tree-width	FPT [10]
clique-width(w)	$n^{O(w)}$ algorithm (Theorem 1)
	no $f(w)n^{o(w)}$ algorithm (Theorem 2)
Distance to cluster graphs	FPT (Theorem 3)
Distance to co-cluster graphs	FPT (Theorem 4)
Distance to threshold graphs	FPT (Theorem 5)
Distance to bipartite graphs	para-NP-hard (Theorem 6)
Distance to split graphs	para-NP-hard (Theorem 7)

parameters reduces the gap between tractable and intractable parameterizations. In Sect. 4, we show that 2-LOAD COLORING is fixed-parameter tractable with respect to the parameters distance to cluster, distance to co-cluster and distance to threshold graphs. Finally in Sect. 5, we show that 2-LOAD COLORING is NP-complete on bipartite graphs and split graphs. Table 1 gives an overview of our results.

2 Preliminaries

In this section, we introduce some basic notation and terminology related to graph theory and parameterized complexity. For $k \in \mathbb{N}$, we use $[k]$ to denote the set $\{1, 2, \dots, k\}$. If $f : A \to B$ is a function and $C \subseteq A$, $f|_C$ denotes the restriction of f to C, that is $f|_C : C \to B$ such that for all $x \in C$, $f|_C(x) = f(x)$ All graphs we consider in this paper are undirected, connected, finite and simple. For a graph $G = (V, E)$, by $V(G)$ and $E(G)$ we denote the vertex set and edge set of G respectively. We use n to denote the number of vertices and m to denote the number of edges of a graph. An edge between vertices x and y is denoted as xy for simplicity. For a subset $X \subseteq V(G)$, the graph $G[X]$ denotes the subgraph of G induced by vertices of X and $E_G[X]$ denote the set of edges having both end vertices in the set X. For subsets $X, Y \subseteq V(G)$, $E_G[X, Y]$ denote the set of edges connecting X and Y.

For a vertex set $X \subseteq V(G)$, we denote $G - X$, the graph obtained from G by deleting all vertices of X and their incident edges. For a vertex $v \in V(G)$, by $N(v)$, we denote the set $\{u \in V(G) \mid vu \in E(G)\}$ and we use $N[v]$ to denote the set $N(v) \cup \{v\}$. The neighbourhood of a vertex set $S \subseteq V(G)$ is $N(S) = (\cup_{v \in V(G)} N(v)) \setminus S$. A vertex is called *universal vertex* if it is adjacent to every other vertex of the graph. For more details on standard graph-theoretic notation and terminology, we refer the reader to the textbook [6].

2.1 Graph Classes

We now define the graph classes which are considered in this paper. A graph is *bipartite* if its vertex set can be partitioned into two disjoint sets such that no two vertices in the same set are adjacent. A *cluster* graph is a disjoint union of complete graphs. A *co-cluster* graph is the complement graph of a cluster graph. A graph is a *split graph* if its vertices can be partitioned into a clique and an independent set. Split graphs are $(C_4, C_5, 2K_2)$-free [7]. A graph is a *threshold graph* if it can be constructed from the one-vertex graph by repeatedly adding either an isolated vertex or a universal vertex. The class of threshold graphs is the intersection of split graphs and cographs [11]. Threshold graphs are $(P_4, C_4, 2K_2)$-free. We denote a split graph (resp. threshold graph) with $G = (C, I)$ where C and I denotes the partition of G into a clique and an independent set.

For a graph class $\mathcal{F}$ the distance to $\mathcal{F}$ of a graph G is the minimum number of vertices that have to be deleted from G in order to obtain a graph in $\mathcal{F}$. The parameters distance to cluster graphs [9], distance to co-cluster graphs, distance to threshold graphs [3] can be computed in FPT time.

2.2 Clique-Width

The *clique-width* of a graph G denoted by $cw(G)$, is defined as the minimum number of labels needed to construct G using the following four operations:

i. *Introducing a vertex.* $\Phi = v(i)$, creates a new vertex v with label i. G_Φ is a graph consisting a single vertex v with label i.
ii. *Disjoint union.* $\Phi = \Phi' \oplus \Phi''$, G_Φ is a disjoint union of labeled graphs $G_{\Phi'}$ and $G_{\Phi''}$
iii. *Introducing edges.* $\Phi = \eta_{i,j}(\Phi')$, connects each vertex with label i to each vertex with label j $(i \neq j)$ in $G_{\Phi'}$.
iv. *Renaming labels.* $\Phi = \rho_{i \to j}(\Phi')$: each vertex of label i is changed to label j in $G_{\Phi'}$.

An expression build from the above four operations using w labels is called as *w-expression*. In otherwords, the *clique-width* of a graph G, is the minimum w for which there exists a w-expression that defines the graph G. A w-expression Ψ is a *nice* w-expression of G, if no edge is introduced twice in Ψ.

3 Graphs of Bounded Clique-Width

3.1 Upper Bound

In this section, we present an algorithm for solving 2-LOAD COLORING which runs in time $n^{O(w)}$ on graphs of clique-width at most w.

Theorem 1. 2-LOAD COLORING *can be solved in time $n^{O(w)}$, where w is the clique-width of the input graph.*

Proof. The algorithm is based on a dynamic programming over the w-expression of the input graph G. We assume that the w-expression Ψ defining G is nice, that is every edge is introduced exactly once in Ψ.

For each subexpression Φ of Ψ,

$$OPT(\Phi, n_{1,r}, n_{1,b}, n_{2,r}, n_{2,b}, \ldots, n_{w,r}, n_{w,b}, k_1)$$

denotes maximum number of blue edges that can be obtained in a 2-coloring $f : V(G_\Phi) \rightarrow \{r, b\}$ of G_Φ with the constraint that number of red edges is at least k_1 in G_Φ and the number of vertices of label i in G_Φ that are colored with a color ℓ in G_Φ is $n_{i,\ell}$, where $\ell \in \{r, b\}, i \in [w]$.

If there are no colorings satisfying the constraint $OPT(\Phi, n_{1,r}, n_{1,b}, \ldots, n_{w,r}, n_{w,b}, k_1)$, then we set its value equal to -∞. Observe that G is a YES-instance of 2-LOAD COLORING if and only if $OPT(\Psi, ., \ldots, ., k_1) \geq k_2$, for some 2-coloring of G_Ψ.

Now we give the details of calculating the values of $OPT(\Phi,)$ at each operation.

1. $\Phi = v(i)$. In this case G_Φ contains one vertex of label i and no edges. Hence $OPT(\Phi, 0, 0, \ldots n_{i,r} = 1, 0, \ldots 0, k_1 = 0) = 0$ and $OPT(\Phi, 0, 0, \ldots 0, n_{i,b} = 1, 0, \ldots 0, 0, k_1 = 0) = 0$. Otherwise $OPT(\Phi, \ldots, k_1) = -\infty$.
2. $\Phi = \rho_{i\rightarrow j}(\Phi')$.
All vertices of label i are relabled to j by $\rho_{i\rightarrow j}$ operation in G_Φ, hence, there are no vertices of label i in G_Φ, so $n_{i,r} = n_{i,b} = 0$. Let c be a coloring corresponding to an entry $OPT(\Phi, n_{1,r}, n_{1,b}, \ldots, n_{w,r}, n_{w,b}, k_1)$. Then c is also a coloring of $G_{\Phi'}$, corresponding to the entry $OPT(\Phi', n'_{1,r}, n'_{1,b}, \ldots, n'_{w,r}, n'_{w,b}, k_1)$, where $n'_{i,\ell} + n'_{j,\ell} = n_{j,\ell}$ for $\ell \in \{r, b\}$ and $n'_{p,\ell} = n_{p,\ell}$ for all $p \in [w] - \{i, j\}$ and $\ell \in \{r, b\}$. The number of red and blue edges in G_Φ is the same as that in $G_{\Phi'}$ with respect to the coloring c. Hence, we have the following relation.

$$OPT(\Phi, n_{1,r}, n_{1,b}, \ldots, n_{w,r}, n_{w,b}, k_1) =$$
$$\begin{cases} \max \left\{ OPT(\Phi', n'_{1,r}, \ldots, n'_{w,b}, k_1) \;\middle|\; \begin{array}{c} n'_{i,\ell} + n'_{j,\ell} = n_{j,\ell}, n_{i,\ell} = 0 \text{ for each } \ell \in \{r, b\} \\ \text{and } n_{p,\ell} = n'_{p,\ell} \text{ for all } p \in [w] - \{i, j\} \\ \text{and } \ell \in \{r, b\} \end{array} \right\} \\ -\infty, \text{ otherwise} \end{cases}$$

3. $\Phi = \Phi' \oplus \Phi''$. As this operation does not add new edges, any coloring c corresponding to $OPT(\Phi, n_{1,r}, n_{1,b}, \ldots, n_{w,r}, n_{w,b}, k_1)$ is split between two colorings $c' = c|_{V(G_{\Phi'})}$ and $c'' = c|_{V(G_{\Phi''})}$ respectively. As G_Φ is the disjoint union of $G_{\Phi'}$ and $G_{\Phi''}$, the number of red and blue edges edges in G_Φ with respect to c is a sum of number red and blue edges with respect to c' and c'' in the graphs $G_{\Phi'}$ and $G_{\Phi''}$.

$$OPT(\Phi, n_{1,r}, n_{1,b}, \ldots, n_{w,r}, n_{w,b}, k_1) =$$
$$\max_{\substack{n'_{i,a} + n''_{i,a} = n_{i,a} \\ k'_1 + k''_1 = k_1}} \{OPT(\Phi', n'_{1,r}, \ldots, n'_{w,b}, k'_1) + OPT(\Phi'', n''_{1,r}, \ldots, n''_{w,b}, k''_1)\}$$

4. $\Phi = \eta_{i,j}(\Phi')$. The graph G_Φ is obtained from $G_{\Phi'}$ by adding the edges between each vertex of label i to each vertex of label j. Any coloring c of G_Φ is also a coloring of $G_{\Phi'}$. As given w-expression is nice, every edge is which is added by this operation was not present in $G_{\Phi'}$. Therefore $\eta_{i,j}$ operation on $G_{\Phi'}$ creates $n_{i,r} \cdot n_{j,r}$ many red edges and $n_{i,b} \cdot n_{j,b}$ blue edges. Hence, we have the following relation

$$OPT(\Phi, n_{1,r}, \ldots, n_{w,b}, k_1) = OPT(\Phi', n_{1,r}, \ldots, n_{w,b}, k_1 - n_{i,r} \cdot n_{j,r}) + n_{i,b} \cdot n_{j,b}$$

We have described the recursive formulas for all possible cases. The correctness of the algorithm follows from the description of the procedure. The number of entries in the OPT table is at most $|\Psi| n^{O(w)}$. We can compute each entry of the OPT table in $n^{O(w)}$ time. The maximum number of blue edges that can be obtained in G_Ψ is equals to $\max_{n_{1,r},\ldots,n_{w,b}} OPT(\Psi, n_{1,r}, n_{1,b}, \ldots, n_{w,r}, n_{w,b}, k_1)$ which can be computed in $n^{O(w)}$ time. This proves that 2-Load Coloring can be solved in time $n^{O(w)}$ on graphs of clique-width at most w. □

3.2 Lower Bound

We now show the lower bound complementing with the corresponding upper bound result of the previous section. To prove our result we give a linear FPT reduction from the Minimum Bisection problem. In the Minimum Bisection problem, we are given a graph G with an even number of vertices and a positive integer k, and the goal is to determine whether there is a partition of $V(G)$ into two sets V_1 and V_2 of equal size such that $|E_G[V_1, V_2]| \leq k$. Fomin et al. [8] showed that Minimum Bisection cannot be solved in time $f(w)n^{o(w)}$ unless ETH fails.

Theorem 2. *The* 2-Load Coloring *problem cannot be solved in time $f(w)n^{o(w)}$ unless ETH fails. Here, w is the clique-width of n vertex input graph.*

Proof. We give a reduction from the Minimum Bisection problem to the 2-Load Coloring problem. Let (G, k) be an instance of Minimum Bisection. We construct a graph H as follows.

1. For every vertex $v \in V(G)$, we introduce two vertices a_v, b_v in H. Let $A = \{a_{v_1}, \ldots, a_{v_n}\}$ and $B = \{b_{v_1}, \ldots, b_{v_n}\}$
2. For every edge $uv \in E(G)$, we add the edges $a_u a_v$ and $b_u b_v$ to H.
3. Finally, for every vertex $a_{v_i} \in A$ and every vertex $b_{v_j} \in B$ add the edge $a_{v_i} b_{v_j}$ to H.

It is easy to see that the graph H has $2n$ vertices and $2m + n^2$ edges and the construction of H can be done in polynomial time. Moreover, if $cw(G) = w$ with w-expression Φ_G, then we can construct a $(w+1)$-expression Φ_H of H by taking two disjoint copies of Φ_G and relabel every vertex in the first copy with the label $w+1$ and every vertex in the second copy with some arbitrary label $\ell \in [w]$ and finally add edges between both the copies using $\eta_{w+1,\ell}$. This shows that $cw(H) \leq w + 1$. Let us set $k_1 = k_2 = m - k + n^2/4$. We now show that (G, k) is a Yes instance of Minimum Bisection if and only if (H, k_1, k_2) is a Yes instance of 2-Load Coloring.

Forward Direction. Let (V_1, V_2) be a partition of $V(G)$ such that $|E_G[V_1, V_2]| \leq k$ and $|V_1| = |V_2|$. We construct a 2-coloring $f : V(H) \rightarrow \{r, b\}$ of H as follows. For each $v \in V(G)$, $f(a_v) = r$ and $f(b_v) = b$ if $v \in V_1$ and $f(a_v) = b$ and $f(b_v) = r$ if $v \in V_2$. Let $A_r = \{a_v : f(a_v) = r\}, A_b = \{a_v : f(a_v) = b\}$ and $B_r = \{b_v : f(b_v) = r\}, B_b = \{b_v : f(b_v) = b\}$. It is easy to see that $|A_r| = |A_b| = |B_r| = |B_b| = n/2$.

$$r_f = |E_H[A_r]| + |E_H[B_r]| + |E_H[A_r, B_r]| = m - k + n^2/4 = k_1$$

Similarly, we can show that $b_f = k_2$. Therefore (H, k_1, k_2) is a YES instance of 2-LOAD COLORING.

Reverse Direction. Let $f : V(H) \rightarrow \{r, b\}$ be a 2-coloring of H such that $r_f = k_1$ and $b_f = k_2$. Let $A_r = \{a_v : f(a_v) = r\}, A_b = \{a_v : f(a_v) = b\}$ and $B_r = \{b_v : f(b_v) = r\}, B_b = \{b_v : f(b_v) = b\}$. Let $V_r = A_r \cup B_r$ and $V_b = A_b \cup B_b$. Then we have $|E_H[V_r, V_b]| = 2m + n^2 - r_f - b_f = 2k + n^2/2$.

Let $V_1 := \{v : f(a_v) = r\}$ and $V_2 := \{v : f(a_v) = b\}$. We show that $|E_G(V_1, V_2)| \leq k$ and $|V_1| = |V_2|$. Let $|A_b| = p$ and $|B_b| = q$ such that $p+q = |V_b|$. Then we have $|A_r| = n - p$ and $|B_r| = n - q$.

We know that

$$|E_H[V_r]| + |E_H[V_b]| = k_1 + k_2 = 2m - 2k + n^2/2$$

$$|E_H[A_r]| + |E_H[B_r]| + |E_H[A_r, B_r]| + |E_H[A_b]| + |E_H[B_b]| + |E_H[A_b, B_b]| = 2m - 2k + n^2/2$$

$$|E_H[A_r]| + |E_H[B_r]| + (n-p)(n-q) + |E_H[A_b]| + |E_H[B_b]| + pq = 2m - 2k + n^2/2$$

After simplifying we get $(n - p)(n - q) + pq = n^2/2$. Which implies $p = q = n/2$, that is $|A_r| = |A_b| = |B_r| = |B_b| = n/2$. Hence $|V_1| = |V_2| = n/2$, that is (V_1, V_2) is a bisection of G.

Finally, we show that $|E_G[V_1, V_2]| \leq k$. We know that $|E_H[V_r, V_b]| = 2k + n^2/2$.

$$|E_H[V_r, V_b]| = |E_H[A_r, A_b]| + |E_H[B_r, B_b]| + |E_H[A_r, B_b]| + |E_H[A_b, B_r]| = 2k + n^2/2$$

By the construction of H we have $|E_H[A_r, A_b]| = |E_H[B_r, B_b]|$. Therefore we get

$$2|E_H[A_r, A_b]| + n^2/4 + n^2/4 = 2k + n^2/2$$

Which implies $|E_H[A_r, A_b]| = k$ and hence $|E_G[V_1, V_2]| = k$. Therefore (G, k) is a YES instance of MINIMUM BISECTION. This concludes the proof. □

4 Parameterized Algorithms

Clique-width of cluster graphs, co-cluster graphs and threshold graphs is at most two. Hence, from the Theorem 1, 2-LOAD COLORING is polynomial time solvable on these graph classes. In this section, we show that 2-LOAD COLORING is FPT parameterized by distance to cluster graphs, distance to co-cluster graphs and distance to threshold graphs.

4.1 Distance to Cluster Graphs

Theorem 3. 2-Load Coloring *is fixed-parameter tractable parameterized by the distance to cluster graphs.*

Proof. Let (G, X, k_1, k_2) be a 2-Load Coloring instance, where $X \subseteq V(G)$ of size k such that $G - X$ is a disjoint union of cliques $C_1, C_2, \ldots, C_\ell$. We first guess the colors of vertices in X in an optimal 2-coloring of G. This can be done in $O(2^k)$ time. Let $h : X \to \{r, b\}$ be such a coloring. Without loss of generality we assume that X is an independent set in G. Otherwise, let e_r and e_b be the number of red and blue edges in the coloring h having both end vertices in the set X. We build the new instance (G', X, k_1', k_2') of 2-Load Coloring, where G' is the graph obtained from G by deleting the edges having both their end vertices in X and $k_1' = k_1 - e_r$ and $k_2' = k_2 - e_b$. It is easy to see that (G, X, k_1, k_2) is a Yes instance of 2-Load Coloring iff there exists a 2-coloring g of G' such that $g|_X = h$ and $r_g \geq k_1'$ and $b_g \geq k_2'$. Hence, we assume that X is an independent set in G.

For each $i \in [\ell]$, let $G_i = G[X \cup C_i]$ be the subgraph of G induced by the vertices of the clique C_i and the set X.

The algorithm is based on dynamic programming technique, which has two phases. In phase-1, given a graph G_i and non-negative integers q and n_i^r, we find a 2-coloring of G_i that maximizes the number of blue edges with the constriant that there are at least q red edges in G_i and n_i^r red vertices in C_i. In phase-2, for each $t \in [\ell]$, and a non-negative integer p, we find a 2-coloring of $\widehat{G_t} = G[C_1 \cup, \ldots, \cup C_t \cup X]$ that maximizes the number of blue edges with the constraint that number of red edges is at least p.

Phase-1. For each $i \in [\ell]$, $q \in [|E(G_i)|] \cup \{0\}$ and $n_i^r \in [|C_i|] \cup \{0\}$, let $b[G_i, n_i^r, q]$ be the maximum number of blue edges that can be attained in a 2-coloring g of G_i satisfying the following constraints.

1. $g|_X = h$.
2. number of red edges in G_i is at least q.
3. number of red vertices in C_i is equal to n_i^r.

If the constraint cannot be satisfied, then we let $b[G_i, n_i^r, q] = -\infty$. From the definition of $b[,]$, we can see that $b[G_i, 0, 0]$ gives the number of blue edges in G_i when all vertices of C_i are colored blue. $b[G_i, 0, q] = -\infty$ for $q > 0$, $b[G_i, n_i^r, 0] = -\infty$ for $n_i^r > 1$. For a given values of i, n_i^r and q, the computation of $b[G_i, n_i^r, q]$ is described as follows.

Let H_i be the graph obtained from G_i by deleting the edges inside the clique C_i. It is easy to see that X is a vertex cover of the graph H_i. If g colors n_i^r vertices red in the clique C_i then we get $\binom{n_i^r}{2}$ red edges and $\binom{n_i^b}{2}$ blue edges inside C_i, where $n_i^r + n_i^b = |C_i|$. Hence we get the following relation.

$$b\Big[G_i, n_i^r, q\Big] = b\Big[H_i, n_i^r, q - \binom{n_i^r}{2}\Big] + \binom{n_i^b}{2}$$

For $v \in H_i - X$, let $r^i(v)$ and $b^i(v)$ denote the number of red and blue vertices in $N(v) \cap X$ respectively. For a vertex $v \in H_i$, we use $H_i - \{v\}$ to denote the graph obtained from H_i by deleting the vertex v and its incident edges. Let $q' = q - \binom{n_i^r}{2}$. Using this notation, we get the following recurrence.

$$b\Big[H_i, n_i^r, q'\Big] = \max\Big\{b\big[H_i - \{v\}, n_i^r - 1, \max\{q' - r^i(v), 0\}\big], b\big[H_i - \{v\}, n_i^r, q'\big] + b^i(v)\Big\}$$

If v is colored red, then we get $r^i(v)$ red edges between v and the neighbors of v in X. If v is colored blue, then we get $b^i(v)$ blue edges between v and neighbors of v in X.

The size of the DP table is at most $O(mn^2)$ and each entry can be computed in $O(n)$ time. Hence the running time of Phase-1 is $O(n^3m)$.

Phase-2. Let $\widehat{G_t}$ be the subgraph of G induced by the cliques $C_1, \cdots, C_t$ and the set X. Let $OPT[t, p]$ be the maximum number of blue edges that can be attained in a 2-coloring f of $\widehat{G_t}$ with the constraint that number of red edges is at least p and $f|_X = h$. If the constraint cannot be satisfied, then we let $OPT[t, p] = -\infty$. From the definition of OPT, we have $OPT[0, 0] = 0$ and $OPT[0, p] = -\infty$ for $p > 0$. For $t > 0$ we have:

$$OPT[t, p] = \max_{\substack{q = 0, \ldots, |E(G_t)| \\ n_t^r = 0, \ldots, |C_t|}} \Big\{OPT\big[t - 1, \max\{p - q, 0\}\big] + b[G_t, n_t^r, q]\Big\}$$

If there are q red edges in G_t and n_t^r red vertices in C_t in the coloring f, then we get $b[G_t, n_t^r, q]$ blue edges in G_t. We consider all possible values for q and n_t^r and pick the values that maximizes the OPT.

Observe that (G, X, k_1, k_2) is a YES instance if and only if $OPT[\ell, k_1] \geq k_2$. There are $O(\ell k_1)$ subproblems, each of which can be solved in $O(n^2m)$ time. As $\ell \leq n, k_1 \leq m$, the running time of phase-2 is $O(\ell k_1 n^2 m) = O(n^3m^2)$. The overall running time of the algorithm is $O(2^k n^3 m^2)$, where $O(2^k)$ is the time required for guessing the coloring of X in an optimal coloring of G. □

4.2 Distance to Co-cluster Graphs

Theorem 4. 2-LOAD COLORING *is fixed-parameter tractable parameterized by the distance to co-cluster graphs.*

Proof. Due to space restriction the proof is presented in the full version of this paper [12].

4.3 Distance to Threshold Graphs

Theorem 5. 2-LOAD COLORING *is fixed-parameter tractable parameterized by the distance to threshold graphs.*

Proof. Let (G, X, k_1, k_2) be a 2-Load Coloring instance, where $X \subseteq V(G)$ of size k such that $G - X$ is a threshold graph. We guess the coloring of X in an optimal 2-coloring of G in $O(2^k)$ time. Let $h : X \rightarrow \{r, b\}$ be such a coloring. We assume that X is an independent set in G, if not we count the number of red edges e_r and blue edges e_b inside X and replace the parameters k_1 and k_2 with $k_1 - e_r$ and $k_2 - e_b$ respectively. Then finally we delete all the edges inside X. For a vertex v in $G - X$, we use $n_X^r(v)$ (resp. $n_X^b(v)$) to denote the number of neighbors of the vertex v in the set X which are colored red (resp. blue). Let $v_1, v_2, \ldots, v_\ell$ be the ordering of the vertices of the threshold graph $G - X$ obtained from its construction (i.e., v_i is added before v_{i+1} to the graph $G - X$).

For $t \in [\ell]$, let $V_t = \{v_1, \ldots, v_t\}$ and G_t be the graph induced by the vertices $V_t \cup X$. Using the definition of threshold graphs, we can see that, for each $t \in [\ell]$, the vertex v_t is either a universal vertex or an isolated vertex in the graph $G_t - X$. We use n_t^r and n_t^b denote the number of red and blue vertices in a 2-coloring of G_t respectively.

Let $OPT[t, n_t^r, p]$ be the maximum number of blue edges that can be obtained in a 2-coloring g of G_t with the constraint that $g|_X = h$ and V_t has n_t^r vertices of color red, G_t has at least p red edges. If the constraint cannot be satisfied, then we let $OPT[t, n_t^r, p] = -\infty$. From the definition of OPT, we have $OPT[0, 0, 0] = 0$, $OPT[0, n_t^r, p] = -\infty$ for $p > 0$ and $OPT[t, 0, p] = -\infty$ for $p > 0$.

For $t > 0$, we have two cases based on whether v_t is a universal vertex or an isolated vertex in the graph $G_t - X$. If v_t is a universal vertex, then it is adjacent to all vertices of $G_t - X$. Hence we get the following relation.

$$OPT[t, n_t^r, p] = \max\{OPT[t-1, n_t^r-1, \max\{p-(n_t^r-1)-n_X^r(v_t), 0\}], OPT[t-1, n_t^r, p]+(n_t^b-1+n_X^b(v_t))\}$$

If v_t is colored red, then we get $(n_t^r - 1)$ red edges between v_t and neighbors of v_t in $G_t - X$ and $n_X^r(v_t)$ red edges between v_t and its neighbors in X. If v_t is colored blue, then we get $n_t^b - 1$ blue edges between v_t and neighbors of v_t in $G_t - X$ and $n_X^b(v_t)$ blue edges between v_t and its neighbors in X.

If v_t is an isolated vertex, then it is not adjacent to any vertex of $G_t - X$. Then we get the following relation.

$$OPT[t, n_t^r, p] = \max\{OPT[t-1, n_t^r-1, \max\{p-n_X^r(v_t), 0\}], OPT[t-1, n_t^r, p]+n_X^b(v_t)\}$$

If v_t is colored red, then we get $n_X^r(v_t)$ red edges between v_t and its neighbors in X. If v_t is colored blue, then we get $n_X^b(v_t)$ blue edges between v_t and its neighbors in X.

Observe that (G, X, k_1, k_2) is a Yes instance of 2-Load Coloring if and only if $OPT[\ell, n_t^r, k_1] \geq k_2$ for some integer n_t^r. There are $O(n^2 k_1)$ subproblems, each of which can be solved in $O(n)$ time. As $k_1 \leq m$, the overall running time of this algorithm is $O(2^k n^3 m)$. □

5 Special Graph Classes

In this section, we show that 2-Load Coloring is NP-complete on bipartite graphs and split graphs.

Theorem 6. 2-LOAD COLORING *is* NP*-complete on bipartite graphs.*

Proof. We give a reduction from 2-LOAD COLORING on general graphs. Let (G, k_1, k_2) be an instance of 2-LOAD COLORING. Without loss of generality, we assume that $k_2 \geq k_1$. We construct a bipartite graph $H = ((X \cup Z) \cup Y), E)$ as follows. For every vertex $v \in V(G)$, we introduce a vertex $x_v \in X$. For every edge $e \in E(G)$, we introduce a vertex $y_e \in Y$, and if $e = uv$, then y_e is adjacent to x_u and x_v in H. For every edge $e \in E(G)$, we also introduce additional m^2 vertices $z_e^1, z_e^2, \ldots, z_e^{m^2}$ in Z. For each $e \in E(G)$ and $i \in [m^2]$ introduce an edge between z_e^i and y_e in H. This completes the construction of the bipartite graph H and clearly can be performed in polynomial time. We set $k_1' = 2k_1 + k_1m^2$, $k_2' = (m + k_2 - k_1) + (m - k_1)m^2$. We argue that (G, k_1, k_2) is a YES instance of 2-LOAD COLORING if and only if (H, k_1', k_2') is a YES instance of 2-LOAD COLORING.

Forward Direction. Let $f : V(G) \to \{r, b\}$ is a 2-coloring of G such that $r_f = k_1$ and $b_f = k_2$. Then we define a coloring $g : V(H) \to \{r, b\}$ of H as follows: $g(x_v) = f(v)$ for all $x_v \in X$. For every edge $e = uv \in E(G)$, if $f(u) = f(v) = r$, then $g(y_e) = f(u)$, else $g(y_e) = b$. For each $e \in E(G)$ and $i \in [m^2]$, $g(z_e^i) = g(y_e)$. Note that for each red (resp blue) edge $e = uv$ in G, $m^2 + 2$ edges (namely $y_e x_v$, $y_e x_u$, $z_e^i x_u$) are red (resp. blue) with respect to g. For each edge $e = uv \in E(G)$ with $f(u) \neq f(v)$ we get $m^2 + 1$ blue edges with respect to g. Therefore, $r_g = (m^2 + 2)k_1 = 2k_1 + m^2k_1$ and $b_g = (m^2 + 2)k_2 + (m - k_1 - k_2)(m^2 + 1) = (m + k_2 - k_1) + (m - k_1)m^2$. Hence (H, k_1', k_2') is a YES instance of 2-LOAD COLORING.

Reverse Direction. Let $g : V(H) \to \{r, b\}$ be a 2-coloring of H such that $r_g = 2k_1 + k_1m^2$ and $b_g = (m + k_2 - k_1) + (m - k_1)m^2$. We assume that $g(z_e^i) = g(y_e)$ for each $e \in E(G)$ and $i \in [m^2]$, otherwise we recolor the vertices z_e^i with the color of y_e, as it only increases the number of red and blue edges. Now, we argue that the 2-coloring g of H when restricted to the vertices of G gives a 2-coloring f of G such that $r_f = k_1$ and $b_f = k_2$.

We partition the vertices of Y in H into four sets based on their coloring in g as follows.

$$P_r = \{y_e : e = uv,\ g(y_e) = g(x_u) = g(x_v) = r\}$$

$$Q_r = \{y_e : e = uv,\ g(y_e) = r \text{ and } g(x_u) \neq g(x_v)\}$$

$$P_b = \{y_e : e = uv,\ g(y_e) = g(x_u) = g(x_v) = b\}$$

$$Q_b = \{y_e : e = uv,\ g(y_e) = b \text{ and } g(x_u) \neq g(x_v)\}$$

Let p_r, q_r,p_b and q_b denote the sizes of the sets P_r, Q_r P_b and Q_b respectively. Using this notation, we have $r_g = 2p_r + q_r + m^2(p_r + q_r)$, $b_g = 2p_b + q_b + m^2(p_b + q_b)$. Also, it is easy to see that $p_r + q_r + p_b + q_b = m$. By using above two equations, we get

$$r_g + b_g = p_r + p_b + m + m^3$$

$p_r + p_b = r_g + b_g - m - m^3 = 2k_1 + k_1m^2 + (m + k_2 - k_1) + (m - k_1)m^2 - m - m^3 = k_1 + k_2$

Claim. $p_r = k_1$ and $p_b = k_2$.

Suppose $p_r = k_1 - \ell$ and $p_b = k_2 + \ell$ for some $\ell \geq 1$. Then

$$r_g = 2p_r + q_r + m^2(p_r + q_r) = 2(k_1 - \ell) + q_r + m^2(k_1 - \ell + q_r)$$

Since $r_g = 2k_1 + k_1m^2$, by substituting in the above equation, we get $q_r = \left(\frac{m^2+2}{m^2+1}\right)\ell \geq \ell + 1$. That implies $q_b \leq m - (k_1 - \ell) - (k_2 + \ell) - (\ell + 1) = (m - k_1 - k_2 - \ell - 1)$. Then

$$\begin{aligned} b_g &= 2p_b + q_b + m^2(p_b + q_b) \\ &= 2(k_2 + \ell) + (m - k_1 - k_2 - \ell - 1) + m^2(k_2 + \ell + (m - k_1 - k_2 - \ell - 1)) \\ &< (m + k_2 - k_1) + m^2(m - k_1) \end{aligned}$$

This is a contradiction as $b_g = (m + k_2 - k_1) + m^2(m - k_1)$. Therefore $p_r = k_1$ and $p_b = k_2$. Define $f : V(G) \to \{r, b\}$ as $f(u) = g(x_u)$. Since $p_r = k_1$ and $p_b = k_2$, we get $r_f \geq k_1$ and $b_f \geq k_2$. Hence, by restricting the coloring g of H to the vertices of G, we get a coloring f of G with at least k_1 red edges and k_2 blues edges. Therefore (G, k_1, k_2) is a YES instance of 2-LOAD COLORING. □

Theorem 7. 2-LOAD COLORING *is* NP*-complete on split graphs.*

Proof. The proof is similar to the case of bipartite graphs except that we add all possible edges between the vertices of Y to make it a clique in the graph H. We set $k_1' = 2k_1 + m^2k_1 + \binom{k_1}{2}$ and $k_2' = (m + k_2 - k_1) + (m - k_1)m^2 + \binom{m-k_1}{2}$. □

6 Conclusion

In this paper, we have studied the parameterized complexity of 2-LOAD COLORING. We showed that 2-LOAD COLORING (a) cannot be solved in time $f(w)n^{o(w)}$, unless ETH fails, (b) can be solved in time $n^{O(w)}$, where w is the clique-width of the graph. We have shown that the problem is FPT parameterized by (a) distance to cluster graphs (b) distance to co-cluster graphs and (c) distance to threshold graphs. We also studied the complexity of the problem on special classes of graphs. We have shown that the problem is NP-complete on bipartite graphs and split graphs. A possible future work would be to study the kernelization complexity of the problem with respect to structural graph parameters.

Acknowledgement. The author acknowledges DST-SERB (SRG/2020/001162) for funding to support this research.

References

1. Ahuja, N., Baltz, A., Doerr, B., Přívětivý, A., Srivastav, A.: On the minimum load coloring problem. J. Discrete Algorithms **5**(3), 533–545 (2007)

2. Barbero, F., Gutin, G., Jones, M., Sheng, B.: Parameterized and approximation algorithms for the load coloring problem. Algorithmica **79**(1), 211–229 (2017)
3. Cai, L.: Fixed-parameter tractability of graph modification problems for hereditary properties. Inf. Process. Lett. **58**(4), 171–176 (1996)
4. Courcelle, B., Olariu, S.: Upper bounds to the clique width of graphs. Discrete Appl. Math. **101**(1), 77–114 (2000)
5. Cygan, M., Fomin, F.V., Kowalik, L., Lokshtanov, D., Marx, D., Pilipczuk, M., Pilipczuk, M., Saurabh, S.: Parameterized Algorithms. Springer, Cham (2015). https://doi.org/10.1007/978-3-319-21275-3_15
6. Diestel, R.: Graph Theory. Graduate Texts in Mathematics (2005)
7. Foldes, S., Hammer, P.: On split graphs and some related questions. In: Problémes Combinatoires et Théorie Des Graphes, pp. 138–140, Colloques internationaux C.N.R.S. 260 (1976)
8. Fomin, F.V., Golovach, P.A., Lokshtanov, D., Saurabh, S.: Algorithmic lower bounds for problems parameterized by clique-width. In: Proceedings of the Twenty-First Annual ACM-SIAM Symposium on Discrete Algorithms, pp. 493–502. SIAM (2010)
9. Guo, J.: A more effective linear kernelization for cluster editing. Theor. Comput. Sci. **410**(8–10), 718–726 (2009)
10. Gutin, G., Jones, M.: Parameterized algorithms for load coloring problem. Inf. Process. Lett. **114**(8), 446–449 (2014)
11. Mahadev, N.V., Peled, U.N.: Threshold Graphs and Related Topics. vol. 56. Elsevier (1995)
12. Reddy, I.V.: On structural parameterizations of load coloring. arXiv preprint arXiv:2010.05186 (2020)

One-Sided Discrete Terrain Guarding and Chordal Graphs

Kasthurirangan Prahlad Narasimhan(✉)

National Institute of Science Education and Research, HBNI, Bhubaneswar, India
kprahlad.narasimhan@niser.ac.in

Abstract. The TERRAIN GUARDING problem, a variant of the famous ART GALLERY problem, has garnered significant attention over the last two decades in Computational Geometry from the viewpoint of complexity and approximability. Both the continuous and discrete versions of the problem were shown to be NP-Hard in [14] and to admit a PTAS [8,15]. The biggest unsolved question regarding this problem is if it is fixed-parameter tractable with respect to the size of the guard set. In this paper, we present two theorems that establish a relationship between a restricted case of the ANNOTATED TERRAIN GUARDING problem and the CLIQUE COVER problem in chordal graphs. These theorems were proved in [11] for a special class of terrains called orthogonal terrains and were used to present a FPT algorithm with respect to the parameter that we require for DISCRETE ORTHOGONAL TERRAIN GUARDING in [2]. We hope that the results obtained in this paper can, in future work, be used to produce such an algorithm for DISCRETE TERRAIN GUARDING.

Keywords: Terrain guarding · Chordal graphs · Visibility graphs

1 Introduction

Let $V = \{v_1, \ldots, v_n\}$ be a finite sequence of three or more points in $\mathbb{R}^2$. The *polygonal chain* defined by V is the curve specified by the line segments connecting v_i and v_{i+1} for all $1 \leq i < n$. In this paper, we additionally assume that polygonal chains are simple curves. For a point v in $\mathbb{R}^2$, we use $x(v)$ and $y(v)$ to denote the x and y coordinates of v. A *1.5-dimensional terrain* (which we will refer to as a *terrain*) is a polygonal chain defined by V where $x(v_i) \leq x(v_j)$ for all i and j such that $1 \leq i < j \leq n$. We also view a terrain T as an undirected graph with vertices V and edges $E = \{(v_i, v_{i+1}) \mid 1 \leq i < n\}$. We say that two points a and b on a terrain T *see* or *guard* each other if no point in the line segment joining these two points lies strictly below the terrain. An example of a terrain is shown in Fig. 1a. Let U be a set of points on the terrain. The *visibility region* of U is defined to be the collection of all points on the terrain which is seen by at least one point of U. We let VIS U denote this set. The encircled vertices in Fig. 1a are precisely the ones that are present in VIS U when $U = \{v_2, v_9\}$.

A. Mudgal and C. R. Subramanian (Eds.): CALDAM 2021, LNCS 12601, pp. 122–134, 2021.
https://doi.org/10.1007/978-3-030-67899-9_10

When U contains a single element, say u, we abuse this notation and write $\text{Vis}\ u$ instead of $\text{Vis}\ U$.

These definitions naturally lead us to the three major versions of the terrain guarding problem. They revolve around finding k-many points (called *guards*) on the terrain to guard a chosen set of points of the terrain. In the CONTINUOUS TERRAIN GUARDING problem, we are required to guard the vertex set of the graph by placing guards anywhere on the terrain. In the DISCRETE TERRAIN GUARDING version, while we are still to guard the vertex set, we can only place guards on the vertices themselves. ANNOTATED TERRAIN GUARDING generalizes the discrete version by restricting the vertices where the guards can be placed to a subset of V while requiring us to guard a given subset of vertices. We will focus on the annotated version of the terrain guarding problem in this paper and define it formally below. This is referenced from [2]. Hereafter, we assume that the number of vertices of a terrain is n.

Problem. ANNOTATED TERRAIN GUARDING: Given a terrain $T(V, E)$, $k \in \mathbb{N}$ and $\mathcal{G}, \mathcal{C} \subseteq V$ decide if there exists a $S \subseteq \mathcal{G}$ with $|S| \leq k$ such that $\text{Vis}\ S \supseteq \mathcal{C}$.

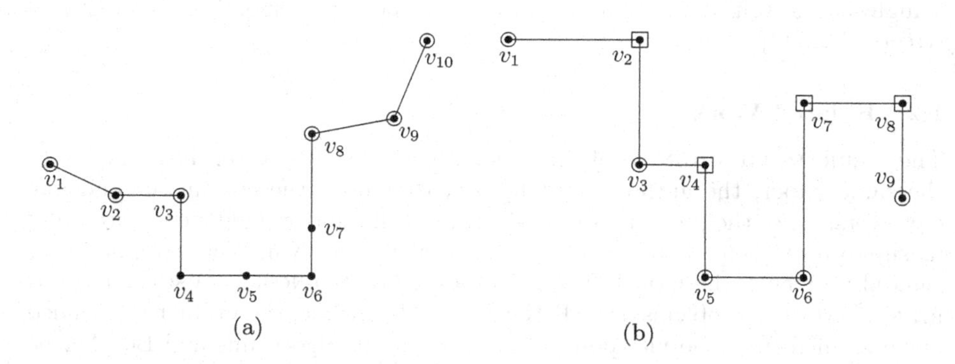

Fig. 1. Examples of terrains where the vertices and edges are marked by small discs and straight lines respectively. In (a), the vertices that are seen by $U = \{v_2, v_9\}$ are encircled. The second figure is an example of an orthogonal terrain.

Note that if $\mathcal{G} = \mathcal{C} = V$ in the annotated version of the problem, then it is exactly the DISCRETE TERRAIN GUARDING problem. We use $(T(V, E), n, k, \mathcal{G}, \mathcal{C})$ to denote an instance of the ANNOTATED TERRAIN GUARDING problem. The *visibility graph* of such an instance, G_T, is defined to be the undirected graph $G_T = (\mathcal{C}, E')$, where $E' = \{(u, v) \in \mathcal{C}^2 \mid \text{there is a } g \in \mathcal{G} \text{ that sees } u \text{ and } v\}$. In some variants of the ART GALLERY problem, vertices in the visibility graph are connected by an edge if those two vertices see each other [16]. Here, however, there exists an edge between two vertices of G_T if there exists an element in the guard set which can see both these vertices.

A subclass of terrains which are of particular interest are orthogonal terrains. In an *orthogonal* (or *rectilinear*) terrain, each edge is either parallel to the x-axis or parallel to the y-axis. Furthermore, each vertex is incident to at most one edge of each type. An example of an orthogonal terrain is given in Fig. 1b.

A graph $G(V, E)$ is *chordal* if for any $V' \subseteq V$, where $|V'| \geq 4$, the subgraph induced by V' is not a cycle. Equivalently, G is chordal if the graph induced by any cycle of length at least 4 is not a cycle. Chordal graphs have been well studied in literature since a lot of the typical NP-Hard graph problems can be solved quickly for this graph class [10]. In particular, there exists a simple polynomial time algorithm which solves the CLIQUE COVER problem in chordal graphs [9]. The CLIQUE COVER problem is defined as follows: given a graph $G(V, E)$ and a $k \in \mathbb{N}$, decide if there exists a collection of k-many cliques of G that covers V. An instance of this problem is denoted by $(G(V, E), n, k)$ where $|V| = n$.

1.1 Motivation

Optimal guarding of terrains arises in the placement of antennas for communication networks. We study this problem in two dimensions to understand better the considerably more difficult problem of guarding terrains in three dimensions. Moreover, 1.5-dimensional terrains arise directly in applications of coverage along a highway as well as in security lamp and camera placement along walls and streets [3,11,14].

1.2 Related Work

The TERRAIN GUARDING problem was stated in 1995 by Chen *et al.* in [4]. In the same paper, the authors hypothesized that both the continuous and discrete versions of the problem are NP-Hard, but did not provide a concrete proof in support of their claim. It was only in 2010 that King and Krohn finally showed that both the CONTINUOUS TERRAIN GUARDING and DISCRETE TERRAIN GUARDING problems are NP-Hard [14]. Meanwhile, the problem continued to be studied from the viewpoint of approximation algorithms and Ben-Moshe *et al.* [3] proposed the first constant-factor approximation for the discrete version of the problem. The factor of approximation was improved over the course of several papers [5,6,13] and finally a PTAS for the discrete version of the problem was given by Krohn *et al.* in 2014 [15]. A PTAS for CONTINUOUS TERRAIN GUARDING was obtained a couple of years later by Friedrichs *et al.* [8].

Thus, we have a satisfactory understanding of the approximability of the terrain guarding problem. In the paper that they proved the NP-Hardness of the terrain guarding problems, King and Krohn stated that the biggest remaining question regarding this problem was its fixed-parameter tractability. Terrain guarding has been shown to have a FPT algorithm with respect to few parameters [1,12] but it is still not known if the problem is fixed-parameter tractable with respect to the number of guards that are required to guard the terrain. In 2018, Ashok *et al.* showed that this is indeed true for the DISCRETE ORTHOGONAL TERRAIN GUARDING problem in [2]. Their algorithm exploited a connection between guarding orthogonal terrains and covering chordal graphs with cliques that was established by Katz and Roisman in Lemmas 3.6 and 3.7 of their paper

[11]. In these lemmas, they considered the visibility graph of the ANNOTATED ORTHOGONAL TERRAIN GUARDING instance $(T(V, E), n, k, \mathscr{R}, \mathscr{C}_l)$ and proved that it is chordal. They then showed that any clique of the visibility graph can been seen by a single guard. In this paper, we will show that these lemmas can be stated and proved for a special case of the annotated version of the terrain guarding problem called the LEFT-SIDED TERRAIN GUARDING problem (we define the problem formally in the next section).

1.3 Results

This paper presents two theorems which prove the equivalence between a restricted case of the LEFT-SIDED TERRAIN GUARDING problem and the CLIQUE COVER problem in chordal graphs. Theorem 3.1 proves that the visibility graph corresponding to an instance of this problem is chordal. Theorem 3.2 builds on top of this and proves that there exists a clique in the visibility graph, if, and only if, there exists a guard that sees all the vertices of that clique. Collating these two theorems gives us the main result of this paper. Lemmas 3.4 and 3.5 show that this paper indeed generalizes results that are known for orthogonal terrains.

Main Result. Let $(T(V, E), n, k, \mathcal{G}, \mathcal{C})$ be a LEFT-SIDED TERRAIN GUARDING instance where $\mathcal{G} \cap \mathcal{C} = \emptyset$ and VIS $\mathcal{G} \supseteq \mathcal{C}$. Then, this is a true instance of the problem if, and only if, $(G_T(\mathcal{C}, E'), |\mathcal{C}|, k)$ is a true instance of the CLIQUE COVER problem where G_T, the visibility graph of T, is a chordal graph.

2 Preliminaries

For points a and b on T, we say a *precedes* b, denoted by $a \prec b$, if a appears on the terrain before b does (the terrain is scanned from left to right). The Order Claim, which was originally stated in [3] and later slightly generalized in [1], lays the foundation for the theorems that follow in the next section.

Lemma 2.1 (Order Claim). *Let a, b, c and d be four points on a terrain $T(V, E)$ such that $a \prec b \prec c \prec d$. If a sees c and b sees d, then, a sees d.*

In an orthogonal terrain $T(V, E)$, a vertex v_i, where $1 < i < n$, is *convex* if $x(v_i) = x(v_{i+1})$ and $y(v_i) < y(v_{i+1})$ or $x(v_i) = x(v_{i-1})$ and $y(v_i) < y(v_{i-1})$ and is *reflex* otherwise. Equivalently, v_i is a convex vertex if the angle formed by the vertices v_{i-1}, v_i and v_{i+1} (measured above the terrain) is convex and is a reflex vertex otherwise. It is a *left* vertex if $x(v_{i-1}) = x(v_i)$ and a *right* vertex if $x(v_i) = x(v_{i+1})$. The set of convex vertices is denoted by $\mathscr{C}$ and the set of reflex vertices is denoted by $\mathscr{R}$. In Fig. 1b, the convex vertices are encircled and the reflex vertices are marked using squares. The set of vertices which are both convex and left are called *left convex* vertices and is denoted by $\mathscr{C}_l$. *Right convex*, *left reflex* and *right reflex* vertices are defined similarly and are denoted by $\mathscr{C}_r$, $\mathscr{R}_l$, and $\mathscr{R}_r$ respectively. Vertices $a \in \mathscr{C}_l$ and $b \in \mathscr{R}_r$ are said to be of the *opposite type* as are vertices $c \in \mathscr{C}_r$ and $d \in \mathscr{R}_l$. v_1 is defined to be of the opposite type as v_2 and v_n is defined to be that of v_{n-1}. In Fig. 1b, $\mathscr{R}_l = \{v_7\}$, $\mathscr{R}_r = \{v_2, v_4, v_8\}$, $\mathscr{C}_l = \{v_1, v_3, v_5, v_9\}$ and $\mathscr{C}_r = \{v_6\}$. Finally, we define a restriction of the annotated version of the terrain guarding problem where we allow the guards to only see in one direction.

Problem. LEFT-SIDED TERRAIN GUARDING: Given a terrain $T(V, E)$, $k \in \mathbb{N}$ and $\mathcal{G}, \mathcal{C} \subseteq V$ decide if there exists a $S \subseteq \mathcal{G}$ with $|S| \leq k$ such that for all $v \in \mathcal{C}$, there is a $g \in \mathcal{G}$ such that $x(v) \leq x(g)$ and g sees v.

Equivalently, in this version of the problem we enforce that the guards of $\mathcal{G}$ can only see to their left. In this case, we say that $\mathcal{G}$ is a set of *left guards*. RIGHT-SIDED TERRAIN GUARDING and *right guards* are defined symmetrically. In the paper that they introduced the terrain guarding problem [4], Chen *et al.* also described the left and right-guarding versions of the problem. They produced an algorithm, which they called *Army-Withdraw*, which ran in linear time to solve these versions. Elbassioni *et al.* [6] constructed a bipartite graph G from a LEFT-SIDED TERRAIN GUARDING instance $(T(V, E), n, k, \mathcal{G}, \mathcal{C})$ where $\mathcal{G} \cap \mathcal{C} = \emptyset$ with the bipartition $(\mathcal{G}, \mathcal{C})$. An element $(g, c) \in \mathcal{G} \times \mathcal{C}$ was an edge of this graph if $x(c) \leq x(g)$ and g sees c. They then proved that the vertex-vertex incidence matrix corresponding to this graph is *totally balanced* an used the properties of such matrices to produce a 4-approximation algorithm for the ANNOTATED TERRAIN GUARDING problem where $\mathcal{G} \cap \mathcal{C} = \emptyset$. The author refers the reader to [7] and [10] for a detailed discussion on totally balanced matrices.

3 Terrains and Chordal Graphs

In this section, we will prove two theorems which will lead us to the main result of this paper. Even though this section deals exclusively with the LEFT-SIDED TERRAIN GUARDING problem, the claims and their proofs apply, by symmetry, to the RIGHT-SIDED TERRAIN GUARDING problem. The first theorem proves that the visibility graph of an LEFT-SIDED TERRAIN GUARDING instance $(T(V, E), n, k, \mathcal{G}, \mathcal{C})$ is chordal. The proof of this theorem considers a cycle C of length k, where $k \geq 4$, in G_T and proves that the subgraph induced by C, denoted by $G_T[C]$, is not a cycle. This is done by an extensive use of Lemma 2.1 on the various cases that arise depending on the positions of the vertices of C and the guards that see them on the terrain.

The second theorem considers a LEFT-SIDED TERRAIN GUARDING instance $(T(V, E), n, k, \mathcal{G}, \mathcal{C})$ where $\mathcal{G}$ and $\mathcal{C}$ are disjoint and VIS $\mathcal{G} \supseteq \mathcal{C}$. It proves that the vertices of any clique of G_T can be seen by a single guard. This proves that k-many guards can see all of $\mathcal{C}$ if, and only if, there exists k-many cliques that cover G_T. This along with the previous theorem directly proves our main result. We prove this theorem using induction over the number of vertices in the clique. In Fig. 8, we provide an example of a terrain where this claim fails if $\mathcal{G} \cap \mathcal{C}$ is non-empty.

Finally, we prove two corollaries of our main result. The first one states that if a LEFT-SIDED TERRAIN GUARDING instance is false, then there exists a small subset of $\mathcal{C}$ (with $k+1$ vertices) that cannot be seen by k-many guards. We prove this by producing an independent set U of size $k+1$ in G_T and observing that if k guards did see all the vertices of U, then U would fail to be an independent set. The second corollary proves that for an orthogonal terrain T, $(T(V,E), n, k, \mathscr{R}, \mathscr{C}_l)$ is a true instance if, and only if, $(G_l(\mathscr{C}_l, E'), |\mathscr{C}_l|, k)$ is a true instance of the CLIQUE COVER problem. This was the result obtained in [11] by Katz and Roisman. This is done by observing that left convex vertices can only see to one side.

Theorem 3.1. *Let $(T(V,E), n, k, \mathcal{G}, \mathcal{C})$ be a* LEFT-SIDED TERRAIN GUARDING *instance. Then, the visibility graph of this instance, say G_T, is chordal.*

Proof. Let $C \subseteq \mathcal{C}$ where $|C| = p \geq 4$ be a cycle in G_T. We prove that $G_T[C]$ is not a cycle. Let $C = \{c_1, c_2 \dots c_p\}$ be the order of the vertices as they appear on the cycle. Also, we assume, without loss in generality, that $c_i \preceq c_1$ for all $c_i \in C$ and that $c_p \prec c_2$. As c_1 and c_p are neighbours in G_T, there is a left guard $g_{1,p}$ which sees both these vertices. Similarly, we have $g_{1,2}$, a left guard, which sees both c_1 and c_2. Note that $c_1 \preceq g_{1,p}$ and $c_1 \preceq g_{1,2}$. If $g_{1,2} = g_{1,p} = g$, then g sees both c_2 and c_p. This implies that c_2 and c_p share an edge in $G_T[C]$. Since $p \geq 4$, (c_2, c_p) is a chord of the cycle. Thus, $G_T[C]$ is not a cycle. We are now left with two cases:

Case 1 ($g_{1,p} \prec g_{1,2}$). This is illustrated in Fig. 2. Here, $c_p \prec c_2 \prec g_{1,p} \prec g_{1,2}$ and c_p sees $g_{1,p}$ while c_2 sees $g_{1,2}$. Thus, by Lemma 2.1, $g_{1,2}$ sees c_p. Since $g_{1,2}$ sees c_2 by construction, there is an edge between c_2 and c_p in G_T. As observed previously, this implies that $G_T[C]$ is not a cycle. Note that $g_{1,p}$ could be c_1.

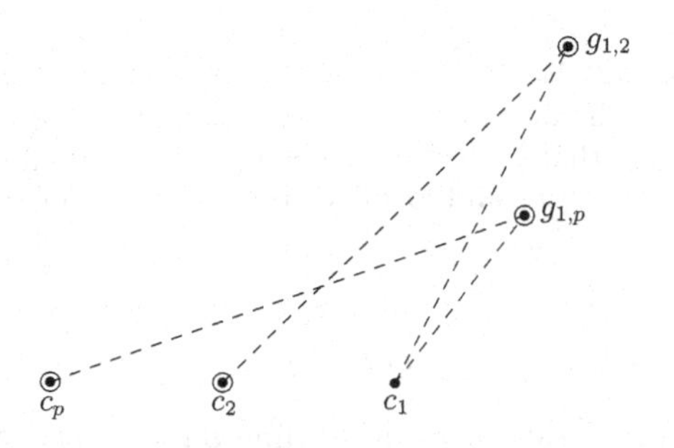

Fig. 2. This figure illustrates Case 1 of Theorem 3.1 where $g_{1,p} \prec g_{1,2}$. Two vertices that see each other are connected by a dashed line. If we substitute a, b, c and d with c_p, c_2, $g_{1,p}$ and $g_{1,2}$ respectively in Lemma 2.1, we get that c_p sees $g_{1,2}$ in this case.

Case 2 ($g_{1,2} \prec g_{1,p}$). In this case, we have two possibilities. The first one is when $c_1 \prec g_{1,2}$ while the second one is where $g_{1,2}$ is c_1. Figure 3a illustrates the first case. Here, $c_2 \prec c_1 \prec g_{1,2} \prec g_{1,p}$ and c_2 sees $g_{1,2}$ while c_1 sees $g_{1,p}$. We infer that $g_{1,p}$ sees c_2 by applying Lemma 2.1 on these vertices. Thus, there is an edge between c_2 and c_p in $G_T[C]$ proving that $G_T[C]$ is not a cycle. Now, assume that $c_1 = g_{1,2}$. This is considered in Fig. 3b. Unfortunately, in this situation, we cannot use Lemma 2.1 directly.

We now consider c_3, the other neighbour of c_2, in C. Note that c_3 exists as $|C| \geq 4$. Since c_2 and c_3 are neighbours in C, and thus in G_T, there exists a left guard, say $g_{2,3}$, which sees both these vertices. If $g_{2,3} = g_{1,p} = g$, then c_3 has an edge with c_1 in G_T. Since the neighbours of c_1 in C are c_2 and c_p, where $p > 3$, (c_1, c_3) is chord of C. This proves that $G_T[C]$ is not a cycle. Thus, we will focus on the situations where $g_{2,3} \neq g_{1,p}$ in the cases that follow.

We will show that if $c_3 \prec c_2$, then $G_T[C]$ is not a cycle. We will then prove that if $c_{j+1} \prec c_2 \prec c_j \prec c_1$ for any $3 \leq j < p$, then $G_T[C]$ is not a cycle. Finally, we prove that these two claims jointly imply that $G_T[C]$ is never a cycle and complete our proof.

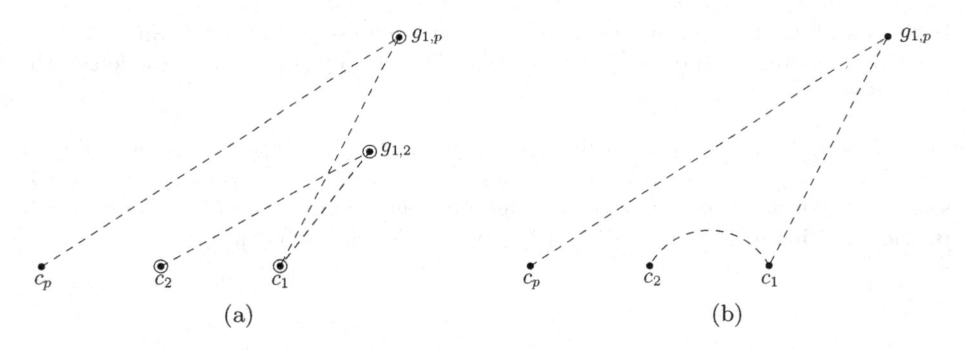

Fig. 3. This figure depicts Case 2 of Theorem 3.1 where $g_{1,2} \prec g_{1,p}$. In (a) $c_1 \prec g_{1,2}$, while in (b) $c_1 = g_{1,2}$. In the first possibility, we can apply Lemma 2.1 on the encircled vertices. Doing so, we get that c_2 sees $g_{1,p}$. We are unable to apply the Lemma 2.1 on (b). The dashed line between c_1 and c_2 in (b) is curved for illustrative purposes.

Claim 1. If $c_3 \prec c_2$, then $G_T[C]$ is not a cycle.

We consider the following two cases depending on the position of c_3: it precedes c_p or it lies between c_p and c_2.

Subcase 2.1 ($c_3 \prec c_p$). We have two possibilities: $g_{2,3} \prec g_{1,p}$ or $g_{1,p} \prec g_{2,3}$. These are shown in Fig. 4a and 4b respectively. Lemma 2.1 guarantees an edge between c_1 and c_3 in the former case and between c_2 and c_p in the latter case in $G_T[C]$ (apply the lemma on c_3, c_p, $g_{2,3}$ and $g_{1,p}$ in the first case and on c_p, c_2, $g_{1,p}$ and

$g_{2,3}$ in the second case). As noted previously, $G_T[C]$ is not a cycle in both these cases. Note that $g_{2,3}$ could be c_1 or c_2 in the first case and $g_{1,p}$ could be c_1 in the second.

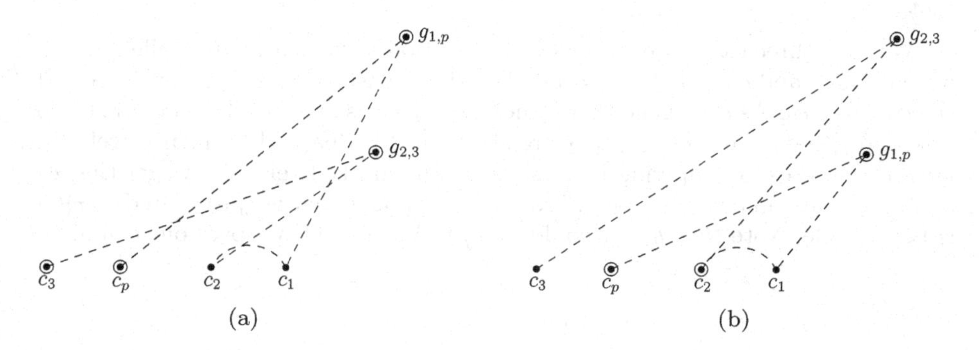

Fig. 4. This figure illustrates Subcase 2.1 of Theorem 3.1 where $c_3 \prec c_p$. In (a) $g_{2,3}$ precedes $g_{1,p}$, while in (b) $g_{1,p}$ precedes $g_{2,3}$. On applying Lemma 2.1 to the marked vertices, we get that c_3 is seen by $g_{1,p}$ in (a) and c_p is seen by $g_{2,3}$ in (b).

Subcase 2.2 ($c_p \prec c_3 \prec c_2$). Depending on the position of $g_{2,3}$, we have three possibilities: $g_{2,3}$ precedes c_1, or it lies between c_1 and $g_{1,p}$, or it lies after $g_{1,p}$. Note that the third case is equivalent to the one in Fig. 4b since the position of c_3 was not used in the proof of the existence of the (c_2, c_p) edge. If $g_{2,3}$ is equal to c_1, then c_1 has an edge with c_3 in $G_T[C]$ since it sees itself as well as c_3. Thus, $G_T[C]$ is not a cycle. This leaves two cases: $g_{2,3} \prec c_1$ and $c_1 \prec g_{2,3} \prec g_{1,p}$. These are shown in Figs. 5a and 5b. On applying Lemma 2.1 on the four encircled vertices in the order they appear on the terrain, we infer that the (c_1, c_3) edge exists in $G_T[C]$ in both these cases proving that it is not a cycle.

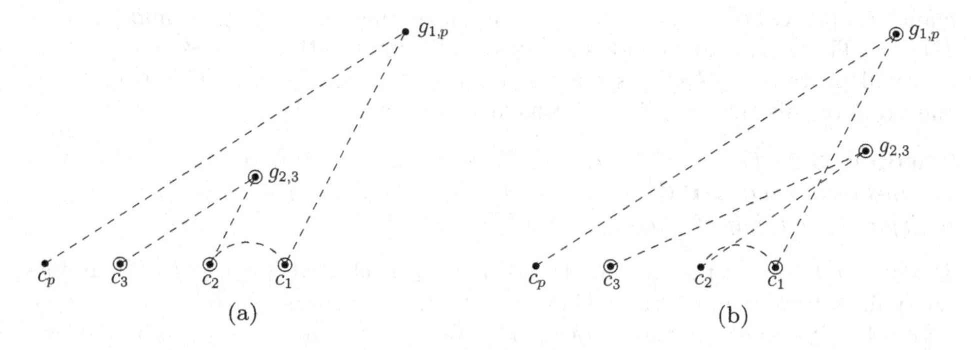

Fig. 5. c_3 lies between c_2 and c_p in both these figures which illustrate Subcase 2.2 of Theorem 3.1. In (a) $g_{2,3} \prec c_1$, while in (b) $c_1 \prec g_{2,3} \prec g_{1,p}$. On applying Lemma 2.1 on the encircled vertices we infer that c_3 is seen by c_1 in (a) and by $g_{1,p}$ in (b).

Thus, we have proven that if c_3 precedes c_2, $G_T[C]$ is not a cycle. We will complete the proof by proving the following claim.

Claim 2. If $c_{j+1} \prec c_2 \prec c_j \prec c_1$ for some j where $3 \leq j < p$, then $G_T[C]$ is not a cycle.

We have three cases depending on the position of the left guard, called $g_{j,j+1}$, which sees c_j and c_{j+1}. If $g_{j,j+1} = c_1$, then there exists a (c_2, c_{j+1}) edge in $G_T[C]$ since c_1 sees c_2. As c_{j+1} is neither c_1 nor c_3, $G_T[C]$ is not a cycle. The other two cases: $g_{j,j+1} \prec c_1$ and $c_1 \prec g_{j,j+1}$ are shown in Figs. 6a and 6b respectively. In both these cases, on applying Lemma 2.1 on the marked vertices, we get that c_2 and c_{j+1} share an edge in $G_T[C]$. From the argument that we just stated, $G_T[C]$ is not a cycle. Note that $g_{j,j+1}$ could be c_j in Fig. 6a. This proves our claim.

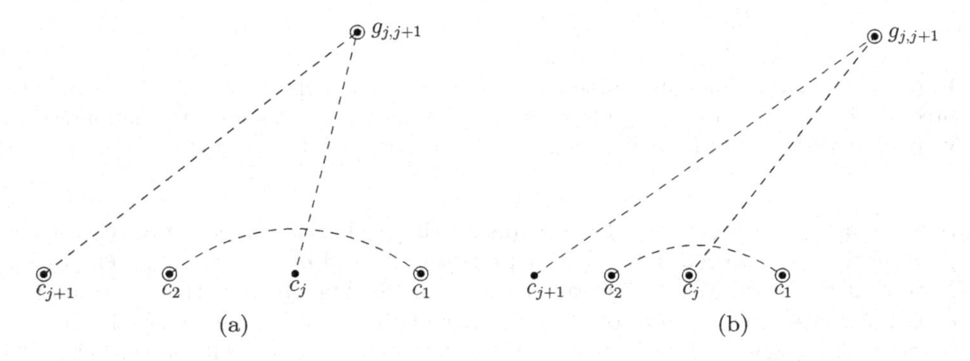

Fig. 6. This figure corresponds to cases that are discussed in Claim 2 of Theorem 3.1. Here, c_j lies between c_2 and c_1. In (a) $g_{j,j+1} \prec c_1$, while in (b) $g_{j,j+1}$ lies after c_1. On applying Lemma 2.1, we get that c_1 sees c_{j+1} in (a) and c_2 sees $g_{j,j+1}$ in (b).

Now, let $D = \{i \mid c_2 \prec c_i \prec c_1\}$. If D is empty, then $c_3 \prec c_2$. By our first claim, $G_T[C]$ is not a cycle. Since we assumed that $c_p \prec c_2$, $j := \max\{i \mid i \in D\} < p$. Thus, c_{j+1} exists and is not in D. This implies that $c_{j+1} \prec c_2 \prec c_j \prec c_1$ and will prove that $G_T[C]$ is not a cycle by the second claim. This completes the proof of this theorem since C was arbitrary cycle of length at least 4. □

Theorem 3.2. *Let $(T(V, E), n, k, \mathcal{G}, \mathcal{C})$ be a* Left-Sided Terrain Guarding *instance where $\mathcal{G} \cap \mathcal{C} = \emptyset$ and* Vis *$\mathcal{G} \supseteq \mathcal{C}$. Then, for $K \subseteq \mathcal{C}$, $G_T[K]$ is a clique if, and only if, there is a $g \in \mathcal{G}$ such that* Vis *$g \supseteq K$.*

Proof. Let K be a set such that there is a $g \in \mathcal{G}$ such that Vis $g \supseteq K$. Then, for any pair of vertices in K there is an edge between them in $G_T[K]$ since there is a guard (g itself) seeing them both. Thus, $G_T[K]$ is a clique. Now, we prove the forward direction of the claim. Assume that $K \subseteq \mathcal{C}$ such that $G_T[K]$ is a clique. We prove that there exists a guard seeing all of K by induction on the number of vertices in K. If $|K| = 1$ or $|K| = 2$, then our claim follows trivially. Assume that our supposition holds for all cliques of size at most p, where $p \geq 2$.

Let $K = \{k_1, k_2, \ldots k_p, k_{p+1}\}$ be a subset of $\mathcal{C}$ such that $G_T[K]$ is a clique. The vertices of K are ordered according to how they appear on the terrain. Let $K' = \{k_2 \ldots k_p, k_{p+1}\}$. Since $G_T[K']$ is a clique of size p, by the induction hypothesis, there is a left guard g_1 such that $\textsc{Vis}\ g_1 \supseteq K'$. Since there is a (k_1, k_{p+1}) edge in $G_T[K]$, there is a left guard, say g_2, which sees them both. If $g_1 = g_2 = g$, then we have $\textsc{Vis}\ g \supseteq K$ proving the supposition. We are now left with two cases:

Case 1 ($g_2 \prec g_1$). This case is shown in Fig. 7a. Here, we observe that $k_1 \prec k_j \prec g_2 \prec g_1$ where k_1 sees g_2 and k_j sees g_1. By Lemma 2.1, g_1 guards k_1 as well. Thus, $\textsc{Vis}\ g_1 \supseteq K$.

Case 2 ($g_1 \prec g_2$). This is illustrated in Fig. 7b. Here, on applying Lemma 2.1 on the marked vertices, we get that g_2 sees k_j for all j where $2 \leq j \leq p$. Thus, $\textsc{Vis}\ g_2 \supseteq K$. Note that we can apply Lemma 2.1 on these four vertices only because $g_1 \neq k_{p+1}$.

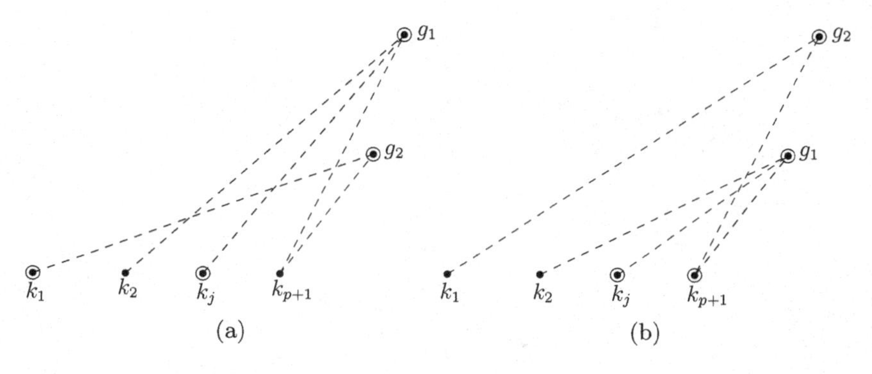

Fig. 7. This figure corresponds to the cases that arise in Theorem 3.2. In (a) $g_2 \prec g_1$, while in (b) $g_1 \prec g_2$. By Lemma 2.1, g_1 also sees k_1 in (a) and g_2 guards all the vertices from k_2 through to k_{p+1} in (b).

This proves our supposition and completes the proof by induction. Note that the situations illustrated in Figs. 7a and 7b are similar to the ones in Figs. 2 and 3a. They are presented in this proof again for clarity. □

This theorem's claim is not true if $\mathcal{G} \cap \mathcal{C} \neq \emptyset$. A non-example is presented in Fig. 8. In the terrain illustrated by that figure, we let $\mathcal{C} = \{v_1, v_3, v_4\}$ and $\mathcal{G} = \{v_4, v_5, v_7\}$ be a set of left guards. v_1 shares an edge with both v_3 and v_4 in G_T since v_7 sees both v_1 and v_3 while v_5 sees both v_1 and v_4. Furthermore, since v_4 sees itself as well as v_3, there is an edge between v_3 and v_4 in G_T. Thus, $G_T[\mathcal{C}]$ is a clique. However, none of the three guards in $\mathcal{G}$ guard all the vertices of $\mathcal{C}$: v_4 does not see v_1, v_5 does not see v_3, and v_7 does not see v_4. It is also clear that Theorem 3.2 fails to hold if $\textsc{Vis}\ \mathcal{G} \supseteq \mathcal{C}$. For example, if $\mathcal{C} = \{v_3\}$ and $\mathcal{G} = \{v_5\}$

in the terrain illustrated in Fig. 8, then the isolated vertex v_3 is a clique but no guard in $\mathcal{G}$ sees it.

Theorem 3.1 and 3.2 are, to the best of the author's knowledge, an addition to existing literature. Combining these two theorems gives us the main result of this paper.

Theorem 3.3. *Let $(T(V,E), n, k, \mathcal{G}, \mathcal{C})$ be a* LEFT-SIDED TERRAIN GUARDING *instance where $\mathcal{G} \cap \mathcal{C} = \emptyset$ and* VIS $\mathcal{G} \supseteq \mathcal{C}$*. Then, this is a true instance of the problem if, and only if, $(G_T(\mathcal{C}, E'), |\mathcal{C}|, k)$ is a true instance of the* CLIQUE COVER *problem where G_T, the visibility graph of T, is a chordal graph.*

As stated in the beginning of this section, the above result also holds for the RIGHT-SIDED TERRAIN GUARDING problem. It is well known that in a chordal graph $G(V,E)$, the minimum number of cliques required to cover V, denoted by $\overline{\chi}(G)$, is equal to the size of a maximum sized independent set of G, denoted by $\alpha(G)$ [10]. The algorithm that solves the CLIQUE COVER problem can be modified slightly to solve the INDEPENDENT SET problem in polynomial time [9]. We use these two properties of chordal graphs in the proof of the lemma that follows.

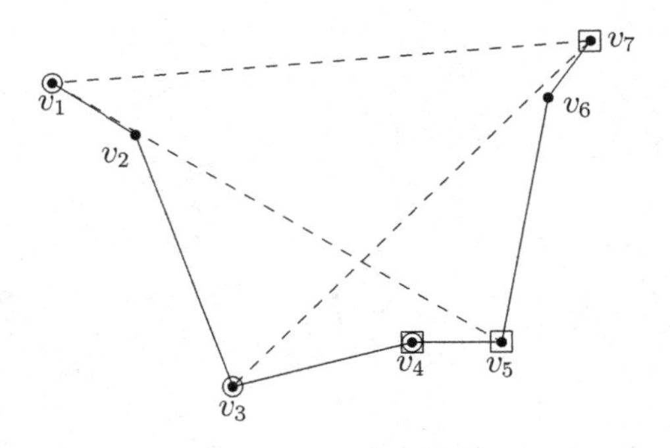

Fig. 8. This terrain presents an example where a clique in G_T is not seen by a single left guard if $\mathcal{G} \cap \mathcal{C} \neq \emptyset$. The vertices of $\mathcal{C}$ and $\mathcal{G}$ have been encircled and marked by squares respectively.

Lemma 3.4. *Let $(T(V,E), n, k, \mathcal{G}, \mathcal{C})$ be a* LEFT-SIDED TERRAIN GUARDING *instance where $\mathcal{G} \cap \mathcal{C} = \emptyset$ and* VIS $\mathcal{G} \supseteq \mathcal{C}$*. One can decide, in polynomial time, if this is a true instance of the problem. If this instance is false, then one can find $U \subseteq \mathcal{C}$ in polynomial time such that $|U| = k+1$ and $(T(V,E), n, k, \mathcal{G}, U)$ is a false instance.*

Proof. By Theorem 3.3, we know that the visibility graph, say G_T, corresponding to $(T(V,E), n, k, \mathcal{G}, \mathcal{C})$ is chordal and that it is a true instance if, and only if, $(G_T(\mathcal{C}, E'), |\mathcal{C}|, k)$ is a true instance of the CLIQUE COVER problem. Since the CLIQUE COVER problem can be solved in polynomial time in chordal graphs, we can decide if $(T(V,E), n, k, \mathcal{G}, \mathcal{C})$ is a true instance of the LEFT-SIDED TERRAIN GUARDING problem in polynomial time.

Now, if $(T(V,E), n, k, \mathcal{G}, \mathcal{C})$ is false, then G_T cannot be covered by k many cliques. Thus, $\overline{\chi}(G_T) > k$. This implies that $\alpha(G_T) > k$. We compute a maximum sized independent set of G_T and let U be a subset of size $k+1$ of this independent set. Since G_T is chordal, this can be done in polynomial time. Clearly, U is an independent set of G_T. If there exists k many guards in $\mathcal{G}$ which guards U, then there must exist one guard which sees at least two vertices of U. By construction of G_T, there must exist an edge between them. This contradicts the fact that U is an independent set of G_T and thus completes the proof of this lemma. □

We note that the above lemma holds for the right-sided version of the terrain guarding problem as well. The lemma stated and proved above generalizes Lemmas 4.8 and 4.9 of [2]. These were used to present a FPT algorithm with respect to the solution size for the DISCRETE ORTHOGONAL TERRAIN GUARDING problem in that paper. We conclude this paper by proving that the following result by Katz and Roisman [11] follows from Theorem 3.3.

Lemma 3.5. *Consider the* ANNOTATED ORTHOGONAL TERRAIN GUARDING *instance* $(T(V,E), n, k, \mathscr{R}, \mathscr{C}_l)$ *and let* G_l *be the visibility graph corresponding to this instance. Then,* G_l *is chordal. Furthermore,* $(T(V,E), n, k, \mathscr{R}, \mathscr{C}_l)$ *is a true instance of the problem if, and only if,* $(G_l(\mathscr{C}_l, E'), |\mathscr{C}_l|, k)$ *is a true instance of the* CLIQUE COVER *problem. The symmetric claim holds for the set of right convex vertices.*

Proof. Note that a vertex $v \in \mathscr{C}_l$ can only see to its right (referring back to Fig. 1b will make this observation straightforward) [11]. Equivalently, a vertex $g \in \mathscr{R}$ which is to guard v needs to look only to its left. Thus, we can consider the guards which are required to guard $\mathscr{C}_l$ to be a set of left guards. Using a symmetric argument, we see that the guard set that is to guard the right convex vertices can be considered to be a set of right guards. Also, we note that VIS $\mathscr{R} \supseteq V$. This implies that $\mathscr{R}$ sees all of $\mathscr{C}_l$ and $\mathscr{C}_r$. Since $\mathscr{C} \cap \mathscr{R} = \emptyset$, we can apply Theorem 3.3 to the LEFT-SIDED TERRAIN GUARDING instance $(T(V,E), n, k, \mathscr{R}, \mathscr{C}_l)$. By a symmetric argument, our claim is also true for the RIGHT-SIDED TERRAIN GUARDING instance $(T(V,E), n, k, \mathscr{R}, \mathscr{C}_r)$. This completes the proof of the lemma. □

Acknowledgments. The author would like to thank Susobhan Bandopadhyay, Dr. Aritra Banik, and Dr. Sushmita Gupta for helpful discussions. He would also like to thank the anonymous reviewers of CALDAM-2021 for their valuable comments.

References

1. Agrawal, A., Kolay, S., Zehavi, M.: Parameter analysis for guarding terrains. In: Albers, S. (ed.) 17th Scandinavian Symposium and Workshops on Algorithm Theory, SWAT 2020, Tórshavn, Faroe Islands, 22–24 June 2020. LIPIcs, vol. 162, pp. 4:1–4:18. Schloss Dagstuhl - Leibniz-Zentrum für Informatik (2020). https://doi.org/10.4230/LIPIcs.SWAT.2020.4

2. Ashok, P., Fomin, F.V., Kolay, S., Saurabh, S., Zehavi, M.: Exact algorithms for terrain guarding. ACM Trans. Algorithms **14**(2), 25:1–25:20 (2018). https://doi.org/10.1145/3186897
3. Ben-Moshe, B., Katz, M.J., Mitchell, J.S.B.: A constant-factor approximation algorithm for optimal 1.5D terrain guarding. SIAM J. Comput. **36**(6), 1631–1647 (2007). https://doi.org/10.1137/S0097539704446384
4. Chen, D.Z., Estivill-Castro, V., Urrutia, J.: Optimal guarding of polygons and monotone chains. In: Proceedings of the 7th Canadian Conference on Computational Geometry, Quebec City, Quebec, Canada, August 1995, pp. 133–138. Carleton University, Ottawa, Canada (1995). http://www.cccg.ca/proceedings/1995/cccg1995_0022.pdf
5. Clarkson, K.L., Varadarajan, K.R.: Improved approximation algorithms for geometric set cover. Discret. Comput. Geom. **37**(1), 43–58 (2007). https://doi.org/10.1007/s00454-006-1273-8
6. Elbassioni, K.M., Krohn, E., Matijevic, D., Mestre, J., Severdija, D.: Improved approximations for guarding 1.5-dimensional terrains. Algorithmica **60**(2), 451–463 (2011). https://doi.org/10.1007/s00453-009-9358-4
7. Farber, M.: Characterizations of strongly chordal graphs. Discret. Math. **43**(2–3), 173–189 (1983). https://doi.org/10.1016/0012-365X(83)90154-1
8. Friedrichs, S., Hemmer, M., King, J., Schmidt, C.: The continuous 1.5D terrain guarding problem: discretization, optimal solutions, and PTAS. J. Comput. Geom. **7**(1), 256–284 (2016). https://doi.org/10.20382/jocg.v7i1a13
9. Gavril, F.: Algorithms for minimum coloring, maximum clique, minimum covering by cliques, and maximum independent set of a chordal graph. SIAM J. Comput. **1**(2), 180–187 (1972). https://doi.org/10.1137/0201013
10. Golumbic, M.C.: Algorithmic graph theory and perfect graphs. Networks **13**(2), 304–305 (1983). https://doi.org/10.1002/net.3230130214
11. Katz, M.J., Roisman, G.S.: On guarding the vertices of rectilinear domains. Comput. Geom. **39**(3), 219–228 (2008). https://doi.org/10.1016/j.comgeo.2007.02.002
12. Khodakarami, F., Didehvar, F., Mohades, A.: A fixed-parameter algorithm for guarding 1.5D terrains. Theor. Comput. Sci. **595**, 130–142 (2015). https://doi.org/10.1016/j.tcs.2015.06.028
13. King, J.: A 4-approximation algorithm for guarding 1.5-dimensional terrains. In: Correa, J.R., Hevia, A., Kiwi, M. (eds.) LATIN 2006. LNCS, vol. 3887, pp. 629–640. Springer, Heidelberg (2006). https://doi.org/10.1007/11682462_58
14. King, J., Krohn, E.: Terrain guarding is NP-hard. In: Charikar, M. (ed.) Proceedings of the Twenty-First Annual ACM-SIAM Symposium on Discrete Algorithms, SODA 2010, Austin, Texas, USA, 17–19 January 2010, pp. 1580–1593. SIAM (2010). https://doi.org/10.1137/1.9781611973075.128
15. Krohn, E., Gibson, M., Kanade, G., Varadarajan, K.R.: Guarding terrains via local search. J. Comput. Geom. **5**(1), 168–178 (2014). https://doi.org/10.20382/jocg.v5i1a9
16. O'Rourke, J.: Art gallery theorems and algorithms. SIAM Rev. **31**(2), 342–343 (1989). https://doi.org/10.1137/1031076

Parameterized Complexity of Locally Minimal Defensive Alliances

Ajinkya Gaikwad, Soumen Maity(✉), and Shuvam Kant Tripathi

Indian Institute of Science Education and Research, Pune, India
soumen@iiserpune.ac.in, {ajinkya.gaikwad, tripathi.shuvamkant}@students.iiserpune.ac.in

Abstract. A defensive alliance in a graph $G = (V, E)$ is a set of vertices S satisfying the condition that every vertex $v \in S$ has at least as many neighbours (including itself) in S as it has in $V \backslash S$. We consider the notion of local minimality in this paper. We are interested in locally minimal defensive alliance of maximum size. This problem is known to be NP-hard but its parameterized complexity remains open until now. We enhance our understanding of the problem from the viewpoint of parameterized complexity. The three main results of the paper are the following: (1) when the input graph happens to be a tree, Locally Minimal Strong Defensive Alliance can be solved in polynomial time, (2) Locally Minimal Defensive Alliance is fixed parameter tractable (FPT) when parametrized by neighbourhood diversity, and (3) Locally Minimal Defensive Alliance can be solved in polynomial time for graphs of bounded treewidth.

Keywords: Parameterized complexity · FPT · Treewidth

1 Introduction

During the last 20 years, the Defensive Alliance problem has been studied extensively. A defensive alliance in an undirected graph is a set of vertices with the property that each vertex has at least as many neighbours in the alliance (including itself) as neighbours outside the alliance. In 2000, Kristiansen, Hedetniemi, and Hedetniemi [10] introduced defensive, offensive, and powerful alliances, and further studied by Shafique [7] and other authors. In this paper, we will focus on defensive alliances. A defensive alliance is *strong* if each vertex has at least as many neighbours in the alliance (not counting itself) as outside the alliance. The theory of alliances in graphs have been studied intensively [2,5,6] both from a combinatorial and from a computational perspective. As

A. Gaikwad—The first author gratefully acknowledges support from the Ministry of Human Resource Development, Government of India, under Prime Minister's Research Fellowship Scheme (No. MRF-192002-211).

S. Maity—The author's research was supported in part by the Science and Engineering Research Board (SERB), Govt. of India, under Sanction Order No. MTR/2018/001025.

A. Mudgal and C. R. Subramanian (Eds.): CALDAM 2021, LNCS 12601, pp. 135–148, 2021.
https://doi.org/10.1007/978-3-030-67899-9_11

mentioned in [1], the focus has been mostly on finding small alliances, although studying large alliances do not only make a lot of sense from the original motivation of these notions, but was actually also delineated in the very first papers on alliances.

Note that defensive alliance is not a hereditary property, that is, a subset of defensive alliance is not necessarily a defensive alliance. Shafique [7] called an alliance a *locally minimal alliance* if the set obtained by removing any vertex of the alliance is not an alliance. Bazgan et al. [1] considered another notion of alliance that they called a *globally minimal alliance* which has the property that no proper subset is an alliance. In this paper we are interested in locally minimal alliances of maximum size. Clearly, the motivation is that big communities where every member still matters somehow are of more interest than really small communities. Also, there is a general mathematical interest in such type of problems, see [13].

2 Basic Notations

Throughout this article, $G = (V, E)$ denotes a finite, simple and undirected graph of order $|V| = n$. The subgraph induced by $S \subseteq V(G)$ is denoted by $G[S]$. For a vertex $v \in V$, we use $N_G(v) = \{u : (u, v) \in E(G)\}$ to denote the (open) neighbourhood of vertex v in G, and $N_G[v] = N_G(v) \cup \{v\}$ to denote the closed neighbourhood of v. The degree $d_G(v)$ of a vertex $v \in V(G)$ is $|N_G(v)|$. For a subset $S \subseteq V(G)$, we define its closed neighbourhood as $N_G[S] = \bigcup_{v \in S} N_G[v]$ and its open neighbourhood as $N_G(S) = N_G[S] \setminus S$. For a non-empty subset $S \subseteq V$ and a vertex $v \in V(G)$, $N_S(v)$ denotes the set of neighbours of v in S, that is, $N_S(v) = \{u \in S : (u, v) \in E(G)\}$. We use $d_S(v) = |N_S(v)|$ to denote the degree of vertex v in $G[S]$. The complement of the vertex set S in V is denoted by S^c.

Definition 1. A non-empty set $S \subseteq V$ is a defensive alliance in G if for each $v \in S$, $|N[v] \cap S| \geq |N(v) \setminus S|$, or equivalently, $d_S(v) + 1 \geq d_{S^c}(v)$.

A vertex $v \in S$ is said to be protected if $d_S(v) + 1 \geq d_{S^c}(v)$. A set $S \subseteq V$ is a defensive alliance if every vertex in S is protected.

Definition 2. A vertex $v \in S$ is said to be *marginally protected* if it becomes unprotected when one of its neighbour in S is moved from S to $V \setminus S$. A vertex $v \in S$ is said to be *strongly protected* if it remains protected even if one of its neighbours is moved from S to $V \setminus S$.

Definition 3. An alliance S is called a *locally minimal alliance* if for any $v \in S$, $S \setminus \{v\}$ is not an alliance.

Definition 4. An alliance S is *globally minimal alliance* or shorter *minimal alliance* if no proper subset is an alliance.

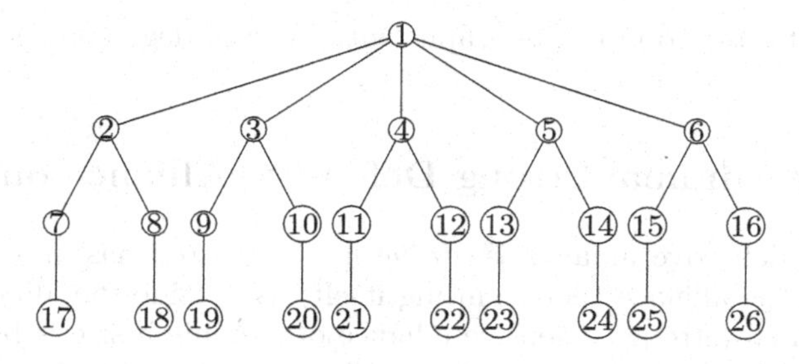

Fig. 1. A graph G with $A_L(G) = 10$ and $A(G) = 3$; $S = \{7, 2, 9, 3, 11, 4, 13, 5, 15, 6\}$ is a locally minimal defensive alliance of size 10 and $\{1, 2, 3\}$ is a globally minimal defensive alliance of size 3.

A defensive alliance S is connected if the subgraph induced by S is connected. An alliance S is called a *connected locally minimal alliance* if for any $v \in S$, $S \setminus \{v\}$ is not a connected alliance. Notice that any globally minimal alliance is also connected. As introduced in [1], we use $A_L(G)$ for the cardinality of the largest locally minimal defensive alliance in a graph G, and $A(G)$ for the cardinality of the largest globally minimal defensive alliance in a graph G (Fig. 1). In this paper, we consider Locally Minimal Defensive Alliance problem under structural parameters. We define the problem as follows:

Locally Minimal Defensive Alliance
Input: An undirected graph $G = (V, E)$ and an integer $k \leq |V(G)|$.
Question: Is there a locally minimal defensive alliance $S \subseteq V(G)$ such that $|S| \geq k$?

Our results are as follows:

- Locally Minimal Strong Defensive Alliance problem is polynomial time solvable on trees.
- Locally Minimal Defensive Alliance problem is FPT when parameterized by neighbourhood diversity of the input graph.
- Locally Minimal Defensive Alliance problem is polynomial time solvable for graphs with bounded treewith.

Known Results: The decision version for several types of alliances have been shown to be NP-complete. Carvajal et al. [3] proved that deciding if a graph contains a strong defensive alliance of size at most k is NP-hard. The defensive alliance problem is NP-complete even when restricted to split, chordal and bipartite graph [8]. Bazgan et al. [1] proved that deciding if a graph contains a locally minimal strong defensive alliance of size at least k is NP-complete, even when restricted to bipartite graphs with average degree less than 3.6; deciding if a graph contains a connected locally minimal strong defensive alliance or a connected locally minimal defensive alliance of size at least k is NP-complete, even when restricted to bipartite graphs with average degree less than $2+\epsilon$, for any $\epsilon > 0$.

3 Locally Minimal Strong Defensive Alliance on Trees

Recall that a defensive alliance is *strong* if each vertex has at least as many neighbours in the alliance (not counting itself) as outside the alliance. Finding a locally minimal (strong) defensive alliance of maximum size is believed to be intractable [1]. However, when the graph happens to be a tree, we solve the problem in polynomial time, using dynamic programming. It may be observed that if S is a locally minimal strong defensive alliance, then for every vertex $v \in S$, at least one of its neighbours in S is marginally protected. A vertex $v \in S$ is said to be *good* if it has at least one marginally protected neighbour in S, otherwise it is called a *bad* vertex. Let v be a non-leaf node with children $v_1, v_2, \ldots, v_d$. Then $v \in S$ is *marginally unprotected* by its children if $\lceil \frac{d+1}{2} \rceil - 1$ of its children are in S; thus the parent of v must be in S for protection of v. Vertex v is *strongly protected* by its children if at least $\lceil \frac{d+1}{2} \rceil + 1$ of its children are in S. We define different possible states of a vertex v as follows:

- 0: vertex v is not in the solution.
- $\hat{1}_b$: vertex v is marginally unprotected by its children and none of its children are marginally protected.
- $\hat{1}_g$: vertex v is marginally unprotected by its children and if v has children then at least one of them is marginally protected.
- 1_{mg}: vertex v is marginally protected by its children and at least one of its children is marginally protected.
- 1_{sb}: vertex v is strongly protected by its children and none of the children is marginally protected.
- 1_{sg}: vertex v is strongly protected by its children and at least one of its children is marginally protected.

Here is the algorithm: Start by rooting the tree at any node v. Each node defines a subtree, the one hanging from it. This immediately suggests subproblems: $A_v(s)$ = the size of the largest locally minimal defensive alliance of the subtree rooted at v and the state of v is s. Our final goal is to compute $\max\left\{A_r(0), A_r(1_{sg}), A_r(1_{mg})\right\}$ where r is the root of T.

Leaf Node: For a leaf node v, we have $A_v(0) = 0$, $A_v(\hat{1}_b) = 1$; $A_v(\hat{1}_g) = A_v(1_{mg}) = A_v(1_{sb}) = A_v(1_{sg}) = -\infty$.

Non-leaf Node: Let v be a non-leaf node with the set $\mathcal{C} = \{v_1, v_2, \ldots, v_d\}$ of children. Suppose we know $A_{v_i}(s)$ for all children v_i of v. How can we compute $A_v(s)$? We now consider the following cases:

Case 1: Let the state of v be 0. Then

$$A_v(0) = \sum_{x \in \mathcal{C}} \max\left\{A_x(0), A_x(1_{sg}), A_x(1_{mg})\right\}$$

Case 2: Let the state of v be $\hat{1}_b$ or 1_{sb}. Let $(v_1, v_2, \ldots, v_d)$ be a descending ordering of $\mathcal{C}$ according to values $\max\{A_{v_i}(1_{sg}), A_{v_i}(1_{sb})\}$, that is,

$$\max\{A_{v_1}(1_{sg}), A_{v_1}(1_{sb})\} \geq \ldots \geq \max\{A_{v_d}(1_{sg}), A_{v_d}(1_{sb})\}.$$

Let $\mathcal{C}_{\lceil \frac{d+1}{2} \rceil - 1} = \{v_1, v_2, \ldots, v_{\lceil \frac{d+1}{2} \rceil - 1}\}$ and $\mathcal{C}_{\lceil \frac{d+1}{2} \rceil + 1} = \{v_1, v_2, \ldots, v_{\lceil \frac{d+1}{2} \rceil + 1}\}$. Then

$$A_v(\hat{1}_b) = 1 + \sum_{x \in \mathcal{C}_{\lceil \frac{d+1}{2} \rceil - 1}} \max\Big\{A_x(1_{sg}), A_x(1_{sb})\Big\} + \sum_{x \in \mathcal{C} \setminus \mathcal{C}_{\lceil \frac{d+1}{2} \rceil - 1}} A_x(0),$$

and

$$\begin{aligned} A_v(1_{sb}) = 1 + & \sum_{x \in \mathcal{C}_{\lceil \frac{d+1}{2} \rceil + 1}} \max\Big\{A_x(1_{sg}), A_x(1_{sb})\Big\} \\ & + \sum_{x \in \mathcal{C} \setminus \mathcal{C}_{\lceil \frac{d+1}{2} \rceil + 1}} \max\Big\{A_x(0), A_x(1_{sg}), A_x(1_{sb})\Big\} \end{aligned}$$

Thus, in this case, v must have at least $\lceil \frac{d+1}{2} \rceil + 1$ non-leaf children, otherwise, $A_v(1_{sb}) = -\infty$.

Case 3: Let the state of v be $\hat{1}_g$, 1_{mg} or 1_{sg}. Let $(v_1, v_2, \ldots, v_d)$ be a descending ordering of $\mathcal{C}$ according to values $\max\{A_{v_i}(\hat{1}_g), A_{v_i}(\hat{1}_b), A_{v_i}(1_{sg}), A_{v_i}(1_{sb})\}$. Let $\mathcal{C}_{k,i}$ be the set of first k children from the ordering $(v_1, v_2, \ldots, v_d)$ except vertex v_i. We have the following recurrence relations:

$$\begin{aligned} A_v(\hat{1}_g) = \max_{v_i \in \mathcal{C}} \Big\{ & 1 + \max\{A_{v_i}(\hat{1}_g), A_{v_i}(\hat{1}_b)\} \\ & + \sum_{x \in \mathcal{C}_{\lceil \frac{d+1}{2} \rceil - 2, i}} \max\{A_x(\hat{1}_g), A_x(\hat{1}_b), A_x(1_{sg}), A_x(1_{sb})\} \\ & + \sum_{x \in \mathcal{C} \setminus \left(\mathcal{C}_{\lceil \frac{d+1}{2} \rceil - 2, i} \cup \{v_i\}\right)} A_x(0) \Big\}, \end{aligned}$$

for $d \geq 2$, and $A_v(\hat{1}_g) = 1$ for $d = 1$. Here $v \in S$ is good and *marginally unprotected* by its children, that is, exactly $\lceil \frac{d+1}{2} \rceil - 1$ of its children are in S and at least one of them is labelled $\hat{1}_g$ or $\hat{1}_b$ so that v is adjacent to at least one marginally protected child. Next, we have

$$\begin{aligned} A_v(1_{mg}) = \max_{v_i \in \mathcal{C}} \Big\{ & 1 + \max\{A_{v_i}(\hat{1}_g), A_{v_i}(\hat{1}_b)\} \\ & + \sum_{x \in \mathcal{C}_{\lceil \frac{d+1}{2} \rceil - 1, i}} \max\{A_x(\hat{1}_g), A_x(\hat{1}_b), A_x(1_{sg}), A_x(1_{sb})\} \\ & + \sum_{x \in \mathcal{C} \setminus \left(\mathcal{C}_{\lceil \frac{d+1}{2} \rceil - 1, i} \cup \{v_i\}\right)} A_x(0) \Big\} \end{aligned}$$

Here $v \in S$ is good and *marginally protected* by its children, that is, exactly $\lceil \frac{d+1}{2} \rceil$ of its children are in S and at least one of them is labelled $\hat{1}_g$ or $\hat{1}_b$ so that v is adjacent to at least one marginally protected child. Finally, we have

$$\begin{aligned} A_v(1_{sg}) = & \max_{\text{non-leaf } v_i \in \mathcal{C}} \Big\{ 1 + \max\{A_{v_i}(\hat{1}_g), A_{v_i}(\hat{1}_b)\} \\ & + \sum_{\text{non-leaf } x \in \mathcal{C}_{\lceil \frac{d+1}{2} \rceil, i}} \max\{A_x(\hat{1}_g), A_x(\hat{1}_b), A_x(1_{sg}), A_x(1_{sb})\} \\ & + \sum_{\text{non-leaf } x \in \mathcal{C} \setminus \left(\mathcal{C}_{\lceil \frac{d+1}{2} \rceil, i} \cup \{v_i\}\right)} \max\{A_x(0), A_x(\hat{1}_g), A_x(\hat{1}_b), A_x(1_{sg})\} \Big\}. \end{aligned}$$

Here $v \in S$ is good and *strongly protected* by its children, that is, at least $\lceil \frac{d+1}{2} \rceil + 1$ of its children are in S and at least one of these $\lceil \frac{d+1}{2} \rceil + 1$ children is labelled $\hat{1}_g$ or $\hat{1}_b$ so that v is adjacent to at least one marginally protected child. It may be noted that if v_i is a leaf node then v_i cannot be in S. The reason is this; v_i's only neighbour is its parent v, which is strongly protected, therefore v_i will never have a marginally protected neighbour. Thus, in this case, v must have at least $\lceil \frac{d+1}{2} \rceil + 1$ non-leaf children, otherwise, $A_v(1_{sg}) = -\infty$. For computation of $A_r(1_{mg})$ and $A_r(1_{sg})$, we replace d by $d - 1$ in the above recurrence relations as the root node r with d children has degree d, whereas other non-leaf node with d children has degree $d + 1$.

The running time of this algorithm is easy to analyze. At each node $v \in V(T)$, we compute $A_v(s)$ where s is a state of v. The time required to get descending ordering of the children of v is $O(d \log d)$, where d is the number of children of vertex v. The number of subproblems is exactly the number of vertices in T. The total running time is therefore equal to $c \sum d_i \log d_i \leq c \log n \sum d_i = cn \log n = O(n \log n)$, where c is a constant.

4 FPT Algorithm Parameterized by Neighbourhood Diversity

In this section, we present an FPT algorithm for LOCALLY MINIMAL DEFENSIVE ALLIANCE problem parameterized by neighbourhood diversity. We say two vertices u and v have the same type if and only if $N(u) \setminus \{v\} = N(v) \setminus \{u\}$. The relation of having the same type is an equivalence relation. The idea of neighbourhood diversity is based on this type structure.

Definition 5. [11] The neighbourhood diversity of a graph $G = (V, E)$, denoted by $\mathtt{nd}(G)$, is the least integer k for which we can partition the set V of vertices into k classes, such that all vertices in each class have the same type.

If neighbourhood diversity of a graph is bounded by an integer k, then there exists a partition $\{C_1, C_2, \ldots, C_k\}$ of $V(G)$ into k type classes. It is known that such a minimum partition can be found in linear time using fast modular decomposition algorithms [14]. Notice that each type class could either be a clique or an independent set by definition. For algorithmic purpose it is often useful to consider a *type graph* H of graph G, where each vertex of H is a type class in G, and two vertices C_i and C_j are adjacent iff there is complete bipartite clique between these type classes in G. It is not difficult to see that there will be either a complete bipartite clique or no edges between any two type classes. The key property of graphs of bounded neighbourhood diversity is that their type graphs have bounded size. In this section, we prove the following theorem:

Theorem 1. *The* LOCALLY MINIMAL DEFENSIVE ALLIANCE *problem is fixed-parameter tractable when parameterized by the neighbourhood diversity.*

Let G be a connected graph such that $\texttt{nd}(G) = k$. Let $C_1, \ldots, C_k$ be the partition of $V(G)$ into sets of type classes. We assume $k \geq 2$ since otherwise the problem becomes trivial. Next we guess $|C_i \cap D|$ and whether the vertices in C_i are marginally or strongly protected, where D is a locally minimal defensive alliance. We make the following guesses:

- *Option 1:* $|C_i \cap D| = 0$.
- *Option 2:* $|C_i \cap D| = 1$ and the vertices in C_i are marginally protected.
- *Option 3:* $|C_i \cap D| = 1$ and the vertices in C_i are strongly protected.
- *Option 4:* $|C_i \cap D| > 1$ and the vertices in C_i are marginally protected.
- *Option 5:* $|C_i \cap D| > 1$ and the vertices in C_i are strongly protected.

There are at most 5^k choices for the tuple $(C_1, C_2, \ldots, C_k)$ as each C_i has 5 options as given above. Finally we reduce the problem of finding a locally minimal defensive alliance of maximum size to an integer linear programming optimization with k variables. Since integer linear programming is fixed parameter tractable when parameterized by the number of variables [12], we conclude that our problem is FPT when parameterized by the neighbourhood diversity.

ILP Formulation: Given a particular choice P of options for $(C_1, C_2, \ldots, C_k)$, our goal here is to find a locally minimal defensive alliance of maximum size. For each C_i, we associate a variable x_i that indicates $|D \cap C_i| = x_i$. Clearly, $x_i = 0$, if C_i is assigned Option 1; $x_i = 1$ if C_i is assigned Option 2 or 3; and $x_i > 1$ if C_i is assigned Option 3 or 4. Because the vertices in C_i have the same neighbourhood, the variables x_i determine D uniquely, up to isomorphism. Let $S_1 = \{C_i \mid x_i = 1\}$, $S_{>1} = \{C_i \mid x_i > 1\}$ and $S = S_1 \cup S_{>1}$. Let $H[S]$ be the subgraph of H induced by S. Now we label the vertices of $H[S]$ as follows: vertex C_i is labelled c_1 if it is a clique and Option 2 is assigned to C_i; vertex C_i is labelled $c_{>1}$ if it is a clique and Option 4 is assigned to C_i; vertex C_i is labelled $\texttt{ind}$ if it is an independent set and Option 2 or 4 is assigned to C_i; vertex C_i is labelled s if it is a clique or an independent set, and Option 3 or 5 is assigned to C_i. To ensure local minimality of defensive alliance, the induced subgraph must satisfy the following conditions:

- Every vertex labelled s in the induced graph must have at least one neighbour labelled $c_1, c_{>1}$ or ind.
- Every vertex labelled c_1 in the induced graph must have at least one neighbour labelled $c_1, c_{>1}$ or ind.
- Every vertex labelled ind in the induced graph must have at least one neighbour labelled c_1, $c_{>1}$ or ind.

Above conditions ensure local minimality of the solution because when we remove a vertex from the solution, we make sure at least one of its neighbours gets unprotected. This happens because every vertex in the solution has at least one neighbour which is marginally protected. If the induced subgraph $H[S]$ satisfies all the above conditions then we proceed for the ILP, otherwise not. Let C be a subset of S consisting of all type classes which are cliques; $I = S \setminus C$ and $R = \{C_1, \dots, C_k\} \setminus S$. Let n_i denote the number of vertices in C_i. We consider two cases:

Case 1: Suppose $v \in C_j$ where $C_j \in I$. Then the degree of v in D satisfies

$$d_D(v) = \sum_{C_i \in N_H(C_j) \cap S} x_i \tag{1}$$

Thus, including itself, v has $1 + \sum_{C_i \in N_H(C_j) \cap S} x_i$ defenders in G. Note that if $C_i \in D$, then only x_i vertices of C_i are in D and the remaining $n_i - x_i$ vertices of C_i are outside D. The degree of v outside D satisfies

$$d_{D^c}(v) = \sum_{C_i \in N_H(C_j) \cap S} (n_i - x_i) + \sum_{C_i \in N_H(C_j) \cap R} n_i \tag{2}$$

Case 2: Suppose $v \in C_j$ where $C_j \in C$. The degree of v in D satisfies

$$d_D(v) = \sum_{C_i \in N_H[C_j] \cap S} x_i \tag{3}$$

The degree of v outside D satisfies

$$d_{D^c}(v) = \sum_{C_i \in N_H[C_j] \cap S} (n_i - x_i) + \sum_{C_i \in N_H[C_j] \cap R} n_i \tag{4}$$

In the following, we present an ILP formulation of locally minimal defensive alliance problem, where a choice of options for $(C_1, \dots, C_k)$ is given:

$$
\begin{aligned}
&\text{Maximize} \sum_{C_i \in S} x_i \\
&\text{Subject to} \\
&1 + \sum_{C_i \in N_H(C_j) \cap S} 2x_i > \sum_{C_i \in N_H(C_j)} n_i, \quad \text{for all } C_j \in I, \text{ labelled } s, \\
&\sum_{C_i \in N_H(C_j) \cap S} 2x_i - \sum_{C_i \in N_H(C_j)} n_i = 0 \text{ or } -1, \quad \text{for all } C_j \in I, \text{ labelled } \texttt{ind}, \\
&\sum_{C_i \in N_H[C_j] \cap S} 2x_i > \sum_{C_i \in N_H[C_j]} n_i, \quad \text{for all } C_j \in C, \text{ labelled } s, \\
&\sum_{C_i \in N_H[C_j] \cap S} 2x_i - \sum_{C_i \in N_H[C_j]} n_i = 0 \text{ or } 1, \quad \text{for all } C_j \in C, \text{ labelled } c_1 \text{ or } c_{>1}, \\
&x_i = 1 \text{ for all } i \,:\, C_i \in S_1; \\
&x_i \in \{2, 3, \ldots, |C_i|\} \text{ for all } i \,:\, C_i \in S_2.
\end{aligned}
$$

Solving the ILP: Lenstra [12] showed that the feasibility version of p-ILP is FPT with running time doubly exponential in p, where p is the number of variables. Later, Kannan [9] proved an algorithm for p-ILP running in time $p^{O(p)}$. In our algorithm, we need the optimization version of p-ILP rather than the feasibility version. We state the minimization version of p-ILP as presented by Fellows et. al. [4].

P-VARIABLE INTEGER LINEAR PROGRAMMING OPTIMIZATION (p-OPT-ILP): Let matrices $A \in Z^{m \times p}$, $b \in Z^{p \times 1}$ and $c \in Z^{1 \times p}$ be given. We want to find a vector $x \in Z^{p \times 1}$ that minimizes the objective function $c \cdot x$ and satisfies the m inequalities, that is, $A \cdot x \geq b$. The number of variables p is the parameter. Then they showed the following:

Lemma 1. [4] p-OPT-ILP can be solved using $O(p^{2.5p+o(p)} \cdot L \cdot log(MN))$ arithmetic operations and space polynomial in L. Here L is the number of bits in the input, N is the maximum absolute value any variable can take, and M is an upper bound on the absolute value of the minimum taken by the objective function.

In the formulation for LOCALLY MINIMAL DEFENSIVE ALLIANCE problem, we have at most k variables. The value of objective function is bounded by n and the value of any variable in the integer linear programming is also bounded by n. The constraints can be represented using $O(k^2 \log n)$ bits. Lemma 1 implies that we can solve the problem with the guess P in FPT time. There are at most 5^k choices for P, and the ILP formula for a guess can be solved in FPT time. Thus Theorem 1 holds.

5 Graphs of Bounded Treewidth

In this section we prove that Locally Minimal Defensive Alliance problem can be solved in polynomial time for graphs of bounded treewidth. In other words, this section presents XP-time algorithm for Locally Minimal Defensive Alliance problem parameterized by treewidth. We now prove the following theorem:

Theorem 2. *Given an n-vertex graph G and its nice tree decomposition T of width at most k, the size of a maximum locally minimal defensive alliance of G can be computed in* $8^k n^{O(2^{k+1})}$ *time.*

Let $(T, \{X_t\}_{t \in V(T)})$ be a nice tree decomposition rooted at node r of the input graph G. For a node t of T, let V_t be the union of all bags present in the subtree of T rooted at t, including X_t. We denote by G_t the subgraph of G induced by V_t. For each node t of T, we construct a table $dp_t(A, \mathbf{x}, a, \alpha, \mathbf{y}, \mathbf{z}, \beta) \in \{\text{true}, \text{false}\}$ where $A \subseteq X_t$; $\mathbf{x}$ and $\mathbf{y}$ are vectors of length n; a, α and β are integers between 0 and n. We set $dp_t(A, \mathbf{x}, a, \alpha, \mathbf{y}, \mathbf{z}, \beta) = \text{true}$ if and only if there exists a set $A_t \subseteq V_t$ such that:

1. $A_t \cap X_t = A$
2. $a = |A_t|$
3. the ith coordinate of vector $\mathbf{x}$ is

$$x(i) = \begin{cases} d_{A_t}(v_i) & \text{for } v_i \in A \\ 0 & \text{otherwise} \end{cases}$$

4. α is the number of vertices $v \in A_t$ that are protected, that is, $d_{A_t}(v) \geq \frac{d_G(v)-1}{2}$.
5. A vertex $v \in A$ is said to be "good" if it has at least one marginally protected neighbour in $A_t \setminus A$. A vertex $v \in A$ is said to be "bad" if it has no marginally protected neighbours in $A_t \setminus A$. Here $\mathbf{y}$ is a vector of length n, and the ith coordinate of vector $\mathbf{y}$ is

$$y(i) = \begin{cases} g & \text{if } v_i \in A \text{ and } v_i \text{ is a good vertex} \\ b & \text{if } v_i \in A \text{ and } v_i \text{ is a bad vertex} \\ 0 & \text{otherwise} \end{cases}$$

6. $\mathbf{z}$ is a 2^k length vector, where the entry $z(S)$ associated with subset $S \subseteq A$ denotes the number of common bad neighbours of S in $A_t \setminus A$. The z vector considers the power set of A in lexicographic order. For example, let $A = \{a, b, c\}$, then $\mathbf{z} = \Big(z(\{a\}), z(\{a, b\}), z(\{a, b, c\}), z(\{a, c\}), z(\{b\}), z(\{b, c\}), z(\{c\})\Big)$.
7. β is the number of good vertices in A_t.

We compute all entries $dp_t(A, \mathbf{x}, a, \alpha, \mathbf{y}, \mathbf{z}, \beta)$ in a bottom-up manner. Since $tw(T) \leq k$, at most $2^k n^k (n+1)^3 2^k n^{2^k} = 4^k n^{O(2^k)}$ records are maintained at each node t. Thus, to prove Theorem 2, it suffices to show that each entry $dp_t(A, \mathbf{x}, a, \alpha, \mathbf{y}, \mathbf{z}, \beta)$ can be computed in $2^k n^{O(2^k)}$ time, assuming that the entries for the children of t are already computed.

Leaf Node: For a leaf node t we have that $X_t = \emptyset$. Thus $dp_t(A, \mathbf{x}, a, \alpha, \mathbf{y}, \mathbf{z}, \beta)$ is true if and only if $A = \emptyset$, $\mathbf{x} = \mathbf{0}$, $a = 0$, $\alpha = 0$, $\mathbf{y} = \mathbf{0}$, $\mathbf{z} = \mathbf{0}$, $\beta = 0$. These conditions can be checked in $O(1)$ time.

Introduce Node: Suppose t is an introduction node with child t' such that $X_t = X_{t'} \cup \{v_i\}$ for some $v_i \notin X_{t'}$. Let A be any subset of X_t. We consider two cases:

Case (i): Let $v_i \notin A$. In this case $dp_t(A, \mathbf{x}, a, \alpha, \mathbf{y}, \mathbf{z}, \beta)$ is true if and only if $dp_{t'}(A, \mathbf{x}, a, \alpha, \mathbf{y}, \mathbf{z}, \beta)$ is true.

Case (ii): Let $v_i \in A$. Here $dp_t(A, \mathbf{x}, a, \alpha, \mathbf{y}, \mathbf{z}, \beta)$ is true if and only if there exist $A', \mathbf{x}', a'$, α', $\mathbf{y}'$, $\mathbf{z}'$, and β' such that $dp_{t'}(A', \mathbf{x}', a', \alpha', \mathbf{y}', \mathbf{z}', \beta')$=true, where

1. $A = A' \cup \{v_i\}$;
2. $x(j) = x'(j) + 1$, if $v_j \in N_A(v_i)$, $x(i) = d_A(v_i)$, and $x(j) = x'(j)$ if $v_j \in A \setminus N_A[v_i]$;
3. $a = a' + 1$;
4. $\alpha = \alpha' + \delta$; here δ is the cardinality of the set
$$\left\{v_j \in A \mid x'(j) < \frac{d_G(v_j) - 1}{2}; x(j) \geq \frac{d_G(v_j) - 1}{2}\right\}.$$
That is, to compute α from α' we need to add the number δ of those vertices not satisfied in $(A', \mathbf{x}', a', \alpha', \mathbf{y}', \mathbf{z}', \beta')$ but satisfied in $(A, \mathbf{x}, a, \alpha, \mathbf{y}, \mathbf{z}, \beta)$.
5. $y(i) = b$ and $y(j) = y'(j)$ for all $j \neq i$.
6. $z(S) = z'(S)$ if $v_i \notin S$; $z(S) = 0$ if $v_i \in S$.
7. $\beta = \beta'$.

For an introduce node t, $dp_t(A, \mathbf{x}, a, \alpha, \mathbf{y}, \mathbf{z}, \beta)$ can be computed in $O(1)$ time. This follows from the fact that there is only one candidate of such tuple $(A', \mathbf{x}', a', \alpha', \mathbf{y}', \mathbf{z}', \beta')$.

Forget Node: Suppose t is a forget node with child t' such that $X_t = X_{t'} \setminus \{v_i\}$ for some $v_i \in X_{t'}$. Let A be any subset of X_t. Here $dp_t(A, \mathbf{x}, a, \alpha, \mathbf{y}, \mathbf{z}, \beta)$ is true if and only if either $dp_{t'}(A, \mathbf{x}, a, \alpha, \mathbf{y}, \mathbf{z}, \beta)$ is true (this corresponds to the case that A_t does not contain v_i) or $dp_{t'}(A', \mathbf{x}', a', \alpha', \mathbf{y}', \mathbf{z}', \beta')$=true for some $A', \mathbf{x}'$, a', α', $\mathbf{y}'$, $\mathbf{z}'$, β' with the following conditions (this corresponds to the case that A_t contains v_i):

1. $A' = A \cup \{v_i\}$;
2. $x(j) = x'(j)$ for all $j \neq i$ and $x(i) = 0$;
3. $a = a'$;
4. $\alpha = \alpha'$;

We now consider four cases:
Case 1: v_i is not marginally protected and v_i is a good vertex.

5. $y(j) = y'(j)$ for all $j \neq i$ and $y(i) = 0$;
6. $z(S) = z'(S)$ for all $S \subseteq A$;
7. $\beta = \beta'$.

Case 2: v_i is not marginally protected and v_i is a bad vertex.

5. $y(j) = y'(j)$ for all $j \neq i$ and $y(i) = 0$;
6. $$z(S) = \begin{cases} z'(S) + 1 & \text{if } S \subseteq N_A(v_i) \\ z'(S) & \text{otherwise} \end{cases}$$
7. $\beta = \beta'$.

Case 3: v_i is marginally protected and v_i is a good vertex.

5. $$y(j) = \begin{cases} g & \text{if } v_j \in N_A(v_i) \\ y'(j) & \text{if } v_j \in A \setminus N_A(v_i) \end{cases}$$
6. $z(S) = z'(S) - z'(S \cup \{v_i\})$ for all $S \subseteq A$;
7. $\beta = \beta' + z'(\{v_i\}) + |\{j \ : \ y'(j) = b; y(j) = g\}|$.

Case 4: v_i is marginally protected and v_i is a bad vertex.

5. $$y(j) = \begin{cases} g & \text{if } v_j \in N_A(v_i) \\ y'(j) & \text{if } v_j \in A \setminus N_A(v_i) \end{cases}$$
6. $$z(S) = \begin{cases} z'(S) - z'(S \cup \{v_i\}) + 1 & \text{if } S \subseteq N_A(v_i) \\ z'(S) - z'(S \cup \{v_i\}) & \text{for all other subsets } S \subseteq A \end{cases}$$
7. $\beta = \beta' + z'(\{v_i\}) + |\{j \ : \ y'(j) = b; y(j) = g\}|$.

For a forget node t, $dp_t(A, \mathbf{x}, a, \alpha, \mathbf{y}, \mathbf{z}, \beta)$ can be computed in $n^{O(2^k)}$ time. This follows from the fact that there are $n^{O(2^k)}$ candidates of such tuple $(A', \mathbf{x}', a', \alpha', \mathbf{z}', \beta')$, and each of them can be checked in $O(1)$ time.

Join Node: Suppose t is a join node with children t_1 and t_2 such that $X_t = X_{t_1} = X_{t_2}$. Let A be any subset of X_t. Then $dp_t(A, \mathbf{x}, a, \alpha, \mathbf{y}, \mathbf{z}, \beta)$ is true if and only if there exist $(A_1, \mathbf{x_1}, a_1, \alpha_1, \mathbf{y_1}, \mathbf{z_1}, \beta_1)$ and $(A_2, \mathbf{x_2}, a_2, \alpha_2, \mathbf{y_2}, \mathbf{z_2}, \beta_2)$ such that $dp_{t_1}(A_1, \mathbf{x_1}, a_1, \alpha_1, \mathbf{y_1}, \mathbf{z_1}, \beta_1) =$ true and $dp_{t_2}(A_2, \mathbf{x_2}, a_2, \alpha_2, \mathbf{y_2}, \mathbf{z_2}, \beta_2) =$ true, where

1. $A = A_1 = A_2$;
2. $x(i) = x_1(i) + x_2(i) - d_A(v_i)$ for all $i \in A$, and $x(i) = 0$ if $i \notin A$;

3. $a = a_1 + a_2 - |A|$;
4. $\alpha = \alpha_1 + \alpha_2 - \gamma + \delta$; γ is the cardinality of the set

$$\left\{v_j \in A \mid x_1(j) \geq \frac{d_G(v_i) - 1}{2};\ x_2(j) \geq \frac{d_G(v_i) - 1}{2}\right\}$$

and δ is the cardinality of the set

$$\left\{v_j \in A \mid x_1(j) < \frac{d_G(v_i) - 1}{2};\ x_2(j) < \frac{d_G(v_i) - 1}{2};\ x(j) \geq \frac{d_G(v_i) - 1}{2}\right\}.$$

To compute α from $\alpha_1 + \alpha_2$, we need to subtract the number of those v_j which are satisfied in both the branches and add the number of vertices v_j not satisfied in either of the branches t_1 and t_1 but satisfied in t.

5.

$$y(j) = \begin{cases} g & \text{if } y_1(j) = g \text{ or } y_2(j) = g \\ b & \text{otherwise} \end{cases}$$

6. $z(S) = z_1(S) + z_2(S)$ for all $S \subseteq A$;
7. $\beta = \beta_1 + \beta_2 - |\left\{j\ :\ y_1(j) = g, y_2(j) = g\right\}|$.

For a join node t, there are n^k possible pairs for $(\mathbf{x_1}, \mathbf{x_2})$ as $\mathbf{x_2}$ is uniquely determined by $\mathbf{x_1}$; $n{+}1$ possible pairs for (a_1, a_2); $n{+}1$ possible pairs for (α_1, α_2); there are 2^k possible pairs for $(\mathbf{y_1}, \mathbf{y_2})$ as $\mathbf{y_2}$ is uniquely determined by $\mathbf{y_1}$; there are n^{2^k} possible pairs for $(\mathbf{z_1}, \mathbf{z_2})$ as $\mathbf{z_2}$ is uniquely determined by $\mathbf{z_1}$; and $n+1$ possible pairs for (β_1, β_2). In total, there are $2^k n^{O(2^k)}$ candidates, and each of them can be checked in $O(1)$ time. Thus, for a join node t, $dp_t(A, \mathbf{x}, a, \alpha, \mathbf{y}, \mathbf{z}, \beta)$ can be computed in $2^k n^{O(2^k)}$ time.

At the root node r, we look at all records such that $dp_r(\emptyset, \mathbf{x}, a, \alpha, \mathbf{y}, \mathbf{z}, \beta) =$ true, and $a = \alpha = \beta$. The size of a maximum locally minimal defensive alliance is the maximum a satisfying $dp_r(\emptyset, \mathbf{x}, a, a, \mathbf{y}, \mathbf{z}, a)=$ true.

6 Conclusion

The main contributions in this paper are that the Locally Minimal Defensive Alliance problem is FPT when parameterized by neighborhood diversity, the problem is polynomial time solvable on trees, and XP in treewidth. We list some nice problems emerge from the results here: is the problem FPT in treewidth, and does it admit a polynomial kernel in neighborhood diversity? Also, noting that the result for neighborhood diversity implies that the problem is FPT in vertex cover, it would be interesting to consider the parameterized complexity with respect to twin cover. The modular width parameter also appears to be a natural parameter to consider here, and since there are graphs with bounded modular-width and unbounded neighborhood diversity; we believe this is also an interesting open problem. The parameterized complexity of the Locally Minimal Defensive Alliance problem remains unsettle when parameterized by other important structural graph parameters like clique-width.

Acknowledgement. We are grateful to the referees for thorough reading and constructive comments that have made the paper better readable.

References

1. Bazgan, C., Fernau, H., Tuza, Z.: Aspects of upper defensive alliances. Discret. Appl. Math. **266**, 111–120 (2019)
2. Cami, A., Balakrishnan, H., Deo, N., Dutton, R.: On the complexity of finding optimal global alliances. J. Comb. Math. Comb. Comput. **58**, 23–31 (2006)
3. Carvajal, R., Matamala, M., Rapaport, I., Schabanel, N.: Small alliances in graphs. In: Kučera, L., Kučera, A. (eds.) MFCS 2007. LNCS, vol. 4708, pp. 218–227. Springer, Heidelberg (2007). https://doi.org/10.1007/978-3-540-74456-6_21
4. Fellows, M.R., Lokshtanov, D., Misra, N., Rosamond, F.A., Saurabh, S.: Graph layout problems parameterized by vertex cover. In: Hong, S.-H., Nagamochi, H., Fukunaga, T. (eds.) ISAAC 2008. LNCS, vol. 5369, pp. 294–305. Springer, Heidelberg (2008). https://doi.org/10.1007/978-3-540-92182-0_28
5. Fernau, H., Rodriguez-Velazquez, J.A.: A survey on alliances and related parameters in graphs. Electron. J. Graph Theory Appl. **2**(1) (2014)
6. Fricke, G., Lawson, L., Haynes, T., Hedetniemi, M., Hedetniemi, S.: A note on defensive alliances in graphs. Bull. Inst. Comb. Appl. **38**, 37–41 (2003)
7. Hassan-Shafique, K.: Partitioning a graph in alliances and its application to data clustering (2004)
8. Jamieson, L.H., Hedetniemi, S.T., McRae, A.A.: The algorithmic complexity of alliances in graphs. J. Comb. Math. Comb. Comput. **68**, 137–150 (2009)
9. Kannan, R.: Minkowski's convex body theorem and integer programming. Math. Oper. Res. **12**(3), 415–440 (1987)
10. Kristiansen, P., Hedetniemi, M., Hedetniemi, S.: Alliances in graphs. J. Comb. Math. Comb. Comput. **48**, 157–177 (2004)
11. Lampis, M.: Algorithmic meta-theorems for restrictions of treewidth. Algorithmica **64**, 19–37 (2012)
12. Lenstra, H.W.: Integer programming with a fixed number of variables. Math. Oper. Res. **8**(4), 538–548 (1983)
13. Manlove, D.: Minimaximal and maximinimal optimisation problems: a partial order-based approach (1998)
14. Tedder, M., Corneil, D., Habib, M., Paul, C.: Simpler linear-time modular decomposition via recursive factorizing permutations. In: Aceto, L., Damgård, I., Goldberg, L.A., Halldórsson, M.M., Ingólfsdóttir, A., Walukiewicz, I. (eds.) ICALP 2008. LNCS, vol. 5125, pp. 634–645. Springer, Heidelberg (2008). https://doi.org/10.1007/978-3-540-70575-8_52

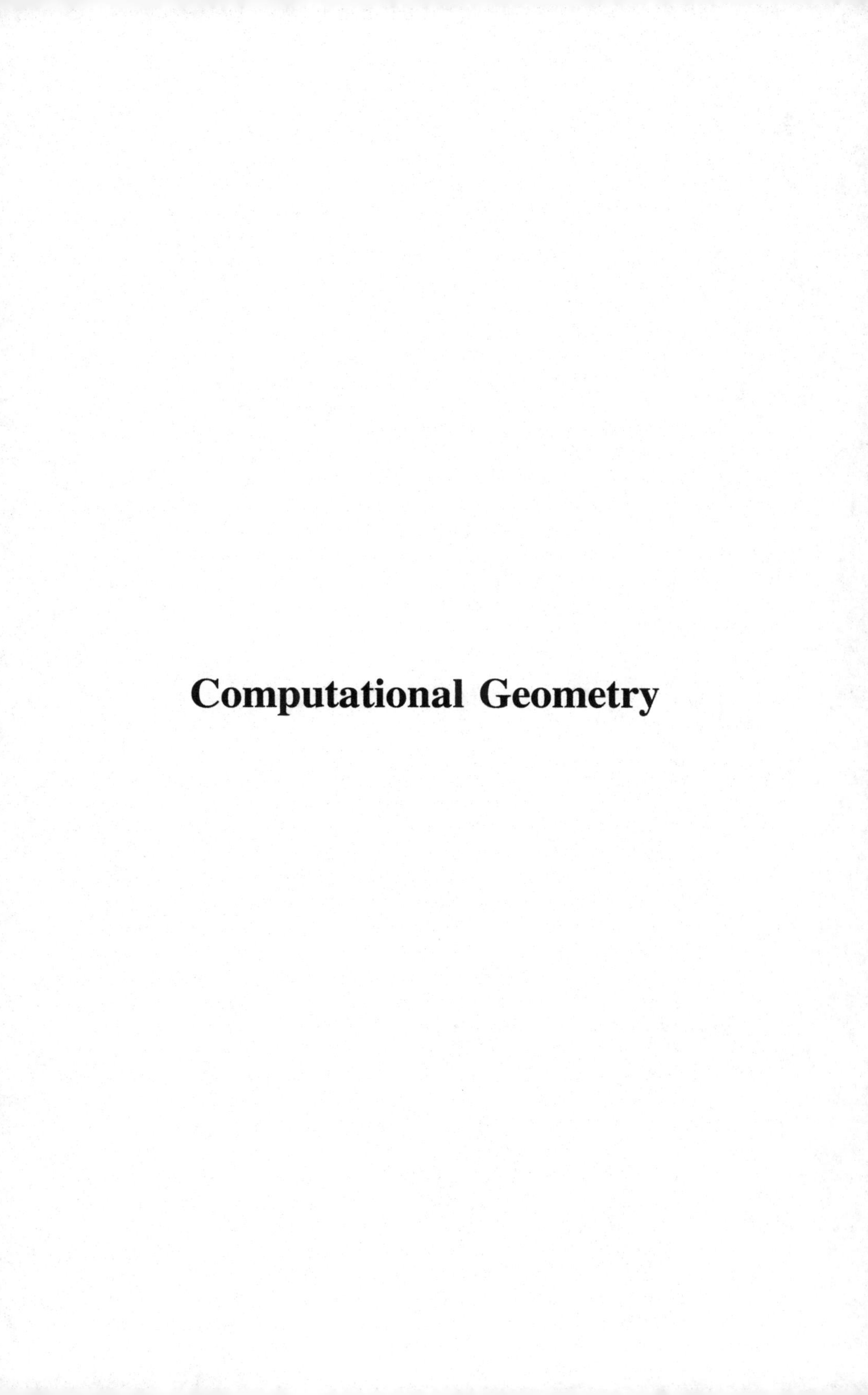

Computational Geometry

New Variants of Perfect Non-crossing Matchings

Ioannis Mantas[1(✉)], Marko Savić[2(✉)], and Hendrik Schrezenmaier[3(✉)]

[1] Faculty of Informatics, Università della Svizzera italiana, Lugano, Switzerland
ioannis.mantas@usi.ch

[2] Department of Mathematics and Informatics, Faculty of Sciences, University of Novi Sad, Novi Sad, Serbia
marko.savic@dmi.uns.ac.rs

[3] Institut für Mathematik, Technische Universität Berlin, Berlin, Germany
schrezen@math.tu-berlin.de

Abstract. Given a set of points in the plane, we are interested in matching them with straight line segments. We focus on perfect (all points are matched) non-crossing (no two edges intersect) matchings. Apart from the well known MinMax variant, where the length of the longest edge is minimized, we extend work by looking into different optimization variants such as MaxMin, MinMin and MaxMax. We consider both the monochromatic and bichromatic versions of these problems and by employing diverse techniques we provide efficient algorithms for various input point configurations.

Keywords: Perfect · Non-crossing · Matchings · Monochromatic · Bichromatic · Bottleneck · MinMax · MaxMin · MaxMax · MinMin

1 Introduction

In the *matching problem*, given is a set of objects, the goal is to partition the set into pairs such that no object belongs in two pairs. This simple problem is a classic in graph theory, which has received a lot of attention, both in an abstract and in a geometric setting. There are plenty of variants of the problem and there is a great plethora of results.

In this paper, we consider the geometric setting where, given a set P of $2n$ points in the plane, the goal is to match points of P with straight line segments, in the sense that each pair of points induces an *edge* of the matching. A matching is *perfect* if it consists of exactly n pairs. A matching is *non-crossing* if all edges induced by the matching are pairwise disjoint. When there are no restrictions on which pairs of points can be matched, the problem is called *monochromatic*. In the *bichromatic* variant, P is partitioned into two sets B and R of blue and red points, respectively, and only points of different colors are allowed to be matched. When $|B| = |R| = n$, the bichromatic point set P is called *balanced*.

A. Mudgal and C. R. Subramanian (Eds.): CALDAM 2021, LNCS 12601, pp. 151–164, 2021.
https://doi.org/10.1007/978-3-030-67899-9_12

1.1 Related Work

Geometric matchings find applications in many diverse fields, with the most famous perhaps being operations research, where it is known as the *assignment problem*. They are useful in the field of shape matching, when shapes are represented by finite point sets, see e.g., [30], and it is a fundamental problem in pattern recognition. Among others, geometric matchings appear in VLSI design problems, see e.g., [11], in computational biology, see e.g., [10], and are used for map construction or comparison algorithms, see e.g., [16].

Requiring the matching to be non-crossing or perfect is rather natural. Given a monochromatic or balanced bichromatic point set, a perfect non-crossing matching always exists and it can be found in $O(n \log n)$ time by recursively computing *ham-sandwich cuts* [20] or by using the algorithm of Hershberger and Suri [18]. Many times though, not any perfect non-crossing matching is sufficient and the interest lies in finding a matching with respect to some optimization criterion.

A well-studied optimization criterion is minimizing the sum of lengths of all edges, which we call the MinSum variant. It is also known as the *Euclidean assignment* or *Euclidean matching* problem. It is interesting, and not difficult to show, that such a matching is always non-crossing. For monochromatic point sets, an $O(n^{1.5} \log n)$-time algorithm was given by Varadarajan [29]. For bichromatic point sets, Kaplan et al. [19] recently presented an $O(n^2 \log^9 n \lambda_6(\log n))$-time algorithm, outperforming previous results [3,28]. When points are in convex position, Marcotte and Suri [22] solved the problem in $O(n \log n)$ time for both the monochromatic and bichromatic settings.

Another popular goal is to minimize the length of the longest edge, which we call the MinMax variant and is also known as the *bottleneck matching*. Given monochromatic points, Abu-Affash et al. [1] showed that finding such a matching is $\mathcal{NP}$-hard. This was accompanied by an $O(n^3)$-time algorithm for points in convex position. Recently this was improved to $O(n^2)$ time by Savić and Stojaković [24]. For bichromatic points, Carlsson et al. [8] proved that finding a MinMax matching is $\mathcal{NP}$-hard. Biniaz et al. [7] gave algorithms with $O(n^3)$-time for points in convex position and $O(n \log n)$-time for points on a circle. These were improved to $O(n^2)$ and $O(n)$, respectively, by Savić and Stojaković [25].

Several other optimzation goals have been studied. In a *fair matching* the goal is to minimize the length difference between the longest and the shortest edge, and in a *minimum deviation matching*, the difference between the length of the shortest edge and the average edge length should be minimized, see [14,15]. Alon et al. [5] studied the MaxSum variant, where the goal is to maximize the sum of edge lengths. They conjectured that the problem is $\mathcal{NP}$-hard, and gave an approximation algorithm, later improved by Dumitrescu and Toth [12].

1.2 Problem Variants Considered and Our Contribution

In this work, we continue exploring similar optimization variants in different settings and give efficient algorithms for constructing optimal matchings. We only deal with perfect non-crossing matchings, so these properties will always be

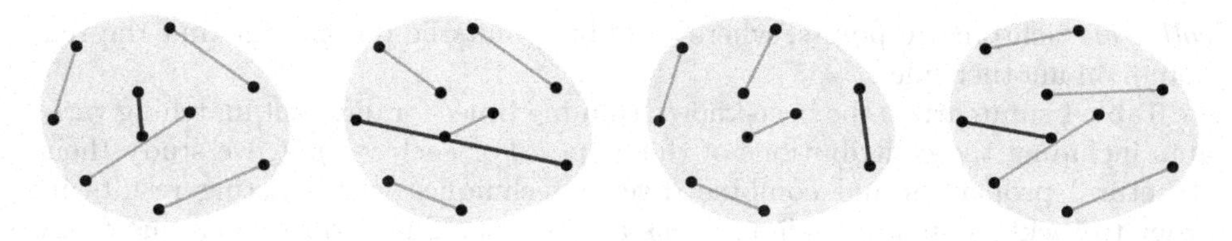

Fig. 1. Optimal MinMin1, MaxMax1, MinMax1, and MaxMin1 matchings of monochromatic points. The edges realizing the values of the matchings are highlighted.

assumed from now on, without further mention. We consider four optimization variants: MinMin where the length of the shortest edge is minimized, MaxMax where the length of the longest edge is maximized, MaxMin where the length of the shortest edge is maximized, and MinMax where the length of the longest edge is minimized. See Fig. 1 for an illustration of these four variants.

To the best of our knowledge, except for MinMax, the other three variants have not been considered before. Studying the MinMin and MaxMax variants is motivated by the analysis of worst-case scenarios for problems where very short or long edges are undesirable, but the selection of edges is not something that we can control. More generally, the values of MinMin and MaxMax serve as lower and upper bounds on the length of any feasible edge and can be helpful in estimating the quality of a matching, with respect to some objective function. The MaxMin variant, similar to MinMax, resembles fair matchings in the sense that all edges have similar lengths, analogously to the variants studied in [14].

We study both the monochromatic and bichromatic versions of these variants in different point configurations. For bichromatic points, we assume that P is balanced. We denote the monochromatic problems with the index 1, e.g., MinMin1, and the bichromatic with the index 2, e.g., MinMin2. In Sect. 2, we consider monochromatic points in general position. In Sect. 3, points are in convex position. In Sect. 4, points lie on a circle. In Sect. 5, we consider *doubly*

Table 1. Summary of results on the optimization of perfect non-crossing matchings. The value of the matching can be obtained in the time not indicated with (*). The time marked with (*) represents the extra time needed to also return a matching. h denotes the size of the convex hull. ε denotes an arbitrarily small positive constant. Results without reference are given in this paper.

Monochromatic	MinMin1	MaxMax1	MinMax1	MaxMin1
General position	$O(nh + n\log n)$, $O(n^{1+\epsilon}+n^{2/3}h^{4/3}\log^3 n)$	$O(nh) + O(n\log n)^*$, $O(n^{1+\epsilon}+n^{2/3}h^{4/3}\log^3 n)$	$\mathcal{NP}$-hard [1]	?
Convex position	$O(n)$	$O(n)$	$O(n^2)$ [24]	$O(n^3)$
Points on circle	$O(n)$	$O(n)$	$O(n)$	$O(n)$
Bichromatic	**MinMin2**	**MaxMax2**	**MinMax2**	**MaxMin2**
General position	?	?	$\mathcal{NP}$-hard [8]	?
Convex position	$O(n)$	$O(n)$	$O(n^2)$ [25]	$O(n^3)$
Points on circle	$O(n)$	$O(n)$	$O(n)$ [25]	$O(n^3)$
Doubly collinear	$O(n)$	$O(1) + O(n)^*$	$O(n^4\log n)$	?

collinear bichromatic points, where the blue points lie on one line and the red points on another line.

Table 1 summarizes the best-known running times for different matching variants including the contributions of this paper. For each variant, we study their structural properties and combine diverse techniques with existing results in order to tackle as many configurations as possible. The various open questions that arise throughout the paper, pave the way for further research in this family of problems. The proofs which are omitted due to lack of space, together with extra details and suggestions for future work, can be found in our full paper [21].

2 Monochromatic Points in General Position

In this section, P is a monochromatic set of points in general position, where we assume that no three points are collinear. We denote by $\mathrm{CH}(P)$ the boundary of the convex hull of P, by h the number of vertices of $\mathrm{CH}(P)$, by $q_1, \ldots, q_h$ the counterclockwise ordering of the vertices along $\mathrm{CH}(P)$, and by $d(v, w)$ the Euclidean distance between two points v and w. We call an edge (v, w) *feasible*, if there exists a matching which contains (v, w), and *infeasible* otherwise.

The following lemma gives us a feasibility criterion for an edge (v, w).

Lemma 1. *An edge (v, w) is infeasible if and only if (1) $v, w \in \mathrm{CH}(P)$ and (2) there is an odd number of points on each side of (v, w).*

Proof. Let l be the line through the points v, w and let A, B be the subdivision of $P \setminus \{v, w\}$ induced by l. For the if-part, let $v, w \in \mathrm{CH}(P)$. Then each edge (a, b) with $a \in A$, $b \in B$ intersects (v, w) and thus cannot be in a matching with (v, w). If further A, B have an odd number of points each, at least one point from each set will not be matched if (v, w) is in the matching. So, (v, w) is infeasible.

For the only-if-part, let (v, w) be infeasible and suppose that (1) or (2) is not fulfilled. If (2) is not fulfilled, then A, B have an even number of points each. Thus, we can find matchings of A and B independently without intersecting (v, w). Hence, (v, w) is a feasible edge, a contradiction. If (1) is not fulfilled, then not both of v, w are in $\mathrm{CH}(P)$. So, l crosses at least one edge (x, y) of $\mathrm{CH}(P)$, with $x \in A, y \in B$, see Fig. 2a. But then, both $A \setminus \{x\}$ and $B \setminus \{y\}$ contain an even number of points. Thus, there exist matchings of $A \setminus \{x\}$ and $B \setminus \{y\}$, which together with (v, w) and (x, y) form a matching of P, a contradiction. □

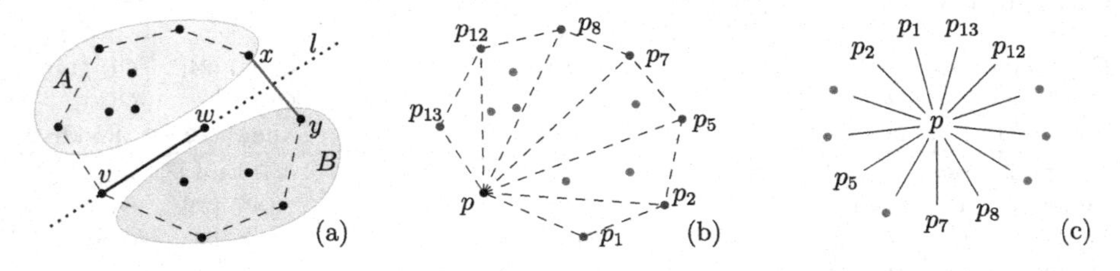

Fig. 2. (a) Illustration for the proof of Lemma 1. (b),(c) The weak radial ordering of p with the points of $P \setminus \mathrm{CH}(P)$ considered as unlabeled points.

2.1 MinMin1 and MaxMax1 Matchings in General Position

The problems MinMin and MaxMax are equivalent to finding the *extremal*, shortest or longest, feasible pair. A main challenge is to check the feasibility of an edge according to Lemma 1. We propose two different approaches.

Using Radial Orderings. The *radial ordering* of a point $p \in P$ is the counter-clockwise circular ordering of the points in $P \setminus p$ by angle around p. The radial orderings of all $p \in P$ can be computed in $O(n^2)$ total time using the dual line arrangement of P, see e.g., [4,6].

Given a subset $A \subseteq P$, we define the *A-weak radial ordering* of a point $p \in P$ as the radial ordering of p where the points from A that occur between two points from $\overline{A} := P \setminus A$ are given as an unordered set, see Figs. 2b and 2c. We are interested in the $\overline{\mathrm{CH}(P)}$-weak radial orderings of the points in CH(P). These are of interest, as they allow us to check the feasibility of all pairs (q_i, q_j) of points $q_i, q_j \in \mathrm{CH}(P)$ in $O(nh)$ total time using Lemma 1.

Lemma 2. *Given a point set P and a subset $A \subseteq P$ with $|P| = n$ and $|A| = k$, the $\overline{A}$-weak radial orderings of all points in A can be computed in $O(nk)$ time.*

Proof. First, we use a point-line duality and compute the dual line arrangement $\mathcal{L}_A$ of A in $O(k^2)$ time [4,6]. We denote the dual line of a point p by l_p. For each edge e of $\mathcal{L}_A$, we initialize a set $X_e := \emptyset$, also in $O(k^2)$ total time. Then, for each point $p \in P \setminus A$, we find the set E_p of edges of $\mathcal{L}_A$ that are intersected by l_p and add p to all sets X_e with $e \in E_p$. Due to the zone theorem [4] this takes $O(k)$ time for each p. Finally, we can read off the weak radial ordering of a point $q \in A$ from $\mathcal{L}_A$ and the sets X_e in the following way: Let $p_1, \ldots, p_{k-1}$ be the ordering of the points in $A \setminus q$ corresponding to the order of intersections of l_q with the other lines in $\mathcal{L}_A$. Further, let e_i be the edge of $\mathcal{L}_A$ between the intersections of l_q with l_{p_i} and $l_{p_{i+1}}$ (with indices understood modulo $k-1$). Then the weak radial ordering of q is $p_1, X_{e_1}, p_2, X_{e_2}, \ldots, X_{e_{k-1}}$. □

We use the feasibility criterion of Lemma 1 and the concept of weak radial orderings to provide algorithms for MinMin1 and MaxMax1.

Theorem 1. *If P is in general position,* MinMin1 *can be solved in $O(nh + n \log n)$ time.*

Proof. We initially construct CH(P) in $O(n \log h)$ time [9]. Then, we compute the $\overline{\mathrm{CH}(P)}$-weak radial orderings of the points in CH(P) in $O(nh)$ total time using Lemma 2. Now we look for the shortest feasible edge.

We first consider edges (v, w) with $v \notin \mathrm{CH}(P)$ and we want to find $m_1 := \min(\{ d(v, w) \colon v \in P \setminus \mathrm{CH}(P), w \in P \})$. By Lemma 1, such edges are always feasible. We can find m_1 in $O(n \log n)$ time using a standard algorithm via a Voronoi diagram. Now we consider edges (v, w) with both $v, w \in \mathrm{CH}(P)$ and we want to find $m_2 := \min(\{ d(v, w) \colon v, w \in \mathrm{CH}(P) \})$. By Lemma 1, an edge (q_i, q_{i+1}) is always feasible and an edge (q_i, q_{j+1}) is feasible if and only if (i) (q_i, q_j) is feasible and there is an odd number of points between q_j, q_{j+1} in the radial ordering of q_i or (ii) (q_i, q_j) is infeasible and there is an even number of points between q_j, q_{j+1} in the radial ordering of q_i. Thus, we can find m_2 in $O(nh)$ time, using weak radial orderings. So, we can find the overall minimum $m_{\mathrm{sol}} = \min(m_1, m_2)$, in $O(n \log n + nh)$ time. □

Observe that using the same algorithm but considering the maximum feasible values for m_1, m_2 and m_{sol}, also solves MAXMAX1 in $O(nh+n \log n)$ time. Using the following lemma we further improve the time complexity to $O(nh)$.

Lemma 3. *If (v, w) is a longest feasible edge, then one of $v, w \in \mathrm{CH}(P)$.*

Theorem 2. *If P is in general position,* MAXMAX1 *can be solved in $O(nh)$ time.*

Proof. The algorithm is similar to the MINMIN1, described in Theorem 1, with two changes: The minimizations of $m_1, m_2, m_{\mathrm{sol}}$ are replaced by maximizations and, to find m_1, we only consider edges (v, w) with $v \in P \setminus \mathrm{CH}(P)$ and $w \in \mathrm{CH}(P)$. This is sufficient, due to Lemma 3, and reduces the time for finding m_1 to $O((n-h)h)$, by simply comparing all $(n-h)h$ edges. Hence, the overall running time is reduced to $O((n-h)h + nh) = O(nh)$. □

Using Halfplane Range Queries. Now we take another approach to decide the feasibility of a pair of points from $\mathrm{CH}(P)$. The task of determining the number of points of a given point set lying on one side of a given straight line is known as *halfplane range query* and has been studied extensively over the last decades, see e.g., [2]. Using these results to check the criterion of Lemma 1, we obtain the following algorithms that are more efficient than those of Theorems 1 and 2, when $h = \Omega(n^c)$ for some constant $c > 0$.

Theorem 3. *Let P be in general position. Then* MINMIN1 *and* MAXMAX1 *can be solved in $O(n^{1+\epsilon} + n^{2/3}h^{4/3}\log^3 n)$ time where $\epsilon > 0$ is an arbitrary constant.*

Proof. We show that the feasibility of all pairs of points of $\mathrm{CH}(P)$ can be decided in the claimed running times. Then, with the aforementioned algorithm and an additional effort of $O(n \log n)$ time, MINMIN1 and MAXMAX1 can be solved.

We distinguish two classes of values of h. Let $h \leq n^{1/4}$. According to [23], halfplane range queries can be answered in $O(n^{1/2})$ time after a preprocessing step, costing $O(n^{1+\epsilon})$ time. We have to do $\binom{h}{2} = O(h^2)$ queries, so the time needed for the queries is $O(h^2 n^{1/2}) = O(n)$. Therefore the preprocessing step dominates the overall time needed, resulting in $O(n^{1+\epsilon})$ total time.

Now let $h \geq n^{1/4}$. We set $m = n^{2/3}h^{4/3}$. Then we have $n \leq m \leq n^2$, which is required by [23] for the following to hold: Halfplane range queries can be answered in $O(\frac{n}{m^{1/2}} \log^3 \frac{m}{n})$ time after a preprocessing step costing $O(n^{1+\epsilon} + m \log^{\epsilon} n)$ time. Thus the time needed for the $O(h^2)$ queries is $O(n^{2/3}h^{4/3}\log^3 n)$ and for the preprocessing is $O(n^{1+\epsilon} + n^{2/3}h^{4/3}\log^{\epsilon} n)$, so $O(n^{1+\epsilon} + n^{2/3}h^{4/3}\log^3 n)$ time overall. Combining the two cases for h, the claim follows. □

3 Points in Convex Position

In this section, we assume that the points in P are in convex position and their counterclockwise ordering, $p_0, \ldots, p_{2n-1}$, is given. For simplicity, we address points by their indices, i.e., we refer to p_i as i. Arithmetic operations with indices are done modulo $2n$. We call edges of the form $(i, i+1)$ *boundary edges* and we call the remaining edges *diagonals.*

We remark that all four optimization variants, for both monochromatic and bichromatic point sets, can be solved in $O(n^3)$ time by a dynamic programming approach. This approach has also been used in [1,7,8] for MinMax problems. We present more efficient algorithms for MinMin and MaxMax.

3.1 MinMin1 and MaxMax1 Matchings in Convex Position

We make use of the following two algorithms. Given two convex polygons P and Q, Toussaint's algorithm [27] finds in $O(|P| + |Q|)$ time the vertices that realize the minimum distance between P and Q. Analogously, Edelsbrunner's algorithm [13] finds in $O(|P| + |Q|)$ time the vertices that realize the maximum distance between P and Q.

Theorem 4. *If P is in convex position,* MinMin1 *and* MaxMax1 *can be solved in $O(n)$ time.*

Proof. A pair (i, j) is feasible if and only if i and j are of different parity. This suggests that we can split P into two sets, P_{even} and P_{odd}, one containing the even and the other containing the odd indices. Then, any edge (v, w) with $v \in P_{\text{even}}$ and $w \in P_{\text{odd}}$ is feasible. Considering P_{odd} and P_{even} as convex polygons, we apply Toussaint's algorithm [27] for MinMin1 or Edelsbrunner's algorithm [13] for MaxMax1. All steps can be done in $O(n)$ time. □

3.2 MinMin2 and MaxMax2 Matchings in Convex Position

We now combine the monochromatic algorithms with the theory of *orbits* [25], a concept which captures well the nature of bichromatic matchings in convex position. More specifically, P is partitioned into orbits, which are balanced sets

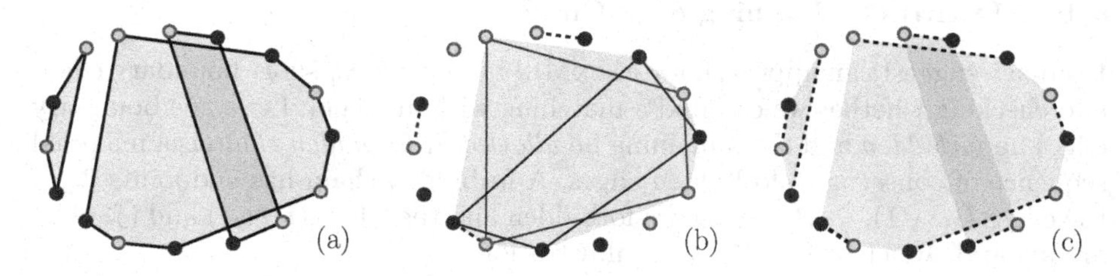

Fig. 3. MinMin2 for P in convex position. (a) Find orbits. (b) Find the shortest edge between the blue and red polygon of an orbit. (c) Extend to a perfect matching. (Color figure online)

of points, and the colors of the points along the boundary of the orbit are alternating, see Fig. 3a. An important property is that a bichromatic edge (b, r) is feasible if and only if b and r are in the same orbit [25].

Theorem 5. *If P is in convex position,* MinMin2 *and* MaxMax2 *can be solved in $O(n)$ time.*

Proof. We first compute all orbits in $O(n)$ time [25]. Due to the alternation of red and blue points along the boundary of the orbits, a single orbit can be considered as a set of points in the monochromatic setting, with respect to the feasibility of the edges. Thus, for MinMin2, we can select for each orbit the shortest edge in $O(n)$ time using Theorem 4. The shortest edge out of all these selected edges is the shortest overall feasible edge of P, see Fig. 3b. Selecting the maximum edges instead solves MaxMax2. □

To construct, in $O(n)$ time, optimal matchings from Theorems 4 and 5 after finding an extremal feasible edge (i, j), we can apply the following lemma to the sets $\{i + 1, \dots, j - 1\}$ and $\{j + 1, \dots, i - 1\}$, see Fig. 3c. The idea is to pair consecutive edges along the boundary of the convex hull or each orbit.

Lemma 4. *If P is in convex position, we can construct an arbitrary matching in $O(n)$ time, both in the monochromatic and bichromatic case.*

4 Points on a Circle

In this section, we assume that all points lie on a circle. Obviously, the points are in convex position, so all the results from Sect. 3 also apply here. We present algorithms with a better time complexity. We use the notation from Sect. 3.

In addition to convexity, the results in this section rely on a property of points lying on a circle, which we call the *decreasing chords property.* A point set has this property if, for any edge (i, j), in at least one of its sides, an edge between any two points on that side is not longer than (i, j) itself, see Fig. 4a.

Due to the decreasing chords property, we can easily infer the following.

Lemma 5. *Any shortest edge of a matching on P is a boundary edge.*

4.1 MaxMin1 Matching on a Circle

Lemma 5 suggests an approach for MaxMin by *forbidding* short boundary edges and checking whether we can find a matching without them. Let some boundary edges be *forbidden* and the remaining be *allowed.* A *forbidden chain* is a maximal sequence of consecutive forbidden edges. A forbidden chain has endpoints i, j if the edges $(i, i+1), \dots, (j-1, j)$ are forbidden and the edges $(i-1, i)$ and $(j, j+1)$ are allowed. Refer to Fig. 4b for an illustration.

Lemma 6. *Given a set of forbidden edges, there exists a matching without forbidden edges if and only the length of a longest forbidden chain is less than n.*

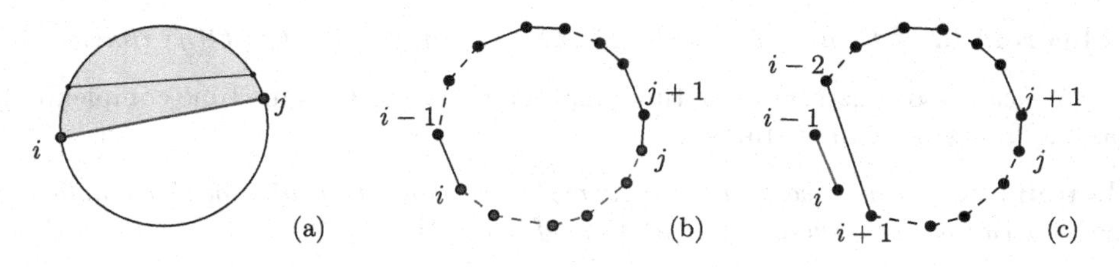

Fig. 4. (a) The decreasing chords property. (b) A forbidden chain with endpoints i, j. (c) Proof of Lemma 6. (Forbidden edges are dashed and allowed edges are solid.)

Proof. If the boundary edges are either all forbidden or all allowed, then the statement trivially holds. So, let us assume that there exists at least one forbidden and at least one allowed boundary edge.

Consider a forbidden chain of length l which has endpoints i and j. First, we assume that $l \geq n$. Then, at least one matched pair (a, b) has both endpoints in $\{i, \ldots, j\}$. Thus, either (a, b) is a forbidden boundary edge or it splits P in a way that all points on one side of the line through (a, b) lie completely in $\{i, \ldots, j\}$. So, there exists a matched boundary edge inside $\{i, \ldots, j\}$ and, thus, a matching without forbidden edges does not exist.

Now let us assume that $l < n$. We construct a matching without forbidden edges using a recursive approach. We match the pair $(i-1, i)$ and consider the set $P' = P \setminus \{i-1, i\}$, see Fig. 4c. In P', $(i-2, i+1)$ is an allowed boundary edge since it is a diagonal in P. We show that P' can be matched by showing that the condition of the lemma holds for P'.

Let i' and j' be the endpoints of a longest forbidden chain in P', going counterclockwise from i' to j', and let l' be its length. If $l' < n-1$, a matching of P' without forbidden edges can be computed recursively. Otherwise, if $l' \geq n-1$, from $l' \leq l < n$ we infer that $l' = l = n-1$. Since $l = l'$, the new longest forbidden chain is disjoint from $\{i+1, \ldots, j\}$, so it is contained in $\{j+1, \ldots, i-2\}$, see Fig. 4c. But since $|P'| = 2n-2$ and $|\{i+1, \ldots, j\}| = n-1$, we have $|\{j+1, \ldots, i-2\}| = n-1$ and thus $l' < n-1$, a contradiction. □

MaxMin is equivalent to finding the largest value μ such that there exists a matching with all edges of length at least μ. By Lemma 5, it suffices to search for μ among the lengths of the boundary edges. By Lemma 6, this means that we need to find the maximal length μ of a boundary edge such that there are no n consecutive boundary edges all shorter than μ. An obvious way to find μ is to employ binary search over the boundary edge lengths and check at each step whether the condition is satisfied or not, which yields an $O(n \log n)$-time algorithm. A faster approach to find μ is as follows. Consider all $2n$ sets of n consecutive boundary edges and associate to each set the longest edge in it. Then, out of the $2n$ longest edges, search for the shortest one. This can be done in $O(n)$ time using a data structure for *range maximum query*, see e.g., [17]. However, our approach fits under the more restricted *sliding window maximum problem*, for which several simple optimal algorithms are known, see e.g., [26].

Theorem 6. *If P lies on a circle,* MaxMin1 *can be solved in $O(n)$ time.*

We can also construct an optimal matching within the same time complexity, as the following lemma states.

Lemma 7. *Given a value $\mu > 0$, a matching consisting of edges of length at least μ can be constructed in $O(n)$ time if it exists.*

4.2 Other Matchings on a Circle

Theorem 7. *If P lies on a circle,* MinMax1 *can be solved in $O(n)$ time.*

Proof. We show that there exists a MinMax1 matching using only boundary edges. Suppose we have a MinMax1 matching M containing a diagonal (i,j). Assume, without loss of generality, that all edges with endpoints in $\{i,\dots,j\}$ are at most as long as (i,j). We construct a new matching M' by taking all matched pairs in M that are outside of $\{i,\dots,j\}$ together with edges $(i,i+1),(i+2,i+3),\dots,(j-1,j)$. The longest edge of M' is not longer than the longest edge of M, proving our claim. There are only two matchings consisting only of boundary edges and in $O(n)$ time we choose the one with the shorter longest edge. □

Points on a circle are in convex position, so, both MinMin1 and MinMin2 can be found in $O(n)$ time using Theorems 4 and 5. Instead, we can do it much simpler by finding the shortest feasible boundary edge. By Lemma 5, the shortest edge of a matching is a boundary edge in both settings. This can then be extended to a perfect matching using Lemma 4.

5 Doubly Collinear Points

In this section, we consider a *doubly collinear* setting. A bichromatic point set P is *doubly collinear* if the blue points lie on a line l_B and the red points lie on a line l_R. We assume that l_B and l_R are not parallel and that the ordering of the points along each line is given. Let $x = l_B \cap l_R$ and assume, for simplicity, that $x \notin P$. Lines l_B and l_R are split at x into two *half-lines*, and the plane is subdivided into four *sectors*. We call a sector *small*, if its angle is acute.

Let $l \in \{l_B, l_R\}$. Then, for two points a, b on l, we denote by (a,b) the open line segment connecting a and b. Further, if $a \neq x$, we denote by $(a_{\rightarrow x \rightarrow})$, ${}_{x\cdots}(a_{\rightarrow}) \subset l$ the open half-lines starting at a that contain x and do not contain x, respectively. If we use square brackets, e.g., in ${}_{x\cdots}[a_{\rightarrow})$, $(a,b]$, or $[a,b]$, the corresponding endpoint is contained in the set.

The following lemma gives us a feasibility criterion for an edge (r,b), see Fig. 5a and 5b, which can be checked in $O(1)$ time. We give a constructive proof that also indicates an algorithm which, given a feasible edge (r,b), returns a matching containing (r,b) in $O(n)$ time.

Lemma 8. *An edge (r,b) is feasible if and only if $|(r,x) \cap P| \leq |(b_{\rightarrow x \rightarrow}) \cap P|$ and $|(b,x) \cap P| \leq |(r_{\rightarrow x \rightarrow}) \cap P|$.*

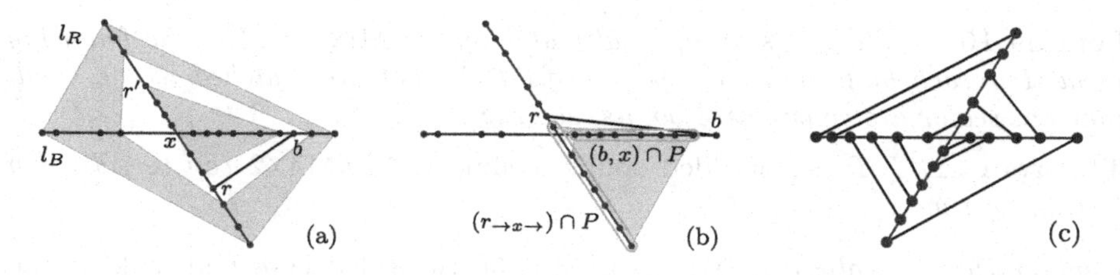

Fig. 5. (a) A feasible edge (r, b). (b) An infeasible edge (r, b). (c) An optimal MINMAX2 matching of the type described in Lemma 10.

5.1 MINMIN2 and MAXMAX2 Matchings on Doubly Collinear Points

Let l'_R and l'_B be a red and a blue half-line, respectively. The following lemma is a consequence of Lemma 8. It allows us to find, for each point in $l'_R \cap P$, the closest point in $l'_B \cap P$ it induces a feasible edge with, in $O(n)$ total time.

Lemma 9. *Let $r \in R$ and $r' \in {}_{x\cdots}(r_{\rightarrow}) \cap P$. Let $b, b' \in l'_B \cap P$ be closest to r and r', respectively, such that (r, b) and (r', b') are feasible. Then $b' \in {}_{x\cdots}[b_{\rightarrow}) \cap P$.*

Theorem 8. *If P is doubly collinear,* MINMIN2 *can be solved in $O(n)$ time.*

We call a point $p \in P$ an *extremal* point if $|{}_{x\cdots}(p_{\rightarrow}) \cap P| = 0$. We show that the longest edge between points in R and B is realized by extremal points, so there are $O(1)$ candidate edges. Hence, we can find the longest feasible edge in $O(1)$ time. This can later be extended in $O(n)$ time to an optimal matching, using the algorithm which appears in the constructive proof of Lemma 8.

Theorem 9. *If P is doubly collinear,* MAXMAX2 *can be solved in $O(1)$ time.*

5.2 MINMAX2 and MAXMIN2 Matchings on Doubly Collinear Points

We start by considering the *one-sided* doubly collinear case, where all red points are on the same side of l_B. Then, we turn to the general (two-sided) case. Finally, for the general case, we present improved results for some special cases.

One-Sided Doubly Collinear. Observe that, in this case, the extremal red point must be matched with one of the two extremal blue points. Thus, using dynamic programming, all four optimization variants can be solved in $O(n^2)$ time.

Theorem 10. *If P is one-sided doubly collinear,* MINMAX2 *and* MAXMIN2 *can be solved in $O(n^2)$ time.*

For MINMAX2, we show that there exists (also in the two-sided case) an optimal matching of a special form, described in the following lemma, allowing us to design a faster algorithm. It can be obtained from an arbitrary optimal matching by applying local changes that do not change the objective value.

Lemma 10. *There exists an optimal matching for* MinMax2 *of the following form. For each half-line l', the points of $l' \cap P$ that are matched in the small incident sector are consecutive points, see Fig. 5c.*

Theorem 11. *If P is* one-sided doubly collinear, MinMax2 *can be solved in $O(n \log n)$ time.*

General Doubly Collinear. We return to the two-sided case and look at the MinMax2 variant. By only considering matchings of the form described in Lemma 10, enumerating all possible choices for the decision which blue point is matched through which sector, and applying Theorem 11 for the two resulting one-sided subproblems, we obtain the following.

Theorem 12. *If P is doubly collinear,* MinMax2 *can be solved in $O(n^4 \log n)$ time.*

Special Angles of Intersection. Let α be the angle of intersection l_B and l_R, with $\alpha \in (0, \frac{\pi}{2}]$. We prove the existence of optimal matchings having a special form, and we then use these to derive improved algorithms for these cases as follows.

Theorem 13. *If $\alpha = \frac{\pi}{2}$,* MinMax2 *and* MaxMin2 *can be solved in $O(n)$ time.*

Theorem 14. *If $\alpha \leq \frac{\pi}{4}$,* MinMax2 *can be solved in $O(n)$ time.*

6 Conclusions and Future Work

We considered new variants for perfect non-crossing matchings. In most MinMin and MaxMax variants, we came up with optimal algorithms by exploiting structural properties of the point sets, combined with existing techniques from diverse problems. On the contrary, the MaxMin variant exhibits a significant difficulty. Designing efficient algorithms even for simple configurations, as cocircular or doubly collinear, is not at all obvious and thus quite interesting on its own. Throughout the paper many open questions have arisen. For instance, regarding convex bichromatic point sets, can orbits help to improve the MaxMin algorithms? Regarding arbitrary point sets, is there a polynomial time feasibility check for a bichromatic edge? Are the MaxMin variants $\mathcal{NP}$-hard as their MinMax counterparts? It would be interesting to see how Table 1 can be filled with improved algorithms or hardness results.

Acknowledgements. M. S. was partially supported by the Ministry of Education, Science and Technological Development, Republic of Serbia, project 174019, and H. S. by the German Research Foundation, DFG grant FE-340/11-1.

Initial discussions took place at the Intensive Research Program in Discrete, Combinatorial and Computational Geometry which took place in Barcelona in 2018. We are grateful to CRM, UAB for hosting the event and to the organizers for providing the platform to meet and collaborate. We would like to thank Carlos Alegría, Carlos Hidalgo Toscano, Oscar Iglesias Valiño, and Leonardo Martínez Sandoval for preliminary discussions, and Carlos Seara for raising a question that motivated this work. Finally, we would like to thank an anonymous reviewer for bringing to our attention the halfplane range queries.

References

1. Abu-Affash, A.K., Carmi, P., Katz, M.J., Trabelsi, Y.: Bottleneck non-crossing matching in the plane. Comput. Geom. **47**(3A), 447–457 (2014)
2. Agarwal, P.K.: Simplex range searching and its variants: a review. In: Loebl, M., Nešetřil, J., Thomas, R. (eds.) A Journey Through Discrete Mathematics, pp. 1–30. Springer, Cham (2017). https://doi.org/10.1007/978-3-319-44479-6_1
3. Agarwal, P.K., Efrat, A., Sharir, M.: Vertical decomposition of shallow levels in 3-dimensional arrangements and its applications. SIAM J. Comput. **29**(3), 912–953 (2000)
4. Agarwal, P.K., Sharir, M.: Arrangements and their applications. In: Handbook of Computational Geometry, chap. 2, pp. 49–119. North-Holland (2000)
5. Alon, N., Rajagopalan, S., Suri, S.: Long non-crossing configurations in the plane. In: Proceedings of the 9th Annual Symposium on Computational Geometry, pp. 257–263 (1993)
6. Asano, T., Ghosh, S.K., Shermer, T.C.: Visibility in the plane. In: Handbook of Computational Geometry, chap. 19, pp. 829–876. North-Holland (2000)
7. Biniaz, A., Maheshwari, A., Smid, M.H.: Bottleneck bichromatic plane matching of points. In: Proceedings of the 26th Canadian Conference on Computational Geometry, pp. 431–435 (2014)
8. Carlsson, J.G., Armbruster, B., Rahul, S., Bellam, H.: A bottleneck matching problem with edge-crossing constraints. Int. J. Comput. Geom. Appl. **25**(4), 245–261 (2015)
9. Chan, T.M.: Optimal output-sensitive convex hull algorithms in two and three dimensions. Discret. Comput. Geom. **16**(4), 361–368 (1996). https://doi.org/10.1007/BF02712873
10. Colannino, J., et al.: An O(n log n)-time algorithm for the restriction scaffold assignment problem. J. Comput. Biol. **13**(4), 979–989 (2006)
11. Cong, J., Kahng, A.B., Robins, G.: Matching-based methods for high-performance clock routing. IEEE Trans. Comput. Aided Des. Integr. Circuits Syst. **12**(8), 1157–1169 (1993)
12. Dumitrescu, A., Tóth, C.D.: Long non-crossing configurations in the plane. Discret. Comput. Geom. **44**, 727–752 (2010)
13. Edelsbrunner, H.: Computing the extreme distances between two convex polygons. J. Algorithms **6**(2), 213–224 (1985)
14. Efrat, A., Itai, A., Katz, M.J.: Geometry helps in bottleneck matching and related problems. Algorithmica **31**(1), 1–28 (2001)
15. Efrat, A., Katz, M.J.: Computing fair and bottleneck matchings in geometric graphs. In: Asano, T., Igarashi, Y., Nagamochi, H., Miyano, S., Suri, S. (eds.) ISAAC 1996. LNCS, vol. 1178, pp. 115–125. Springer, Heidelberg (1996). https://doi.org/10.1007/BFb0009487
16. Eppstein, D., van Kreveld, M., Speckmann, B., Staals, F.: Improved grid map layout by point set matching. Int. J. Comput. Geom. Appl. **25**(02), 101–122 (2015)
17. Fischer, J., Heun, V.: Space-efficient preprocessing schemes for range minimum queries on static arrays. SIAM J. Comput. **40**(2), 465–492 (2011)
18. Hershberger, J., Suri, S.: Applications of a semi-dynamic convex hull algorithm. BIT Numer. Math. **32**(2), 249–267 (1992)
19. Kaplan, H., Mulzer, W., Roditty, L., Seiferth, P., Sharir, M.: Dynamic planar Voronoi diagrams for general distance functions and their algorithmic applications. Discret. Comput. Geom. **64**(3), 838–904 (2020)

20. Lo, C.-Y., Matoušek, J., Steiger, W.: Algorithms for ham-sandwich cuts. Discret. Comput. Geom. **11**(4), 433–452 (1994). https://doi.org/10.1007/BF02574017
21. Mantas, I., Savić, M., Schrezenmaier, H.: New variants of perfect non-crossing matchings. arXiv preprint arXiv:2001.03252 (2020)
22. Marcotte, O., Suri, S.: Fast matching algorithms for points on a polygon. SIAM J. Comput. **20**(3), 405–422 (1991)
23. Matoušek, J.: Range searching with efficient hierarchical cuttings. Discret. Comput. Geom. **10**(2), 157–182 (1993)
24. Savić, M., Stojaković, M.: Faster bottleneck non-crossing matchings of points in convex position. Comput. Geom. **65**, 27–34 (2017)
25. Savić, M., Stojaković, M.: Bottleneck bichromatic non-crossing matchings using orbits (2018). arxiv.org/abs/1802.06301
26. Tangwongsan, K., Hirzel, M., Schneider, S.: Low-latency sliding-window aggregation in worst-case constant time. In: Proceedings of the 11th ACM International Conference on Distributed and Event-Based Systems, pp. 66–77 (2017)
27. Toussaint, G.T.: An optimal algorithm for computing the minimum vertex distance between two crossing convex polygons. Computing **32**(4), 357–364 (1984)
28. Vaidya, P.M.: Geometry helps in matching. SIAM J. Comput. **18**(6), 1201–1225 (1989)
29. Varadarajan, K.R.: A divide-and-conquer algorithm for min-cost perfect matching in the plane. In: Proceedings of the 39th Symposium on Foundations of Computer Science, pp. 320–329 (1998)
30. Veltkamp, R.C., Hagedoorn, M.: State of the art in shape matching. In: Lew, M.S. (ed.) Principles of Visual Information Retrieval. ACVPR, pp. 87–119. Springer, London (2001). https://doi.org/10.1007/978-1-4471-3702-3_4

Cause I'm a Genial Imprecise Point: Outlier Detection for Uncertain Data

Vahideh Keikha[1(✉)], Hamidreza Keikha[2], and Ali Mohades[3]

[1] The Czech Academy of Sciences, Institute of Computer Science, Pod Vodárenskou věží, Prague, Czech Republic
keikha@cs.cas.cz

[2] Department of Mathematics and Computer Science, Sistan and Baluchestan University, Zahedan, Iran
keikha.eng@gmail.com

[3] Department of Mathematics and Computer Science, Amirkabir University of Technology, Tehran, Iran
mohades@aut.ac.ir

Abstract. In this paper, we introduce the outlier detection problem in a set of uncertain points. We study two variants of the problems based upon the definition of the outlier. For a given positive integer $k(< n)$ and a set $\Re$ of n regions as the imprecise points, the first type of the outlier detection problem that we study is to locate $n - k$ points on distinct regions, such that the size of the smallest axis-aligned bounding box (AABB), the diameter or the smallest enclosing circle (SEC) of the resulting points gets minimized. The uncertainty regions we study are squares or disks, and the excluded k regions are considered as outliers.

We also study the covering versions in which the objectives of the SEC and the AABB problems are to find the smallest circle or axis-aligned bounding box, respectively, that covers the area of at least $n - k$ regions.

In the second-type of outliers, the outliers are those k regions that mostly reduce the *uncertainty-induced gap* between the lower bound and the upper bound on the size of the output. We give polynomial time algorithms for several variants of the mentioned problems, ranging in running time from $O(n \log n)$ to $O(n^{5.5} \log n)$.

1 Introduction

Geometric modeling is a vigorous fit for facing many of the world's challenges. However, the evident uncertainty due to a variety of reasons alters the efficiency of the geometric algorithms. Such imprecision can occur at different occasions of the data collection process, e.g., one may gather data by sampling or one may not have enough space to store all the numerical information without rounding. In geometric problems, it has often been assumed that the input points have precise coordinates, however, real datasets are mostly uncertain or incomplete. Also, some of the collected data might be irrelevant and situated away from the other data. Motivated by these, we introduce the "outlier detection problem in

A. Mudgal and C. R. Subramanian (Eds.): CALDAM 2021, LNCS 12601, pp. 165–178, 2021.
https://doi.org/10.1007/978-3-030-67899-9_13

a set of geometric imprecise data". To the best of our knowledge, this problem is not studied yet to make comparisons with the related works.

Data Uncertainty. If the exact coordinates of a point are not known, we refer to this point as an *uncertain point.* We assume an uncertain point is modeled by a square (or disk) according to the *region-based* model of uncertainty, in which the point may lie at any place within the region with the same probability [13]. Let $\mathfrak{R}$ be a set of n squares. Then $\mathfrak{R}$ can introduce infinitely many *placements*, where each placement consists of n points from distinct squares, and each point is allowed to translate within the square the point lies on. Also, each different placement of points can have different measures, which implies that there exists a *range* in which the size of the various *descriptors* of the points bounce. Such descriptors are most conveniently defined by the smallest enclosing circle (SEC), axis-aligned bounding box (AABB), the diameter, or the convex hull. We refer to the largest value of the descriptor as the *upper bound* and to the smallest one as the *lower bound.*

Calculating the upper/lower bound on the area of AABB of a set of n squares (or disks) is to find a placement of points such that the area of AABB has its maximum/minimum value. For squares or disks, the choices of the optimal solutions lie on the boundary of the axis-extreme regions and cost $O(n)$ time [12]. In the SEC problem, computing the upper/lower bound is defined as computing a placement of points such that the radius of the SEC has its maximum/minimum value. Solving the maximization problem on both the disks and the squares takes $O(n)$ time. The minimization problem for disks is LP-type, and for squares (or any set of convex bodies) can be modeled as finding the minimum of a convex function, and they both take $O(n)$ time [12]. In the diameter problem, computing the upper/lower bound is defined as computing a placement of points such that the maximum (minimum) pairwise distances has its maximum (minimum) value. On a set of n squares, both the minimization and the maximization problems cost $O(n \log n)$ time [12]. On a set of disks, the largest possible diameter can be computed in $O(n \log n)$ time, but for the smallest possible diameter problem, only approximation algorithms are known [11,12]. These problems are introduced by Löffler and van Kreveld [12]. For arbitrary values of k, these algorithms may suggest exponential time solutions to the problem of computing a subset of size $n - k$ ($n - k$-subset) with the optimized cost function among all $n - k$-subsets.

Outlier Removal. There are lots of studies in the field of outlier detection when the input is a set of points. Several problems are studied under different assumptions for the number of outliers, where it varies from one to any value in the order of the input; see, e.g., [4] and the references therein. In the literature, most of the studies aim to remove the outliers to reduce the most the size of a specific measure, prior to any analyzes. Let P be a set of n points. For arbitrary values of k, the problem of choosing a subset of size $n - k$ of P such that this set has the smallest AABB can be solved in $O(n^{5/2})$ time [2,3,10]. Very recently, this running time is improved to $O(n^2 \log n)$ by Chan and Har-Peled [5].

The problem of computing a set of $n - k$ points with the smallest possible diameter can be solved in $O(n \log n + k^2 n \log^2 k)$ [1,8] time. Note that removing

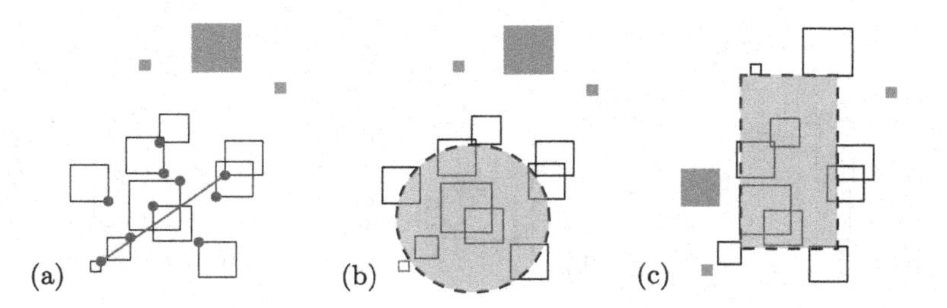

Fig. 1. Problem definitions on a set of squares in the minimization version, with $k = 3$ outliers. The optimal solutions are shown in purple, and the shaded squares are the first-type outliers. (a) Minimizing the smallest possible diameter. (b) Minimizing the SEC. (c) Minimizing the smallest area AABB. (Color figure online)

even one outlier from a set of n points takes $\Omega(n \log n)$ time [3] since one of the vertices of the diameter determines the outlier. Using such vertex, the diameter would be computed in linear time. We refer the interested readers to the references in [2,5] for related studies.

For a set P of n points and an integer $k < n$, the problem of computing the SEC of only $n - k$ points is referred to as $n - k$-enclosing problem. It is already shown that the $n-k$-enclosing circle of a set of points is centered at a vertex or an edge of a Voronoi diagram of order $n-k-1$, and can be computed in $O(n^2 \log n)$ time [7]. They also provided a $O(nk \log^2 n)$ time algorithm with $O(nk)$ space, and an algorithm with $O(nk \log^2 n \log n/k)$ time and $O(n \log n)$ space [7]. Har-peled and Mazumdar [9] presented a $O(n(n-k))$ expected time randomized algorithm for computing the $n - k$-enclosing circle of a set of n points in the plane. They also provided several approximation algorithms with $O(n)$ time for constant approximation factors, and with $O(n + n\ min(1/(n-k)\epsilon^3) \log^2(\frac{1}{\epsilon}, n-k))$ time for a $(1 + \epsilon)$ approximation factor. In $\mathbb{R}^d$, if d is fixed, the $n - k$-enclosing circle problem can be solved in polynomial time, otherwise the problem becomes NP-hard [15]. The authors also introduced a PTAS with running time $O(n^{\frac{2}{\epsilon^2}+1}d)$.

Contribution

Problem Definition. We study two variants of the problems based upon the definition of the outlier. For a set $\mathfrak{R}$ of n imprecise points, the first type of problem we study is to locate $n - k$ (for a given k) points on distinct squares (or disks), such that the diameter, the radius of the smallest enclosing circle, or the area of the smallest axis-aligned bounding box of the resulting points gets minimized. The excluded k squares are considered as the *outliers.*

In the covering version, we study the case where the radius of the SEC or the area of the AABB needs to get minimized but they should cover the area of at least $n - k$ squares (against the previous version, that stabbing $n - k$ regions suffices). In the maximization version of the diameter problem, the selected $n - k$ points have the largest possible diameter among all possible choices. More formally we study the following problems:

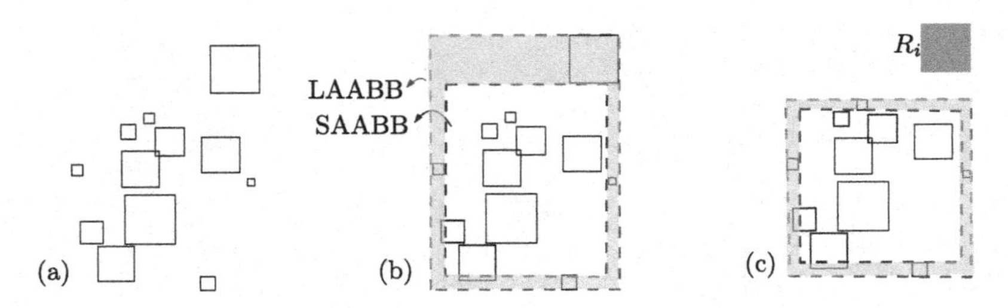

Fig. 2. (a) A set $\mathfrak{R}$ of uncertain points. (b) There exists a gap between $\mathfrak{U}(\alpha_{\mathfrak{R}})$ (shown in red) and $\mathfrak{L}(\alpha_{\mathfrak{R}})$ (shown in blue). (c) For $k = 1$, the outlier which narrows the uncertainty gap is R_i, i.e. the uncertainty gap of $\mathfrak{R}' = \mathfrak{R} \setminus R_i$ is minimized among all choices of $\mathfrak{R}'$. (Color figure online)

Problem 1. Let $\mathfrak{R} = \{R_1, \ldots, R_n\}$ be a set of n squares (or disks), and let $k < n$ be a positive integer. The objective is to find a set P of placement of points on $n - k$ distinct squares such that the diameter, the radius of the SEC, or the area of the AABB of those points is minimized among any other choices of P.

Problem 2. Let $\mathfrak{R} = \{R_1, \ldots, R_n\}$ be a set of n squares (or disks), and let $k < n$ be a positive integer. The objective is to find a circle with the smallest radius (or a smallest area AABB) such that the circle (or the AABB) covers the area of at least $n - k$ squares. In the diameter problem, we need to select $n - k$ points on distinct squares such that the selected points realizes the largest possible diameter among all possible choices.

We refer to the deleted k squares as the *first-type outliers.* An example of the first-type outliers in the minimization problems is illustrated in Fig. 1.

For a measure α on a set $\mathfrak{R}$ of imprecise points, let $\mathfrak{U}(\alpha_{\mathfrak{R}})$ and $\mathfrak{L}(\alpha_{\mathfrak{R}})$, respectively, denote the upper and the lower bound on the size of α for any placement of points on $\mathfrak{R}$. See Fig. 2(b) for an illustration. We call the range $\mathfrak{U}(\alpha_{\mathfrak{R}}) - \mathfrak{L}(\alpha_{\mathfrak{R}})$ *uncertainty-induced-gap* or simply *uncertainty gap.*

In the second-type outlier detection problem, we should identify k outliers based upon this observation that the value of the uncertainty gap (for a given measure, the difference between the size of the largest and the smallest value) gets minimized. The *second-type outliers* are those k regions that narrow this gap the most; in the sense that the smaller uncertainty gap results in the smaller range of a specific measure on different placements.

To define the second-type outliers more formally, let $\mathfrak{R}^* \subseteq \mathfrak{R}$ be a set of $n - k$ squares such that $\mathfrak{R} \setminus \mathfrak{R}^*$ determines the set of the second-type outliers. Then $\mathfrak{U}(\alpha_{\mathfrak{R}^*}) - \mathfrak{L}(\alpha_{\mathfrak{R}^*})$ has the smallest possible value among all possible choices of $\mathfrak{R}^*$. See Fig. 2 as an example, in which the difference between the largest- and the smallest-area AABB determines the uncertainty gap. Indeed, in the second-type outliers, we look for a set of $n - k$ uncertain points, in which, for a specific extent measure, any of the placement of the points on those regions have almost the same extension. In other words, the different placement of the points on $\mathfrak{R}^*$ have

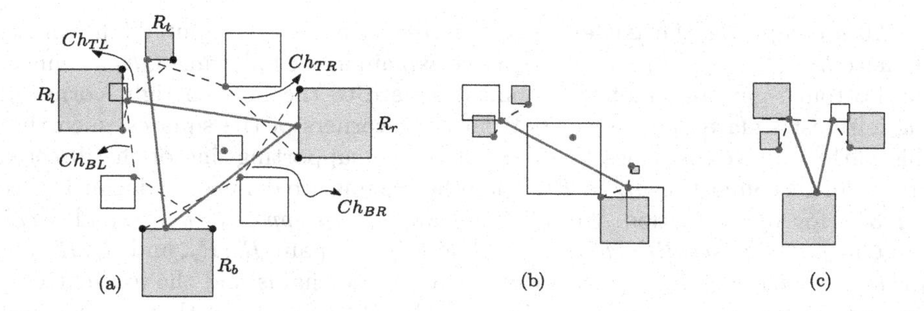

Fig. 3. (a) The axis-extreme squares (shaded squares), and convex chains. Case (iii) of the smallest diameter with two vertices at R_b and R_r, and two vertices at Ch_{TL} and Ch_{TR}. (b) case (i). (c) case (ii). In each case, the optimal placement is shown in purple, and the diameter is denoted by solid line segment (in purple). (Color figure online)

the most similarity among other choices of $\Re^*$ in some sense. As a motivation, one may use this definition to find a cluster of size $n-k$ on $\Re$.

Our Results. Let $\Re$ be a set of n squares/disks, and let $k < n$ be a positive integer. We show that the problem of locating $n-k$ points on distinct squares such that the size of the diameter of the selected points gets minimized takes $O(n^{5.5} \log n)$ time (Sect. 2), locating $n-k$ points on distinct disks (resp. squares) such that the radius of the SEC of the selected points gets minimized takes $O(n^2 \log n)$ (resp. $O(n^3 \log n)$) time (Sect. 3). For large values of k, we introduce a divide-and-conquer algorithm with running time $O(n \log n)$. The problem of locating $n-k$ points on distinct squares such that the area of the bounding box of the selected points gets minimized takes $O(n + k^2(n-k))$ time (Sect. 4). In each particular problem, we also study the maximization/covering version and the problem of computing the second-type outliers.

2 Minimum Diameter

Let $\Re$ be a set of squares. The problem of computing *the smallest diameter* of $\Re$, denoted by d^*, is computing a placement of n points on distinct squares such that this placement has the smallest possible diameter among all possible choices [12]. Our objective is to find a subset $\Re^* \subseteq \Re$ of $n-k$ squares such that the smallest diameter of $\Re^*$ has the smallest possible value among all choices of $\Re^*$.

Löffler and van Krevel [12] proved that in the smallest diameter problem on the squares, except for the four *axis-extreme squares*, we can discretize the problem on the corners. In other words, they proved that the optimal placement of any of the squares except for at most four squares always lies at a corner. They defined the topmost axis-extreme square R_t as a square with the topmost bottom-side. The left, bottom and right axis-extreme squares R_l, R_b, R_r are defined analogously. See Fig. 3(a) as an illustration. Also, for R_t, the bottom side is the candidate of the placement of a vertex of the smallest diameter [12], etc.

After computing the axis-extreme squares, the authors computed four convex chains $Ch_{TL}, Ch_{TR}, Ch_{BL}$ and T_{BR} as we explain in the following: Ch_{TL} connect the bottom right corner of the topmost square to the bottom right corner of the leftmost square via a set of bottom right corners of the squares, such that this path is convex and lies to the right of the supporting line of the directed line segment connecting R_t to R_l. The other chains are defined analogously. See Fig. 3(a) for an illustration. We call Ch_{TL} and Ch_{BR} *opposite chains*, and Ch_{TL} and Ch_{TR} are *consecutive chains*, etc. Similarly, the pairs R_b, R_r, and R_r, R_t, etc. are *consecutive extreme squares*. Computing these chains and the axis-extreme squares takes at most $O(n \log n)$ time [12]. In [12] it is proven that the smallest diameter d^* may occur at multiple pairs of squares at the same time, but it can have at most three different configurations:

Case (i): d^* has two vertices at two opposite chains, see Fig. 3(b).
Case (ii): d^* has one vertex at an axis-extreme square, in which this vertex is in balance between two other vertices on two consecutive chains (or one chain), in which case, the smallest diameter occurs at two pairs at the same time, such that if we move the vertex on the axis-extreme square in either directions, it becomes further from one of the vertices at a convex chain, see Fig. 3(c). We explain this configuration for R_b. It is already proved that for R_b, a candidate of the solution only lies on the top side, and the discrete set of candidate points on the top side of R_b must be computed according to the vertices on Ch_{TL} and Ch_{TR}, which takes $O(n \log n)$ time.
Case (iii): d^* has two adjacent vertices on two consecutive axis-extreme squares, such that they are connected to each other, and each of which is connected to some vertices on a convex chain, in which case, the smallest diameter occurs at three pairs at the same time; see Fig. 3(a). In this configuration, again, the discrete set of candidate points of two axis-extreme squares can be computed in $O(n \log n)$ time [12]. The symmetric configurations of each case will also be treated at the same time.

Discrete Set of Points on Axis-Extreme Squares. For every possible placement of a point on an axis-extreme square, there is one vertex on one of the chains that are furthest away from it, and this determines a candidate for the smallest possible diameter; as the axis-extreme point moves over its edge, this furthest vertex can move only in restricted ways [12], which makes this possible to compute a discrete set of points on a side of an axis-extreme square, such that each of this set can have the complexity $O(n)$ and can be computed in $O(n \log n)$ time [12] (Lemma 4).

Algorithm. Observe that for the set $\Re^*$ of $n-k$ squares of the smallest diameter, we still have four axis-extreme squares. In each step of our algorithm, we first fix four squares R_t, R_l, R_b and R_r as the axis-extreme squares of $\Re^*$, and ignore all the squares that their right (resp. top) sides lie to the left (resp. bellow) of the right (resp. top) side of R_l (R_b). Similarly, we also delete all the squares which lie on an axis-extreme position with respect to R_r and R_t; see the hatched

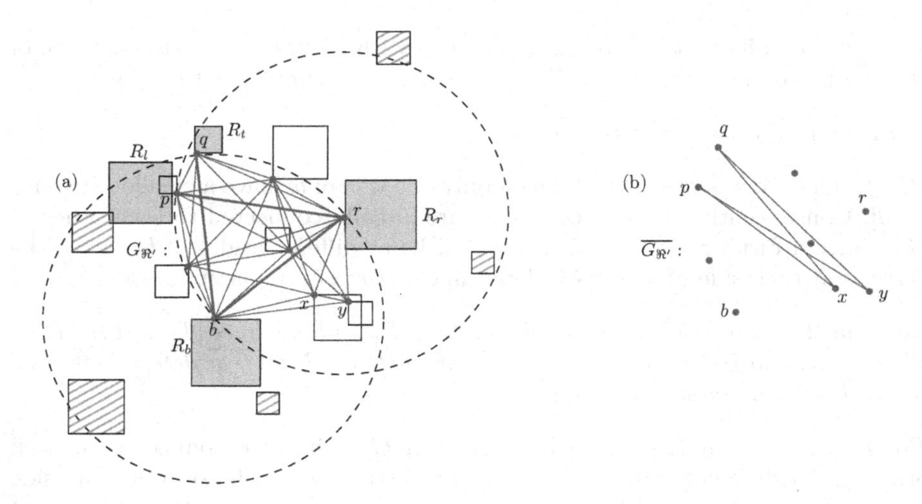

Fig. 4. (a) The graph $G_{\Re'}$ for a selection of R_l, R_t, R_r and R_b as the axis-extreme squares. (b) $\overline{G_{\Re'}}$ is a bipartite graph.

squares in Fig. 4. Let $\Re' \subseteq \Re$ denote the set which has R_t, R_l, R_b and R_r as the axis-extreme squares. We solve the minimum diameter problem of each selection of $\Re'$ independently, and we describe later that how one can determine whether is $\Re'$ a proper $n-k$-subset or not.

First consider the handling of case (iii) on $\Re'$, in which the optimal solution of the smallest diameter of $n-k$ squares has two vertices at R_b and R_r. Note that the symmetric possible cases on $\Re'$ can be treated similarly. Since we have fixed R_t, R_l, R_b and R_r, we can compute $Ch_{TL}, Ch_{TR}, Ch_{BL}$ and Ch_{BR} on $\Re'$, and also the discrete set of points determining the candidate of the vertices of the smallest diameter on R_b and R_r (using Lemma 4 [12]). Computing the chains and the discrete sets takes at most $O(n \log n)$ time [12].

Since we are in case (iii), two vertices at R_b and R_r must be connected directly. Let d denote the computed minimum diameter of $\Re'$, determined by a segment that is connecting a pair $b \in R_b$ and $r \in R_r$, and also by two other connected pairs, each of which is connecting to b or r, as illustrated in Fig. 3(a) and Fig. 4(a). Consider two circles C_b and C_r with radius $|d|$, centered at b and r, respectively. If our selection is a valid choice for the axis-extreme squares of $\Re^*$, there must be at least $n-k$ distinct squares intersecting $C_b \cap C_r$, where each of the pairwise smallest distances is at most $|d|$. Observe that the intersecting squares by $C_b \cap C_r$ contribute a corner on $C_b \cap C_r$. If a square contributes more than one corner, we consider the closest corner to each of b or r, such that for each square, we only consider one corner. Note that it is possible to distinguish between different corners of the squares since we already know that the diameter is at most d. See Fig. 3 as an illustration. Let P' denote the set that contains b, r and all the (possibly selected) corners lying on $C_b \cap C_r$. We make a graph $G_{\Re'}$

on P', such all its vertices (lying on $C_b \cap C_r$) with a smaller pairwise distance of $|d|$, will be connected together. Let $\overline{G_{\Re'}}$ denote the complement of $G_{\Re'}$.

Lemma 1. *$\overline{G_{\Re'}}$ is a bipartite graph.*

Proof. The vertices of P' which lie on only one side of br have a smaller distance of $|d|$. Consequently, the vertices which are already connected to each other at $\overline{G_{\Re'}}$, have a further distance than $|d|$, and lie at different sides of br. Thus the vertices at each side of br in $\overline{G_{\Re'}}$ determine a part of a bipartite graph. □

Lemma 2. *For a selection of $b \in R_b$ and $r \in R_r$ with distance $|d|$, and $R_b, R_r \in \Re'$, a maximum independent set of size (at least) $n - k$ on $\overline{G_{\Re'}}$ determines a set of $n - k$ squares in which the diameter is at most $|d|$.*

Proof. Lemma 1 implies that the vertices of $\overline{G_{\Re'}}$ that are connected to each other lie on different sides of br since the vertices which lie on only one side has a distance smaller than $|d|$, and they are not connected to each other at $\overline{G_{\Re'}}$. Such vertices would be determined by a maximum independent set on $\overline{G_{\Re'}}$. If the maximum independent set has a size at least $n - k$, the pairwise distances of the candidates of this number of squares are at most $|d|$. The lemma follows. □

It is shown that an independent set of maximum size of a bipartite graph of n vertices can be computed in $O(n^{1.5} \log n)$ time [1], in which a similar idea is used for computing a subset with the smallest possible diameter. Thereupon, for a fixed R_l, R_t, R_b and R_r, determining whether they introduce a valid instance can be done in $O(n^{1.5} \log n)$ time, and the smallest possible value of d among all possible configurations of case (iii) of $\Re$ (by considering the freedom of the axis-extreme squares) takes $O(n^{5.5} \log n)$ time.

In case (iii), in each selection of the extreme squares, 3 edges determine d simultaneously, and we considered the existence of a solution of size $n - k$ around br. Let rp and qb denote the two other pairs. It is required to repeat above procedure also for rp and qb; see Fig. 4(a).

For each selection of $\Re'$, we also consider the solutions of case (i) and case (ii), and remember the minimum d for which there are $n - k$ squares to realize a diameter of size at most d. To consider the solution of case (i) on $\Re'$, we consider all possible distinct pairs of the bottom left corners and the upper right corners on Ch_{TL} and Ch_{BR}, respectively, as the candidates to determine d, and construct the graph $G_{\Re'}$, as we discussed in case (iii). In case (ii), d will be determined by a vertex on the discrete set of R_b and one vertex at Ch_{TL} and one vertex at Ch_{TR} (or both vertices at one of these chains). The symmetric configurations have the same statement.

Observe that case (i) and case (ii) also have at most $O(n^2)$ different candidates for d, and can be treated with the same time cost of case(iii).

Theorem 1. *Let $\Re$ be a set of n squares. The problem of computing a subset of $n - k$ squares with the minimum diameter can be solved in $O(n^{5.5} \log n)$ time.*

Maximization Problem. In the maximization version, the solution always occurs at the corners of the squares and there is no need to compute the convex chains [12]. We adjust our algorithm as explained below. First, fix four axis extreme squares as before, and compute the set $\Re'$. We then solve the maximum diameter problem on $\Re'$ [12] and compute d. Then check whether $\overline{G_{\Re'}}$ has an independent of size $n-k$, if so, we update the maximum solution that we have computed so far. It follows that the maximization version can also be solved in $O(n^{5.5}\log n)$ time. As we discussed in the introduction, this algorithm does not work on disks since we cannot compute the discrete sets on disks.

Second-Type Outliers. In the second-type outlier detection problem, a candidate of the optimal solution has two properties: (1) There exists a set $\Re'$ for which the corresponding $\overline{G_{\Re'}}$ graphs on both the maximum and the minimum possible diameter of $\Re'$ has an independent set of size (at least) $n-k$ and (2) the difference between the values of d for the maximum and the minimum possible diameter of $\Re'$ has the smallest possible value among all other choices of $\Re'$.

To compute a set $\Re'$ of these properties, for each selection of $\Re'$, we solve both the maximization and the minimization problems at the same time. If $\Re'$ justifies both conditions, we keep it for the comparison with the other candidates.

Theorem 2. *Let $\Re$ be a set of n squares. For a given k, the problem of computing a subset of size $n-k$ that minimizes the uncertainty gap can be solved in $O(n^{5.5}\log n)$ time.*

3 Smallest Enclosing Circle with Outliers

In this section, we compute the smallest circle which is intersecting or covering at least $n-k$ regions of $\Re$. We first discuss the stabbing problem on a set of disks, and then we generalize the idea to the squares. We also introduce a time and space efficient divide-and-conquer algorithm for large values of k.

$\Re$ Is a Set of Disks. The objective is to find a subset $\Re' \subseteq \Re$, such that $\Re'$ intersects at least $n-k$ disks and has the smallest possible radius among all possible choices of $\Re'$. Observe that for a set of unit disks of radius r, solving the problem on the center of the disks suffices. It is because the distance of the center of the smallest $n-k$-enclosing circle (which is realized by $\Re'$) to the center of at least one disk of $\Re'$ is at least r, otherwise, we still can reduce its radius. This property holds for all possible $n-k$ minimal enclosing circles of $\Re$. Consequently, if we first solve the $n-k$-enclosing circles of the centers of the disks of $\Re$ and reduce the radius of the computed optimal solution by r, the resulting circle is the SEC of $n-k$ disks. From now on, we assume the disks have different sizes.

Our algorithm is based on the parametric search. Suppose we know the radius of the smallest circle that intersects at least $n-k$ squares equals r'. We enlarge any disk R_i of radius r_i to have the radius r_i+r'. Observe that there is an intersection point p on one of the enlarged disks R_i', so that p has a *ply* $n-k$, where the ply of a point is the number of distinct regions containing p. Here we mean p lies in the common intersection of $n-k$ disks. The intersection points of the enlarged disks should be determined by at least two disks. Consequently, we discretize the possible values of r' to a set with complexity $O(n^2)$.

Algorithm. Let X denote the set of all candidates for r', and let c_i denote the center of R_i. For any two disks R_i and R_j with $j \neq i$, let $r' = |c_i c_j|$. If $r' \geq (c_i - c_j - r_i - r_j)/2$ which means if we increase their radius by summing with r', R_i and R_j intersect, we add r' to X. After sorting the elements of X, we binary search on X to find the smallest r' for which there is a point on the boundary of a disk with ply $n-k$. In each iteration, we construct the arrangement of the enlarged disks by an additive radius r', and we consider the intersection points of the boundary of all the disks to find a point with ply $n-k$. The optimal solution would be determined by the minimum value of r' for which there exists a point with ply $n-k$ in the arrangement of the enlarged disks. The complexity of the vertices of the arrangement is $O(n^2)$, and a binary search for finding the smallest r' takes $O(n^2 \log n)$ time and $O(n^2)$ space. Notice that we can extend our algorithm to a set of squares, but for finding the exact location of the center of $n-k$-enclosing disk, we need to solve a linear programming problem at each iteration of the binary search, which increases the running time to $O(n^3 \log n)$. We remark that the discussed complexities can be improved slightly by using the introduced oracles in [7] for the same problem on a set of points.

Theorem 3. *Let $\Re$ be a set of disks (resp. squares). The problem of finding a subset of size $n-k$ of $\Re$ for which the smallest enclosing circle intersects at least $n-k$ regions and has the smallest possible radius can be solved in $O(n^2 \log n)$ (resp. $O(n^3 \log n)$) time and $O(n^2)$ space.*

A Divide-and-Conquer Algorithm for Large Values of k

We design a time and space efficient algorithm for large values of k, that is based on the divide-and-conquer technique and costs $O(n)$ space and $O(n \log n)$ time. We first describe the case where $k = n-3$. Observe that we have a naive $O(n^3)$ algorithm for this problem. In each iteration of the algorithm, we decompose the set of squares into two subsets with equal (or almost equal) size by a vertical line ℓ. Let l^* and r^* denote the radius of the optimal solution of the left and the right subsets of the squares, respectively, as illustrated in Fig. 5. Consider two strips of length $2\times \min(l^*, r^*)$ to the left and to the right of ℓ. We decompose each strip to a set of cells of side length $2\times \min(l^*, r^*)$. Observe that each cell intersects at most 5 squares since otherwise, we can construct a circle which is intersecting three squares with a radius smaller than $\min(l^*, r^*)$ (on one side of ℓ). To find a solution that is smaller than $\min(l^*, r^*)$, we need to consider at most 5 adjacent cells of each cell of the strip. Let m denote the radius of the solution on any 6 adjacent cells of the strip. Observe that m is an SEC with two or three points on its boundary, with at least one point on each side of ℓ. At most 30 squares can intersect 6 adjacent cells. Therefore, computing m takes $O(n)$ time, if we sort the squares (based on their bottom sides) according to y-coordinates,

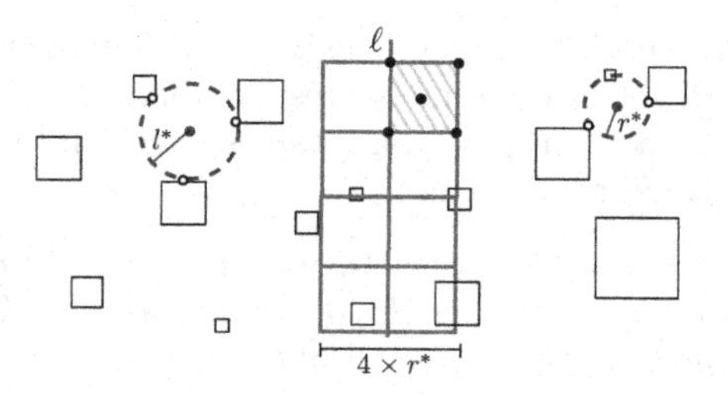

Fig. 5. For $k = n - 3$, there could be at most 5 squares intersecting with a cell of the strip; see the hatched cell. Here, the intersecting squares are points (degenerate squares).

prior to the algorithm. This iteration returns $\min(l^*, r^*, m)$. If a sub-problem has only two or three squares, we simply return the smallest stabbing circle. Then $T(n) = 2T(n/2) + O(n)$, and we conclude that the algorithm runs in $O(n \log n)$ time and $O(n)$ space. Observe that this algorithm carries over to disks and to any $k \in O(n)$, e.g., if $k = n - 4$, each cell cannot intersect more than 8 squares.

Largest Covering Circle with Outliers. We show that computing a smallest $n - k$-enclosing circle which covers the area of at least $n - k$ disks is LP-type. Observe that this problem can be formulated as below

$$Min \;\; r$$

$$s.t. \;\; ||x - c_i|| + r_i \leq r \;\; for \;\; i = 1, \ldots, n,$$

where c_i and r_i are the center and the radius of $R_i \in \Re$, and x and r are the center and the radius of the $n - k$-enclosing circle. According to Theorem 1.2. of [14], above program can be reformulated to satisfy only k constraints in $O(nk^d)$ time, where d equals the geometric dimension of the original problem (here $d = 2$), and this would be performed by finding the optimal solution of $O(k^2)$ independent LP-type problems. Therefore, this problem can be solved in $O(nk^2)$ time. Moreover, this implies that computing k second-type outliers takes $O(nk^2 + n^2 \log n)$ time.

Theorem 4. *The problem of computing the smallest circle which covers the area of at least $n - k$ disks can be solved in $O(nk^2)$ time.*

4 Bounding Box

In the case where there exists a point in the common intersection of $n - k$ squares, the optimal solution to Problem 1 is a single point. This case can be distinguished in $O(n \log n)$ time. From now on, suppose this is not the case. We first look at the problem in $\mathbb{R}^1$. First let $I = \{I_1, \ldots, I_n\}$ be a set of n uncertain points modeled as intervals. For a given k, the objective is to locate $n - k$ points on $n-k$ intervals, so that the length of the interval that intersects $n-k$ intervals gets minimized. For a set $\Re$ of squares (resp. intervals), let $\Re^*_{n-k}$ (resp. I^*_{n-k}) denote the smallest area bounding box (resp. shortest interval) that intersects at least $n-k$ squares (resp. intervals).

Lemma 3. *Let I be a set of n intervals, and let $k < n$ be any positive integer. The interval of minimum length which covers at least $n-k$ points from $n-k$ distinct intervals selects its left (resp. right) endpoint on the rightmost (rest. leftmost) endpoint of an interval in I.*

The lemma can be easily proved by contradiction.

Algorithm. Lemma 3 implies that we can discretize the problem on the endpoints. Then a sliding window idea can be applied to solve the problem. Suppose we have sorted and constructed two sorted lists $\overleftarrow{I}$ (increasing order) and $\overrightarrow{I}$ (decreasing order) of the endpoints of k distinct intervals, from the beginning and the end of I. For a point $p_i \in \overleftarrow{I}$, the candidate point $q_j \in \overrightarrow{I}$ has index $n-k-i$. The algorithm and computing the partially ordered sets take $O(n + k \log n)$ time [6].

2-d Case. Now let $\Re = \{R_1, \ldots, R_n\}$ be a set of squares, the objective is to locate $n-k$ points on $n-k$ distinct squares such that the area of the smallest AABB of at least $n-k$ squares gets minimized. It can be shown that Lemma 3 carries over to the squares, i.e., only the sides of the squares need to be considered.

Observation 1. *For any value of k, $\Re^*_{n-k}$ is entirely located within $\Re^*_n$.*

The same argument also holds for I^*_{n-k} and I^*_n. Similar to the 1-d case, a sliding window algorithm solves the problem, in which we try all possible ways to delete k squares from four directions. We start from $\Re^*_n$, and try to shrink it optimally.

The first candidate of being an outlier from the top direction is the square with the topmost bottom side, since removing any other square does not influence the top side. Also, any other potential outlier from the other 3 directions has the same property. We first compute the k-th square with topmost bottom side in linear time. Let $R_{t'}$ denote it. Then we compute all the squares which have a higher bottom side than $R_{t'}$, and compute a (decreasing) sorted list of them. We similarly compute three other sorted lists of distinct sides of the squares in other directions, each of them consists of k squares. This takes $O(n + k \log n)$ time.

W.l.o.g, in each step of the algorithm we first try removing some squares from the left and right, and then from top and bottom. If we need to remove m squares from left and right, it can be done in $m+1$ different ways. Then the remaining $k-m$ squares can be removed in $k-m+1$ different ways. Checking whether each candidate square is already removed or not can be easily done by assigning a true/false flag to each square. Considering the freedom of the choices in each step, the optimal solution can be found in $O(n + k^2(n-k))$ time.

Theorem 5. *For a given set* $\Re$ *of* n *squares and a positive integer* $k < n$*, the problem of choosing* $n - k$ *squares such that the smallest AABB of those squares has the smallest area can be solved in* $O(n + k^2(n - k))$ *time.*

The Covering Bounding Box. In both 1-d and 2-d cases, we again discretize the problem on the endpoints of the intervals or the sides of the squares, and adjust the minimization algorithms for the covering versions. In particular, in $\mathbb{R}^1$, the interval of minimum length that covers at least $n - k$ distinct intervals completely selects its left (resp. right) endpoint on the leftmost (rest. rightmost) endpoint of an interval in I, etc. All the extensions to the disks are obvious.

Second-Type Outliers. In the second-type outlier detection problem, (1) we should find a set $\Re'$ for which $\Re'$ is intersecting with at least $n - k$ squares and (2) the difference between the area of the smallest-area intersecting and the smallest-area covering AABB of $\Re'$ has the smallest possible value. Since both the minimization and the covering problems can be solved in $O(n + k^2(n - k))$ time, the second-type outliers of size k can be detected at the same time.

Acknowledgment. V. Keikha was supported by the Czech Science Foundation, grant number GJ19-06792Y, and with institutional support RVO: 67985807.

References

1. Aggarwal, A., Imai, H., Katoh, N., Suri, S.: Finding k points with minimum diameter and related problems. J. Algorithms **12**(1), 38–56 (1991)
2. Ahn, H.-K., et al.: Covering points by disjoint boxes with outliers. Comput. Geom. **44**(3), 178–190 (2011)
3. Atanassov, R., et al.: Algorithms for optimal outlier removal. J. Discret. Algorithms **7**(2), 239–248 (2009)
4. Bae, S.W.: Computing a minimum-width square or rectangular annulus with outliers. Comput. Geom. **76**, 33–45 (2019)
5. Chan, T.M., Har-Peled, S.: Smallest k-enclosing rectangle revisited. Discret. Computat. Geom. 1–23 (2020)
6. Daskalakis, C., Karp, R.M., Mossel, E., Riesenfeld, S.J., Verbin, E.: Sorting and selection in posets. SIAM J. Comput. **40**(3), 597–622 (2011)
7. Efrat, A., Sharir, M., Ziv, A.: Computing the smallest k-enclosing circle and related problems. Comput. Geom. **4**(3), 119–136 (1994)
8. Eppstein, D., Erickson, J.: Iterated nearest neighbors and finding minimal polytopes. Discret. Comput. Geom. **11**(3), 321–350 (1994). https://doi.org/10.1007/BF02574012
9. Har-Peled, S., Mazumdar, S.: Fast algorithms for computing the smallest k-enclosing circle. Algorithmica **41**(3), 147–157 (2005)
10. Kaplan, H., Roy, S., Sharir, M.: Finding axis-parallel rectangles of fixed perimeter or area containing the largest number of points. Comput. Geom. **81**, 1–11 (2019)
11. Keikha, V., Löffler, M., Mohades, A.: A fully polynomial time approximation scheme for the smallest diameter of imprecise points. Theor. Comput. Sci. **814**, 259–270 (2020)

12. Löffler, M., van Kreveld, M.: Largest bounding box, smallest diameter, and related problems on imprecise points. Comput. Geom. **43**(4), 419–433 (2010). Special Issue: 10th Workshop on Algorithms and Data Structures (WADS 2007)
13. Löffler, M.: Data imprecision in computational geometry. Ph.D. thesis, Utrecht University (2009)
14. Matoušek, J.: On geometric optimization with few violated constraints. Discret. Comput. Geom. **14**(4), 365–384 (1995). https://doi.org/10.1007/BF02570713
15. Shenmaier, V.: The problem of a minimal ball enclosing k points. J. Appl. Ind. Math. **7**(3), 444–448 (2013)

A Worst-Case Optimal Algorithm to Compute the Minkowski Sum of Convex Polytopes

Sandip Das[1], Subhadeep Ranjan Dev[1(✉)], and Swami Sarvottamananda[2]

[1] Indian Statistical Institute, Kolkata, Kolkata, India
sandipdas@isical.ac.in, info.subhadeep@gmail.com
[2] Ramakrishna Mission Vivekananda Educational and Research Institute, Howrah, Howrah, India
sarvottamananda@rkmvu.ac.in

Abstract. We propose algorithms to compute the Minkowski sum of a set of convex polytopes in $\mathbb{R}^d$. The input and output of the proposed algorithms are the face lattices of the input and output polytopes respectively. We first present the algorithm for the Minkowski sum of two convex polytopes. The time complexity of this algorithm is $O(d^\omega nm)$ where n and m are the face lattice sizes of the two input polytopes and ω is the matrix multiplication exponent ($\omega \sim 2.373$). Our algorithm for two summands is worst-case optimal for fixed d. We generalize this algorithm for r convex polytopes, say P_i, $1 \leq i \leq r$. The time complexity of this generalization is $O(\min\{d^\omega NM, d^\omega r \prod |P_i|\})$ where $N = \sum |P_i|$ is the total size of the face lattices of the r input polytopes and M is the size of the face lattice of their Minkowski sum $P_1 \oplus \cdots \oplus P_r$. Our algorithm for multiple summands is worst-case optimal for fixed $d \geq 3$ and $r < d$.

1 Introduction

The Minkowski sum, initially defined by Hermann Minkowski [1864–1909], is an important concept, useful in computational geometry, computer graphics, robot motion planning, assembly planning, computer-aided design, computer-aided manufacturing and various other fields. The abundant literature from the late nineteenth century on the subject corroborates the importance and applicability of the concept. Recently Das et al. [10] implicitly used the Minkowski sums to compute the diameter, the width, the minimum enclosing/stabbing sphere, the maximum inscribed sphere and the minimum enclosing/stabbing cylinder for various types of inputs, some involving convex polygons/polytopes, for polyhedral distance functions [9]. The present algorithm was motivated by it.

The Minkowski Sum of two or more convex polytopes in $\mathbb{R}^d$ can be computed efficiently by a simple application of the optimal convex hull algorithm by Chazelle [6] or the output sensitive algorithm by Seidel [20]. The time complexities are $O(n^{\lfloor d/2 \rfloor})$ and $O(n^2 + h \log h)$ respectively for fixed d where n is the total number of Cartesian sums of the vertices of the input polytopes and

A. Mudgal and C. R. Subramanian (Eds.): CALDAM 2021, LNCS 12601, pp. 179–195, 2021.
https://doi.org/10.1007/978-3-030-67899-9_14

h is the size of the resultant Minkowski sum. For the sub-linear output size, an improved algorithm by Chan [5] can be used instead. The Minkowski sum computing algorithm using the results by Chazelle is a worst-case optimal algorithm, only when the summands and the sum, all are asymptotically worst-cases for the convex hulls as well as for the Minkowski sum. The Minkowski sum of two convex polygons in $\mathbb{R}^2$ can be computed optimally in linear time by a well known algorithm [4]. Bekker et al. [3] gave an optimal output sensitive algorithm, linear in the output size, to compute the Minkowski sum of two convex polyhedra in $\mathbb{R}^3$. In the case of $\mathbb{R}^d$, Fukuda et al. [14] presented an alternative non-optimal output sensitive algorithm for enumerating the faces in the Minkowski sum of convex polytopes. Their polynomial-time per-face algorithm did not improve upon the method using Seidel's $O(n^2 + h \log h)$-time or Chan's sub-linear sized convex hull algorithm for output sensitive computation. Agarwal et al. [2], and later Fogel et al. [12], devised the algorithms to compute the Minkowski Sum of non-convex polygons in $\mathbb{R}^2$ and non-convex polyhedra in $\mathbb{R}^3$, respectively, using techniques of decomposition of non-convex polygons and polyhedra, respectively, by Chazelle et al. [7,8].

In addition to the algorithmic results above, there are also several combinatorial results on the Minkowski sum of convex polytopes in $\mathbb{R}^d$. Grunbaum [16] proved that in the worst case, for $d \geq 3$, the number of faces of the Minkowski sum of two convex polytopes is the product of the number of faces of its two summands. This important fact contributes for the optimality of our proposed worst-case optimal algorithm. Adiprasito [1], Gritzmann et al. [15], Karavelas et al. [17], Weibel [21] and others derived various upper and lower bounds for the Minkowski sums.

The different approaches to compute the Minkowski sum of convex polytopes are to compute convex hulls [5,6,20], to compute the boundary faces of the resultant polytope using LPP [13,14] or to use convolutions [3,4,19,22]. The convolution method is well understood in the case of $\mathbb{R}^2$ and $\mathbb{R}^3$ but not in higher dimensions. In this paper, we compute the Minkowski sum of convex polytopes in $\mathbb{R}^d$ using their face lattices [5,6,11,20], defined in Sect. 2, as input and output for traversal. Our traversal of the Minkowski sum is similar to the convolution method.

Our contributions in this is paper are two worst-case optimal algorithms for fixed d to compute the face lattices of Minkowski sums for two and multiple convex polytopes, respectively, in $\mathbb{R}^d$. We are able to achieve worst-case optimal bounds in both the algorithms by devising efficient criteria to check the boundary faces of the Minkowski sum. For this check, we first characterize the faces and the incidences of the Minkowski sum in terms of the face lattice by providing algorithm specific necessary and sufficient conditions for faces (Lemma 1, Lemma 2 and Lemma 3) and for incidences (Lemma 6). A novel use of Lemma 4 is crucial to avoid duplicate faces. We use these characterizations to effectively remove the duplicate or invalid faces and incidences, on the one hand, and identify the valid faces and incidences of the Minkowski sum uniquely, on the other. Secondly, we augment the face lattices in a novel way. This allows us to apply the sufficiency conditions in $O(d^\omega)$ time per candidate face of the Minkowski sum (Lemma 2(ii) and Lemma 3(ii)). The quantity ω, $\omega \sim 2.373$, in d^ω is the matrix multiplication exponent [18] and d^ω is the complexity of computing the rank of the $d \times 2d$ matrices in the two algorithms. The other significant points to note in this paper are an efficient face lattice traversal method independent of adjacency information and an effective table search for finding duplicates or identifying new faces (Sect. 4.2).

A brief summary of time complexities in our algorithms is as follows. Let P and Q be two convex polytopes in $\mathbb{R}^d$ with face lattice sizes n and m, respectively. Our algorithm for two summands computes the face lattice of the Minkowski sum $P \oplus Q$ of P and Q in $O(d^\omega nm)$ time (Theorem 2). This running time is worst-case optimal for any fixed $d \geq 3$. For $d \leq 2$, a trivial modification to our algorithm suffices to compute the face lattice of $P \oplus Q$ in $O(n + m)$ time. Furthermore, let $P_1, P_2, \ldots, P_r$ be r convex polytopes in $\mathbb{R}^d$ with respective face lattice sizes $n_1, n_2, \ldots, n_r$. Let $N = \sum_{i=1}^{r} n_i$ and let M denote the size of the face lattice of the Minkowski sum, $\bigoplus_{i=1}^{r} P_i$, of the r polytopes. We generalize our previous algorithm to compute the face lattice of the Minkowski sum $\bigoplus_{i=1}^{r} P_i$ of r summands in time $O\left(\min\{d^\omega NM, d^\omega r \prod_{i=1}^{r} n_i\}\right)$ (Theorem 3). This running time is optimal for fixed d when $r < d$ and is our primary contribution in this paper. Note that the size of the Minkowski sum for $r \geq d$ is bounded by the sum of sizes of $(d-1)$-subset Minkowski sum among r polytopes.

The remaining paper is organized as follows. In Sect. 2 we provide a few preliminary definitions and notations used throughout the paper. In Sect. 3 we describe an augmentation to the input face lattices and develop some basic methods needed by our algorithms. Next, in Sect. 4 we present our first algorithm to compute the Minkowski sum of two convex polytopes in $\mathbb{R}^d$. In the same section, we also present various geometric results on the face lattice of convex polytopes and their Minkowski sum that prove the correctness and complexity of our algorithm. In Sect. 5 we generalize the method used in our first algorithm to efficiently compute the Minkowski sum of multiple convex polytopes in $\mathbb{R}^d$.

2 Preliminaries

Let S_1 and S_2 be two sets of points in $\mathbb{R}^d$. The *Minkowski sum* of these two sets, denoted by $S_1 \oplus S_2$, is defined as the set $\{p \mid p = p_1 + p_2, \forall p_1 \in S_1$ and $\forall p_2 \in S_2\}$ where $p_1 + p_2$ denotes the sum of the positional vectors of p_1 and p_2. This definition is equally applicable to the Minkowski sum of convex polytopes or, as a matter of fact, to any geometric objects as they can be equivalently seen as sets of points. We assume, for the sake of simplicity, that the origin of reference lies strictly in the interior of all the input convex polytopes, i.e, not on the boundary or in the exterior. However, note that the illustrations in the paper, for the sake of clarity, are in general positions.

It is a well-known fact that the Minkowski sum of two convex sets is also convex. Extending this to polytopes we have the following observation (see Fig. 1 for an example).

Observation 1. The Minkowski sum of any two convex polytopes is also a convex polytope.

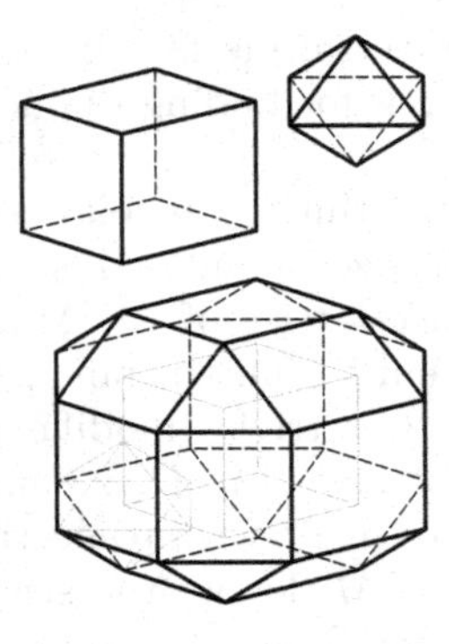

Fig. 1. Minkowski sum of two convex polytopes in $\mathbb{R}^3$.

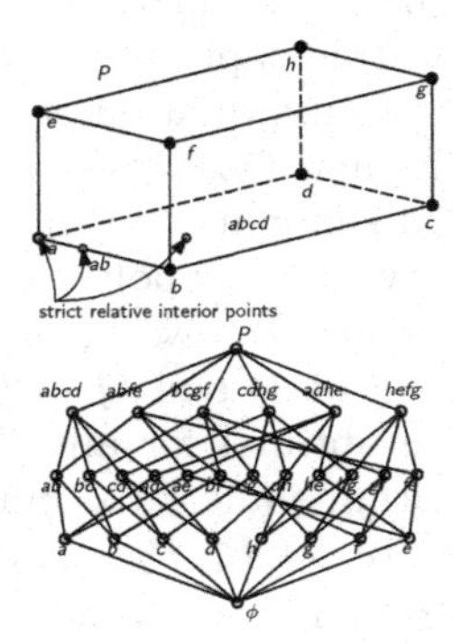

Fig. 2. Face lattice $\mathscr{L}(P)$ as a Hasse diagram of the cuboid P. For example, $ab \prec abcd$ as well as $ab \succ abcd$.

Let S be a set of points in $\mathbb{R}^d$. The *affine space* of S, denoted by *affine*(S), is the affine combinations of the points in S, i.e., the set $\{p \mid p = \sum_{i=1}^{k} \lambda_i p_i, \forall i, 1 \leq i \leq k, p_i \in S, \lambda_i \in \mathbb{R}$ and $\sum_{i=1}^{k} \lambda_i = 1\}$. Every affine space A has an associated *affine dimension* which is the cardinality of its basis and is denoted by *dim*(A). In our algorithms we represent an affine space A by a point in A and any of its several bases consisting of *dim*(A) number of basis vectors.

We represent a *hyperplane* h in $\mathbb{R}^d$ by an equation of the form $\vec{x} \cdot \vec{n} = c$, where $\vec{n}$ is the normal to h and c is a fixed constant. The two open halfspaces determined by h are $\vec{x} \cdot \vec{n} > c$ and $\vec{x} \cdot \vec{n} < c$ and are denoted by h^+ and h^- respectively. We say that a hyperplane h is a *supporting hyperplane* of a polytope P if $P \cap h \neq \varnothing$ and $P \cap h^+ = \varnothing$, i.e., P touches h and its interior lies completely in h^- because of our assumption that origin lies in the interior.

A *nontrivial face* of a polytope P is the intersection of P with a supporting hyperplane h of P. The only two *trivial faces* of P are the face P itself and the null face ϕ. Observe that the faces of any convex polytope are also convex. Moreover, for a face f and a supporting hyperplane h of P if $f = P \cap h$ we call h the *supporting hyperplane* of P *on* f. We have the following observation.

Observation 2. Every supporting hyperplane h of a d dimensional polytope P corresponds to exactly one face of P and every nontrivial face f of P has one supporting hyperplane if f is a $d-1$ dimensional face or infinite supporting hyperplanes otherwise.

We call a $(k-1)$-dimensional face, say f, of a k-dimensional face F a *facet* of F. We denote this by $f \prec F$ and $F \succ f$. A face f is called a *subface* of a face

F and F a *superface* of f if there exists a sequence of faces $f_1, f_2, \ldots, f_i$ (i can be 0) such that $f \prec f_1 \prec f_2 \cdots \prec f_i \prec F$.

Every face f of P has a corresponding affine space *affine*(f) and a corresponding affine dimension *dim*(f). For brevity, we use the same terms that we used for point sets earlier. In subsequent sections, we represent an affine space A, with $dim(A) = k$, by the tuple $\langle \vec{p}, \vec{u_1}, \vec{u_2}, ..., \vec{u_k} \rangle$, if the affine space A corresponds to the parametric equation $\vec{x}(\vec{t}) = \vec{p} + \sum_{i=1}^{k} t_i \cdot \vec{u_i}$ where $\vec{p}$ is the vector representation of a point in A, the set $\{\vec{u_1}, \vec{u_2}, \ldots, \vec{u_k}\}$ is a vector basis of A, the vector $\vec{t} = (t_1, t_2, \ldots, t_k)$ is the parameter with $\vec{t} \in \mathbb{R}^k$ and $\vec{x}(\vec{t})$'s are points of the affine space A.

We assume that the input consists of the face lattices of summand convex polytopes and either $\mathcal{H}$-representations or the $\mathcal{V}$-representations of the summands. We briefly describe the face lattice of any convex polytopes and a linear-time augmentation of any face lattice, that we need for the algorithm, in the next section.

3 Augmented Face Lattice of Convex Polytopes

3.1 Face Lattice of a Convex Polytope

Let P be a convex polytope of affine dimension δ, where $\delta \leq d$, i.e. the polytope P may be in a subspace of $\mathbb{R}^d$. We first define the face lattice [5,6,11,20] of any convex polytope for reference.

Definition 1. *The* face lattice, *denoted by $\mathscr{L}(P)$, of a convex polytope P is a Hasse diagram where the nodes are faces of P including a unique null face ϕ and the polytope P. Moreover, the reflexive, anti-symmetric and transitive binary relation $\prec^*$ or, equivalently, $\succ^*$, the transitive closure of $\prec$ or $\succ$, respectively, is used as the partial order relation for the Hasse diagram.*

For our purpose, we represent the Hasse diagram of the face lattice as a layered graph in the algorithm, with the nodes arranged in layers according to the dimension k, $-1 \leq k \leq \delta$, of their faces. See Fig. 2 for the face lattice of a cuboid in $\mathbb{R}^3$. Though we need additional description of the faces in the nodes in many applications, in this paper, we only need either the $\mathcal{H}$-representation of the polytope stored in $(\delta - 1)$-dimensional faces, i.e. facets, or the $\mathcal{V}$-representation of the polytope stored in 0-dimensional faces, i.e. vertices. Any representation that allows us to augment the face lattice described below is acceptable. We briefly describe the data structure representing $\mathscr{L}(P)$ below.

3.2 Data Structure for a Face Lattice

For convenience, we denote the set of all faces of the polytope P by $\mathscr{F}(P)$ and its subsets consisting of faces of dimension k by $\mathscr{F}_k(P)$, $-1 \leq k \leq \delta$. The face lattice of P has nodes arranged in $(\delta+2)$ layers where the nodes in the k-th layer represent the faces in $\mathscr{F}_k(P)$. First, we denote the face corresponding to a node x

by *face*(x), and, vice versa, the node corresponding to a face f by *node*(f). In this paper, due to one-to-one mapping, the nodes and their corresponding faces refer to each other uniquely and we use them interchangeably in the discussions. Next, for the sake of brevity, *dim*(x) and *affine*(x) for a node x denote the dimension and affine space respectively of the face *face*(x). We store *dim*(x) and the tuple representing *affine*(x) in each node x, except in the node of the null face ϕ, for which *dim*$(x) = -1$. Each node x also maintains two lists, $super(x)$ and $sub(x)$ where the list $super(x)$ is the list of all nodes x' such that *face*$(x') \succ$ *face*(x) and $sub(x)$ is the list of all nodes x' such that *face*$(x') \prec$ *face*(x). We also store a strict relatively interior point of the face of the node, denoted by *point*(x), in each node x, except the node of null face ϕ. By *relative interior*, we mean that the point is an interior, i.e. non-boundary, point of face *face*(x) relative to its affine space and by *strict*, we mean that the point is not a relative interior point of any of the proper subfaces of *face*(x). See Fig. 2 for illustration of a strict relatively interior point. In the discussion below, we show how we augment the face lattice $\mathscr{L}(P)$ to include *point*(x) and *affine*(x) in each node x, in time linear in the size of the face lattice. We assume that the face lattice inherently has *dim*(x), $sup(x)$ and $super(x)$ stored in its each node x.

3.3 Augmenting a Face Lattice

In literature, the $\mathcal{V}$-representation of the convex polytope P is the minimal set of points whose convex hull is the polytope P. An alternative representation is the $\mathcal{H}$-representation of P which is the set of minimal halfspaces whose intersection is the polytope P. Both the representations equivalently and uniquely describe the polytope P for our purpose. The conversion from one representation to another requires a simple top-down or bottom-up traversal of the face lattice. We can do this in linear time. Thus, we can augment the face lattice of the polytope P by the method given below, if the input contains either the $\mathcal{H}$-representation or the $\mathcal{V}$-representation of the polytope P.

In the following discussion, we assume that the $\mathcal{V}$-representation of the convex polytope P, equivalently the coordinates of all the vertices in $\mathscr{F}_0(P)$, is provided. We do a bottom-up traversal on $\mathscr{L}(P)$ to augment the face lattice with *point*(x) for each node x. Initially, for each node x corresponding to vertices, i.e. the faces in $\mathscr{F}_0(P)$, we set *point*(x) to be the coordinates of the vertex determined by the $\mathcal{V}$-representation. Then at intermediate steps, at each node x with *dim*$(x) > 0$ we arbitrarily select two nodes y and y' in the list $sub(x)$ and set *point*(x) to be the middle point $(point(y) + point(y'))/2$. We can easily show that *point*(x) is a strict relative interior point of face *face*(x). It is interior because of convexity, relative because of being an affine combination of points in subfaces and strict because the affine spaces of distinct subfaces are distinct.

Next, we show how we augment in linear time the face lattice to include the affine spaces of all the nodes required by our algorithm.

As mentioned earlier the parametric equation of the affine space of a k-dimensional face f is represented by the $(k+1)$ tuple $\langle \vec{p}, \vec{u_1}, \vec{u_2}, \ldots, \vec{u_k} \rangle$. In this representation, $\vec{p}$ is the vector representing a strict relative interior point of f and the set $\{\vec{u_1}, \vec{u_2}, \ldots, \vec{u_k}\}$ is a basis of *affine*(f). We show below how we compute the tuple $\langle \vec{p}, \vec{u_1}, \vec{u_2}, \ldots, \vec{u_k} \rangle$ representing the affine space *affine*(x) for any node x in $\mathscr{L}(P)$. We first make the following observation.

Observation 3. If f and g are two faces of P such that $f \succ g$ then any point in the strict relative interior of the face f is *affinely independent* of the points in the face g, i.e., no strict relative interior point of the face f is an affine combination of points in the face g.

In short, the observation above is the consequence of the fact that the affine dimension increases as we add the strict relative interior points of the superfaces. Thus, in the pseudocode below, we compute the affine space of each of the subfaces of *face*(x) in $\mathscr{L}(P)$ recursively, in linear time.

```
procedure AugmentAffine(x)
  if affine(x) is unset then
    if x is a vertex then
      Set affine(x) for the vertex appropriately;
    else
      foreach y ∈ sub(x) do
        AugmentAffine(y)
      Choose a node y arbitrarily from sub(x);
      Set affine(x) as affine(y) with additional basis vector
      point(y) − point(x);
```

The correctness of the recursive algorithm follows from the observation above. We, therefore, have the following theorem.

Theorem 1. *We can augment the face lattice $\mathscr{L}(P)$ of the convex polytope P with affine dimensions, affine spaces and strict relative interior points in linear time for either the $\mathcal{V}$-representation or the $\mathcal{H}$-representation of the polytope P. The space required is $O(d^2 \cdot |\mathscr{L}(P)|)$.*

Once we compute the affine spaces of the faces of the summand convex polytopes, we can readily compute the affine dimension and the affine space of the Minkowski sum of the faces, which may be less than the sum of individual dimensions of the summand faces. This is required to determine if the sum is a valid face and then, if it is, to determine the appropriate layer of the face lattice of the sum. We show next how we compute the affine dimension.

3.4 Determination of the Dimension of the Minkowski Sum of Faces

Let P and Q be two convex polytopes with face lattices $\mathscr{L}(P)$ and $\mathscr{L}(Q)$. Let x be a node in $\mathscr{L}(P)$ and x' a node in $\mathscr{L}(Q)$. We show in pseudocode below, how we determine the affine dimension of *face*$(x) \oplus$ *face*(x'). We note that *face*$(x) \oplus$ *face*(x') may not be a valid face of $P \oplus Q$. It may be a partial face or it may partially lie inside P.

We also note that for any two faces f in P and g in Q the affine space *affine*$(f \oplus g)$ = *affine*$(f) \oplus$ *affine*(g). This follows from the definition of the affine spaces and the Minkowski sum. So, briefly, we consider the combined set of affinely independent vectors in *affine*(x) and *affine*(x') and return the resulting rank as the dimension of $f \oplus g$. The function RANK( ) in our algorithm computes the rank of a matrix using the fastest rank determination algorithm. We give below the pseudocode to determine the dimension of $f \oplus g$ where $f =$ *face*(x) and $g =$ *face*(x') in $\mathscr{L}(P)$ and $\mathscr{L}(Q)$ respectively.

function DIM(x, x')
  $\langle \vec{p}, \vec{u_1}, \vec{u_2}, \ldots, \vec{u_k} \rangle \leftarrow$ *affine*(x);
  $\langle \vec{p'}, \vec{u'_1}, \vec{u'_2}, \ldots, \vec{u'_{k'}} \rangle \leftarrow$ *affine*(x');
  $r \leftarrow$ RANK$[\vec{u_1}\ \vec{u_2}\ \ldots\ \vec{u_k}\ \vec{u'_1}\ \vec{u'_2}\ \ldots\ \vec{u'_{k'}}]$;
  return r;

The runtime complexity of the function DIM is d^ω where ω is the matrix multiplication exponent [18]. We will use this time complexity later. The correctness of the function above follows from the fact that *affine*$(A \oplus B)$ = *affine*$(A) \oplus$ *affine*(B) for any two sets of points A and B. We use the methods described above in the subsequent sections to compute the Minkowski sum of convex polytopes.

4 The Minkowski Sum of Two Convex Polytopes

Let P and Q be two convex polytopes in $\mathbb{R}^d$ with the face lattice representations $\mathscr{L}(P)$ and $\mathscr{L}(Q)$. The face lattice sizes are n and m respectively. In this section, we present an $O(d^\omega nm)$-time and $O(d^2nm)$-space algorithm to compute the Minkowski sum $P \oplus Q$. In Sect. 4.1 we first present some important characterizations for the Minkowski sum $P \oplus Q$ necessary for our algorithm. We then present the main algorithm of the algorithm in Sect. 4.2.

In this section, we assume unless mentioned otherwise, that the affine dimension of the polytope $P \oplus Q$, i.e., *dim*$(P \oplus Q)$, is d. Otherwise, if *dim*$(P \oplus Q) < d$, we compute the Minkowski sum in lower affine dimension by projecting the input to the affine space of the polytope $P \oplus Q$.

4.1 Necessary and Sufficient Conditions for the Faces of $P \oplus Q$

First we provide a necessary condition for the faces of the polytope $P \oplus Q$. The following lemma is a well known fact so the proof is omitted.

Lemma 1 (Necessary Condition for a face F in $P \oplus Q$). *Let P and Q be convex polytopes. If F is any face of $P \oplus Q$ then there exist a unique face f in P and a unique face g in Q such that $F = f \oplus g$.*

As a consequence of Lemma 1, all the faces of the polytope $P \oplus Q$ are of the form $f \oplus g$ where f is a face of the polytope P and g is a face of the polytope Q. However, not all possibilities of $f \oplus g$ are faces of the polytope $P \oplus Q$. To remove the invalid possibilities we need sufficient conditions. By definition of a face, for $f \oplus g$ to be a face of $P \oplus Q$, a sufficient condition is that there must be a supporting hyperplane h of $f \oplus g$. We state the condition below.

Observation 4 (Sufficient condition). Let f and g be faces of convex polytopes P and Q respectively. The Minkowski sum $f \oplus g$ is a face of polytope $P \oplus Q$, if there exists a supporting hyperplane h such that $(P \oplus Q) \cap h^+ = \varnothing$ and $(P \oplus Q) \cap h = f \oplus g$.

This condition, however, can not be efficiently checked. So it is not practical, We need a computationally efficient sufficient condition. Therefore we present alternative conditions. First, we suggest a sufficient condition for $f \oplus g$ to be a facet of polytope $P \oplus Q$ and then we generalize the condition for other lower dimensional faces. For facets of $P \oplus Q$, the condition is simplified because there is only one choice of supporting hyperplane h, $h = \mathit{affine}(f \oplus g)$, since the affine dimension will be $d - 1$ because of our assumption, i.e., the affine space will be the supporting hyperplane because it happens to be the only choice. Now, the hyperplane h will be a supporting hyperplane of $f \oplus g$, if, equivalently (1) every point of the polytope $P \oplus Q$, other than of $f \oplus g$, is in h^-, (2) every vertex in the polytope $P \oplus Q$, other than the vertices of $f \oplus g$, is in h^-, since due to convexity every point in $P \oplus Q$ is a convex combination of its vertices, or (3) every strictly relative interior point of possible $f_i \oplus g$ and $f \oplus g_j$ are in h^- where f_i's and g_j's are the immediate superfaces of f and g respectively. The criterion in (3) can be computed efficiently and the equivalence of the three criteria can be proved. Summarily, the idea of the proof is that if $f \oplus g$ is not a facet then at least one of the several $f \oplus g_j$'s and $f_i \oplus g$'s will have its interior outside h^-. We present the sufficient condition of (3) in the following lemma. We omit the proof because of space constraint.

Lemma 2 (Sufficient condition for a facet $f \oplus g$ in $P \oplus Q$). *Let f be a face of a convex polytope P and g be a face of a convex polytope Q. The Minkowski sum $f \oplus g$ is a facet of polytope $P \oplus Q$, if*

(i) $\mathit{dim}(f \oplus g) = d - 1$, *and*
(ii) $\forall f'$, $f' \succ f$, $\mathit{point}(f') + \mathit{point}(g) \in h^-$ *and* $\forall g'$, $g' \succ g$, $\mathit{point}(f) + \mathit{point}(g') \in h^-$ *where h is* $\mathit{affine}(f \oplus g)$.

Once we have a sufficient condition for the facets, we can get sufficient conditions for the other faces by treating them as the facets of any of their immediate superfaces. We can effectively and efficiently do this by checking the conditions relative to the affine space of the immediate superfaces. So, if we are able to compute all the facets of $P \oplus Q$ correctly, then we can also compute the $(d-2)$-faces of the polytope $P \oplus Q$ by checking the sufficient conditions, then we can continue to $(d-3)$-faces, and so on. This will allow us to check sufficient conditions for the faces of all affine dimensions. We show later how we do this efficiently. We present the generalized sufficient condition in the lemma below. As mentioned earlier, the face $f \oplus g$, in the lemma must be part of a known face of $P \oplus Q$ to be its facet. The proof, which is similar to that of Lemma 2 is omitted.

Lemma 3 (Sufficient condition for a face $f \oplus g$ in $P \oplus Q$). *Let $F \oplus G$ be a face of polytope $P \oplus Q$ where F and G are faces of convex polytopes P and Q respectively. Let f and g be any faces of P and Q, such that $f \prec F$ and $g \prec G$ respectively and let $dim(f \oplus g) = dim(F \oplus G) - 1 = k$. Then $f \oplus g$ is a facet of $F \oplus G$, and hence is a face of $P \oplus Q$, if,*

(i) $dim(f \oplus g) = k$, and
(ii) $\forall f'$, $F \succ \cdots \succ f' \succ f$, $point(f') + point(g) \in A^-$ and $\forall g'$, $G \succ \cdots \succ g' \succ g$, $point(f) + point(g') \in A^-$ where A is the hyperplane, $A = affine(f \oplus g)$, in the relative subspace $affine(F \oplus G)$, i.e., in $\mathbb{R}^{k+1}$.

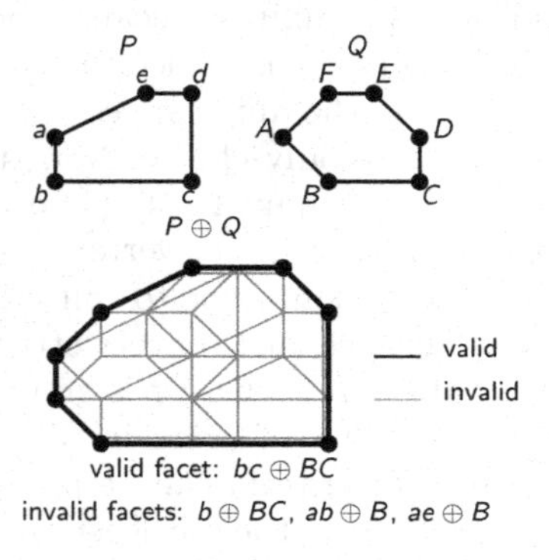

Fig. 3. Invalid sums $f \oplus g$ can be partial facets, partially on the boundary or completely in the interior.

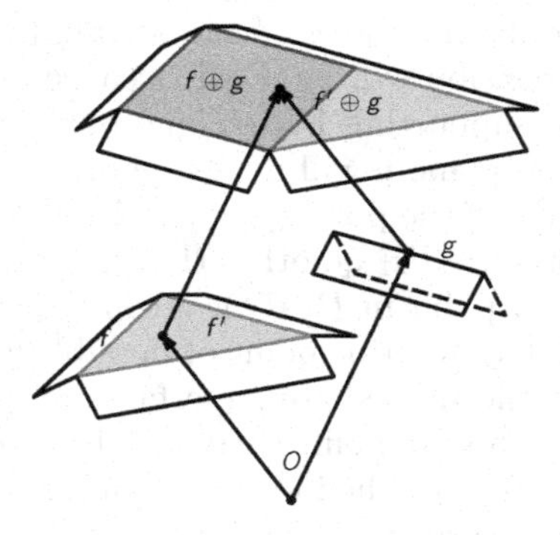

Fig. 4. Invalid partial facet $f \oplus g$ not satisfying the sufficiency condition, i.e. $point(f' \oplus g) \notin h^-$.

The necessity and sufficiency lemmas presented above, unfortunately, do not remove the possibility that same face of the polytope $P \oplus Q$ is computed multiple times. We need an efficient method to identify duplicates. Even after identifying duplicates correctly, we also need an efficient method to add missing incidences at a later stage correctly. First we note, owing to Lemma 1, that multiple $f_i \oplus g_i$'s can not give the same face of $P \oplus Q$. Second, if $f_i \oplus g_i$'s refer to partial faces on the boundary of $P \oplus Q$ with same affine space then the suprema of f_i's and g_i's will be faces F and G respectively, such that $F \oplus G$ is a face of $P \oplus Q$ with the same affine space. This allows us to add a node corresponding only to the Minkowski sum of suprema, $F \oplus G$, when we do a top down computation. We present this property in the following lemma needed in the algorithm to halt the top down processing (Figs. 3 and 4).

Lemma 4. *Let f be a face of P and g be a face of Q such that affine$(f \oplus g) = A$. Let h be a supporting hyperplane on A, i.e., $(P \oplus Q) \cap h \subseteq A$, and $(P \oplus Q) \cap h^+ = \varnothing$. Then there exist suprema faces F and G of P and Q respectively, such that $f \prec \cdots \prec F$, $g \prec \cdots \prec G$ respectively, affine$(F \oplus G) = A$ and $F \oplus G$ is a face of $P \oplus Q$.*

The proof of Lemma 4 is omitted. An immediate consequence of the lemma is all $f_i \oplus g_i$'s that account for partial faces will have a proper face $F \oplus G$ which is unique and therefore in a top down computation, once we compute a proper face of $P \oplus Q$, we neither encounter any partial faces nor we compute the same face twice. This is the consequence of the following lemma which we state without the proof.

Lemma 5. *Let f and f' be two faces of P and g and g' be two faces of Q. If $f \oplus g$ and $f' \oplus g'$ correspond to the same face of $P \oplus Q$ then $f \equiv f'$ and $g \equiv g'$.*

Consequently, if we compute a suprema face of two summands, we are sure that there are no other summands for the same suprema face. Thus there would be no partial faces, no invalid faces and no duplicate faces. The only task that remains in the present section is the characterization of incidences of the computed faces. The incidences of the facets to P computed by the application of Lemma 2 are obvious. Following Lemma 1 and Lemma 3, we only need to implement the remaining incidences by the application necessary and sufficient condition for the facets of the faces of the Minkowski sum $P \oplus Q$ in the algorithm. We present the necessary and sufficient condition for these incidences in the following lemma.

Lemma 6 (Necessary and Sufficient Condition for an Incidence $f \oplus f' \prec g \oplus g'$ in $P \oplus Q$). *Let f and g be faces of P and f' and g' be faces of Q where P and Q are convex polytopes. Then $f \oplus f' \prec g \oplus g'$ iff*

(i) $f \oplus f'$ and $g \oplus g'$ are faces of $P \oplus Q$,
(ii) dim$(f \oplus f') =$ dim$(g \oplus g') - 1$, and
(iii) $f \prec \cdots \prec g$ and $f' \prec \cdots \prec g'$

The above lemma can be proved by analyzing the supporting hyperplanes of faces $f \oplus f'$ and $g \oplus g'$. The immediate consequence of this lemma is that whenever we compute a facet of a face of $P \oplus Q$ we can immediately add an incidence relation whether it is newly computed or not. The lemma also ensures that we do not miss any incidences during the execution of the algorithm.

We describe the algorithm to compute the Minkowski sum in the next section.

4.2 The Minkowski Sum Algorithm

To simplify the discussion we treat the nodes in the face lattices as the corresponding faces. Our algorithm computes the faces of $P \oplus Q$ and the incidences layer by layer in a top-down manner. At each layer k, $-1 \le k \le d$, we maintain a list of the computed nodes in $\mathscr{F}_k(P \oplus Q)$. For each computed node z in

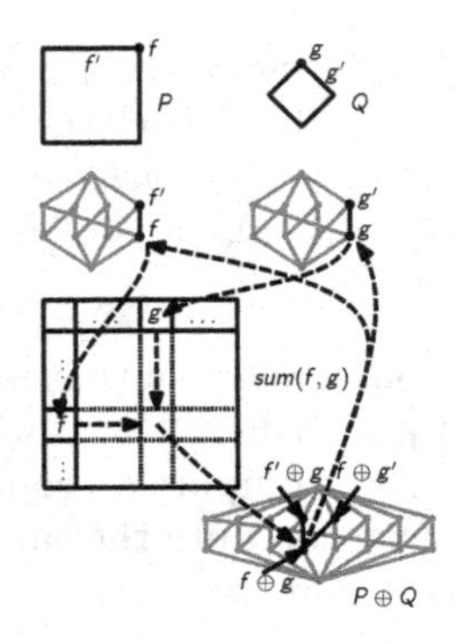

Fig. 5. Identifying replication using the table *sum*.

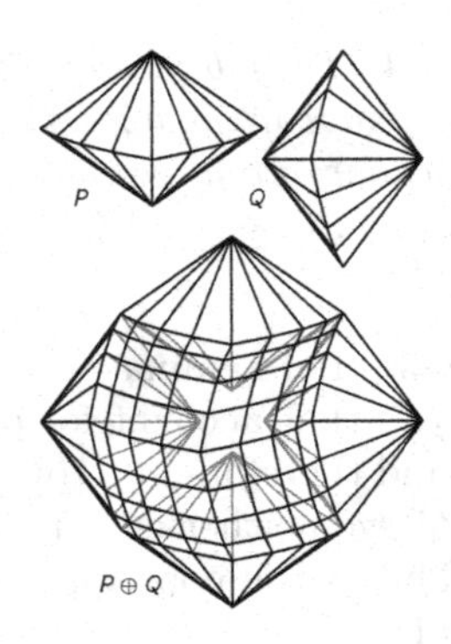

Fig. 6. Worst case: Even in $\mathbb{R}^3$, the number of faces, i.e., sum of vertices, edges and facets, in $P \oplus Q$ can be $\Theta(nm)$.

$\mathscr{F}(P \oplus Q)$, we store its summand nodes x and y from $\mathscr{L}(P)$ and $\mathscr{L}(Q)$ respectively. We create a 2-dimensional table *sum* that takes as indices two nodes x in $\mathscr{L}(P)$ and y in $\mathscr{L}(Q)$ such that the table entry maps to the node z. Other table entries will be empty. This table is updated whenever a new node is added to $\mathscr{L}(P \oplus Q)$. See Fig. 5. Internally, we create unique consecutive indices for the nodes.

We describe the computation of $P \oplus Q$ below. Initially, the set $\mathscr{F}_d(P \oplus Q)$ contains a single node representing the trivial face $P \oplus Q$. We compute the set $\mathscr{F}_k(P \oplus Q)$ inductively from the already computed set $\mathscr{F}_{k+1}(P \oplus Q)$, from $k = d - 1$ down to 0, as follows. For each node $z \in \mathscr{F}_{k+1}(P \oplus Q)$ we look for the facets of z among the candidate Minkowski sums involving the subfaces x_i's and y_j's of the summands x and y of the sum z. A sum $x_i \oplus y_j$ is a candidate sum if $dim(x_i \oplus y_j)$ is $dim(z) - 1$. To verify whether a candidate sum, i.e $x_i \oplus y_j$, is a new valid facet of z we need to check two conditions: (1) whether the sum satisfies Lemma 3 and (2) a node corresponding to the sum does not already exist in $\mathscr{L}(P \oplus Q)$. Condition (1) depends only on the immediate superfaces of x_i and y_j and the affine space of z. This can be readily checked since we have all necessary information with us from the previous iteration. Condition (2) can be verified by checking the table $sum(x, y)$. Once the sum of $x_i \oplus y_j$ is determined to be a valid immediate subface of z we create a node z' in $\mathscr{L}(P \oplus Q)$ corresponding to the sum of x_i and y_j along with proper incidences and update the table *sum* as per Lemma 6. During traversal if we reach either a $k-1$ dimensional candidate node or a non-facet k dimensional candidate node we do not traverse to further subfaces of x' and y' as outlined by Lemma 4. This concludes our algorithm. We have the following theorem as a result.

Theorem 2. *The Minkowski sum $P \oplus Q$ of the convex polytopes P and Q in $\mathbb{R}^d$ of face lattice sizes n and m respectively can be computed in $O(d^{\omega} nm)$-time* *and $O(d^2nm)$-space. Furthermore, the algorithm is worst-case optimal in n and m, i.e., for fixed d, when $|P \oplus Q| = O(nm)$.*

Proof. **Correctness:** A direct consequence of Lemma 1 is that in order to compute the faces of $P \oplus Q$ we need only check the faces of type $f \oplus g$ where f is a face of P and g a face of Q. Not all such faces are valid and Lemma 3 ensures that we choose only the correct faces of $P \oplus Q$. Furthermore Lemma 4 and 5 guarantee that no duplicate nodes are generated by our algorithm.

Time Complexity: The run-time complexity is $O(d^{\omega}nm)$ because our algorithm traverses each pair of incidences of the face lattices $\mathscr{L}(P)$ and $\mathscr{L}(Q)$ only once. The expression d^{ω} is due to the time required to compute rank in the necessary and sufficient condition.

Space Complexity: The $O(d^2nm)$ space requirement is due to the extra storage for face-lattice augmentation when we store d-dimensional vectors in the nodes.

Optimality: We assume that d is fixed in the following discussion. We show that the worst case space and time complexities are tight. See Fig. 6 for a worst-case construction in $\mathbb{R}^3$. For $d > 3$, we describe a d-dimensional construction in Sect. 4.4. Since there are $\Theta(nm)$ vertices in the Minkowski Sum in the construction, any reasonable representation of the polytope, $\mathcal{V}$-representation, $\mathcal{H}$-representation, face enumeration, facet enumeration, facial structure, face lattice, etc.; all will have $\Omega(nm)$ length. This is a lower bound for any Minkowski Sum algorithm for these constructions because no algorithm can report the vertices in less time. So the complexity of any algorithm that constructs the Minkowski sum for this example will be $\Omega(nm)$. Since our algorithm runs in $\Theta(nm)$, i.e., linear in output length, therefore the lower bound is tight and our algorithm is worst-case optimal. □

4.3 Minkowski Sum from Other Input Representations

V-representation or H-representation. We note that instead of the face lattices of the convex polytopes P and Q, if only the $\mathcal{V}$-representations or $\mathcal{H}$-representations of P and Q are provided, then we can compute the Minkowski Sum $P \oplus Q$ by first computing the face lattices of P and Q using the best known algorithm for convex hull construction before applying our techniques.

This may lead to a better performing algorithm than existing techniques in many cases except, for example, when Chan's and Seidel's output sensitive convex hull applications perform better. For example, if we use the best of Chazelle's, Seidel's or Chan's algorithm to compute the convex hull, the time complexity of our algorithm would be $O(nm + T(\mathcal{L}(P)) + T(\mathcal{L}(P)))$ instead of the current $O(T(\mathcal{L}(P \oplus Q)))$ for fixed d where $T(\mathcal{L}) = \min\{T_{bc}(\mathcal{L}), T_{rs}(\mathcal{L}), T_{tc}(\mathcal{L})\}$ and T_{bc}, T_{rs} and T_{tc} are the time complexities of the convex hull algorithms of Chazelle, Seidel and Chan, respectively, to compute the final convex hull face lattice $\mathcal{L}$. For illustration, this is always better than Chazelle's convex hull application and our $O(nm + (n + m)\log(n + m)$-time, surprisingly, fares better compared to Seidel's output sensitive $O((\dots)^2 + h \log h)$ whenever $h \log h$, $h = |P \oplus Q|$, is $\omega(nm)$. □

Face Enumeration Suppose we are provided an enumeration of all the faces (or only facets and vertices) of the input Polytopes P and Q. In this case, there are algorithms in the literature to compute face lattices from the face enumerations in $O(F \log F)$ time using a dictionary [5,20], where $F = |\mathcal{F}|$, $\mathcal{F}$ is the set of faces of the input convex polytope. We expect that we should have at least the affine space of each face in the input, or a way to compute it efficiently, in order to compute the incidences of the face lattice. For the case of enumeration of only facets and vertices, we need more data on facets and vertices. Thus if the faces of input polytopes P and Q are enumerated instead of face lattice as input, then the resulting complexity of our algorithm which computes the face lattice from the face enumeration first will be $O(nm + (n+m)\log(n+m))$-time in the worst case. □

From the above discussion, it can be seen that our algorithm fares quite well even in the case when the input is provided in common representations. Summarily, only a term $(n+m)\log(n+m)$ is added (which is the time complexity for sort). In the next section, we give an example to prove the worst-case optimality of the algorithm.

4.4 An Example to Prove Worst-Case Optimality

Let us consider the convex polytopes (in fact, polygons placed orthogonally) P_i, $1 \leq i \leq d-1$ in d-dim that is the convex hull of n_i points with the coordinates $(x_1 = 0, \ldots, x_{i-1} = 0, x_i = h(1-t^2), x_{i+1} = 0, \ldots, x_{d-1} = 0, x_d = t)$ for n_i distinct values of t, $-1 \leq t \leq 1$. In short, except for coordinates x_i and x_d, other coordinates are 0. We construct convex polytopes $P = P_1 \oplus \cdots \oplus P_k$ and $Q = P_{k+1} \oplus \cdots \oplus P_{d-1}$. Note that the origin is on the bases of polytopes P and Q. In this section, we consider d as fixed and express the complexities independent of d.

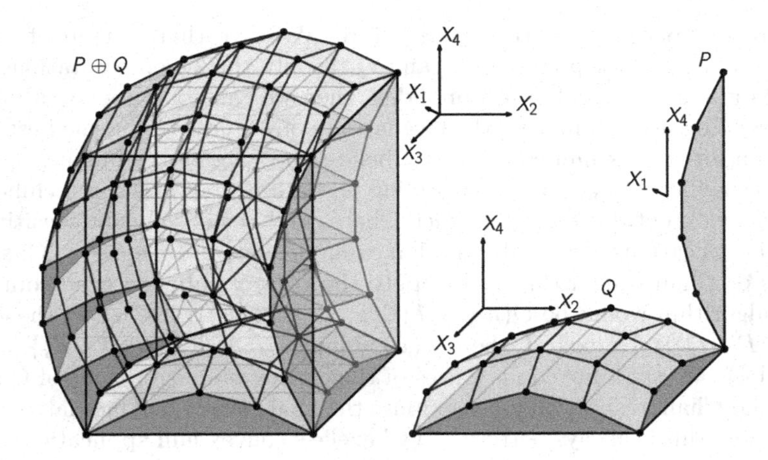

Fig. 7. Construction of the worst-case polytope $P \oplus Q$ in $\mathbb{R}^4$ to prove the optimality

We draw the above example in $\mathbb{R}^4$, where $P = P_1$ and $Q = P_2 \oplus P_3$. See Fig. 7 for illustration. If we wish to construct d-dimensional input polytopes instead of k or $d-k-1$ dimensional polytopes then we may consider the polytopes $P \oplus \epsilon$-cube and $Q \oplus \epsilon$-cube instead, where the ϵ-cube is an axis-parallel d-dimensional hypercube of ϵ size.

The number of vertices, as well as number of faces and the size of face-lattice of P is $\Theta(n_1 n_2 \ldots n_k)$. Let the size of the face lattice be n. The number of vertices, as well as number of faces and the size of face-lattice of Q is $\Theta(n_{k+1} n_{k+2} \ldots n_{d-1})$. Let the size of face lattice be m. The number of vertices, as well as number of faces and the size of face-lattice of $P \oplus Q$ is $\Theta(n_1 n_2 \ldots n_{d-1})$. This is $\Theta(nm)$, since every point $u \oplus v$, where u is a vertex of P and v is a vertex of Q, is a vertex of $P \oplus Q$ (i.e. on the boundary of the convex hull).

Our algorithm computes the Minkowski sum for this example in $\Theta(nm)$, whereas the best known output sensitive algorithm, i.e. Seidel's, is $\Theta((nm)^2 + h \log h)$. Since the output size is $\Theta(nm)$ for this worst-case, our algorithm is linear in the output size and therefore optimal. This completes the argument for worst-case optimality.

The algorithm to compute the Minkowski sum of two convex polytopes can be used to compute the Minkowski sum of multiple convex polytopes. We present the algorithm and analyze its complexity in the next section.

5 The Minkowski Sum of Multiple Convex Polytopes

Let P_1, P_2, ..., P_r be r convex polytopes of the face lattice sizes n_1, n_2, ..., n_r, respectively. We can compute the Minkowski sum $\bigoplus_{i=1}^{n} P_i$ in $O(d^\omega \sum_{j=1}^{r} \prod_{i=1}^{j} n_i) = O(d^\omega \cdot r \cdot \prod_{i=1}^{r} n_i))$ in the worst case by the method presented in Sect. 4 after sorting the polytopes in ascending order of the sizes of face lattices. The worst-case complexity of the resulting algorithm is tight for $r < d$. For $r \geq d$, however, it is known that the worst-case space complexity of the face lattice of the resulting polytope does not increase with r as an exponent [21]. Thus the time complexity has an upper bound. We describe this bound from the citation for reference. Let $M = O(\sum_{S \in \mathcal{C}} |\bigoplus_{P \in S} P|)$ where $\mathcal{C}$ is the set of all $d-1$ combinations of r polytopes. By Stirling's approximation, this quantity is asymptotically $\approx (er/(d-1))^{d-1}$ times maximum size of $(d-1)$-subset Minkowski sum among r polytopes. The size of the Minkowski sum for any r, $r \geq d$ is bounded by M. Consequently, we give the running time for the computation of the Minkowski sum of r convex polytopes in the theorem below.

Theorem 3. *The Minkowski sum $\bigoplus_{i=1}^{r} P_i$ of r convex polytopes P_i's of face lattice sizes n_i's, $1 \leq i \leq r$, can be computed in time $O(d^\omega min\{NM, r \prod_{i=1}^{r} n_i\})$, where N is the total input size $\sum_{i=1}^{r} n_i$ and M is the face lattice size of the resultant Minkowski sum. The space complexity of the algorithm is $O(d^2 M)$. Moreover, the algorithm is worst-case optimal for fixed for $d > 3$ and $r < d$.*

Proof. The proof follows from the observation that the size of the face lattice of the sum is always larger than the size of the face lattice of any of the summands. The time complexity for $r \geq d$ is $O(\sum_{i=1}^{r} d^{\omega} n_i M) = O(d^{\omega} NM)$ by the application of Theorem 2. For $r < d$ too, the expression for time-complexity, $O(d^{\omega} r \prod_{i=1}^{r} n_i))$, which might be larger than the tighter complexity of $O(d^{\omega} \sum_{i=1}^{r} n_i M)$, follows directly from Theorem 2. Thus the time-complexity is $O(d^{\omega} min\{NM, r \prod_{i=1}^{r} n_i\})$. The correctness of the algorithm follows from the correctness of the algorithm of previous section. The space requirement is due to augmented face lattice of the output polytope in the intermediate steps. The argument for worst-case optimality for fixed d is same as in the proof of Theorem 2. □

6 Concluding Remarks and Acknowledgments

In the paper, we presented a worst-case optimal algorithm to compute the Minkowski sum with better bounds than the currently known algorithms. We acknowledge anonymous referees for providing constructive suggestions and suggesting improvements in the paper.

References

1. Adiprasito, K.A., Sanyal, R.: Relative Stanley-Reisner theory and upper bound theorems for Minkowski sums. Publications mathématiques de l'IHÉS **124**(1), 99–163 (2016)
2. Agarwal, P.K., Flato, E., Halperin, D.: Polygon decomposition for efficient construction of Minkowski sums. Comput. Geom. **21**(1–2), 39–61 (2002)
3. Bekker, H., Roerdink, J.B.T.M.: An efficient algorithm to calculate the Minkowski sum of convex 3D polyhedra. In: Alexandrov, V.N., Dongarra, J.J., Juliano, B.A., Renner, R.S., Tan, C.J.K. (eds.) ICCS 2001. LNCS, vol. 2073, pp. 619–628. Springer, Heidelberg (2001). https://doi.org/10.1007/3-540-45545-0_71
4. de Berg, M., Cheong, O., van Kreveld, M., Overmars, M.: Computational Geometry: Algorithms and Applications, 3rd edn. Springer, Heidelberg (2008). https://doi.org/10.1007/978-3-540-77974-2
5. Chan, T.M.: Output-sensitive results on convex hulls, extreme points, and related problems. Discret. Comput. Geom. **16**(4), 369–387 (1996). https://doi.org/10.1007/BF02712874
6. Chazelle, B.: An optimal convex hull algorithm in any fixed dimension. Discret. Comput. Geom. **10**(4), 377–409 (1993). https://doi.org/10.1007/BF02573985
7. Chazelle, B.M.: Convex decompositions of polyhedra. In: Proceedings STOC, pp. 70–79 (1981)
8. Chazelle, B.M., Dobkin, D.P.: Optimal convex decompositions. In: Toussaint, G.T. (ed.) Computational Geometry. Machine Intelligence and Pattern Recognition, vol. 2, pp. 63–133 (1985)
9. Chew, L.P., Scot Drysdale, R.L.: Voronoi diagrams based on convex distance functions. In: Proceedings SoCG, pp. 235–244 (1985)
10. Das, S., Nandy, A., Sarvottamananda, S.: Radius, diameter, incenter, circumcenter, width and minimum enclosing cylinder for some polyhedral distance functions. Discret. Appl. Math. (2020, in press)

11. Edelsbrunner, H.: Algorithms in Combinatorial Geometry. EATCS Monographs on Theoretical Computer Science. Springer, Heidelberg (1987). https://doi.org/10.1007/978-3-642-61568-9
12. Fogel, E., Halperin, D.: Exact Minkowski sums of convex polyhedra. In: Proceedings SoCG, pp. 382–383 (2005)
13. Fukuda, K.: From the zonotope construction to the Minkowski addition of convex polytopes. J. Symb. Comput. **38**(4), 1261–1272 (2004)
14. Fukuda, K., Weibel, C.: Computing all faces of the Minkowski sum of V-polytopes. In: Proceedings of the 17th CCCG, pp. 253–256 (2005)
15. Gritzmann, P., Sturmfels, B.: Minkowski addition of polytopes: computational complexity and applications to Gröbner basis. SIAM J. Discret. Math. **6**(2), 246–269 (1993)
16. Grünbaum, B., Kaibel, V., Klee, V., Ziegler, G.M.: Convex Polytopes. Graduate Texts in Mathematics. Springer, Heidelberg (2003). https://doi.org/10.1007/978-1-4613-0019-9
17. Karavelas, M.I., Tzanaki, E.: The maximum number of faces of the Minkowski sum of two convex polytopes. In: Proceedings SODA, pp. 11–28 (2012)
18. Le Gall, F.: Powers of tensors and fast matrix multiplication. In: Proceedings ISSAC, pp. 296–303 (2014)
19. Ramkumar, G.D.: An algorithm to compute the Minkowski sum outer-face of two simple polygons. In: Proceedings SoCG, pp. 234–241 (1996)
20. Seidel, R.: Constructing higher-dimensional convex hulls at logarithmic cost per face. In: Proceedings STOC, pp. 404–413 (1986)
21. Weibel, C.: Maximal F-vectors of Minkowski sums of large numbers of polytopes. Discret. Comput. Geom. **47**(3), 519–537 (2012)
22. Wein, R.: Exact and efficient construction of planar Minkowski sums using the convolution method. In: Azar, Y., Erlebach, T. (eds.) ESA 2006. LNCS, vol. 4168, pp. 829–840. Springer, Heidelberg (2006). https://doi.org/10.1007/11841036_73

On the Intersections of Non-homotopic Loops

Václav Blažej[1(✉)], Michal Opler[2], Matas Šileikis[3], and Pavel Valtr[4]

[1] Faculty of Information Technology, Czech Technical University in Prague, Prague, Czech Republic
vaclav.blazej@fit.cvut.cz

[2] Computer Science Institute, Charles University, Prague, Czech Republic
opler@iuuk.mff.cuni.cz

[3] The Czech Academy of Sciences, Institute of Computer Science, Prague, Czech Republic
matas.sileikis@gmail.com

[4] Department of Applied Mathematics, Faculty of Mathematics and Physics, Charles University, Prague, Czech Republic

Abstract. Let $V = \{v_1, \ldots, v_n\}$ be a set of n points in the plane and let $x \in V$. An x-*loop* is a continuous closed curve not containing any point of V, except of passing exactly once through the point x. We say that two x-loops are *non-homotopic* if they cannot be transformed continuously into each other without passing through a point of V. For $n = 2$, we give an upper bound $2^{O(k)}$ on the maximum size of a family of pairwise non-homotopic x-loops such that every loop has fewer than k self-intersections and any two loops have fewer than k intersections. This result is inspired by a very recent result of Pach, Tardos, and Tóth who proved the upper bounds 2^{16k^4} for the slightly different scenario when $x \notin V$.

Keywords: Graph drawing · Non-homotopic loops · Curve intersections · Plane

1 Introduction

The so-called *crossing lemma*, which was proved independently by Ajtai, Chvátal, Newborn, Szemerédi [1] and by Leighton [2], bounds the number of crossings in any planar drawing of any graph with n vertices and $m \geq 4n$ edges. The crossing lemma has many applications in discrete and computational geometry and other fields. Very recently, Pach, Tardos, and Tóth [3] proved an

This research was initiated during the workshop KAMAK 2020 in Kytlice in Sept. 20–25, 2020. MŠ was supported by the Czech Science Foundation, grant number GJ20-27757Y, with institutional support RVO: 67985807. VB acknowledges the support of the OP VVV MEYS funded project CZ.02.1.01/0.0/0.0/16_019/0000765 "Research Center for Informatics". PV and MO were supported by project 18-19158S of the Czech Science Foundation.

A. Mudgal and C. R. Subramanian (Eds.): CALDAM 2021, LNCS 12601, pp. 196–205, 2021.
https://doi.org/10.1007/978-3-030-67899-9_15

interesting natural modification of the crossing lemma for multigraphs with non-homotopic edges. In the proof of their result, Pach, Tardos, and Tóth [3] applied a bound on the maximum size of certain collections of so-called non-homotopic loops. In this paper, we show that their bound can be significantly improved for a closely related problem.

For an integer $n \geq 1$, let $V_n = \{v_1, \ldots, v_n\}$ be a set of n distinct points in the plane $\mathbb{R}^2$. Given $x \in \mathbb{R}^2$, an *x-loop* is a continuous function $f : [0,1] \to \mathbb{R}^2$ such that $f(0) = f(1) = x$ and $f(t) \notin V_n$ for $t \in (0,1)$. We will only consider x-loops that do not pass through x, that is $f(t) = x$ only for $t \in \{0,1\}$. When x is clear from the context we will also call an x-loop simply a *loop*. Two loops f_0, f_1 are *homotopic* (with respect to V_n), denoted $f_0 \sim f_1$, if there is a continuous function $H : [0,1]^2 \to \mathbb{R}^2$ (a *homotopy*) such that

$$H(0,t) = f_0(t) \quad \text{and} \quad H(1,t) = f_1(t) \quad \text{for all } t \in [0,1],$$

$$H(s,0) = H(s,1) = x \quad \text{for all } s \in [0,1].$$

and

$$H(s,t) \notin V_n \quad \text{for all } s,t \in (0,1).$$

A self-intersection of a loop f corresponds to a pair $\{t,u\} \subset (0,1)$ of distinct numbers such that $f(t) = f(u)$, while an intersection of two loops f_1, f_2 corresponds to an *ordered* pair $t, u \in (0,1)$ such that $f_1(t) = f_2(u)$.

Given integers $n, k \geq 1$ and $x \in V_n$, let $g(n,k)$ be the largest number of pairwise non-homotopic loops such that every loop has fewer than k self-intersections and any two loops have fewer than k intersections.

Pach, Tardos and Tóth [3] considered the same quantity, but for x outside of V_n (they also added a restriction that no loop passes through x, which holds trivially in our setting with $x \in V_n$). Although the two settings seem to be very similar, we were able to obtain an upper bound on $g(2,k)$ which is significantly smaller than the upper bound on $f(2,k)$ obtained by Pach, Tardos and Tóth [3]. In the setting of Pach, Tardos and Tóth [3] with $x \notin V_n$, the largest number of pairwise non-homotopic loops so that every loop has fewer than k self-intersections and any two loops have fewer than k intersections is denoted by $f(n,k)$. The two aforementioned quantities are related by the following inequalities.

Proposition 1. *For every $n, k \geq 1$ we have*

$$g(n,k) \leq f(n,k) \leq g(n+1,k). \tag{1}$$

Proposition 1 is proved in Sect. 5.

Pach, Tardos and Tóth [3] showed that for $n \geq 2$

$$f(n,k) \leq 2^{(2k)^{2n}} \tag{2}$$

and

$$f(n,k) \geq \begin{cases} 2^{\sqrt{nk}/3}, & n \leq 2k, \\ (n/k)^{k-1}, & n \geq 2k. \end{cases}$$

Also in [3] it was proved that if $n = 1$, then there are at most $2k + 1$ non-homotopic loops with fewer than k self-intersections (that is, if we do not bound the number of intersections) implying $f(1, k) \leq 2k + 1$.

In our main result we focus on the function g in case $n = 2$. Inequalities (1) and (2) imply that $g(2, k) \leq 2^{16k^4}$. The following theorem improves this bound significantly.

Theorem 1. *Let $n = 2$ and $x \in V_2$. For any k, the size of any collection of non-homotopic x-loops with fewer than k self-intersections is at most $2^{O(k)}$. In particular*

$$g(2, k) \leq 2^{12k}.$$

We believe that the bound in Theorem 1 can be further improved by reducing the exponent to $O(\sqrt{k} \log k)$. We plan to address this in a follow-up paper.

2 Setup and Notation

Depending on the context, we will treat $S := \mathbb{R}^2 \setminus V_n$ either as the plane with n points removed, or as a sphere with $n + 1$ points removed (where n of these points is the set $V_n = \{v_1, \ldots, v_n\}$ and the last one, denoted by v_0, corresponds to the "point at infinity"). We refer to the points v_i as *obstacles.*

For convenience, we will always assume the following properties of a finite collection of loops:

1. the set of points of intersections and the set of points of self-intersections are disjoint,
2. every (self-)intersection is simple (that is, no point in S belongs to more than two loops and no loop passes through the same point more than twice),
3. every intersection between two loops is a *crossing*, that is, one loop "passes to the other side" of the other loop (otherwise an intersection is called a *touching*).

Assumptions 1–3 can be attained by infinitesimal perturbations without creating any new intersections or self-intersections.

Given a drawing of the x-loops satisfying the above conditions, we choose a closed curve on the sphere without self-intersections which goes through the points $v_0, \ldots, v_n$ in this order (if $x \notin V_n$, we choose this curve so that it avoids x). We call this loop the *equator.* Removing the equator from the sphere, we obtain two connected sets, which we arbitrarily name the *top half* and the *bottom half.* We refer to the $n + 1$ sets into which the equator is split by excluding points v_i as *gaps.* We label the gaps by elements of $A_n := \{0, \ldots, n\}$, assigning label i to the gap between v_i and v_{i+1}, with indices counted modulo $n + 1$.

By a careful choice of the equator, we can assume the following conditions:

4. every x-loop in the collection intersects the equator a finite number of times,
5. each of these intersections (except for, possibly, the intersection at x) is a crossing (as opposed to a touching),
6. no point of self-intersection or intersection (other than x) lies on the equator.

Part of a given loop f between a pair of distinct intersections with the equator (inclusively) is called a *segment* (it is a restriction of f to a closed subinterval of $[0, 1]$). Whenever the two intersections defining a segment are consecutive (along the loop f), the segment is called an *arc*. If some arc intersects itself, we can remove the part of the arc between these self-intersections without changing the homotopy class of the loop; this trivially does not increase the number of (self-) intersections; therefore we can make yet another assumption about the family of x-loops and the equator:

7. there are no self-crossings within any arc (i.e., between consecutive crossings of the equator).

Given a x-loop ℓ, we list the labels of gaps in the order they are crossed by the loop. This way we obtain a word w over alphabet A_n. We say that ℓ *induces* w. The empty word corresponds to a trivial loop.

A segment that intersects the equator k times (including the beginning and the end) consists of $k-1$ arcs, which we order in a natural way so that the segment traverses the arcs in the increasing order. A segment is called a *downsegment* if its first arc is contained in the bottom half, and an *upsegment* otherwise. If a segment does not start nor end at x, by listing the labels of gaps that the segment intersects, we obtain a word w in alphabet A_n, and call such a segment a *w-segment* (or, more specifically, *w-downsegment* or *w-upsegment*, if we want to specify the location of the first arc). For example, a loop with the first arc in the top half that induces the word 01201 has a 01-downsegment and a 01-upsegment, as well as a 012-downsegment but no 012-upsegment.

3 General n

In this section we state and prove several facts that are valid for general n, including all prerequisites for the proof of Theorem 1.

We start with a simple proposition which allows bounding the number of non-homotopic loops in terms of the different words that they induce.

Proposition 2. *Let $x \in V_n$ and suppose that two x-loops ℓ_1 and ℓ_2 start with an arc which belongs to the same half of the sphere. If they induce the same word of length m, they are homotopic.*

Proof. For $i \in \{1, 2\}$ and $k \in \{1, \dots, m\}$, suppose that the kth arc of ℓ_i ends at point p_i^k (with p_1^k and p_2^k lying in the same gap). Let γ^k be a loop which first goes along the first k arcs of ℓ_2, then goes along a gap from p_2^k to p_1^k and then continues to x along the last $m+1-k$ arcs of ℓ_1. Since the curved quadrilateral consisting of kth arcs of the loops and the two parts of two gaps does not wind around any obstacle, we have $\gamma^{k-1} \sim \gamma^k$ for $k = 2, \dots, m$. By a similar argument, we have $\ell_1 \sim \gamma^1$ and $\ell_2 \sim \gamma^m$. The proposition follows.

Lemma 1. *Let letters $a, b, c \in A_n$ be distinct. If f_1, f_2 are two, not necessarily distinct, x-loops, then any abc-downsegment in f_1 and any abc-upsegment in f_2 intersect.*

Proof. Choose a cyclic orientation of the equator such that gaps appear in the order a, b, c. Let s_1 be an *abc*-downsegment of f_1, and s_2 be an *abc*-upsegment of f_2 and, for contradiction, suppose s_1, s_2 do not intersect.

By removing the point of intersection with s_1 from the gap b, we obtain two disjoint sets, and name them B' and B'' so that — in the same orientation of the equator — gap a is followed by set B' followed by B'' followed by gap c. The *bc*-arc of s_1 partitions the top half into two connected components. Note that gap a and B'' belong to different components. Since the *ab*-arc of s_2 belongs to the top half, s_2 must intersect the gap b in B'. Similarly, *ab*-arc of s_1 partitions the bottom half into two components, so that B' and c belong to different components. Since the *bc*-arc of s_2 traverses the bottom half, s_2 must intersect the gap b in B''. Since B' and B'' are disjoint, we obtain a contradiction.

Corollary 1. *Given three distinct letters $a, b, c \in A_n$, suppose that loop f_1 has i disjoint abc-downsegments and loop f_2 has j disjoint abc-upsegments. Then the number of intersections between f_1 and f_2 is at least $i \cdot j$. In particular, if $f_1 = f_2$, the number of self-intersections is at least $i \cdot j$.*

In the following lemma we chose $x = v_1$ for simplicity, since in the application we can choose $x = v_1$ without loss of generality. The proof of Lemma 2 shows that if $x = v_i$, then the set $\{2, \ldots, n\}$ should be replaced by the set of gaps not incident to x, that is $\{0, \ldots, n\} \setminus \{i-1, i\}$.

Lemma 2. *Let $n \geq 1$ and assume $x = v_1$. Any family of non-homotopic x-loops can be redrawn without increasing the number of self-intersections and intersections so that every x-loop induces a word such that (i) no two consecutive letters are equal and (ii) the first and the last letter belongs to $\{2, \ldots, n\}$.*

Proof. By an *ear* we mean a segment inducing a word *aa* for any letter a. Taking into account that the gaps 0 and 1 are incident to $x = v_1$, by *x-ear* we mean a segment, which corresponds to a letter 0 or 1 at the start or end of the word: that is, an x-ear has x as one of its endpoints and a crossing of the gap 0 or the gap 1 as its other endpoint. We will remove ears in the first step (thus deleting the consecutive pairs of equal letters) and x-ears in the second step (thus deleting the wrong letters at the beginning or the end).

For the first step, we choose an ear in some loop (between two points of some gap a) and denote its endpoints by u and v. By *uv-gap* denote the set of points in the gap a strictly between u and v. An ear is minimal if there is no other ear with both endpoints in the *uv*-gap. We remove ears one by one, always picking a minimal ear.

The chosen ear partitions one of the halves of the sphere into two simply connected sets, one of which, that we denote by P, contains the *uv*-gap in its boundary.

We remove the chosen ear by continuously transforming it to a path which closely follows the uv-gap inside the other half of the sphere, as shown on Fig. 1. By choosing the new path sufficiently close to the equator we can make sure that if a new (self-)intersection with some loop ℓ appears, then by tracing ℓ from that (self-)intersection in a certain direction we cross the uv-gap, thus entering the set P.

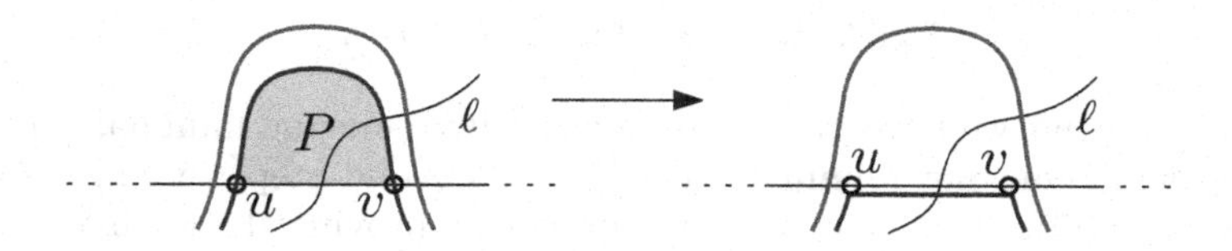

Fig. 1. Removal of a minimal ear

Since $x \notin P$, by tracing ℓ further we must leave P. This cannot happen by crossing the uv-gap again, since that would contradict the fact that we picked a minimal ear. Hence we leave P by crossing the original path of ear. This gives a way to assign, for each newly created intersection with ℓ, a unique intersection with ℓ that was removed, showing that the transformation of the ear does not increase the total number of intersections with ℓ. In particular the number of self-intersections does not increase since we can choose ℓ to be the loop containing the ear in question.

The second step, removing x-ears, is similar to the first one, except that we have to deal with the endpoint x separately. Let v be the point where an x-ear crosses a gap a incident to x (either 0 or 1). Similarly as for ears, by xv-gap we mean the points of gap a strictly between x and v. An x-ear is minimal, if no other x-ear crosses gap a through the xv-gap. We will remove the x-ears one by one, always picking a minimal x-ear.

Since the x-ear is contained in one of the halves of the sphere, it partitions it into two simply connected sets, one of which, that we denote by P, has the xv-gap in its boundary. We remove the x-ear by continuously transforming it into a path that closely follows the xv-gap in the opposite half of the sphere, as shown on Fig. 2. By choosing the new path sufficiently close to the equator, we can make sure that if a new (self-)intersection with some loop ℓ appears, then by tracing ℓ from that (self-)intersection in a certain direction we cross the xv-gap, thus entering set P. Tracing ℓ further we must eventually leave the set P, since $x \notin P$. This cannot happen by crossing the xv-gap again, since that would contradict the fact that we removed all ears in the first step. It also cannot happen by crossing x, since this would contradict that we chose a minimal x-ear. Hence we leave P by crossing the original path of the x-ear, which determines an intersection with the loop ℓ that was removed by transforming the x-ear.

Similarly as in the first step this assigns a unique removed intersection with ℓ to each new intersection with ℓ, showing that removal of a minimal x-ear does not increase the number of (self-)intersections.

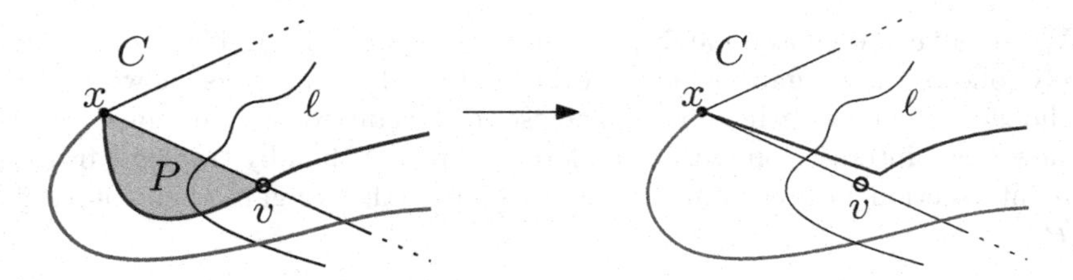

Fig. 2. Removal of a minimal x-ear

We recap what we have proved: by repeatedly removing minimal ears in the first step and removing minimal x-ears in the second step we end up with a drawing which does not have any ears nor x-ears, proving the lemma.

Note that the following lemma holds for $n = 1$ vacuously, since neither of the two conditions can be satisfied.

Lemma 3. *Let $n \geq 1$ and assume $x \in V_n$. Suppose that a, b are adjacent distinct gaps, i.e., $b = a + 1$ or $b = a - 1$ modulo $n + 1$. Let ℓ be an x-loop that induces a word in which no two consecutive letters are equal. Let s be an aba-segment in ℓ. If either (i) x is not a shared endpoint of a and b or (ii) ℓ crosses gaps different from a, b before and after the segment s, then s intersects some other segment of ℓ.*

Proof. Let y be the obstacle incident to gaps a and b. Let u and v be the endpoints of the segment s (thus distinct and both in the gap a) and label them so that u is closer to y than v. Let yu-gap and uv-gap denote part of the a-gap between respective points (or obstacle). The union of s and the uv-gap forms a closed curve without self-intersections (neither the ab arc of segment s nor the ba arc has any self-intersections and the arcs are on the opposite halves of the sphere),which divides the sphere into two parts. Let P be the part of the divided sphere which contains vertex y. As P contains y and its border intersects gap a and gap b, we can partition P by the equator into parts P_1 and P_2, where P_1 denotes the part incident to v, as depicted on Fig. 3.

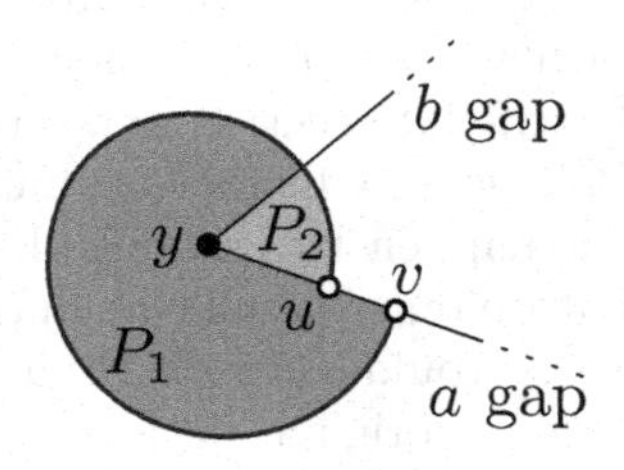

Fig. 3. An *aba*-segment forces a self-intersection if $x \neq y$ or the loop crosses a gap different from a, b both before and after the segment

Choose an orientation of ℓ so that v precedes u. After passing through u, the loop enters into P_1. Loop ℓ must eventually leave P due to the assumption of the lemma: in the case (i) this is because $x \neq y$ and therefore $x \notin P$ while in the case (ii) it has to reach some gap $c \notin \{a, b\}$ which does not intersect P. Before it leaves the set P, it may cross the equator several times, but only through gaps a and b. Since by assumption ℓ does not cross the same gap twice in a row, the location of the loop before leaving set P is determined by the last crossed gap: it always enters set P_1 after crossing the gap a and enters set P_2 after passing through the gap b.

If the loop leaves P through s, we obtain the desired self-intersection. Otherwise ℓ leaves P through the uv-gap (it cannot leave through point u, since by assumption self-intersections do not occur on the equator). As leaving P through the uv-gap is only possible from set P_1, we obtain that ℓ crosses gap a twice in a row (once to enter P_1 and then to leave through the uv-gap), contradicting the assumption.

4 Case $n = 2$

From now on, we focus on the task at hand and assume that $n = 2$ (meaning that we have exactly 3 obstacles on a sphere, one of which is x), in which case words use letters $0, 1$ and 2.

Proof (of Theorem 1)
Without loss of generality assume that $x = v_1$ and fix an x-loop ℓ. By Lemma 2, we can assume that ℓ induces a word w starting and ending in 2 and with no two consecutive equal letters. Lemma 3 implies that for any two distinct letters a, b in w every aba-segment participates in a self-intersection: for $aba \in \{121, 212, 020, 202\}$ this is because x is not incident to both a and b, while for $aba \in \{010, 101\}$ this is because the word induced by the loop starts and ends in 2.

Since every self-intersection is simple, it occurs in at most two disjoint segments. We claim that the word w induced by ℓ has fewer than $12k$ letters. For contradiction, assume the contrary and partition the first $12k$ letters of w into $2k$ disjoint subwords of length 6. Each of these subwords either contains an aba subword or the word is of the form $abcabc$. Segments in the form $abcabc$ contain an abc-upsegment and an abc-downsegment (for the same word abc) which by Lemma 1 forces a self-intersection. Segments which contain aba subword participate in an intersection. As each intersection may cause at most two participations of disjoint segments, it follows that ℓ has at least k self-intersections, giving a contradiction.

It is easy to see that there are fewer than 2^{12k-1} permitted words with fewer than $12k$ letters. By Proposition 2 at most two non-homotopic x-loops induce the same word. Therefore we conclude that $g(2, k) \leq 2^{12k}$.

5 Proof of Proposition 1

We denote $V_n = \{v_1, \ldots, v_n\}$ the set of points removed from the plane. To see the first inequality in (1), let $x \in V_n$, and fix a family of x-loops that attains the maximum $g(n,k)$. Since the number of (self-)intersections is finite, by continuity of loops, there is a circle centered at x such that (i) each loop intersects it at exactly two distinct points, (ii) inside the circle there are no intersections (other than those at x) and no self-intersections, and (iii) inside the circle there are no points of V_n (other than x). Property (iii) implies that we can homotopically transform the loops inside the circle so that between the circle and x they form straight lines.

Pick a point x' on the circle so that neither x' nor its antipodal point lie on any x-loop. Denoting the points where a loop ℓ crosses the circle by p_l while 'departing' and q_l while 'returning', replace each ℓ by an x'-loop ℓ' in which $p_\ell x$ and $q_\ell x$ are replaced by straight segments $p_\ell x'$ and $q_\ell x'$. Since the pairs p_ℓ and q_ℓ are pairwise disjoint, this does not create additional intersections, see Fig. 4.

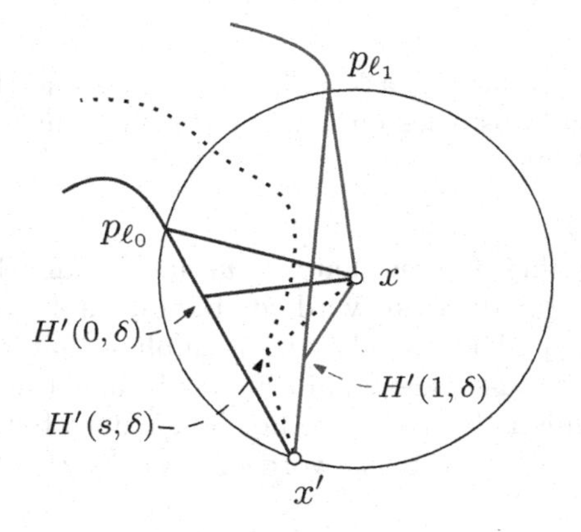

Fig. 4. Constructing homotopy for modified loops, showing only the 'departing' ends of the loops, i.e., for the argument t close to 0.

It remains to check that no two of the resulting x'-loops are homotopic (with respect to V_n). Assuming for contradiction that H' is a homotopy between x'-loops $\ell'_0 = H'(0,\cdot)$ and $\ell'_1 = H'(1,\cdot)$, we will construct a homotopy between original loops ℓ_0 and ℓ_1. Pick $\varepsilon > 0$ such that there are no obstacles in a ball of radius ε around x'. Elementary analysis implies that there is $\delta = \delta(\varepsilon)$ such that

$$\max_{s\in[0,1]} |H'(s,t) - x'| < \varepsilon, \quad t \in [0,\delta] \cup [1-\delta, 1]. \tag{3}$$

Fix $s \in [0,1]$ and define a function $H(s,t)$ as follows. Set $H(s,t) = H'(s,t)$ for $t \in [\delta, 1-\delta]$. On the interval $[0,\delta]$ connects x and $H'(s,\delta)$ by a straight line, that is, set $H(s,t) = x(1-t) + tH'(s,\delta)$, and symmetrically on $[1-\delta,1]$ connects x

and $H'(s, 1-\delta)$ by a straight line, that is, set $H(s,t) = xt + (1-t)H'(s, 1-\delta)$. By (3) and choice of ε none of these two segments hits any obstacle other than x.

Assuming δ is small enough we can make sure that for $i \in \{0,1\}$, $H'(i,\delta)$ and $H'(i, 1-\delta)$ lie inside the circle on the straight segments of the loop ℓ_i'. It is easy to see that H is a homotopy with respect to V_n, so $H(0,\cdot) \sim H(1,\cdot)$. By replacing the two initial straight segments of $H(0,\cdot)$ (namely x to $H'(0,\delta)$ and $H'(0,\delta)$ to p_{ℓ_0}) by the segment from x to p_{l_0} (and similarly the final two segments at the other end of the loop) we obtain the loop ℓ_0. The three segments form a triangle with no elements of V_n inside it (and similarly for the triangle at the other end), which implies that $H(0,\cdot) \sim \ell_0$. By the same argument $H(1,\cdot) \sim \ell_1$. Recalling that $H(0,\cdot) \sim H(1,\cdot)$, we obtain $\ell_0 \sim \ell_1$, a contradiction.

To see the second inequality in (1), we choose a family of x-loops that attains the maximum $f(n,k)$. Since none of the x-loops passes through x, they are also x-loops with respect to $V_{n+1} := V_n \cup \{x\}$. To show that this family of x-loops gives a lower bound to $g(n+1,k)$, we observe that if two x-loops f_0, f_1 are non-homotopic with respect to V_n, then they are non-homotopic with respect to V_{n+1}. Indeed, assuming for contradiction that that there is a homotopy H between f_0 and f_1 satisfying $H(s,t) \notin V_{n+1}$ for all $s,t \in (0,1)$, it trivially satisfies $H(s,t) \notin V_n$ for all $s,t \in (0,1)$, and thus $f_0 \sim f_1$ with respect to V_{n-1}, giving a contradiction.

This completes the proof of Proposition 1.

Acknowledgement. This research was initiated during the workshop KAMAK 2020 organized by two departments (KAM and IÚUK) of the Faculty of Mathematics and Physics, Charles University, in Kytlice in September 20–25, 2020. We thank the organizers and the participants of the workshop for creating a stimulating atmosphere and for helpful discussions.

References

1. Ajtai, M., Chvátal, V., Newborn, M.M., Szemerédi, E.: Crossing-free subgraphs. In: Theory and Practice of Combinatorics. North-Holland Mathematics Studies, vol. 60, pp. 9–12. North-Holland, Amsterdam (1982). https://doi.org/10.1016/S0304-0208(08)73484-4
2. Leighton, F.T.: Complexity Issues in VLSI. Foundations of Computing. MIT Press, Cambridge (1983)
3. Pach, J., Tardos, G., Tóth, G.: Crossings between non-homotopic edges (2020). arXiv:2006.14908. To appear in LNCS, Springer, Proc. of Graph Drawing 2020

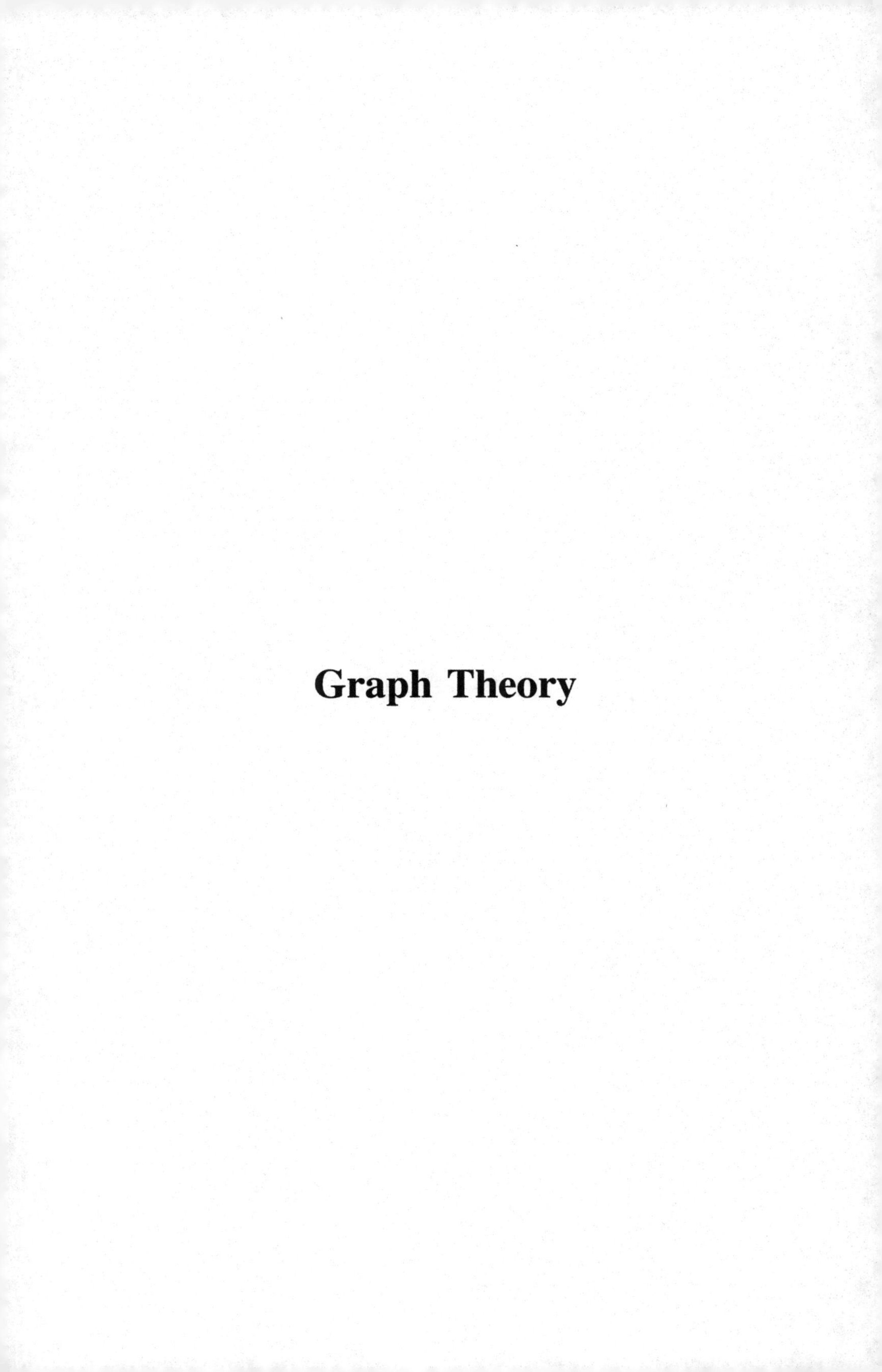

Graph Theory

On cd-Coloring of Trees and Co-bipartite Graphs

M. A. Shalu and V. K. Kirubakaran(✉)

IIITDM Kancheepuram, Chennai, India
{shalu,mat19d002}@iiitdm.ac.in

Abstract. A k-class domination coloring (k-cd-coloring) is a partition of the vertex set of a graph G into k independent sets $V_1, \ldots, V_k$, where each V_i is dominated by some vertex u_i of G. The least integer k such that G admits a k-cd-coloring is called the cd-chromatic number, $\chi_{cd}(G)$, of G. A subset S of the vertex set of a graph G is called a subclique in G if $d_G(x, y) \neq 2$ for every $x, y \in S$. The cardinality of a maximum subclique in G is called the subclique number, $\omega_s(G)$, of G.

In this paper, we present algorithms to compute an optimal cd-coloring and a maximum subclique of (i) trees with time complexity $O(n)$ and (ii) co-bipartite graphs with time complexity $O(n^{2.5})$. This improves $O(n^3)$ algorithms by Shalu et al. [2017, 2020]. In addition, we prove tight upper bounds for the subclique number of the class of (i) P_5-free graphs and (ii) double-split graphs.

1 Introduction

Mathematical models for many real life problems demand optimum (i) sharing (vertex coloring) and (ii) monitoring the usage (domination) of the limited available resources [1,2,9]. The class domination coloring, a combination of vertex coloring and domination [7,13,14], is one of the attempts to address such problems. For other variations and details visit [3,6,10–12]. A k-vertex coloring $V_1, \ldots, V_k$ of a graph G is called a *k-cd-coloring* if for every i, $1 \leq i \leq k$, there exists a vertex u_i such that $V_i \subseteq N[u_i]$. The *cd-chromatic number* of a graph G is defined as $\chi_{cd}(G) = \min\{k : G \text{ admits a } k\text{-cd-coloring}\}$. We observe that two vertices x, y of a graph G receive same color in a cd-coloring of G only if the length of a shortest path between x and y in G is two; i.e., $d_G(x, y) = 2$. By this observation, Shalu et al. [11] introduced a lower bound for the cd-chromatic number of a graph called the subclique number. A subset S of the vertex set of a graph G is called a *subclique* in G if $d_G(x, y) \neq 2$ for every $x, y \in S$. The *subclique number* of a graph G is defined as $\omega_s(G) = \max\{|S| : S \text{ is a subclique in } G\}$. Clearly, $\omega_s(G) \leq \chi_{cd}(G)$. In this paper, we explore the following computational problems.

CD-COLORABILITY
Instance : A graph G and a positive integer k.
Question : Is G k-cd-colorable?

M. A. Shalu—Supported by SERB (DST), MATRICS scheme MTR/2018/000086.

A. Mudgal and C. R. Subramanian (Eds.): CALDAM 2021, LNCS 12601, pp. 209–221, 2021.
https://doi.org/10.1007/978-3-030-67899-9_16

SUBCLIQUE	
Instance	: A graph G and a positive integer k.
Question	: Does G contain a subclique of size k?

The problem CD-COLORABILITY is proved to be NP-complete for several classes of graphs such as bipartite graphs [7] and chordal graphs [12]. It is also proved that testing whether a graph is k-cd-colorable is NP-complete for $k \geq 4$ and is polynomial time solvable for $k \leq 3$. Shalu and Sandhya [10] gave an $O(n^5)$ time algorithm to check whether a graph is 3-cd-colorable where n is the cardinality of the vertex set. Kiruthika et al. [6] gave an $O(2^n n^4 log\, n)$ time algorithm to find the cd-chromatic number of a graph with n vertices and addressed FPT version of the problem CD-COLORABILITY.

Shalu et al. [11] proved that the problem SUBCLIQUE is NP-complete for bipartite graphs, chordal graphs and P_5-free graphs. They also proved that (i) an optimal cd-coloring and (ii) a maximum subclique in trees and co-bipartite graphs can be found in $O(n^3)$ time [11,12]. We improve the complexity of these results.

In this paper, we present (i) a linear time algorithm to find a maximum subclique and an optimal cd-coloring of trees in Sect. 3, and (ii) an $O(n^{2.5})$ algorithm to find a maximum subclique and an optimal cd-coloring of a co-bipartite graphs with n vertices in Sect. 4. In addition, we give a tight upper bound for the problem SUBCLIQUE on P_5-free graphs in Sect. 5, and we show that the subclique number of a double-split graph is equal to its clique number in Sect. 6.

2 Preliminaries

We follow West [15] for terminologies and notation. We denote (i) by $x_1 \ldots x_k$ a path with vertex set $\{x_1, \ldots, x_k\}$ and edge set $\{x_i x_{i+1} : 1 \leq i \leq k-1\}$, and (ii) by $(x_1, \ldots, x_k)$ a cycle with vertex set $\{x_1, \ldots, x_k\}$ and edges $x_1x_2, \ldots, x_{k-1}x_k, x_kx_1$. A clique (An independent set) C of a graph G is a collection of vertices such that any two vertices (no two vertices) in C are adjacent. The cardinality of a maximum clique (maximum independent set) is called clique number (independence number), $\omega(G)$ ($\alpha(G)$), of G. For $X \subseteq V(G)$, $G[X]$ denotes the subgraph induced by the vertices of X in G. If we say $\{x_1, \ldots, x_k\}$ induces a path P_k, then it means that $G[\{x_1, \ldots, x_k\}]$ is the path $x_1 \ldots x_k$. For two vertices $x, y \in G$, $d_G(x, y)$ denotes the length of a shortest path joining x and y in G. For a graph H, G is H-free if no induced subgraph of G is isomorphic to H. For a vertex $x \in V(G)$, $N(x) = \{y \in V(G) : xy \in E(G)\}$, $N[x] = \{x\} \cup N(x)$ and $A(x) = V(G) \setminus N[x]$.

3 Trees

In this section, we present a linear time algorithm to find a maximum subclique and an optimal cd-coloring of trees, and thereby we show that the cd-chromatic number and the subclique number of a tree are equal.

Algorithm 1: To compute a maximum subclique and an optimal cd-coloring of a tree

Input : A tree T

Output : a subclique S, a collection $\{V_u : u \in S\}$, and a collection of singleton sets $\{D_u : u \in S\}$ where $V(T) = \cup_{u \in S} V_u$ is a $|S|$-cd-coloring of T, and each color class V_u is dominated by the vertex in D_u.

Initialization: $S = \emptyset$, $V_u = D_u = \emptyset$ for all $u \in V(T)$, and all vertices of T are unmarked.

```
while T ≇ K_1 do
    Choose a leaf vertex u of T
    if u is unmarked then
        S = S ∪ {u}
        D_u = N(u)
        V_u = {u} ∪ {v ∈ V(T) : d_T(u, v) = 2 and v is unmarked}
        Mark all vertices in V_u
    end
    T = T \ {u}
end
if T ≅ K_1 and u is unmarked then
    S = S ∪ {u}
    D_u = {u}
    V_u = {u}
    Mark u
end
```

We present an example of the execution of Algorithm 1 at the end of this section.

Theorem 1. *For every tree T, $\omega_s(T) = \chi_{cd}(T)$. Besides, when a tree T is given as input, Algorithm 1 outputs a maximum subclique and an optimal cd-coloring of T.*

Proof. For an input tree T of Algorithm 1, let S, $\{V_u : u \in S\}$, and $\{D_u : u \in S\}$ be the sets output by the algorithm.

Claim 1: For all $u \in S$, either (i) $D_u = V_u = \{u\}$ or (ii) there exists a vertex $v \in V(T)$ distinct from u such that $D_u = \{v\}$ and $V_u \subseteq N(v)$.
If u is added to S at the end of the algorithm, then $D_u = V_u = \{u\}$. If not, u is added to S at i^{th} iteration where $i < n$. At i^{th} iteration, u is a leaf vertex, and hence u has only one neighbor, say v. Therefore, $D_u = \{v\}$. This implies that, during i^{th} iteration, every vertex at distance two from u is a neighbor of v. Thus, $V_u \subseteq N(v)$. This proves Claim 1.
Note that, in both cases, the vertex in D_u dominates V_u.

Claim 2: S is a subclique in T.
If not, there exist $u, v \in S$ such that $d_G(u, v) = 2$. W.l.o.g., assume that u is

added to S before v. Let i be the iteration in which u is added to S. Then, during i^{th} iteration, the vertex v is marked. Therefore, v will not be added to S during any succeeding iteration, a contradiction.

Claim 3: V_u is an independent set for all $u \in S$.
If not, for some $u \in S$, there exist $x, y \in V_u$ such that $xy \in E(T)$. Since V_u contains at least two vertices, $V_u \neq \{u\}$. Then, by Claim 1, there exists a vertex v of T distinct from u such that $D_u = \{v\}$ and $V_u \subseteq N(v)$. In particular, $xv, yv \in E(T)$. Thus, (x, v, y) is a cycle in T, a contradiction.

Claim 4: $\bigcup_{u \in S} V_u$ is a cd-coloring of T.
A vertex of T is added to a set V_u for some $u \in S$, only when the vertex is unmarked. In addition, every vertex is marked soon after it is added to some V_u. This implies that every vertex in $V(T)$ belongs to at most one V_u. Since every vertex in T is marked only when it is added to some V_u and also every vertex in T is marked by the end of the algorithm. This implies that every vertex of T belongs to V_u for some $u \in S$. Thus, $\cup_{u \in S} V_u$ is a partition of $V(T)$. By Claim 3, each V_u is an independent set. Hence, $V(T) = \cup_{u \in S} V_u$ is a coloring.

By Claim 1, the vertices in V_u are dominated by the vertex in D_u. Thus, the coloring $V(T) = \cup_{u \in S} V_u$ is a $|S|$-cd-coloring of T. Therefore, $\chi_{cd}(T) \leq |S| \leq \omega_s(T)$, and we know that, $\omega_s(T) \leq \chi_{cd}(T)$. This implies that $\omega_s(T) = \chi_{cd}(T) = |S|$.
Hence, the set S output by the algorithm is a maximum subclique of T. Similarly, the partition $V(T) = \cup_{u \in S} V_u$ is an optimal cd-coloring of T where each V_u is dominated by the vertex in D_u. □

Corollary 1. *Algorithm 1 runs in $O(n)$ time where $n = |V(T)|$.*

Proof. Our analysis is based on the assumption that the tree is represented using adjacency list. To compute the running time of Algorithm 1, we consider each step in the algorithm, and determine the time complexity of the step as well as the number of times the step is executed. The step of choosing a leaf vertex from T requires $O(1)$ time, and is repeated $n-1$ times. Similarly, deleting a leaf vertex takes constant time, and is repeated $n-1$ times. The steps in which we add a new vertex to the sets S and D_u also requires only constant time, and are repeated at most n times. In addition, adding vertices to V_u and marking them takes $O(deg(v))$ time where $D_u = \{v\}$. Whenever a vertex v is added to D_u during some iteration, each of its neighbors are marked at the end of the iteration. Hence, every vertex u' added to S during any succeeding iteration is not a neighbor of v. Thus, for any two vertices u, w added to S during iterations of the while loop, vertices in D_u and D_w are distinct. And the steps after all iterations of the while loop takes only $O(1)$ time. Thus, the time complexity of the algorithm is at most $(n-1) \cdot O(1) + n \cdot O(1) + \sum_{v \in V(T)} O(deg(v)) + O(1) = O(n)$. □

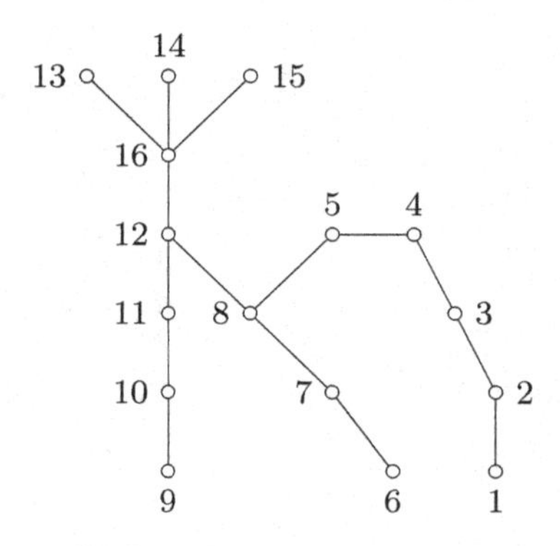

Fig. 1. Tree T

Consider the tree T in Fig. 1, and assume that the algorithm chooses the least labelled leaf during every iteration. Then, the output of the algorithm will be a subclique $S = \{1, 2, 5, 6, 9, 10, 13, 16\}$, $\{V_u : u \in S\}$ and $\{D_u : u \in S\}$ where $V_1 = \{1, 3\}$, $V_2 = \{2, 4\}$, $V_5 = \{5, 7, 12\}$, $V_6 = \{6, 8\}$, $V_9 = \{9, 11\}$, $V_{10} = \{10\}$, $V_{13} = \{13, 14, 15\}$, $V_{16} = \{16\}$, $D_1 = \{2\}$, $D_2 = \{3\}$, $D_5 = \{8\}$, $D_6 = \{7\}$, $D_9 = \{10\}$, $D_{10} = \{11\}$, $D_{13} = \{16\}$ and $D_{16} = \{16\}$. Observe that $\{V_1, V_2, V_5, V_6, V_9, V_{10}, V_{13}, V_{16}\}$ is an 8-cd-coloring of T where vertices in V_u are dominated by the vertex in D_u for every $u \in S$. Thus, $\chi_{cd}(T) = \omega_s(T) = 8$.

4 Co-bipartite Graphs

Let $G(X \cup Y, E)$ be a co-bipartite graph where the vertex set is partitioned into cliques X and Y. In this section, we show that the cd-chromatic number of a co-bipartite graph G is equal to its subclique number .

Lemma 1. *Let $G(X \cup Y, E)$ be a co-bipartite graph. Then, the set $A = \{x \in X : N(x) \cap Y = \emptyset\} \cup \{y \in Y : N(y) \cap X = \emptyset\}$ is either a clique or a maximal subclique in G(see Fig. 2).*

Proof. If either $A \cap X = \emptyset$ or $A \cap Y = \emptyset$, then A is a clique. Assume that both $A \cap X$ and $A \cap Y$ are non-empty.

Claim: A is a maximal subclique in G.
Clearly, $A \cap X$ and $A \cap Y$ are cliques. If $d_G(x, y) = 2$, for some $x \in A \cap X$ and $y \in A \cap Y$, then there exists a path xwy in G for some $w \in V(G)$. W.l.o.g., assume that $w \in Y$. Then $N(x) \cap Y \neq \emptyset$ and $x \in A$, a contradiction. Therefore, A is a subclique of G. Next, we prove that the subclique A is maximal. On the contrary, assume that A is a proper subset of some subclique, say S in G. Let $x \in (S \setminus A)$. W.l.o.g., assume that $x \in X$. Since $x \notin A$, it has a neighbor $y \in Y$ (by the definition of A). Let $y^* \in A \cap Y$. Clearly, $xy^* \notin E(G)$. This implies that $\{x, y, y^*\}$ induces a P_3 in G. Thus, $d_G(x, y^*) = 2$, a contradiction because S is a subclique. Therefore, A is a maximal subclique of G. □

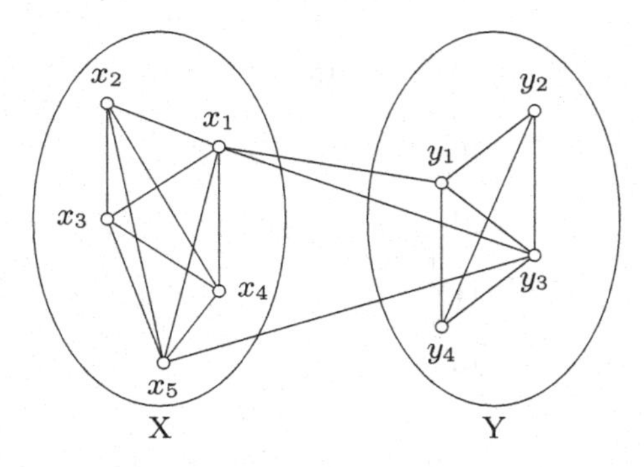

Fig. 2. A co-bipartite graph $G(X \cup Y, E)$. For this graph G, $A = \{x_2, x_3, x_4, y_2, y_4\}$.

Theorem 2. *Let $G(X \cup Y, E)$ be a co-bipartite graph, and let $A = \{x \in X : N(x) \cap Y = \emptyset\} \cup \{y \in Y : N(y) \cap X = \emptyset\}$. Then, any subclique of G is either a clique or a subset of A. Hence, $\omega_s(G) = \max\{\omega(G), |A|\}$.*

Proof. Let S be a subclique of G. If S is a subset of A, we are done. Otherwise, there exists a vertex $x \in S \setminus A$. W.l.o.g. assume that $x \in X$. By the definition of A, x has a neighbor $y \in Y$.

Claim: S is a clique in G.
On the contrary, assume that there exist vertices $x' \in S \cap X$ and $y' \in S \cap Y$, such that $x'y' \notin E(G)$. If $xy' \in E(G)$, then $x \neq x'$ and thus $\{x', x, y'\} \subseteq S$ induces a P_3 in G; since $d_G(x', y') = 2$ and S is a subclique, we have a contradiction. If $xy' \notin E(G)$. Then $y \neq y'$ and thus $\{x, y, y'\}$ induces a P_3 in G; since $d_G(x, y') = 2$ and S is a subclique, we have a contradiction.

Thus, every subclique in G is either a clique or a subset of A, and hence $\omega_s(G) = \max\{\omega(G), |A|\}$. □

Remark: A maximum clique of a co-bipartite graph G can be found in $O(n^{2.5})$, since a maximum independent set in the bipartite graph G^c can be found in $O(n^{2.5})$ [8]. Also, the set $A \subseteq V(G)$ in Theorem 2 can be found in $O(n^2)$ time. Thus, a maximum subclique of a co-bipartite graph can be found in $O(n^{2.5})$ time.

In order to find the cd-chromatic number of a co-bipartite graph, we use the concept of system of distinct representatives (SDR) [15].

System of Distinct Representatives (SDR)
Let $\mathcal{D} = \{D_1, \ldots, D_n\}$ be a collection of subsets of a set Y. An *SDR* for $\mathcal{D}$ is a set of distinct elements $d_1, \ldots, d_n$ of Y such that $d_i \in D_i$.

Note: $\mathcal{D}$ has an *SDR* if and only if $|\bigcup_{i \in B} D_i| \geq |B|$ for every $B \subseteq \{1, \ldots, n\}$ [15].

Theorem 3. *If $G(X \cup Y, E)$ is a co-bipartite graph, then $\chi_{cd}(G) = \omega_s(G)$.*

Proof.
Case 1: G is disconnected.

Then the vertex set of G itself is a subclique of G (union of cliques X and Y), and thus, $\chi_{cd}(G) = |V(G)| = \omega_s(G)$ (since $|V(G)| = \omega_s(G) \leq \chi_{cd}(G) \leq |V(G)|$).

Case 2: G is connected.
We know that, $\omega_s(G) \geq \max\{|X|, |Y|\}$. Let S be a maximum subclique of G. Let $X_1 = X \cap S$, $X_2 = X \setminus S$, $Y_1 = Y \cap S$, and $Y_2 = Y \setminus S$. Since $|Y_1| + |X_1| = |S| = \omega_s(G) \geq |X|$, we have $|Y_1| \geq |X| - |X_1|$, and thus $|Y_1| \geq |X_2|$. Similarly, $|X_1| \geq |Y_2|$.

Let us label the vertices in X_1 as $x_{(1,1)}, \ldots, x_{(1,|X_1|)}$, the vertices in X_2 as $x_{(2,1)}, \ldots, x_{(2,|X_2|)}$, the vertices in Y_1 as $y_{(1,1)}, \ldots, y_{(1,|Y_1|)}$, and the vertices in Y_2 as $y_{(2,1)}, \ldots, y_{(2,|Y_2|)}$.

Let us construct a one to one function $f : X_2 \to Y_1$, such that $d_G(x, f(x)) = 2$ for every $x \in X_2$. To this end, we define $D_i = \{y \in Y_1 : d_G(x_{(2,i)}, y) = 2\}$ for $1 \leq i \leq |X_2|$. Let $\mathcal{D} = \{D_1, \ldots, D_{|X_2|}\}$.

Claim 1: $\mathcal{D} = \{D_i : 1 \leq i \leq |X_2|\}$ has an *SDR*.
If not, there exists a set $\{i_1, \ldots, i_k\} \subseteq \{1, \ldots, |X_2|\}$ such that $|\bigcup_{j=1}^{k} D_{i_j}| < k$ (by Note). Clearly, $S \setminus \bigcup_{j=1}^{k} D_{i_j}$ is a subclique and $\{x_{(2,i_1)}, \ldots, x_{(2,i_k)}\}$ is a clique. Since no vertex of $S \setminus \bigcup_{j=1}^{k} D_{i_j}$ is at distance two from each vertex in $\{x_{(2,i_1)}, \ldots, x_{(2,i_k)}\}$, $S' = (S \setminus \{\bigcup_{j=1}^{k} D_{i_j}\}) \cup \{x_{(2,i_1)}, \ldots, x_{(2,i_k)}\}$ is a subclique in G. The cardinality of S' is at least $|S| + 1$, a contradiction because S is a maximum subclique. This proves Claim 1.

By Claim 1, $\mathcal{D} = \{D_1, \ldots, D_{|X_2|}\}$ has an *SDR*. W.l.o.g., assume that $y_{(1,1)}, \ldots, y_{(1,|X_2|)}$, are the distinct elements of Y_1 such that $y_{(1,i)} \in D_i$ for $1 \leq i \leq |X_2|$.

Next, we define a function $f : X_2 \to Y_1$ as $f(x_{(2,i)}) = y_{(1,i)}$ for $1 \leq i \leq |X_2|$.

Clearly, f is a one-one function, thanks to SDR. Also, by the definition of D_i, $d_G(x_{(2,i)}, y_{(1,i)}) = 2$ for $1 \leq i \leq |X_2|$.

Similarly, we define a one-one function $g : Y_2 \to X_1$ such that $d_G(y, g(y)) = 2$ for every $y \in Y_2$. W.l.o.g., assume that $g(y_{(2,j)}) = x_{(1,j)}$ for $1 \leq j \leq |Y_2|$. Let

$$U_i = \begin{cases} \{x_{(2,i)}, y_{(1,i)}\} & for\ 1 \leq i \leq |X_2| \\ \{y_{(1,i)}\} & for\ |X_2| + 1 \leq i \leq |Y_1|, \end{cases}$$

and let

$$V_j = \begin{cases} \{x_{(1,j)}, y_{(2,j)}\} & for\ 1 \leq j \leq |Y_2| \\ \{x_{(1,j)}\} & for\ |Y_2| + 1 \leq j \leq |X_1|. \end{cases}$$

Then, the partition $V(G) = (U_1 \cup \ldots \cup U_{|Y_1|}) \cup (V_1 \cup \ldots \cup V_{|X_1|})$ is an $\omega_s(G)$-coloring of G. For $1 \leq i \leq |Y_1|$, U_i is either a singleton set or a pair of vertices at distance two from each other, and hence dominated by some vertex in G. Similarly, each $V_j (1 \leq j \leq |X_1|)$ is dominated by some vertex in G. Therefore, the above coloring is an $\omega_s(G)$-cd-coloring of G. This implies that $\chi_{cd}(G) \leq \omega_s(G)$. We know that, $\omega_s(G) \leq \chi_{cd}(G)$. Hence, $\chi_{cd}(G) = \omega_s(G)$. □

Corollary 2. *An optimal cd-coloring of a co-bipartite graph G with n vertices can be found in $O(n^{2.5})$ time.*

Proof. We construct a graph G^* associated with a graph G as $V(G^*) = V(G)$ and $E(G^*) = \{xy : d_G(x,y) = 2\}$ [11]. The sets $\{U_i = \{x_{(2,i)}, y_{(1,i)}\} : 1 \leq i \leq |X_2|\}$ in Theorem 3 corresponds to the matching $\{x_{(2,i)}y_{(1,i)} : 1 \leq i \leq |X_2|\}$ in $G^*[X_2 \cup Y_1]$ (a bipartite graph with partitions X_2 and Y_1). Since the matching saturates every vertex in X_2, it is a maximum matching of $G^*[X_2 \cup Y_1]$. Similarly, the sets $\{V_j = \{x_{(1,j)}, y_{(2,j)}\} : 1 \leq j \leq |Y_2|\}$ corresponds to the matching $\{x_{(1,j)}y_{(2,j)} : 1 \leq j \leq |Y_2|\}$ in $G^*[X_1 \cup Y_2]$ (a bipartite graph with partitions X_1 and Y_2). Since the matching saturates every vertex in Y_2, it is a maximum matching of $G^*[X_1 \cup Y_2]$. We know that a maximum matching in a bipartite graph can be found in $O(n^{2.5})$ time [8]. Also, G^* can be computed in $O(n^2)$ time by removing the edges between the vertices of A (defined in Theorem 2) from G^C. Thus, the problem is solvable in $O(n^{2.5})$ time. □

5 P_5-free Graphs

In this section we give a tight upper bound for subclique number on P_5-free graphs. Shalu et al. [11] proved that SUBCLIQUE problem is NP-complete for the class of P_5-free graphs (a superclass of co-bipartite graphs). We use the following observation in the proof of our theorem on P_5-free graphs.

Observation 1. *If G is a graph with a universal vertex, then every subclique is a clique, and thus $\omega_s(G) = \omega(G)$.* □

Theorem 4. *Every connected P_5-free graph has a dominating P_3 or a dominating clique [4].* □

Theorem 5. *Let G be a non trivial connected P_5-free graph. Then*

(i) $\omega_s(G) \leq \omega(G)(\omega(G) - 1)$, if G has a dominating clique and
(ii) $\omega_s(G) = \omega(G)$, if G has a dominating P_3 and no subgraph K_2 of the P_3 dominates G. In addition, the above upper bounds are tight.

Proof. Let S be a subclique of a non trivial connected P_5-free graph G.

Part 1: Suppose that G has a dominating clique, say $C = \{x_1, \ldots, x_k\}$. We consider two cases.

Claim 1: If $C \cap S \neq \emptyset$, then $|S| \leq \omega(G)$.
W.l.o.g., assume that $x_i \in S$ for some $i \in \{1, \ldots, k\}$. We prove that $S \subseteq N[x_i]$. If not, there exists a vertex $y \in S$ such that $yx_i \notin E(G)$ and let x_j be the vertex in $C \setminus \{x_i\}$ that dominates y, then $\{x_i, x_j, y\}$ induces a P_3 in G. Hence $d_G(y, x_i) = 2$ and $y, x_i \in S$, a contradiction. Therefore $S \subseteq N[x_i]$. Note that S is a subclique in $G[N[x_i]]$, a graph with universal vertex x_i. Thus by Observation 1, S is a clique and $|S| \leq \omega(G)$.
Claim 2: If $C \cap S = \emptyset$, then $|S| \leq |C|(\omega(G) - 1)$.
Clearly, $S \subseteq \bigcup_{i=1}^{k} N(x_i)$ and $S = \bigcup_{i=1}^{k} (N(x_i) \cap S)$. Note that $N(x_i) \cap S$ is a clique of size at most $\omega(G) - 1$, because $N(x_i) \cap S$ is a subclique in $G[N[x_i]]$, and any maximum clique in $G[N[x_i]]$ contains x_i. Thus, $|S| = |\bigcup_{i=1}^{k} (N(x_i) \cap S)| \leq \sum_{i=1}^{k} |N(x_i) \cap S| \leq k(\omega(G) - 1)$ and $k \leq \omega(G)$.

By Claims 1 and 2, $\omega_s(G) \leq \omega(G)(\omega(G) - 1)$ which completes the proof of the Part 1 and an example showing this bound is tight is given in Fig. 3. Note that $\{2, 3, 4, 6, 7, 8, 10, 11, 12, 14, 15, 16\}$ forms a subclique of size 12.

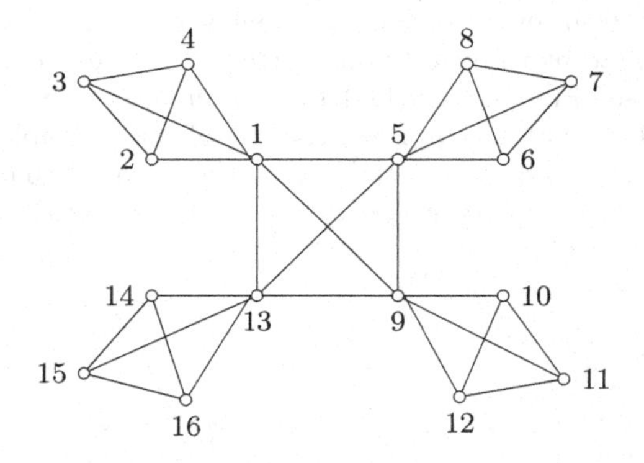

Fig. 3. A P_5-free graph G with $\omega(G) = 4$ and $\omega_s(G) = 12$.

Part 2: Suppose that G has a dominating P_3 and no subgraph K_2 of the P_3 dominates G. We prove that $\omega_s(G) = \omega(G)$. We reserve the rest of the proof for the longer version of this paper. □

Note that Theorems 4 and 5 proves Corollary 3, and Part 1 of Theorem 5 proves Corollary 4.

Corollary 3. *If G is a non trivial connected P_5-free graph, then $\omega_s(G) \leq \omega(G)(\omega(G) - 1)$.*

Corollary 4. *If G is a non trivial graph with a dominating clique, then $\omega_s(G) \leq \omega(G)(\omega(G) - 1)$.*

Remark: The difference between subclique number and clique number is arbitrarily large even for P_6-free trees. An example is shown in Fig. 4.

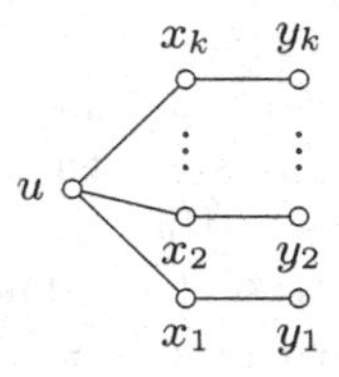

Fig. 4. A P_6-free tree, T with $\omega(T) = 2$, and $S = \{x_1, y_1, y_2, \ldots y_k\}$ is a subclique of size $k + 1$.

6 Double-Split Graphs

In this section, we prove that the subclique number of a double-split graph is equal to its clique number. The class of double split graphs play a key role in the proof of Strong Perfect Graph Theorem [5]. A graph $G(A \cup B, E)$ is a double split graph if (i) $G[A] \cong pK_2$ for some positive integer p (i.e., $G[A]$ is a disjoint union of edges), (ii) $G[B] \cong (mK_2)^c$ for some positive integer m (i.e., $G[B]$ is the complement of a disjoint union of edges), (iii) every vertex $u \in B$ is adjacent to exactly one vertex from every pair of adjacent vertices $x, y \in A$, and (iv) every vertex $x \in A$ is adjacent to exactly one vertex from every pair of non-adjacent vertices $u, v \in B$ (see Fig. 5). Clearly, $|A| = 2p$ and $|B| = 2m$.

Shalu et al. [12] proved that if G is a double-split graph with $p \geq 2$ and $m \geq 2$, then $m + 1 \leq \chi_{cd}(G) \leq m + 2$. Also, they proved that $\chi_{cd}(G) = m + 1$ if and only if there are two vertices $u, v \in B$ such that $N(u) \cap A = N(v) \cap A$.

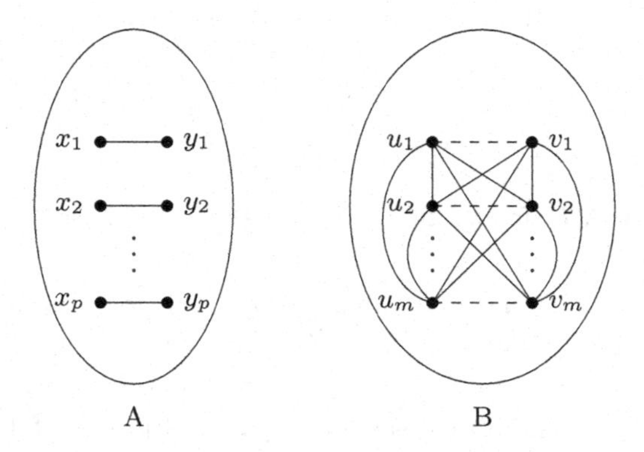

Fig. 5. Structure of a double-split graph. Edges between A and B are not displayed. Dashed lines join non-adjacent vertices in B.

Theorem 6. *Let $G(A \cup B, E)$ be a double-split graph. Then, $\omega_s(G) = \omega(G) = m + 1$.*

Proof. Let S be a subclique in G. We prove that $|S| \leq m+1$. First, we establish the following claims.

Claim 1: If $x, x' \in S \cap A$, then x and x' do not have a common neighbor in B. If not, there exists a vertex $u \in B$ such that ux and ux' are edges in G. By condition (iii) (of the definition of double-split graph), u is adjacent to exactly one end vertex of every edge in A. Since u is adjacent to both x and x', xx' is not an edge in G. Hence, $d_G(x, x') = 2$, a contradiction because S is a subclique.

Claim 2: S can contain at most two vertices from A.
On the contrary, assume that $x, y, z \in S \cap A$. Since $B \neq \emptyset$, there exist $u, v \in B$ such that $uv \notin E(G)$. Then, by condition (iv), x is adjacent to either u or v. W.l.o.g., assume that x is adjacent to u. Since $x, y \in S \cap A$, $uy \notin E(G)$ by Claim 1. Again, by condition (iv), vy is an edge in G. Now, by Claim 1, z is neither adjacent to u nor adjacent to v, a contradiction to condition (iv).

Claim 3: Every vertex in $S \cap B$ is adjacent to each vertex in $S \cap A$.
If not, there exist $x \in S \cap A$ and $u \in S \cap B$ such that $ux \notin E(G)$. Let y be the neighbor of x in A. Then, by condition (iii), $uy \in E(G)$. This implies that $d_G(x, u) = 2$, a contradiction to the subclique property of S. Hence, every vertex in $S \cap B$ is adjacent to each vertex in $S \cap A$.

Claim 4: If $S \cap B \neq \emptyset$, then $|S \cap A| \leq 1$.
On the contrary, assume that $S \cap B \neq \emptyset$, and there are two vertices $x, z \in S \cap A$. Let $u \in S \cap B$. Then by Claim 3, $ux, uz \in E(G)$, a contradiction to Claim 1.

Claim 5: If $m = 1$, then $B = \{u_1, v_1\}$ is a subclique in G.
Clearly, $u_1v_1 \notin E(G)$. By condition (iv), u_1 and v_1 do not have a common neighbor in A. This implies that $d_G(u_1, v_1) \neq 2$. Hence, B is a subclique of size $2 (= m+1)$.

Claim 6: For $m > 1$, S can contain at most m vertices of B.
Clearly, for every non-adjacent pair $u, v \in B$, there is a vertex $u' \in B$ such that $u'u, u'v \in E(G)$. This implies that $d_G(u, v) = 2$. Thus, S can contain at most one vertex from every non-adjacent pair in B. Since there are exactly m such disjoint non-adjacent pairs in B, $|S \cap B| \leq m$.

Next, we prove that $|S| \leq m+1$. We consider the following three cases.
(1) $S \cap A = \emptyset$ implies $S \subseteq B$, and hence $|S| \leq m+1$, by Claims 5 and 6.
(2) $S \cap B = \emptyset$ implies $S \subseteq A$, and hence $|S| \leq 2 \leq m+1$, by Claim 2.
(3) $S \cap B \neq \emptyset$ and $S \cap A \neq \emptyset$ implies $|S| \leq m+1$ (Proof of *(3)*: Let $x \in S \cap A$. Then, by Claims 3 and 4, it is evident that $S \subseteq (N(x) \cap B) \cup \{x\}$. By condition (iv), $|N(x) \cap B| = m$, and hence $|S| \leq m+1$).

By above three cases, $|S| \leq m + 1$. Since S is a arbitrary sublcique and $|S| \leq m + 1$, $\omega_s(G) \leq m + 1$. We know that every clique is also a subclique. Thus, $\omega(G) \leq \omega_s(G)$. If $x \in A$, then $(N(x) \cap B) \cup \{x\}$ is a clique in G. This implies that $\omega(G) \geq m + 1$. Since $m + 1 \leq \omega(G) \leq \omega_s(G) \leq m + 1$, we have $\omega_s(G) = \omega(G) = m + 1$. □

7 Conclusion

We present algorithms to compute an optimal cd-coloring and a maximum subclique of (i) trees with time complexity $O(n)$ and (ii) co-bipartite graphs with time complexity $O(n^{2.5})$, and thereby improve the known $O(n^3)$ algorithms. If G is a non-trivial connected P_5-free graph, then $\omega_s(G) \leq \omega(G)(\omega(G) - 1)$. Note that such an upper bound (for subclique number) doesn't exist even for the class of P_6-free tress, since for every integer $k \geq 1$ there exists a P_6-free tree with $\omega = 2$ and $\omega_s = k + 1$ (see Fig. 4).

References

1. Amin, S.M., Wollenberg, B.F.: Towards a smart grid. IEEE Power Energ. Mag. **3**, 34–41 (2005). https://doi.org/10.1109/MPAE.2005.1507024
2. Androutsellis-Theotokis, S., Spinellis, D.: A survey of peer-to-peer content distribution technologies. ACM Comput. Surv. **36**, 335–371 (2004). https://doi.org/10.1145/1041680.1041681
3. Arumugam, S., Chandrasekar, K.R., Misra, N., Philip, G., Saurabh, S.: Algorithmic aspects of dominator colorings in graphs. Comb. Algorithms **7056**, 19–30 (2011). https://doi.org/10.1007/978-3-642-25011-8_2
4. Bocsó, D., Tuza, Z.: Dominating cliques in P_5-free graphs. Periodica Mathematica Hungarica **21**, 303–308 (1990). https://doi.org/10.1007/BF02352694
5. Chudnovsky, M., Robertson, N., Seymour, P., Thomas, R.: The strong perfect graph theorem. Ann. Math. **164**, 51–229 (2006). https://doi.org/10.4007/annals.2006.164.51
6. Krithika, R., Rai, A., Saurabh, S., Tale, P.: Parameterized and exact algorithms for class domination coloring. In: Steffen, B., Baier, C., van den Brand, M., Eder, J., Hinchey, M., Margaria, T. (eds.) SOFSEM 2017. LNCS, vol. 10139, pp. 336–349. Springer, Cham (2017). https://doi.org/10.1007/978-3-319-51963-0_26
7. Merouane, H.B., Haddad, M., Chellali, M., Kheddouci, H.: Dominated colorings of graphs. Graphs Comb. **31**(3), 713–727 (2014). https://doi.org/10.1007/s00373-014-1407-3
8. Micali, S., Vazirani, V.V.: An $O(\sqrt{|V|}|E|)$ algorithm for finding maximum matching in general graphs. In: Proceedings of 21st IEEE Symposium on Foundations of Computer Science, pp. 17–27 (1980). https://doi.org/10.1109/SFCS.1980.12
9. Monti, A., Ponci, F., Benigni, A., Liu, J.: Distributed intelligence for smart grid control. In: International School on Nonsinusoidal currents and Compensation, Lagow, Poland (2010). https://doi.org/10.1109/ISNCC.2010.5524469
10. Shalu, M.A., Sandhya, T.P.: The cd-coloring of graphs. In: Govindarajan, S., Maheshwari, A. (eds.) CALDAM 2016. LNCS, vol. 9602, pp. 337–348. Springer, Cham (2016). https://doi.org/10.1007/978-3-319-29221-2_29

11. Shalu, M.A., Vijayakumar, S., Sandhya, T.P.: A lower bound of the cd-chromatic number and its complexity. In: Gaur, D., Narayanaswamy, N.S. (eds.) CALDAM 2017. LNCS, vol. 10156, pp. 344–355. Springer, Cham (2017). https://doi.org/10.1007/978-3-319-53007-9_30
12. Shalu, M.A., Vijayakumar, S., Sandhya, T.P.: On complexity of cd-coloring of graphs. Discret. Appl. Math. **280**, 171–185 (2020). https://doi.org/10.1016/j.dam.2018.03.004
13. Swaminathan, V., Sundareswaran, R.: Color class domination in graphs. In: Mathematical and Experimental Physics. Narosa Publishing House (2010)
14. Venkatakrishnan, Y.B., Swaminathan, V.: Color class domination number of middle and central graph of $K_{1,n}, C_n, P_n$. Adv. Model. Optim. **12**, 233–237 (2010)
15. West, D.B.: Introduction to Graph Theory, 2nd edn. Pearson, London (2018)

Cut Vertex Transit Functions of Hypergraphs

Manoj Changat[1(✉)], Ferdoos Hossein Nezhad[1], and Peter F. Stadler[2,3,4,5,6]

[1] Department of Futures Studies, University of Kerala, Trivandrum 695 581, India
mchangat@keralauniversity.ac.in, ferdows.h.n@gmail.com
[2] Bioinformatics Group, Department of Computer Science, and Interdisciplinary Center for Bioinformatics, Universität Leipzig, Härtelstrasse 16-18, 04107 Leipzig, Germany
studla@bioinf.uni-leipzig.de
[3] Max Planck Institute for Mathematics in the Sciences, Leipzig, Germany
[4] Institute for Theoretical Chemistry, University of Vienna, Vienna, Austria
[5] Facultad de Ciencias, Universidad Nacional de Colombia, Bogotá, Colombia
[6] Santa Fe Institute, Santa Fe, NM, USA

Abstract. We study the cut vertex transit function of a connected graph G and discuss its betweenness properties. We show that the cut vertex transit function can be realized as the interval function of a block graph and derive an axiomatic characterization of the cut vertex transit function. We then consider a natural generalization to hypergraphs and identify necessary conditions.

Keywords: Transit function · Convexity · Cut vertices · Block graphs

1 Introduction

A *transit function* R defined on a non-empty set V is a function $R : V \times V \to 2^V$ satisfying the three axioms

(t1) $x \in R(x, y)$ for all $x, y \in V$,
(t2) $R(x, y) = R(y, x)$ for all $x, y \in V$,
(t3) $R(x, x) = \{x\}$ for all $x \in V$.

Transit functions on discrete structures were introduced by H.M. Mulder [13] to generalize the concept of betweenness in an axiomatic way. Intuitively, the *transit sets* $R(x, y)$ can be interpreted as an interval delimited by x and y. Transit functions captured attention in particular on discrete sets endowed with some additional structure, such as graphs, partially ordered sets, hypergraphs, etc. Several types of interval functions that can be defined in terms of paths were studied in some detail. Most of the literature concerns the shortest path transit function

$$I(u, v) := \{w \in V | w \text{ lies on a shortest } uv\text{-path}\}. \quad (1)$$

A. Mudgal and C. R. Subramanian (Eds.): CALDAM 2021, LNCS 12601, pp. 222–233, 2021.
https://doi.org/10.1007/978-3-030-67899-9_17

on a connected graph G, see e.g. [12,14,16,17]. As alternatives in particular the induced path [4,5,8,11,18] and the all-paths transit functions [3] have been considered. P. Duchet [10] considered the following notion of betweenness for both graph and hypergraphs:

$$C(u, v) := \{w \in V | w \text{ lies on every } uv\text{-path}\}. \tag{2}$$

On graphs, $C(u, v) = \{u, v\}$ whenever u and v are located in the same block, and $C(u, v) = V$ if u and v are located in different connected components. For a connected graph G we therefore have the equivalent definition [13]

$$C(u, v) = \{u, v\} \cup \{w | w \text{ is a cut vertex between } u \text{ and } v\}. \tag{3}$$

Therefore, C is usually called the *cut vertex transit function* of a graph G. A similar relation with cut vertices can be found for hypergraphs, see Sect. 3 below.

2 Cut Vertex Transit Functions of Graphs

Notation and Terminology

Let $G = (V, E)$ be a finite, simple graph with vertex set V and edge-set E. Two graphs $G = (V, E)$ and $H = (W, F)$ are *isomorphic* if and only if there is a bijection f from V to W such that for adjacent vertices $u, v \in V$ the images $f(u), f(v)$ are adjacent vertices in H. Given G and a vertex $v \in V$, we write $G - v$ for the graph obtained by removing v and all its incident edges. We say that v is a cut vertex in a connected graph G if G has at least one edge and $G - v$ is disconnected.

A graph is 2-connected if it contains no cut vertex. A block of G is a maximal 2-connected subgraph. A clique is a complete subgraph. A graph is a *block graph* if all its blocks are cliques. The *block closure* G^* of a connected graph G is the graph obtained from G by joining two vertices whenever they are in the same block of G. Thus G^* is the block graph.

Let R be a transit function on V. The *underlying graph* G_R of R has vertex set V and $uv \in E$ is an edge of G_R if and only if $R(u, v) = \{u, v\}$. Note that if R is a transit function on G, then G_R need not be isomorphic with G, see [13] for counterexamples. The *transit graph* G^t_R of a transit function R on V is defined as the graph with vertex set V and $uv \in E$ is an edge of G^t_R if there is no $x \neq u, v$ such that $R(u, x) \cap R(x, v) = \{x\}$.

For any arbitrary transit function R, we define

$$R(u, v, w) := R(u, v) \cap R(v, w) \cap R(w, u). \tag{4}$$

The cardinality $|R(u, v, w)|$ for several of path-based transit function characterizes interesting graph classes. For instance, in terms of the shortest path function, the graphs for which $|I(u, v, w)| > 0$ are the modular graphs and $|I(u, v, w)| = 1$ characterizes median graphs [14]. For the induced path function, $|J(u, v, w)| > 0$ determines the triangle-free graphs [2], and $|J(u, v, w)| = 1$ identifies the svelte graphs [11]. For the all-path functions, $|A(u,v,w)| > 0$ characterizes the connected graphs and $|A(u,v,w)| = 1$ determines the trees.

The following betweenness axioms were considered by H.M. Mulder in [12] and [11], see also [13]:

(b1) $x \in R(u,v)$, $x \neq v \Rightarrow v \notin R(u,x)$,
(b2) $x \in R(u,v) \Rightarrow R(x,v) \subseteq R(u,v)$,
(b3) if $x \in R(u,v)$ and $y \in R(u,x)$, then $x \in R(y,v)$ for all u,v,x,y.
(b4) if $x \in R(u,v)$, then $R(u,x) \cap R(x,v) = \{x\}$ for all u,v,x,
(m) $x,y \in R(u,v) \Rightarrow R(x,y) \subseteq R(u,v)$.

Properties of C on Graphs

We briefly review the relationships between the underlying graph G_C, the graph G and transit graph G^t_C of the cut vertex transit function C of G.

Proposition 1. (Prop. 9 in [13]) *Let C be a cut vertex transit function of a connected graph G. Then the underlying graph G_C, the block closure G^* of G and transit graph G^t_C of C are isomorphic.*

Proposition 2. *The cut vertex transit function C of a connected graph G satisfies axioms* (b1), (b2), (b3), (b4), *and* (m).

Proof. Propositions 8 and 10 in [13] show that C satisfies (b1), (b2) and (m). The statement that C satisfies (b4) is equivalent to the observation that "$F_C = C$" in [13]. To see that (b3) is satisfied, consider four vertices $x \neq u, x \neq v, y \neq x$ and $y \neq u$ such that $x \in C(u,v)$ and $y \in C(u,x)$. By construction, x is a cut vertex separating u and v, and y is a cut vertex separating u and x. That is, y lies within every ux-path and x within every uv-path in G. Since G is connected, x also lies within every yv-path in G. Hence the cut vertex x separates vertices y and v in G. That is, $x \in C(y,v)$. □

As shown in [5], the underlying graph G_R is connected if R satisfies axioms (b1) and (b2). Since C satisfies this condition, the underlying graph G_C of the cut vertex transit function C of G is connected whenever G is connected.

Proposition 3. (Prop. 11 and 12 in [13]) *Let G be a connected graph. Then for any three vertices of G holds $|C(u,v,w)| \leq 1$. G is a tree if and only if any three distinct vertices u,v,w of G satisfy $|C(u,v,w)| = 1$.*

Lemma 1. *Let R be a transit function satisfying axioms* (b2) *and* $|R(u,v,w)| \leq 1$, *then R satisfies axiom* (b1) *and* (b4) *on V, and G_R is connected.*

Proof. Let $x \in R(u,v)$, suppose R does not satisfy axiom (b4). Then there exists at least one $y \neq x$, such that $y \in R(u,x) \cap R(x,v)$. Since $x \in R(u,v)$, by (b2), $R(u,x) \subseteq R(u,v)$ and $R(x,v) \subseteq R(u,v)$. Therefore $y \in R(u,v)$. Hence $|R(u,x,v)| > 1$, which violates $|R(u,v,w)| \leq 1$. Therefore R satisfies axiom (b4). Since R satisfies (t1), (t2), and (b4), R also satisfies (b1). Since R satisfies (b1) and (b2), G_R is connected. □

An axiomatic characterization of the cut vertex transit function of a graph can be obtained starting from the following simple observation:

Proposition 4. *If G is a block graph, then the shortest path transit function and the cut vertex transit function coincide.*

Proof. Let G be a block graph. If u, v are in the same block, then $I(u, v) = \{u, v\}$ since every block is a clique. Thus $I(u, v) = C(u, v)$. If u and v are in distinct blocks, then there is a unique sequence of blocks between them, which pair-wisely intersect in cut-vertices. Since two consecutive cut vertices are contained in the same block, they are adjacent in G, and thus the sequence of cut vertices form the unique shortest path connecting u and v in G. Therefore $I(u, v) = C(u, v)$.

We note in passing that Proposition 4 can also be obtained using the observation of [15] that G is a block graph if and only the shortest path between any two vertices is unique, and noting that all inner vertices of a shortest path in a block graph are cut vertices since it contains at most one edge from each block. The second ingredient is a characterization of the shortest path transit function, usually called the interval function as defined in (1), for block graphs.

Proposition 5. (Thm. 6 in [1]) *R is the interval function of a block graph G, i.e., $R = I_{G_R}$, if and only if R satisfies the axioms* (t1), (t2), (b1), (b2), *and the additional axiom*

(U*) $R(u, x) \cap R(x, v) = \{x\}$ *implies* $R(u, v) \subseteq R(u, x) \cup R(x, v)$, *for all* $u, v, x \in V$.

Taken together, Propositions 4 and 5 imply

Corollary 1. *A transit function R is the cut vertex transit function of the graph G_R if and only if R satisfies axioms* (t1), (t2), (b1), (b2), *and* (U*).

An alternative characterization can be obtained using $R(u, v, w)$.

Theorem 1. *Let $R : V \times V \to 2^V$ be a function on V. Then R is the cut vertex transit function of the graph G_R if and only if R satisfies the axioms* (t1), (t2), (b2), (U*), *and $|R(u, v, w)| \leq 1$ for all $u, v, w \in V$.*

Proof. Suppose R satisfies the axioms (t1), (t2), (b2), (U*), and $|R(u, v, w)| \leq 1$ for all $u, v, w \in V$. By Lemma 1, R also satisfies (b1), and hence Corollary 1 implies that R is the cut vertex transit function of the block graph G_R. Conversely, let R by the cut vertex transit function of a graph. Therefore, $|R(u, v, w)| \leq 1$ for any $u, v, w \in V$ by Proposition 3. On the other hand, R is the interval function of a block graph by Proposition 4, and hence satisfies in particular (t1), (t2), (b2), and (U*). □

We conclude this section with the following remark which can be deduced from the results of this section.

Corollary 2. *The cut vertex transit function C of a connected graph G coincides with the interval function of its block closure G^*.*

3 Cut Vertex Transit Function of Hypergraphs

3.1 Notation and Terminology

A hypergraph H consists of a set V of vertices and a set $E \subseteq 2^V$ of non-empty edges. A path in a hypergraph H is an alternating sequence of hyperedges $x_1e_1x_2e_2x_3 \ldots x_{k-1}e_{k-1}x_ke_kx_{k+1}$ such that $x_1 \in e_1$, $x_i \in e_{i-1} \cap e_i$, and $x_{k+1} \in e_k$ and for $i \neq j$ we have $x_i \neq x_j$ and $e_i \neq e_j$. Every edge e_i in a path thus contains at least two vertices. A hypergraph is connected if every pair of vertices is connected by a path. A path in H is called simple if $e_i \cap e_j = \emptyset$ for $j \neq i, i\pm 1$. A cycle is a simple path, say $x_1e_1x_2e_2x_3 \ldots x_{k-1}e_{k-1}x_ke_kx_{k+1}$, in which $x_1 = x_{k+1}$. We will need the following simple observation.

Lemma 2. *Let P by a uv-path in H. Then there is a simple uv-path composed of a subset of the hyperedges of P.*

In order to determine $C(u, v)$, therefore, it suffices to consider only simple uv-path. Now let P by a simple uv-path and consider a vertex $x \in P$. We note that x is either contained in a single edge, say e_j, or in the intersection of two consecutive edges $e_i \cap e_{i+1}$. In either case, P can be subdivided into a ux-path P_1 and a xv-path P_2. In the first case, P_1 and P_2 share e_j, while in the second case they have no edge in common. Both paths P_1 and P_2 are again simple.

Paths can also be concatenated, provided they do not contain the same edge. Let P_1 and P_2 be a ux-path and an xv-path, respectively. Then their concatenation $P_1P_2 = ue_1 \ldots e_jxe'_1 \ldots e'_kv$ is again a path provided no edges appear twice as we traverse from u to v. We also define a concatenation in which the last edge of P_1 and the first edge of P_2 coincides: For $P_1 = ue_1 \ldots x_je^*x$ and $P_1 = xe^*x'_1e'_2 \ldots e'_kv$ we set $P_1 \bullet P_2 := ue_1 \ldots x_je^*x'_1e'_2 \ldots e'_kv$. Note that although x does not appear explicitly in $P_1 \bullet P_2$, it is still contained in e^*. Note, however, that the concatenated uv-path does not necessarily contain a simple path that still contains x.

The *strong vertex deletion* removes with a vertex y also all edges from H that contain y. As in the graph case we write $H - y$ for the resulting hypergraph. A *strong cut vertex* is a vertex whose strong deletion renders H disconnected [9]. That is, x is a strong cut vertex in H if and only if there are two distinct vertices $u \neq x$ and $v \neq x$ in H such that every uv-path contains an edge containing x. In this case, we say that *x separates u und v*.

Before we proceed, we consider the concatenation of simple paths at a strong cut vertex.

Lemma 3. *Let x be a strong cut vertex in H separating u and v. Let P_1 and P_2 be simple ux- and xv-paths, respectively. Then there are simple uv-paths P' and P'' that contain the edges of P_1 and P_2, respectively.*

Proof. First, we observe that the concatenations P_1P_2 or $P_1 \bullet P_2$ (in the case of equal end edges) are again paths in this case since all edges of P_1 except for the last one, e_1^*, are contained in the component of $H - x$ that contains u, all edges of P_2 except for the first one, e_2^* are contained in the component of $H - x$ that contains v. In particular, therefore $e_i \cap e_j = \emptyset$ for all edges $e_i \neq e_1^*$ in P_1 and $e_j \neq e_2^*$ in P_2. Thus, if $e_1^* = e_2^*$, then the concatenation $P_1 \bullet P_2$ is again simple, and the assertion follows. If $e_1^* \neq e_2^*$, then we have to consider the following case: (i) If P_1P_2 is simple, then the assertion follows trivially. (ii) otherwise, there is a minimal i such that there is a $x' \in e_i \cap e_2^*$ (in which case $ux_1 \ldots x_ie_ix'P_2$ is a simple uv-path), or there is a maximal j such that $x'' \in e_2^* \cap e_j'$ (in which case $P_1x''e_j' \ldots e_k'v$ is a simple xv-path).

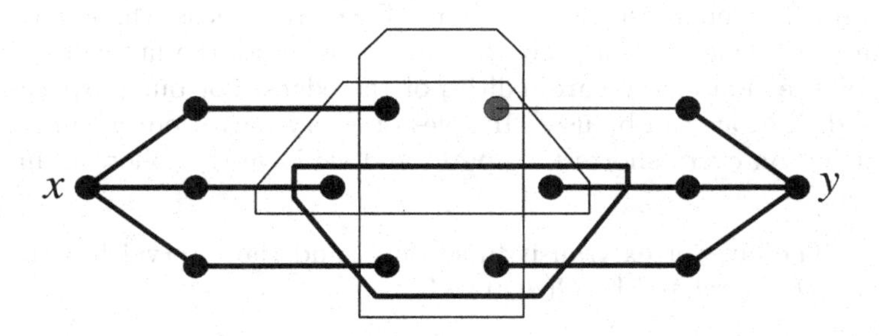

Fig. 1. The absence of strong cut vertices does not imply the existence of two vertex disjoint paths. Every pair of xy-paths shares at least two vertices in the intersection of two of hyperedges with rank 4. Removal of all edges incident to any given vertex, however, still leave an xy-path behind, i.e., there is no strong cut vertex separating x and y.

Very little seems to be known about path-related transit functions on hypergraphs. For our purposes, one of the difficulties seems to be rooted in the much looser connection between alternative paths and cut vertices compared to the special case of graphs, see also [9].

Remark 1. As in graphs, the existence of two edge-disjoint paths is necessary – but not sufficient – to exclude strong cut vertices. We say that two xy-path P' and P'' are vertex-disjoint if their vertex sets only share the endpoints, i.e., $\bigcup_{e \in P'} \cap \bigcup_{e \in P''} = \{x, y\}$. In contrast to graphs, the existence of two vertex disjoint paths is necessary but not sufficient in hypergraphs to rule out strong cut vertices. In the example of Fig. 1, any two xy-paths share (at least) a pair of vertices located in the hyperedges of rank 4, i.e., there is no pair of vertex-disjoint xy-paths. Nevertheless, no vertex is contained in all xy-paths, and hence there is no strong cut vertex separating x and y.

Properties of C on Hypergraphs

As in the case of graphs, we consider the transit function defined in Eq. (2), i.e., for all $u \neq v$ we have $x \in C(u, v)$ if every uv-path in the hypergraph contains an edge that contains x. That is, $x \in C(u, v)$ if and only if x is a strong cut vertex separating u and v.

Since the definition of uv-paths is symmetric and u and v are contained in every uv-path, it is clear that C satisfies (t1) and (t2). By convention we set $C(x, x) = \{x\}$ for all x, i.e., C is a well-defined transit function. Note that $C(u, v)$ in particular contains all vertices in the intersection of all edges in uv-paths that contain u. For instance, if every edge containing u also contains $w \neq u$, then $w \in C(u, v)$ for all v.

Similar to the case of graphs, the interval function $I(u, v)$ of a hypergraph H is defined as the function that returns, for every pair of vertices u, v of H, the set of all vertices lying on shortest uv-paths in H. Note that the exact definition depends on what one means by "shortest". There are at least three natural ways to measure the "length" of a path: its number of edges, the number of included vertices, or the sum of the cardinalities of the edges. For our purposes, any of the above definitions can be used. If x lies on every path from u to v, then x in particular lies on every shortest uv-path, and we have the following immediate observation.

Remark 2. The cut vertex transit function C and the interval function I of a hypergraph H are related by $C(u, v) \subseteq I(u, v)$.

The characterization of cut vertex transit functions of graphs benefits greatly from the existence of the block closure G^* and the convenient properties of block graphs. As a generalization we define the hypergraph H^* obtained from H by adding the hyperedge $\{u, v\}$ whenever $C(u, v) = \{u, v\}$. The inclusion of these edges clearly does not change the cut vertex transit function, i.e., $C = C^*$. Its underlying graph G_{C^*} is of course a block graph. In contrast to the graph case, G^*_C is not connected in general, even if H is connected. For example, if $e = \{x, u, y\}$ is the only edge containing u, then $\{x, u, y\} \in C(u, v)$ for all v, and thus u is an isolated vertex in G_{C^*}. It is tempting to speculate that the correspondence between C and the interval function I^* of the block closure generalizes to hypergraphs. However, the example in Fig. 2 shows that this is not the case. Note that the example is independent of the choice of the exact length function for paths.

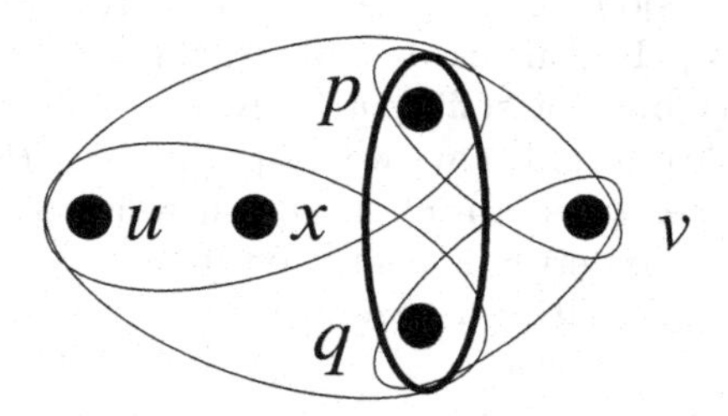

Fig. 2. Construction of the block closure of H with $V = \{p, q, u, v, x\}$ and edges $\{u, x, p\}$, $\{u, x, q\}$, $\{p, v\}$, and $\{q, v\}$ adds only the edge $\{p, q\}$. We have $C^*(u, v) = C(u, v) = \{u, x, v\}$ and $I(u, v) = I^*(u, v) = V$, since both H and H^* contain the two equivalent shortest uv-paths $P_1 = (u\{u, x, p\}p\{p, v\}v)$ and $P_2 = (u\{u, x, q\}q\{q, v\}v)$. Note that this remains true for any "reasonable" notion of "shortest paths".

We start our investigation of the cut vertex transit function C with a simple observation

Lemma 4. *The cut vertex transit function of a hypergraph satisfies axiom* (b2).

Proof. From $x \in C(u,v)$, we know that every uv-path contains an edge containing x, and thus an xv-subpath. By assumption, $y \in C(x,v)$, i.e., every xv-path contains an edge containing y. Thus every uv-path contains such an edge. Consequently we have $y \in C(u,v)$ and $C(x,v) \subseteq C(u,v)$.

Next we investigate to what extent C captures betweenness properties between cut vertices.

Lemma 5. *Let x and y be two distinct strong cut vertices separating u and v. Then y is a strong cut vertex separating u and x or x and v.*

Proof. By assumption every (simple) uv-path contains an edge containing x and an edge containing y. By Lemma 3 the concatenation of any simple ux-path and any simple xv-path always contains y. Now suppose there are two simple uv-paths P_1 and P_2 such that y appears only in edges of the ux-subpath of P_1 and only in edges of the xv-subpath of P_2. Since P_1 and P_2 are simple, we can concatenate the xv-subpath of P_1 and the ux-subpath of P_2 to obtain a simple path that contains no edge containing u, a contradiction. Hence y is contained in every ux-path or in every xv-path.

Consider the following axiom for general transit functions, which also appeared in the characterization of hierarchies [7].

(h") $x \in R(u,v)$ implies $R(u,v) = R(u,x) \cup R(x,v)$.

We can now restate Lemma 5 as

Corollary 3. *The cut vertex transit function C of a hypergraph satisfies axiom* (h").

Proof. Lemma 5 translates to: "$x, y \in C(u,v)$ implies $y \in C(u,x)$ or $y \in C(x,v)$ for all $u, v \in V$." Thus $C(u,v) \subseteq C(u,x) \cup C(x,v)$. Property (b2) ensures that $C(u,x) \cup C(x,v) \subseteq C(u,v)$.

Note that axiom (h") trivially implies axiom (b2).

Lemma 6. *Let x and y be two distinct strong cut vertices in H such that x separates u and v, and y separates u and x. Then x separates y and v.*

Proof. We first observe that both x and y are contained in every uv-path. We again invoke Lemma 3 to argue that every simple uv-path can be subdivided into a ux- and xv-path, with the property that every ux-path contains an edge containing y. Furthermore, if every xv-path also contains y, then x and y are always contained in a common edge of every uv-path. In the latter case x of course always contained in the yv-subpath. Now suppose there is a xv-path that does not contain y. We can then subdivide every simple uv-path into an ux-path and an xv-path, and further subdivide the ux-path into a uy-path and yx-path. If x and y always appear in the same edge along P, we can argue as above. Otherwise, x is always contained in the yv-subpath of every uv-path. Thus x always separates y and v. □

Somewhat surprisingly, this result is no longer true when some of the vertices u, v, x, and y are not assumed to be pairwisely distinct. To see this, consider the hypergraph H_3 consisting of three vertices and the single edge $e = V = \{u, x, v\}$. We have $C(u, x) = C(x, v) = C(u, v) = \{u, x, v\}$. Setting $y = v$, however, we would claim $x \in C(y, v) = C(v, v)$, contradicting (t3). In particular, therefore, C does not satisfy (b3) but only the following, slightly weaker axiom:

(b3') If x, y, u, v distinct, $x \in R(u, v)$, and $y \in R(u, x)$ then $x \in R(y, v)$.

The hypergraph H_3 also serves as a counter-example for (b1): We have $x \in C(u, v)$ but $C(u, x) = C(x, v) = C(u, v)$. Hence we do not seem to have an analog of axiom (b1). By the same argument, (b4) does not hold.

The following axiom is motivated by the idea that, given three distinct vertices, at least one of the paths connecting them must pass through each cut vertex that separates any two of them.

(X') If $x \notin R(u, v)$ then $R(u, v) \subseteq R(u, x) \cup R(x, v)$ for $u, v \in V$.

Lemma 7. *The cut vertex transit function of a hypergraph satisfies axiom* (X').

Proof. Assume, for contradiction, that there is $y \in C(u, v)$ such that $y \notin C(u, x)$ and $y \notin C(x, u)$. Then there is a ux- and xv-path, neither of which contains an edge containing y. Their concatenation, therefore, contains an uv-path without an edge containing y, and thus $y \notin C(u, v)$, a contradiction. The assertion follows immediately.

In fact, axiom (X') can be viewed as the following axiom (X) ("triangle inequality")

(X) For all $u, v, x \in V$ holds $R(u, v) \subseteq R(u, x) \cup R(x, v)$.

Corollary 4. *The cut vertex transit function of a hypergraph satisfies axiom* (X).

Proof. Let $u, v, x \in V$ If $x = u$ or $x = v$ or $u = v$ the assertion holds trivially; hence assume that u, v, and x are pairwise disjoint. If $x \in C(u, v)$ then (h") implies $C(u, x) \cup C(x, v) = C(u, v)$. If $x \notin C(u, v)$, the assertion follows immediately from (X').

Axiom (U*) is a trivial consequence of (X).

We note in passing that (X) is satisfied also by the transit functions of hierarchies, which satisfy (m), (h") and (h"') [$x \notin R(u, v)$ implies $R(u, x) = R(x, v)$] [7]. To see this, we note that (h"') implies $v \in R(u, x)$ and $u \in R(x, v)$, which together with (m) implies $R(u, v) \subseteq R(u, x) \cup R(x, v)$, i.e., (X').

Finally we consider the axiom

(m') For all $u, v, x, y \in V$ with $R(u, v) \cap R(x, y) \neq \emptyset$ there are $p, q \in V$ such that $R(u, v) \cap R(x, y) = R(p, q)$.

Property (m') is a key ingredient towards constructing convexities from transit sets: it states that the intersection of two transit sets is either empty or a transit set [6]. We have

Lemma 8. *The cut vertex transit function C of a hypergraph satisfies* (m').

Proof. First note that if $C(u, v) \cap C(x, y) = \{p\}$, the assertion holds because $C(p, p) = \{p\}$ by definition. This covers in particular the cases $u = v$ or $x = y$. For $\{u, v\} = \{x, y\}$ there is nothing to show. Thus we consider $u \neq v$, $x \neq y$, $\{u, v\} \neq \{x, y\}$, and assume that there are two distinct vertices $p', q' \in C(u, v) \cap C(x, y)$. For contradiction, suppose that (m') does not hold. That is, for every choice of p', q', there is a vertex $r \in C(u, v) \cap C(x, y) \setminus C(p', q')$. On the other hand, every uv-path and every xy path contains an rp'-path that contains an edge containing q' or an rq'-path that contains an edge containing p'; otherwise, at least one of p', q', r' would not be contained in $C(u, v) \cap C(x, y)$. Thus we have $C(p', q') \subsetneq C(p', r) \subseteq C(u, v) \cap C(x, y)$ or $C(p', q') \subsetneq C(r, q') \subseteq C(u, v) \cap C(x, y)$. That is, the initial choice $C(p', q')$ can be replaced by a strict super-set that contains r. Repeating the argument therefore produces arbitrarily large super-sets of $C(p', q')$ that are still contained in $C(u, v) \cap C(x, y)$, a contradiction. Thus there is an inclusion-maximal choice p, q for p', q' such that $C(p, q) = C(u, v) \cap C(x, y)$. □

Note that in graphs the choice of p, q is unique due to the uniqueness of the shortest paths in the block closure [15] and the fact the end points of a path are uniquely defined. The latter is no longer true in hypergraphs since there may be a vertex $p' \neq p$ that is contained in every choice for the first edges in a pq-path, in which case $C(p', q) = C(p, q)$. For the same reason $|C(u, v, w)| \leq 1$ fails for hypergraphs.
We summarize our observations as

Theorem 2. *The cut vertex transit function C of a hypergraph satisfies* (t1), (t2), (b2), (b3'), (m), (h"), (X'), (X), *and* (m').

Proof. The general axioms (t1) and (t2) are the basic transit axioms already discussed at the beginning of this section. Lemma 4 establishes (b2). Axiom (b3') is simple rewording of Lemma 6. Axiom (h") holds due to Corollary 3. It is shown in [7] that (h") implies (m). Axioms (X) and (U*) are shown in Lemma 7 and Corollary4, resp., and (m') holds because of Lemma 8.

Note that (m) and thus also (b2) are implied by (h") and thus will not be needed in a characterization that makes use of axiom (h"). Similarly, (U*) is redundant.

The results above provide some insights into the cut vertex transit function C of hypergraphs. Even though C has some very strong properties (axioms), they do not appear to provide a complete characterization. It remains an interesting problem, therefore, whether an axiomatic characterization exists for the cut vertex transit function of hypergraphs and if so, whether it will be similar in spirit to characterization of the cut vertex transit function of graphs.

Acknowledgments. This research work was performed while MC was visiting the Max Plank Institute for Mathematics in the Sciences (MPI-MIS), Leipzig and Leipzig University's Interdisciplinary Center for Bioinformatics (IZBI). MC acknowledges the financial support of the MPI-MIS, the hospitality of the IZBI, and the Commission for Developing Countries of the International Mathematical Union (CDC-IMU) for providing the individual travel fellowship supporting the research visit to Leipzig. This work was supported in part by SERB-DST, Ministry of Science and Technology, Govt. of India, under the MATRICS scheme for the research grant titled "Axiomatics of Betweenness in Discrete Structures" (File: MTR/2017/000238).

References

1. Balakrishnan, K., Changat, M., Lakshmikuttyamma, A.K., Mathew, J., Mulder, H.M., Narasimha-Shenoi, P.G., Narayanan, N.: Axiomatic characterization of the interval function of a block graph. Discret. Math. **338**, 885–894 (2015). https://doi.org/10.1016/j.disc.2015.01.004
2. Changat, M., Hossein Nezhad, F., Mohandas, S., Mulder, H.M., Narasimha-Shenoi, P.G., Stadler, P.F.: Interval function, induced path transit function, modular, geodetic and block graphs and axiomatic characterizations (2019, in preparation)
3. Changat, M., Klavžar, S., Mulder, H.M.: The all-paths transit function of a graph. Czechoslovak Math. J. **51**, 439–448 (2001). https://doi.org/10.1023/A:1013715518448
4. Changat, M., Mathew, J.: Induced path transit function, monotone and Peano axioms. Discret. Math. **286**, 185–194 (2004). https://doi.org/10.1016/j.disc.2004.02.017
5. Changat, M., Mathews, J., Mulder, H.M.: The induced path function, monotonicity and betweenness. Discret. Appl. Math. **158**, 426–433 (2010). https://doi.org/10.1016/j.dam.2009.10.004
6. Changat, M., Narasimha-Shenoi, P.G., Stadler, P.F.: Axiomatic characterization of transit functions of weak hierarchies. Art Discret. Appl. Math, **2**, P1.01 (2019). https://doi.org/10.26493/2590-9770.1260.989
7. Changat, M., Nezhad, F.H., Stadler, P.F.: Axiomatic characterization of transit functions of hierarchies. Ars Math. Contemp. **14**, 117–128 (2018)
8. Changat, M., Peterin, I., Ramachandran, A., Tepeh, A.: The induced path transit function and the Pasch axiom. Bull. Malaysian Math. Sci. Soc. **39**(1), 123–134 (2015). https://doi.org/10.1007/s40840-015-0285-z
9. Dewar, M., Pike, D., Proos, J.: Connectivity in hypergraphs. Canad. Math. Bull. **61**, 252–271 (2018). https://doi.org/10.4153/CMB-2018-005-9
10. Duchet, P.: Classical perfect graphs: an introduction with emphasis on triangulated and interval graphs. Ann. Discret. Math. **21**, 67–96 (1984). https://doi.org/10.1016/S0304-0208(08)72924-4
11. Morgana, M.A., Mulder, H.M.: The induced path convexity, betweenness and svelte graphs. Discret. Math. **254**, 349–370 (2002). https://doi.org/10.1016/S0012-365X(01)00296-5
12. Mulder, H.M.: The Interval function of a Graph, MC Tract, vol. 132. Mathematisch Centrum, Amsterdam (1980)

13. Mulder, H.M.: Transit functions on graphs (and posets). In: Changat, M., Klavžar, S., Mulder, H.M., Vijayakumar, A. (eds.) Convexity in Discrete Structures. Ramanujan Lecture Notes Series, vol. 5, pp. 117–130. International Press, Boston (2008)
14. Mulder, H.M., Nebeský, L.: Axiomatic characterization of the interval function of a graph. Eur. J. Comb. **30**, 1172–1185 (2009). https://doi.org/10.1016/j.ejc.2008.09.007
15. Mulder, H.M.: An observation on block graphs. Bull. Inst. Comb. Appl. **77**, 57–58 (2016)
16. Nebeský, L.: A characterization of the interval function of a connected graph. Czech. Math. J. **44**, 173–178 (1994). https://doi.org/10.21136/CMJ.1994.128449
17. Nebeský, L.: Characterization of the interval function of a (finite or infinite) connected graph. Czech. Math. J. **51**, 635–642 (2001). https://doi.org/10.1023/A:1013744324808
18. Nebeský, L.: The induced paths in a connected graph and a ternary relation determined by them. Math. Bohem. **127**, 397–408 (2002). https://doi.org/10.21136/MB.2002.134072

Lexicographic Product of Digraphs and Related Boundary-Type Sets

Manoj Changat[1], Prasanth G. Narasimha-Shenoi[2], and Mary Shalet Thottungal Joseph[2](✉)

[1] Department of Futures Studies, University of Kerala, Trivandrum, India
mchangat@gmail.com
[2] Department of Mathematics, Government College Chittur, Palakkad, India
prasanthgns@gmail.com, mary_shallet@yahoo.co.in

Abstract. Let $D = (V, E)$ be a digraph and $u, v \in V(D)$. The metric, *maximum distance* is defined by $md(u, v) = \max\{\overrightarrow{d}(u, v), \overrightarrow{d}(v, u)\}$ where $\overrightarrow{d}(u, v)$ denote the length of a shortest directed $u - v$ path in D. The relationship between the boundary-type sets of the lexicographic product of two digraphs and its factor graphs have been studied in this article.

Keywords: Maximum distance · Boundary-type sets · Strongly connected digraph · Lexicographic product · DDLE digraph

Subject Classification (2020): 05C12 · 05C20 · 05C76

1 Introduction

The operations on digraphs that result in a bigger digraph have been of great research interest. There exist many digraph products for which the vertex set is the Cartesian product of vertex sets of its factors. They differ by the definitions of the edge sets. Among them, four are called standard products. These are the Cartesian product, the strong product, the direct product, and the lexicographic product. For a rich bibliography about them, see [10].

The study of the boundary vertex sets of a graph, namely the *boundary, contour, eccentricity*, and *peripheral sets* was initiated in [5,7]. They find applications in various contexts like facility location [8] and rebuilding in graphs [3].

Minimizing the distance between nodes in the digraph sense is equivalent to minimizing the distance in either direction. Thus the metric, *maximum distance* (see [9]) $md(u, v)$ for $u, v \in V(D)$ is the most suitable in these networks. The concept of boundary type vertices defined in graphs using the usual distance can be extended to digraphs using the metric maximum distance.

The study of lexicographic product of two graphs was initiated by Frank Harary in 1959. In [12], Harary defined a binary operation on graphs, which was called composition, such that the group of the composition of two graphs is permutationally equivalent to the composition of their groups.

A. Mudgal and C. R. Subramanian (Eds.): CALDAM 2021, LNCS 12601, pp. 234–246, 2021.
https://doi.org/10.1007/978-3-030-67899-9_18

In this article, the four boundary-type sets of lexicographic product of two directed graphs is investigated. A similar attempt was made in [6] to find the boundary-type sets of the Cartesian product of two directed graphs.

The article is organised as follows. In Sect. 2, all the basic definitions are provided. In Sect. 3, the definition of a digraph satisfying the DDLE property is introduced. The four boundary-type sets of the lexicographic product $D_1 \circ D_2$ are considered in the following cases. Subsection 3.1 deals with the case when D_1 is a digraph satisfying the DDLE property. Subsections 3.2 and 3.3 deal with the cases when D_1 is a directed cycle and a symmetric digraph, respectively.

2 Preliminaries

A *digraph* D consists of a non-empty finite set $V(D)$ of elements called vertices and a finite set $E(D)$ of ordered pairs of distinct vertices called arcs or edges [1]. The first vertex of the ordered pair is the tail of the edge, and the second is the head; together they are the endvertices. The underlying graph UD of a digraph D is the simple graph with the vertex set $V(D)$ and the unordered pair $(x, y) \in E(UD)$ if and only if either $(x, y) \in E(D)$ or $(y, x) \in E(D)$.

The following definitions are from [2]. A *directed walk* is an alternating sequence $W = x_1 a_1 x_2 a_2 x_3 \ldots x_{k-1} a_{k-1} x_k$ of vertices x_i and arcs a_j from D such that x_i and x_{i+1} are the tail and head of a_i, respectively, for every $i \in [k-1]$. If the vertices of W are distinct, then W is a *directed path* (*dipath*). If the vertices $x_1, \ldots, x_{k-1}$ are distinct, $k \geq 3$ and $x_1 = x_k$, then W is a *directed cycle* (*dicycle*). A pair of opposite arcs forms a directed cycle of length 2.

In this article, a path will always mean a 'directed path'. A digraph is *strongly connected* or *strong* if, for each ordered pair u, v of vertices, there is a path from u to v. The *length* of a path is the number of edges in the path. Let u and v be vertices of a strongly connected digraph D. A shortest directed $u-v$ path is also called a directed $u-v$ *geodesic*. The number of edges in a directed $u-v$ geodesic is called the directed distance $\vec{d}(u, v)$. But this distance is not a metric because $\vec{d}(u, v) \neq \vec{d}(v, u)$ is possible. So in [9], Chartrand and Tian introduced two other distances in a strong digraph, namely the maximum distance $md(u, v) = \max\{\vec{d}(u, v), \vec{d}(v, u)\}$ and sum distance $sd(u, v) = \vec{d}(u, v) + \vec{d}(v, u)$, both of which are metrics.

This article deals with the distance md. The *m-eccentricity* of a vertex v, the *m-radius*, and the *m-diameter* of a digraph D are also defined in [9]. We denote them respectively as $\mathrm{mecc}(v)$, $\mathrm{mrad}(D)$, and $\mathrm{mdiam}(D)$. Thus, $\mathrm{mecc}(v) = \max\limits_{u \in V(D)} \{md(v, u)\}$, $\mathrm{mrad}(D) = \min\limits_{v \in V(D)} \{\mathrm{mecc}(v)\}$, and $\mathrm{mdiam}(D) = \max\limits_{v \in V(D)} \{\mathrm{mecc}(v)\}$.

In a strongly connected digraph, the m-distance between every pair of vertices and the m-eccentricity of every vertex is finite.

The concept of neighborhood of a vertex v in a digraph D is as follows: $N_D^+(v) = \{u \in V : (v, u) \in E\}$, $N_D^-(v) = \{w \in V : (w, v) \in E\}$. The sets

$N_D^+(v)$, $N_D^-(v)$ and $N_D(v) = N_D^+(v) \bigcup N_D^-(v)$ are called the out-neighborhood, the in-neighborhood and the neighborhood of v, respectively.

Most of the following definitions are analogous to the definitions in [7]. Let $D = (V, E)$ be a strong digraph and $u, v \in V$. The vertex v is said to be an *m-boundary vertex* of u if no neighbor of v is further away from u than v. A vertex v is called an *m-boundary vertex* of D if it is the m-boundary vertex of some vertex $u \in V$. The *m-boundary* $m\partial(D)$ of D is the set of all of its m-boundary vertices; that is, $m\partial(D) = \{v \in V : \exists u \in V \text{ such that } \forall w \in N(v),\ md(u, w) \leq md(u, v)\}$. The vertex v is called an *m-eccentric vertex* of u if no vertex in V is further away from u than v. Then $md(u, v) = \text{mecc}(u)$. A vertex v is called an *m-eccentric vertex* of D if it is the m-eccentric vertex of some vertex $u \in V$. The *m-eccentricity* $\text{mEcc}(D)$ of D is the set of all of its m-eccentric vertices; that is, $\text{mEcc}(D) = \{v \in V : \exists u \in V \text{ such that } \text{mecc}(u) = md(u, v)\}$. A vertex $v \in V$ is called an *m-peripheral vertex* of D if no vertex in V has an m-eccentricity greater than $\text{mecc}(v)$; that is, if the m-eccentricity of v is equal to the m-diameter of D. The *m-periphery* $\text{mPer}(D)$ of D is the set of all of its m-peripheral vertices; that is, $\text{mPer}(D) = \{v \in V : \text{mecc}(u) \leq \text{mecc}(v), \forall u \in V\} = \{v \in V : \text{mecc}(v) = \text{mdiam}(D)\}$. A vertex $v \in V$ is called an *m-contour vertex* of D if no neighbor vertex of v has an m-eccentricity greater than $\text{mecc}(v)$. The following definition is from [5]. The *m-contour* $\text{mCt}(D)$ of D is the set of all of its contour vertices; that is, $\text{mCt}(D) = \{v \in V : \text{mecc}(u) \leq \text{mecc}(v), \forall u \in N(v)\}$.

From the definitions, it follows that

1. $\text{mPer}(D) \subseteq \text{mCt}(D) \cap \text{mEcc}(D)$,
2. $\text{mEcc}(D) \cup \text{mCt}(D) \subseteq m\partial(D)$.

3 Lexicographic Product of Directed Graphs

The lexicographic product of two digraphs D_1 and D_2 is the digraph $D_1 \circ D_2$, having the vertex set $V(D_1) \times V(D_2)$ and with arc set defined as follows. A vertex (u_i, v_r) is adjacent to (u_j, v_s) in $D_1 \circ D_2$ if either

1. $(u_i, u_j) \in E(D_1)$, or
2. $u_i = u_j$, $(v_r, v_s) \in E(D_2)$.

The lexicographic product of two graphs is not commutative [10]. The distance between two vertices (u_i, v_r) and (u_j, v_s) in the lexicographic product $G \circ H$ of a connected graph G and a graph H is obtained from [10] as:

$$d_{G \circ H}((u_i, v_r), (u_j, v_s)) = \begin{cases} d_G(u_i, u_j) & \text{if } u_i \neq u_j \\ \min\{2, d_H(v_r, v_s)\} & \text{if } u_i = u_j \end{cases}$$

Prior to the definition of the distance between two vertices in the lexicographic product of two digraphs, several other notions from [11] need to be introduced. Let D be a digraph. Given a vertex x of a digraph D, $\xi_D(x)$ is the length of a shortest dicycle in D containing x, or infinity if no such dicycle exists. In this article, $\xi_D(x)$ will be called the dicycle distance of x in D.

Consider two digraphs D_1 and D_2 with vertex sets $V(D_1) = \{u_1, u_2, \ldots, u_n\}$ and $V(D_2) = \{v_1, v_2, \ldots, v_m\}$, respectively. Let $(u_i, v_r), (u_j, v_s) \in V(D_1 \circ D_2)$. The formula for directed distance $\overrightarrow{d}_{D_1 \circ D_2}((u_i, v_r), (u_j, v_s))$ is obtained from [11] as follows:

$$\overrightarrow{d}_{D_1 \circ D_2}((u_i, v_r), (u_j, v_s)) = \begin{cases} \overrightarrow{d}_{D_1}(u_i, u_j) & \text{if } u_i \neq u_j \\ \min\{\xi_{D_1}(u_i), \overrightarrow{d}_{D_2}(v_r, v_s)\} & \text{if } u_i = u_j. \end{cases}$$

Thus

$$\begin{aligned} md_{D_1 \circ D_2}((u_i, v_r), (u_j, v_s)) &= \max\{\overrightarrow{d}_{D_1 \circ D_2}((u_i, v_r), (u_j, v_s)), \overrightarrow{d}_{D_1 \circ D_2}((u_j, v_s), (u_i, v_r))\} \\ &= \begin{cases} \max\{\overrightarrow{d}_{D_1}(u_i, u_j), \overrightarrow{d}_{D_1}(u_j, u_i)\} & \text{if } u_i \neq u_j \\ \min\{\xi_{D_1}(u_i), \max\{\overrightarrow{d}_{D_2}(v_r, v_s), \overrightarrow{d}_{D_2}(v_s, v_r)\}\} & \text{if } u_i = u_j. \end{cases} \\ &= \begin{cases} md_{D_1}(u_i, u_j) & \text{if } u_i \neq u_j \\ \min\{\xi_{D_1}(u_i), md_{D_2}(v_r, v_s)\} & \text{if } u_i = u_j. \end{cases} \end{aligned}$$

Hence it follows that

$$\text{mecc}_{D_1 \circ D_2}(u_i, v_r) = \begin{cases} \min\{\text{mecc}_{D_2}(v_r), 2\} & \text{if } \text{mecc}_{D_1}(u_i) = 1 \\ \max\{\text{mecc}_{D_1}(u_i), \min\{\xi_{D_1}(u_i), \text{mecc}_{D_2}(v_r)\}\} & \text{if } \text{mecc}_{D_1}(u_i) \geq 2. \end{cases}$$

As the digraphs under consideration are clear from the vertex labelling, we may denote $md_{D_1}(u_i, u_j)$ by $md(u_i, u_j)$, $md_{D_1 \circ D_2}((u_i, v_r), (u_j, v_s))$ by $md((u_i, v_r), (u_j, v_s))$, and $\text{mecc}_{D_1 \circ D_2}(u_i, v_r)$ by $\text{mecc}(u_i, v_r)$.

For every vertex x of a strongly connected digraph D with at least two vertices, there exists a dicycle in D containing x. So $\xi_D(x)$ is finite for every vertex x in D. Also, $\text{mecc}_D(x)$ is finite for every vertex x of a strongly connected digraph D. The following definition is introduced to begin with the study of the boundary vertices of the lexicographic product of two digraphs.

Definition 1. *A strong digraph D is said to satisfy the* ***dicycle distance less than eccentricity property*** *or in short the* ***DDLE property****, if for every vertex $x \in V(D)$, $\text{mecc}(x) > \xi_D(x)$. A digraph D which satisfy the DDLE property is called a* ***DDLE digraph****.*

If D_1 is a DDLE digraph, then $\text{mecc}(u_i) > 2$ for all $u_i \in V(D_1)$, and hence $\text{mecc}(u_i, v_r) = \text{mecc}(u_i) > \xi_{D_1}(u_i)$ for all $(u_i, v_r) \in V(D_1 \circ D_2)$. Let G and H be the underlying graphs of the digraphs D_1 and D_2, respectively. Since $N_{G \circ H}(u_i, v_r) = (N_G(u_i) \times V(H)) \bigcup (\{u_i\} \times N_H(v_r))$ [10], and since the neighbors of (u_i, v_r) are exactly the same in $D_1 \circ D_2$ and its underlying graph $G \circ H$, $N_{D_1 \circ D_2}(u_i, v_r) = (N_{D_1}(u_i) \times V(D_2)) \bigcup (\{u_i\} \times N_{D_2}(v_r))$. The lexicographic product of two digraphs D_1 and D_2 is strongly connected if and only if D_1 is strongly connected [10]. Examples of digraphs that do not satisfy the DDLE property are the directed cycles C_n and the complete graphs K_n, $n \geq 2$.

3.1 $D_1 \circ D_2$, D_1 Is a *DDLE* Digraph

As D_1 is a DDLE digraph, $\text{mecc}(u_i) > \xi_{D_1}(u_i)$ for all $u_i \in V(D_1)$. Also since $\text{mecc}(u_i) > 2$, $\text{mecc}(u_i, v_r) = \max\{\text{mecc}(u_i), \min\{\xi_{D_1}(u_i),$

$\text{mecc}(v_r)\}\} = \text{mecc}(u_i)$ for all $(u_i, v_r) \in V(D_1 \circ D_2)$. Thus if D_1 is a DDLE digraph, then we get the following results regarding the m-periphery, m-contour and m-eccentricity sets of $D_1 \circ D_2$. Since the DDLE property is related to the m-eccentricity of a vertex, nothing could be inferred about the m-boundary set $m\partial(D_1 \circ D_2)$.

Proposition 1. *Let D_1 be a strongly connected DDLE digraph and D_2 be an arbitrary digraph. Then*

1. $\text{Per}(D_1 \circ D_2) = \text{mPer}(D_1) \times V(D_2)$,
2. $\text{Ct}(D_1 \circ D_2) = \text{mCt}(D_1) \times V(D_2)$,
3. $\text{Ecc}(D_1 \circ D_2) = \text{mEcc}(D_1) \times V(D_2)$.

Proof. Given D_1 is a DDLE digraph. Let $u_i \in V(D_1)$ and $v_r \in V(D_2)$. Then, $\text{mecc}(u_i, v_r) = \text{mecc}(u_i)$.

$$
\begin{aligned}
1.\ (u_i, v_r) \in \text{mPer}(D_1 \circ D_2) &\iff \text{mecc}(u_i, v_r) \geq \text{mecc}(u_k, v_q) \\
&\qquad \text{for all } (u_k, v_q) \in V(D_1 \circ D_2) \\
&\iff \text{mecc}(u_i) \geq \text{mecc}(u_k) \text{ for all } u_k \in V(D_1) \\
&\iff u_i \in \text{mPer}(D_1).
\end{aligned}
$$

Thus, $\text{mPer}(D_1 \circ D_2) = \text{mPer}(D_1) \times V(D_2)$.

2. $N_{D_1 \circ D_2}(u_i, v_r) = (N_{D_1}(u_i) \times V(D_2)) \bigcup (\{u_i\} \times N_{D_2}(v_r))$.

$$
\begin{aligned}
(u_i, v_r) \in \text{mCt}(D_1 \circ D_2) &\iff \text{mecc}(u_i, v_r) \geq \text{mecc}(u_k, v_q) \text{ for all } (u_k, v_q) \in N(u_i, v_r) \\
&\iff \text{mecc}(u_i) \geq \text{mecc}(u_k) \text{ for all } u_k \in N(u_i) \\
&\iff u_i \in \text{mCt}(D_1).
\end{aligned}
$$

Hence it follows that $\text{mCt}(D_1 \circ D_2) = \text{mCt}(D_1) \times V(D_2)$.

3. $(u_i, v_r) \in \text{mEcc}(D_1 \circ D_2)$ if and only if there exists $(u_j, v_s) \in V(D_1 \circ D_2)$ such that $\text{mecc}(u_j, v_s) = \text{mecc}(u_j) = md((u_j, v_s), (u_i, v_r))$. Since D_1 is a DDLE digraph, $\text{mecc}(u_j, v_s) = \text{mecc}(u_j)$, and $u_j \neq u_i$. Thus, $md((u_j, v_s), (u_i, v_r)) = md(u_j, u_i)$. Hence it follows that $(u_i, v_r) \in \text{mEcc}(D_1 \circ D_2)$ if and only if there exists a vertex u_j in D_1 such that $\text{mecc}(u_j) = md(u_j, u_i)$; if and only if $u_i \in \text{mEcc}(D_1)$. Thus, $\text{mEcc}(D_1 \circ D_2) = \text{mEcc}(D_1) \times V(D_2)$. □

A digraph D is said to be symmetric if $(u, v) \in E(D)$ if and only if $(v, u) \in E(D)$, and so the maximum distance md is the usual distance d and likewise m-eccentricity is the usual eccentricity and so on. Thus the prefix m can be avoided for boundary-type sets also. If D is a connected symmetric digraph, then $\xi_D(x) = 2$ for all $x \in V(D)$. Hence the DDLE property for D is $\text{ecc}(x) > 2$ for all $x \in V(D)$; that is, $\text{rad}(D) > 2$.

Thus, an immediate corollary follows from Proposition 1.

Corollary 1. *Let D_1 be a connected symmetric digraph with* $\text{rad}(D_1) > 2$ *and D_2 be an arbitrary digraph. Then*

1. $\text{mPer}(D_1 \circ D_2) = \text{Per}(D_1) \times V(D_2)$,
2. $\text{mCt}(D_1 \circ D_2) = \text{Ct}(D_1) \times V(D_2)$,
3. $\text{mEcc}(D_1 \circ D_2) = \text{Ecc}(D_1) \times V(D_2)$.

3.2 $C_n \circ D_2$, C_n Is a Dicycle

Proposition 2. *Let C_n be the dicycle on n vertices and D_2 be an arbitrary digraph.*

1. *If* $\text{mrad}(D_2) \geq n$ *or* $\text{mdiam}(D_2) < n$, *then* $m\partial(C_n \circ D_2) = \text{mCt}(C_n \circ D_2) = \text{mEcc}(C_n \circ D_2) = \text{mPer}(C_n \circ D_2) = V(C_n) \times V(D_2)$.
2. *If* $\text{mrad}(D_2) < n$ *and* $\text{mdiam}(D_2) \geq n$, *then* $\text{mPer}(C_n \circ D_2) = \text{mCt}(C_n \circ D_2) = V(C_n) \times [\bigcup_{\text{mecc}(v_r) \geq n} v_r]$, *and* $\text{mEcc}(C_n \circ D_2) = m\partial(C_n \circ D_2) = V(C_n) \times V(D_2)$.

Proof. $\text{mecc}(u_i) = n - 1$ and $\xi_{C_n}(u_i) = n$ for all $u_i \in C_n$. Hence

$$\text{mecc}_{C_n \circ D_2}(u_i, v_r) = \begin{cases} n-1 & \text{if } \text{mecc}_{D_2}(v_r) \leq n-1 \\ n & \text{if } \text{mecc}_{D_2}(v_r) \geq n. \end{cases}$$

1. If $\text{mrad}(D_2) \geq n$, then $\text{mecc}(u_i, v_r) = n$ for all $(u_i, v_r) \in V(C_n \circ D_2)$. If $\text{mdiam}(D_2) < n$, then $\text{mecc}(u_i, v_r) = n - 1$ for all $(u_i, v_r) \in V(C_n \circ D_2)$. So in both the cases, $m\partial(C_n \circ D_2) = \text{mCt}(C_n \circ D_2) = \text{mEcc}(C_n \circ D_2) = \text{mPer}(C_n \circ D_2) = V(C_n) \times V(D_2)$.
2. If $\text{mrad}(D_2) < n$ and $\text{mdiam}(D_2) \geq n$, then $\text{mPer}(C_n \circ D_2)$ consists of all those vertices (u_i, v_r) such that $\text{mecc}(u_i, v_r) = n$. Hence $\text{mPer}(C_n \circ D_2) = V(C_n) \times [\bigcup_{\text{mecc}(v_r) \geq n} v_r]$. Since $\text{mPer}(C_n \circ D_2) \subseteq \text{mCt}(C_n \circ D_2)$, $V(C_n) \times [\bigcup_{\text{mecc}(v_r) \geq n} v_r] \subseteq \text{mCt}(C_n \circ D_2)$. If $v_r \in V(D_2)$ is such that $\text{mecc}(v_r) < n$, then $\text{mecc}(u_i, v_r) = n - 1$ for all $u_i \in V(C_n)$. $N(u_i) \times V(D_2) \subseteq N(u_i, v_r)$. Since $\text{mdiam}(D_2) \geq n$, there exists a vertex $(u_k, v_q) \in N(u_i, v_r)$ such that $\text{mecc}(u_k, v_q) = n$. Hence if $\text{mecc}(v_r) < n$, then $(u_i, v_r) \notin \text{mCt}(C_n \circ D_2)$. Hence $\text{mCt}(C_n \circ D_2) = V(C_n) \times [\bigcup_{\text{mecc}(v_r) \geq n} v_r]$.

Let $u_i \in V(C_n)$ and $v_r \in V(D_2)$. If $\text{mecc}(v_r) < n$, then $\text{mecc}(u_i, v_r) = n - 1$ and there exists $u_j \neq u_i$ such that $md_{C_n \circ D_2}((u_j, v_r), (u_i, v_r)) = n - 1 = \text{mecc}(u_j, v_r)$ and hence (u_i, v_r) is an eccentric vertex of (u_j, v_r). If $\text{mecc}(v_r) \geq n$, then $\text{mecc}(u_i, v_r) = n$ and there exists a vertex $v_s \in V(D_2)$ such that $md(v_s, v_r) \geq n$ and so $\text{mecc}(u_i, v_s) = n$. Thus $\text{mecc}(u_i, v_s) = md_{C_n \circ D_2}((u_i, v_s), (u_i, v_r)) = n$ and hence (u_i, v_r) is an eccentric vertex of (u_i, v_s). Hence $\text{mEcc}(C_n \circ D_2) = V(C_n) \times V(D_2)$. Since $\text{mEcc}(C_n \circ D_2) \subseteq m\partial(C_n \circ D_2)$, $m\partial(C_n \circ D_2) = V(C_n) \times V(D_2)$. □

If D_1 is not a DDLE digraph, then the boundary-type sets are no longer characterized by the m-radius or m-diameter of the two digraphs. This is because $\text{mecc}_{D_1 \circ D_2}(u_i, v_r)$ depends on $\xi_{D_1}(u_i)$, in addition to $\text{mecc}_{D_1}(u_i)$ and $\text{mecc}_{D_2}(v_r)$.

To see this, consider the digraphs D_1 and D_1' in Fig. 1. The eccentricity of each vertex is displayed near the vertex. Here, $\text{mdiam}(D_1) = \text{mdiam}(D_1') = 2$. Let D_2 be the symmetric dipath P_4 with labels v_1, v_2, v_3, v_4 in order. Then, $\text{mecc}_{D_2}(v_1) = \text{mecc}_{D_2}(v_4) = 3$ and $\text{mecc}_{D_2}(v_2) = \text{mecc}_{D_2}(v_3) = 2$. In D_1, $\xi_{D_1}(u_1) = \xi_{D_1}(u_3) = 3$ and $\xi_{D_1}(u_2) = \xi_{D_1}(u_4) = 2$. Hence in $D_1 \circ D_2$, $\text{mecc}(u_1, v_1) = \text{mecc}(u_3, v_1) = \text{mecc}(u_1, v_4) = \text{mecc}(u_3, v_4) = 3$

and the m-eccentricity of all the other vertices is 2. Thus, $\text{mPer}(D_1 \circ D_2) = \{(u_1, v_1), (u_3, v_1), (u_1, v_4), (u_3, v_4)\}$. In D_1', the dicycle distance of every vertex is 2. Thus, the m-eccentrity of every vertex in $D_1' \circ D_2$ is 2 and hence $\text{mPer}(D_1' \circ D_2) = V(D_1) \times V(D_2)$. So, the remaining discussion is restricted to the case when D_1 is a symmetric digraph.

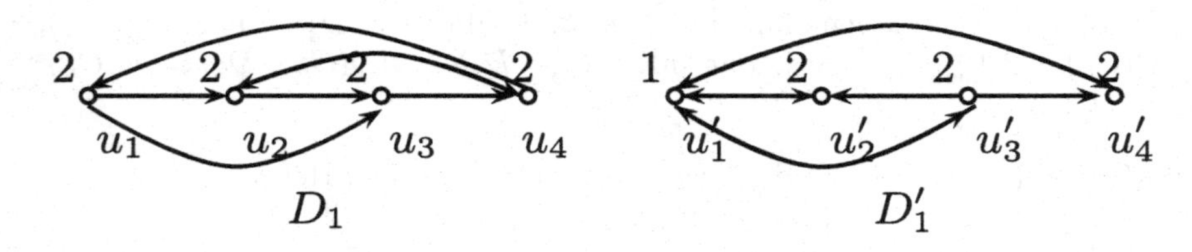

Fig. 1. Digraphs D_1 and D_1'

3.3 $D_1 \circ D_2$, D_1 Is a Symmetric Digraph

Suppose that D_1 is a symmetric digraph. Then $\xi_{D_1}(u_i) = 2$ for all $u_i \in V(D_1)$. Thus, $md_{D_1 \circ D_2}((u_i, v_r), (u_j, v_s)) = \begin{cases} md_{D_1}(u_i, u_j) & \text{if } u_i \neq u_j \\ \min\{2, md_{D_2}(v_r, v_s)\} & \text{if } u_i = u_j \end{cases}$

and $\text{mecc}_{D_1 \circ D_2}(u_i, v_r) = \begin{cases} \text{ecc}_{D_1}(u_i) & \text{if } \text{ecc}_{D_1}(u_i) \geq 2 \\ \min\{\text{mecc}_{D_2}(v_r), 2\} & \text{if } \text{ecc}_{D_1}(u_i) = 1. \end{cases}$

The distance between two vertices in $D_1 \circ D_2$, when D_1 is a symmetric digraph is the same as in the case of the lexicographic product of two symmetric digraphs. So all the results for the boundary type sets are also the same. The description of the boundary-type sets of the lexicographic product of two symmetric digraphs is given in [4]. In this article, the results for the boundary-type sets are proved, when D_1 is a symmetric digraph and D_2 is an arbitrary digraph. In correspondence with the notation in [4], the notation $mZ(D)$ is used to denote the set of vertices with minimum m-eccentricity in the digraph D. If all the vertices have the same m-eccentricity, then we take $mZ(D) = \emptyset$. Also, $\text{mdiam}(D) \leq 2\ \text{mrad}(D)$, for all digraphs D [9].

If $D_1 = K_n$, $n \geq 2$, then $\text{ecc}(u_i) = 1$ for all $u_i \in V(D_1)$.

Hence $md_{D_1 \circ D_2}((u_i, v_r), (u_j, v_s)) = \begin{cases} 1 & \text{if } u_i \neq u_j \\ \min\{2, md_{D_2}(v_r, v_s)\} & \text{if } u_i = u_j \end{cases}$

and $\text{mecc}_{D_1 \circ D_2}(u_i, v_r) = \min\{\text{mecc}_{D_2}(v_r), 2\}$.

Proposition 3. *Let K_n be the complete symmetric digraph on n vertices and D_2 be an arbitrary digraph. Then*

1. $\text{mPer}(K_n \circ D_2) = \text{mCt}(K_n \circ D_2) = \begin{cases} V(K_n) \times V(D_2) & \textit{if } \text{mrad}(D_2) \geq 2 \textit{ or } D_2 = K_m \\ V(K_n) \times [V(D_2) \backslash mZ(D_2)] & \textit{if } \text{mrad}(D_2) = 1 \textit{ and } D_2 \neq K_m \end{cases}$
2. $\text{Ecc}(K_n \circ D_2) = V(K_n) \times V(D_2)$.

Proof.

1. If $\text{mrad}(D_2) \geq 2$, then $\text{mecc}(u_i, v_r) = 2$ for all $(u_i, v_r) \in V(K_n \circ D_2)$. If $D_2 = K_m$, then $\text{mecc}(u_i, v_r) = 1$ for all $(u_i, v_r) \in V(K_n \circ D_2)$. Thus in both the cases, $\text{mPer}(K_n \circ D_2) = \text{mCt}(K_n \circ D_2) = V(K_n) \times V(D_2)$.

If $\text{mrad}(D_2) = 1$ and $D_2 \neq K_m$, then $\text{mecc}(u_i, v_r) = 1$ for all $v_r \in mZ(D_2)$ and $\text{mecc}(u_i, v_r) = 2$ for all $v_r \in V(D_2) \backslash mZ(D_2)$. Thus, $\text{mPer}(K_n \circ D_2) = V(K_n) \times (V(D_2) \backslash mZ(D_2))$. If $\text{mecc}(u_i, v_r) = 1$, then $(u_i, v_r) \notin \text{mCt}(K_n \circ D_2)$, because there exists $v_q \in N(v_r)$ such that $\text{mecc}(v_q) = 2$ and hence $\text{mecc}(u_k, v_q) = 2$, where $(u_k, v_q) \in N(u_i, v_r)$. Since every vertex with eccentricity 2 is in $\text{mPer}(K_n \circ D_2)$, it follows that $\text{mPer}(K_n \circ D_2) = \text{mCt}(K_n \circ D_2) = V(K_n) \times (V(D_2) \backslash mZ(D_2))$.

2. First, suppose that $\text{mrad}(D_2) = 1$. If $v_s \in mZ(D_2)$, then $\text{mecc}(v_s) = 1$, and hence $\text{mecc}(u_i, v_s) = 1 = md_{K_n \circ D_2}((u_i, v_s), (u_i, v_r))$ for all $(u_i, v_r) \in V(K_n \circ D_2)$.

Now, if $\text{mrad}(D_2) \geq 2$, then for all $v_s \in V(D_2)$, $\text{mecc}(v_s) \geq 2$, and hence $\text{mecc}(u_i, v_s) = 2 = d_{K_n \circ D_2}((u_i, v_s), (u_i, v_r))$ for all $(u_i, v_r) \in V(K_n \circ D_2)$.
Hence in both the cases, $\text{Ecc}(K_n \circ D_2) = V(K_n) \times V(D_2)$. □

From Corollary 1, if $\text{mrad}(D_1) > 2$, then $\text{mPer}(D_1 \circ D_2) = \text{mPer}(D_1) \times V(D_2)$. The next two propositions discuss the relation between $\text{mPer}(D_1 \circ D_2)$ and $\text{mdiam}(D_1)$. $\text{mdiam}(D_1) = 1$ is the case when D_1 is a complete symmetric digraph.

Proposition 4. *Let D_1 be a connected symmetric digraph with* $\text{diam}(D_1) \geq 3$, *and D_2 be an arbitrary digraph. Then* $\text{mPer}(D_1 \circ D_2) = \text{Per}(D_1) \times V(D_2)$.

Proof. If $\text{diam}(D_1) = n \geq 3$, then $\text{mdiam}(D_1 \circ D_2) = n$, since $\text{mecc}(u_i, v_r) = n$ for all vertices (u_i, v_r) in $D_1 \circ D_2$ such that $u_i \in \text{Per}(D_1), v_r \in V(D_2)$ and $\text{mecc}(u_i, v_r) < n$ for the remaining vertices. Hence $\text{mPer}(D_1 \circ D_2) = \text{Per}(D_1) \times V(D_2)$. □

Proposition 5. *Let D_1 be a connected symmetric digraph with* $\text{diam}(D_1) = 2$, *and D_2 be an arbitrary digraph. Then*

$$\text{mPer}(D_1 \circ D_2) = \begin{cases} V(D_1) \times V(D_2) & \textit{if } \text{mrad}(D_2) \geq 2 \\ [\text{Per}(D_1) \times V(D_2)] \cup [V(D_1) \times (V(D_2) \backslash mZ(D_2))] & \textit{if } \text{mrad}(D_2) = 1. \end{cases}$$

Proof. If $\text{diam}(D_1) = 2$ and $\text{mrad}(D_2) \geq 2$, then $\text{mecc}(u_i, v_r) = 2$ for all $(u_i, v_r) \in V(D_1 \circ D_2)$. Hence in this case, $\text{mPer}(D_1 \circ D_2) = V(D_1) \times V(D_2)$.

If $\text{diam}(D_1) = 2$ and $\text{mrad}(D_2) = 1$, then

$$\begin{aligned}(u_i, v_r) \in \text{mPer}(D_1 \circ D_2) &\iff \text{mecc}(u_i, v_r) = 2\\ &\iff \text{either } \text{ecc}(u_i) = 2 \text{ or } \text{mecc}(v_r) = 2\\ &\iff \text{either } u_i \in \text{Per}(D_1) \text{ or } v_s \in V(D_2)\backslash mZ(D_2).\end{aligned}$$

Hence $\text{mPer}(D_1 \circ D_2) = [\text{Per}(D_1) \times V(D_2)] \bigcup [V(D_1) \times (V(D_2)\backslash mZ(D_2))]$. □

The m-contour and m-eccentricity sets of $D_1 \circ D_2$ depends on the m-radii of both D_1 and D_2, unless $\text{mrad}(\text{D}_1) \geq 3$.

Proposition 6. *Let D_1 be a connected symmetric digraph different from K_n and D_2 be an arbitrary digraph. Then*

$$\text{mCt}(D_1 \circ D_2) = \begin{cases} \text{Ct}(D_1) \times V(D_2) & \text{if } \text{rad}(D_1) \geq 2 \\ V(D_1) \times V(D_2) & \text{if } \text{rad}(D_1) = 1 \text{ and } \text{mrad}(D_2) \geq 2 \\ [(\text{Ct}(D_1)\backslash Z(D_1)) \times V(D_2)] \cup [Z(D_1) \times (V(D_2)\backslash mZ(D_2))] & \text{if } \text{rad}(D_1) = 1 \text{ and } \text{mrad}(D_2) = 1. \end{cases}$$

Proof. Let $u_i \in V(D_1)$ and $v_r \in V(D_2)$.

Suppose that $\text{rad}(D_1) \geq 2$. By Corollary 1, $\text{mCt}(D_1 \circ D_2) = \text{Ct}(D_1) \times V(D_2)$, if $\text{rad}(D_1) > 2$. Let $\text{rad}(D_1) = 2$. Then $\text{mecc}(u_i, v_r) = \text{ecc}(u_i)$ for all $(u_i, v_r) \in V(D_1 \circ D_2)$. $N_{D_1 \circ D_2}(u_i, v_r) = (N_{D_1}(u_i) \times V(D_2)) \bigcup (\{u_i\} \times N_{D_2}(v_r))$. If $u_i \in \text{Ct}(D_1)$, then $\text{ecc}(u_i) \geq \text{ecc}(u_k)$ for all $u_k \in N(u_i)$. Thus, $\text{mecc}(u_i, v_r) \geq \text{mecc}(u_k, v_s)$ for all $(u_k, v_s) \in N(u_i, v_r)$. Hence $\text{Ct}(D_1) \times V(D_2) \subseteq \text{mCt}(D_1 \circ D_2)$. If $u_i \notin \text{Ct}(D_1)$, then there exists $u_q \in N(u_i)$ such that $\text{ecc}(u_q) > \text{ecc}(u_i)$. Hence there exists $(u_q, v_s) \in N(u_i, v_r)$ such that $\text{mecc}(u_i, v_r) < \text{mecc}(u_q, v_s)$ and thus $(u_i, v_r) \notin \text{mCt}(D_1 \circ D_2)$. Hence $\text{mCt}(D_1 \circ D_2) = \text{Ct}(D_1) \times V(D_2)$.

Suppose that $\text{rad}(D_1) = 1$ and $\text{mrad}(D_2) \geq 2$. Thus, $\text{diam}(D_1) \leq 2$, and $\text{ecc}_{D_1}(u_i) = 1$ or 2. Since $mecc(v_r) \geq 2$ for all $v_r \in V(D_2)$, $\text{mecc}(u_i, v_r) = 2$ for all $(u_i, v_r) \in V(D_1 \circ D_2)$. Hence in this case, $\text{mCt}(D_1 \circ D_2) = V(D_1) \times V(D_2)$.

Consider the case $\text{rad}(D_1) = \text{mrad}(D_2) = 1$. We have $\text{mecc}(u_i, v_r) = 1$ or 2 for all $(u_i, v_r) \in V(D_1 \circ D_2)$. Let $u_i \in V(D_1)$ and $v_r \in V(D_2)$. If $\text{mecc}(u_i, v_r) = 2$, then, $(u_i, v_r) \in \text{mCt}(D_1 \circ D_2)$. $\text{mecc}(u_i, v_r) = 2$ if either ($\text{ecc}(u_i) = 2$ and $\text{mecc}(v_r) \geq 1$), or ($\text{ecc}(u_i) = 1$ and $\text{mecc}(v_r) = 2$). The first possibility is $u_i \in \text{Ct}(D_1)\backslash Z(D_1)$ and $v_r \in V(D_2)$. The second possibility is $u_i \in Z(D_1)$ and $v_r \in V(D_2)\backslash mZ(D_2)$. $\text{mecc}(u_i, v_r) = 1$ if and only if $\text{ecc}(u_i) = \text{mecc}(v_r) = 1$. As $D_1 \neq K_n$ and $\text{ecc}(u_i) = 1$, there is at least one $u_k \in N(u_i)$ such that $\text{ecc}(u_k) = 2$. Hence $\text{mecc}(u_k, v_r) = 2$ and since $(u_k, v_r) \in N(u_i, v_r)$, it follows that $(u_i, v_r) \notin \text{mCt}(D_1 \circ D_2)$. Hence in this case, $\text{mCt}(D_1 \circ D_2) = [(\text{Ct}(D_1)\backslash Z(D_1)) \times V(D_2)] \bigcup [Z(D_1) \times (V(D_2)\backslash mZ(D_2))]$. □

Lemma 1. *Let D_1 be a connected symmetric digraph and D_2 be an arbitrary digraph. Then* $\text{mEcc}(D_1 \circ D_2) \subseteq \text{Ecc}(D_1) \times V(D_2)$.

Proof. Let $u_i \in V(D_1)$ and $v_r \in V(D_2)$. Whenever $u_i \in \mathrm{Ecc}(D_1)$, there exists a vertex $u_j \in V(D_1)$ such that $\mathrm{ecc}(u_j) = d(u_j, u_i)$. If $\mathrm{rad}(D_1) \geq 2$, then in $D_1 \circ D_2$, $\mathrm{mecc}(u_j, v_r) = \mathrm{ecc}(u_j) = d(u_j, u_i) = md((u_j, v_r), (u_i, v_r))$ and hence $(u_i, v_r) \in \mathrm{mEcc}(D_1 \circ D_2)$. If $\mathrm{rad}(D_1) = 1$ and if $u_i \in \mathrm{Ecc}(D_1)$, let $u_j \in V(D_1)$ be such that $\mathrm{ecc}(u_j) = 1 = d(u_j, u_i)$. Then, there are two cases.

If $\mathrm{mecc}(v_r) = 1$, then $\mathrm{mecc}(u_j, v_r) = md((u_j, v_r), (u_i, v_r)) = 1$ and so $(u_i, v_r) \in \mathrm{mEcc}(D_1 \circ D_2)$. If $\mathrm{mecc}(v_r) \geq 2$, then there exists a vertex $v_s \in V(D_2)$ such that in $D_1 \circ D_2$, $\mathrm{mecc}(u_i, v_s) = md((u_i, v_s), (u_i, v_r)) = 2$ and so $(u_i, v_r) \in \mathrm{mEcc}(D_1 \circ D_2)$. So in both the cases, $\mathrm{Ecc}(D_1) \times V(D_2) \subseteq \mathrm{mEcc}(D_1 \circ D_2)$. □

Proposition 7. *Let D_1 be a connected symmetric digraph with* $\mathrm{rad}(D_1) = 2$, *and D_2 be an arbitrary digraph. Then*

$$\mathrm{mEcc}(D_1 \circ D_2) = \begin{cases} [\mathrm{Ecc}(D_1) \times V(D_2)] \bigcup [Z(D_1) \times (V(D_2) \backslash Z(D_2))] & \text{if } \mathrm{mrad}(D_2) = 1 \\ [\mathrm{Ecc}(D_1) \cup Z(D_1)] \times V(D_2) & \text{if } \mathrm{mrad}(D_2) \geq 2. \end{cases}$$

Proof. By Lemma 1, $\mathrm{Ecc}(D_1) \times V(D_2) \subseteq \mathrm{mEcc}(D_1 \circ D_2)$. Now, it is enough to find the vertices $(u_i, v_r) \in \mathrm{mEcc}(D_1 \circ D_2)$ such that $u_i \notin \mathrm{Ecc}(D_1)$.
First, suppose that $\mathrm{mrad}(D_2) = 1$. If $u_i \in Z(D_1)$, then $\mathrm{ecc}(u_i) = 2$ and hence $\mathrm{mecc}(u_i, v_r) = 2$ for all $v_r \in V(D_2)$. If $v_r \notin mZ(D_2)$, then $\mathrm{mecc}(v_r) \geq 2$ and so there exists a vertex $v_s \in V(D_2)$ such that $md(v_s, v_r) \geq 2$ and hence $md_{D_1 \circ D_2}((u_i, v_s), (u_i, v_r)) = 2$. But, if $v_r \in mZ(D_2)$, $md(v_s, v_r) = 1$ for all $v_s \in V(D_2)$. Hence there exists no vertex (u_i, v_s) in $D_1 \circ D_2$ such that $md_{D_1 \circ D_2}((u_i, v_s), (u_i, v_r)) = 2$. Thus if $u_i \notin \mathrm{Ecc}(D_1)$, then $(u_i, v_r) \in \mathrm{mEcc}(D_1 \circ D_2)$ if and only if $u_i \in Z(D_1)$ and $v_r \notin Z(D_2)$. Hence $\mathrm{Ecc}(D_1 \circ D_2) = [\mathrm{Ecc}(D_1) \times V(D_2)] \bigcup [Z(D_1) \times (V(D_2) \backslash mZ(D_2))]$.

Next, suppose that $\mathrm{mrad}(D_2) \geq 2$. Let $v_r \in V(D_2)$. Since $\mathrm{rad}(D_1) = 2$ and $\mathrm{mrad}(D_2) \geq 2$, $\mathrm{mecc}(u_i, v_s) = \begin{cases} \mathrm{ecc}(u_i) & \text{if } u_i \notin Z(D_1) \\ 2 & \text{if } u_i \in Z(D_1). \end{cases}$
$(u_i, v_r) \in \mathrm{mEcc}(D_1 \circ D_2)$ if and only if there exists a vertex $(u_j, v_s) \in V(D_1 \circ D_2))$ such that $\mathrm{mecc}(u_j, v_s) = md((u_j, v_s), (u_i, v_r))$. If $v_s \in V(D_2)$ is such that $md(v_s, v_r) \geq 2$, then $\mathrm{mecc}_{D_1 \circ D_2}(u_i, v_s) = 2 = md_{D_1 \circ D_2}((u_i, v_s), (u_i, v_r))$ for all $u_i \in Z(D_1)$. Thus besides $\mathrm{Ecc}(D_1) \times V(D_2)$, all elements (u_i, v_r), where $u_i \in Z(D_1)$ are also eccentric vertices in $D_1 \circ D_2$. For vertices u_i such that $u_i \notin \mathrm{Ecc}(D_1)$ and $u_i \notin Z(D_1)$, $\mathrm{ecc}(u_i) > 2$ and there exist no vertex $u_j \in V(D_1)$ such that $\mathrm{ecc}(u_j) = d(u_j, u_i)$. Hence in $D_1 \circ D_2$, there exist no vertex (u_j, v_s) such that $\mathrm{mecc}_{D_1 \circ D_2}(u_j, v_s) = md_{D_1 \circ D_2}((u_j, v_s), (u_i, v_r))$. That is, (u_i, v_r) is not an eccentric vertex of any vertex (u_j, v_s) in $D_1 \circ D_2$. Thus $\mathrm{mEcc}(D_1 \circ D_2) = (\mathrm{Ecc}(D_1) \bigcup Z(D_1)) \times V(D_2)$. □

Proposition 8. *Let D_1 be a connected symmetric digraph with* $\text{rad}(D_1) = 1$ *different from K_n and D_2 be an arbitrary digraph. Then* $\text{mEcc}(D_1 \circ D_2)$

$$= \begin{cases} [\text{Ecc}(D_1) \times V(D_2)] \cup [Z(D_1) \times (V(D_2) \backslash mZ(D_2))] & \textit{if } \text{mrad}(D_2) = 1 \textit{ and} \\ & |Z(D_1)| = |mZ(D_2)| = 1 \\ \text{Ecc}(D_1) \times V(D_2) & \textit{if } \text{mrad}(D_2) = 1 \textit{ and} \\ & |Z(D_1)| \geq 2 \\ [\text{Ecc}(D_1) \cup Z(D_1)] \times V(D_2) & \textit{if } \text{mrad}(D_2) \geq 2 \textit{ or } \text{mrad}(D_2) = 1, \\ & |Z(D_1)| = 1, \textit{ and } |mZ(D_2)| \geq 2. \end{cases}$$

Proof. By Lemma 1, $\text{Ecc}(D_1) \times V(D_2) \subseteq \text{mEcc}(D_1 \circ D_2)$.
Thus, it is enough to find the remaining vertices in $\text{Ecc}(D_1 \circ D_2)$ in each case.

Consider the case: $\text{rad}(D_2) = 1$ and $|Z(D_1)| = |mZ(D_2)| = 1$. Let $v_r \in V(D_2)$.

$$\text{mecc}_{D_1 \circ D_2}(u_i, v_s) = \begin{cases} \text{ecc}_{D_1}(u_i) & \text{if } u_i \notin Z(D_1) \\ \min\{\text{mecc}_{D_2}(v_s), 2\} & \text{if } u_i \in Z(D_1). \end{cases}$$

Thus if $u_i \notin \text{Ecc}(D_1)$, then $(u_i, v_r) \in \text{mEcc}(D_1 \circ D_2)$ only if $u_i \in Z(D_1)$ and $v_r \notin mZ(D_2)$. For, if $v_s \in V(D_2)$ is such that $md(v_r, v_s) = 2$, then $\text{mecc}_{D_1 \circ D_2}(u_i, v_s) = 2$. Thus $\text{mecc}_{D_1 \circ D_2}(u_i, v_s) = md_{D_1 \circ D_2}((u_i, v_s), (u_i, v_r)) = 2$ for $u_i \in Z(D_1)$ and $v_r \notin mZ(D_2)$, and there exists only one vertex in $D_1 \circ D_2$ having eccentricity 1 (the single vertex in $Z(D_1) \times Z(D_2)$). This vertex cannot be the eccentric vertex of any vertex in $D_1 \circ D_2$. Hence $\text{mEcc}(D_1 \circ D_2) = [\text{Ecc}(D_1) \times V(D_2)] \bigcup [Z(D_1) \times (V(D_2) \backslash mZ(D_2))]$.

Now consider the case: $\text{mrad}(D_2) = 1$ and $|Z(D_1)| \geq 2$. If $u_i \in Z(D_1)$, then there exists another vertex $u_j \in Z(D_1)$ such that $\text{ecc}(u_j) = d(u_j, u_i) = 1$. Hence $u_i \in \text{Ecc}(D_1)$. Thus, $\text{mEcc}(D_1 \circ D_2) = \text{Ecc}(D_1) \times V(D_2)$.

Next, consider the case: $\text{mrad}(D_2) = 1$, $|Z(D_1)| = 1$ and $|mZ(D_2)| \geq 2$. For each $v_r \in Z(D_2)$, there exists another vertex $v_s \in mZ(D_2)$ such that $\text{mecc}(v_s) = md(v_s, v_r) = 1$. Thus, if $u_i \in Z(D_1)$, then $\text{mecc}(u_i, v_s) = md((u_i, v_s), (u_i, v_r)) = 1$. That is, every vertex (u_i, v_r) such that $u_i \in Z(D_1)$, $v_r \in mZ(D_2)$ is an eccentric vertex in $D_1 \circ D_2$. But all vertices in $Z(D_1) \times (V(D_2) \backslash mZ(D_2))$ are in $\text{mEcc}(D_1 \circ D_2)$, as proved in the first case. Since no other vertex could be an eccentric vertex in $\text{mEcc}(D_1 \circ D_2)$, it follows that $\text{mEcc}(D_1 \circ D_2) = [\text{Ecc}(D_1) \bigcup Z(D_1)] \times V(D_2)$.

Finally, consider the case: $\text{mrad}(D_2) \geq 2$.

Then, $\text{mecc}_{D_1 \circ D_2}(u_i, v_s) = \begin{cases} \text{ecc}_{D_1}(u_i) & \text{if } u_i \notin Z(D_1) \\ 2 & \text{if } u_i \in Z(D_1). \end{cases}$

Hence if $u_i \in Z(D_1)$, then $\text{mecc}_{D_1 \circ D_2}(u_i, v_s) = md_{D_1 \circ D_2}((u_i, v_s), (u_i, v_r)) = 2$ for all $v_r \in V(D_2)$. So the vertices $(u_i, v_r) \in Z(D_1) \times V(D_2)$ are the other vertices in $\text{Ecc}(D_1 \circ D_2)$. Thus, $\text{mEcc}(D_1 \circ D_2) = [\text{Ecc}(D_1) \bigcup Z(D_1)] \times V(D_2)$. □

Proposition 9. *Let D_1 be a connected symmetric digraph and D_2 be an arbitrary digraph. Then*

1. $m\partial(D_1 \circ D_2) = V(D_1) \times V(D_2)$, *if* $\text{mrad}(D_2) \geq 2$ *or* $\text{mrad}(D_2) = 1$ *and* $|mZ(D_2)| \geq 2$.
2. $m\partial(D_1 \circ D_2) = [V(D_1) \times (V(D_2) \backslash mZ(D_2))] \bigcup [\partial(D_1) \times mZ(D_2)]$, *if* $\text{mrad}(D_2) = 1$ *and* $|mZ(D_2)| = 1$.

Proof. Let $u_i \in V(D_1)$ and $v_r \in V(D_2)$.

1. Suppose that $\text{mrad}(D_2) \geq 2$.

   Then $\text{mecc}_{D_1 \circ D_2}(u_i, v_r) = \begin{cases} \text{ecc}_{D_1}(u_i) & \text{if } \text{ecc}_{D_1}(u_i) \geq 2 \\ 2 & \text{if } \text{ecc}_{D_1}(u_i) = 1 \end{cases}$

   Since $md_{D_1 \circ D_2} d((u_j, v_s), (u_i, v_r)) = \begin{cases} d_{D_1}(u_j, u_i) & \text{if } u_i \neq u_j \\ 1 & \text{if } u_i = u_j \text{ and } md(v_s, v_r) = 1 \\ 2 & \text{if } u_i = u_j \text{ and } md(v_s, v_r) > 1 \end{cases}$

   If v_s is such that $md_{D_2}(v_s, v_r) > 1$, then $md((u_i, v_s), (u_i, v_r)) = 2$. If $(u_k, v_q) \in N(u_i, v_r)$, then $md((u_i, v_s), (u_k, v_q)) = \begin{cases} 1 & \text{if } u_k \in N_{D_1}(u_i) \\ 1 & \text{if } u_k = u_i \text{ and } md(v_s, v_q) = 1 \\ 2 & \text{if } u_k = u_i \text{ and } md(v_s, v_q) > 1 \end{cases}$

   as $N(u_i, v_r) = [N(u_i) \times V(D_2)] \bigcup [\{u_i\} \times N(v_r)]$. Hence (u_i, v_r) is a boundary vertex of (u_i, v_s). Thus, $V(D_1) \times V(D_2) \subseteq m\partial(D_1 \circ D_2)$.

   Now suppose that $\text{mrad}(D_2) = 1$ and $|mZ(D_2)| \geq 2$. Let $v_a \in mZ(D_2)$. So $md(v_a, v_r) = 1$ for all $v_r \in V(D_2)$. Then $md((u_i, v_a), (u_i, v_r)) = 1$, and $md((u_i, v_a), (u_k, v_q)) = 1$ for all $(u_k, v_q) \in N(u_i, v_r)$, since $d(u_i, u_k) = 1$ for all $u_k \in N(u_i)$. Since $|mZ(D_2)| \geq 2$, there exist $v_b \in mZ(D_2)$ such that $md(v_b, v_a) = 1$. Thus we get $md((u_i, v_b), (u_i, v_a)) = 1$, and also $md_{D_1 \circ D_2}((u_i, v_b), (u_k, v_q)) = 1$ for all $(u_k, v_q) \in N(u_i, v_a)$. So again in this case, $V(D_1) \times V(D_2) \subseteq m\partial(D_1 \circ D_2)$.

2. Suppose that $\text{mrad}(D_2) = 1$ and $|mZ(D_2)| = 1$. So there is only one vertex v_a such that $md_{D_2}(v_a, v_r) = 1$ for all $v_r \in V(D_2)$. Hence every vertex $(u_i, v_r) \in V(D_1 \circ D_2)$ is a boundary vertex of (u_i, v_a). Also, (u_i, v_a) is not a boundary vertex in $D_1 \circ D_2$ unless u_i is a boundary vertex in D_1. For, if $u_i \in \partial(D_1)$, then u_i is a boundary vertex of $u_j \in V(D_1)$, and hence (u_i, v_a) is a boundary vertex of (u_j, v_a) in $D_1 \circ D_2$. So in this case, $m\partial(D_1 \circ D_2) = [V(D_1) \times (V(D_2) \backslash mZ(D_2))] \bigcup [\partial(D_1) \times mZ(D_2)]$. □

Acknowledgements. Prasanth G. Narasimha-Shenoi and Mary Shalet Thottungal Joseph are supported by Science and Engineering Research Board, a statutory body of Government of India under their Extra Mural Research Funding No. EMR/2015/002183. Also, their research was partially supported by Kerala State Council for Science Technology and Environment of Government of Kerala under their SARD project grant Council(P) No. 436/2014/KSCSTE. Prasanth G. Narasimha-Shenoi is also supported by Science and Engineering Research Board, under their MATRICS Scheme No. MTR/2018/000012.

The authors thank the anonymous referees for their valuable comments which helped in improving the article.

References

1. Bang-Jensen, J., Gutin, G.Z.: Digraphs: Theory, Algorithms and Applications. Springer, Heidelberg (2008). https://doi.org/10.1007/978-1-84800-998-1
2. Bang-Jensen, J., Gutin, G.Z.: Classes of Directed Graphs. Springer, Heidelberg (2018). https://doi.org/10.1007/978-3-319-71840-8
3. Cáceres, J., Hernando, C., Mora, M., Pelayo, I.M., Puertas, M.L., Seara, C.: On geodetic sets formed by boundary vertices. Discret. Math. **306**(2), 188–198 (2006)
4. Cáceres, J., Hernando Martín, M.d.C., Mora Giné, M., Pelayo Melero, I.M., Puertas González, M.L.: Boundary-type sets and product operators in graphs. In: VII Jornadas de Matemática Discreta y Algorítmica, pp. 187–194 (2010)
5. Cáceres, J., Márquez, A., Oellermann, O.R., Puertas, M.L.: Rebuilding convex sets in graphs. Discret. Math. **297**(1), 26–37 (2005)
6. Changat, M., Narasimha-Shenoi, P.G., Thottungal Joseph, M.S., Kumar, R.: Boundary vertices of cartesian product of directed graphs. Int. J. Appl. Comput. Math. **5**(1), 1–19 (2019). https://doi.org/10.1007/s40819-019-0604-4
7. Chartrand, G., Erwin, D., Johns, G.L., Zhang, P.: Boundary vertices in graphs. Discret. Math. **263**(1), 25–34 (2003)
8. Chartrand, G., Gu, W., Schultz, M., Winters, S.J.: Eccentric graphs. Netw.: Int. J. **34**(2), 115–121 (1999)
9. Chartrand, G., Tian, S.: Distance in digraphs. Comput. Math. Appl. **34**(11), 15–23 (1997)
10. Hammack, R., Imrich, W., Klavžar, S.: Handbook of Product Graphs. CRC Press, Boco Raton (2011)
11. Hammack, R.H.: Digraphs products. In: Bang-Jensen, J., Gutin, G. (eds.) Classes of Directed Graphs. SMM, pp. 467–515. Springer, Cham (2018). https://doi.org/10.1007/978-3-319-71840-8_10
12. Harary, F., et al.: On the group of the composition of two graphs. Duke Math. J. **26**(1), 29–34 (1959)

The Connected Domination Number of Grids

Adarsh Srinivasan[1(✉)] and N. S. Narayanaswamy[2]

[1] Indian Institute of Science Education and Research (IISER) Pune, Pune, India
adarsh.srinivasan@students.iiserpune.ac.in
[2] Indian Institute of Technology (IIT) Madras, Chennai, India
swamy@cse.iitm.ac.in

Abstract. Closed form expressions for the domination number of an $n \times m$ grid have attracted significant attention, and an exact expression has been obtained in 2011 [7]. In this paper, we present our results on obtaining new lower bounds on the connected domination number of an $n \times m$ grid. The problem has been solved for grids with up to 4 rows and with 6 rows and the best currently known lower bound for arbitrary m, n is $\lceil \frac{mn}{3} \rceil$ [11]. Fujie [4] came up with a general construction for a connected dominating set of an $n \times m$ grid. In this paper, we investigate whether this construction is indeed optimum. We prove a new lower bound of $\left\lceil \frac{mn+2\left\lceil \frac{\min\{m,n\}}{3} \right\rceil}{3} \right\rceil$ for arbitrary $m, n \geq 4$.

Keywords: Connected dominating set · Maximum leaf spanning tree · Grid graph · Connected domination number

1 Introduction

In this paper, we study the Minimum Connected Dominating Set (Min-CDS) problem in grid graphs. Given a connected graph $G = (V, E)$, a *connected dominating set* (CDS) is a subset S of V which induces a connected subgraph in G such that every vertex of G is either in S or adjacent to a vertex in S. The Min-CDS problem asks for a CDS of minimum size. This is a well studied problem in combinatorial optimisation. The *connected domination number* of G is the minimum size of a connected dominating set of G. This problem is equivalent to the Maximum Leaf Spanning Tree (MLST) problem, which is the problem of finding a spanning tree of G with maximum number of leaves. A graph has a spanning tree with k leaves if and only if it has a connected dominating set of size $|V| - k$. These problems are known to be NP-complete [6, ND2, Appendix 2], and have been widely studied and have applications in areas such as networking, circuit layout, etc (See [15] for example). A common theme in the study of any NP-complete problem is to consider the problem in special classes of inputs with

This research was supported by the first author's INSPIRE fellowship from Department of Science and Technology (DST), Govt. of India.

A. Mudgal and C. R. Subramanian (Eds.): CALDAM 2021, LNCS 12601, pp. 247–258, 2021.
https://doi.org/10.1007/978-3-030-67899-9_19

more structure than the general case and try to understand whether the problem remains NP-complete or admits a polynomial time solution. MIN-CDS is known to be NP-complete when it is restricted to planar bipartite graphs of maximum degree 4 [9]. It is also known to be NP-complete for unit disk graphs [10] and subgraphs of grid graphs [2, Theorem 6.1]. When viewed in terms of approximation algorithms, the minimum connected dominating set and maximum leaf spanning tree problems are not equivalent. The MLST problem is MAX-SNP-hard which makes a Polynomial Time Approximation Scheme (PTAS) unlikely [5], but linear time 3-approximation algorithms [12] and 2-approximation algorithms [13] exist. A PTAS exists for the MIN-CDS problem on unit disk graphs [3,8]. The complexity of this problem for complete grid graphs remains unknown.

By comparison, the computation of the domination number of an $n \times m$ grid graph is a well studied problem. Chang [1] devoted their PhD thesis to calculating the domination number of grids and Gonçalves et al. [7] solved this problem in 2011 by proving it to be $\left\lfloor \frac{(n+2)(m+2)}{5} \right\rfloor - 4$ for $16 \leq m \leq n$. Hence it is natural to ask the following questions about the connected domination number of grid graphs:

- Can we come up with a closed form expression for the connected domination number of an $n \times m$ grid?
- Can we design an algorithm that takes n, m as input, with run-time polynomial in n, m that outputs the domination number of an $n \times m$ grid?

An answer for the first question would imply an answer for the second one, but not vice versa. A partial answer can be obtained by showing lower and upper bounds on the connected domination number. Upper bounds can be obtained using constructions or heuristic algorithms. Fujie [4] came up with a general construction of a spanning tree of a grid with a large number of leaves which leads to an upper bound on the connected domination number. Li and Tolouse [11] determined the optimum maximum leaf spanning tree for grid graphs with up to 4 rows and with 6 rows. The only known general lower bound is $\lceil \frac{mn}{3} \rceil$. This was obtained by Li and Tolouse using an easy counting argument and by Fujie using a mathematical programming approach.

In this paper, we come up with improved lower bounds on the connected domination number of a grid. We show a lower bound of $\left\lceil \frac{mn + 2\lceil \frac{\min\{m,n\}}{3} \rceil}{3} \right\rceil$ for arbitrary $m, n \geq 4$ (Theorem 3). To our knowledge, this is the first non-trivial result of this kind. Our proof also leads to some insight on the structure of the optimum connected dominating set of a grid.

1.1 Preliminaries and Terminology

We first introduce the definitions and notations that we will use in the rest of the paper. For all definitions and notations not defined here, we refer to [14]. Let $G = (V, E)$ be a connected graph. A leaf refers to a vertex of degree 1 in G. The open neighbourhood of a subset S of V in G is defined to be the set of all vertices adjacent to a vertex in S which are not in S and denoted by $N_G(S)$. The closed neighbourhood of S in G, $N_G[S]$ is defined to be $N_G(S) \cup S$. $G[S]$ denotes the subgraph of G induced by S. A set S is called a connected dominating set of G if $N_G[S] = V$ and $G[S]$ is a connected subgraph of G. The size of the minimum connected dominating set of G is called its connected domination number and is denoted by $\gamma_c(G)$. The maximum leaf number of G is the number of leaves in the maximum leaf spanning tree of G. The connected domination number and the maximum leaf number add up to $|V|$. A connected dominating set of G can be obtained by deleting the leaves of a spanning tree of G.

The notation $[i]$ denotes the set $\{1, 2, \ldots, i\}$. The $n \times m$ grid graph $G_{n,m}$ is the graph with the vertex set $[n] \times [m]$ with two vertices (i_1, j_1) and (i_2, j_2) being adjacent if and only if $|i_i - i_2| + |j_1 - j_2| = 1$. It can also be defined as a unit disk graph in which the disks have the integer points mentioned as centers and radius 1/2. For the reminder of the paper, we just use G instead of $G_{n,m}$ without any ambiguity. We assume whenever necessary that G is embedded in a larger grid graph. Specifically, we embed G in G' which is a grid graph with two additional rows and columns. The vertex set of G' is $\{0, 1, \ldots, n+1\} \times \{0, 1, \ldots, m+1\}$ with the same incidence relation as G.

S is a connected dominating set of G. l is the number of leaves in the graph $G[S]$. It has no relation to the number of leaves in the corresponding spanning tree of G. For any $v \in G$ we define the *loss function* of that vertex to be $\ell(v) = |N[v] \cap S| - 1$. The loss function of the set S is defined to be $\ell(S) = \sum_{v \in N_{G'}[S]} \ell(v)$. The *boundary* of $G_{n,m}$ is defined to be the set of points in $G_{n,m}$ which have three neighbours or less in $G_{n,m}$ (excluding the points themselves). The *excess function* $e(S)$ is defined to be the number of points in S present in the boundary of G. These definitions are inspired by similar definitions in [7].

2 Bounds on the Connected Domination Number

2.1 Known Upper Bounds

Upper bounds for $\gamma_c(G)$ can be easily obtained by constructing spanning trees for G with a large number of leaves, which leads to an upper bound for the maximum leaf number of G, and a corresponding lower bound on the connected domination number. Fujie gave a construction of a spanning tree with a large number of leaves [4, Lemma 2]. We reproduce their construction here:

Let D_1 be a CDS of $G_{n,m}$ with the following vertices-

$$\begin{aligned}
&(1,2), (2,2), \ldots, (n,2)\\
&(1,m-1), (2,m-1), \ldots (n,m-1)\\
&(2,3), (2,4), \ldots, (2,m-2)\\
&(i, 3k+2) \text{ for } i = 3,4,\ldots,n, k = 1,2,\ldots, \left\lfloor \frac{m-4}{3} \right\rfloor
\end{aligned}$$

Let D_2 be a CDS with the following vertices-

$$(2,1),(2,2),\ldots,(2,m)$$
$$(n-1,1),(n-1,2),\ldots(n-1,m)$$
$$(3,2),(4,2),\ldots,(m-2,2)$$
$$(3k+2,i) \text{ for } k=1,2,\ldots,\left\lfloor\frac{n-4}{3}\right\rfloor, i=3,4,\ldots,m$$

Hence, we have the following upper bound on $\gamma_c(G)$:

$$\gamma_c(G) \le \min\left\{2n+(m-4)+\left\lfloor\frac{m-4}{3}\right\rfloor(n-2), 2m+(n-4)+\left\lfloor\frac{n-4}{3}\right\rfloor(m-2)\right\} \tag{1}$$

Figure 1 describes an example for the constructions for the graph $G_{7,11}$.

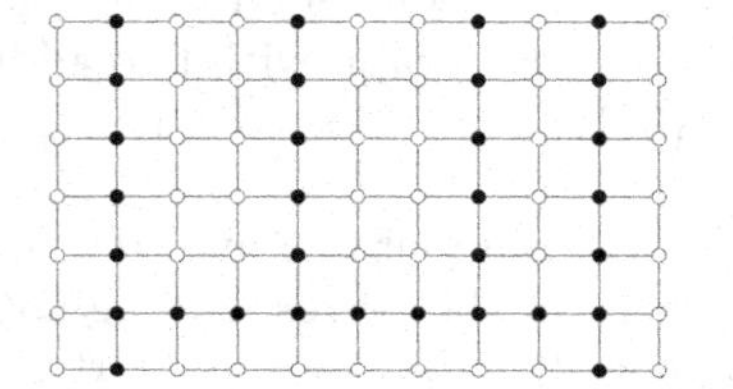
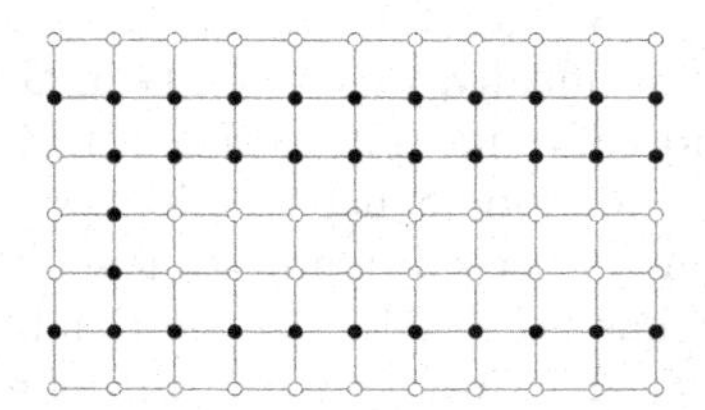

Fig. 1. D_1 and D_2 for $G_{7,11}$ (Black vertices present in CDS)

2.2 New Lower Bounds

Proving lower bounds would require combinatorial arguments, which will be the main contribution of this paper. In [7] the authors introduced a combinatorial parameter called the loss function to prove lower bounds on the domination number of a grid. The loss and excess functions which we have defined are inspired by that definition. Using these parameters, we prove a sequence of lower bounds on $|S|$, each an improvement on the previous one, culminating in Theorem 3.

Our approach is to obtain lower bounds in $\ell(S)$ and $e(S)$ parametrized by l and then combine them to obtain absolute lower bounds on $\ell(S)+e(S)$. This will, in turn lead to lower bounds on $|S|$.

$G[S]$ can be divided into a number of horizontal and vertical line segments, with each vertical line segment connected to at least one horizontal line segment and vice versa as the graph is connected. The vertices, called *joins* where a horizontal line segment meets a vertical line segment can be of degree 2, 3 or 4. When the join is of degree 2, we refer to the horizontal line segment and vertical line segment that meet at the vertex as a *bend.*

Let d_3 and d_4 be the number of vertices in $G[S]$ with degree 3 and 4 respectively and d_2 be the number of bends. Note that here d_2 counts only those degree 2 vertices which form a bend. We make the following observation on the number of leaves in $G[S]$:

Lemma 1. *For any CDS S of G: $l \leq d_3 + 2d_4 + 2$.*

Proof. The well known handshake lemma states that for a graph $G = (V, E)$ $\sum_{v \in V} \delta(v) = 2|E|$. We apply this for $G[S]$. The number of vertices of degree 2 is $|S| - d_3 - d_4 - l$ and as $G[S]$ is connected, the number of edges in $G[S]$ is at least $|S| - 1$.

$$3d_3 + 4d_4 + 2(|S| - d_3 - d_4 - l) + l = 2|E| \geq 2(|S| - 1)$$
$$\implies l \leq d_3 + 2d_4 + 2$$

□

We now relate the parameters $\ell(S)$ and $e(S)$ to the size of a connected dominating set:

Lemma 2. *For any $m, n \geq 3$, $G_{n,m}$ has a minimum CDS that does not contain any corner of $G_{n,m}$.*

Proof. Consider a CDS S of $G_{n,m}$ which contains the corner point $(1, 1)$. As $G[S]$ is a connected subgraph, S must contain either $(1, 2)$ or $(2, 1)$. There exists a maximal horizontal or vertical line segment in $G[S]$ containing $(1, 1)$. Assume $G[S]$ contains the path $(1, 1), (1, 2), \ldots, (1, k)$ As $G[S]$ is connected, one of these points must contain a neighbour in S. Let $(1, i)$ be the first such point with a neighbour $(2, i)$. Now, for all $j < i$, we replace $(1, j)$ in S with $(2, j)$ to obtain a new CDS of G with the same number of points. As $n, m \geq 4$, $(2, 1)$ which replaces $(1, 1)$ is not a corner point. We perform a similar procedure with the path $(1, 1), (2, 1), \ldots, (k, 1)$ if S does not contain the point $(1, 2)$, and repeat this for all four corner points of G to obtain a new CDS for $G_{n,m}$ with the same number of points as S. □

Lemma 3. *For a CDS S of G: $nm = 5|S| - \ell(S) - e(S)$.*

Proof. For a set S which dominates G, consider the set $N_{G'}[S]$, which is its closed neighbourhood in G'. Any point in S dominates 5 points including itself, and for each $v \in N_{G'}[S]$, the number of points which dominate v is $1 + \ell(v)$. Hence, $|N_{G'}[S]| = 5|S| - \ell(S)$.

As no point in S is a corner point of G (see previous lemma), every point in the boundary of G dominates exactly one point in G' outside G. Thus the number of points in $N_{G'}[S]$ outside G is $e(S)$. Hence, $mn = |N_{G'}[S]| - e(S)$ and this proves the lemma. □

Lemma 4. $\ell(S) \geq 2|S| - l + d_2 + 3d_3 + 6d_4$ *and* $e(S) \geq 4$ *if* $m, n \geq 4$.

Proof. Consider the four corners of the grid, the points $(1, 1), (n, 1), (1, m)$ and (n, m). These points have to be dominated by a point in S, and all their neighbours in G are in the boundary of G. As both n and m are greater than or equal to 4, two corner points cannot be dominated by the same point in S, and hence $e(S) \geq 4$.

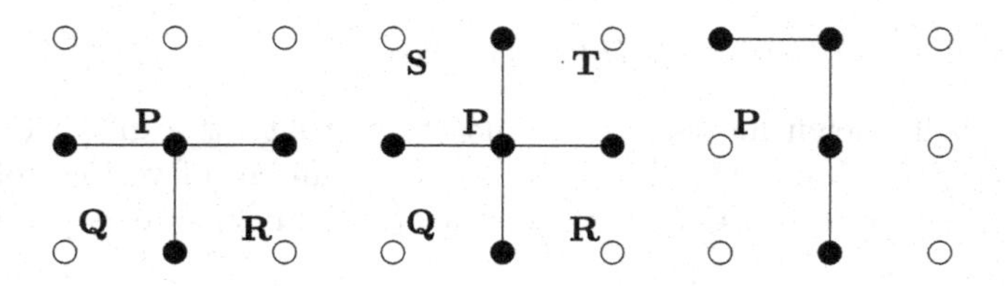

Fig. 2. The three different types of joins

For any point v in S which is not a leaf in $G[S]$, $\ell(v) \geq 2$. This is because it has at least 2 neighbours in S. If v is a leaf, $\ell(v) = 1$. Hence, if l is the number of leaves of S, $\ell(S) \geq 2|S| - l$. In addition to this, consider a vertex of degree 3 as in Fig. 2. The loss function of the point P is at least 3 and the loss function of Q and R is at least 1. For the vertex of degree 4 shown, the loss function of P is at least 4 and the loss functions of Q, R, S and T are each at least 1. For the bend shown, the loss function of the point P is at least 1. Hence, $\ell(S) \geq 2|S| - l + d_2 + 3d_3 + 6d_4$. □

Putting these observations together, we get our first bound on $e(S) + \ell(S)$:

Lemma 5 (Parametrized bound 1). *Consider any CDS S for G, with $G[S]$ having l leaves. Then $\ell(S) + e(S) \geq 2|S| + 2l - 2$.*

Proof. This follows from the fact that $\ell(S) + e(S) \geq 2|S| - l + 3d_3 + 6d_4 + 4$ (Lemma 4) and $d_3 + 2d_4 \geq l - 2$ (Lemma 1). □

From this bound it is easy to derive the already known lower bound of $\lceil \frac{mn}{3} \rceil$ for $\gamma_c(G)$:

Theorem 1 (Bound 1). *For any CDS S of G:*

$$|S| \geq \left\lceil \frac{mn}{3} \right\rceil$$

Proof. As S must have at least 2 leaves, $\ell(S) + e(S) \geq 2|S| + 2$. We can now use this in Lemma 3:

$$mn \leq 5|S| - 2|S| - 2 = 3|S| - 2$$
$$\implies |S| \geq \left\lceil \frac{mn}{3} \right\rceil$$

□

We have not used any structural information on $G[S]$ yet. Specifically, we have not used the fact that it might contain bends. Next, we use the fact that the connected dominating set must contain a certain minimum number of joins to show a new bound on $\ell(S) + e(S)$. We first prove the following simple lemma on the structure of $G[S]$. We say that the horizontal line segments *span* the height of the graph if the subgraph induced by their closed neighbourhood contains a point from every row of G and we say that the vertical line segments *span* the width of the graph if the subgraph induced by their closed neighbourhood contains a point from every column of G.

Lemma 6. *$G[S]$ either has at least $\lceil \frac{n}{3} \rceil$ horizontal line segments which span the width of G or $\lceil \frac{m}{3} \rceil$ vertical line segments which span the height of G.*

Proof. Any horizontal line segment dominates an area that spans at most three rows. Hence, if S has less than $\lceil \frac{n}{3} \rceil$ horizontal line segments, there exists at least one row which is not dominated by any of the points in the horizontal line segments. The rows not dominated by horizontal line segments must be dominated by the vertical line segments. Any vertical line segment can dominate an area which spans at most 3 columns and hence there must be at least $\lceil \frac{m}{3} \rceil$ vertical line segments in S. Similarly, if S has less than $\lceil \frac{m}{3} \rceil$ vertical line segments, it must have at least $\lceil \frac{n}{3} \rceil$ horizontal line segments. □

We use this lemma to prove the following bound on the number of joins in $G[S]$.

Lemma 7. *Let d_2, d_3 and d_4 denote the number of bends, joins of degree 3 and joins of degree 4 respectively.*

$$d_2 + d_3 + d_4 \geq \left\lceil \frac{\min\{m, n\}}{3} \right\rceil$$

Proof. Either the horizontal line segments dominate an area that spans the entire height of the graph or the vertical line segments dominate an area that spans the width of the graph. We assume the latter without loss of generality. Each of these vertical line segments must be connected to a point in the previous or next column in the grid. Hence, each of these must contain either a bend, a degree 3 join or a degree 4 join. From the previous lemma, we know that there are at least $\lceil \frac{m}{3} \rceil$ vertical line segments and the result follows. □

We use these two lemmas to obtain two new lower bounds on $\ell(S) + e(S)$ parametrized by the number of leaves, which we can then combine to obtain an improved lower bound for $|S|$.

Lemma 8 (parametrized bound 2). *Consider any CDS S for G, with $G[S]$ having l leaves. Then $\ell(S) + e(S) \geq 2|S| + \left\lceil \frac{\min\{m,n\}}{3} \right\rceil + l$.*

Proof. We know that $e(S) \geq 4$ and $\ell(S) \geq 2|S| - l + 3d_3 + 6d_4 + d_2$, and because $d_2 + d_3 + d_4 \geq \left\lceil \frac{\min\{m,n\}}{3} \right\rceil$,

$$\ell(S) \geq 2|S| - l + \left\lceil \frac{\min\{m, n\}}{3} \right\rceil + 2d_3 + 5d_4 \geq 2|S| - l + \left\lceil \frac{\min\{m, n\}}{3} \right\rceil + 2(d_3 + 2d_4 + 2) - 4$$

We now use the fact that $l \leq d_3 + 2d_4 + 2$ to prove the lemma □

Lemma 9 (parametrized bound 3). *Consider any CDS S for G, with $G[S]$ having l leaves. Then $\ell(S) + e(S) \geq 2|S| + 2\left\lceil \frac{\min\{m,n\}}{3} \right\rceil + 2 - l$.*

Proof. We can assume without loss of generality that the vertical lines span the width of the grid. Every vertical line must contain a join. Hence, at least $\lceil \frac{m}{3} \rceil - d_3 - d_4$ of them must have one or more bends. A vertical line with only one bend must also contain a leaf. This implies that the number of bends is at least $2\left(\left\lceil \frac{\min\{m,n\}}{3} \right\rceil - d_3 - d_4\right) - l$. We use this in our estimation of $\ell(S)$:

$$\begin{aligned}
\ell(S) &\geq 2|S| - l + d_2 + 3d_3 + 6d_4 \\
&\geq 2|S| - l + 3d_3 + 6d_4 + 2\left(\left\lceil \frac{\min\{m,n\}}{3} \right\rceil - d_3 - d_4\right) - l \\
&\geq 2|S| + 2\left\lceil \frac{\min\{m,n\}}{3} \right\rceil + d_3 + 2d_4 + 2 - 2l - 2 \\
&\geq 2|S| + 2\left\lceil \frac{\min\{m,n\}}{3} \right\rceil - l - 2
\end{aligned}$$

Using the fact that $e(S) \geq 4$, the result follows. □

Combining the previous two parametrized bounds leads to the following lower bound on $|S|$ which is an improvement over the currently known bound of $|S| \geq \lceil \frac{mn}{3} \rceil$:

Theorem 2 (Bound 2). *For a CDS S of G:*

$$|S| \geq \left\lceil \frac{mn + \left\lceil \frac{3}{2} \left\lceil \frac{\min\{m,n\}}{3} \right\rceil \right\rceil + 1}{3} \right\rceil$$

Proof. The lower bound in Lemma 8 increases with l and the bound in Lemma 9 decreases with l. As they both lower bound $\ell(S) + e(S)$, $\ell(S) + e(S)$ is always greater than or equal to $2|S| + \left\lceil \frac{3}{2} \left\lceil \frac{\min\{m,n\}}{3} \right\rceil \right\rceil + 1$. From Lemma 3,

$$mn = 5|S| - \ell(S) - e(S) \geq 5|S| - \left(2|S| + \left\lceil \frac{3}{2} \left\lceil \frac{\min\{m,n\}}{3} \right\rceil \right\rceil + 1\right)$$

This means that $3|S| \geq mn + \left\lceil \frac{3}{2} \left\lceil \frac{\min\{m,n\}}{3} \right\rceil \right\rceil + 1$ which proves the theorem. □

This bound can be further improved by counting the number of bends in $G[S]$ more carefully. In the proof of Lemma 9, we used the fact that the number of bends is at least $2\left(\left\lceil \frac{\min\{m,n\}}{3} \right\rceil - d_3 - d_4\right) - l$. In the following lemma, we improve on that:

Lemma 10. *Consider a CDS S of G. The number of bends in S is at least* $2\left(\left\lceil \frac{\min\{m,n\}}{3} \right\rceil - l + 1\right)$.

Proof. We first assume that $G[S]$ has no vertices of degree 4. If it does, we can just treat a vertex of degree 4 as two vertices of degree 3. Hence, the number of vertices of degree 3 has to be at least $l - 2$. As before, we can assume without loss of generality that the vertical lines dominate an area that spans the width of the grid and that there are $\lceil \frac{m}{3} \rceil$ vertical lines. Some of these lines have one or more degree 3 vertices and some of them have bends.

Observe that a vertical line with only one join must contain a leaf. Every join in a vertical line must be paired with another join or be paired with a leaf as shown in Fig. 3.

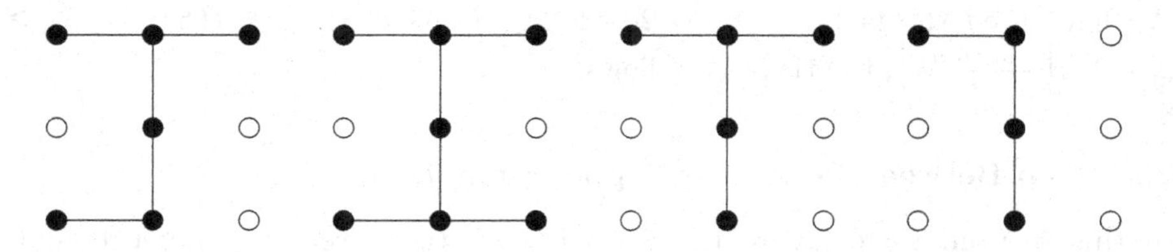

Fig. 3. Joins and leaves

Out of all the vertices of degree 3, let t_1 be the number of vertices not paired with a bend or a degree 3 vertex, t_2 be the number of bend-degree 3 vertex pairs, and t_3 be the number of degree 3 vertex- degree 3 vertex pairs. Hence, it is clear that out of the vertical lines, at most $(l - 2) - t_3$ contain vertices of degree 3. Hence there are at least $\left(\lceil \frac{m}{3} \rceil - (l - 2) + t_3\right)$ vertical lines without vertices of degree 3 and these vertical lines contain at most $l - t_1$ leaves. Consider such a vertical line. It can have only one bend if and only if it has a leaf and hence there can be at most $(l - t_1)$ of such columns. We have already counted t_2 bends. Hence, we can bound the number of bends:

$$\begin{aligned} d_2 &\geq 2\left(\left\lceil \frac{m}{3} \right\rceil - (l - 2) + t_3 - (l - t_1)\right) + t_2 + (l - t_1) \\ &= 2\left\lceil \frac{m}{3} \right\rceil - 3l + 4 + t_1 + t_2 + 2t_3 \end{aligned}$$

$t_1 + t_2 + 2t_3 \geq l - 2$ as every vertex of degree 3 belongs in at least one of the three categories mentioned and the lemma follows. □

We now have the necessary material to prove the main result of our paper.

Theorem 3 (Main Theorem). *For a CDS S of G:*

$$|S| \geq \left\lceil \frac{mn + 2\left\lceil \frac{\min\{m,n\}}{3} \right\rceil}{3} \right\rceil$$

Proof.

$$\begin{aligned}\ell(S) &\geq 2|S| - l + d_2 + 3d_3 + 6d_4\\ &\geq 2|S| - l + 3d_3 + 6d_4 + 6 + 2\left(\left\lceil \frac{\min\{m,n\}}{3} \right\rceil - l + 1\right) - 6\\ &\geq 2|S| + 2\left\lceil \frac{\min\{m,n\}}{3} \right\rceil - 4\end{aligned}$$

We have used the fact that $d_3 + 2d_4 + 2 \geq l$. As $e(S) \geq 4$, $\ell(S) + e(S) \geq 2|S| + 2\left\lceil \frac{\min\{m,n\}}{3} \right\rceil$, the theorem follows. □

2.3 Gap Between Lower and Upper Bounds

In this section, we compare the gap between the lower and upper bounds obtained. To do that, we have to consider this case by case, for reminders n and m leave on division by 3. We assume $m \leq n$. Let L denote the lower bound obtained in Theorem 3. We let $L = \frac{mn}{3} + \frac{2m}{9}$, omitting the ceiling functions as they would only increase L by at most 2. Out of the two CDS's we constructed in Sect. 2.1, D_1 and D_2, the upper bound is given by the construction of smaller size. If m is divisible by 3, $|D_1| \leq |D_2|$, and the gap between the lower and upper bounds is $\frac{m}{9}$. If m is not divisible by 3 and n is however, then $|D_2| \leq |D_1|$ and the gap is $\frac{n}{3} - \frac{2m}{9}$. Similarly, we can analyse the other cases using Table 1 and Table 2. The new lower bound is closest to the constructions in the case that m is divisible by 3.

Table 1. Gaps between $|D_1|$ and L

m mod 3	$\|D_1\|$	$\|D_1\| - \frac{mn}{3}$	$\|D_1\| - L$
0	$\frac{mn}{3} + \frac{m}{3}$	$\frac{m}{3}$	$\frac{m}{9}$
1	$\frac{mn}{3} + \frac{m}{3} + \frac{2n}{3} - \frac{4}{3}$	$\frac{m}{3} + \frac{2n}{3} - \frac{4}{3}$	$\frac{m}{9} + \frac{2n}{3} - \frac{4}{3}$
2	$\frac{mn}{3} + \frac{m}{3} + \frac{n}{3} - \frac{2}{3}$	$\frac{m}{3} + \frac{n}{3} - \frac{2}{3}$	$\frac{m}{9} + \frac{n}{3} - \frac{2}{3}$

Table 2. Gaps between $|D_2|$ and L

n mod 3	$\|D_2\|$	$\|D_2\| - \frac{mn}{3}$	$\|D_2\| - L$
0	$\frac{mn}{3} + \frac{n}{3}$	$\frac{n}{3}$	$\frac{n}{3} - \frac{2m}{9}$
1	$\frac{mn}{3} + \frac{n}{3} + \frac{2m}{3} - \frac{4}{3}$	$\frac{n}{3} + \frac{2m}{3} - \frac{4}{3}$	$\frac{n}{3} + \frac{4m}{9} - \frac{4}{3}$
2	$\frac{mn}{3} + \frac{m}{3} + \frac{n}{3} - \frac{2}{3}$	$\frac{m}{3} + \frac{n}{3} - \frac{2}{3}$	$\frac{m}{9} + \frac{n}{3} - \frac{2}{3}$

3 Conclusions and Further Research

In this paper, we come up with improved lower bounds on the connected domination number of a grid. The question of finding a closed form expression however, remains open. We have broadly used the following approach to prove lower bounds on $|S|$. Using the fact that G is a grid graph, we obtained some structural results for any connected set that dominates G, which lead to lower bounds on the number of bends, vertices of degree 3 and vertices of degree 4 in $G[S]$. We then used Lemma 4 to get lower bounds on $|S|$. This approach however, does not capture the full picture. Consider Fujie's construction detailed in Sect. 2.1. There are no bends or vertices of degree 4 and the number of vertices of degree 3 is $\lceil \frac{m}{3} \rceil$ or $\lceil \frac{n}{3} \rceil$, and our techniques have already accounted for this. There is still a gap between our lower bound and this upper bound because $e(S) = \lceil \frac{m}{3} \rceil + 2$ and $\lceil \frac{n}{3} \rceil + 2$ for these constructions, while we have used a lower bound of 4 for $e(S)$. While this is the best possible lower bound for $e(S)$ separately, it might be possible to obtain better lower bounds for $\ell(S) + e(S)$ by trying to lower bound the sum of two quantities, rather than lower bound each quantity separately as we have done.

An approach to this problem which we have not pursued here would be to design an algorithm or an approximation algorithm that returns the size of the minimum connected dominating set of an $n \times m$ grid in time polynomial in n and m. It is important to note that an approximation algorithm would lead to an *upper bound* on the connected domination number of the grid, while our work has focused on lower bounds. The constructions described in Sect. 2.1 for example, would lead to a trivial approximation algorithm. The gap between our current lower and upper bounds is linear in m and n, which means that this would be asymptotically better than any $(1 + \epsilon)$-approximation algorithm (the input size is $O(mn)$). A non-constructive approach to obtaining upper bounds, for example using the probabilistic method could also be tried.

Our approach has been completely analytical. In [7] the authors used a computational approach to answer the analogous questions about the domination number of grids, using dynamic programming algorithms to calculate the minimum value of a similar loss function near the boundary of the grid. The connected dominating set problem is one of a more 'global' nature than the dominating set problem, as it involves connectivity as a constraint. This entails a very different set of challenges and a computational approach to the problem would likely require new techniques.

References

1. Chang, T.: Domination numbers of grid graphs. Ph.D. thesis, University of South Florida (1992)
2. Clark, B.N., Colbourn, C.J., Johnson, D.S.: Unit disk graphs. Discret. Math. **86**(1), 165–177 (1990). https://doi.org/10.1016/0012-365X(90)90358-O

3. Du, H., Ye, Q., Zhong, J., Wang, Y., Lee, W., Park, H.: Polynomial-time approximation scheme for minimum connected dominating set under routing cost constraint in wireless sensor networks. Theoret. Comput. Sci. **447**, 38–43 (2012). https://doi.org/10.1016/j.tcs.2011.10.010
4. Fujie, T.: An exact algorithm for the maximum leaf spanning tree problem. Comput. Oper. Res. **30**(13), 1931–1944 (2003). https://doi.org/10.1016/S0305-0548(02)00117-X
5. Galbiati, G., Maffioli, F., Morzenti, A.: A short note on the approximability of the maximum leaves spanning tree problem. Inf. Process. Lett. **52**(1), 45–49 (1994). https://doi.org/10.1016/0020-0190(94)90139-2
6. Garey, M.R., Johnson, D.S.: Computers and Intractability: A Guide to the Theory of NP-Completeness. W. H. Freeman & Co., New York (1979)
7. Gonçalves, D., Pinlou, A., Rao, M., Thomassé, S.: The domination number of grids. SIAM J. Discret. Math. **25**(3), 1443–1453 (2011). https://doi.org/10.1137/11082574
8. Hunt III, H.B., Marathe, M.V., Radhakrishnan, V., Ravi, S.S., Rosenkrantz, D.J., Stearns, R.E.: NC-approximation schemes for NP-and PSPACE-hard problems for geometric graphs. J. Algorithms **26**(2), 238–274 (1998). https://doi.org/10.1006/jagm.1997.0903
9. Li, B.P., Toulouse, M.: Variations of the maximum leaf spanning tree problem for bipartite graphs. Inf. Process. Lett. **97**(4), 129–132 (2006). https://doi.org/10.1016/j.ipl.2005.10.011
10. Lichtenstein, D.: Planar formulae and their uses. SIAM J. Comput. **11**(2), 329–343 (1982). https://doi.org/10.1137/0211025
11. Lie, P., Toulouse, M.: Maximum leaf spanning tree problem for grid graphs. JCMCC. J. Comb. Math. Comb. Comput. **73** (2010). http://www.cs.umanitoba.ca/~lipakc/gridgraph-aug6-08.pdf
12. Lu, H.I., Ravi, R.: Approximating maximum leaf spanning trees in almost linear time. J. Algorithms **29**(1), 132–141 (1998). https://doi.org/10.1006/jagm.1998.0944
13. Solis-Oba, R., Bonsma, P., Lowski, S.: A 2-approximation algorithm for finding a spanning tree with maximum number of leaves. Algorithmica **77**(2), 374–388 (2015). https://doi.org/10.1007/s00453-015-0080-0
14. West, D.B.: Introduction to Graph Theory, 2nd edn. Prentice Hall, Upper Saddle River (2000)
15. Wu, J., Li, H.: On calculating connected dominating set for efficient routing in ad hoc wireless networks. In: Proceedings of the 3rd International Workshop on Discrete Algorithms and Methods for Mobile Computing and Communications, DIALM 1999, pp. 7–14. Association for Computing Machinery, New York (1999). https://doi.org/10.1145/313239.313261

On Degree Sequences and Eccentricities in Pseudoline Arrangement Graphs

Sandip Das[1], Siddani Bhaskara Rao[2], and Uma kant Sahoo[1(✉)]

[1] Indian Statistical Institute, Kolkata, India
umakant.iitkgp@gmail.com
[2] CRRAO Advanced Institute of Mathematics, Statistics and Computer Science, Hyderabad, India

Abstract. A pseudoline arrangement graph is a planar graph induced by an embedding of a psuedoline arrangement. We give a simple criterion based on the degree sequence that says when a degree sequence will have a pseudoline arrangement graph as one of its realizations. We then study the eccentricities of vertices in such graphs.

1 Introduction

A *pseudoline* is a curve that extends to infinity on both ends. An *arrangement* $\mathcal{A}(L)$ of pseudolines in the Euclidean plane $\mathbb{R}^2$ is a collection L of (at least three) pairwise intersecting pseudolines. A pair of pseudolines intersect exactly once, where they *cross* each other. It is *simple* if no three pseudolines meet at a point. This implies that all the intersection points are distinct. In this article, we shall consider only simple arrangements, and for the sake of convenience, we omit the word *simple.*

The class of *pseudoline arrangement graphs* $\mathcal{G}_{\mathcal{L}}$ are graphs induced by simple arrangements $\mathcal{A}(L)$, for any set of pseudolines L, whose vertices are intersection points of pseudolines in L, and there is an edge between two vertices if they appear on one of the pseudolines, say $l \in L$, with no other vertices in the part of l between the two vertices. The realization of a pseudoline arrangement graph G_L by pseudolines in L is its *pseudoline arrangement realization*, denoted $R(G_L)$. To obtain a pseudoline arrangement realization, we delete the two infinite segments of each pseudoline from the corresponding pseudoline arrangement. Hence it differs from the pseudoline arrangement (also note that $R(G_L)$ has just one unbounded face. We have analogous definitions by replacing pseudolines with lines.

Arrangements are basic objects of interest in both Discrete and Computational Geometry. Two of the main reasons driving the study of pseudoline arrangements are (1) the numerous problems and conjectures collected in the

S. Das and U. K. Sahoo—Authors are partially supported by the IFCAM project MA/IFCAM/18/39.

A. Mudgal and C. R. Subramanian (Eds.): CALDAM 2021, LNCS 12601, pp. 259–271, 2021.
https://doi.org/10.1007/978-3-030-67899-9_20

survey book by Grünbaum [16], and (2) a consequence of the topological representation theorem of Folkman and Lawrence [11] that gives a geometric interpretation of oriented matroids of rank three in terms of pseudoline arrangements (see [1, Chapter 6] for detailed discussions). For further details on line and pseudoline arrangements, see surveys by Erdős and Purdy [5], and by Felsner and Goodman [8]; also see the book by Felsner [7, Chapter 5 and 6]. We present a brief relevant review in Sect. 1.3.

Pseudoline arrangements are a natural generalization of line arrangements, preserving their basic topological and combinatorial properties. It is well-known that pseudoline arrangements strictly contain line arrangements (see [7,16]). In this article, we study the corresponding *graph realization problem* on pseudoline arrangement graphs, and the *eccentricities* of its vertices. These results are explored in Sects. 2 and 3, respectively. For graph-theoretic terms, refer to the standard text by West [24].

Next we describe the first problem addressed in this article, that is, the graph realization problem concerning pseudoline arrangement graphs.

1.1 Degree Sequences and Graph Realization Problem

The *degree sequence* of a graph is the non-increasing list of degrees of its vertices. A graph with degree sequence π is said to be a *realization* of π. Given an arbitrary finite sequence of non-increasing numbers π, the *graph realization problem* asks whether there is a graph that realizes π. The Erdős-Gallai theorem [6] and the Havel-Hakimi algorithm [17,19] (strengthening of the former) are the two popular methods of solving the graph realization problem. We first address the following problem:

Pseudoline Arrangement Graph Realization Problem: Given a sequence of finite numbers, whether there is a pseudoline arrangement graph that realizes it.

The following theorem solves the pseudoline arrangement graph realization problem. If the answer to this problem is yes, then we construct a pseudoline arrangement realization. This construction and the proof is given in Sect. 2.

A vertex with degree i is a *i-vertex*, for $2 \leq i \leq 4$; and let d_i denote the number of i-vertices. Let $\langle a^d \rangle$ denote the sequence $\langle a, \ldots, a \rangle$ of length d.

Theorem 1. *A finite non-increasing sequence of positive numbers π is a degree sequence of a pseudoline arrangement graph if and only if it satisfies the following two conditions.*

1. $\pi = \langle 4^{d_4}, 3^{d_3}, 2^{d_2} \rangle$ *with* $3 \leq d_2 \leq n$, $d_3 = 2(n - d_2)$ *and* $d_4 = n(n-5)/2 + d_2$ *for some integer* $n \geq 3$.
2. *If* $d_2 = n$, *then* n *is odd.*

There are stronger characterizations based on degree sequences for other graph classes. A graph class $\mathcal{G}$ has a *degree sequence characterization* if one can recognize whether a graph $G \in \mathcal{G}$ or not, solely based on its degree sequence. Hence to recognize whether $G \in \mathcal{G}$ or not, one needs to check whether all the conditions of the degree sequence characterizations are met by the degree sequence of G; often leading to linear-time recognition algorithms [3,18,23]. However, for the following reason, we cannot infer anything about the recognition of pseudoline arrangement graphs from Theorem 1.

A 2–*switch* operation is the replacement of a pair of edges xy and zw in a simple graph by the edges yz and wx, given that yz and wx were not edges in the graph originally. Performing a 2–switch operation in a graph does not change its degree sequence. The class of pseudoline arrangement graphs $\mathcal{G}_\mathcal{L}$ is not *closed* under the 2–*switch* operation, that is, after performing a 2–*switch* operation in $G_L \in \mathcal{G}_\mathcal{L}$, the resulting graph may not be in $\mathcal{G}_\mathcal{L}$. This kills all the hope for obtaining a degree sequence characterization for pseudoline arrangement graphs. Thus in this "sense", Theorem 1 is the best one can hope for.

We further want to highlight that Theorem 1 also implies that the class of pseudoline arrangement graphs cannot have a *forbidden graph characterization*, that is, a characterization for recognizing a graph class by specifying a list of graphs that are forbidden to exist as (or precisely, be isomorphic to) an induced subgraph of any graph in the class. A result of Greenwell et al. [15] says that a graph class has a forbidden graph characterization if and only if it is closed under taking induced subgraphs. The pseudoline arrangement graphs are not closed under vertex deletions. Indeed, Theorem 1 implies that deleting any vertex in a pseudoline arrangement graph does not result in a pseudoline arrangement graph. Alternately, consider the pseudoline arrangement graph on four pseudolines: on deleting a vertex with degree two, we do not get a pseudoline arrangement graph. Hence pseudoline arrangement graphs cannot have a forbidden graph characterization.

As mentioned earlier, pseudoline arrangements are a natural generalization of line arrangements. One more important advantage in studying pseudoline arrangements is that it is possible to decide efficiently whether a combinatorial structure represents an arrangement of psuedolines, whereas the corresponding question for line arrangements is NP-complete [7, Chapter 6]. In our context of graphs, Bose et al. [4] proved that recognizing line arrangement graphs is NP-hard. In this light, it makes more sense to study the line arrangement graph realization problem. But observe that Theorem 1 also addresses the corresponding problem for lines (see Remark 3). Hence we stick with studying pseudolines.

Next we study the eccentricities of vertices in pseudoline arrangement graphs.

1.2 Eccentricities in Pseudoline Arrangement Graphs

The *distance* $d(u, v)$ between two vertices u, v of a graph G is the length of the shortest path between them. The *eccentricity* $e(u)$ of a vertex u is the maximum distance of a vertex in $V(G)$ from u. A vertex v is an *eccentric vertex* of u if $d(u, v) = e(u)$. The *diameter* $d(G)$ of G is the maximum eccentricity of any vertex in $V(G)$. The *radius* $r(G)$ of G is the minimum eccentricity of any vertex in $V(G)$.

In studying the eccentricities of vertices, one of our main aims is to find the diameter and radius of pseudoline arrangement graphs. To this end, one needs to find properties of shortest paths and eccentric vertices. Hence we begin with some basic observations regarding them in Sect. 3.1. They vary from the restrictions on the shortest paths between two vertices, to the existence of particular types of eccentric vertices. These observations are also of independent interest. Using these observations, or otherwise, we find the diameter of pseudoline arrangement graphs.

Proposition 1. *The diameter of a pseudoline arrangement graph on n lines is $n-2$.*

Surprisingly, the diameter of a pseudoline arrangement graph is independent of the graph, and depends only on the number of pseudolines in its realization.

In the following theorem, we characterize the vertices of a pseudoline arrangement graph that have the maximum eccentricity, that is, equal to the diameter of the graph.

Theorem 2. *For a vertex v in a pseudoline arrangement graph G on n pseudolines, its eccentricity $e(v)$ is $n-2$ if and only if v lies in the outer face of its realization $R(G)$.*

Hence Theorem 2 fixes the vertices that occur in the outer face of every realization of a pseudoline arrangement graph.

Unlike the diameter, one can see that the radius of a pseudoline arrangement graph will depend on the graph structure. However, we are still to prove any non-trivial bounds on the radius. We hope Theorem 2 to be a starting point for such a result. We keep this line of questioning for further study.

Organization: In the remaining part of this section,we give a brief review, and then introduce some required tools and definitions. In Sect. 2, we prove Theorem 1. In Sect. 3, we present our results on eccentricity of vertices in pseudoline arrangement graphs leading to Proposition 1 and Theorem 2. Due to space constraints, the proofs of these results are deferred to the full version. We conclude with some remarks in Sect. 4.

1.3 A Brief Review on Pseudoline Arrangement Graphs

In this survey, we focus on pseudoline arrangement graphs, and hence many other results concerning arrangements are not mentioned. They can be found in the survey articles and books on this topic by Grünbaum [16], by Erdős and Purdy [5], by Felsner [7], and by Felsner and Goodman [8]. Grünbaum [16] (in the 1970s) was the first to collect relevant results and posed many problems and conjectures on arrangements of lines as well as pseudolines. The chapter by Erdős and Purdy [5] also addresses various aspects of arrangements. The Chapter 5 in Felsner's book [7] contains results on line arrangements, whereas pseudoline arrangements are studied in Chapter 6. The survey by Felsner and Goodman [8] is the most recent (2017).

Bose et al. [4] (in 2003) proved that recognizing line arrangement graphs is NP-hard. Amongst other results, they gave examples of non-Hamiltonian line arrangement graphs. Felsner et al. [9] studied connectivity, Hamiltonian path and Hamiltonian cycle decomposition, 4-edge and 3-vertex coloring for arrangement graphs of pseudolines, and pseudocircles on spheres. Triangles in line arrangement realizations have been studied by Melchior [22], Levi [21], Füredi and Palsti [12], Felsner and Kriegel [10] and others. Bose et al. [2] studied some hypergraph-theoretic properties of line arrangements.

1.4 Tools, Definitions and Notations

Tools Used: We need the following widely used notions and constructions for pseudoline arrangements. Definitions and constructions are explained in detail in [7, Chapter 6]. We give a succinct description. For a fixed unbounded region f of a pseudoline arrangement $\mathcal{A}$ on n pseudolines, there is always an unbounded region f^* that is *separated* from f by all pseudolines. Note that the boundaries of f and f^* have two pseudolines in common. Fix points $x \in f$ and $x^* \in f^*$. We *topologically sweep* the arrangement to form an aesthetic arrangement of polylines (pseudolines made up of line segments) called the *wiring diagram* [13] corresponding to the sweep. This process uses *allowable sequences* [14], which we do not describe here.

Consider the internally disjoint oriented x^*, x-curves that do not contain any vertex of the arrangement and that crosses each pseudoline exactly once. A *topological sweep* of the arrangement is a sequence $c_0, \dots, c_r$ of such oriented x^*, x-curves with $r = \binom{n}{2}$ such that there is exactly one vertex between the curves c_i and c_{i+1}. Here c_0 is the oriented x^*, x-curve such that all the vertices in $\mathcal{A}$ lie to the right of c_0 (with respect to the orientation of c_0). Label the pseudolines from 1 to n such that c_0 intersects the pseudolines in increasing order. Next we form the wiring diagram corresponding to this topological sweep.

Fix n horizontal wires. We confine the pseudolines to these wires, except for the parts where they cross each other. Corresponding to the topological sweep we have a sequence $p_0, \dots, p_r$ of vertical lines with p_i to the left of p_j, for $i < j$. The ordering of polylines in which p_0 intersects from bottom to top is $1, 2, \dots, n$. Between p_i and p_{i+1}, we allow only the two pseudolines that form the vertex between c_i and c_{i+1} to intersect. Hence the ordering of the polylines intersecting p_i from bottom to top is the same as the ordering of pseudolines intersecting c_i from x^* to x. We call this as the *wiring diagram* corresponding to the topological sweep. See Fig. 1 for an illustration.

Definitions and Notations: Since we are dealing with pseudolines, which are topological analogues of lines, we shall come across terms like *pseudohalfplane*, *pseudoquadrant*, *pseudotriangle*, *pseudopolygon* etc., in our arguments; the prefix *pseudo* denotes the topological analogue of the following term.

Let G_L be a pseudoline arrangement graph with realization $R(G_L)$. The *span* of each pseudoline is the part of pseudoline drawn in $R(G_L)$, that is, the part of the pseudoline between its end vertices. A *path* P is a sequence of distinct vertices, such that consecutive vertices are adjacent. The *length* of a path P, denoted $|P|$, is the number of edges in P. The length of the shortest u, v-path is

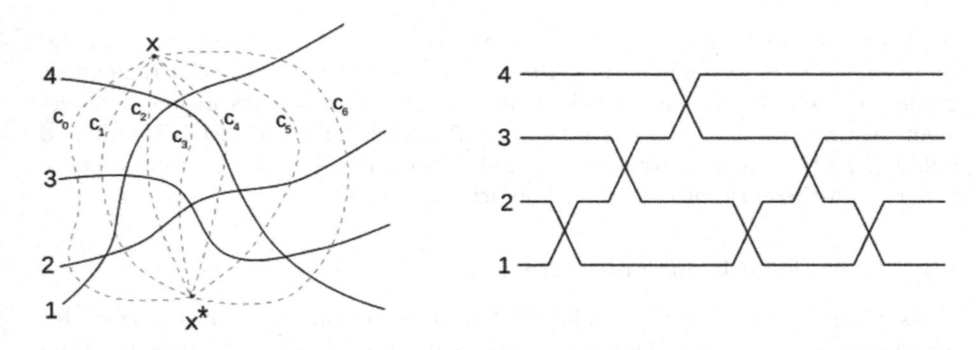

Fig. 1. Topological sweep of a pseudoline arrangement and its wiring diagram.

$d(u,v)$. For vertices u and v in the path P in G_L, let $uv|_P$ denote the u,v-path in P, with length $d(u,v)_P$. For vertices u and v in line $l \in L$, let $uv|_l$ denote the u,v-path in l, with length $d(u,v)_l$.

For points c and d (may not be in $V(G_L)$) that are on different pseudolines, let $\overline{cd}$ denote the line segment between c and d. If vertex $u \in V(G_L)$ is an intersection point of two pseudolines l_1 and l_2, then we say $u = l_1 \cap l_2$. For vertices $u, v \in V(G_L)$ not lying on a pseudoline l, we say l *separates* u and v if they lie on different pseudohalfplanes bounded by l.

Relevant definitions specific to the proofs are given in their respective sections.

2 Pseudoline Arrangement Graph Realization Problem

Now we solve the pseudoline arrangement graph realization problem, that is, we prove Theorem 1. For the sake of the reader, we restate Theorem 1.

Theorem 1. *A finite non-increasing sequence of positive numbers π is a degree sequence of a pseudoline arrangement graph if and only if it satisfies the following two conditions.*

1. $\pi = \langle 4^{d_4}, 3^{d_3}, 2^{d_2} \rangle$ with $3 \leq d_2 \leq n$, $d_3 = 2(n - d_2)$ and $d_4 = n(n-5)/2 + d_2$ for some integer $n \geq 3$.
2. If $d_2 = n$, then n is odd.

Remark 1. We study degree sequences in pseudoline arrangement graphs on the Euclidean plane as opposed to the *real projective plane*, where the corresponding problem is not interesting ($\pi = \langle 4^{\binom{n}{2}} \rangle$). It further indicates the importance of the second condition in Theorem 1 for completing the characterization, as most of the first condition is relatively intuitive.

The rest of this section is devoted to the proof of Theorem 1. In Sect. 2.1, we prove the necessity of Theorem 1. In Sect. 2.2, we define some constructions and operations that we need to prove the sufficiency of Theorem 1, which is presented in Sect. 2.3.

2.1 Proof of Necessity of Theorem 1

Suppose the degree sequence π has a pseudoline arrangement realization on $n \geq 3$ pseudolines. Since every pair of pseudolines intersect, $d_2 + d_3 + d_4 = n(n-1)/2$. Each of the end vertices of every pseudoline is either a 2–vertex or a 3–vertex: each 2–vertex is an end vertex of two pseudolines, and each 3–vertex is an end vertex of one pseudoline. Thus $2d_2 + d_3 = 2n$. From both these equations, $d_3 = 2(n - d_2)$, $d_4 = n(n-5)/2 + d_2$ and $d_2 \leq n$.

We claim that $d_2 \geq 3$. First we extend the realization on n pseudolines to an arrangement on n pseudolines. Next extend the arrangement to a pseudoline arrangement on $n+1$ pseudolines in the real projective plane by adding an imaginary pseudoline l at infinity. A standard result by Levi [21] (also see [7, Prop. 5.13]) shows that every such pseudoline is incident to at least three triangles. In particular, l is incident to at least three triangles in this arrangement in the real projective plane. Each of these triangles corresponds to unbounded regions with exactly two pseudolines in its boundary in the arrangement in the Euclidean plane. The intersection point of these two pseudolines corresponds to a 2-vertex in the realization. Hence $d_2 \geq 3$.

If $d_2 = n$ then $d_3 = 0$, that is, each of the end vertex of every pseudoline is a 2–vertex. In such an arrangement consider any pseudoline l. Let u and v be the end vertices of l. Pseudoline l divides the plane into two open pseudohalfplanes, denoted l^+ and l^-. The other two pseudolines from u and v meet in one of the pseudohalfplanes, say (without loss of generality) l^+. Let the number of 2–vertices in l^+ be k. Thus the number of 2–vertices in l^- is $n-k-2$. All other pseudolines, except the three incident at u or v, cross the part of l between u and v. We double count such crossings: for all the 2-vertices in l^- there are $2(n-k-2)$ such crossings, and for all the 2–vertices in l^+ there are $2k-2$ such crossings. Thus $2k - 2 = 2(n-k-2)$. This implies $n = 2k+1$. Hence if $d_2 = n$ then n is odd. This completes the necessity part of the proof.

Remark 2. One can derive the first condition of Theorem 1 from the argument involving the projective plane in proving $d_2 \geq 3$. However, we want to highlight the approach using the two equations, as they hold for graphs induced by a more general arrangement of simple finite curves, which is helpful in alternately proving a result of Kostochka and Nešetřil [20]. An alternate proof of $d_2 \geq 3$ using *wiring diagrams* is given in Sect. 4.1.

To prove the sufficiency of Theorem 1, we give a line arrangement realization having the given degree sequence. First we need the following construction and operations.

2.2 Constructions and Operations

Star Construction: For odd n, place n vertices uniformly on a circle and join each vertex to its opposite two farthest vertices. Resulting is a line arrangement realization on n lines called a *star construction on n vertices*. The center of the circle is called the *center of the star construction.* The rest of the vertices are 4–vertices that lie within the circle. The star construction on n vertices has degree sequence $\langle 4^{n(n-3)/2}, 2^n \rangle$.

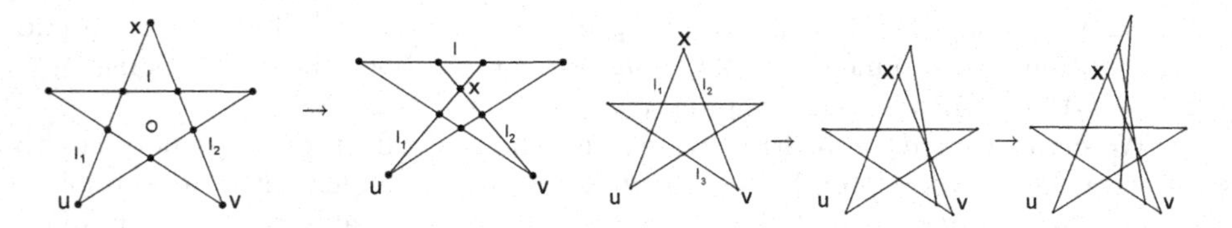

Fig. 2. Pull operation

Fig. 3. Two line operations on the 2-vertex x

Pull Operation: Consider a 2–vertex x in a star construction on at least 5 vertices with O as the center of the star construction. Let $x = l_1 \cap l_2$ with the other end vertices of l_1 and l_2 as u and v respectively. Let l be the first line crossed while moving from x to O along the line segment $\overline{xO}$. Rotate l_1 and l_2 about u and v such that $x = l_1 \cap l_2$ comes closer to O till x crosses l, while keeping the slope of $\overline{xO}$ fixed (see Fig. 2). Now x becomes a 4–vertex and two new 3–vertices are created at the expense of two 4–vertices in the star construction. In this operation, the number of 2–vertices decreases by one. Hence the degree sequence changes from $\langle 4^{d_4}, 2^{d_2} \rangle$ to $\langle 4^{d_4-1}, 3^2, 2^{d_2-1} \rangle$.

Line Operation: In the target degree sequence, if d_2 is odd, then consider a star construction. Choose a 2–vertex $x = l_1 \cap l_2$ with the other end vertices of l_1 and l_2 as u and v respectively, which are also 2-vertices. Let l_3 be the other line intersecting l_2 at v. Take a point on the line l_1 that is close to x and just outside the span of l_1 in the realization. Also, take a point in the realization on the line l_3 and close to v. Joining these two points we add a new line l to the realization that intersects all the span of other lines except the span of l_1 in the realization of the star construction (as shown in Fig. 3). In the new realization, $l_1 \cap l$ forms a new 2–vertex, making x a 3–vertex. Thus the number of 2–vertices is unaffected. The other end vertex of l is also a 3–vertex, increasing number of 3–vertices by 2. The rest of the new vertices introduced are 4–vertices.

By doing k line operations on x, we add k such new lines to the realization close to x. On constructing such new lines we make sure that their intersection points with l_3 (as well as with l_1), in order of their addition, form a monotonic sequence of points in l_3 (and in l_1). It ensures that the new line added also intersects the previously added lines before reaching its end vertex, which is a 3–vertex (refer Fig. 3). Upon performing k line operations on a star construction on d_2 vertices, the number of 2–vertices remains unchanged; the number of 3–vertices increases by $2k$; and the number of 4–vertices increases by

$\binom{d_2+k}{2} - \binom{d_2}{2} - 2k = k(d_2 + \frac{k-5}{2})$. Hence the degree sequence changes from $\langle 4^{d_4}, 2^{d_2}\rangle$ to $\langle 4^{d_4+k(d_2+\frac{k-5}{2})}, 3^{2k}, 2^{d_2}\rangle$.

In the target sequence, if d_2 is even, then we first consider a star construction on $d_2 + 1$ with one pull operation on it. This realization has d_2 2–vertices. We can also perform line operations on this realization. Repeating the above calculations, upon performing k line operations on a star construction on d_2+1 vertices with a pull operation, the number of 2–vertices remains unchanged; the number of 3–vertices increases by $2k$; and the number of 4–vertices increases by $\binom{d_2+1+k}{2} - \binom{d_2+1}{2} - 2k = k(d_2 + \frac{k-3}{2})$. Hence the degree sequence changes from $\langle 4^{d_4}, 3^2, 2^{d_2}\rangle$ to $\langle 4^{d_4+k(d_2+\frac{k-3}{2})}, 3^{2+2k}, 2^{d_2}\rangle$.

Now we are ready to prove the sufficiency part of Theorem 1.

2.3 Proof of Sufficiency of Theorem 1

Let π be a degree sequence satisfying the properties given in Theorem 1 for some value of n. We give an algorithm to draw a line arrangement realization with degree sequence π.

Algorithm: For odd d_2, we do a star construction on d_2 vertices. If $d_2 = n$, then n is odd, and we have the required line arrangement realization; else do $n - d_2$ line operations on a 2–vertex of the star to get the required realization. For even d_2, we do a star construction on $d_2 + 1$ vertices and then do a pull operation on one of the 2–vertices, resulting in d_2 2–vertices. If $d_2 = n - 1$, then we have the required line arrangement realization; else if $n > d_2+1$, then we do $n-d_2-1$ line operations on a 2–vertex of the star construction to get the required realization. Hence we have the required line arrangement realization, with degree sequence π.

Correctness: For odd d_2, a star construction on d_2 vertices results in the degree sequence $\pi_s = \langle 4^{d_2(d_2-3)/2}, 2^{d_2}\rangle$. If $d_2 = n$, then $\pi_s = \pi$. If $d_2 < n$, then performing $n - d_2$ line operations increases d_4 by $(n - d_2)(d_2 + \frac{n-d_2-5}{2}) = \frac{n^2}{2} - \frac{d_2^2}{2} - \frac{5n}{2} + \frac{5d_2}{2}$; increases d_3 by $2(n - d_2)$; and d_2 remains same. This results in the degree sequence $\langle 4^{n(n-5)/2+d_2}, 3^{2(n-d_2)}, 2^{d_2}\rangle$.

For even d_2, a pull operation on a star construction on $d_2 + 1 \geq 5$ vertices results in degree sequence $\pi_{sp} = \langle 4^{(d_2+1)(d_2-2)/2-1}, 3^2, 2^{d_2}\rangle$. If $d_2 = n - 1$, then $\pi = \langle 4^{n(n-3)/2-1}, 3^2, 2^{n-1}\rangle = \langle 4^{n(n-5)/2}, 3^2, 2^{n-1}\rangle$. Hence $\pi_{sp} = \pi$. If $d_2 < n-1$, then performing $n - d_2 - 1$ line operations increases d_4 by $(n - d_2 - 1)(d_2 + \frac{n-d_2-4}{2}) = \frac{n^2}{2} - \frac{d_2^2}{2} - \frac{5n}{2} + \frac{3d_2}{2} + 2$; increases d_3 by $2(n - d_2 - 1)$; and d_2 remains same. This results in the degree sequence $\langle 4^{n(n-5)/2+d_2}, 3^{2(n-d_2)}, 2^{d_2}\rangle$.

So the resulting realization given by the algorithm has degree sequence π. This completes the proof of the sufficiency of Theorem 1 and hence the proof of Theorem 1. □

Remark 3. We constructed a line arrangement realization in the proof of the sufficiency of Theorem 1, and the proof of necessity also goes through for line arrangements. Hence Theorem 1 also solves the line arrangement graph realization problem.

Next we study the eccentricities of vertices in pseudoline arrangement graphs.

3 Eccentricities in Pseudoline Arrangement Graphs

To find the eccentricity of a vertex, we shall find one of its eccentric vertices. For this purpose, we derive some basic results on the shortest paths and eccentric vertices in pseudoline arrangement graphs, which are of independent interest. To this end, we need the following definitions.

Consider two vertices u and v in a pseudoline arrangement that do not lie on the same pseudoline. Let $u = l_1 \cap l_2$ and $v = l_3 \cap l_4$. For each vertex in the pseudoline arrangement, the two intersecting pseudolines divide the Euclidean plane into four pseudoquadrants. Let Q_v denote the pseudoquadrant defined by l_1 and l_2 that contains vertex v. Similarly, let Q_u denote the pseudoquadrant defined by l_3 and l_4 that contains vertex u. Let $Q_{uv} = Q_u \cap Q_v$.

Due to space constraints, the proofs of our results in this section are deferred to the full version.

3.1 Basic Results on Eccentricities

Our first observation says that the shortest path between two vertices in a pseudoline is the path between them on the pseudoline.

Proposition 2. *For vertices u and v on pseudoline l, the shortest u, v-path completely lies on l, and this path is unique.*

Next we study the shortest paths between any two vertices in a pseudoline arrangement graph. The following is a consequence of Proposition 2.

Proposition 3. *For any two vertices u and v, a shortest u, v-path of length k has vertices on $k + 2$ pseudolines.*

Our next proposition is the analogue of Proposition 2, for vertices that do not lie on a pseudoline.

Proposition 4. *For vertices u and v not on the same pseudoline, the shortest path between them completely lies in Q_{uv}.*

As a corollary of our next result, it follows that one of the eccentric vertices of any vertex lies on the outer face.

Proposition 5. *For a vertex $u \in V(G_L)$, there exists an eccentric vertex of u that is a 2–vertex or 3–vertex.*

Corollary 1. *For a vertex $u \in V(G_L)$, there exists an eccentric vertex of u that lies in the outer face of $R(G_L)$.*

Next we present the main results of this section.

3.2 Diameter and Characterization of Vertices with Maximum Eccentricity

Finding the diameter of a pseudoline arrangement graph is a straightforward implication of Proposition 3. However, one can also prove it without using Proposition 3. For the sake of the reader, we restate Proposition 1.

Proposition 6. *The diameter of a pseudoline arrangement graph on n lines is $n-2$.*

The context of the proof of Proposition 3, implies the following remark.

Remark 4. If $d(u,v) = n-2-i$, then every shortest u,v-path has vertices on $n-i$ pseudolines. In particular, if $d(u,v) = n-2$, then any shortest u,v-path has vertices on all the n pseudolines.

For vertices u and v, any shortest u,v-path has a vertex on every pseudoline that separates u and v. So the number of such separating pseudolines lower bounds $d(u,v)$.

Remark 5. For vertices u and v in a pseudoline arrangement graph,

$$d(u,v) \geq \begin{cases} \#\text{separating pseudolines} + 1 & \text{if } u \text{ and } v \text{ lie on the same pseudoline;} \\ \#\text{separating pseudolines} + 2, & \text{if } u \text{ and } v \text{ do not lie on the same} \\ & \text{pseudoline.} \end{cases}$$

Theorem 2 characterizes the vertices of a pseudoline arrangement graph that have the maximum eccentricity, that is, equal to its diameter. For the sake of the reader, we restate Theorem 2.

Theorem 2. *For a vertex v in a pseudoline arrangement graph G on n pseudolines, its eccentricity $e(v)$ is $n-2$ if and only if v lies in the outer face of its realization $R(G)$.*

As a direct consequence of Theorem 2, we can also find eccentricities of some vertices in the next layer. Let G_L be a pseudoline arrangement graph having vertices $V_{out} \subset V(G_L)$ on the outer face of its realization $R(G_L)$. The 1–*layer* of $R(G_L)$ is the outer face of the realization upon removing all vertices in V_{out} and their incident edges.

Theorem 2 implies that any interior vertex has eccentricity less than $n-2$. Notice that it is possible for vertices in the 1–layer of $R(G_L)$ to have no neighbors on the outer face. However, for vertices in the 1–layer which have a neighbor on the outer face, we have the following corollary.

Corollary 2. *Let u_1 be a vertex in the 1–layer of $R(G_L)$. If u_1 has a neighbor u in the outer face of $R(G_L)$, then $e(u_1) = n-3$.*

4 Final Remarks

4.1 Alternate Proof of $d_2 \geq 3$ in Theorem 1 Using Wiring Diagrams

Observe that the leftmost and rightmost intersection point in the wiring diagram of a pseudoline arrangement is always a 2-vertex; so $d_2 \geq 2$. Next we consider a 'restricted wiring diagram' in which there is just one intersection point between the top two levels. In this case, observe that such an intersection point is also a 2-vertex. This 2-vertex is different from the leftmost and the rightmost intersection point in the wiring diagram (else one of the pseudolines has just one intersection point in it; a contradiction). Thus for pseudoline arrangements with a wiring diagram that has only one intersection point between the top two levels, $d_2 \geq 3$. We claim that all pseudoline arrangements have such a restricted wiring diagram. To show this, we need to carefully set up the topological sweep that fixes the wiring diagram. Choose the sweep lines to originate from an unbounded face that is bounded by two pseudolines in the pseudoline arrangement (this always exists as $d_2 \geq 2$). Perform the topological sweep to form the required restricted wiring diagram. See Fig. 1 for an illustration.

Acknowledgement. The authors thank Prof. Douglas B. West for his encouragement to pursue the line arrangement graph realization problem, which was the starting point of this work. The authors also thank Dibyayan Chakraborty for suggesting to pursue the questions on eccentricity. The authors also thank the reviewers of both of this version and the earlier drafts for greatly enhancing both the content and the presentation of this paper. The first and third authors are partially supported by IFCAM project Applications of graph homomorphisms (MA/IFCAM/18/39).

References

1. Björner, A., Las Vergnas, M., Sturmfels, B., White, N., Ziegler, G.M.: Oriented Matroids, 2nd ed. Encyclopedia of Mathematics, vol. 46. Cambridge University Press, New York (1999)
2. Bose, P., et al.: Coloring and guarding arrangements. Discret. Math. Theor. Comput. Sci. **15**, 139–154 (2013)
3. Bose, P., et al.: A characterization of the degree sequences of 2-trees. J. Graph Theory **58**, 191–209 (2008)
4. Bose, P., Everett, H., Wismath, S.: Properties of arrangement graphs. Int. J. Comput. Geom. Appl. **13**(6), 447–462 (2003)
5. Erdős, P., Purdy, G.: Extremal problems in combinatorial geometry. In: Graham, R.L., et al. (eds.) Handbook of Combinatorics, vol. I, pp. 809–874. Elsevier, Amsterdam (1995)
6. Erdős, P., Gallai, T.: Graphs with prescribed degrees of vertices. Mat. Lapok **11**, 264–274 (1960)
7. Felsner, S.: Geometric Graphs and Arrangements. Advanced Lectures in Mathematics. Springer, Heidelberg (2004). https://doi.org/10.1007/978-3-322-80303-0
8. Felsner, S., Goodman, J.E.: Pseudoline arrangements, Chapter 5. In: Goodman, J.E., et al. (eds.) Handbook of Discrete and Computational Geometry, 3rd edn. CRC Press, Boco Raton (2017)

9. Felsner, S., Hurtado, F., Noy, M., Streinu, I.: Hamiltonicity and colorings of arrangement graphs. Discret. Appl. Math. **154**(17), 2470–2483 (2006). Extended abstract in Proceedings of SODA 2000, 155–164
10. Felsner, S., Krieger, K.: Triangles in Euclidean arrangements. In: Hromkovič, J., Sýkora, O. (eds.) WG 1998. LNCS, vol. 1517, pp. 137–148. Springer, Heidelberg (1998). https://doi.org/10.1007/10692760_12
11. Folkman, J., Lawrence, J.: Oriented matroids. J. Comb. Theory Ser. B **25**, 199–236 (1978)
12. Füredi, Z., Palasti, I.: Arrangements of lines with a large number of triangles. Proc. Am. Math. Soc. **92**, 561–566 (1984)
13. Goodman, J.E.: Proof of a conjecture of Burr, Grünbaum and Sloane. Discret. Math. **32**, 27–35 (1980)
14. Goodman, J.E., Pollack, R.: Semispaces of configurations, cell complexes of arrangements. J. Comb. Theory Ser. A **37**, 257–293 (1984)
15. Greenwell, D.L., Hemminger, R.L., Kleitman, J.: Forbidden subgraphs. In: Proceedings of 4th South East Conference of Graph Theory and Computing, Florida Atlantic University, pp. 389–394 (1973)
16. Grünbaum, B.: Arrangements and Spreads. Regional Conference Series in Mathematics. Amer. Math. Soc. Providence, RI (1972)
17. Hakimi, S.L.: On realizability of a set of integers as degrees of the vertices of a linear graph. Int. J. Soc. Ind. Appl. Math. **10**, 496–506 (1962)
18. Hammer, P.L., Simeone, B.: The splittance of a graph. Combinatorica **1**, 275–284 (1981). https://doi.org/10.1007/BF02579333
19. Havel, V.: A remark on the existence of finite graphs. Časopis Pěst Mat **80**, 477–480 (1955)
20. Kostochka, A.V., Nešetřil, J.: Coloring relatives of intervals on the plane, I: chromatic number versus girth. Eur. J. Comb. **19**, 103–110 (1998)
21. Levi, F.: Die Teilung der projektiven Ebene durch Gerade oder Pseudogerade, Ber. Math.-Phys. Kl. Sächs. Akad. Wiss. Leipzig, **78**, 256–267 (1926)
22. Melchior, E.: Über Vielseite der projektiven Ebene. Deutsche Mathematik **5**, 461–475 (1940)
23. Merris, R.: Split graphs. Eur. J. Comb. **24**(4), 413–430 (2003)
24. West, D.B.: Introduction to Graph Theory, 2nd edn. Pearson Education, London (2002)

Cops and Robber on Butterflies and Solid Grids

Sheikh Shakil Akhtar, Sandip Das, and Harmender Gahlawat(✉)

Indian Statistical Institute, Kolkata, India
harmendergahlawat@gmail.com

Abstract. Cops and robber is a well-studied two player pursuit-evasion game played on a graph. In this game, a set of cops, controlled by the first player, tries to capture the position of a robber, controlled by the second player. The *cop number* of a graph is the minimum number of cops required to capture the robber in the graph. We show that the cop number for butterfly networks and for solid grids is two.

1 Introduction and Results

Cops and robber is a two player pursuit-evasion game played on a graph. The first player controls a set of agents, called *cops*, and the second player controls a single agent, called the *robber*. The game starts with the cops occupying some vertices of the graph and then the robber occupies a vertex of the graph. More than one agent may simultaneously occupy the same vertex of the graph. In each subsequent round, first each cop and then the robber make a move. In a move, an agent either moves to an adjacent vertex (along an edge) or stays on the same vertex. If at any point in the game one of the cops occupies the same vertex as the robber, then we call it a *capture*. If the cops can capture the robber in a graph, then the cops *win* and if the robber can evade the capture indefinitely, then the robber *wins*. The *cop number* of a graph G, written as $c(G)$, is the minimum number of cops required to capture a robber in G. Furthermore, for a family $\mathcal{F}$ of graphs, $c(\mathcal{F}) = \max\{c(G)|G \in \mathcal{F}\}$. Classically, the game is a perfect information game in the sense that all the agents know each other's position.

The game of cops and robber was independently introduced by Quilliot [19] and by Nowakowski and Winkler [18]. Both of these papers considered the game with a single cop and a robber, and both characterized the graphs where a single cop can win. Later, Aigner and Fromme [1] generalized the game to multiple cops and introduced the concept of the cop number. They also showed that the cop number for planar graphs is 3 and for every n, there exists a graph G such that $c(G) \geq n$. Since then, cops and robber game has been studied extensively and has applications in artificial intelligence, graph searching, game development, etc. [2,11,12] as well as significant implications in theory [20]. For a detailed survey of the subject, see the book by Bonato and Nowakowski [5].

Berarducci and Inrigila [6] gave a backtracking algorithm that decides whether the cop number of a graph is at most k in $O(n^k)$ time; so for a fixed

A. Mudgal and C. R. Subramanian (Eds.): CALDAM 2021, LNCS 12601, pp. 272–281, 2021.
https://doi.org/10.1007/978-3-030-67899-9_21

k, this is a polynomial-time algorithm. Goldstein and Reingold [9] proved that, deciding if k cops can capture a robber in a graph is EXPTIME-complete if k is not fixed and either the initial positions are given or the graph is directed. Fomin et al. [8] proved that determining the cop number of a graph is NP-hard. Later Kinnersley [13] proved that determining the cop number of a graph or digraph is EXPTIME-complete.

Butterfly networks (defined later) are extensively studied interconnection networks and have applications in parallel computing [14]. We study the game of cops and robber on butterfly networks and have the following theorem.

Theorem 1. *Cop number for finite butterfly networks is two.*

We also show that on a k-dimensional butterfly network, two cops can capture the robber in $O(k^2)$ cop moves.

Cops and robber game has been studied on grids and it is known that the cop number for grids is 2 [16]. Several variations of cops and robber game also have been studied on grids [3,4,7], and the classical cops and robber game has been studied on various kinds of grids [15,17]. Continuing this, we study the game of cops and robber on *solid grids*, which are subgraphs of grids such that each internal face has unit area (formally defined later). We have the following theorem.

Theorem 2. *Cop number for solid grids is two.*

1.1 Organisation

In Sect. 2 we define the important definitions and tools that we will use for our proofs. In Sect. 3 we study the game on butterfly networks and prove Theorem 1. In Sect. 4 we study the game on solid grids and prove Theorem 2. In Sect. 5 we suggest some future directions.

2 Definitions

All the graphs considered in this paper are finite, connected, and simple. We will denote the robber as $\mathcal{R}$. Let G be a graph and H be a subgraph of G. Then $G - H$ refers to the graph induced by vertices that are in G but not in H.

Let H be a subgraph of G. We say that $\mathcal{R}$ is *restricted* to H, if $\mathcal{R}$ cannot leave the vertices of H without getting captured. We also say that H is the *robber territory*.

Let H be a subgraph of G. A cop $\mathcal{C}$ *guards* H if the robber cannot enter the vertices of H without getting captured by $\mathcal{C}$ in the next cop move.

For a graph G, *capture time* using k cops, is the number of cop moves to ensure the capture (of robber) using k cops.

A k-dimensional butterfly network, is a graph, consisting of $2^k(k+1)$ vertices arranged in $k+1$ columns and 2^k rows. The 2^k rows are coded in k-bit binary from $00\dots00$ to $11\dots11$ and the $k+1$ columns are coded in decimal from 0 to k. These columns are also referred to as *levels*. A vertex in i-th row and j-th column is denoted by (i, j). There exists an edge between two vertices (i, j) and (i', j') if (1) $j' = j + 1$ and (2) either $i' = i$, referred as *straight edge*, or the binary representations of i and i' differ exactly in the j'-th least significant bit, referred as *cross edge*. See Fig. 1 for an illustration.

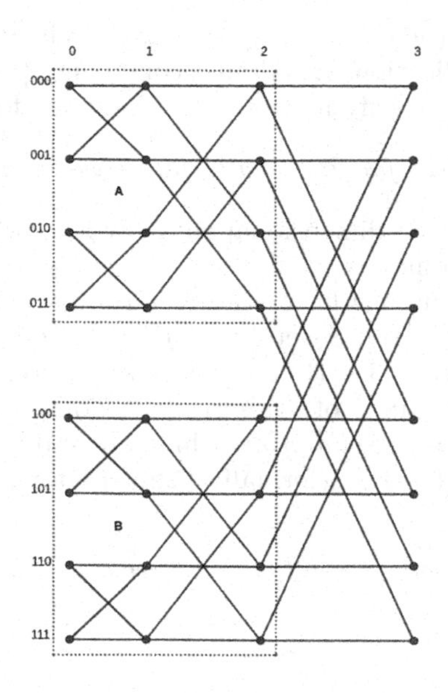

Fig. 1. A 3-dimensional butterfly network.

We also define the *recursive* definition of a k-dimensional butterfly network, which we will use in our algorithm. For that purpose, take two $(k-1)$-dimensional butterfly networks A and B. Now, for all the vertices of A, append 0 as the most significant bit and for all the vertices of B, append 1 as the most significant bit. Levels of all the vertices remain as they were in A and B. Next, add level k with 2^k vertices and add edges between vertices of level k and $k-1$ as per the rules defined for butterfly networks in the previous definition. See Fig. 1 for an illustration. Observe that all paths from vertices of A to vertices of B go through the vertices of the new level (level k).

Let u and v be two vertices of a graph G. Then $N(u)$ denote the open neighbourhood of vertex u, that is, $N(u) = \{x \mid (x, u) \text{ is an edge in } G\}$. Closed neighbourhood of a vertex u, $N[u] = N(u) \cup \{u\}$. A vertex u is a *corner vertex*, if there exists a vertex v such that $N[u] \subseteq N[v]$. We also say that u is a corner of v. If a graph G has no corner vertex, then $c(G) > 1$ [1,18]. Aigner and Fromme [1] proved the following result (we restate the result to suit our definitions), which we will use.

Result 1. *(Aigner and Fromme [1]) Let P be a shortest path between two vertices u and v of a graph G. Then one cop can guard P after a finite number of moves.*

3 Butterfly Networks

In this section, we give a strategy to capture the robber in a k-dimensional butterfly network using two cops in $O(k^2)$ moves. We need the following definitions.

We refer the vertex in level 0 and row $00\ldots00$ as the *start vertex* of a k-dimensional butterfly network. An agent, cop or robber, makes a *forward move* if the level of vertex occupied by the agent increases, and makes a *backward move* if the level of the vertex occupied by the agent decreases.

Note 1. When we say l-th least significant bits or l significant bits, it means the usual for $l > 0$. For $l = 0$, assume that the l least significant bits or the l-th least significant bit of all the tuples are the same.

Vertex (i', j) is said to be an *image* of (i, j), if the j least significant bits of i and i' are the same. For example, each vertex is an image of itself, all the vertices of level 0 are images of each other, and the vertices of level k do not have any image other than themselves. A cop $\mathcal{C}$ *captures an image* of $\mathcal{R}$ if $\mathcal{R}$ is at a vertex (i, j) and $\mathcal{C}$ is at an image of (i, j).

A path is a *monotone path* if all its vertices are from different levels and the first vertex is from level 0. A cop $\mathcal{C}$ *guards* level l if the robber cannot enter the vertices of level l without getting captured by $\mathcal{C}$ in the subsequent cop move.

Recall that we refer $(00\ldots00, 0)$ as the start vertex. We have the following lemma.

Lemma 1. *Let (i, j) be a vertex of a butterfly network. Then there exists an image (i', j) of (i, j) such that there exists a monotone path from the start vertex to (i', j).*

Proof. We follow a simple strategy to create such a monotone path P. We start our path P from the *start vertex*. When we move from level l to $l + 1$, we take a straight edge if the $l + 1$-th least significant bit of i is 0 and take a cross edge otherwise. This way when we reach level j, at a vertex (i', j), the j least significant bits of i and i' will be the same and each vertex of path P is in a different level. Hence this path P is a monotone path from start to an image (i', j) of vertex (i, j). □

Lemma 2. *In a k-dimensional butterfly network, one cop can capture an image of the robber in at most k steps.*

Proof. Cop $\mathcal{C}$ begins at the start vertex. Let $\mathcal{R}$ be at a vertex (i, j). First, $\mathcal{C}$ will find a monotone path P, from the start vertex to an image of $\mathcal{R}$ (by Lemma 1). The cop will update the monotone path P dynamically following the moves of $\mathcal{R}$ and will move forward through P until it reaches an image of $\mathcal{R}$.

Let $\mathcal{R}$ be at vertex (r, l). We will maintain the invariant that the last vertex of P is (p, l) such that (p, l) is an image of $\mathcal{R}$. Also in each cop move, $\mathcal{C}$ will make a forward move on P. If at any point $\mathcal{C}$ and $\mathcal{R}$ are at the same level, then observe that $\mathcal{C}$ has captured an image of $\mathcal{R}$.

If $\mathcal{R}$ moves forward to increase the level from vertex (r, l) to $(r', l+1)$, then r and r' differ in at most one bit (that is the $(l+1)$-th least significant bit). Before this move let the last vertex of P be (p, l). Since l least significant bits of p and r are the same, l least significant bits of p and r' are also the same. So if the $l+1$-th bit of p and r' is the same, then we extend our path P using a straight edge, else we extend P using a cross edge.

If $\mathcal{R}$ moves backward, then we truncate our path by one vertex. Suppose $\mathcal{R}$ moves from (r, l) to $(r', l-1)$. Here r and r' differ in at most one bit, that is, the l-th bit. Hence the first $l-1$ bits of r, r', p and p's neighbour in level $l-1$ are the same. So our invariant holds if we just remove the last vertex from our path P.

Since in each cop move $\mathcal{C}$ is strictly increasing its level in a monotone path, in at most k moves, both $\mathcal{C}$ and $\mathcal{R}$ will be in the same level. So $\mathcal{C}$ captures an image of $\mathcal{R}$ in at most k steps. □

Lemma 3. *In a k-dimensional butterfly network, one cop can guard level k in at most k steps.*

Proof. In the level k of a k-dimensional butterfly network, each vertex has only itself as an image. Thus, if $\mathcal{R}$ is in level k and $\mathcal{C}$ captures an image of $\mathcal{R}$, then $\mathcal{C}$ captures $\mathcal{R}$. Hence, if $\mathcal{C}$ can ensure that after each cop move $\mathcal{C}$ has captured an image of the robber $\mathcal{R}$, then $\mathcal{R}$ cannot enter level k without being captured by $\mathcal{C}$.

Cop $\mathcal{C}$ starts by capturing an image of the robber (by Lemma 2). In Lemma 2 we are maintaining a dynamic path P such that its last vertex (p, l) is an image of $\mathcal{R}$. When $\mathcal{C}$ captures an image of $\mathcal{R}$, cop $\mathcal{C}$ is on the last vertex (p, l) of P. We keep maintaining this dynamic path P as we did in Lemma 2 and $\mathcal{C}$ will remain on last vertex of P. Let (p, l) be the last vertex of P and after the move of $\mathcal{R}$, the new last vertex of P is (p', l'). Since vertices (p, l) and (p', l') are adjacent, $\mathcal{C}$ can and will move to (p', l').

Thus, once $\mathcal{C}$ captures an image of $\mathcal{R}$ in a k-dimensional butterfly network, $\mathcal{R}$ cannot enter level k. Hence $\mathcal{C}$ guards level k when $\mathcal{C}$ captures an image of $\mathcal{R}$. □

The following lemma is central to our strategy to capture $\mathcal{R}$ using two cops.

Lemma 4. *Let $\mathcal{R}$ be restricted to levels from 0 to x, for $1 \leq x \leq k$, in a k-dimensional butterfly network. Then after a finite number of moves, one cop, say $\mathcal{C}$, can restrict the robber to levels from 0 to $x-1$.*

Proof. If $x = k$, then $\mathcal{C}$ can restrict $\mathcal{R}$ to levels from 0 to $x-1$ simply by guarding level k (using Lemma 3).

If $x < k$, then we consider the recursive definition of butterfly networks. If we consider the levels from 0 to x of a k-dimensional butterfly network and consider only the x least significant bits of binary codes of rows, then we have 2^{k-x} butterfly networks of dimension x. (If $v = (i, j)$ was a vertex of original network, then here we consider the vertex v as (i', j), where i' is a x bit binary tuple containing x least significant bits of i.)

Now observe that, if $\mathcal{R}$ on a vertex of one of these x-dimensional butterfly networks, say A, then $\mathcal{R}$ cannot leave A without entering the level $x+1$ (as all these x-dimensional butterfly networks are connected only through vertices of level $x+1$). Now $\mathcal{C}$ will consider only the x least significant bits of the butterfly network A and follow the strategy from Lemma 3 to guard level x (here the start vertex becomes the start vertex of A). Once $\mathcal{C}$ guards level x, the robber $\mathcal{R}$ cannot enter level x and hence is restricted to levels from 0 to $x-1$. □

Theorem 1. *Cop number for finite butterfly networks is two.*

Proof. We give a cop strategy to capture $\mathcal{R}$ in a k-dimensional butterfly network using two cops. In this strategy, cops keep restricting the robber territory level by level, finally restricting $\mathcal{R}$ to level 0, where it cannot move. Then the cops capture $\mathcal{R}$.

Initially, $\mathcal{R}$ is restricted to levels from 0 to k. Using Lemma 4, one cop restricts $\mathcal{R}$ to levels from 0 to $k-1$. While this cop guards $\mathcal{R}$, other cop moves and restricts $\mathcal{R}$ to levels from 0 to $k-2$ using Lemma 4. Once $\mathcal{R}$ is restricted to levels from 0 to $k-2$ by the second cop, the first cop guarding level k can be freed. (Cops can do so because if $\mathcal{R}$ cannot enter level $k-1$, it cannot enter level k.) This way whenever cops restrict $\mathcal{R}$ using a new cop, the previous cop gets free and restricts $\mathcal{R}$ further to smaller levels.

The cops, subsequently, restrict $\mathcal{R}$ to level 0 where one cop is ensuring the guard position. Now the second cop moves and captures $\mathcal{R}$. Hence two cops are sufficient to capture the robber in a butterfly network.

To show that two cops are necessary, we prove a stronger result that all k-dimensional butterfly networks, for $k > 0$, have cop number greater than 1. We prove this by proving that k-dimensional butterfly networks, for $k > 0$, do not have a corner vertex. For contradiction, suppose that u and v are two vertices of a k-dimensional butterfly network such that u is a corner of v; so $N[u] \subseteq N[v]$. Thus u and v must be adjacent and hence must be in different but consecutive levels. Now u has two neighbours in the level of v and one of them is v. Let the other neighbour be x. If $N[u] \subset N[v]$, then $x \in N[v]$. This is a contradiction as x and v are in the same level. Hence there is no corner in a butterfly network. So two cops are necessary to capture a robber in a butterfly network.

Hence the cop number for butterfly networks is 2. □

Since the cops capture $\mathcal{R}$ by restricting $\mathcal{R}$ to smaller levels in each iteration and each iteration takes $O(k)$ time for a k-dimensional butterfly network (having $2^k(k+1)$ vertices), we have the following corollary.

Corollary 1. *Capture time for a k-dimensional butterfly network using two cops is $O(k^2)$.*

4 Solid Grids

In this section, we consider the game of cops and robber on solid grids. An $m \times n$ grid is a set of points in two dimensions with integer coordinates (i, j) where $0 \leq i < m$ and $0 \leq j < n$, where each point represents a vertex and there is an edge between two vertices if and only if the Euclidean distance between the points representing these vertices 1. A graph is a solid grid if it has an embedding such that it is a subset of a grid and all the internal faces have unit area.

We consider a grid representation of the solid grid graph. In this representation rows and columns are clearly defined.

A *column path* is a path which has all its vertices from the same column, say c_i, and both endpoints of this path have exactly one neighbour in c_i, each. A column c_i may have multiple column paths. A column path P in column c_i is a *boundary column path* if vertices of P have neighbours only in P and in either column c_{i+1} or in c_{i-1}. See Fig. 2 for an illustration. Two column paths P and P' are *adjacent* if some vertex $p \in P$ and $p' \in P'$ have an edge.

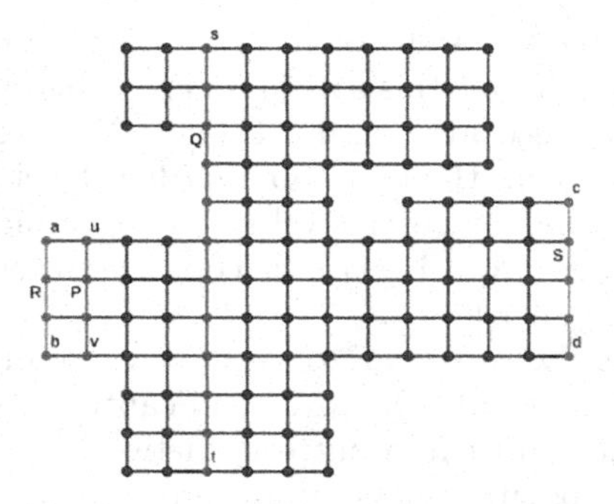

Fig. 2. A solid grid. Here *P, Q, R* and *S* are some of the column paths, of which *R* and *S* are boundary paths.

Let P be a column path of a solid grid graph G, and P have endpoints u and v. It is easy to see that:

1. P is a shortest u, v-path.
2. If P is not a boundary column path, then $G - P$ has at least two connected components.

We have the following lemma.

Lemma 5. *Let P be a column path in solid grid G and let S be one of the components of $G - P$. Then S has a unique column path P' adjacent to P.*

Proof. We will prove this by contradiction. Let P_1 and P_2 be two paths of component S that are adjacent to P, such that bottom most vertex of P_1 is in a higher row that top most vertex of P_2. Let the bottom most vertex of P_1 be u

and top most vertex of P_2 be v. Also let the neighbours of u and v in P be u' and v' respectively. Note that (u, v) can not be an edge, by definition of column paths.

Let $v', x_1, \ldots, x_k, u'$ be the path between u' and v' in P and $u, y_1, \ldots, y_j, v$ be the shortest path between u and v in S. Then $u, y_1, \ldots, y_j, v, v', x_1, \ldots, x_k, u', u$ is an internal face of the solid grid and has area more than 1 (since (u, v) is not an edge). Since it is not possible in a solid grid, this leads to a contradiction.

Hence S can have only one column path P' that is adjacent to P and this proves our claim.

□

From Lemma 5, we have the following observation, which is central to our strategy to capture $\mathcal{R}$ using two cops.

Observation 1. *Let P be a column path in solid grid G and let $\mathcal{R}$ be in one of the components of $G - P$, say S. Then if a cop is guarding the column path P' of S, that is adjacent to P, then $\mathcal{R}$ cannot leave the component S without being captured.*

Now we prove the following theorem.

Theorem 2. *Cop number for solid grids is two.*

Proof. We give a cop strategy to capture $\mathcal{R}$ using two cops. In this strategy, cops will reduce the robber territory after every finite number of steps, subsequently capturing the robber. Let the two cops be $\mathcal{C}_1$ and $\mathcal{C}_2$. Cops follow the following strategy.

1. $\mathcal{C}_1$ begins by guarding a column path P. If P is a boundary path, then $\mathcal{R}$ is restricted to $G - P$, else $\mathcal{R}$ is restricted to one of the connected components of $G - P$, say S.
2. Now, cops find the column path P' in S that is adjacent to P, and $\mathcal{C}_2$ guards P'. This restricts $\mathcal{R}$ to S and hence we can free $\mathcal{C}_1$, which was guarding P earlier.
   $\mathcal{R}$ is now restricted to S and a column path P' is guarded by $\mathcal{C}_2$. This further restricts $\mathcal{R}$ to either $S - P'$ (if P' is a boundary column path of S) or to one of the connected components of $S - P'$. Let the connected component $\mathcal{R}$ is restricted to be S'.
   This situation is same as situation in the end of step 1. So, we rename $\mathcal{C}_2$ as $\mathcal{C}_1$, $\mathcal{C}_1$ as $\mathcal{C}_2$, P' as P and S' as S, and repeat step 2.

Here cops reduce the robber territory in each step. Subsequently, the robber will be restricted to a single column and then $\mathcal{C}_2$ will capture $\mathcal{R}$.

To see the two cops are necessary to capture a robber in some solid grids, we can see that a cycle of 4 vertices, which is a solid grid, has cop number 2. □

5 Conclusion

In this paper, we studied the game of cops and robbers on butterfly networks and solid grids. We showed that the cop number for both of them is 2.

For butterfly networks, in each iteration of the algorithm, a cop guards a level of the network. Conventionally, in the cops and robber game on a graph G, a set of cops guard a connected subgraph H of G, and cops stay on the vertices of H. In our strategy, a cop guards a subgraph of the butterfly network that is an independent set, and the cop never enters that subset until it can capture $\mathcal{R}$. We believe that this way of *guarding a disconnected subgraph from a distance* can be useful in finding the cop number of other graph classes.

We gave an asymptotic bound on the capture time of butterfly networks using two cops. It might be interesting to find the exact bounds on capture time of butterfly networks (assuming optimal play from the robber). Moreover, Luccio and Pagli [15] studied the cops and robber game on grids and studied if increasing the cops can decrease the capture time. For a graph, they defined the *work* W_k as $k \cdot capture(k)$, where $capture(k)$ is the number of moves required by k cops to capture the robber. Then they defined the *speedup* using $j > i$ cops as W_i/W_j. Since butterfly networks have an inherent structure to support parallel computations, a natural question is whether more cops can work simultaneously to give a speedup greater than one.

It is well-known that the cop number of grids is 2. We extended this result to solid grids. A superclass of solid grids is the *Partial grids*, which are subgraphs of grids. The cop number of partial grids is still not known. Cop number of partial grids is lower bounded by solid grids which is 2 and upper bounded by planar graphs which is 3 [1]. This motivates the question, whether the cop number for partial grids is 2 or 3.

Acknowledgements. We would like to thank the IFCAM project Applications of graph homomorphisms (MA/IFCAM/18/39). We would also like to thank Uma kant Sahoo for positive discussions.

References

1. Aigner, M., Fromme, M.: A game of cops and robbers. Discrete Appl. Math. **8**, 1–12 (1984)
2. Alspach, B.: Sweeping and searching in graphs: a brief survey. Matematiche **59**, 5–37 (2006)
3. Balister, P., Bollobas, B., Narayanan, B., Shaw, A.: Catching a fast robber on the grid. J. Combin. Theory Ser. A **152**, 341–352 (2017)
4. Bonato, A., Inerney, F.M.: The game of wall cops and robbers. In: Senthilkumar, M., Ramasamy, V., Sheen, S., Veeramani, C., Bonato, A., Batten, L. (eds.) Computational Intelligence, Cyber Security and Computational Models. AISC, vol. 412, pp. 3–13. Springer, Singapore (2016). https://doi.org/10.1007/978-981-10-0251-9_1
5. Bonato, A., Nowakowski, R.: The Game of Cops and Robbers on Graphs. American Mathematical Society (2011)

6. Berarducci, A., Intrigila, B.: On the cop number of a graph. Adv. Appl. Math. **14**, 389–403 (1993)
7. Das, S., Gahlawat, H.: Variations of cops and robbers game on grids. In: Panda, B.S., Goswami, P.P. (eds.) CALDAM 2018. LNCS, vol. 10743, pp. 249–259. Springer, Cham (2018). https://doi.org/10.1007/978-3-319-74180-2_21
8. Fomin, F., Golovach, P., Kratochvil, J., Nisse, N., Suchan, K.: Pursuing a fast robber on a graph. Theor. Comput. Sci. **411**, 1167–1181 (2010)
9. Goldstein, A.S., Reingold, E.M.: The complexity of pursuit on a graph. Theor. Comput. Sci. **143**, 93–112 (1995)
10. Hosseini, S.A., Mohar, B.: Game of cops and robbers in oriented quotients of the integer grid. Discrete Math. **341**(2), 439–450 (2018)
11. Isaza, A., Lu, J., Bulitko, V., Greiner, R.: A cover-based approach to multi-agent moving target pursuit. In: Proceedings of The 4th Conference on Artificial Intelligence and Interactive Digital Entertainment (2008)
12. Isler, V., Kannan, S., Khanna, S.: Randomized pursuit-evasion with local visibility. SIAM J. Discrete Math. **1**, 26–41 (2006)
13. Kinnersley, W.B.: Cops and robbers is EXPTIME-complete. J. Combin. Theory Ser. B **111**, 201–220 (2015)
14. Leighton, F.: Introduction to Parallel Algorithms and Architectures. Morgan Kaufmann, Burlington (1992)
15. Luccio, F., Pagli, L.: Cops and robber on grids and tori. ArXiv e-prints. arXiv:1708.08255 (2017)
16. Maamoun, M., Meyniel, H.: On a game of policemen and robber. Discrete Appl. Math. **17**, 307–309 (1987)
17. Mehrabian, A.: The capture time of grids. Discrete Math. **311**, 102–105 (2011)
18. Nowakowski, R., Winkler, P.: Vertex-to-vertex pursuit in a graph. Discrete Math. **43**, 253–259 (1983)
19. Quilliot, A.: Thése d'Etat. Ph.D. thesis, Université de Paris VI (1983)
20. Seymour, P.D., Thomas, R.: Graph searching and a min-max theorem for treewidth. J. Combin. Theory Ser. B **58**, 22–33 (1993)

b-Coloring of Some Powers of Hypercubes

P. Francis[1], S. Francis Raj[2], and M. Gokulnath[2](✉)

[1] Department of Computer Science, Indian Institute of Technology Palakkad, Palakkad 678557, India
pfrancis@iitpkd.ac.in
[2] Department of Mathematics, Pondicherry University, Puducherry 605014, India
francisraj_s@yahoo.com, gokulnath.math@gmail.com

Abstract. The b-chromatic number $b(G)$ of a graph G is the maximum k for which G has a proper vertex coloring using k colors such that each color class contains at least one vertex adjacent to a vertex of every other color classes. In this paper, we mainly investigate on one of the open problems given in [1]. As a consequence, we obtain an upper bound for the b-chromatic number of some powers of hypercube. This turns out to be the improvement of the existing bounds.

Keywords: b-coloring · b-chromatic number · Hypercubes · Powers of graphs

2000 AMS Subject Classification: 05C15

1 Introduction

All graphs considered in this paper are simple, finite and undirected. Let G be a graph with vertex set $V(G)$ and edge set $E(G)$. A b-coloring of a graph G using k colors is a proper coloring of the vertices of G using k colors in which each color class has a color dominating vertex, that is, a vertex which has a neighbor in each of the other color classes. The b-chromatic number, $b(G)$ of G is the largest k such that G has a b-coloring using k colors. The concept of b-coloring was introduced by Irving and Manlove [2] in analogy to the achromatic number of a graph G.

It is clear from the definition of $b(G)$ that $\chi(G) \leq b(G) \leq \Delta(G) + 1$, where $\chi(G)$ and $\Delta(G)$ denote the chromatic number of G and the maximum degree of G respectively. The pth power of a graph G denoted by G^p is a graph whose vertex set $V(G^p) = V(G)$ and edge set $E(G^p) = \{xy : d_G(x, y) \leq p\}$, where $d_G(x, y)$ denotes the distance between x and y in G.

Let $[n] = \{1, 2, \ldots, n\}$ and $2^{[n]} = \{A : A \subseteq [n]\}$, the power set of $[n]$. Let us define the hypercube Q_n in a slightly different way. For $n \in \mathbb{N}$, $V(Q_n) = 2^{[n]}$ and for $x, y \in V(Q_n)$, $xy \in E(Q_n)$ if and only if $|x \triangle y| = 1$. For $0 \leq i \leq n$, let $V_i = \{A \in 2^{[n]} : |A| = i\}$. Let us define the simplicial ordering on the elements of $V(Q_n)$ in the following way. For two distinct vertices $x, y \in V(Q_n)$, we say that

A. Mudgal and C. R. Subramanian (Eds.): CALDAM 2021, LNCS 12601, pp. 282–287, 2021.
https://doi.org/10.1007/978-3-030-67899-9_22

x precedes y, denoted by $x < y$, if $|x| < |y|$ or $|x| = |y|$ and $\min\{x \triangle y\} \in x$. Let $\mathcal{I}_m$ denote the first m elements of $2^{[n]}$ in the simplicial ordering. Sometimes, we refer to $\mathcal{I}_m$ as an initial segment of size m in the hypercube. Also, let us define the simplicial ordering of sets consisting of vertices of $V(Q_n)$ in the following way. For $A, B \subseteq V(Q_n)$, we say that A precedes B, denoted by $A < B$, if $|A| < |B|$ or $|A| = |B|$ and $\min\{A \triangle B\} \in A$, where $\min\{A \triangle B\}$ is the first set in the simplicial ordering of $A \triangle B$. For $A \subseteq 2^{[n]}$ and $p \in [n]$, let us define $C^p[A] = \{y \in 2^{[n]} : |x \triangle y| \leq p \text{ for every } x \in A\}$ and let $C^p(A) = C^p[A] \backslash A$.

Let us recall some of the definitions due to Tsukerman [3] which are required for this paper. For $A \subseteq 2^{[n]}$ and $i \in [n]$, the i-sections of A are given by $A_{i-} = \{x \in 2^{[n] \backslash \{i\}} : x \in A\}$ and $A_{i+} = \{x \in 2^{[n] \backslash \{i\}} : x \cup \{i\} \in A\}$. Clearly, $A = A_{i-} \cup (A_{i+} + \{i\})$ where $A + \{i\} = \{x \cup \{i\} : x \in A\}$. Let us define i-compression of A, denoted by $S_i(A)$, as follows: $S_i(A)_{i-} = \mathcal{I}_{|A_{i-}|} \subseteq 2^{[n] \backslash \{i\}}$, $S_i(A)_{i+} = \mathcal{I}_{|A_{i+}|} \subseteq 2^{[n] \backslash \{i\}}$ and $S_i(A) = S_i(A)_{i-} \cup (S_i(A)_{i+} + \{i\}) = \mathcal{I}_{|A_{i-}|} \cup (\mathcal{I}_{|A_{i+}|} + \{i\})$. Thus either $S_i(A) = A$ or $S_i(A) < A$ in the simplicial ordering of sets. We say that A is i-compressed if and only if $A = S_i(A)$.

For notation and terminologies not mentioned in this paper, see [4].

Let us recall the clique number and bounds for the b-chromatic number of powers of hypercubes given in [1].

Theorem 1. *[1] (i) For $n \geq 3$ and $1 \leq p \leq n-1$, the clique size of Q_n^p is*

$$\omega(Q_n^p) = \begin{cases} \sum_{i=0}^{\frac{p}{2}} \binom{n}{i} & \text{if } p \text{ is even} \\ 2 \sum_{i=0}^{\frac{p-1}{2}} \binom{n-1}{i} & \text{if } p \text{ is odd.} \end{cases}$$

(ii) For all $n \geq 2$ and $\lfloor \frac{n}{2} \rfloor < p < n-1$, the b-chromatic number of Q_n^p is
$2^{n-1} \leq b(Q_n^p) \leq 2^{n-1} + \left\lfloor \frac{\omega(Q_n^p)}{2} \right\rfloor$.

In [1], the authors have also obtained the maximum number of common neighbors for a clique of size 2 in Q_n^p. Also, for a clique of larger size, they expected that the number of common neighbors will be maximum if the vertices of the clique in Q_n^p are chosen as an initial segment in the simplicial order. This has been mentioned as an open problem.

Problem 1. [1] let $\mathcal{F} \subseteq 2^{[n]}$ such that for all $A, B \in \mathcal{F}$, $|A \triangle B| \leq p$. Suppose $|\mathcal{F}| = m$, for some $2 \leq m \leq 2^{n-1}$, then for what kind of $\mathcal{F}$ will the $|C^p[\mathcal{F}]|$ be maximum?

In Sect. 2, we have answered Problem 1 and we have shown that $|C^p[\mathcal{F}]|$ is maximum if $\mathcal{F}$ is chosen as an initial segment in the simplicial ordering. As a consequence, for $\lfloor \frac{n}{2} \rfloor < p < n-1$, we have obtained an upper bound, better than the existing bound given in Theorem 1, for the b-chromatic number of the powers of hypercube.

2 Bounds for the b-Chromatic Number of Some Powers of Hypercube

Let us start Sect. 2 by answering Problem 1. First let us observe that the set of all common neighbors of an initial segment of a power of hypercube is again an initial segment of the power of hypercube.

Lemma 1. *For any $n, p \in \mathbb{N}$ and $a \in [2^n]$, there exists an integer $b \in [2^n]$ such that $C^p[\mathcal{I}_a] = \mathcal{I}_b$.*

Proof. It is enough to prove that for any $x, y \in 2^{[n]}$ such that if $x \notin C^p[\mathcal{I}_a]$ and $x < y$, then $y \notin C^p[\mathcal{I}_a]$. Since $x \notin C^p[\mathcal{I}_a]$, there exists a set $z \in \mathcal{I}_a$ such that $|x \triangle z| \geq p + 1$. Let us first prove that $y \notin C^p[\mathcal{I}_a]$ for the case $x \cap z = \emptyset$.

Case 1: $x \cap z = \emptyset$.

Clearly, $z \subseteq x^c$ and $|x \triangle z| = |x| + |z| \leq n$. Let us assume that the elements in all the sets of $2^{[n]}$ are arranged in ascending order. It is easy to observe that $y^c < x^c$. Also, in simplicial ordering, for any integer $q < |y^c|$, the set containing the first q elements of y^c is equal or comes before than the set containing the first q elements of x^c. Let $k = \max\{0, (|x| + |z| - |y|)\}$ and let us define a set $z' \in 2^{[n]}$ which contains the first k elements in y^c. Since $|y^c| = n - |y| \geq |x| + |z| - |y|$, z' is a well defined set. If $z' = \emptyset$, then $z' \in \mathcal{I}_a$. If $z' \neq \emptyset$, then $|z'| = |x| + |z| - |y| = |z| - (|y| - |x|) \leq |z|$. Since z is the set containing $|z|$ elements of x^c, we have either $z' = z$ or $z' < z$. In both cases, $z' \in \mathcal{I}_a$. Since $z' \subseteq y^c$, we have $z' \cap y = \emptyset$ and $|z' \triangle y| = |z'| + |y| \geq |x| + |z| - |y| + |y| = |x| + |z| \geq p + 1$. Thus $y \notin C^p[\mathcal{I}_a]$.

Case 2: $x \cap z \neq \emptyset$.

Let us define $z_0 = z \backslash (x \cap z)$. Clearly, $z_0 < z$ and $|x \triangle z_0| > |x \triangle z| \geq p + 1$. Thus $z_0 \in \mathcal{I}_a$ such that $x \cap z_0 = \emptyset$ and $|x \triangle z_0| \geq p + 1$. By using Case 1, we have $y \notin C^p[\mathcal{I}_a]$. □

Let us recall a result due to Tsukerman [3] which is useful to prove our next result.

Theorem 2. *[3] for $B \subseteq 2^{[n]}$, if B is i-compressed for each $i \in [n]$, but not an initial segment, then $|B| = 2^{n-1}$ and B is of the following form*

$$B = \begin{cases} \mathcal{I}_\ell \backslash \{\{\frac{n+3}{2}, \frac{n+5}{2}, \ldots, n\}\} & \text{where } n \text{ is odd and } \ell = \sum_{i=0}^{\frac{n-1}{2}} \binom{n}{i} + 1 \\ \mathcal{I}_{\ell'} \backslash \{\{1, \frac{n}{2}+2, \frac{n}{2}+3, \ldots, n\}\} & \text{where } n \text{ is even and } \ell' = \sum_{i=0}^{\frac{n}{2}-1} \binom{n}{i} + \binom{n-1}{n/2} + 1. \end{cases}$$

By using Lemma 1 and Theorem 2 and with some involved arguments, we obtain a positive answer to Problem 1.

Theorem 3. *For $n, p \in \mathbb{N}$, if $A \subseteq 2^{[n]}$ such that for all $x, y \in A, |x \triangle y| \leq p$, then $|C^p[A]| \leq |C^p[\mathcal{I}_{|A|}]|$.*

Finally, let us establish an improved upper bound for the b-chromatic number of some power of hypercubes. Before doing it, let us observe a few results which will help us get the upper bound. Let us start by giving a relationship between the b-chromatic number and the number of common neighbors of a clique in powers of hypercubes.

Lemma 2. *For $n, p \in \mathbb{N}$ and $\ell \leq 2^{n-1}$, if $b(Q_n^p) \geq 2^{n-1} + \ell$, then there exists a clique of size 2ℓ, say $A_{2\ell}$, such that $|C^p(A_{2\ell})| \geq 2^{n-1} - \ell$.*

Next, we have given an upper bound for the number of common neighbors for some particular initial segments of some powers of hypercubes.

Lemma 3. *For n being odd, $\frac{n+1}{2} \leq p \leq n-2$, if $r = \sum_{i=0}^{p-\frac{n-1}{2}} \binom{n}{i}$ and $s = \binom{p}{p-\frac{n-1}{2}}$, then $|C^p(\mathcal{I}_{(r-s+1)})| < 2^{n-1} - (r - 2s + 1)$. For n being even, $\frac{n}{2} + 1 \leq p \leq n-2$, if $r' = \sum_{i=0}^{p-\frac{n}{2}} \binom{n}{i} + \binom{n-1}{p-\frac{n}{2}}$ and $s' = \binom{p-1}{p-\frac{n}{2}}$, then $|C^p(\mathcal{I}_{(r'-s'+1)})| < 2^{n-1} - (r' - 2s' + 1)$.*

In Lemma 4 and Lemma 5, we have established upper bounds for the b-chromatic number of some particular powers of hypercubes.

Lemma 4. *$b(Q_7^5) \leq 2^6 + 9$*

Proof. Suppose $b(Q_7^5) \geq 2^6 + 10$, by Lemma 2, there exists A_{20} such that $|C^5(A_{20})| \geq 2^6 - 10$. By Theorem 3, $|C^5(\mathcal{I}_{20})| \geq |C^5(A_{20})| \geq 2^6 - 10$. Let us find $C^5[\mathcal{I}_{20}]$ in Q_7^5. $\mathcal{I}_{20} = \{\emptyset, \{1\}, \{2\}, \ldots, \{7\}, \{1,2\}, \{1,3\}, \ldots, \{1,7\}, \{2,3\}, \{2,4\}, \ldots, \{2,7\}, \{3,4\}\}$. So $C^5[\mathcal{I}_{20}]$ will contain all the vertices in V_0, V_1, V_2, V_3 and $\{1,2,3,4\}, \{1,2,3,5\}, \{1,2,3,6\}, \{1,2,3,7\}, \{1,2,4,5\}, \{1,2,4,6\}, \{1,2,4,7\}$ in V_4. Therefore, $|C^5(\mathcal{I}_{20})| = 2^6 + 7 - 20 = 2^6 - 13 < 2^6 - 10$, a contradiction. □

Lemma 5. *For n being odd and $n \geq 5$, $b(Q_n^{\frac{n+1}{2}}) \leq 2^{n-1} + \lfloor \frac{n+1}{4} \rfloor$.*

Proof. On contrary, if $b(Q_n^{\frac{n+1}{2}}) \geq 2^{n-1} + \lfloor \frac{n+1}{4} \rfloor + 1$, then by using Lemma 2, there exists a clique of size $2\left(\lfloor \frac{n+1}{4} \rfloor + 1\right)$, say $A_{2(\lfloor \frac{n+1}{4} \rfloor + 1)}$ such that $\left|C^{\frac{n+1}{2}}\left(A_{2(\lfloor \frac{n+1}{4} \rfloor + 1)}\right)\right| \geq 2^{n-1} - \left(\lfloor \frac{n+1}{4} \rfloor + 1\right)$. Since $2\left(\lfloor \frac{n+1}{4} \rfloor + 1\right) \geq \frac{n+1}{2} + 1$, $\left|C^{\frac{n+1}{2}}\left(\mathcal{I}_{\frac{n+1}{2}+1}\right)\right| \geq \left|C^{\frac{n+1}{2}}\left(\mathcal{I}_{2(\lfloor \frac{n+1}{4} \rfloor + 1)}\right)\right|$. One can easily observe that $\left|C^{\frac{n+1}{2}}\left(\mathcal{I}_{\frac{n+1}{2}+1}\right)\right| = 2^{n-1} - \frac{n+1}{2} - 1 + 1 = 2^{n-1} - \frac{n+1}{2}$. By using Theorem 3, we have $\left|C^{\frac{n+1}{2}}\left(\mathcal{I}_{2(\lfloor \frac{n+1}{4} \rfloor + 1)}\right)\right| \geq \left|C^{\frac{n+1}{2}}\left(A_{2(\lfloor \frac{n+1}{4} \rfloor + 1)}\right)\right|$. Thus $2^{n-1} - \frac{n+1}{2} \geq 2^{n-1} - \left(\lfloor \frac{n+1}{4} \rfloor + 1\right)$ which implies, $n \leq 3$, a contradiction. □

Without much difficulty, one can observe Lemma 6.

Lemma 6. *For $n \geq 9$, $\lfloor \frac{n}{2} \rfloor + 3 \leq p \leq n-2$, let $r = \sum_{i=0}^{p-\lfloor \frac{n}{2} \rfloor} \binom{n}{i}$ and $s = \binom{p}{p-\lfloor \frac{n}{2} \rfloor}$, then $r \geq 3s$.*

Now with the help of Lemmas 1, 2, 3, 4, 5, 6 and Theorem 3, we have established an upper bound for the b-chromatic number of some powers of hypercubes which turns out to be an improvement of Theorem 1.

Theorem 4. *For n being odd and $n \geq 5$, $\frac{n+1}{2} \leq p \leq n-2$, if $r = \sum_{i=0}^{p-\frac{n-1}{2}} \binom{n}{i}$ and $s = \binom{p}{p-\frac{n-1}{2}}$, then $b(Q_n^p) \leq 2^{n-1} + \left\lfloor \frac{r-s}{2} \right\rfloor$. For n being even and $n \geq 6$, $\frac{n}{2}+1 \leq p \leq n-2$, if $r' = \sum_{i=0}^{p-\frac{n}{2}} \binom{n}{i} + \binom{n-1}{p-\frac{n}{2}}$ and $s' = \binom{p-1}{p-\frac{n}{2}}$, then $b(Q_n^p) \leq 2^{n-1} + \left\lfloor \frac{r'-s'}{2} \right\rfloor$.*

Proof. Let us start with n being odd. By using Lemma 4 and Lemma 5, for n being odd, it is enough to assume that $n \geq 9$ and $\frac{n+3}{2} \leq p \leq n-2$. Suppose $b(Q_n^p) \geq 2^{n-1} + \left\lfloor \frac{r-s}{2} \right\rfloor + 1$, then by using Lemma 2, there exists a clique of size $2\left(\left\lfloor \frac{r-s}{2} \right\rfloor + 1\right)$, say $A_{2(\lfloor \frac{r-s}{2} \rfloor+1)}$ such that $\left|C^p\left(A_{2(\lfloor \frac{r-s}{2} \rfloor+1)}\right)\right| \geq 2^{n-1} - \left(\left\lfloor \frac{r-s}{2} \right\rfloor + 1\right)$. Since $2\left(\left\lfloor \frac{r-s}{2} \right\rfloor + 1\right) \geq r-s+1$, $|C^p(\mathcal{I}_{r-s+1})| \geq \left|C^p\left(\mathcal{I}_{2(\lfloor \frac{r-s}{2} \rfloor+1)}\right)\right|$. By combining these results with Theorem 3 and Lemma 3, we can show that $2^{n-1} - (r-2s+1) \geq 2^{n-1} - \left(\left\lfloor \frac{r-s}{2} \right\rfloor + 1\right)$ and this would imply that $r < 3s$. When n is odd and $p = \frac{n+3}{2}$, $r < 3s$ will yield the following.

$$
\begin{aligned}
1 + n + \tfrac{n(n-1)}{2} &< \tfrac{3\left(\frac{n+3}{2}\right)\left(\frac{n+1}{2}\right)}{2} \\
n^2 - 8n - 1 &< 0 \\
n^2 - 8n - 9 &< 0 \\
(n+1)(n-9) &< 0
\end{aligned}
$$

Thus $n < 9$, a contradiction to the assumption that $n \geq 9$. Therefore, $\frac{n+5}{2} = \frac{n-1}{2} + 3 \leq p \leq n-2$.

Next let us consider n to be even with $n \geq 6$ and $\frac{n}{2}+1 \leq p \leq n-2$. Suppose $b(Q_n^p) \geq 2^{n-1} + \left\lfloor \frac{r'-s'}{2} \right\rfloor + 1$, then as done in the odd case we can show that $r' < 3s'$. But for n being even and $n \geq 6$, if $p = \frac{n}{2}+1$, then $r' = 2n$ which is greater than $3s' = \frac{3n}{2}$. Also, when $n \geq 8$ and $p = \frac{n}{2}+2$, $r' < 3s'$ implies the following.

$$
\begin{aligned}
1 + n + \tfrac{n(n-1)}{2} + \tfrac{(n-1)(n-2)}{2} &< \tfrac{3\left(\frac{n}{2}+1\right)\left(\frac{n}{2}\right)}{2} \\
5n^2 - 14n + 16 &< 0 \\
5n^2 - 14n + 8 &< 0 \\
(5n-4)(n-2) &< 0
\end{aligned}
$$

Thus $n < 2$, a contradiction to the assumption that $n \geq 8$. So when n is even, $n \geq 8$ and $\frac{n}{2}+3 \leq p \leq n-2$.

Let us introduce a new variable q in the following way, in order to prove that $r \geq 3s$ in both parity of n. Let $q = p - \left\lfloor \frac{n}{2} \right\rfloor$ and we rewrite r, r', s, s' as $r = \sum_{i=0}^{q} \binom{n}{i}$, $r' = \sum_{i=0}^{q} \binom{n}{i} + \binom{n-1}{q}$, $s = \binom{q+\lfloor \frac{n}{2} \rfloor}{q}$ and $s' = \binom{q+\lfloor \frac{n}{2} \rfloor-1}{q}$. It is clear that $r' \geq r$ and $s \geq s'$. Since $n \geq 9$ and $q \geq 3$, by using Lemma 6, we get that $r \geq 3s$. This will also imply that $r' \geq 3s'$, a contradiction to both $r < 3s$ and $r' < 3s'$ in the odd and even case respectively. □

Acknowledgment. For the first author, this research was supported by Post Doctoral Fellowship, Indian Institute of Technology, Palakkad. And for the second author, this research was supported by SERB DST, Government of India, File no: EMR/2016/007339. Also, for the third author, this research was supported by the UGC-Basic Scientific Research, Government of India, Student id: gokulnath.res@pondiuni.edu.in.

References

1. Francis, P., Francis Raj, S.: On b-coloring of powers of hypercubes. Discrete Appl. Math. **225**, 74–86 (2017)
2. Irving, R.W., Manlove, D.F.: The b-chromatic number of a graph. Discrete Appl. Math. **91**(1–3), 127–141 (1999)
3. Tsukerman, E.: Isoperimetric inequalities and the Alexandrov theorem. Master's thesis, Stanford University (2013)
4. West, D.B.: Introduction to Graph Theory. Prentice-Hall of India Private Limited (2005)

Chromatic Bounds for the Subclasses of pK_2-Free Graphs

Athmakoori Prashant(✉) and M. Gokulnath

Department of Mathematics, Pondicherry University, Puducherry 605014, India
11994prashant@gmail.com, gokulnath.math@gmail.com

Abstract. In this paper, we study the chromatic number for graphs with forbidden induced subgraphs. We improve the existing χ-binding functions for some subclasses of $2K_2$-free graphs, namely $\{2K_2, H\}$-free graphs where $H \in \{HVN, K_5 - e, K_1 + C_4\}$. In addition, for $p \geq 3$, we find the polynomial χ-binding functions for several subclasses of pK_2-free graphs, namely $\{pK_2, H\}$-free graphs where $H \in \{HVN, gem, diamond, K_5 - e, dart, C_4, K_1 + C_4, \overline{P_5}\}$.

Keywords: Coloring · Chromatic number · χ-binding funtion · $2K_2$-free graphs · pK_2-free graphs

2000 AMS Subject Classification: 05C15, 05C75

1 Introduction

All graphs considered in this paper are simple, finite and undirected. Let G be a graph with vertex set $V(G)$ and edge set $E(G)$. For any positive integer k, a proper k-coloring of a graph G is a mapping c: $V(G) \rightarrow \{1, 2, \ldots, k\}$ such that for any two adjacent vertices $u, v \in V(G)$, $c(u) \neq c(v)$. If a graph G admits a proper k-coloring then G is said to be k-colorable. The chromatic number, $\chi(G)$, of a graph G is the smallest k such that G is k-colorable. In this paper, P_n, C_n and K_n respectively denotes the path, the cycle and the complete graph on n vertices. The neighborhood $N(x)$ of a vertex x is $\{u \colon ux \in E(G)\}$ and for $S \subseteq V(G)$, we denote the neighborhood S by $N(S)$ is $\cup_{v \in S} N(v)$. Also for $S, T \subseteq V(G)$, we define $N_T(S) = N(S) \cap T$. For $S, T \subseteq V(G)$, let $\langle S \rangle$ denote the subgraph induced by S in G and let $[S, T]$ denote the set of all edges with one end in S and the other end in T. If every vertex in S is adjacent with every vertex in T, then $[S, T]$ is said to be complete. For any graph G, let $\overline{G}$ denotes the complement of G.

Let $\mathcal{F}$ be a family of graphs. We say that G is $\mathcal{F}$-free if it contains no induced subgraph which is isomorphic to a graph in $\mathcal{F}$. For two vertex-disjoint graphs G_1 and G_2, the join of G_1 and G_2, denoted by $G_1 + G_2$, is the graph whose vertex set $V(G_1 + G_2) = V(G_1) \cup V(G_2)$ and the edge set $E(G_1 + G_2) = E(G_1) \cup E(G_2) \cup \{xy : x \in V(G_1),\ y \in V(G_2)\}$. In this paper, we write $H \sqsubseteq G$

A. Mudgal and C. R. Subramanian (Eds.): CALDAM 2021, LNCS 12601, pp. 288–293, 2021.
https://doi.org/10.1007/978-3-030-67899-9_23

whenever H is an induced subgraph of G. A clique (independent set) in a graph G is a set of pairwise adjacent (non-adjacent) vertices. The size of a largest clique (independent set) in G is called the clique number (independence number) of G, and is denoted by $\omega(G)(\alpha(G))$.

A graph G is called perfect if $\chi(H) = \omega(H)$, for every induced subgraph H of G. A hereditary class $\mathcal{G}$ of graphs is said to be χ-bounded [7] if there is a function f (called a χ-binding function) such that $\chi(G) \leq f(\omega(G))$, for every $G \in \mathcal{G}$. We say that the χ-binding function f is special linear if $f(x) = x + c$ where c is a constant. If $c = 1$, then this special upper bound is called the Vizing bound for the chromatic number. There has been extensive research done on χ-binding function for various graph classes. See for instance, [8,10,11].

Throughout this paper, we use a particular partition of the vertex set of a graph G as defined initially by Wagon in [12] and improved by Bharathi et al. in [1] as follows. Let $A = \{v_1, v_2, \ldots, v_\omega\}$ be a maximum clique of size $\omega(G)$, in short we say ω. Let us define the lexicographic ordering on the set $L = \{(i,j) : 1 \leq i < j \leq \omega\}$ in the following way. For two distinct elements $(i_1, j_1), (i_2, j_2) \in L$, we say that (i_1, j_1) precedes (i_2, j_2), denoted by $(i_1, j_1) <_L (i_2, j_2)$ if either $i_1 < i_2$ or $i_1 = i_2$ and $j_1 < j_2$. For every $(i,j) \in L$, let $C_{i,j} = \{v \in V(G)\backslash A : v \notin N(v_i) \cup N(v_j)\}\backslash \{\bigcup_{(i',j') <_L (i,j)} C_{i',j'}\}$. Note that, for any $k \in \{1, 2, \ldots, j-1\}\backslash\{i\}$, $[v_k, C_{i,j}]$ is complete. Hence $\omega(\langle C_{i,j}\rangle) \leq \omega(G) - j + 2$.

For $1 \leq i \leq \omega$, let us define $I_i = \{v \in V(G)\backslash A : v \in N(a), \text{for any } a \in A\backslash\{v_i\}\}$. Since A is a maximum clique, for $1 \leq i \leq \omega$, I_i is an independent set and for any $x \in I_i$, $xv_i \notin E(G)$. Clearly, each vertex in $V(G)\backslash A$ is non-adjacent to at least one vertex in A. Hence those vertices will be contained either in I_i for some integer i, $1 \leq i \leq \omega$, or in $C_{i,j}$ for some $(i,j) \in L$. Thus $V(G) = A \cup \left(\bigcup_{1 \leq i \leq \omega} I_i\right) \cup \left(\bigcup_{(i,j) \in L} C_{i,j}\right)$.

In [12], Wagon showed that the class of pK_2-free graphs admit an $O(\omega^{2p-2})$ χ-binding function for all $p \in \mathbb{N}$. In particular, he showed that the χ-binding function for $2K_2$-free graphs is $\binom{\omega+1}{2}$. In [9], Karthick and Mishra proved that the families of $\{2K_2, H\}$-free graphs, where $H \in \{HVN, diamond, gem, K_1 + C_4, \overline{P_5}, \overline{P_2 \cup P_3}, K_5 - e\}$ admit special linear χ-binding functions. In this paper, we improve the bounds and show that the family of $\{2K_2, HVN\}$-free graphs and $\{2K_2, K_1 + C_4\}$-free graphs admit the vizing bound when $\omega \geq 4$ and $\omega \geq 3$ respectively. Also, we prove that the family of $\{2K_2, K_5 - e\}$-free graphs are ω-colorable when $\omega \geq 5$. Further, we prove that the families of $\{pK_2, H\}$-free graphs, where $H \in \{HVN, gem, diamond, K_5 - e\}$ admit linear χ-binding function and the family of $\{pK_2, dart\}$-free graphs admit a quadratic χ-binding function. In addition, we show that the families of $\{pK_2, H\}$-free graphs, where $H \in \{C_4, K_1 + C_4, \overline{P_5}\}$ admit $O(\omega^{p-1})$ χ-binding function. Some graphs that are considered as a forbidden induced subgraphs in this paper are shown in Fig. 1.

Notations and terminologies not mentioned here are as in [13].

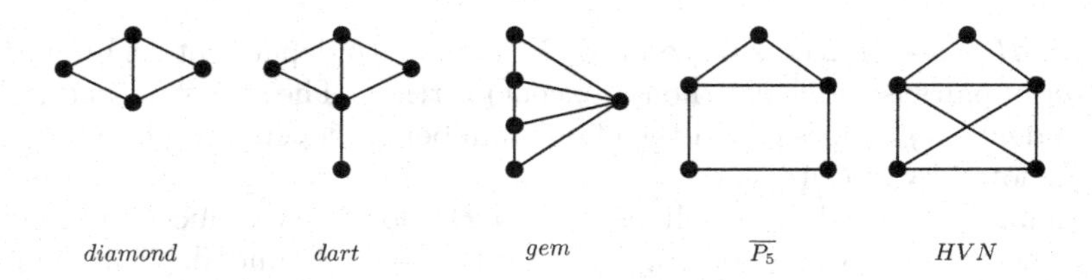

Fig. 1. Some special graphs

2 Coloring of Some Classes of $2K_2$-Free Graphs and pK_2-Free Graphs

Let us start this section by recalling some χ-binding results due to Wagon [12], Gaspers and Huang [6] and Brandt [3].

Theorem 1. *[12] If G is a $2K_2$-free graph, then $\chi(G) \leq \binom{\omega(G)+1}{2}$.*

Theorem 2. *[6] If G is a $2K_2$-free graph such that $\omega(G) \leq 3$, then $\chi(G) \leq 4$.*

Theorem 3. *[3] For $p \geq 3$, if G is a pK_2-free graph such that $\omega(G) = 2$, then $\chi(G) \leq 2p - 2$.*

In [9], Karthick and Mishra have showed that if G is a $\{2K_2, HVN\}$-free graph, then $\chi(G) \leq \omega(G) + 3$. In Theorem 4, we reduce the bound to $\omega(G) + 1$ for $\omega \geq 4$.

Theorem 4. *If G is a $\{2K_2, HVN\}$-free graph such that $\omega \geq 4$, then $\chi(G) \leq \omega(G) + 1$.*

Proof. Let G be a $\{2K_2, HVN\}$- free graph such that $\omega \geq 4$. Let $j \geq 4$. We shall first show that $C_{i,j} = \emptyset$. If there exists a vertex $a \in C_{i,j}$, then there exist at least two integers $s, q \in \{1, 2, \ldots, j\} \backslash \{i, j\}$ such that $av_s, av_q \in E(G)$ and $\langle\{a, v_s, v_q, v_i, v_j\}\rangle \cong HVN$, a contradiction. Thus $V(G) = A \cup C_{1,2} \cup C_{1,3} \cup C_{2,3} \cup \left(\bigcup_{1 \leq i \leq \omega} I_i\right)$. By using similar arguments, we can show that $N_A(C_{1,3}) = \{v_2\}$ and $N_A(C_{2,3}) = \{v_1\}$. Moreover, we claim that $[C_{1,3}, I_k] = \emptyset$ for $k \neq 2$ and $[C_{2,3}, I_\ell] = \emptyset$ for $\ell \neq 1$. On contrary, for $k \neq 2$, there exists $a \in C_{1,3}$ and $b \in I_k$ such that $ab \in E(G)$. Since $\omega(G) \geq 4$ and $N_A(C_{1,3}) = \{v_2\}$, there exist at least two integers $s, q \in \{1, 2, \ldots, \omega\} \backslash \{2, k\}$ such that $av_s, av_q \notin E(G)$ and thus $\langle\{a, v_2, b, v_s, v_q\}\rangle \cong HVN$, a contradiction. Similarly, for $\ell \neq 1$, we can show that $[C_{2,3}, I_\ell] = \emptyset$.

Let the set of colors be $\{1, 2, \ldots, \omega + 1\}$. For $1 \leq i \leq \omega$, let us assign the color i to the vertex v_i and to all the vertices of I_i and assign the colors $\omega + 1$, 1 and 2 to the vertices of $C_{1,2}$, $C_{1,3}$ and $C_{2,3}$ respectively. Clearly, this is a proper coloring of G and thus $\chi(G) \leq \omega(G) + 1$. □

Note that, all those properties mentioned in Theorem 4 are due to the fact that G is a HVN-free and hence those properties are also valid in Theorem 5.

For $p \geq 2$, let us define a sequence of functions $f_p^1 : \mathbb{N} \to \mathbb{N}$ to serve as a χ-binding function for $\{pK_2, HVN\}$-free graphs as follows. For $p, s \geq 2$, $t \geq 3$, $m \geq 4$, $f_p^1(1) = 1$, $f_2^1(s) = s + 1$, $f_t^1(2) = 2t - 2$, $f_t^1(3) = f_{t-1}^1(3) + 2f_{t-1}^1(2) + 3$, $f_t^1(m) = f_{t-1}^1(m) + 2f_{t-1}^1(m-1)$.

Theorem 5. *For $p \geq 2$, if G is a $\{pK_2, HVN\}$-free graph, then $\chi(G) \leq f_p^1(\omega(G))$.*

Proof. When $\omega = 1$, the result is obvious. Let us assume that $\omega \geq 2$. Let us prove the result by induction on p. For $p = 2$, by using Theorem 1, Theorem 2 and Theorem 4, the result holds. By induction hypothesis, for $s \geq 2$, let us assume that if G' is an $\{sK_2, HVN\}$-free graph, then $\chi(G') \leq f_s^1(\omega(G'))$.

Let G be an $\{(s+1)K_2, HVN\}$-free graph. By using Theorem 3, the result is true for $\omega(G) = 2$. Let us consider $\omega \geq 3$. As observed in Theorem 4, we have $V(G) = A \cup C_{1,2} \cup C_{1,3} \cup C_{2,3} \cup \left(\bigcup_{1 \leq i \leq \omega} I_i \right)$. For $1 \leq i \leq \omega$, assign the color i to the vertex v_i and to the vertices of I_i. Hence $A \cup \left(\bigcup_{1 \leq i \leq \omega} I_i \right)$ can be colored with $\omega(G)$ colors. Clearly, for $(i,j) \in L$, each $\langle C_{i,j} \rangle$ is $\{sK_2, HVN\}$-free and $\omega(\langle C_{1,3} \rangle) \leq \omega(G) - 1$ and $\omega(\langle C_{2,3} \rangle) \leq \omega(G) - 1$.

Let us first consider $\omega(G) = 3$. By using induction hypothesis, $C_{1,2}$ can be colored using $f_s^1(3)$ colors, $C_{1,3}$ and $C_{2,3}$ can be colored with $f_s^1(2)$ colors separately. Thus $V(G)$ can be colored with at most $3 + f_s^1(3) + 2f_s^1(2) = f_{s+1}^1(3)$ colors. When $\omega \geq 4$, by using similar strategies as done for $\omega = 3$, with a little more involvement we can show that $V(G)$ can be colored with at most $f_{s+1}^1(\omega)$ colors. □

Let us recall a result in chromatic number of $\{2K_2, gem\}$-free graphs.

Theorem 6. *[4] Let G be a $\{2K_2, gem\}$-free graph, then $\chi(G) \leq \max\{3, \omega(G)\}$.*

For $p \geq 2$, let us define a sequence of functions $f_p^2 : \mathbb{N} \to \mathbb{N}$ to serve as a χ-binding function for $\{pK_2, gem\}$-free graphs as follows. For $p \geq 2$, $t, s \geq 3$, $f_p^2(1) = 1$, $f_2^2(2) = 3$, $f_2^2(t) = t$, $f_t^2(2) = 2t - 2$, $f_t^2(s) = f_{t-1}^2(s) + 2s - 2$.

Theorem 7. *For $p \geq 2$, if G is a $\{pK_2, gem\}$-free graph, then $\chi(G) \leq f_p^2(\omega(G))$.*

Since *diamond* is an induced subgraph of *gem*, for $p \geq 2$, $f_p^2(\omega(G))$ will be the χ-binding function for $\{pK_2, diamond\}$-free graphs as well.

Corollary 1. *For $p \geq 2$, if G is a $\{pK_2, diamond\}$-free graph, then $\chi(G) \leq f_p^2(\omega(G))$.*

In [9], Karthick and Mishra have showed that if G is a $\{2K_2, K_5 - e\}$-free graph, then $\chi(G) \leq \omega(G) + 4$. In Theorem 8, we reduce the bound to 6 when $\omega(G) = 4$ and we prove that G is $\omega(G)$-colorable, for all $\omega(G) \geq 5$.

Theorem 8. *If G is a $\{2K_2, K_5 - e\}$-free graph, then*
$$\chi(G) \leq \begin{cases} 6 & \text{for } \omega(G) = 4 \\ \omega(G) & \text{for } \omega(G) \geq 5. \end{cases}$$

Next, for $p \geq 2$, let us define a sequence of functions $f_p^3 : \mathbb{N} \to \mathbb{N}$ to serve as a χ-binding function for $\{pK_2, K_5 - e\}$-free graphs as follows. For $p \geq 2$, $t \geq 3$, $m \geq 4$, $s \geq 5$, $f_p^3(1) = 1$, $f_2^3(2) = 3$, $f_2^3(3) = 4$, $f_2^3(4) = 6$, $f_2^3(s) = s$, $f_t^3(2) = 2t - 2$, $f_t^3(3) = f_t^1(3)$, $f_t^3(m) = f_{t-1}^3(m) + 2f_{t-1}^2(m-1) + 3m - 6$.

Theorem 9. *If G is a $\{pK_2, K_5 - e\}$-free graph, then $\chi(G) \leq f_p^3(\omega(G))$.*

Next, for $p \geq 2$, let us define a sequence of functions $f_p^4 : \mathbb{N} \to \mathbb{N}$ to serve as a χ-binding function for $\{pK_2, dart\}$-free graphs as follows. For $p \geq 2$, $m, t \geq 3$, $s \geq 4$, $f_p^4(1) = 1$, $f_2^4(2) = 3$, $f_2^4(3) = 4$, $f_2^4(s) = \binom{s+1}{2}$, $f_t^4(2) = 2t - 2$, $f_t^4(m) = f_{t-1}^4(m) + \binom{m+1}{2} + m - 2$.

Theorem 10. *If G is a $\{pK_2, dart\}$-free graph, then $\chi(G) \leq f_p^4(\omega(G))$.*

Next, let us recall a result due to Blazsik et al. in [2].

Theorem 11. *[2] If G is a $\{2K_2, C_4\}$- free graph, then $\chi(G) \leq \omega(G) + 1$.*

For $p \geq 2$, let us define a sequence of functions $f_p^5 : \mathbb{N} \to \mathbb{N}$ to serve as a χ-binding function for $\{pK_2, C_4\}$-free graphs as follows. For $p, s \geq 2$, $m, t \geq 3$, $f_p^5(1) = 1$, $f_2^5(s) = s + 1$, $f_t^5(2) = 2t - 2$, $f_t^5(m) = \sum\limits_{i=2}^{m} f_{t-1}^5(i)$.

Theorem 12. *If G is a $\{pK_2, C_4\}$-free graph, then $\chi(G) \leq f_p^5(\omega(G))$.*

In [9], Karthick and Mishra have showed that if G is a $\{2K_2, K_1+C_4\}$-free graph, then $\chi(G) \leq \omega(G) + 5$. In Theorem 13, we improve the bound to $\omega(G) + 1$.

Theorem 13. *If G is a $\{2K_2, K_1 + C_4\}$-free graph such that $\omega(G) \geq 3$, then $\chi(G) \leq \omega(G) + 1$.*

For $p \geq 2$, let us define a sequence of functions $f_p^6 : \mathbb{N} \to \mathbb{N}$ to serve as a χ-binding function for $\{pK_2, K_1+C_4\}$-free graphs as follows. For $p, s \geq 2$, $m, t \geq 3$, $f_p^6(1) = 1$, $f_2^6(s) = s+1$, $f_t^6(2) = 2t-2$, $f_t^6(m) = \left(\sum\limits_{i=2}^{m} f_{t-1}^6(i)\right) + f_{t-1}^6(m-1) + 1$.

Theorem 14. *If G is a $\{pK_2, K_1 + C_4\}$-free graph, then $\chi(G) \leq f_p^6(\omega(G))$.*

Next, let us recall a result due to Fouquet et al. in [5].

Theorem 15. *[5] If G is a $\{2K_2, \overline{P_5}\}$-free graph, then $\chi(G) \leq \frac{3}{2}\omega(G)$.*

Next, for $p \geq 2$, let us define a sequence of functions $f_p^7 : \mathbb{N} \to \mathbb{N}$ to serve as a χ-binding function for $\{pK_2, \overline{P_5}\}$-free graphs as follows. For $p \geq 2$, $m, t \geq 3$ $s \geq 4$, $f_p^7(1) = 1$, $f_2^7(2) = 3$, $f_2^7(3) = 4$, $f_2^7(s) = \frac{3}{2}s$, $f_t^7(2) = 2t - 2$, $f_t^7(m) = m + \sum\limits_{i=2}^{m} f_{t-1}^7(i)$.

Theorem 16. *If G is a $\{pK_2, \overline{P_5}\}$-free graph, then $\chi(G) \leq f_p^7(\omega(G))$.*

Acknowledgment. For the first author, this research was supported by the Council of Scientific and Industrial Research, Government of India, File No: 09/559(0133)/2019-EMR-I. And for the second author, this research was supported by the UGC-Basic Scientific Research, Government of India, Student id: gokulnath.res@pondiuni.edu.in.

References

1. Bharathi, A.P., Choudum, S.A.: Colouring of $(P_3 \cup P_2)$-free graphs. Graphs Comb. **34**(1), 97–107 (2018)
2. Blázsik, Z., Hujter, M., Pluhár, A., Tuza, Z.: Graphs with no induced C_4 and $2K_2$. Discrete Math. **115**(1–3), 51–55 (1993)
3. Brandt, S.: Triangle-free graphs and forbidden subgraphs. Discrete Appl. Math. **120**(1–3), 25–33 (2002)
4. Brause, C., Randerath, B., Schiermeyer, I., Vumar, E.: On the chromatic number of $2K_2$-free graphs. Discrete Appl. Math. **253**, 14–24 (2019)
5. Fouquet, J.L., Giakoumakis, V., Maire, F., Thuillier, H.: On graphs without P_5 and $\overline{P_5}$. Discrete Math. **146**(1–3), 33–44 (1995)
6. Gaspers, S., Huang, S.: $(2P_2, K_4)$-free graphs are 4-colorable. SIAM J. Discrete Math. **33**(2), 1095–1120 (2019)
7. Gyárfás, A.: Problems from the world surrounding perfect graphs. Zastosowania Matematyki Applicationes Mathematicae **19**(3–4), 413–441 (1987)
8. Karthick, T., Maffray, F.: Vizing bound for the chromatic number on some graph classes. Graphs Comb. **32**(4), 1447–1460 (2016)
9. Karthick, T., Mishra, S.: Chromatic bounds for some classes of $2K_2$-free graphs. Discrete Math. **341**(11), 3079–3088 (2018)
10. Randerath, B., Schiermeyer, I.: Vertex colouring and forbidden subgraphs-a survey. Graphs Comb. **20**(1), 1–40 (2004)
11. Schiermeyer, I., Randerath, B.: Polynomial χ-binding functions and forbidden induced subgraphs: a survey. Graphs Comb. **35**(1), 1–31 (2019)
12. Wagon, S.: A bound on the chromatic number of graphs without certain induced subgraphs. J. Comb. Theory Ser. B **29**(3), 345–346 (1980)
13. West, D.B.: Introduction to Graph Theory. Prentice-Hall of India Private Limited (2005)

Axiomatic Characterization of the Median Function of a Block Graph

Manoj Changat[1(✉)], Nella Jeena Jacob[1], and Prasanth G. Narasimha-Shenoi[2]

[1] Department of Futures Studies, University of Kerala, Thiruvananthapuram 695581, India
mchangat@keralauniversity.ac.in, nellajeenaj@gmail.com

[2] Department of Mathematics, Government College Chittur, Palakkad, India
prasanthgns@gmail.com

Abstract. A median of a profile of vertices (a sequence of vertices) on a connected graph is a vertex that minimizes the sum of the distances to the elements in the profile. The median function has as output the set of medians of a profile. Median function is an important consensus function for the location of a desirable facility in a network. The axiomatic characterization of the median function is studied by several authors on special classes of graphs like trees and median graphs. In this paper, we determine the median sets of all types of profiles and obtain an axiomatic characterization for the median function on block graphs, an immediate generalization of trees.

Keywords: Profiles · Block graph · Median sets · Median function · Axiomatic characterization

1 Introduction

Consensus is an important concept where the aggregation of processed data is desired. Consensus is a collective opinion or a general understanding among a group of agents or clients. Consensus functions are introduced to model the problem of achieving consensus amongst agents or clients in a rational way. The input of a consensus function is information on the clients and the output concerns the issue on which consensus should be reached. The problems of consensus can also be treated in location theory as finding an optimal facility among possible locations. Facility location problems in the discrete case deal with functions that find an appropriate location for a common facility or resource in a discrete structure like a network, or a graph or an ordered set [10].

An arbitrary consensus function on a graph $G = (V, E)$ can be defined on the vertex set V which returns for every sequence of vertices (profile) of V a non empty subset of V. Axiomatic studies in this area resulted in better understanding of the process of location and consensus. An interesting problem in

A. Mudgal and C. R. Subramanian (Eds.): CALDAM 2021, LNCS 12601, pp. 294–308, 2021.
https://doi.org/10.1007/978-3-030-67899-9_24

consensus theory is to find a set of axioms that characterizes a given consensus function. The axiomatic study of social decision problems can be traced to 1950 with the seminal work of Kenneth Arrow [1], followed by Kemeny and Snell where, perhaps for the first time, the axiomatic approach was joined with consensus terminology [8]. Axiomatic characterization of the mean function is due to Holzman in [7] and that of the median function by Vohra in [22] on the continuous variant of a tree, where internal points of edges are also allowed as locations. The discrete case of trees was first attempted by McMorris, Mulder and Roberts [13], where they also characterized the median function on cube-free median graphs using three simple and appealing axioms, known as anonymity (A), betweenness (B) and consistency (C). Anonymity describes the property that the output does not rely upon the ordering of the elements in the profile. Betweenness implies that anything between two vertices has equal importance and consistency means that if two profiles agree on some output, then the aggregation of the two profiles agrees on that output. The A, B, C axioms are known as *universal axioms* for the median function as these axioms are satisfied by the median function on any connected graph. The median function has been characterized on hypercubes and median graphs using only the A, B, C axioms [13,18,19]. It is expected that the axioms depend on the consensus function at hand and on the structure of the graph on which the function is studied as we will see in this paper. The axiomatic characterization of other consensus functions on trees studied were the mean function [11], the ℓ_p-function (discrete case) [12], the center function [15,20]. So far, in most of the consensus functions mentioned above, a characterization is only obtained on trees.

A *block graph* is a graph in which every block (maximal 2-connected subgraph) is a complete graph. Since trees are special cases of block graphs, they are natural extensions to trees to consider for the axiomatic characterization of the median function. In this paper motivated by the studies on the median function on trees, we attempt for an axiomatic characterization of the median function on block graphs.

The paper is organized as follows. In the remaining part of this section, we fix the notations and terminology, in Sect. 2, we present a consensus strategy known as plurality strategy for computing median sets of profiles. This strategy will produce median sets on graphs having connected median sets. In Sect. 4, some basic facts about the block graphs is discussed and determine the structure of median sets in block graphs. Based on the structure of the median sets, we formulate the required axioms for the median function of block graphs and describe the axiomatic characterization.

In this paper we consider only finite simple connected graphs $G = (V, E)$ with vertex set V and edge set E. The distance $d(u, v)$ between u and v in G is the length of a shortest u, v-path. The interval $I(u, v)$ between two vertices u and v in G consists of all vertices on shortest u, v-paths, that is: $I(u, v) = \{x : d(u, x) + d(x, v) = d(u, v)\}$. A profile π of length $k = |\pi|$ on G is a nonempty sequence $\pi = (x_1, x_2, \ldots, x_k)$ of vertices of V with repetitions allowed. We define V^* to be the set of all profiles of finite length on V and $x_1, x_2, \ldots, x_k$ are known as the elements of the profile. A vertex of π is a vertex that occurs as an element in π. By $\{\pi\}$ we denote the set of all vertices of π. Note that a vertex x may occur more than once as element in π. If we say that x is an element of π, then we mean an element in a certain position, say $x = x_j$ in the jth position. A subprofile of π is just a subsequence of π. For convenience we also allow the empty subprofile. Let $\pi = (x_1, x_2, \ldots, x_k)$ and $\rho = (y_1, y_2, \ldots, y_\ell)$ be two profiles. The profile $(x_1, x_2, \ldots, x_k, y_1, y_2, \ldots, y_\ell)$ is called a *concatenation* of π and ρ, and is denoted by $\pi\rho$. Note that in most cases we have $\pi\rho \neq \rho\pi$. By $x \in \pi$, we mean that x is a vertex of the profile π. A *consensus function* on G is a function $L : V^* \to 2^V - \emptyset$ that gives a nonempty subset of V as output for each profile on G. For convenience, we write $L(x_1, x_2, \ldots, x_k)$ instead of $L((x_1, x_2, \ldots, x_k))$, for any function L defined on profiles, but will keep the brackets where needed. The remoteness of a vertex v to a profile π is defined as $D(v, \pi) = \sum_{i=1}^{k} d(v, x_i)$. A vertex minimizing $D(v, \pi)$ is called a *median* of the profile. The set of all medians of π is the median set of π and is denoted by $Med(\pi)$.

We can also consider Med as a function from V^* to $2^V - \emptyset$, and then Med function becomes the median function of G. Note that $Med(x) = \{x\}$, and $Med(x, y) = I(x, y)$. Also, if $I(u, v) \cap I(v, w) \cap I(w, u) \neq \emptyset$, then $Med(u, v, w) = I(u, v) \cap I(v, w) \cap I(w, u)$. The median function has been studied extensively on median graphs. A *median graph* is defined by the property that $|I(u, v) \cap I(v, w) \cap I(w, u)| = 1$, for any three vertices u, v, w. Equivalently, a median graph is a graph such that any profile of length 3 has a unique median. See e.g. [9,17,21] for structure theory and [13,14,19], for the axiomatic characterizations for the median function on median graphs, using the A, B, C axioms mentioned above. Formally, the A, B, C axioms are defined as follows.

(A) Anonymity: $L(\pi) = L(x_{\chi(1)}, x_{\chi(2)}, \ldots, x_{\chi(k)})$, for any profile $\pi = (x_1, x_2, \ldots, x_k)$ on V and for any permutation χ of $\{1, 2, \ldots, k\}$.

(B) Betweenness: $L(u, v) = I(u, v)$, for all u, v in V.

(C) Consistency: If $L(\pi) \cap L(\rho) \neq \emptyset$, for profiles π and ρ, then $L(\pi\rho) = L(\pi) \cap L(\rho)$.

We may observe that the first and third axioms are defined without any reference to the distance function. It can be seen that the median function Med satisfies axioms $(A), (B), (C)$ on any connected graph.

2 Plurality Strategy and Median Sets in Graphs

We present the plurality strategy on profiles of a graph G and describe the graphs where median sets can be computed using this strategy.

For any three vertices, u, v, w in a graph G, if $d(w, u) < d(w, v)$, then we say that w is closer to u than to v. For a profile π and an arbitrary edge uv of a graph G, we denote π_{uv} as the subprofile of π consisting of all elements in π closer to u than to v and π_{vu} as the subprofile of π consisting of all elements in π closer to v

than to u. We call the pairs of subprofiles, (π_{uv}, π_{vu}) as a *split.* If $|\pi_{uv}| = |\pi_{vu}|$, then we call (π_{uv}, π_{vu}) as a balanced split of π. We call (π_{uv}, π_{vu}) an unbalanced split with respect to π if $|\pi_{uv}| > |\pi_{vu}|$ or $|\pi_{vu}| > |\pi_{uv}|$. If (π_{uv}, π_{vu}) is an unbalanced split with $|\pi_{uv}| > |\pi_{vu}|$, then π_{uv} will be in the *majority side* of the split. The *majority strategy* on graphs is introduced by Mulder in 1997 [16], which can be adopted on a profile of vertices of a graph. The procedure is to start from any initial vertex of the graph and always moving along edges to a side where there is a majority of the profile lies and finally get stuck or moves along a set of vertices and cannot move out of those vertices, which is called the outcome of the majority strategy. It is proved in [16] that in the case of median graphs, the outcome is precisely the median set of the profile independent of the starting vertex. In this strategy, we do not compute the distance, but only the number of vertices on each side of the edge that we move along.

Motivated by the majority strategy in graphs, another consensus strategy called *plurality strategy* was introduced in [2]. It is defined below.

Plurality Strategy
Plurality strategy on a profile π for a graph G is the following.

1 Start at an initial vertex v.
2 If we are in v, and w is a neighbor of v with $|\pi_{vw}| \leq |\pi_{wv}|$, then we move to w.
3 We move only to a vertex already visited if there is no alternative.
4 We stop when
  (a) We are stuck at a vertex v
  or
  (b) We have visited vertices at least twice, and, for each vertex v visited at least twice and for each neighbor w of v, either w is also visited at least twice or $|\pi_{vw}| > |\pi_{wv}|$.
5 Median set is the set of vertices where we get stuck or visited at least twice.

Example 1. Consider the block graph G shown in Fig. 1:

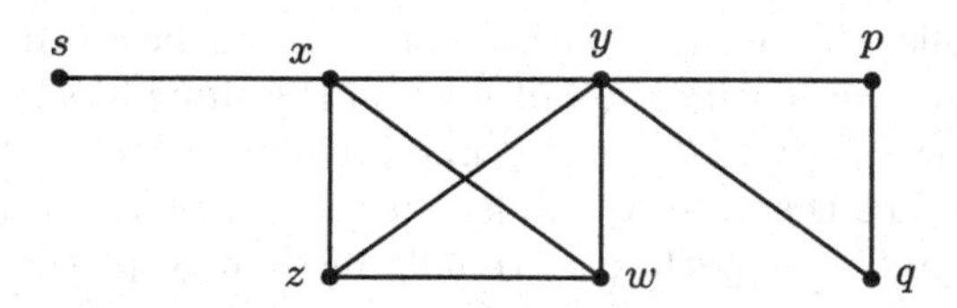

Fig. 1. Illustration of plurality strategy

Consider the profile $\pi = (s, x, w, p, p, q)$. For starting the plurality strategy from s, we compute, $\pi_{sx} = \{s\}, \pi_{xs} = \{x, w, p, p, q\}, \pi_{xz} = \{s, x\}, \pi_{xw} = \{x, s\}$, $\pi_{wx} = \{w\}, \pi_{yx} = \{p, p, q\}, \pi_{xy} = \{x, s\}, \pi_{yp} = \{s, x, w\}, \pi_{py} = \{p, p\}, \pi_{yq} = \{s, x, w\}, \pi_{qy} = \{q\}$. Thus, $|\pi_{sx}| = 1, |\pi_{xs}| = 5, |\pi_{xy}| = 2, |\pi_{yp}| = 3, |\pi_{py}| = 2, |\pi_{yq}| = 3, |\pi_{qy}| = 1$, So if we apply the plurality strategy starting from s, we move to x, then to y and get stuck at y. So the outcome of plurality strategy starting from y for profile π is y which is the median of π. Again, for $\pi = (s, s, s, p, q, q)$, for starting the plurality strategy from z, we compute, $|\pi_{zx}| = 0, |\pi_{xz}| = 3, |\pi_{xs}| = 3, |\pi_{sx}| = 3, |\pi_{xy}| = 3, |\pi_{yx}| = 3, |\pi_{yp}| = 3, |\pi_{py}| = 1, |\pi_{yq}| = 3, |\pi_{qy}| = 2$. Thus from z, we move to x, then to s, then again to x, then to y. We cannot move to p or q, so we back track from y to x to s. Thus we move around the path s, x, y, which will be the outcome of the plurality strategy, which is also the median of π. So, we get the path or interval $I(s, y)$ as the median of π.

We have the following results from [2].

Lemma 1. *[2] Let G be a connected graph and π a profile on G. Plurality strategy makes a move from vertex u to vertex v if and only if $D(v, \pi) \leq D(u, \pi)$.*

We have an immediate remark for Plurality strategy.

Remark 1. $D(v, \pi) \leq D(u, \pi)$ if and only if $|\pi_{vu}| \leq |\pi_{uv}|$ which is the condition for plurality strategy to make a move from v to u.

The following theorem is proved in [2].

Theorem 1. *[2] Following are equivalent for a connected graph G.*

1. *Plurality strategy produces $Med(\pi)$ starting from an arbitrary vertex for all profiles π.*
2. *$Med(\pi)$ is connected for all profiles π.*

3 Median Sets in Block Graphs

First we describe the preliminary notions and some basic facts about the block graphs for our purpose. In the rest of this paper, we consider non-trivial block graphs having at least two blocks. It may be noted that all internal vertices of the unique induced path (same as shortest path) are cut vertices in a block graph. Another fact is that for any three vertices u, v, w in a block graph $G : I(u, v) \cap I(u, w) = \{u\}$ implies that either all the vertices u, v, w lie on the same block or u lies on the shortest v, w- path or u and v lie on the same block. In each of these cases, it can be noted that $d(v, w) \geq \max(d(u, v), d(u, w))$. This observation together with the fact that the intervals $I(u, v)$ in a block graph is the unique shortest u, v-path in G (the paths are trivially median graphs) implies that every block graph is a quasi-median graph. Quasi-median graphs are non-bipartite generalizations of median graphs and there are several characterizations of these graphs, for e.g., [4,6]. The definition of a quasi-median graph which is more relevant in the case of a block graph is the following. A graph G is a *quasi-median graph* if every interval in G induces a median graph and for any three vertices u, v, w: $I(u, v) \cap I(u, w) = \{u\}$ implies that $d(v, w) \geq \max(d(u, v), d(u, w))$. In fact the family of block graphs is one of the maximal subclass of the family of quasi-median graphs.

It may be observed that a quasi-median graph have all median sets connected [3] and hence the plurality strategy will produce median sets for all profiles in a quasi-median graph. Since block graphs are special class of quasi-median graphs, have median sets connected for all profiles according to Theorem 1.

In this section we determine the structure of median sets for all profiles on block graphs. Let π be a profile on G. Note that a vertex may occur more than once as an element in π. A vertex with highest occurrence in π is called a *plurality vertex* of π. We denote the set of plurality vertices of π by $Pl(\pi)$. For a profile $\pi = (x_1, x_2, \ldots, x_i, \ldots, x_k)$, by $\pi - x_i$, we mean the subprofile of π obtained by deleting the vertex x_i from the ith position of π. The vertex x_i that occurs in other positions in π, remain to be in $\pi - x_i$ in their respective positions. A profile π is a *balanced profile*, if there exists an edge uv in G such that $\pi = (\pi_{uv}, \pi_{vu})$ and $|\pi_{uv}| = |\pi_{vu}|$. A profile π is a *unbalanced profile*, if π is not a balanced profile. That is, there doesn't exists an edge uv in G such that $\pi = (\pi_{uv}, \pi_{vu})$ and $|\pi_{uv}| = |\pi_{vu}|$. A profile π is called a *complete profile* if $\{\pi\}$ induces a clique in G. That is, all the vertices of a complete profile lies on a block B of G. A profile π is called a *partially complete* profile, if π has a complete subprofile π_1 consisting of at least half of $|\pi|$. It is to be noted that throughout this section, we use the fact that any profile π (balanced or unbalanced) in a block graph G can be represented as the concatenation of the subprofiles π_{uv} and π_{vu}; that is, $\pi = (\pi_{uv}, \pi_{vu})$, for some edge uv in G. For profile $\pi = (\pi_{uv}, \pi_{vu})$ and a vertex z, we use the terminology that z *is closer to* π_{uv} to mean that z is closer to vertices in π_{uv} than the vertices in π_{vu}.

Theorem 2. *Let G be a block graph. The median sets of G are singleton set $\{v\}, v \in V(G)$, a set of vertices in a path or a block or a proper subset of a block.*

Proof. Let π be a profile containing vertices from G.

Case 1: Let $\pi = (v)$, then $D(v, \pi) = 0 < D(x, \pi)$ for any $x \in V(G)$ different than v. Hence $\{v\}$, where $v \in V$ are all the median sets. If π contains more than one vertex, we proceed as follows.

Case 2: Let $\pi = (u, v)$. Then by betweenness axiom we will get the output for the profile of length 2. Hence $Med(\pi)$ is the set of all vertices in the u, v-path. i.e. $Med(\pi) = I(u, v)$ for all u, $v \in V$.

Case 3: Let π be a complete profile. Suppose that every vertex of π occurs only once in π. Let u be a vertex in π. Let v be a neighbour of u. There are two possibilities for v. Either $v \notin \{\pi\}$ or $v \in \{\pi\}$. If $v \notin \{\pi\}$, then it is clear that π_{uv} will be $\{u\}$, while π_{vu} will be the empty set. Also, if $v \in \{\pi\}$, then π_{uv} is $\{u\}$ and π_{vu} is $\{v\}$. Hence by the plurality strategy starting from the vertex u, we can move to all the vertices in π and nowhere else. Therefore $Med(\pi) = Pl(\{\pi\})$. If $Pl(\{\pi\}) \neq \emptyset$, let $u \in Pl(\{\pi\})$. Then, for any neighbour v of u, π_{uv} will be the profile consisting u repeating according to the number of times u occurs in π, where as π_{vu} will be the empty set if $v \notin \{\pi\}$ and π_{vu} will be $\{v\}$, if v is profile vertex, but not a plurality vertex of π. If $v \in Pl(\{\pi\})$, then

π_{vu} will be the profile consisting v repeating according to the number of times v occurs in π. Then it clear that π_{uv} and π_{vu} have the same cardinality and hence $Med(\pi) = Pl(\{\pi\})$. Hence, for a complete profile π, $Med(\pi) = Pl(\{\pi\})$, if $Pl(\{\pi\}) \neq \emptyset$ and $Med(\pi) = \{\pi\}$, if $Pl(\{\pi\}) = \emptyset$. Similarly, we can prove that if π is a partially complete profile, then $Med(\pi) = Pl(\{\pi_1\})$, if $Pl(\{\pi_1\}) \neq \emptyset$ and $Med(\pi) = \{\pi_1\}$, if $Pl(\{\pi_1\}) = \emptyset$.

Case 4: $\pi = (\pi_{uv}, \pi_{vu})$ is a balanced profile.

Subcase 4.1: $u, v \notin \pi$.
Let u_1 be a vertex adjacent to u and closer to π_{uv}. Then it is clear that $|\pi_{u_1u}| \leq |\pi_{uv}|$. If $|\pi_{u_1u}| < |\pi_{uv}|$, then some vertex or vertices in π_{uv} branches out from vertex u, otherwise (if u_1 is closer to π_{uv}) $|\pi_{u_1u}| = |\pi_{uv}|$ and we can move to u_1 by the plurality strategy. Using the same strategy, we continue to move to vertices $u_i, \ldots, u_k = x$, $i = 1, \ldots, k$, where $u, u_1, \ldots, u_k = x$ form a path, say $P_{u,x}$ in G. It is clear that $|\pi_{u_iu_{i+1}}| = |\pi_{u_{i+1}u_{i+2}}| = |\pi_{uv}|$, for $i = 1, \ldots, k-2$. Let z be the first vertex adjacent to x and closer to π_{uv} such that $|\pi_{zx}| < |\pi_{xz}| = |\pi_{uv}|$. Then some vertex or vertices in π_{uv} branches out from vertex x. This implies that removing x will disconnect the vertices in π_{uv} or in other words x is a cut vertex separating π_{uv} and π_{vu}.

Similarly following the same strategy to the side of π_{vu}, we obtain a path $v, v_1, \ldots, v_\ell = y$, say $P_{v,y}$ such that $|\pi_{v_iv_{i+1}}| = |\pi_{v_{i+1}v_{i+2}}| = |\pi_{vu}|$, for $i = 1, \ldots, \ell - 2$. If we move to a vertex w adjacent to y and closer to π_{vu}, then $|\pi_{yw}| > |\pi_{wy}|$. This imply that y is a vertex that separates π_{uv} and π_{vu} and closest to π_{vu}. So the only alternative is to retrace the path $P_{v,y}$ we took from y to v and then from v to x through the path u,$P_{u,x}$ and thus we have visited vertices in the shortest x, y-path obtained by concatenating the paths $P_{u,x}$ and $P_{v,y}$, at least twice. Therefore, by plurality strategy the median set, $Med(\pi) = I(x, y)$, where x is the cut vertex that separates π_{uv} and π_{vu} and closest to π_{uv} and y is the cut vertex that separates π_{uv} and π_{vu} and closest to π_{vu}.

Subcase 4.2: $\pi = (\pi_{uv}, \pi_{vu})$ is a balanced profile and either u, v or both u, v belongs to π.
In this case, applying the plurality strategy similar to that of Subcase 4.1, we can prove that $Med(\pi) = I(x, y)$, where there are three cases for x, y according as $u \in \pi$, $v \in \pi$ or both $u, v \in \pi$. If $u \in \pi$ and $v \notin \pi$, then, it is clear that $x = u$ and y is a cut-vertex that separates π_{uv} and π_{vu} and closest to π_{vu}. If $v \in \pi$ and $u \notin \pi$, then, it is clear that x is a cut-vertex that separates π_{uv} and π_{vu} and closest to π_{uv} and $y = v$. If $u, v \in \pi$, then $x = u$ and $y = v$.

Case 5: π is an unbalanced profile, which is not a complete profile nor a partially complete profile. (Note that a complete profile or a partially complete profile is an unbalanced profile).
Therefore there exists an edge uv such that $\pi = (\pi_{uv}, \pi_{vu})$ and either $|\pi_{uv}| > |\pi_{vu}|$ or $|\pi_{vu}| > |\pi_{uv}|$. Assume that $|\pi_{uv}| > |\pi_{vu}|$. Therefore, median set $Med(\pi)$ lies to the side of π_{uv}. By plurality strategy we make a move from v to u as $|\pi_{uv}| > |\pi_{vu}|$. We may continue to move from u through vertices of the path

$v, u, u_1 \ldots u_k = z$ to z by plurality strategy until $|\pi_{u_1u}| \geq |\pi_{uu_1}| \geq \ldots \geq |\pi_{u_{i+1}u_i}| \geq |\pi_{u_iu_{i+1}}|$, for $i = 1, \ldots, k-1$. There are two cases.

Subcase 5.1: $|\pi_{u_1u}| > |\pi_{uu_1}| > \ldots > |\pi_{u_iu_{i+1}}| > |\pi_{u_{i+1}u_i}|$, for $i = 1, \ldots, k-1$ and $|\pi_{z'z}| < |\pi_{zz'}|$ for all $z' \in N(z) \setminus \{u_{k-1}\}$.
Then, we are stuck at z and hence by plurality strategy, $\text{Med}(\pi) = \{x\}$.

Subcase 5.2: $|\pi_{u_{i+1}u_i}| = |\pi_{u_iu_{i+1}}|$, for some $i = 1, \ldots, k-1$.
Let r be the first index among $i = 1, \ldots, k$ such that $|\pi_{u_{r+1}u_r}| = |\pi_{u_ru_{r+1}}|$ so that $|\pi_{u_1u}| > |\pi_{uu_1}| > \ldots > |\pi_{u_ru_{r-1}}| > |\pi_{u_{r-1}u_r}|$ and $|\pi_{u_{j+1}u_j}| = |\pi_{u_ju_{j+1}}|$, for $j = r, \ldots, k-1$. This implies that vertices in π_{uv} branches out from u_r so that vertices in the subprofiles $\pi_{u_{j+1}u_j}$ and $\pi_{u_ju_{j+1}}$ are equidistant from u_r, for $j = r, \ldots, k-1$. This is possible if and only if the vertices u_j, for $j = r, \ldots, k$ are mutually adjacent so that they form a clique and hence lie on some block. Therefore, we move around the vertices, u_r, for $j = r, \ldots, k-1$ any number of times and hence by plurality strategy, $\text{Med}(\pi) = \{u_r, \ldots, u_k\}$.

So, considering all the cases, we can summarize that the median sets of a block graph G are the singleton set $\{v\}, v \in V(G)$, a set of vertices in a path or a block or a proper subset of a block. □

Remark 2. The median sets of a block graph for all profiles π are as follows:

1. $\pi = (x)$, then $Med(\pi) = \{x\}$.
2. $\pi = (x, y)$, then $Med(\pi) = I(x, y)$.
3. $\pi = (\pi_{uv}, \pi_{vu})$ is a balanced profile. If $u, v \notin \pi$, then $Med(\pi) = I(x, y)$ such that x and y are vertices that separates π_{uv} and π_{vu}, where x is a vertex closest to π_{uv} and y is a vertex closest to π_{vu}.
4. $\pi = (\pi_{uv}, \pi_{vu})$ is a balanced profile. If $u \in \pi$, then $Med(\pi) = I(u, y)$ such that y is a vertex that separates π_{uv} and π_{vu} closest to π_{vu}.
5. $\pi = (\pi_{uv}, \pi_{vu})$ is a balanced profile. If $v \in \pi$, then $Med(\pi) = I(x, v)$ such that x is a vertex that separates π_{uv} and π_{vu} and closest to π_{uv}.
6. $\pi = (\pi_{uv}, \pi_{vu})$ is a balanced profile. If $u, v \in \pi$, then $Med(\pi) = I(u, v)$.
7. $\pi = (\pi_{uv}, \pi_{vu})$ is an unbalanced profile with $|\pi_{uv}| < |\pi_{vu}|$. Then
   (a) $Med(\pi) = \{x\}$ such that x is a vertex that separates π_{uv} and π_{vu} which is closest to π_{uv}.
   (b) $Med(\pi) = Pl(\pi_1)$, when π is a complete profile or a partially complete profile provided $Pl(\pi_1) \neq \emptyset$ and $Med(\pi) = \{\pi\}$, if $Pl(\pi_1) = \emptyset$.
   (c) $Med(\pi) = V(K)$, where K is a clique in G (K can be the entire block of G).

We conclude this section with two Lemmata for an unbalanced profile. These two results will give us the median sets of profiles π from the median sets of vertex deleted subprofiles $\pi - x$.

Lemma 2. *Let π be an unbalanced profile in a block graph G with $|\pi| = n$. If $\bigcap_{i=1}^{n} Med(\pi - x_i) \neq \emptyset$, then $Med(\pi) = \bigcap_{i=1}^{n} Med(\pi - x_i)$.*

Proof. Let G be a block graph and $\pi = (\pi_{uv}, \pi_{vu})$ with $|\pi_{uv}| > |\pi_{vu}|$ be an unbalanced profile having $\bigcap_{i=1}^{k} Med(\pi - x_i) \neq \emptyset$. Let $\pi = (x_1, x_2, \ldots, x_n)$, $\pi_{uv} = (x_1, x_2, \ldots, x_k)$ and $\pi_{vu} = (x_{k+1}, x_{k+2}, \ldots, x_n)$. Let $x \in \bigcap_{i=1}^{n} Med(\pi - x_i)$. Assume that $x \notin Med(\pi)$. Therefore there exists some $y \in V(G)$ such that

$$\sum_{i=1}^{n} d(y, x_i) < \sum_{i=1}^{n} d(x, x_i) \tag{1}$$

Since G is a block graph, there will be a unique x, y- shortest path. Since $x \in Med(\pi - x_i)$ for all $x_i \in \pi$, we have that x is closer to π_{uv} than to π_{vu}. Therefore all the y, x_i- shortest paths will pass through x for all $i = 1, 2, \ldots, k$. Therefore all the x, x_i - shortest paths will pass through y for all $i = k+1, k+2, \ldots, n$. So,

$$\begin{aligned}
\sum_{i=1}^{n} d(x, x_i) &= \sum_{i=1}^{k} d(x, x_i) + \sum_{i=k+1}^{n} d(x, x_i) \\
&= \sum_{i=1}^{k} d(x, x_i) + \sum_{i=k+1}^{n} (d(x, y) + d(y, x_i)) \\
&< \sum_{i=1}^{k} (d(x, x_i) + d(x, y)) + \sum_{i=k+1}^{n} (d(x_i, y)) \qquad (2) \\
&\text{(as } |\pi_{uv}| > |\pi_{vu}| \text{ we need at least one more } d(x, y))
\end{aligned}$$

Equations 1 and 2 contradicts each other. Hence $x \in Med(\pi)$.
Now let $x \in Med(\pi)$.

$$\begin{aligned}
x \in Med(\pi) &\iff D(x, \pi) \text{ is minimum} \\
&\iff \sum_{i=1}^{n} d(x, x_i) \text{ is minimum} \\
&\iff \sum_{j \neq i} d(x, x_j) \text{ is minimum} \\
&\iff x \in Med(\pi - x_i), \forall x_i \in \pi \\
&\iff x \in \bigcap_{i=1}^{n} Med(\pi - x_i)
\end{aligned}$$

Hence the lemma. □

Next lemma enable us to get the median sets of profiles π when the median sets of vertex deleted profiles $\pi - x$ have an empty intersection.

Lemma 3. *Let π be an unbalanced profile in a block graph G with $|\pi| = n$ and $\bigcap_{i=1}^{n} Med(\pi - x_i) = \emptyset$, then $\bigcup_{i=1}^{n} Med(\pi - x_i) = Med(\pi)$.*

Proof. Let π be an unbalanced profile with $\bigcap_{i=1}^{n} Med(\pi - x_i) = \emptyset$. Also let $\pi = (\pi_{uv}, \pi_{vu})$ with $|\pi_{uv}| > |\pi_{vu}|$. Let $x \in \bigcup_{i=1}^{n} Med(\pi - x_i)$. Therefore $x \in Med(\pi - x_i)$ for some $x_i \in \pi$. Without lose of generality assume that $x_i = x_1$. Also since $\bigcap_{i=1}^{n} Med(\pi - x_i) = \emptyset$, there exists an x_j so that $x \notin Med(\pi - x_j)$. Again without lose of generality we can assume $x_j = x_2$. Now, let us examine all possible cases for x_1, x_2 with respect to their occurrence in π_{uv} and π_{vu}.

Case 1: $x_1, x_2 \in \pi_1$.
Since $x \in Med(\pi - x_1)$ and Since $|\pi_{uv}| > |\pi_{vu}|$ and $x_1, x_2 \in \pi_{uv}$, either $|\pi_{uv} - x_1| = |\pi_{uv} - x_2| > |\pi_{vu}|$ or $|\pi_{uv} - x_1| = |\pi_{uv} - x_2| = |\pi_{vu}|$. If $|\pi_{uv} - x_1| = |\pi_{uv} - x_2| > |\pi_{vu}|$, then since $x \in Med(\pi - x_1)$, it follows by plurality strategy that $x \in Med(\pi - x_2)$, a contradiction. If $|\pi_{uv} - x_1| = |\pi_{uv} - x_2| = |\pi_{vu}|$, then the profiles $(\pi_{uv} - x_1, \pi_{vu})$ and $(\pi_{uv} - x_2, \pi_{vu})$ are balanced profiles having the same median sets and since $x \in Med(\pi - x_1)$, it follows by plurality strategy again that $x \in Med(\pi - x_2)$, a contradiction. This implies that it is impossible to have both $x_1, x_2 \in \pi_{uv}$. That is, if $x_1 \in \pi_1$, then $x_2 \notin \pi_1$ and vice versa.

Case 2: $x_1 \in \pi_{uv}$ and $x_2 \in \pi_{vu}$.
Here $|\pi_{uv}| > |\pi_{vu}|$ will imply that $|\pi_{uv} - x_1| > |\pi_{vu} - x_2|$. So x will also be closer to $\pi_{uv} - x_1$ than $\pi_{vu} - x_2$. Since $x \in Med(\pi - x_1)$, it follows by plurality strategy that $x \in Med(\pi)$.

Case 3: $x_1 \in \pi_{vu}$ and $x_2 \in \pi_{vu}$.
The median set of π will be closer to π_{uv} than π_{vu} since $|\pi_{uv}| > |\pi_{vu}|$. Now, $|\pi_{uv} - x_1| > |\pi_{vu} - x_2|$ and so by plurality strategy from v towards u, we come to a vertex x in $Med(\pi - x_2)$ whenever $x \in Med(\pi - x_1)$. So $x \in Med(\pi - x_1)$ implies that $x \in Med(\pi - x_2)$, which is a contradiction. So if $x_1 \in \pi_{vu}$, then $x_2 \notin \pi_{vu}$ and vice versa.

Case 4: $x_1 \in \pi_{vu}$ and $x_2 \in \pi_{uv}$.
This case is similar to that of Case 3 and here also, we infer that $x \in Med(\pi - x_2)$, a contradiction. So this case also will not happen.

In all the above cases, we have proved that except Case 2, none of the other case will happen. since $\bigcap_{i=1}^{n} Med(\pi - x_i) = \emptyset$, and in case 2, we have proved that $x \in Med(\pi)$. Hence $\bigcup_{i=1}^{n} Med(\pi - x_i) \subseteq Med(\pi)$.

Conversely suppose $x \in Med(\pi)$. Since $|\pi_{uv}| > |\pi_{vu}|$, choosing any vertex $x_i \in \pi_{vu}$, we can see that x lies closer to π_{uv} than $\pi_{vu} - x_i$, so that $x \in Med(\pi - x_i)$. So $Med(\pi) \subseteq \bigcup_{i=1}^{n} Med(\pi - x_i)$. Hence $\bigcup_{i=1}^{n} Med(\pi - x_i) = Med(\pi)$. $\square$

Remark 3. It is observed that in the proof Lemma 3, we do not use the assumption that G is a block graph, but we have used the plurality strategy in the proof to obtain the median vertex of profiles. So, it may be true that Lemma 3, works for graphs with connected medians as the plurality strategy produces medians of any profile for graphs with connected medians. The same may be true for Lemma 2.

4 Axiomatic Characterization of the Median Function

In this section, we formulate the axioms for the characterization of median function in a block graph. We require three specialized axioms along with the universal axioms (A) and (C). The consistency axiom can be viewed as a means to determine the output of a profile by the output of subprofiles: if we can write $\pi = \pi_1\pi_2$ as the concatenation of the subprofiles π_1 and π_2, and it so happens that the outputs of these subprofiles intersect, then we know the output of π. There are two axioms for an unbalanced profile, where knowledge on the output of vertex deleted subprofiles of π will give the output of π.

Let $G = (V, E)$ be a block graph and let L be a consensus function on G. We formulate the following axioms for a consensus function L for profiles $\pi = (\pi_{uv}, \pi_{vu})$.

Axiom (BB): Betweenness of a Balanced profile
$\pi = (\pi_{uv}, \pi_{vu})$ is a balanced profile. If $u, v \notin \pi$, then $L(\pi) = I(x, y)$ such that x and y are vertices that separates π_{uv} and π_{vu} where x is a vertex closest to π_{uv} and y is a vertex closest to π_{vu}.
$\pi = (\pi_{uv}, \pi_{vu})$ is a balanced profile. If $u \in \pi$, then $L(\pi) = I(u, y)$ such that y is a cut-vertex that separates π_{uv} and π_{vu} closest to π_{vu}.
$\pi = (\pi_{uv}, \pi_{vu})$ is a balanced profile. If $v \in \pi$, then $L(\pi) = I(x, v)$ such that x is a cut-vertex that separates π_{uv} and π_{vu} closest to π_{uv}.
$\pi = (\pi_{uv}, \pi_{vu})$ is a balanced profile. If $u, v \in \pi$, then $L(\pi) = I(u, v)$.

Axiom (IU): Inconclusiveness of an unbalanced profile If π is an unbalanced profile, then

$$L(\pi) = \bigcup_{i=1}^{n} L(\pi - x_i), \text{ if } \bigcap_{i=1}^{n} L(\pi - x_i) = \emptyset$$

Axiom (CU): Conclusiveness of an unbalanced profile If π is an unbalanced profile, then

$$L(\pi) = \bigcap_{i=1}^{n} L(\pi - x_i), \text{ if } \bigcap_{i=1}^{n} L(\pi - x_i) \neq \emptyset$$

Remark 4. It is straightforward to observe that Axiom (BB) implies the betweenness axiom (B) as the profile $\pi = (x, y)$ is a balanced profile with respect to any edge uv on the shortest x, y-path in G. Another observation is that a complete or a partially complete profile π is an unbalanced profile which satisfies axiom (IU), if $|Pl(\pi)| = 1$ and satisfies axiom (CU), otherwise.

Now we are in position to utilize the previous Lemmata to prove the main result: that is, a characterization for the median function of a block graph.

Theorem 3. *Let L be a consensus function on a block graph G. Then L is precisely the median function Med of G if and only if L satisfies (A), (C), (BB), (IU) and (CU).*

Proof. If $L = Med$. The median function Med obviously satisfies (A) and (C). By Theorem 2, the median function Med satisfies axiom (BB). By Lemma 3, the function Med satisfies axiom (IU) and by Lemma 2, the function Med satisfies axiom (CU). Therefore the function L satisfies all the axioms (BB), (IU) and (CU). So we only have to prove the converse part.

Conversely, suppose that L is a consensus function on a block graph G satisfying the five axioms (A),(C),(BB), (IU) and (CU). We have to show that for every profile π, the consensus function $L(\pi)$ returns the median set $Med(\pi)$ and hence L is the median function Med. We use induction on the length of the profile to prove the converse. Let π be an arbitrary profile of length k, we split the profile into different cases.

Case 1: π is a balanced profile.
Let $\pi = (\pi_{uv}, \pi_{vu})$, where $|\pi_{uv}| = |\pi_{vu}| = 1$. Let $\pi = (x, y)$, $x \neq y$. Since L satisfies (BB) and hence axiom (B), we have $L(x, y) = I(x, y)$. Also we know that Med satisfies (B) and thus $Med(x, y) = I(x, y)$. Therefore $L(x, y) = Med(x, y)$. Thus the theorem is true for $|\pi_{uv}| = |\pi_{vu}| = 1$.

Now let $\pi = (\pi_{uv}, \pi_{vu})$ be an arbitrary profile so that $|\pi_{uv}| = |\pi_{vu}| \geq 2$. Let $\pi'_{uv} = (\pi_{uv} - x_i)$ and $\pi'_{vu} = (\pi_{vu} - x_j)$. Clearly, $|\pi'_{uv}| < |\pi_{uv}|$ and $|\pi'_{vu}| < |\pi_{vu}|$. Also $|\pi'_{vu}| = |\pi'_{vu}|$. Let $\pi' = (\pi'_{uv}, \pi'_{vu})$, then by the induction hypothesis $L(\pi') = Med(\pi') = I(x', y')$, where x' and y' are vertices defined by the following cases.

Subcase 1.1: If $u, v \notin \pi'$ and so $u, v \notin \pi$. Then by Remark 2, x' is a vertex separating π'_{uv} and π'_{vu} and closer to π'_{uv} and y' is a vertex separating π'_{uv} and π'_{vu} and closer to π'_{vu}.

Subcase 1.2: Either $u \in \pi'$ or $v \in \pi'$. If $u \in \pi'$, then by Remark 2, $Med(\pi') = I(x', y')$ such that $x' = u$ and y' is a vertex that separates π'_{uv} and π'_{vu} closest to π'_{vu}. If $v \in \pi'$, then $Med(\pi') = I(x', y')$ such that x' is a vertex that separates π'_{uv} and π'_{vu} closest to π'_{uv} and $y' = v$.

Subcase 1.3: $u, v \in \pi'$. Then by Remark 2, $Med(\pi') = I(x', y')$, where $x' = u$ and $y' = v$.
We have different possibilities for x' and y' and we use axiom (A) in several places for rearranging the profile vertices.
If $x_i = x$ and $x_j = y \Rightarrow x' = x'_i$ and $y' = x'_j \Rightarrow Med(x'_i, x'_j) = I(x'_i, x'_j)$.
If $x_i \neq x$ and $x_j = y \Rightarrow x' = x$ and $y' = x'_j \Rightarrow Med(x, x'_j) = I(x, x'_j)$.
If $x_i = x$ and $x_j \neq y \Rightarrow x' = x'_i$ and $y' = y \Rightarrow Med(x'_i, y) = I(x'_i, y)$.
If $x_i \neq x$ and $x_j \neq y \Rightarrow x' = x$ and $y' = y \Rightarrow Med(x, y) = I(x, y)$.
Clearly, in each of these cases $Med(\pi')$ contain the vertices in $I(x, y)$.
Let $\pi = (\pi', \rho')$ where $\pi' = (\pi'_{uv}, \pi'_{vu})$ and $\rho' = (x_i, x_j)$. By betweenness axiom (B), we have $L(x_i, x_j) = I(x'_i, x'_j)$. Thus by consistency (C) and induction

hypothesis, we get $L(\pi) = L(\pi'\rho') = L(\pi') \cap L(\rho') = Med(\pi') \cap L(x_i, x_j) \neq \phi$. since Med satisfies consistency, we have $Med(\pi', \rho') = Med(\pi)$. Thus,for a balanced profile $L(\pi) = Med(\pi)$.

Case 2: $\pi = (\pi_{uv}, \pi_{vu})$ with $|\pi_{uv}| > |\pi_{vu}|$ is an unbalanced profile. Let $\pi = (x_1, x_2, \ldots, x_n)$
By induction hypothesis, $L(\pi - x_i) = Med(\pi - x_i)$.
Suppose that $\bigcap_{i=1}^{n} Med(\pi - x_i) = \emptyset$, then by axiom (IU), we have $L(\pi) = \bigcup_{i=1}^{n} L(\pi - x_i)$ and by using Lemma 3, $Med(\pi) = \bigcup_{i=1}^{n} Med(\pi - x_i)$. Thus, $L(\pi) = Med(\pi)$. Now suppose that $\bigcap_{i=1}^{n} Med(\pi - x_i) \neq \emptyset$, then by axiom (CU), $L(\pi) = \bigcap_{i=1}^{n} L(\pi - x_i)$. By Lemma 2, $Med(\pi) = \bigcap_{i=1}^{n} Med(\pi - x_i)$. Hence $L(\pi) = Med(\pi)$, in this case also. Thus for all unbalanced profiles π, $L(\pi) = Med(\pi)$. Since we have proved that $L(\pi) = Med(\pi)$, for all profiles π, the proof is complete. □

4.1 Independence of the Axioms

To prove the independence of the axioms, we consider the following examples.

Example 2 (Consensus function $L \neq Med$, satisfies $(A), (BB)$ and (CU), but not (IU) and (C)). Consider a simple block graph G consisting of two triangles T_1 and T_2 joined by an edge xy, where $T_1 = (x_1, x_2, x)$ and $T_2 = (y_1, y_2, y)$. Let $\pi = (\pi_{uv}, \pi_{vu})$ (for some edge uv in G) be any profile such that $\pi_{uv} \subseteq V(T_1)$ and $\pi_{vu} \subseteq V(T_2)$. Define, $L(\pi) = \begin{cases} Med(\pi) & \text{if } \pi \text{is a balanced profile,} \\ \{y\} & \text{if } |\pi_{vu}| > |\pi_{uv}|, \\ \{x\} & \text{if } |\pi_{uv}| > |\pi_{vu}|. \end{cases}$
$L(\pi)$ doesn't satisfy axiom (IU), for consider the profile $\pi = (y_1, y_2, y)$. Here π is an unbalanced profile with $L(\pi) = \{y\}$, $L(\pi - y_1) = \{y, y_2\}$, $L(\pi - y_2) = \{y, y_1\}$, $L(\pi - y) = \{y_1, y_2\}$ and $\bigcap_i L(\pi - x_i) = \emptyset$, but $\bigcup_i L(\pi - x_i) = \{y_1, y_2, y\} \neq L(\pi)$. Similarly L doesn't satisfy (C), for consider the profiles $\pi = (x_1, y_1)$ and $\rho = (y, y_1)$. Here $L(\pi) \cap L(\rho) = \{x_1, x, y, y_1\} \cap \{y, y_1\} = \{y, y_1\} \neq L(\pi\rho) = \{y\}$. (BB) holds for L, as L coincides with Med, for balanced profiles. To prove that L satisfies (CU), consider an unbalanced profile $\pi = (x_1, x_2, \ldots, x_n)$. Suppose $|\pi_{uv}| > |\pi_{vu}|$, then n should be odd, and assume that $|\pi_{uv}| = |\pi_{vu} - \ell|$, $\ell \geq 1$. We prove the case when $\ell = 1$ as the case $\ell > 1$ is similar. Therefore, let $n = 2k + 1$ and w.l.o.g., we can consider the profile π as $\pi = (x_1, x_2, \ldots, x_k, x_{k+1}, y_1, \ldots, y_k)$, where $\{x_1, x_2, \ldots, x_k, x_{k+1}\} \subseteq V(T_1)$ and $\{y_1, \ldots, y_k\} \subseteq V(T_2)$. Let $\bigcap_{i=1}^{n} L(\pi - x_i) \neq \emptyset$. This implies that $\bigcap_{i=1}^{n} L(\pi - x_i) = \{x_i\}$. By definition of $L(\pi) = \{x\}$ and thus (CU) holds for L.

Example 3 (Consensus function $L \neq Med$, satisfies $(A), (C), (CU)$ and (IU), but not (BB)). Let G be a block graph with vertex set V. Define $L(\pi) = V$, for all profiles π. It is trivial that L satisfies $(A), (C), (IU)$. L satisfies (CU) trivially as $\bigcap_{i=1}^{k} Med(\pi - x_i) \neq \emptyset$, for all profiles π. It is clear that L does not satisfy (BB).

Example 4 (Consensus function $\mathbf{L \neq Med}$, satisfies $(A), (C), (BB)$ and (IU), but not (CU) and (C)). Consider the graph G consisting of a complete graph K_n together with a pendent edge xy attached to a vertex x in K_n. Clearly G is a block graph. Define L as follows.

$$L(\pi) = \begin{cases} \{\pi\} & \text{for } \pi \text{ with } \{\pi\} \subseteq V(K_n) \\ Med(\pi) & \text{if } y \in \pi. \end{cases}$$

L satisfies $(A), (BB)$ and (IU). (BB) holds for L, since balanced profiles π with $\{\pi\} \subseteq V(K_n)$ are of the form (x^ℓ, y^ℓ), for some $\ell \geq 1$ and $L(\pi) = Med(\pi)$, for profiles π containing vertex y. In both cases L satisfies (BB). (IU) holds as the only unbalanced profiles $\pi = (x_1, \ldots, x_k)$ with $\bigcap_{i=1}^{k} L(\pi - x_i) = \emptyset$ with $\{\pi\} \subseteq V(K_n)$, are those profiles with each $x_i \in \pi$ occurs only once. Then $L(\pi) = \{x_1, \ldots, x_k\} = \bigcup_{i=1}^{k} L(\pi - x_i)$. For unbalanced profiles π containing vertex y, $L(\pi) = Med(\pi)$ and so we can prove that L satisfies (IU), in this case. Clearly L doesn't satisfy (CU), as for the unbalanced profile $\pi = (x_1, x_1, x_2)$, $\bigcap_{i=1}^{k} L(\pi - x_i) = \{x_1\} \neq L(\pi) = \{\pi\}$. For profiles $\pi = (x_1, x_1, x_2)$ and $\rho = (x_1)$, $L(\pi) \cap L(\rho) = \{x_1\} \neq L(\pi\rho) = \{x_1, x_2\}$ and so L doesn't satisfy (C) also.

We cannot construct examples of consensus functions L on a block graph that satisfy the axioms $(A), (BB)$ (CU) and (IU), but not (C); that satisfy $(A), (C), (BB), (CU)$, but not (IU); that satisfy $(A), (C), (BB), (IU)$, but not (CU); and $(C), (BB), (CU), (IU)$, but not (A). It is already established that for any consensus function, it is a hard problem to establish the independence of the anonymity axiom (A), see [5]. From the construction of the examples for independence of the axioms, we infer that for an arbitrary consensus function on a block graph, the axioms $(C), (CU)$ and (IU) may be related, and it may be an interesting problem to check whether there exists such a relation.

References

1. Arrow, K.J.: Social Choice and Individual Values, vol. 12. Yale University Press, London (2012)
2. Balakrishnan, K., Changat, M., Mulder, H.M.: The plurality strategy on graphs. Australas. J. Comb. **46**, 191–202 (2010)
3. Bandelt, H.J., Chepoi, V.: Graphs with connected medians. SIAM J. Discrete Math. **15**(2), 268–282 (2002)
4. Bandelt, H.J., Mulder, H.M., Wilkeit, E.: Quasi-median graphs and algebras. J. Graph Theory **18**(7), 681–703 (1994)

5. Changat, M., Lekha, D.S., Mohandas, S., Mulder, H.M., Subhamathi, A.R.: Axiomatic characterization of the median and antimedian function on a complete graph minus a matching. Discrete Appl. Math. **228**, 50–59 (2017)
6. Hagauer, J.: Skeletons, recognition algorithm and distance matrix of quasi-median graphs. Int. J. Comput. Math. **55**(3–4), 155–171 (1995)
7. Holzman, R.: An axiomatic approach to location on networks. Math. Oper. Res. **15**(3), 553–563 (1990)
8. Kemeny, J.G., Snell, L.: Preference ranking: an axiomatic approach. Math. Models Soc. Sci. 9–23 (1962)
9. Klavzar, S., Mulder, H.M.: Median graphs: characterizations, location theory and related structures. J. Comb. Math. Comb. Comput. **30**, 103–128 (1999)
10. McMorris, F., Neumann, D.: Consensus functions defined on trees. Math. Soc. Sci. **4**(2), 131–136 (1983)
11. McMorris, F.R., Mulder, H.M., Ortega, O.: Axiomatic characterization of the mean function on trees. Discrete Math. Algorithms Appl. **2**(03), 313–329 (2010)
12. McMorris, F.R., Mulder, H.M., Ortega, O.: The ℓ_p-function on trees. Networks **60**(2), 94–102 (2012)
13. McMorris, F.R., Mulder, H.M., Roberts, F.S.: The median procedure on median graphs. Discrete Appl. Math. **84**(1–3), 165–181 (1998)
14. McMorris, F.R., Mulder, H.M., Vohra, R.V.: Axiomatic characterization of location functions. In: Advances in Interdisciplinary Applied Discrete Mathematics, pp. 71–91. World Scientific (2011)
15. McMorris, F.R., Roberts, F.S., Wang, C.: The center function on trees. Netw.: Int. J. **38**(2), 84–87 (2001)
16. Mulder, H.M.: The majority strategy on graphs. Discrete Appl. Math. **80**(1), 97–105 (1997)
17. Mulder, H.M.: Median graphs.: a structure theory. In: Advances in Interdisciplinary Applied Discrete Mathematics, pp. 93–125. World Scientific (2011)
18. Mulder, H.M., Novick, B.: An axiomatization of the median procedure on the n-cube. Discrete Appl. Math. **159**(9), 939–944 (2011)
19. Mulder, H.M., Novick, B.: A tight axiomatization of the median procedure on median graphs. Discrete Appl. Math. **161**(6), 838–846 (2013)
20. Mulder, H.M., Pelsmajer, M.J., Reid, K.: Axiomization of the center function on trees. Australas. J. Comb. **41**, 223–226 (2008)
21. Mulder, H.: The interval function of a graph, math. Centre Tracts **132** (1980)
22. Vohra, R.: An axiomatic characterization of some locations in trees. Eur. J. Oper. Res. **90**(1), 78–84 (1996)

On Coupon Coloring of Cartesian Product of Some Graphs

P. Francis(✉) and Deepak Rajendraprasad

Department of Computer Science, Indian Institute of Technology, Palakkad 678557, India
{pfrancis,deepak}@iitpkd.ac.in

Abstract. Let G be a graph with no isolated vertices. A *k-coupon coloring* of G is an assignment of k colors to the vertices of G such that every vertex contains vertices of all k colors in its neighborhood. The *coupon chromatic number* of G, denoted $\chi_c(G)$, is the maximum k for which a k-coupon coloring exists. In this paper, we present an upper bound for the coupon chromatic number of Cartesian product of graphs G and H in terms of $|V(G)|$ and $|V(H)|$. Further, we prove that if G and H are bipartite graphs then $G \square H$ has a coupon coloring with $2\min\{\chi_c(G), \chi_c(H)\}$ colors. As consequences, for any positive integer n, we obtain the coupon chromatic number and total domination number of n-dimensional torus $\overset{n}{\underset{i=1}{\square}} C_{k_i}$ with some suitable conditions to each k_i, which turns out to be a generalization of the result due to S. Gravier [*Total domination number of grid graphs*, Discrete Appl. Math. 121 (2002) 119–128]. Finally, for any $r \geq 0, d \geq 2$, we obtain the coupon coloring for the Hamming graph $K_d \square K_d \square \cdots \square K_d$ (2^r times).

Keywords: Product graphs · Coupon chromatic number · Total domination number · Torus

2000 AMS Subject Classification: 05C15, 05C63

1 Introduction

All graphs considered in this paper are simple and undirected without an isolated vertex. Let C_n and K_n respectively denote the cycle and the complete graph on n vertices. Let $\delta(G)$ denote the minimum degree of the graph G. For any two vertices $x, y \in V(G)$, let $d(x, y)$ be the length of a shortest path between x and y. The neighborhood $N(x)$ of a vertex x is $\{u \colon ux \in E(G)\}$ and for $S \subseteq V(G)$, we denote the neighborhood S by $N(S)$ is $\cup_{v \in S} N(v)$. Let $\mathbb{Z}_n$ be the set of all integers modulo n.

The *Cartesian product* of two graphs G and H, denoted $G \square H$, is a graph whose vertex set is $V(G) \times V(H) = \{(x, y) \colon x \in V(G) \text{ and} y \in V(H)\}$ and two vertices (x_1, y_1) and (x_2, y_2) of $G \square H$ are adjacent if and only if either $x_1 = x_2$

A. Mudgal and C. R. Subramanian (Eds.): CALDAM 2021, LNCS 12601, pp. 309–316, 2021.
https://doi.org/10.1007/978-3-030-67899-9_25

and $y_1y_2 \in E(H)$ or $y_1 = y_2$ and $x_1x_2 \in E(G)$. For each vertex $u \in V(G)$, $\langle \{u\} \times V(H) \rangle$ is isomorphic to H and it is denoted by H_u and for each vertex $v \in V(H)$, $\langle V(G) \times \{v\} \rangle$ is isomorphic to G and it is denoted by G_v. For $d \geq 2$, let $\mathop{\Box}\limits_{i=1}^{d} G_i$ denotes $G_1 \Box G_2 \Box \cdots \Box G_d$. For $r_i \geq 2$, we call $\mathop{\Box}\limits_{i=1}^{d} C_{r_i}$ a d-dimensional *torus graph.*

The *Hamming Graph*, denoted $H_{n,q}$, has the vertex set $(x_1, x_2, \ldots, x_n)$, for $1 \leq i \leq n$, $x_i \in \{0, 1, \ldots, q-1\}$ and two q-ary n-tuples are adjacent if and only if they differ in exactly one coordinate. The special case $H_{n,2}$ is a hypercube of dimension n, denoted as Q_n.

Let $D \subset V(G)$, if $N(D) \supseteq V(G) \backslash D$ then D is a dominating set for $V(G)$ and if $N(D) = V(G)$ then D is a total dominating set for $V(G)$. The *domination* (*total domination*) number of a graph G is the smallest cardinality among all the dominating (total dominating) sets and is denoted $\gamma(G)$ ($\gamma_t(G)$). The *domatic* (*total domatic*) number of G is the maximum number of classes of a partition of $V(G)$ such that each class is a dominating (total dominating) set and is denoted $d(G)$ ($d_t(G)$). The concept of total domatic number was introduced by Cockayne et al. in [5] and investigated further in [2,14,15]. There has been a lot of papers on total domination in graphs. See for instance, [3,7,11,13] and a survey of selected topics by Henning in [10].

A new vertex coloring, namely coupon coloring, was introduced by Chen et al. in [4] which relates the coloring and domination parameters. The concepts of coupon chromatic number and total domatic number are the same. Note that in any coupon coloring of G, each color class must be a total dominating set of G. The motivation for the study of coupon coloring and its applications were mentioned by Chen et al. in [4]. This concept was further studied in [1,6,12].

It is clear from the definition of coupon coloring, $1 \leq \chi_c(G) \leq \delta(G)$, for any graph G without an isolated vertex. In this paper, we characterize some families of graph which attain the upper bound. In this direction, first we present an upper bound for the coupon chromatic number of $G \Box H$ in terms of $|V(G)|$ and $|V(H)|$. Further, we prove that if G and H are bipartite graphs then $G \Box H$ has a coupon coloring with $2\min\{\chi_c(G), \chi_c(H)\}$ colors. As consequences, when n is a power of 2, we obtain the coupon chromatic number of n-dimensional hypercube Q_n is n and torus $\mathop{\Box}\limits_{i=1}^{n} C_{r_i}$ is $2n$ such that each $r_i \equiv 0 (\mathrm{mod}\ 4)$. Also, for any positive integer d, we obtain the coupon chromatic number of d-dimensional torus $\mathop{\Box}\limits_{i=1}^{d} C_{k_i}$ is $2d$, with some suitable conditions to each k_i's as according to d is odd or even. In addition, we obtain the total domination number of d-dimensional torus as mentioned above to be $(\prod\limits_{i=1}^{d} k_i)/2d$, which turns out to be a generalization of the result due to S. Gravier in [7]. Finally, for any $r \geq 0, k \geq 1$, we obtain the coupon coloring for the Hamming graph $H_{2^r,k}$.

2 Coupon Coloring of Cartesian Product of Graphs

Let us start this section by considering some observations on the existence of an r-coupon coloring of an r-regular graphs.

Proposition 1. *Let G be an r-regular graph, $r \geq 1$, and there exists an r-coupon coloring for G, say $f\colon V(G) \to [r] = \{1, 2, \ldots, r\}$. Then*
(i) Each color class of f has the same size $\frac{|V(G)|}{r}$ [15].
(ii) r divides $|V(G)|$ and r^2 divides $|E(G)|$.
(iii) Each color class of f has an even number of vertices and G contains a perfect matching.
(iv) Each color class of f is a minimal total dominating set.

Now, let us consider the relation between the domatic number of a graph G and coupon chromatic number of its Cartesian product with some graph H.

Proposition 2. *Let G and H be any two graphs without an isolated vertex, if G has a domatic number r, then $\chi_c(G \square H) \geq r$.*

Next, we present an upper bound for the coupon chromatic number of Cartesian product of graphs.

Theorem 1. *For any two graphs G and H with no isolated vertices, we have $2 \leq \chi_c(G \square H) \leq \max\{|V(G)|, |V(H)|\}$.*

Proof. Let G and H be any graphs without an isolated vertex of order m and n respectively. Without loss of generality, let $n \geq m$. Let us consider the coloring of $G \square H$ as filling the cells of $m \times n$ grid with colors. For a cell (i, j), $1 \leq i \leq m$, $1 \leq j \leq n$, call the set of cells in the ith row and j^{th} column except the cell itself as a cross-hair at (i, j). There are mn cross-hairs, one corresponding to each cell of the grid. Each cross-hair has $m + n - 2$ cells. Suppose the coloring is a k-coupon coloring, then each cross-hair contains every other colors at least once.

Claim. In any coupon coloring of $G \square H$, each color should appear in at least m cells.

Suppose a color c_1 appears less than m times, then there exists a row i in the grid which does not contain c_1. Similarly, there is a column j which does not contain c_1 in the grid. In this case, the cross-hair at (i, j) does not contain c_1, and hence the coloring is not a coupon coloring. Thus the claim holds. Since each color should appears in at least m cells and there are mn cells in the grid, the maximum possible value of k in any coupon k-coloring is n. Thus $\chi_c(G \square H) \leq n = \max\{|V(G)|, |V(H)|\}$.

It is easy to get the lower bound for $\chi_c(G \square H)$.

The lower bound given in Theorem 1 is tight for graphs with minimum degree 1 and the tightness of an upper bound which is follow from Corollary 1.

Corollary 1. *Let m, n be any two integers greater than* 1 *such that $m \leq n$ and let G be any graph of order m with no isolated vertices, we have $\chi_c(G \square K_n) = n$. In particular, $\chi_c(K_m \square K_n) = n$.*

An upper bound given in Theorem 1 is not tight for cycles of length larger than the size of the complete graphs.

Proposition 3. *Let m, n be any two integers at least* 3*, we have $\chi_c(C_m \square K_n) = n$.*

By Proposition 2 and the domatic number of C_{3k} is 3, we obtain Corollary 2.

Corollary 2. *For any positive integer k, if G be any graph with minimum degree* 1*, then $\chi_c(G \square C_{3k}) = 3$.*

3 Coupon Coloring of Bipartite Graphs

In this section, we examine the families of bipartite graphs G and H such that the coupon chromatic number of $G \square H$ attains its maximum possible value, namely, $\delta(G \square H)$. Let us start considering a lower bound for the coupon chromatic number of Cartesian product of any two bipartite graphs in terms of its coupon chromatic numbers. The bound given in Theorem 2 has been applied multiple times in this paper.

Theorem 2. *If G and H are bipartite graphs without isolated vertex, then $\chi_c(G \square H) \geq 2 \min\{\chi_c(G), \chi_c(H)\}$.*

Proof. Let G and H be the bipartite graphs of order m and n respectively, and let $[X, Y]$ and $[X', Y']$ be their bipartition. Let $u_0, u_1, \ldots, u_{m-1}$ and $v_0, v_1, \ldots, v_{n-1}$ be the vertices of G and H. Note that, any graph G' is l-coupon colorable for all l such that $1 \leq l \leq \chi_c(G')$. Let g and h be the k-coupon coloring of G and H, where $k = \min\{\chi_c(G), \chi_c(H)\}$ and let the k colors be $\{0, \ldots, k-1\}$. Now consider the coloring f for the graph $G \square H$. For $0 \leq i \leq m-1$ and $0 \leq j \leq n-1$,

$$f((u_i, v_j)) = \begin{cases} (2g(u_i) + 2h(v_j)) (\text{mod } 2k) & \text{if } u_i \in X \\ (2g(u_i) + 2h(v_j) + 1) (\text{mod } 2k) & \text{if } u_i \in Y. \end{cases}$$

By the coloring f, for any vertex $u_i \in X$ and $v_j \in V(H)$, the vertex (u_i, v_j) receives the colors from the neighbors of v_j in H_{u_i} are $\{[2g(u_i) + 2s](\text{mod } 2k) : s \in \{0, 1, \ldots, k-1\}\} = \{0, 2, \ldots, 2k-2\}$ and (u_i, v_j) receives the colors from the neighbors of u_i in G_{v_j} are $\{[2t + 2h(v_j) + 1](\text{mod } 2k) : t \in \{0, 1, \ldots, k-1\}\} = \{1, 3, \ldots, 2k-1\}$. Similarly, for any vertex $u_i \in Y$ and $v_j \in V(H)$, the vertex (u_i, v_j) receives the colors $\{1, 3, \ldots, 2k-1\}$ from the neighbors of v_j in H_{u_i} and the colors $\{0, 2, \ldots, 2k-2\}$ from the neighbors of u_i in G_{v_j}. Thus, each vertex (u_i, v_j) in $G \square H$ receives all the colors $\{0, 1, \ldots, 2k-1\}$ in its neighbors and f is a coupon coloring using $2k$ colors. Hence $\chi_c(G \square H) \geq 2 \min\{\chi_c(G), \chi_c(H)\}$. If at least one of the graphs G, H is disconnected, then apply the same technique to each component of $G \square H$ separately.

The bound given in Theorem 2 is tight, which follows by taking $G \cong K_2$ and $H \cong C_{4n}$. Also, these graphs G and H violates the bound $\chi_c(G\Box H) \geq 2\max\{\chi_c(G), \chi_c(H)\}$. One of the simplest examples such that the strict inequality holds in Theorem 2 is $G \cong H \cong C_6$, by Proposition 2, we have $\chi_c(C_6\Box C_6) \geq 3$ but $\chi_c(C_6) = 1$. Theorem 2 is not true for all graphs in general. For $G \cong H \cong K_2\Box K_3$, we have $\chi_c(G) = 3$ and $\chi_c(G\Box H) < 6 = 2\min\{\chi_c(G), \chi_c(H)\}$.

As a consequences of Theorem 2 and (iii) of Proposition 1, we obtain the following Corollaries. Also, the result proved for 2^r-dimensional hypercube in [4,9] follows from the fact that $\mathop{\Box}\limits_{i=1}^{2^r} K_2 \cong Q_{2^r}$.

Corollary 3. *For any two positive integers d, r, if G_i is a bipartite graph such that $\chi_c(G_i) = \delta(G_i) = d$, $1 \leq i \leq 2^r$, then $\chi_c(\mathop{\Box}\limits_{i=1}^{2^r} G_i) = 2^r d$.*

Corollary 4. *Let r be any positive integer and G_i be any graph with no isolated vertices, for $1 \leq i \leq 2^r$, we have $\chi_c(\mathop{\Box}\limits_{i=1}^{2^r} G_i) \geq 2^r$. If each G_i contains a leaf, then $\chi_c(\mathop{\Box}\limits_{i=1}^{2^r} G_i) = 2^r$. In particular, if T_i is a tree, then $\chi_c(\mathop{\Box}\limits_{i=1}^{2^r} T_i) = 2^r$ and $\chi_c(Q_{2^r}) = 2^r$.*

Corollary 5. *If n is a positive integer not a power of 2, then $\chi_c(Q_n) < n$. Also, for any positive integer r, we have $\chi_c(Q_{2^r+1}) = 2^r$.*

Corollary 6. *Let $r, k_1, \ldots, k_{2^r}$ be any positive integers such that $k_i \equiv 0 (\text{mod } 4)$, $1 \leq i \leq 2^r$, we have $\chi_c(\mathop{\Box}\limits_{i=1}^{2^r} C_{k_i}) = 2^{r+1}$.*

From Corollary 6, we find the coupon chromatic number of torus graph of dimension 2^r for every $r \geq 1$ such that the length of the each cycle is congruent to $0(\text{mod } 4)$. In the remaining part of this section, we try to generalize this result to a larger collection of tori. Suppose we consider d-dimensional torus graph for any $d \geq 2$. In [7], Gravier independently obtained the total domination number for some torus by using the concept of tiling.

Theorem 3. *[7] Let $d, k_1, k_2, \ldots, k_d$ be any positive integers such that $d \geq 2$ and $k_i \equiv 0(\text{mod } 4d)$, for $1 \leq i \leq d$, we have $\gamma_t(\mathop{\Box}\limits_{i=1}^{d} C_{k_i}) = (\prod\limits_{i=1}^{d} k_i)/2d$. Moreover, if d is even and for any positive integer k_i, $1 \leq i \leq d$ such that $k_i \equiv 0(\text{mod } 2d)$, then this equality still holds.*

We use the same proof technique as mentioned by Gravier et al., in [8] and we obtain coupon chromatic number and total domination number for some torus. Let us first consider the odd dimensional torus.

Theorem 4. *Let $d, k_1, k_2, \ldots, k_d$ be any positive integers. If d is odd, at least two k_i's are congruent to $0(\text{mod } 4d)$ and remaining k_i's are congruent to $0(\text{mod } 4)$, then $\chi_c(\mathop{\Box}\limits_{i=1}^{d} C_{k_i}) = 2d$.*

Proof. Let $d, k_1, k_2, \ldots, k_d$ be any positive integers and let $G \cong \mathop{\square}\limits_{i=1}^{d} C_{k_i}$. Since $\delta(G) = 2d$, it is enough to give $2d$ coupon coloring for G. For $1 \leq i \leq d$ and d is odd, by assumption there are at least two of k_i's are congruent to $0 (\text{mod } 4d)$ and the remaining k_i's are congruent to $0 (\text{mod } 4)$. The graph $\mathop{\square}\limits_{i=1}^{d} C_{k_i}$ is transitive, without loss of generality, let us take the two of k_i's which are congruent to $0 (\text{mod } 4d)$ be k_{d-1} and k_d. Let us consider the vertices of G be $\{(x_1, x_2, \ldots, x_d) : x_i \in \{0, 1, \ldots, k_i - 1\}, 1 \leq i \leq d\}$. Now, let us consider the subset $\mathcal{F}$ of $V(G)$ formed by the vertices of the type

$$x = \left(x_1, \ldots, x_{d-2}, (2a + \epsilon)(\text{mod } k_{d-1}), (2a + \epsilon + 2db + \sum_{i=1}^{d-2} (2i + 1)x_i)(\text{mod } k_d) \right)$$

where $x_i \in \{0, 1, \ldots, k_i - 1\}$ for $1 \leq i \leq d - 2$, $a \in \mathbb{Z}_{\frac{k_{d-1}}{2}}$ $b \in \mathbb{Z}_{\frac{k_d}{2d}}$ and

$$\epsilon = \begin{cases} 0 \text{ if } \sum_{i=1}^{d-2} x_i \equiv 0 \text{ or } 1 \ (\text{mod } 4) \\ 1 \text{ if } \sum_{i=1}^{d-2} x_i \equiv 2 \text{ or } 3 \ (\text{mod } 4). \end{cases}$$

Let us consider a bicoloring (black and white) of $V(G)$. A vertex is black if and only if $\sum_{i=1}^{d} x_i$ is even. Let $\mathcal{B}$ and $\mathcal{W}$ be the set of black and white vertices of $V(G)$ respectively. Since the sum of the co-ordinates of x equal to $2(2a + \epsilon + db + \sum_{i=1}^{d-2} (i + 1)x_i)$, every vertex $x \in \mathcal{F}$ is black. In G, black and white vertices are equal in number which equals to $(\prod_{i=1}^{d} k_i)/2$.

Claim. $N(\mathcal{F}) = \mathcal{W}$
Let $\mathcal{T}$ be the d-dimensional torus which is isomorphic to $C_4 \square \cdots \square C_4 \square C_{4d} \square C_{4d}$. Note that the set $\mathcal{F}$ is periodic on the torus $\mathcal{T}$. Since $N(\mathcal{F}) \subseteq \mathcal{W}$, we can prove the claim by showing that $|N(\mathcal{F})| = |\mathcal{W}| = (\prod_{i=1}^{d} k_i)/2$. It is easy to check that $|\mathcal{F} \cap V(\mathcal{T})| = (4^{d-2})(\frac{4d}{2})(\frac{4d}{2d})$ and $|\mathcal{F}| = ((4^{d-2})(\frac{4d}{2})(\frac{4d}{2d})) \left((\prod_{i=1}^{d-2} \frac{k_i}{4})(\frac{k_{d-1}}{4d})(\frac{k_d}{4d}) \right) = \left((\prod_{i=1}^{d-2} k_i)(\frac{k_{d-1}}{2})(\frac{k_d}{2d}) \right) = \left(\prod_{i=1}^{d} k_i \right) / 4d = \frac{|\mathcal{W}|}{2d}$. Also, for any vertex $u \in V(G)$, we have $|N(u)| = 2d$. For any two distinct vertices $x, y \in \mathcal{F}$, it is enough to show that $N(x) \cap N(y) = \emptyset$, that is $d(x, y) > 2$. Let $x = (x_1, x_2, \ldots, x_d)$ and $y = (y_1, y_2, \ldots, y_d)$ be any two distinct vertices in $\mathcal{F}$, and let r be the distance between $(x_1, x_2, \ldots, x_{d-2})$ and $(y_1, y_2, \ldots, y_{d-2})$, that is $r = \sum_{i=1}^{d-2} |x_i - y_i|$. If $r > 2$ then the claim holds, so let us assume that $r \leq 2$.

The proof of claim is similar to the proof given by Gravier et al., in [8] and we have $N(\mathcal{F}) = \mathcal{W}$. Now, let us consider the set

$$\mathcal{F}' = \{x + (0, \ldots, 0, 1, 0) \colon \text{ for all } x \in \mathcal{F}\}.$$

Clearly, every vertex $y \in \mathcal{F}'$ is white. Similarly, we can prove that $N(\mathcal{F}') = \mathcal{B}$. Also, $N(\mathcal{F} \cup \mathcal{F}') = \mathcal{W} \cup \mathcal{B} = V(G)$ and thus $\mathcal{F} \cup \mathcal{F}'$ is a total dominating set for $V(G)$.

For $0 \leq i \leq 2d-1$, let $\mathcal{P}_i = \{y + (0, \ldots, 0, 0, i)\colon \text{ for all } y \in (\mathcal{F} \cup \mathcal{F}')\}$. For $i \neq j$, $\mathcal{P}_i$ and $\mathcal{P}_j$ are disjoint and each $\mathcal{P}_i$ is a total dominating set which partition $V(G)$. Let g be the coloring such that $g(x) = i$, for all $x \in \mathcal{P}_i$. Clearly, g is a coupon coloring for G and thus $\chi_c(\mathop{\square}\limits_{i=1}^{d} C_{k_i}) = 2d$.

Next, let us obtain the coupon chromatic number for an even dimensional torus by the proof techniques similar to the previous theorem.

Theorem 5. *Let $d, k_1, k_2, \ldots, k_d$ be any positive integers. If d is even, at least two k_i's are congruent to $0 (\text{mod } 2d)$ and remaining k_i's are congruent to $0 (\text{mod } 4)$, then $\chi_c(\mathop{\square}\limits_{i=1}^{d} C_{k_i}) = 2d$.*

As a consequence of Theorem 5, we obtain the coupon chromatic number of generalized even dimensional torus in Corollary 7.

Corollary 7. *Let $d, d_1, k, p, q, k_1, k_2, \ldots, k_d$ be any positive integers such that $d = 2^p k$, $d_1 = 2^q k$ and $2 \nmid k$, $1 \leq q \leq p$. If at least 2^{p-q+1} number of k_i's are congruent to $0 (\text{mod } 2d_1)$ and remaining k_i's are congruent to $0 (\text{mod } 4)$, then $\chi_c(\mathop{\square}\limits_{i=1}^{d} C_{k_i}) = 2d$.*

Corollary 8 is an immediate consequence of Theorem 4, Corollary 7 and (iv) of Proposition 1 which turns to be a generalization of the result due to S. Gravier in [7].

Corollary 8. *Let $d, d_1, k, p, q, k_1, k_2, \ldots, k_d$ be any positive integers. If d is odd and at least two k_i's are congruent to $0 (\text{mod } 4d)$ and remaining k_i's are congruent to $0 (\text{mod } 4)$, then $\gamma_t(\mathop{\square}\limits_{i=1}^{d} C_{k_i}) = (\prod_{i=1}^{d} k_i)/2d$. Let $d = 2^p k$, $d_1 = 2^q k$ and $2 \nmid k$, for $1 \leq q \leq p$. If at least 2^{p-q+1} number of k_i's are congruent to $0 (\text{mod } 2d_1)$ and remaining k_i's are congruent to $0 (\text{mod } 4)$, then $\gamma_t(\mathop{\square}\limits_{i=1}^{d} C_{k_i}) = (\prod_{i=1}^{d} k_i)/2d$.*

Note that, the family of torus given in Corollary 8 contains the family of a torus as mentioned in Theorem 3. For example, the integer 24 can be expressed as $2^0 \times 24, 2^1 \times 12, 2^2 \times 6$ and $2^3 \times 3$. Corollary 8 finds γ_t for the graphs $\mathop{\square}\limits_{i=1}^{24} C_{p_i}$ where at least two $p_i \equiv 0 (\text{mod } 48)$, $\mathop{\square}\limits_{i=1}^{24} C_{q_i}$ where at least four $q_i \equiv 0 (\text{mod } 24)$ and $\mathop{\square}\limits_{i=1}^{24} C_{r_i}$ where at least eight $r_i \equiv 0 (\text{mod } 12)$ and the remaining p_i, q_i, r_i are congruent to $0 (\text{mod } 4)$ but Theorem 3 only finds γ_t for the graph $\mathop{\square}\limits_{i=1}^{24} C_{s_i}$, where each $s_i \equiv 0 (\text{mod } 48)$.

Finally, we obtain the coupon coloring for the Hamming graph $H_{2^k, r}$.

Theorem 6. *For all $k \geq 0$ and $r \geq 1$, we have $\chi_c(H_{2^k, r}) \geq 2^k \lfloor \frac{r}{2} \rfloor$.*

The proof is obtained by repeated application of Theorem 2 to the complete bipartite subgraph $K_{\lceil \frac{r}{2} \rceil, \lfloor \frac{r}{2} \rfloor}$ of K_r.

Note that an equality holds for the graph $H_{1,k}$, $H_{2,2k}$, and the problem remains open for $H_{2^r,k}$, $r \geq 2$ except $H_{2^r,2}$.

Acknowledgment. For the first author, this research was supported by Post Doctoral Fellowship, Indian Institute of Technology, Palakkad.

References

1. Akbari, S., Motiei, M., Mozaffari, S., Yazdanbod, S.: Cubics graphs with total domatic number at least two. Discussiones Math. Graph Theory **38**, 75–82 (2018)
2. Aram, H., Sheikholeslami, S., Volkmann, L.: On the total domatic number of regular graphs. Trans. Comb. **1**(1), 45–51 (2012)
3. Brešar, B., Hartinger, T.R., Kos, T., Milanič, M.: On total domination in the Cartesian products of graphs. Discussiones Math. Graph Theory **38**, 963–976 (2018)
4. Chen, B., Kim, J.H., Tait, M., Verstraete, J.: On coupon coloring of graphs. Discrete Appl. Math **193**, 94–101 (2015)
5. Cockayne, E.J., Dawes, R.M., Hedetniemi, S.T.: Total domination in graphs. Networks **10**, 211–219 (1980)
6. Goddard, W., Henning, M.A.: Thoroughly dispersed colorings. J. Graph Theory **88**, 174–191 (2018)
7. Gravier, S.: Total domination number of grid graphs. Discrete Appl. Math. **121**, 119–128 (2002)
8. Gravier, S., Mollard, M., Payan, C.: Variations on tilings in the Manhattan metric. Geom. Dedicata. **76**, 265–273 (1999)
9. Hahn, G., Kratochvíl, J., Širáň, J., Sotteau, D.: On the injective chromatic number of graphs. Discrete Math. **256**, 179–192 (2002)
10. Henning, M.A.: A survey of selected recent results on total domination in graphs. Discrete Math. **309**, 32–63 (2009)
11. Henning, M.A., Rall, D.F.: On the total domination number of Cartesian products of graph. Graphs Comb. **21**, 63–69 (2005)
12. Nagy, Z.L.: Coupon-coloring and total domination in Hamiltonian planar triangulations. Graphs Comb. **34**, 1385–1394 (2018)
13. Klavžar, S., Seifter, N.: Dominating Cartesian products of cycles. Discrete Appl. Math. **59**, 129–136 (1995)
14. Zelinka, B.: Total domatic number and degrees of vertices of a graph. Math. Slovaca **39**, 7–11 (1989)
15. Zelinka, B.: Regular totally domatically full graphs. Discrete Math. **86**, 71–79 (1990)

On the Connectivity and the Diameter of Betweenness-Uniform Graphs

David Hartman[1,2], Aneta Pokorná[1(✉)], and Pavel Valtr[3]

[1] Computer Science Institute of Charles University, Faculty of Mathematics and Physics, Charles University, 118 00 Prague, Czech Republic
pokorna@iuuk.mff.cuni.cz

[2] The Institute of Computer Science of the Czech Academy of Sciences, Prague, Czech Republic

[3] Department of Applied Mathematics, Faculty of Mathematics and Physics, Charles University, 118 00 Prague, Czech Republic
https://iuuk.mff.cuni.cz
https://kam.mff.cuni.cz

Abstract. Betweenness centrality is a centrality measure based on the overall amount of shortest paths passing through a given vertex. A graph is *betweenness-uniform* if all its vertices have the same betweenness centrality. We study the properties of betweenness-uniform graphs. In particular, we show that every connected betweenness-uniform graph is either a cycle or a 3-connected graph. Also, we show that betweenness uniform graphs of high maximal degree have small diameter.

Keywords: Betweenness centrality · Betweenness-uniform · Connectivity · Distance

1 Introduction and Definitions

There are many complex networks that play a key role in our society. Well-known examples include the Internet, systems of roads or railroads, electricity networks or social networks. In such networks, it is often the case that information, people or goods travel between different parts of the network, usually using shortest paths between points. From this point of view, points with high throughput are the most important, valuable and often also the most vulnerable parts of the network. Evaluating importance of nodes via their ability to provide information transfer might help in various application areas such as the human brain [10] or in construction of utilized algorithms such as community detection algorithms [8].

A network can be viewed as a graph G with vertex set $V(G)$ of size n and edge set $E(G)$ that has maximal degree $\Delta(G)$ and minimal degree $\delta(G)$. A graph G has vertex connectivity $\kappa(G)$, if $|V(G)| > \kappa(G)$ and remains connected after the removal of less than $\kappa(G)$ vertices. We say that a graph is k-connected, if it has vertex connectivity k. For a fixed vertex x, $N(x)$ stands for the set of all vertices adjacent to x. For two vertices x, y, the length of the shortest xy-path is

A. Mudgal and C. R. Subramanian (Eds.): CALDAM 2021, LNCS 12601, pp. 317–330, 2021.
https://doi.org/10.1007/978-3-030-67899-9_26

their distance $d(x, y)$. Diameter $d(G)$ of a graph G is then $\max_{x,y \in V(G)} d(x, y)$. We denote the set $\{1, \ldots, k\}$ by $[k]$.

A network centrality measure is a tool helping us to assess how important are nodes in the network. For a connected graph, the *betweenness centrality* is the following centrality measure evaluating the importance of a vertex x based on the amount of shortest paths going through it:

$$B(x) := \sum_{\{u,v\} \in \binom{V(G) \setminus \{x\}}{2}} \frac{\sigma_{u,v}(x)}{\sigma_{u,v}},$$

where $\sigma_{u,v}$ denotes the number of shortest paths between u and v and $\sigma_{u,v}(x)$ denotes the number of shortest paths between u and v passing through x [6].

Note that we count over each (unordered) pair $\{u, v\}$ only once. It would be possible to count each pair both as uv and as vu. In such case we would obtain the betweenness value that is two times larger than in the unordered version. Similarly, we can define the betweenness centrality for an edge e in a connected graph:

$$B(e) := \sum_{\{u,v\} \in \binom{V(G)}{2}} \frac{\sigma_{u,v}(e)}{\sigma_{u,v}}$$

where $\sigma_{u,v}(e)$ is the number of shortest paths between u and v passing through edge e. Note that $B(e) \geq 1$ for every $e \in E(G)$, as the edge always forms the shortest path between its endpoints.

There is a close relationship between betweenness centrality of edges and vertices. By summing up the edge betweenness of all edges incident with a vertex x we obtain the adjusted betweenness centrality $B_a(x)$ of this vertex. Relation between normal and adjusted betweenness of a vertex is as follows

$$B(x) = \frac{B_a(x) - n + 1}{2} \tag{1}$$

as has been shown by Caporossi, Paiva, Vukicevic and Segatto [3].

Betweenness centrality is frequently used in applications, even to identify influential patients in the transmission of infection of SARS-CoV-2 [16]. It is often studied from the algorithmic point of view [4,13]. Betweenness centrality and its variants are also studied from the graph-theoretical perspective [1,2,9, 12,15,19]. In this paper we focus on graphs having the same betweenness on all vertices.

A *betweenness-uniform graph* is a graph, in which all vertices have the same value of betweenness centrality. Thus, betweenness-uniform graphs are graphs with all vertices being equally important in terms of the (weighted) number of shortest paths on which they are lying. Networks having this property (or being close to it), are more robust and resistant to attacks, which causes betweenness-uniformity to be a promising feature for infrastructural applications. Moreover, betweenness-uniform graphs are also interesting from theoretical point of view. When studying the distribution of betweenness in a graph, betweenness-uniform graphs are one of the two possible extremal cases and have been already studied by Gago, Hurajová-Coroničová and Madaras [7,11]. The other extremal case are graphs where each vertex has a unique value of betweenness, which were studied by Florez, Narayan, Lopez, Wickus and Worrell [14].

The class of betweenness-uniform graphs includes all vertex-transitive graphs, which are graphs with the property that for every pair of vertices there exists an automorphism, which maps one onto the other. It is easy to see that vertex-transitive graphs are betweenness-uniform. Similarly, edge-transitive graphs are graphs with the property that for each pair of edges there exists an automorphism of the graph mapping one edge onto the other. Pokorná [18] showed that edge-transitive graphs are betweenness-uniform if and only if they are regular. There are also betweenness-uniform graphs which are neither vertex- nor edge-transitive. A construction of Gago, Hurajová-Corončová and Madaras [7] shows that, for n large enough, there are superpolynomially many of these graphs of order n. Also, all distance-regular graphs are betweenness-uniform [7]. Apart from the above mentioned results, not much is known about characterisation of betweenness-uniform graphs.

In this paper we prove two conjectures stated by Hurajová-Coroničová and Madaras [11]. The first one is about the connectivity of betweenness-uniform graphs. Having a connected betweenness-uniform graph, it is not too hard to show that there cannot be any vertex cut of size one. Consider connected components $C_0, \dots, C_p$ created by removing this cut vertex v. When we consider a vertex $a \in C_i$ for some $i \in \{0, 1, \dots, p\}$, only pairs of vertices from $V(C_i) \setminus \{a\}$ contribute to the betweenness of this vertex. On the other hand, all pairs of vertices $\{a, b\}$ such that $a \in C_i$ and $b \in C_j$ for $i \neq j$ contribute to betweenness of the vertex v. Using these two observations, along with some general bounds, we get the following property.

Theorem 1 (Gago, Hurajová-Coroničová and Madaras, 2013 [7]**).** *Any connected betweenness-uniform graph is* 2*-connected.*

As we have mentioned above, vertex-transitive graphs are betweenness-uniform. Thus, all cycles are betweenness-uniform. In this paper we show that cycles are the only betweenness-uniform graphs which are not 3-connected, as has been conjectured by Hurajová-Coroničová and Madaras [11].

Theorem 2. *If G is a connected betweenness-uniform graph then it is a cycle or a* 3*-connected graph.*

A variant of this theorem has already been proved for vertex-transitive graphs and for regular edge-transitive graphs [18]. Note that there exists a betweenness-uniform graph, which is 3-connected and is not vertex-transitive, see Fig. 1. This implies that we cannot generalize this result to claim that all betweenness-uniform graphs are either vertex-transitive or 4-connected.

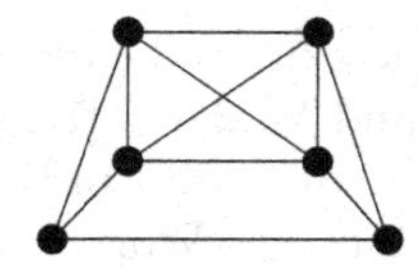

Fig. 1. Example of a 3-connected betweenness-uniform graph, which is not vertex transitive.

The second conjecture of Hurajová-Coroničová and Madaras [11] proven in the following theorem gives a relation between the maximum degree and the diameter of a betweenness-uniform graph.

Theorem 3. *If G is betweenness-uniform graph and $\Delta(G) = n-k$, then $d(G) \leq k$.*

In fact, the bound in Theorem 3 can be still significantly improved to $d(G) \leq \left\lfloor \frac{k+5}{2} \right\rfloor$; see Corollary 12 from Sect. 3.

2 Proof of Theorem 2

Before we start with the proof, we introduce some definitions and notation. *Betweenness centrality* of a vertex $u \in V(G)$ *induced by a subset of vertices* $\emptyset \neq S \subseteq V(G)$ of a graph G is defined as

$$B_S(u) := \sum_{\{x,y\} \in \binom{S \setminus \{u\}}{2}} \frac{\sigma_{x,y}(u)}{\sigma_{x,y}}$$

Average betweenness of $\emptyset \neq U \subseteq V(G)$ in G is

$$\bar{B}(U) := \frac{\sum_{u \in U} B(u)}{|U|}.$$

Average betweenness of U *induced* by $\emptyset \neq S \subseteq V(G)$ is defined analogically as

$$\bar{B}_S(U) := \frac{\sum_{u \in U} B_S(u)}{|U|}.$$

There is a relationship between average distance in a graph and the average betweenness of it's vertices.

Lemma 4 (Comellas, Gago, 2007 [5]**).** *For a graph G of order n,*

$$\bar{B}(V(G)) = \frac{(n-1)}{2} \cdot \left(\frac{\sum_{(u,v) \in V(G)^2} d(u,v)}{n(n-1)} - 1 \right)$$

Let G be a 2-connected graph, which is not 3-connected. We are going to show that it is not betweenness-uniform, unless it is isomorphic to a cycle. The following lemma gives a bound on average distance to a vertex in a 2-connected graph.

Lemma 5 (Plesník, 1984, [17]). *Let G be a 2-connected graph on n vertices and $u \in V(G)$. Then*

$$\frac{\sum_{v \in V(G)} d(v,u)}{n} \leq \left\lfloor \frac{n}{4} \right\rfloor$$

and equality is obtained for G isomorphic to a cycle.

Let us consider a 2-connected graph G. We denote $\{p,q\}$ to be the cut of size two minimizing the size of the smallest connected component of $G - \{p,q\}$, which is denoted by K. Let $k := |K|$ and K^+ be the subgraph of G induced by $V(K) \cup \{p,q\}$.

Observation 6. *Either we have $k = 1$ or both p,q have at least two neighbours in K.*

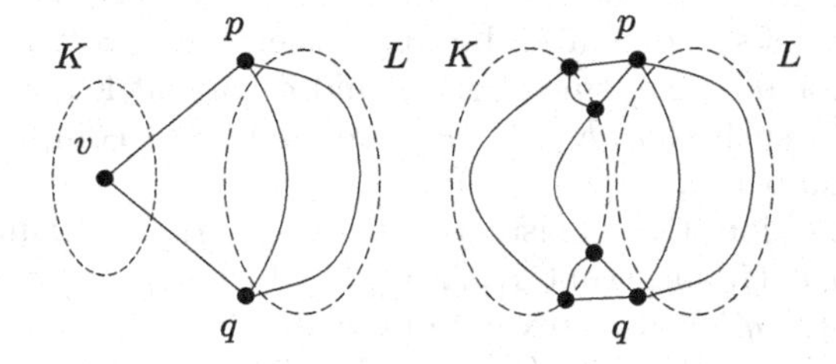

Fig. 2. Examples of the two possible situations from Observation 6.

There might exist one or more connected components in a graph $L = G - \{\{p,q\} \cup V(K)\}$ denoted by $L_1, \ldots, L_j$. We denote $\ell := |L|$. See Fig. 2 for notation and the two cases of Observation 6.

Throughout the proof, we use a notion based on a trivial observation below.

Observation 7. *In a betweenness-uniform graph,*

$$\bar{B}(S) = \bar{B}(R)$$

for any $\emptyset \neq R, S \subseteq V(G)$.

A discrepancy between the average betweenness of the vertices of the cut and of the vertices in component K is defined as

$$\text{disc} := \bar{B}(\{p,q\}) - \bar{B}(V(K)).$$

Let us define $\mathrm{disc}_S := \bar{B}_S(\{p,q\}) - \bar{B}_S(V(K))$ for $\emptyset \neq S \subseteq V(G)$. We split the discrepancy according to which pairs of vertices contribute to it. Namely,

$$\mathrm{disc} = \mathrm{disc}_{\binom{V(K^+)}{2}} + \mathrm{disc}_{\binom{V(L)}{2}} + \mathrm{disc}_{V(K^+)\times V(L)}$$

where $\binom{V(K^+)}{2}$, resp. $\binom{V(L)}{2}$, denotes pairs of vertices with both vertices taken from $V(K^+)$, resp. $V(L)$, and $V(K^+)\times V(L)$ denotes pairs with one vertex from $V(K^+)$ and the second from $V(L)$.

Using Observation 6, we are going to show that if $|N(p)\cap V(K)| \geq 2$ and $|N(q)\cap V(K)| \geq 2$, then the discrepancy is always strictly positive and that discrepancy is zero in the case $k = 1$ if and only if G is isomorphic to a cycle.

2.1 Vertices of the Cut Have at Least Two Neighbours in K

We start with the case $|N(p)\cap V(K)|, |N(q)\cap V(K)| \geq 2$. In this case we have the following useful observations.

Observation 8. *Connected component K^+ is 2-connected.*

Proof. Let us assume that there is a vertex cut x of size one in K^+. Observe that x is also a cut of size one in K, because otherwise x would separate K from $\{p,q\}$, which is not possible, since both p and q have at least two neighbors in K. Let $x \in V(K)$ such that $K_1,\ldots,K_j$ are connected components of $K - x$. Two situations can occur.

In the first situation, there exists a vertex of $\{p,q\}$, for example p, for which there exists K_i, $i \in [j]$ such that $K_i \cap N(p) = \emptyset$. Let $p' := x$ and $q' := q$. We can observe that $\{p',q'\}$ is a vertex cut of G with the property that the smallest component $K' = K_i$ of $G - \{p',q'\}$ is smaller than K, which is a contradiction with the choice of $\{p,q\}$.

In the second situation, both p and q have at least one neighbour in each component of $K - x$. In this case it is clear that $K^+ - x$ is connected.

Counting $\mathrm{disc}_{\binom{V(L)}{2}}$. We take any pair of vertices ℓ_1, ℓ_2 from $V(L)$ and examine how the shortest path between them influences the discrepancy. Basically, there are three different types of shortest paths between ℓ_1 and ℓ_2.

1. The shortest path between ℓ_1 and ℓ_2 passes only through vertices of $V(L)$. In this case, the shortest path does not influence the discrepancy.
2. The shortest path between ℓ_1 and ℓ_2 passes through K, especially it enters K by one cut vertex and leaves through the second cut vertex. This adds one to $\bar{B}(\{p,q\})$ and at most one to $\bar{B}(V(K))$.
3. The shortest path between ℓ_1 and ℓ_2 passes through p or q without visiting component K. This adds something to $\bar{B}(\{p,q\})$ and nothing to $\bar{B}(V(K))$.

Overall, we get that $\mathrm{disc}_{\binom{V(L)}{2}} \geq 0$.

Counting $\mathbf{disc}_{\binom{V(K^+)}{2}}$**.** Using Lemma 5 and Observation 8 we obtain that the sum of the lengths of all shortest paths from a fixed vertex in K^+ is $\frac{(k+2)^2}{4}$. Moreover, by multiplying the sum of shortest paths from a fixed vertex by the number of vertices in K^+, we obtain that the sum of all shortest paths in K^+ is at most $\frac{(k+2)^3}{4}$.

To obtain an upper bound on $\bar{B}_{\binom{V(K^+)}{2}}(V(K))$, we use the relation from Lemma 4, where we use the sum of distances in K^+, but use k for number of vertices. This corresponds to dividing all the contributions of shortest paths in K^+ only to vertices of K. Note that some of the shortest paths might pass though p or q, but this can only decrease the average betweenness of K. As a result,

$$\bar{B}_{\binom{V(K^+)}{2}}(V(K)) \leq \frac{(k-1)}{2}\left(\frac{\frac{(k+2)^3}{4}}{k(k-1)} - 1\right) = \frac{k^2}{8} + \frac{1}{k} + \frac{3}{2}$$

Finally, we assume that $\bar{B}_{\binom{V(K^+)}{2}}(\{p,q\}) = 0$ to obtain a a lower bound on the discrepancy. Then we get that

$$\mathrm{disc}_{\binom{V(K^+)}{2}} \geq 0 - \left(\frac{k^2}{8} + \frac{1}{k} + \frac{3}{2}\right) = -\frac{k^2}{8} - \frac{1}{k} - \frac{3}{2}.$$

Counting $\mathbf{disc}_{V(K^+)\times V(L)}$**.** Clearly, each path from K to L passes through at least one vertex of the cut $\{p,q\}$, adding at least $\frac{1}{2}$ to $\bar{B}_{V(K^+)\times V(L)}(\{p,q\})$. As a result, $\bar{B}_{V(K^+)\times V(L)}(\{p,q\}) \geq \frac{k\ell}{2}$.

Now we show an upper bound on the average betweenness of K. Take any $x \in V(L)$ and suppose $d(x,p) < d(x,q)$. The contribution of x to $\bar{B}(V(K))$ is maximized, when all paths from K^+ to x pass through p. Otherwise, there exists $y \in V(K)$ such that the shortest path between x and y passes through q. This means that $d(y,q) + d(x,q) \leq d(y,p) + d(x,p)$. Together with $d(x,q) \geq d(x,p)$, we get $d(y,q) \leq d(y,p)$, so the path passes through smaller or the same number of vertices of K, then it would if it went through p. From now on, we assume that for each $x \in V(L)$ there exists $r \in \{p,q\}$ such that all paths from K^+ to x are passing through r and we denote $s := \{p,q\} \setminus r$.

We can use Lemma 5 and the fact that K^+ is 2-connected to obtain that $\sum_{v\in V(K^+)} d(v,r) = \sum_{v\in V(K)\cup\{s\}} d(v,r) \leq \frac{(k+2)^2}{4}$. This corresponds to a sum of distances travelled inside K by all paths from $V(K)$ to a fixed $x \in V(L)$. Note that for any $v \in V(K)$, vr-path of length d contributes $d-1$ to the betweenness of K and thus

$$\bar{B}_{V(K^+)\times V(L)}(V(K)) \leq \frac{\ell}{k}\Big(\sum_{v\in V(K)\cup\{s\}} (d(v,r)-1)\Big) \leq \frac{\ell}{k}\Big(\frac{(k+2)^2}{4} - (k+1)\Big) = \frac{k\ell}{4}$$

When we take the lower bound on $\bar{B}_{V(K^+)\times V(L)}(\{p,q\})$ and upper bound on $\bar{B}_{V(K^+)\times V(L)}(K)$ we can bound the discrepancy

$$\mathrm{disc}_{V(K^+)\times V(L)} \geq \frac{k\ell}{2} - \left(\frac{k\ell}{4}\right) = \frac{k\ell}{4}.$$

Together we obtain

$$\mathrm{disc} = \mathrm{disc}_{\binom{V(K^+)}{2}} + \mathrm{disc}_{\binom{V(L)}{2}} + \mathrm{disc}_{V(K^+)\times V(L)} \geq -\frac{k^2}{8} - \frac{1}{k} - \frac{3}{2} + 0 + \frac{k\ell}{4} > 0$$

which holds for $\ell \geq k \geq 4$. It remains to discuss the cases $\ell \geq k = 2$ and $\ell = k = 3$.

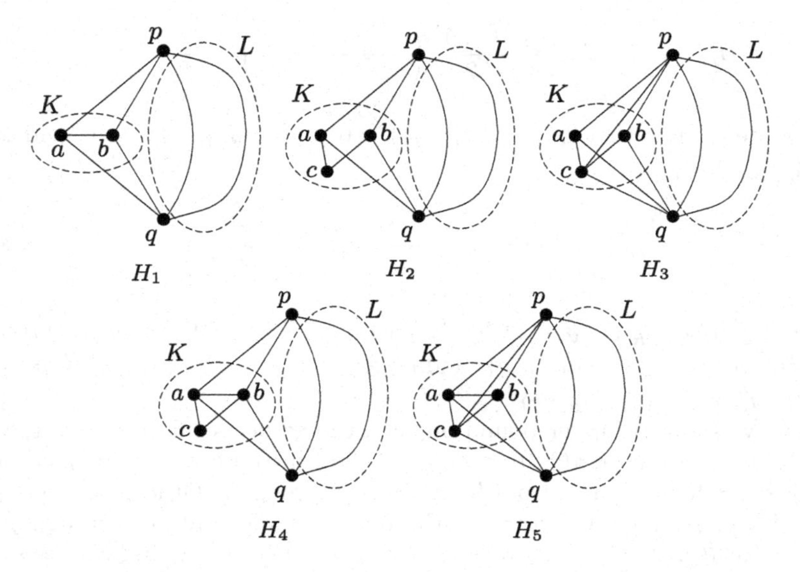

Fig. 3. The five classes of graphs for which we do not obtain positive discrepancy by the general computation.

Let $k = 2$ and $V(K) = \{a, b\}$. Observe that a and b are adjacent, because otherwise $K' = G[\{a\}]$ is a component of $G - \{p, q\}$ smaller than K, which is a contradiction with K being minimal. This implies that G corresponds to a graph with the same K^+ as in graph H_1 shown in Fig. 3.

Let $k = \ell = 3$. Clearly, K is connected. Observation 8 also implies that if there is a vertex cut x in K, then p and q are adjacent to all components of $K - x$. This leaves us with the four classes of graphs H_2, H_3, H_4, H_5 from Fig. 3. The discrepancy in these classes of graphs is positive, which can be shown by counting discrepancy similarly as above together with using the knowledge of the structure of K^+. We omit these details for the sake of brevity.

We have seen that discrepancy is always greater than zero if $|N(p) \cap K| \geq 2$ and $|N(q) \cap K| \geq 2$, implying G is not betweenness-uniform in this case. From this fact and Observation 6, we see that the size of the minimal connected component K must be one if the 2-connected graph G is betweenness-uniform.

2.2 Vertices of Degree Two

Here we consider the case that $|K| = 1$ which means that G contains a vertex v of degree 2. If L contains more connected components $L_1, \ldots, L_j$, we consider a graph $G_i := G[\{p, q, v\} \cup V(L_i)]$ for each of the components separately. Note that each component L_i of L is connected, so G_i is a 2-connected graph. We use $L := L_i$ for simplicity in the text below.

Let p, q be the two neighbors of v. Let K be the one-vertex graph with vertex v, let $K^+ := G[\{p, v, q\}]$ and let $L := G[V \setminus \{p, q, v\}]$. Observe that if G contains the edge pq, then there is no shortest path passing through v and thus $B(v) = 0$. As a consequence, G is isomorphic to K_3, because the only betweenness-uniform graphs with betweenness value 0 are complete graphs [18]. Therefore, we assume that G does not contain the edge pq from now on.

Throughout this section, we decompose discrepancy as follows:

$$\text{disc} = \text{disc}_{\binom{V(L)}{2}} + \text{disc}_{K^+ \times V(L) \cup \binom{K^+}{2}}$$

Observe that $\text{disc}_{\binom{V(L)}{2}} \geq 0$ because every (shortest) path between two vertices of L going through the vertex v of K must contain both p and q. We now focus on paths with at least one end-vertex in the set $\{p, q, v\}$.

For a vertex w in G, set $\alpha(w) := d(w, p) - d(w, q)$. Let P be the shortest path from p to q in $G - v$, and let λ be the number its vertices different from p and q. Along the path P from p to q, the function α consecutively gets values $-\lambda - 1, -\lambda + 1, \ldots, \lambda - 1, \lambda + 1$. It follows that, depending on the parity of λ, one of the following two cases happens:

Case 1: There are two consecutive vertices x, y on P such that

$$\alpha(x) = -1, \ \alpha(y) = 1.$$

Case 2: There are three consecutive vertices x, y, z on P such that

$$\alpha(x) = -2, \ \alpha(y) = 0, \ \alpha(z) = 2.$$

Now, our first aim is to show that in both of the two cases the vertex v together with one of the vertices x, y (and z) forms a 2-cut in G such that after removing the vertices of the 2-cut from G, we obtain two components of equal or almost equal size. Before considering Cases 1 and 2 separately, we prove the following proposition.

Proposition 9. *Let $w \in V(L)$.*

(i) If $|\alpha(w)| \leq 1$, then

$$disc_{K+\times\{w\}} = \frac{1}{2}.$$

(ii) If $\alpha(w) = -2$, then

$$disc_{K+\times\{w\}} = \frac{1}{2}\left(1 - \frac{\sigma_{w,p}}{\sigma_{w,q}}\right) \in \left(0, \frac{1}{2}\right).$$

(iii) If $\alpha(w) = 2$, then

$$disc_{K+\times\{w\}} = \frac{1}{2}\left(1 - \frac{\sigma_{w,q}}{\sigma_{w,p}}\right) \in \left(0, \frac{1}{2}\right).$$

(iv) If $|\alpha(w)| \geq 3$, then

$$disc_{K+\times\{w\}} = 0.$$

Proof.

(i) If $|\alpha(w)| \leq 1$, then all shortest paths from w to p avoid q and v, all shortest paths from w to q avoid p and v, and every shortest path from w to v goes through exactly one of the vertices p and q. Part (i) of the proposition follows.

(ii) If $\alpha(w) = -2$, then every shortest path from w to p avoids q and v. Every shortest path from w to v visits p and avoids q. There are two types of shortest paths between w and q. First type passes through both p and v, second type avoids both of them. As $d(w,p) + 2 = d(w,q)$, there are $\sigma_{w,p}$ paths passing through p and v. As a result, exactly $\sigma_{w,p}$ of the $\sigma_{w,q}$ shortest paths from w to q visit p and v, all the other shortest paths from w to q avoid both p and v. Part (ii) follows.

(iii) Analogous to the proof of part (ii), with the roles of p and q exchanged.

(iv) If $\alpha(w) \leq -3$ then all shortest paths from w to p visit none of the vertices v and q, all shortest paths from w to v visit p and do not visit q, and all shortest paths from w to q visit both p and v. Part (iv) then follows. The case $\alpha(w) \geq 3$ is analogous, with the roles of p and q exchanged.

Suppose now that Case 1 holds. Due to Proposition 9 (i), $\text{disc}_{K+\times\{x,y\}} = 1/2 + 1/2 = 1$. Further, we have $\text{disc}_{\{p\}\times\{q\}} = -1$, since the only shortest path between p and q goes through v. Proposition 9 now implies that if disc $= 0$ then every vertex w in $V(L) \setminus \{x, y\}$ satisfies $|\alpha(w)| \geq 3$. Since the function α differs by at most two on any pair of neighbors in L, $\{v, x\}$ and $\{v, y\}$ are 2-cuts of G. Let L_p and L_q be the two connected components of the graph $G - \{x, y, v\}$, where $p \in L_p$ and $q \in L_q$. We have $\alpha(w) \leq -3$ for every vertex w of L_p, and $\alpha(w) \geq 3$ for every vertex w of L_q. It follows that L_p and L_q have the same number of vertices, since otherwise if $|V(L_p)| > |V(L_q)|$, say, then $B(p) > B(q)$.

If G is not a cycle, then, without loss of generality, the connected graph $H := G[V(L_p) \cup \{p, x\}]$ is not a path from p to x. If H is 2-connected, we can continue in the same way as in the second of the two cases described in Observation 6. Otherwise there is a cut vertex t of H. Let H_p and H_x be the two connected components of $G-t$, where $p \in H_p$ and $x \in H_x$. Since H is not a path from p to x, at least one of the two subgraphs of G induced by $V(H_p) \cup \{t\}$ and by $V(H_x) \cup \{t\}$, respectively, is not a path from p to t and from x to t, respectively. We consider such a subgraph and check again if it is 2-connected. Continuing this process, due to the 2-connectivity of G, we end up with a 2-cut of G which cuts off a 2-connected component, which is a subgraph of H. Then we can continue as in the second of the two cases described in Observation 6. This ends the proof in Case 1.

Suppose now that Case 2 happens. We have $\mathrm{disc}_{\{p\}\times\{q\}} \geq -1$.

According to Proposition 9 (i), $\mathrm{disc}_{K^+\times\{y\}} = 1/2$. According to Proposition 9 (ii),(iii),

$$\mathrm{disc}_{K^+\times\{x,z\}} = \frac{1}{2}\left(1 - \frac{\sigma_{x,p}}{\sigma_{x,q}}\right) + \frac{1}{2}\left(1 - \frac{\sigma_{z,q}}{\sigma_{z,p}}\right),$$

which is positive. It now follows from Proposition 9 (i) that $|\alpha(w)| \geq 2$ for every vertex $w \in V(L), w \neq y$, since otherwise disc > 0.

We have $\sigma_{x,q} \geq \sigma_{x,p} + \sigma_{z,q}$. Similarly, $\sigma_{z,p} \geq \sigma_{x,p} + \sigma_{z,q}$. Thus,

$$\mathrm{disc}_{K^+\times\{x,z\}} = \frac{1}{2}\left(1 - \frac{\sigma_{x,p}}{\sigma_{x,q}}\right) + \frac{1}{2}\left(1 - \frac{\sigma_{z,q}}{\sigma_{z,p}}\right) \geq$$

$$\geq \frac{1}{2}\left(1 - \frac{\sigma_{x,p}}{\sigma_{x,p} + \sigma_{z,q}}\right) + \frac{1}{2}\left(1 - \frac{\sigma_{z,q}}{\sigma_{x,p} + \sigma_{z,q}}\right) = \frac{1}{2}.$$

Proposition 9 now implies that if disc $= 0$ then every vertex w in $V(L) \setminus \{x, y, z\}$ satisfies $|\alpha(w)| \geq 3$. It follows that $\{v, x\}, \{v, y\}, \{v, z\}$ are 2-cuts in G.

Let L_p and L_q be the two connected components of the graph $G - \{v, y\}$, where $p \in V(L_p)$ and $q \in V(L_q)$. We have $\alpha(w) \leq -3$ for every vertex $w \in V(L_p) \setminus \{p, x\}$, and $\alpha(w) \geq 3$ for every vertex $w \in V(L_q) \setminus \{q, z\}$.

Since G is not a cycle, at least one of the following two conditions is satisfied:

(C1) L_p is not a path from p to x, or
(C2) L_q is not a path from q to z.

If both (C1) and (C2) hold, then without loss of generality, we suppose that $|V(L_p)| \leq |V(L_q)|$. Then we consider the graph $H' := G[V(L_p) \cup \{v, y\}]$ and proceed on H' in the same way as we proceeded on H in Case 1.

Suppose now without loss of generality that (C1) holds and (C2) does not hold. Then there is a single shortest path from z to q. It follows that $|V(L_p)| \leq |V(L_q)|$, since otherwise $B(p) > B(q)$. We now can again consider the graph $H' := G[V(L_p) \cup \{v, y\}]$ and proceed on it in the same way as we proceeded on H in Case 1. This finishes the proof in Case 2.

2.3 More Components

If $G - (\{p, q\} \cup V(K))$ had only one connected component L, we are finished. Otherwise, we know that $L = \bigcup_{i=1}^{j} L_i$ and any $G_i := G[\{p, q, v\} \cup V(L_i)]$ has either positive discrepancy, or it is isomorphic to a cycle. We can observe that for G_i and G_j with positive discrepancy,

$$G_{i+j} := G[V(K) \cup \{p, q\} \cup V(L_i) \cup V(L_j)]$$

has also positive discrepancy. This follows from the fact that whenever we obtain positive discrepancy, it is due to the second of the two cases in Observation 8, which has been solved in Subsect. 2.1. From there, it is clear that discrepancy rises with growing difference between k and ℓ. The only remaining case is that each L_i is isomorphic to a path between p and q. Suppose $\ell_i \in N(p) \cap V(L_i)$ and $\ell_j \in N(p) \cap V(L_j)$ for any two connected components L_i, L_j of L. Then the shortest path between ℓ_i and ℓ_j passes through p and avoids K, making $\mathrm{disc}_{\binom{V(L)}{2}} > 0$, which leads to disc > 0.

By the results above, any 2-connected graph has either disc > 0, implying it is not betweenness-uniform, or it has disc $= 0$ and is isomorphic to a cycle. This finishes the proof of Theorem 2.

3 Relation Between Maximal Degree and Diameter of Betweenness-Uniform Graphs

In this section we prove a conjecture of Hurajová-Coroničová and Madaras [11] saying that betweenness-uniform graphs with high maximal degree have small diameter.

Conjecture 10 ([11]). If G is a betweenness-uniform graph and $\Delta(G) = n - k$, then $d(G) \leq k$.

In a previous article by Gago, Hurajová-Coroničová and Madaras [7], this conjecture has been proven for $k = 1$ and $k = 2$ and later Hurajová-Coroničová and Madaras [11] proved the conjecture for $k = 3$ by showing that a betweenness-uniform graph with $\Delta(G) = n - 3$ has $d(G) = 2$ for $n \geq 4$.

Theorem 11. *Let G be 2-connected graph with $\Delta(G) = n - k$. Then $d(G) \leq \lfloor \frac{k+5}{2} \rfloor$.*

Proof. Let u, v be vertices of maximal distance, which implies $d(u, v) = d(G)$, and y be a vertex with $\deg(y) = n - k$. Due to G being 2-connected, there exist at least 2 vertex disjoint paths P_1, P_2 between u and v. Suppose $P_i = x_0 x_1 \cdots x_j$ of length j for some $i \in \{1, 2\}$ contains at least three vertices $y_1, \ldots y_\ell$ of $N(y)$. Clearly, by exchanging the sequence $y_1 \cdots y_\ell$ by sequence $y_1 y y_\ell$ we create a new path P_i' of length at most j. The paths P_1' and P_2' have the following properties:

- each of them contains at most three vertices from $\{y \cup N(y)\}$
- each of them has length at least $d(G)$
- each vertex from $V(G) \setminus \{y, N(y)\}$ is contained in at most one of these paths

Using these properties we can observe that each P_i' contains at least $d(G) - 4$ vertices of $V(G) - \{u, v, y, N(y)\}$, because it contains at least $d(G) + 1$ vertices, which means at least $d(G) - 1$ vertices from $V(G) - \{u, v\}$ and at most three vertices from $\{y \cup N(y)\}$. By summing vertices of $V(G) \setminus \{u, v, y, N(y)\}$ on these paths with the size of $\{u, v, y, N(y)\}$ we obtain a lover bound on the number of vertices,

$$n \geq 2(d(G) - 4) + 3 + (n - k)$$

giving $d(G) \leq \frac{k+5}{2}$. Since $d(G)$ is an integer, we get $d(G) \leq \lfloor \frac{k+5}{2} \rfloor$.

By combining Theorem 1 and Theorem 11 above, we obtain the following corollary, which implies Theorem 3 for $k \geq 1$.

Corollary 12. *Let G be betweenness-uniform graph with $\Delta(G) = n - k$. Then $d(G) \leq \lfloor \frac{k+5}{2} \rfloor$.*

Using our Theorem 2, we can obtain an even better bound.

Acknowledgments. David Hartman and Aneta Pokorná were partially supported by ERC Synergy grant DYNASNET grant agreement no. 810115. Pavel Valtr was partially supported by the H2020-MSCA-RISE project CoSP- GA No. 823748.

References

1. Akgün, M.K., Tural, M.K.: k-step betweenness centrality. Comput. Math. Organ. Theory **26**(1), 55–87 (2019). https://doi.org/10.1007/s10588-019-09301-9
2. Barthelemy, M.: Betweenness Centrality, pp. 51–73. Springer, Cham (2018). https://doi.org/10.1007/978-3-319-20565-6_4
3. Caporossi, G., Paiva, M., Vukicevic, D., Segatto, M.: Centrality and betweenness: vertex and edge decomposition of the Wiener index. MATCH - Commun. Math. Comput. Chem. **68** (2012)
4. Chehreghani, M.H.: An efficient algorithm for approximate betweenness centrality computation. Comput. J. **57**(9), 1371–1382 (2014). https://doi.org/10.1093/comjnl/bxu003
5. Comellas, F., Gago, S.: Spectral bounds for the betweenness of a graph. Linear Algebra Appl. **423**(1), 74–80 (2007). https://doi.org/10.1016/j.laa.2006.08.027. Special Issue devoted to papers presented at the Aveiro Workshop on Graph Spectra
6. Freeman, L.C.: A set of measures of centrality based on betweenness. Sociometry **40**, 35–41 (1977). https://doi.org/10.2307/3033543
7. Gago, S., Hurajová, J.C., Madaras, T.: On betweenness-uniform graphs. Czechoslovak Math. J. **63**(3), 629–642 (2013). https://doi.org/10.1007/s10587-013-0044-y
8. Girvan, M., Newman, M.: Community structure in social and biological networks. Proc. Natl. Acad. Sci. U.S.A. **99**(12), 7821–7826 (2002). https://doi.org/10.1073/pnas.122653799
9. Govorč in, J., Škrekovski, R., Vukašinović, V., Vukičević, D.: A measure for a balanced workload and its extremal values. Discrete Appl. Math. **200**, 59–66 (2016). https://doi.org/10.1016/j.dam.2015.07.006

10. Hagmann, P., et al.: Mapping the structural core of human cerebral cortex. PLoS Biol. **6**(7), 1479–1493 (2008). https://doi.org/10.1371/journal.pbio.0060159
11. Coroničová Hurajová, J., Madaras, T.: More on betweenness-uniform graphs. Czechoslovak Math. J. **68**(2), 293–306 (2018). https://doi.org/10.21136/CMJ.2018.0087-16
12. Kumar, R.S., Balakrishnan, K., Jathavedan, M.: Betweenness centrality in some classes of graphs. Int. J. Comb. (2014). https://doi.org/10.1155/2014/241723
13. Lee, M.J., Choi, S., Chung, C.W.: Efficient algorithms for updating betweenness centrality in fully dynamic graphs. Inf. Sci. **326**, 278–296 (2016). https://doi.org/10.1016/j.ins.2015.07.053
14. Lopez, R., Worrell, J., Wickus, H., Florez, R., Narayan, D.A.: Towards a characterization of graphs with distinct betweenness centralities. Austr. J. Comb. **68**(2), 285–303 (2017)
15. Majstorović, S., Caporossi, G.: Bounds and relations involving adjusted centrality of the vertices of a tree. Graphs Comb. **31**(6), 2319–2334 (2014). https://doi.org/10.1007/s00373-014-1498-x
16. Nagarajan, K., Muniyandi, M., Palani, B., Sellappan, S.: Social network analysis methods for exploring SARS-CoV-2 contact tracing data. BMC Med. Res. Methodol. **20**(1) (2020). https://doi.org/10.1186/s12874-020-01119-3
17. Plesník, J.: On the sum of all distances in a graph or digraph. J. Graph Theory **8**(1), 1–21 (1984). https://onlinelibrary.wiley.com/doi/abs/10.1002/jgt.3190080102
18. Pokorná, A.: Characteristics of network centralities. Master's thesis, Charles University, Prague, Czech Republic (2020). https://is.cuni.cz/webapps/zzp/detail/215181/
19. Unnithan, S., Balakrishnan, K.: Betweenness centrality in convex amalgamation of graphs. J. Algebra Comb. Discrete Struct. Appl. **6**, 21–38 (2019). https://doi.org/10.13069/jacodesmath.508983

Combinatorics and Algorithms

On Algorithms to Find p-ordering

Aditya Gulati(✉), Sayak Chakrabarti, and Rajat Mittal

IIT Kanpur, Kanpur, India
{aditg,sayak,rmittal}@iitk.ac.in

Abstract. The concept of *p-ordering* for a prime p was introduced by Manjul Bhargava (in his PhD thesis) to develop a generalized factorial function over an arbitrary subset of integers. This notion of p-ordering provides a representation of polynomials modulo prime powers, and has been used to prove properties of roots sets modulo prime powers. We focus on the complexity of finding a p-ordering given a prime p, an exponent k and a subset of integers modulo p^k.

Our first algorithm gives a p-ordering for a set of size n in time $\widetilde{\mathcal{O}}(nk \log p)$, where set is considered modulo p^k. The subsets modulo p^k can be represented concisely using the notion of representative roots (Panayi, PhD Thesis, 1995; Dwivedi et al., ISSAC, 2019); a natural question is, can we find a p-ordering more efficiently given this succinct representation. Our second algorithm achieves precisely that, we give a p-ordering in time $\widetilde{\mathcal{O}}(d^2 k \log p + nk \log p + nd)$, where d is the size of the succinct representation and n is the required length of the p-ordering. Another contribution is to compute the structure of roots sets for prime powers p^k, when k is small. The number of root sets have been given before (Dearden and Metzger, Eur. J. Comb., 1997; Maulick, J. Comb. Theory, Ser. A, 2001), we explicitly describe all the root sets for $k \leq 4$.

Keywords: Root-sets · p-ordering · Polynomials · Prime powers

1 Introduction

Polynomials over finite fields have played a crucial role in computer science with impact on diverse areas like error correcting codes [8,14,22,23], cryptography [11,17,20], computational number theory [1,2] and computer algebra [16,24]. Mathematicians have studied almost all aspects of these polynomials; factorization of polynomials, roots of a polynomial and polynomials being irreducible or not are some of the natural questions in this area. There is lot of structure over finite field; we can deterministically count roots and find if a polynomials is irreducible in polynomial time [18]. Not just that, we also have efficient randomized algorithms for the problem of factorizing polynomials over finite fields [3,9].

The situation changes drastically if we look at rings instead of fields. Focusing our attention on the case of numbers modulo a prime power (ring $\mathbb{Z}_{p^k}$, for a

The full version is available at https://arxiv.org/abs/2011.10978.

A. Mudgal and C. R. Subramanian (Eds.): CALDAM 2021, LNCS 12601, pp. 333–345, 2021.
https://doi.org/10.1007/978-3-030-67899-9_27

prime p and a natural number $k \geq 2$) instead of numbers modulo a prime (field $\mathbb{F}_p$), we don't even have unique factorization and the fact that the number of roots are bounded by the degree of the polynomial. Still, there has been some interesting work in last few decades. Maulik [19] showed bound on number of roots sets, sets which are roots for a polynomial modulo a prime power. There has been some recent works giving a randomized algorithm for root finding [4] and a deterministic algorithm for root counting [10,13].

The concept of *p-ordering* and *p-sequences* for a prime p, introduced by Bhargava [5], is an important tool in studying the properties of roots sets and polynomials over powers of prime p [5,7,19]. Bhargava's main motivation to introduce p-ordering was to generalize the concept of factorials ($n!$ for $n \geq 0 \in \mathbb{Z}$) from the set of integers to any subset of integers. He was able to show that many number-theoretic properties of this factorial function (like the product of any k consecutive non-negative integers is divisible by $k!$) remain preserved even with the generalized definition for a subset of integers [6].

For polynomials, Bhargava generalized Polya's theorem, showing that the GCD of the outputs of a degree k polynomial on a subset S of integers divides the analogous factorial of k for S. He also gave a characterization of polynomials which are integer valued on a subset of integers. Both results use a convenient basis, using p-ordering, for representing polynomials on a subset of integers.

A similar convenient basis can be obtained for subsets of $\mathbb{Z}_{p^k}$. An interesting problem for polynomials over rings, of the kind $\mathbb{Z}_{p^k}$, is to find the allowed *root sets* (Definition 4). Maulik [19] was able to use this representation of polynomials over $\mathbb{Z}_{p^k}$ (from p-ordering) to give asymptotic estimates on the number of root sets modulo a prime power p^k; he also gave a recursive formula for root counting.

Our Contributions. While a lot of work has been done on studying the properties of p-orderings [7,15,19], there's effectively no work on finding the complexity of the problem: given a subset of numbers modulo a prime power, find a p-ordering. Our main contribution is to look at the computational complexity of finding p-ordering in different settings. We also classify and count the root-sets for $\mathbb{Z}_{p^k}$, when $k \leq 4$, by looking at their symmetric structure.

- *p-ordering for a general set:* Suppose, we want to find the p-ordering of a set $S \subseteq \mathbb{Z}_{p^k}$ so that $|S| = n$. A naive approach gives a $\widetilde{\mathcal{O}}(n^3 k \log(p))$ time algorithm. We exploit the recursive structure of p-orderings and optimize the resulting algorithm using data structures. These optimizations allow us to give an algorithm that works in $\widetilde{\mathcal{O}}(nk \log(p))$ time. The details of the algorithm, its correctness and time complexity is given in Sect. 3.
- *p-ordering for a subset in representative root representation:* A polynomial of degree d in $\mathbb{Z}_{p^k}$ can have exponentially many roots, but they can have at most d *representative roots* [4,13,21] giving a succinct representation. In general, any efficient algorithm for root finding or factorization should output the roots in form of such representative roots (the complete set of roots could be exponentially large). The natural question is, can we have an efficient algorithm for finding a p-ordering where the complexity scales according to the number of representative roots and not the size of the complete

set. We answer this in affirmative, and provide an algorithm which works in $\widetilde{\mathcal{O}}(d^2 k \log p + nk \log p + nd)$ time, where d is the number of representative roots and n is the length of p-ordering. The details of this algorithm and its analysis are presented in Sect. 4.

- *Roots sets for small powers:* A polynomial in $\mathbb{Z}_{p^k}$, even with small degree, can have exponentially large number of roots. Still, not all subsets of $\mathbb{Z}_{p^k}$ are a root-set for some polynomial. The number of root-sets for the first few values of k were calculated numerically by Dearden and Metzger [12]. Building on previous work, Maulik [19] produced an upper bound on the number of root-sets for any p and k. He also gave a recursive formula for the exact number of root-sets using the symmetries in their structure. We look at the structure of these root sets and completely classify all possible root-sets for $k \leq 4$ (we describe these in Sect. 5).

2 Preliminaries

The proofs for this section can be found in the full version.

Notation and Time Complexity: Our primary goal is to find a p-ordering of a given set $S \subseteq \mathbb{Z}_{p^k}$, for a given prime p and an integer $k > 0$. Since the input size is polynomial in $|S|, \log p, k$; an efficient algorithm should run in time polynomial in these parameters. For the sake of clarity, $\log k$ factors will be ignored from complexity calculations; this omission will be expressed by using notation $\widetilde{\mathcal{O}}$ instead of $\mathcal{O}$ in time complexity. We also use $[n]$ for the set $\{0, 1, \ldots, n-1\}$.

Definition 1. *For any ring S with the usual operations $+$ and $*$, we define*

$$S + a := \{x + a \mid x \in S\} \quad \text{and} \quad a * S := \{a * x \mid x \in S\}$$

p-ordering and p-sequence: We begin by defining the valuation of an integer modulo a prime p.

Definition 2. *Let p be a prime and $a \neq 0$ be an integer. The* valuation *of the integer a modulo p, denoted $v_p(a)$, is the integer v such that $p^v \mid a$ but $p^{v+1} \nmid a$. We also define $w_p(a) := p^{v_p(a)}$.*

If $a = 0$ then both, $v_p(a)$ and $w_p(a)$, are defined to be ∞.

Bhargava [5] introduced the concept of p-ordering for any subset of a Dedekind domain. We restrict to the rings of the form $\mathbb{Z}_{p^k}$ (similar to $\mathbb{Z}$) which has been explained in [6]. p-ordering was introduced in a series of papers to generalise the factorial function.

Definition 3 ([5])**.** *p-ordering on a subset S of $\mathbb{Z}_{p^k}$ is defined inductively.*

1. *Choose any element $a_0 \in S$ as the first element of the sequence.*
2. *Given an ordering $a_0, a_1, \ldots a_{i-1}$ up to $i-1$, choose $a_i \in S \backslash \{a_0, a_1 \ldots a_{i-1}\}$ which minimizes $v_p((a_i - a_0)(a_i - a_1) \ldots (a_i - a_{i-1}))$.*

The i-th element of the associated *p-sequence* for a p-ordering $a_0, a_1, \ldots a_n$ is defined by

$$v_p(S,i) = \begin{cases} 1 & i = 0, \\ w_p((a_i - a_0)(a_i - a_1)\ldots(a_i - a_{i-1})) & i > 0. \end{cases}$$

In the $(i+1)$-th step, let $x \in S \setminus \{a_0, a_1, ..., a_{i-1}\}$ then the value $v_p((x - a_0)(x - a_1)\ldots(x - a_{i-1}))$ is denoted as the *p-value* of x at that step. If the step is clear from context, we call the p-value of that element at that step as its p-value.

To take an example, $S = \{1, 3, 4, 6, 9, 10\} \in \mathbb{Z}_{3^3}$ has $(4, 6, 1, 9, 3, 10)$ and $(3, 10, 6, 4, 9, 1)$ as two valid 3-orderings. The 3-sequence associated with both these 3-orderings is $(1, 1, 3, 3, 9, 27)$. At the first glance it is not clear why associated p-sequences are same. In fact, Bhargava proved the following theorem.

Theorem 1 ([5]). *For any two p-orderings of a subset $S \subseteq \mathbb{Z}$ and a prime p, the associated p-sequences are same.*

We notice few more facts about p-ordering.

Observation 2. *Let S be a subset of integers, let $(a_0, a_1, a_2, ...)$ be a p-ordering on S, then*

1. *For any $x \in \mathbb{Z}$, $(a_0 + x, a_1 + x, a_2 + x, ...)$ is a p-ordering on $S + x$.*
2. *For any $x \in \mathbb{Z}$, $(x * a_0, x * a_1, x * a_2, ...)$ is a p-ordering on $x * S$.*

Observation 3. *Let S be a subset of integers, let $(a_0, a_1, a_2, ...)$ be a p-ordering on S. Then, for any $x \in \mathbb{Z}$*

1. *$v_p(x * S, k) = v_p(S, k) + k \cdot w_p(x)$.*
2. *$v_p(S + x, k) = v_p(S, k)$.*

Theorem 4 ([19]). *Let S be a subset of integers, let $S_j = \{s \in S \mid s \equiv j \pmod{p}\}$ for $j = 0, 1, ..., p-1$, then for any $x \in \mathbb{Z}$, s.t. $x \equiv j \pmod{p}$,*

$$w_p\left(\prod_{a_i \in S}(x - a_i)\right) = w_p\left(\prod_{a_i \in S_j}(x - a_i)\right). \quad (1)$$

Root Sets and Representative Roots:

Definition 4. *A given set S is called a* root set *in a ring R if there is a polynomial $f(x) \in R[x]$, whose roots in R are exactly the elements of S.*

It is non-trivial to check if a subset is a root set. For example, $\{0, 3\}$ is not a root set in $\mathbb{Z}_{3^2}$, whereas a large set $\{0, 3, 6, 9, 12, 15, 18, 21, 24\} \in \mathbb{Z}_{3^3}$ is a root set for a small degree polynomial x^3.

The notion of representative roots in the ring $\mathbb{Z}_{p^k}$ has been used to concisely represent roots of a polynomial [4,13,21].

Definition 5. *The representative root* $(a + p^i*)$ *is a subset of* $\mathbb{Z}_{p^k}$,

$$a + p^i* := \{a + p^i y \mid y \in \mathbb{Z}_{p^{k-i-1}}\} \quad (2)$$

For example, the set $\{1, 26, 51, 76, 101\} \in \mathbb{Z}_{5^3}$ can be represented as $1+25*$. It gives a powerful way to represent big subsets concisely; a polynomial of degree d in $\mathbb{Z}_{p^k}$ can have at most d *representative roots* [4,13,21] (but exponentially many roots). *Extending* a set $S = \{r_1, \cdots, r_l\}$ of representative roots corresponds to the subset $\bigcup_{i=1}^{l} r_i \subseteq \mathbb{Z}_{p^k}$. Conversely, we show that an $S \subseteq \mathbb{Z}_{p^k}$ can be uniquely represented by representative roots.

Definition 6. *Let* $S \subseteq \mathbb{Z}_{p^k}$, *then the set of representative roots* $S^{rep} = \{r_1, ..., r_l\}$ *(for* $r_i = \beta_i + p^{k_i}*$, *for some* $\beta_i \in \mathbb{Z}_{p^k}$ *and* $k_i \in [k]$*) is said to be a minimal root set representation of* S *if*

1. $S = \bigcup_{i=1}^{l} r_i$,
2. $\nexists\, r_i, r_j \in S^{rep} : r_i \subseteq r_j$,
3. $\forall i : \bigcup_{b \in [p]} \left(r_i + p^{k_i - 1} \cdot b\right) \nsubseteq S$

Theorem 5. *Given any set* $S \subseteq \mathbb{Z}_{p^k}$, *the minimal root set representation of* S *is unique.*

We note a few observations about representative roots.

Observation 6. *Given any two representative roots* $A_1 = \beta_1 + p^{k_1}*$ *and* $A_2 = \beta_2 + p^{k_2}*$, *then either* $A_1 \subseteq A_2$ *or* $A_2 \subseteq A_1$ *or* $A_1 \cap A_2 = \emptyset$.

Observation 7. *Let* $a_1 \in \beta_1 + p^{k_1}*$ *and* $a_2 \in \beta_2 + p^{k_2}*$ *be any 2 elements of the representative roots* $\beta_1 + p^{k_1}*$ *and* $\beta_2 + p^{k_2}*$ *respectively, for* $\beta \neq \alpha_2$, *then,*

$$w_p(a_1 - a_2) = w_p(\beta_1 - \beta_2).$$

Observation 8. *Let* $(a_0, a_1, ...)$ *be a p-ordering on* $\mathbb{Z}_{p^k}$, *then* $(\beta + a_0 * p^j, \beta + a_1 * p^j, \beta + a_2 * p^j, ...)$ *is a p-ordering on* $\beta + p^j*$.

Proof. A simple proof of this theorem follows from Observation 2 and the fact that $1, 2, 3, \ldots$ form an obvious p-ordering in $\mathbb{Z}_{p^k}$. □

Observation 9. *Let* $k < p$, $S_j := \{s \in \mathbb{Z}_{p^k} \mid s \equiv j \bmod p\}$ *and* $f(x)$ *be a polynomial in* $\mathbb{Z}_{p^k}[x]$ *with root-set* A. *If there exist* $\alpha_1, ..., \alpha_k \in A \cap S_j$ *such that* $\alpha_i \not\equiv \alpha_j \bmod p^2$ *for all* (i, j) *pairs, then* $S_j \subseteq A$.

3 Algorithm to Find p-ordering on a Given Set

The naive algorithm for finding the p-ordering, from its definition, has time complexity $\widetilde{\mathcal{O}}(n^3 k \log(p))$. The main result of this section is the following theorem.

Theorem 10. *Given a set $S \subseteq \mathbb{Z}$, a prime p and an integer k such that each element of S is less than p^k, we can find a p-ordering on this set in $\widetilde{\mathcal{O}}(nk \log p)$ time.*

The proof of Theorem 10 follows from the construction of Algorithm 1.

Outline of Algorithm 1. We use the recursive structure of p-ordering given by Maulik [19]. Crucially, to find the p-value of an element a at each step, we only need to look at elements congruent modulo p to a (Theorem 4).

Suppose S_j is the set of elements of S congruent to $j \bmod p$. By the observation above, our algorithm constructs the p-ordering of set S by merging the p-ordering of S_j's. Given a p-ordering up to some step, the next element for the p-ordering of S is computed by just comparing the first elements in p-ordering of S_j's (not present in the already computed p-ordering). The p-ordering of translated S_j's is computed recursively (Observation 2).

While merging the p-orderings on each of the S_i's, at each step we need to extract and remove the element with the minimum p-value over all S_j's and replace it with the next element from the p-ordering on the same set S_j. Naively, it would need to find the minimum over all p number of elements taking $\widetilde{\mathcal{O}}(p)$ time. Instead, we use min heap data structure, using only $\widetilde{\mathcal{O}}(\log p)$ time for extraction and insertion of elements.

Each node of the min-heap(H) contains the value of the element (given as input) as *value* and a key, p-value. For every element there is another key *set*, which stores the index of the subsets $S_0, S_1, \ldots S_{p-1}$ to which it belongs. The heap is sorted in terms of the p-value and whenever we do any operation like extracting the minimum, or adding to the heap, we consider ordering given by increasing value of the p-value attribute of the nodes.

3.1 Proof of Correctness

To prove the correctness of Algorithm 1, we need two results: MERGE() procedure works and valuation is computed correctly in the algorithm.

Theorem 11 (Correctness of MERGE()). *In Algorithm 1, given S be a subset of integers, let for $k \in \{0, 1, ..., p-1\}$, $S_k = \{s \in S \mid s \equiv k \pmod p\}$, then given a p-ordering on each of the S_k's, MERGE($S_0, S_1, ..., S_{p-1}$) gives a valid p-ordering on S.*

Proof Outline. We start with p-orderings on each of the non-empty sets $(S_0, S_1, ..., S_{p-1})$, and create a heap taking the first element of each of these p-ordering. At each successive step, we pick the element in the heap with minimum p-value to add to the p-ordering, and insert the next element from the corresponding S_j to the heap.

Algorithm 1. Find p-ordering

1: **procedure** MERGE($S_0, S_1, ..., S_{p-1}$)
2: $S \leftarrow [\,]$
3: **for** $i \in [0, 1, ..., p-1]$ **do**
4: **for** $j \in [0, ..., len(S_i) - 1]$ **do**
5: $S_i[j].set \leftarrow i$
6: $i_0, i_1, i_2, ...i_{p-1} \leftarrow (0, 0, ..., 0)$
7: $H \leftarrow$ CREATE_MIN_HEAP($node = \{S_0[i_0], ..., S_{p-1}[i_{p-1}]\}, key = p\_value$)
8: **while** H.IsEmpty()!=true **do**
9: $a \leftarrow$ EXTRACT_MIN(H)
10: $j \leftarrow a.set$
11: **if** $i_j < len(S_j)$ **then**
12: $i_j \leftarrow i_j + 1$
13: INSERT($H, S_j[i_j]$)
14: $S \leftarrow a$
15: **return** S
16: **procedure** FIND_p-ORDERING(S)
17: **if** length(S)==1 **then**
18: $S[0].p\_value \leftarrow 1$
19: **Return** S
20: $S_0, S_1, ..., S_{p-1} \leftarrow ([\,], [\,], ..., [\,])$
21: **for** $i \in S$ **do**
22: $S_{i.value \bmod p}.append(i)$
23: **for** $i \in [0, 1, ..., p-1]$ **do**
24: $S_i \leftarrow$ FIND_p-ORDERING($(S_i - i)/p$)
25: **for** $j \in [0, ..., len(S_i) - 1]$ **do**
26: $S_i[j].value \leftarrow p * S_i[j].value + i$
27: $S_i[j].p\_value \leftarrow S_i[j].p\_value + j$
28: $S \leftarrow$ MERGE($S_0, S_1, ..., S_{p-1}$)
29: **return** S

In Algorithm 1, we use a sorted list $\mathcal{I}$ of non-empty S_i's, and only iterate over $\mathcal{I}$ in steps 3-5, 23-28. Hence, decreasing the time complexities of these loops. We can create/update the list $\mathcal{I}$ in the loop at steps 21-22.

We know that the valuation of any element in the combined p-ordering is going to be equal to their valuation in the p-ordering over the set S_j containing them (by Theorem 4). If we show that at each step the element chosen has the least valuation out of all the elements left MERGE() works correctly. We prove this by getting a contradiction. If any element other than the ones obtained from the heap is selected, we show that the p-value will be greater than what we get from MERGE(). The details of the proof can be found in the full version. □

Theorem 12 (Correctness of valuations). *In Algorithm 1, let S be a subset of integers, then* FIND_p-ORDERING(S) *gives a valid p-values for all elements of S.*

Proof Outline. The proof requires two parts: MERGE() preserves valuation and changes in the valuation due to *translation* does not induce errors.

- To prove that MERGE() preserves valuation, we make use of the fact that the combined p-ordering after merge has the individual p-orderings as a sub-sequence. This is true as when an element is added to the p-ordering, the power of p contributed to the p-sequence is only due to the elements in that sequence which are congruent to this new element modulo p. So minimizing the valuation over the entire sequence is same as minimizing the valuation with respect to p in the sub-sequence which are elements congruent to one number modulo p. And now minimizing the power of p in this sub-sequence is same as finding a p-ordering on it. Hence, the valuation of each element in the combined p-ordering is going to be equal to the valuation in the individual p-ordering (by Theorem 4). Hence, MERGE() preserves valuations.
- We show that the change in valuations due to translation (Step 24) are corrected (Steps 26–27). This is easy to show by just updating the valuation according to Observation 3.

Hence, valuations are correct maintained throughout the algorithm. The details of the proof can be found in the full version. □

Using the above two theorems, we prove the correctness of Algorithm 1.

Proof of Correctness of Algorithm 1. For the base case, if S is a singleton, then the p-ordering over it is just a single element which is also what FIND_p-ORDERING(S) gives. Let FIND_p-ORDERING() works for $|S| < k$, if we show it works for $|S| = k$, then by induction, FIND_p-ORDERING() works for sets of arbitrary sizes.

Let $|S| = k$, then when we break the set into $S_0, S_1, ..., S_{p-1}$ (Steps 20–22), either all element belong in a single S_i or get distributed into multiple sets. If all of the elements are congruent to each other modulo p, then we apply the recursive step until we reach a point when there is at least one element which is not congruent to all the other elements in S, i.e. S breaks into subsets (Step 21). This follows from the p-adic expansion being unique and not all elements being same in our input set S. Since, by Observation 2, we know that the p-ordering on reduced elements is preserved, we will get the correct p-ordering on the original set. Hence, we only need to prove this for the later case.

Since all the element of the set S_i follow $\forall y \in S_i$, $y \equiv i \bmod p$, hence $\forall y \in S_i$, $p \mid (y - i)$, this implies $(S_i - i)/p \subset \mathbb{Z}$. Hence, FIND_$p$-ORDERING($(S_i - i)/p$) gives a p-ordering on $(S_i - i)/p$ with the correct valuations associated with each element (Theorem 12).

From Observation 2, we know that if $(a_0, a_1, ...)$ is a p-ordering on some set A, then $(p * a_0 + i, p * a_1 + i, ...)$ is a p-ordering on $p * A + i$. Since, FIND_p-ORDERING($(S_i - i)/p$) is a p-ordering on $(S_i - i)/p$, then $p *$ FIND_p-ORDERING($(S_i - i)/p$) $+ x$ is a p-ordering on S_i (Steps 26–27).

Next, since we have valid p-orderings on $S_0, S_1, ..., S_{p-1}$, MERGE($S_0, S_1, ..., S_{p-1}$) returns a valid p-ordering on S (Theorem 11). By induction, our algorithm returns a valid p-ordering on any subset of integers. □

3.2 Time Complexity

Theorem 13. *Given a set $S \subset \mathbb{Z}$ of size n and a prime p, such that for all elements $a \in S$, $a < p^k$ for some k, Algorithm 1 returns a p-ordering on S in $\widetilde{\mathcal{O}}(nk \log p)$ time.*

Proof. We break the complexity analysis into 2 parts, the time complexity for merging the subsets S_i's and the time complexity due the to recursive step.

Time Complexity of MERGE$(S_0, S_1, ..., S_{p-1})$ *in Algorithm* 1. Let $|S_0| + |S_1| + ... + |S_{p-1}| = m$. Then, the time complexity of making the heap (Step 7) is $\widetilde{\mathcal{O}}(\min(m, p))$ (the size of the heap). Next, the construction of common p-ordering(Steps 8–14) takes $\widetilde{\mathcal{O}}(m \log p)$ time, this is because extraction of an element and addition of an element are both bound by $\widetilde{\mathcal{O}}(\log p)$ and the runs a total of m times. Hence, the total time complexity of MERGE$(S_0, S_1, ..., S_{p-1})$ is $\widetilde{\mathcal{O}}(\min(m, p) + m \log p) = \widetilde{\mathcal{O}}(m \log p)$ time.

Time Complexity of Algorithm 1. Let $|S| = n$ and $S \subset \mathbb{Z}_{p^k}$. Then the recursion depth of FIND_p-ORDERING(S) is bound by k. Now at each depth, all the elements are distributed into multiple heaps(of sizes $m_1, m_2, ..., m_q$). Hence, the sum of sizes of all smaller sets at a given depth $\sum_{i=1}^{q} m_i < n$. Hence, the time to run any depth is $\sum_{i=1}^{q} \widetilde{\mathcal{O}}(m_i \log p) = \widetilde{\mathcal{O}}(n \log p)$. Hence, total time complexity for k depth is $\widetilde{\mathcal{O}}(nk \log p)$. □

The proof of Theorem 10 follows from the proof of correctness of Algorithm 1 and time complexity obtained from Theorem 13.

4 Algorithm to Find p-ordering on a Set of Representative Roots

The notion of representative roots (Definition 5) allows us to represent an exponentially large subset of $\mathbb{Z}_{p^k}$ succinctly. Further imposing a few simple conditions on this representation, namely the minimal representation (Definition 6), our subset is represented in a unique way (Theorem 5). A natural question arises, can we efficiently find a p-ordering given a set in terms of representative roots? For example, given a set $\{1, 2, 4, 7, 10, 11, 13, 16, 19, 20, 22, 25\}$ and prime $p = 3$, we can write this set as an union of root sets modulo 3^3 as $\{1 + 3*, 2 + 3^2*\}$. A 3-ordering on this set is $\{1, \mathbf{2}, 4, 7, \mathbf{11}, 10, \mathbf{20}, 13, 16, 19, 22, 25\}$. In this section we give an efficient algorithm to find a p-ordering of a given length n on a set expressed in minimal representation.

Theorem 14. *Given a set $S \subset \mathbb{Z}_{p^k}$, for a prime p and an integer k, that can be represented in terms of d representative roots, we can efficiently find a p-ordering of length n for S in $\widetilde{\mathcal{O}}(d^2 k \log p + nk \log p + np)$ time.*

Outline of the Algorithm. The important observation is, we already have a natural p-ordering defined on a representative root (Observation 8). Since a p-ordering on each representative root is already known, we just need to find a way to merge them. Merging was easy in Algorithm 1 because progress in any one of the p-ordering of an S_j did not affect the p-value of an element outside S_j. However, in this case the exact increase in the p-value is known by Observation 7, and is precisely equal to $v_p(\beta_i - \beta_j)$.

We are given with a set S containing d representative roots. It is of the form $\{\beta_i + p^{\ell_i} * | i = 1, 2 \ldots d\}$. We will use the notation $S[i].value$ for β_i and $S[i].exponent$ for ℓ_i in the pseudocode. Further, we can assume that the representative roots are disjoint. If they are not, one representative root will be contained in another (Observation 6), all such occurrences can be deleted in $\widetilde{\mathcal{O}}(d^2 k \log p)$ time.

In our algorithm, we maintain an array of size d to store the valuations that we would get whenever we add the next element from a representative root. To update the i-th value of this array when an element from the j-th representative root is added, we simply add the value $v_p(\beta_i - \beta_j)$ $(i \neq j)$. Hence, at each step we find the minimum value in this array (in $\widetilde{\mathcal{O}}(\mathrm{d})$) and add it to the combined p-ordering (in $\widetilde{\mathcal{O}}(1)$) and we update all the d values in this array (in $\widetilde{\mathcal{O}}(\mathrm{d})$). We repeat this process till we get the p-ordering of the desired length.

With the above intuition in mind, we develop Algorithm 2 to find the p-ordering of length n on a subset S of $\mathbb{Z}_{p^k}$ given in representative root representation.

4.1 Proof of Correctness

To prove the correctness of this algorithm, we first prove that valuations are correctly maintained.

Theorem 15. *In Algorithm 2, FIND_p-ORDERING(S, n) maintains the correct valuations on the set S of representative roots in valuations at every iteration of the loop.*

Proof Outline. All elements have 0 valuation at the beginning (Step 17). Also, adding an element from the i-th representative root increases the valuation of the j-th representative root by $corr(i, j)$ (Step 28) for $i \neq j$ (Observation 7). The increase for the next element of i is exponent times the increase in p-sequence of $\mathbb{Z}_{p^k}$ (Step 26) (Observation 8). So, we correctly update the valuations array in each iteration. Please see full version for a proof. □

Proof of Correctness of Algorithm 2. By the definition of p-ordering we know that at each iteration if we choose the element with the least valuation then we get a valid p-ordering. By Theorem 15, we know that *valuations* array has the correct next valuations. Hence, to find the representative root with the least valuation, we find the index of the minimum element in *valuations*.

Algorithm 2. Find p-ordering from minimal notation

```
 1: procedure CORRELATE(S)
 2:     Corr ← [0]_{len(S)×len(S)}
 3:     Corr ← [0]_{len(S)×len(S)}
 4:     for j ∈ [1, ..., len(S)] do
 5:         for k ∈ [1, ..., len(S)] do
 6:             Corr[j][k] ← v_p(S[j].value − S[k].value)
 7:     return Corr
 8: procedure p-EXPONENT_INCREASE(n)
 9:     v_p(1) ← 1
10:     for j ∈ [1, ..., n] do
11:         v_p((j + 1)!) ← v_p(j + 1) * v_p(j!)
12:         p_exponent[j] ← v_p((j + 1)!) − v_p(j!)
13:     return p_exponent
14: procedure FIND_p-ORDERING(S, n)
15:     corr ← CORRELATE(S)
16:     increase ← p-EXPONENT_INCREASE(n)
17:     valuations ← [0]_{|S|}
18:     p_ordering ← {}
19:     i_1, i_2 ... i_{|S|} ← 0
20:     for i ∈ {1, 2, ... n} do
21:         min ← min{valuations}
22:         index ← argmin{valuations}
23:         p_ordering.append(S[index].value + p^{S[index].exponent} * i_index)
24:         for j ∈ [1, ..., len(S)] do
25:             if j = index then
26:                 valuations[j] ← valuations[j] + S[index].exponent * increase[i_j]
27:             else
28:                 valuations[j] ← valuations[j] + corr(index, j)
29:         i_index ← i_index + 1
30:     return p_ordering
```

To add the next value, by Observation 8, we find the next element from the p-ordering on the representative root. Hence, the element added has the least valuation. Hence, FIND_p-ORDERING(S, n) returns the correct p-ordering. □

4.2 Time Complexity

Theorem 16. *Given a set $S \subset \mathbb{Z}_{p^k}$, for a prime p and an integer k, that can be represented in terms of d representative roots, Algorithm 2 finds a p-ordering of length n for S in $\widetilde{\mathcal{O}}(d^2 k \log p + nk \log p + np)$ time.*

Proof. Let S contains d representative roots of $\mathbb{Z}_{p^k}$ and we want to find the p-ordering up to length n, then, CORRELATE(S) runs a double loop, each of size d, and each iteration takes $\widetilde{\mathcal{O}}(k \log p)$, hence, CORRELATE$(S)$ takes $\widetilde{\mathcal{O}}(d^2 k \log p)$. p-EXPONENT_INCREASE(n) runs a single loop of size n where each iteration takes $\widetilde{\mathcal{O}}(k \log p)$ time, hence, it takes $\widetilde{\mathcal{O}}(nk \log p)$. Then main loop run a loop of size n, inside this loop we do $\mathcal{O}(d)$ operations on elements of size $\log k$, hence, it takes $\widetilde{\mathcal{O}}(nd)$ time. Hence, in total, our algorithm takes $\widetilde{\mathcal{O}}(d^2 k \log p + nk \log p + nd)$ time. □

Now, the proof of Theorem 14 follows directly from the proof of correctness of Algorithm 2 and the time complexity analysis shown in Theorem 16.

5 Structure of Root Sets for a Given k

We know that $\mathbb{Z}_{p^k}$ is not a unique factorization domain. In fact, even small degree polynomials can have exponentially large number of roots as seen in Sect. 2. Interestingly, not all subsets of $\mathbb{Z}_{p^k}$ can be a root set (Definition 4). Distinctly knowing the description of root sets can help us decide if a given set is a root set. In this section, we discuss and distinctly describe all the root sets in Z_{p^2}, Z_{p^3} and Z_{p^4}.

Dearden and Metzger [12] showed that R is a root-set iff $R_j = \{r \in R \mid r \equiv j \pmod p\}$ is also a root-set for all $j \in [p]$. For example, we know that $R = \{1, 4, 5, 7, 9, 10, 13, 14, 16, 19, 22, 23, 25\}$ is the root set in $\mathbb{Z}_{3^3}$ for the polynomial $f(x) = (x-1)^3(x-5)^2(x-9)$, then $R_0 = \{9\}$, $R_1 = \{1, 4, 7, 10, 13, 16, 19, 22, 25\}$, and $R_2 = \{5, 14, 23\}$ are also root sets.

The number and structure of R_j is symmetric for all j. Let N_{p^k} be the number of possible R_j's, then total number of possible root-sets become $(N_{p^k})^p$ [12].

Let $S_j = \{s \in \mathbb{Z}_{p^k} \mid s \equiv j \pmod p\}$, we take the following approach to find all possible root-sets R_j's. Given an R_j, define $R = \{(r-j)/p : r \in R_j\}$ to be the translated copy. We show that if R contains at least k many distinct residue classes mod$p$, then $R_j = S_j$ (Observation 9). We exhaustively cover all the other cases, when R contains less than k residue classes (possible because k is small). For example, in $\mathbb{Z}_{p^3}$, we find that

$$R_j = \begin{cases} j + p \cdot *, \\ (j + p \cdot \alpha_1 + p^2 *) \cup (j + p \cdot \alpha_2 + p^2 *), \text{ for } \alpha_1 \neq \alpha_2 \in [p], \\ j + p \cdot \alpha + p^2 *, \text{ for } \alpha \in [p], \\ j + p \cdot \alpha_1 + p^2 \cdot \alpha_2, \text{ for } \alpha_1, \alpha_2 \in [p], \\ \emptyset. \end{cases}$$

Please see full version for the classification of root sets ($k \leq 4$) and its proof.

Acknowledgements. We would like to thank Naman Jain for helpful discussions. R.M. would like to thank Department of Science and Technology, India for support through grant DST/INSPIRE/04/2014/001799.

References

1. Adleman, L., Lenstra, H.: Finding irreducible polynomials over finite fields. In: Proceedings of 18th Annual ACM Symposium on Theory of Computing (STOC), pp. 350–355 (1986). https://doi.org/10.1145/12130.12166

2. Agrawal, M., Kayal, N., Saxena, N.: Primes is in p. Ann. Math. 781–793 (2004)
3. Berlekamp, E.: Factoring polynomials over large finite fields. Math. Comput. **24**, 713–735 (1970). https://doi.org/10.1090/S0025-5718-1970-0276200-X
4. Berthomieu, J., Lecerf, G., Quintin, G.: Polynomial root finding over local rings and application to error correcting codes. Appl. Algebra Eng. Commun. Comput. **24**(6), 413–443 (2013). https://doi.org/10.1007/s00200-013-0200-5
5. Bhargava, M.: P-orderings and polynomial functions on arbitrary subsets of dedekind rings. Journal Fur Die Reine Und Angewandte Mathematik - J. REINE ANGEW Math. 101–128 (1997). https://doi.org/10.1515/crll.1997.490.101
6. Bhargava, M.: The factorial function and generalizations. Am. Math. Mon. **107** (2000). https://doi.org/10.2307/2695734
7. Bhargava, M.: On p-orderings, rings of integer values functions, and ultrametric analysis. J. Am. Math. Soc. **22**(4), 963–993 (2009)
8. Bose, R., Ray-Chaudhuri, D.: On a class of error correcting binary group codes *. Inf. Control **3**, 68–79 (1960). https://doi.org/10.1016/S0019-9958(60)90287-4
9. Cantor, D., Zassenhaus, H.: A new algorithm for factoring polynomials over finite fields. Math. Comput. **36** (1981). https://doi.org/10.2307/2007663
10. Cheng, Q., Gao, S., Rojas, J.M., Wan, D.: Counting roots for polynomials modulo prime powers. Open Book Ser. **2**(1), 191–205 (2019)
11. Chor, B., Rivest, R.: A knapsack type public key cryptosystem based on arithmetic in finite fields. IEEE Trans. Inf. Theory **34** (2001). https://doi.org/10.1109/18.21214
12. Dearden, B., Metzger, J.: Roots of polynomials modulo prime powers. Eur. J. Comb. **18**, 601–606 (1997). https://doi.org/10.1006/eujc.1996.0124
13. Dwivedi, A., Mittal, R., Saxena, N.: Efficiently factoring polynomials modulo p^4. In: International Symposium on Symbolic and Algebraic Computation (ISSAC), pp. 139–146 (2019). https://doi.org/10.1145/3326229.3326233
14. Hocquenghem, A.: Codes correcteurs d'erreurs. Chiffres, Revue de l'Association Française de Calcul 2 (1959)
15. Johnson, K.: P-orderings of finite subsets of dedekind domains. J. Algebraic Combinatorics **30**, 233–253 (2009)
16. Lenstra, A., Lenstra, H., Lovász, L.: Factoring polynomials with rational coefficients. Mathematische Annalen **261** (1982). https://doi.org/10.1007/BF01457454
17. Lenstra, H.W.: On the Chor—Rivest knapsack cryptosystem. J. Cryptol. **3**(3), 149–155 (1991). https://doi.org/10.1007/BF00196908
18. Lidl, R., Niederreiter, H.: Finite Fields, vol. 20. Cambridge University Press, Cambridge (1997)
19. Maulik, D.: Root sets of polynomials modulo prime powers. J. Comb. Theory, Ser. A **93**, 125–140 (2001). https://doi.org/10.1006/jcta.2000.3069
20. Odlyzko, A.M.: Discrete logarithms in finite fields and their cryptographic significance. In: Beth, T., Cot, N., Ingemarsson, I. (eds.) EUROCRYPT 1984. LNCS, vol. 209, pp. 224–314. Springer, Heidelberg (1985). https://doi.org/10.1007/3-540-39757-4_20
21. Panayi, P.N.: Computation of Leopoldt's P-adic regulator. Ph.D. thesis, University of East Anglia (1995)
22. Reed, I., Solomon, G.: Polynomial codes over certain finite fields. J. Soc. Ind. Appl. Math. **8**, 300–304 (1960). https://doi.org/10.2307/2098968
23. Sudan, M.: Decoding reed solomon codes beyond the error-correction bound. J. Complexity **13**, 180–193 (1997). https://doi.org/10.1006/jcom.1997.0439
24. Zassenhaus, H.: On hensel factorization ii. J. Number Theory **1**, 291–311 (1969). https://doi.org/10.1016/0022-314X(69)90047-X

Experimental Evaluation of a Local Search Approximation Algorithm for the Multiway Cut Problem

Andrew Bloch-Hansen[1], Nasim Samei[2], and Roberto Solis-Oba[1](✉)

[1] Department of Computer Science, The University of Western Ontario, London, ON N6A 5B7, Canada
{ablochha,rsolisob}@uwo.ca

[2] IST Austria, Am Campus 1, 3400 Klosterneuburg, Austria
nasim.samei@ist.ac.at

Abstract. In the multiway cut problem we are given a weighted undirected graph $G = (V, E)$ and a set $T \subseteq V$ of k terminals. The goal is to find a minimum weight set of edges $E' \subseteq E$ with the property that by removing E' from G all the terminals become disconnected. In this paper we present a simple local search approximation algorithm for the multiway cut problem with approximation ratio $2 - \frac{2}{k}$. We present an experimental evaluation of the performance of our local search algorithm and show that it greatly outperforms the isolation heuristic of Dalhaus et al. and it has similar performance as the much more complex algorithms of Calinescu et al., Sharma and Vondrak, and Buchbinder et al. which have the currently best known approximation ratios for this problem.

Keywords: Multiway cut · Approximation algorithms · Experimental evaluation

1 Introduction

Given an undirected graph $G = (V, E)$ with non-negative weights or costs on the edges and a set $T = \{t_1, t_2, ..., t_k\} \subseteq V$ of k terminals, a multiway cut is a set $E' \subseteq E$ of edges whose removal from G separates all the terminals from each other. In the multiway cut problem the goal is to find a multiway cut with minimum weight. For the case when $k = 2$ the multiway cut problem reduces to the well known minimum s-t cut problem. For $k \geq 3$ Dahlhaus et al. [8] proved that the multiway cut problem is MAX SNP-hard.

The multiway cut problem has a variety of applications including task scheduling in multi-processors systems [14], task allocation in parallel computing systems [11], labelling the pixels of an image [3], integrated circuit layout design [1,12] and combinatorial optimization [10].

Dahlhaus et al. [8] presented the first approximation algorithm for the multiway cut problem, called the isolation heuristic, which has approximation ratio $2 - \frac{2}{k}$. Calinescu et al [6] utilized an elegant geometric relaxation algorithm and

A. Mudgal and C. R. Subramanian (Eds.): CALDAM 2021, LNCS 12601, pp. 346–358, 2021.
https://doi.org/10.1007/978-3-030-67899-9_28

improved the approximation ratio to $1.5 - \frac{1}{k}$. Karger et al. [9] used a similar geometric relaxation technique to design a $\frac{12}{11}$-approximation algorithm for the problem for the case when $k = 3$ and for general k they improved the approximation ratio to 1.3438. Independently, Cunningham and Tang [7] designed an approximation algorithm with the same approximation ratio $\frac{12}{11}$ for the case $k = 3$. Sharma and Vondrak [13] designed a better algorithm with approximation ratio 1.2965 based on the linear programming relaxation used in [6]. Buchbinder et al. [4,5] almost matched Sharma and Vondrak's approximation ratio, but using a much simpler algorithm. These last two algorithms have the currently best known approximation ratio for the multiway cut problem.

In this paper we present a simple local search algorithm for the multiway cut problem with approximation ratio $2 - \frac{2}{k}$. Our algorithm is a simplification of the local search algorithm in [3] for an energy minimization problem in image processing which is also a 2-approximation algorithm for the multiway cut problem. We not only improve the approximation ratio to $2 - \frac{2}{k}$ but we show that this ratio is tight. Due to space limitations the tight example is not included.

A modification of our local search algorithm can be used on two variations of the multiway cut problem: 1) when some set of nodes needs to be in the same partition, and 2) when certain nodes can only be in some partitions. These results are not included because of the space limitations.

The main contribution of this paper is an experimental comparison of our local search algorithm with four other approximation algorithms for the multiway cut problem: Dahlhaus et al. [8] isolation heuristic, the algorithm of Calinescu et al. [6], the algorithm of Sharma and Vondrak [13], and the algorithm of Buchbinder et al. [4,5]. Even though our algorithm does not have the best known approximation ratio for the problem, our results show that in practice our algorithm performs much better than the isolation heuristic [8] even though they have the same theoretical worst-case approximation ratio. Furthermore, our algorithm has comparable performance, with respect to running time and the values of the solutions that it produces, to the three currently best known algorithms for the multiway cut problem: the algorithm of Calinescu et al. [6], the algorithm of Sharma and Vondrak [13], and the algorithm of Buchbinder et al. [4,5].

Our results show that local search algorithms are easy to implement, they are fast and produce high quality solutions. We hope that our research will help increase the use of local search algorithms in practice.

2 Our Local Search Algorithm

As mentioned in the previous section our local search algorithm is a simplification of the algorithm in [3]. Here for convenience we give a brief explanation of the algorithm. Given a graph $G = (V, E)$ and a set L of labels, a labelling function f assigns a label to each node of G. A relabel operation $R\langle A, \alpha, f\rangle$ modifies a labelling function f by changing the labels of all the nodes in a given set $A \subseteq V$ to some label α while keeping the other labels unchanged.

We can formulate the multiway cut problem as a labelling problem as follows: Given a graph $G = (V, E)$, a set of k terminals $T = \{t_1, t_2, ..., t_k\} \subseteq V$ and a set of k labels $L = \{\alpha_1, \alpha_2, ..., \alpha_k\}$, where each terminal t_i has a fixed label α_i, the goal is to label the nodes in $V - T$ in such a way as to minimize the total cost of the set E' of edges whose endpoints have different labels. Note that E' is a multiway cut for G and therefore finding a solution for this labelling problem is equivalent to finding a minimum weight multiway cut.

The *neighborhood function* for a given labelling function f and label α_i, $i = 1, 2, ..., k$ is defined as follows: $\mathcal{N}_i(f) = \{$all the labellings f' that can be obtained by $R\langle A, \alpha_i, f\rangle$, for all possible sets $A \subseteq V\}$. Our local search algorithm for the multiway cut problem is as follows.

Algorithm. MultiwayCut ($G = (V, E)$, L, T)
In: Graph $G = (V, E)$, set L of k labels, and set $T \subseteq V$ of k terminals
Out: Labelling of a local optimum solution for the multiway cut problem
$f \leftarrow$ any labelling function that assigns to each terminal a different label
success $\leftarrow 1$
while *success* $= 1$
    success $\leftarrow 0$
    for $i \leftarrow 1$ **to** k **do**
        Compute a minimum cost labelling $f' \in \mathcal{N}_i(f)$
        if cost of labelling f' < cost of labelling f **then**
            $f \leftarrow f'$
            success $\leftarrow 1$
        end if
    end for
end while
Output f

Note that algorithm MultiwayCut might not run in time that is polynomial in the size of the input as each iteration of the while loop might only give a marginal improvement in the cost of the solution, so it might need a very large number of iterations to find a local optimum solution. We proceed as in [2] to ensure a polynomial running time: Replace the condition of the if statement as follows

if cost of labelling $f' < (1 - \epsilon) \times$ cost of labelling f **then**

where ϵ is a positive value. With this change the maximum number of iterations of the outer loop is $O((log\, n + log(c_{max})/\epsilon)$, where n is the number of vertices and c_{max} is the largest edge cost. Each iteration of the loop needs polynomial time, so the running time of the algorithm is polynomial. This change causes the approximation ratio of the algorithm to be at most $\frac{1}{1-\epsilon}(2 - \frac{2}{k})$.

3 Finding a Minimum Cost Relabel Operation

A minimum cost labelling f' obtainable from f through a relabel $R\langle A, \alpha, f\rangle$ is obtained by computing a minimum cut in the following graph $G_\alpha = (V_\alpha, E_\alpha)$:

- $V_\alpha = \{\{\alpha, \overline{\alpha}\}, V, \{\bigcup_{(p,q)\in E} a_{pq}\}\}$, where α is a source node, $\bar{\alpha}$ is a sink node and a_{pq} are auxiliary nodes.
- $E_\alpha = \bigcup_{p\in V}\{(p,\alpha),(p,\overline{\alpha})\} \bigcup_{\substack{(p,q)\in E \\ f(p)\neq f(q)}} \varepsilon_{\{p,q\}} \bigcup_{\substack{(p,q)\in E \\ f(p)=f(q)}} (p,q)$

The edges have assigned weights as shown in Table 1, where P_α is the set of nodes labelled α in G.

Table 1. Weights for the edges in G_α.

Edge	Weight	Edge	Weight
(α, t_i), $t_i \in T$	∞ if $t_i \notin P_\alpha$ 0 if $t_i \in P_\alpha$	(p, a_{pq}), $f(p) \neq f(q)$	0 if $p \in P_\alpha$ $cost(p,q)$ if $p \notin P_\alpha$
$(\overline{\alpha}, t_i)$, $t_i \in T$	0 if $t_i \notin P_\alpha$ ∞ if $t_i \in P_\alpha$	(a_{pq}, q), $f(p) \neq f(q)$	0 if $q \in P_\alpha$ $cost(p,q)$ if $q \notin P_\alpha$
(α, p), $p \in V - T$	0	(p,q), $f(p) = f(q)$	$cost(p,q)$
$(\overline{\alpha}, p)$, $p \notin P_\alpha$	0	$(\overline{\alpha}, a_{pq})$, $f(p) \neq f(q)$	$cost(p,q)$
$(\overline{\alpha}, p)$, $p \in P_\alpha$	∞		

Due to space limitation the proofs of the following lemmas are not included.

Lemma 1. *Each minimal cut C of G_α of bounded cost separating α from $\bar{\alpha}$ defines a labelling f' for G that can be obtained from f through a relabel operation and its cost is at most the cost of C. This labelling f' is defined as follows: $f'(p) = \alpha$ if $(p,\alpha) \in C$ and $f'(p) = f(p)$ if $(p,\overline{\alpha}) \in C$, $\forall p \in V$.*

Lemma 2. *A labelling f' obtained from f by a relabel operation $R(A,\alpha,f)$ defines a minimal cut of G_α separating α from $\bar{\alpha}$ of cost equal to the cost of f'.*

Theorem 1. *There is a polynomial time algorithm that given a labelling function f for a graph $G = (V,E)$ and a label α it computes a minimum cost labelling f' that can be obtained from f through a single relabel operation $R\langle A,\alpha,f\rangle$.*

Proof. The algorithm builds the graph G_α, computes a minimum cut C' of G_α separating α from $\bar{\alpha}$ and outputs the labelling f' defined by C' as described in Lemma 1. To see that f' is a minimum cost labelling obtained from f through a relabel operation, assume that there is a labelling f'' obtained from f through a relabel operation of cost $cost(f'') < cost(f')$. By Lemma 2, f'' defines a cut C'' of G_α separating α from $\bar{\alpha}$ of cost $cost(C'') = cost(f'')$. But then by Lemma 1, $cost(C'') = cost(f'') < cost(f') \leq cost(C')$ contradicting C' is minimum. □

4 Analysis of MultiwayCut for the 3-Way Cut Problem

Since the analysis of our algorithm is a bit complex, we only sketch the analysis for the case when $k = 3$. The analysis for the case when $k > 3$ is not included because of the space limitations. Let α_1, α_2 α_3 be the 3 labels, let $\hat{f}$ be the labelling function computed by MultiwayCut and f^* be the labelling function of a global optimal solution. We define partitions $\hat{A}_1$, $\hat{A}_2$, $\hat{A}_3$ in the local optimal solution and A_1^*, A_2^*, A_3^* in the global optima solution as follows: $\hat{A}_i = \{v \in V|\hat{f}(v) = \alpha_i\}$ and $A_i^* = \{v \in V|f^*(v) = \alpha_i\}$, for all $1 \leq i \leq 3$.

Note that if we perform a relabel operation on the local optimum solution, we produce a new solution of the same or larger cost. We bound the cost of the local optimal solution interms of the cost of the global optimum solution using this local optimality property. Let $\hat{S}$ and S^* be the sets of edges crossing the partitions of the local optimal solution and of the global optimal solution, respectively. Let $\hat{S}_p = \hat{S} - S^*$ and $S_p^* = S^* - \hat{S}$.

Let $\hat{B}_{\alpha_i\alpha_j} = \{(v, u) \in E|\hat{f}(v) = \alpha_i, \hat{f}(u) = \alpha_j\}$ and $B^*{}_{\alpha_i\alpha_j} = \{(v, u) \in E|f^*(v) = \alpha_i, f^*(u) = \alpha_j\}$, for all $1 \leq i < j \leq 3$. Let P and Q be sets of nodes and B be a set of edges. We define $(B|P) = \{(u, v) \in B|u, v \in P\}$ and $(B|P : Q) = \{(u, v) \in B|\ u \in P, v \in Q\}$. The cost $C(B)$ of set B is the total cost of the edges in B. We will compute two different bounds for the value of the local optimum solution that will yield the approximation ratio of our algorithm.

First Bound. Let $A_{i_s} = A_i^* \cap \hat{A}_i$ and $A_{i_s}{}^c = \hat{A}_i - A_{i_s}$ for $i = 1, 2, 3$. We first perform the relabel operation $R\langle A_{1_s}{}^c, \alpha_2, \hat{f}\rangle$ which changes the label that $\hat{f}$ assigns to nodes in $A_{1_s}{}^c$ to α_2: Thus, (1) it decreases the contribution to the cost of the solution made by the set Δ_1 of edges with one endpoint in $A_{1_s}{}^c$ and one endpoint in $\hat{A}_2$, and (2) it increases the contribution to the cost of the solution made by the set Δ_2 of edges with one endpoint in $A_{1_s}{}^c$ and the other in A_{1_s}.

Note that $A_{1_s}{}^c = \hat{A}_1 - A_{1_s} = \hat{A}_1 \cap (A_2^* \cup A_3^*)$ and so $\Delta_1 = (\hat{B}_{\alpha_1\alpha_2}|A_2^* \cup A_3^* : V)$ and $\Delta_2 = (B^*_{\alpha_1\alpha_3} \cup B^*_{\alpha_1\alpha_2}|\hat{A}_1)$. Observe that set $\Delta = (\hat{B}_{\alpha_1\alpha_2}|A_2^*) \cup (\hat{B}_{\alpha_1\alpha_2}|A_3^*) \subseteq \Delta_1$ and sets $(\hat{B}_{\alpha_1\alpha_2}|A_2^*)$ and $(\hat{B}_{\alpha_1\alpha_2}|A_3^*)$ are disjoint. Then, after performing the relabel operation $R\langle A_{1_s}{}^c, \alpha_2, \hat{f}\rangle$, by the local optimality condition we can show that

$$C(\hat{B}_{\alpha_1\alpha_2}|A_2^*) + C(\hat{B}_{\alpha_1\alpha_2}|A_3^*) \leq C(B^*_{\alpha_1\alpha_3} \cup B^*_{\alpha_1\alpha_2}|\hat{A}_1). \quad (1)$$

By performing $R\langle A_{1_s}{}^c, \alpha_3, \hat{f}\rangle$, $R\langle A_{2_s}{}^c, \alpha_1, \hat{f}\rangle$, $R\langle A_{2_s}{}^c, \alpha_3, \hat{f}\rangle$, $R\langle A_{3_s}{}^c, \alpha_1, \hat{f}\rangle$ and $R\langle A_{3_s}{}^c, \alpha_2, \hat{f}\rangle$ and adding the corresponding inequalities we can show that:

$$C(\hat{S}_p) \leq 2\left[C(S_p^*) - \left(C(B^*_{\alpha_2\alpha_3}|\hat{A}_1) + C(B^*_{\alpha_1\alpha_3}|\hat{A}_2) + C(B^*_{\alpha_1\alpha_2}|\hat{A}_3)\right)\right]. \quad (2)$$

Adding $C(\hat{S} \cap S^*)$ to the left hand side and $2C(\hat{S} \cap S^*)$ to the right hand side of this last inequality we get

$$C(\hat{S}) \leq 2\left[C(S^*) - \left(C(B^*_{\alpha_2\alpha_3}|\hat{A}_1) + C(B^*_{\alpha_1\alpha_3}|\hat{A}_2) + C(B^*_{\alpha_1\alpha_2}|\hat{A}_3)\right)\right]. \quad (3)$$

Second Bound. First, we perform the relabel operation $R\langle A_1^*, \alpha_1, \hat{f}\rangle$. This operation decreases the contribution to the cost of the solution made by $\Theta_1 = (\hat{B}_{\alpha_1\alpha_2} \cup \hat{B}_{\alpha_1\alpha_3} \cup \hat{B}_{\alpha_2\alpha_3}|A_1^*) = (\hat{S}|A_1^*)$ and by $\Theta_2 = (\hat{S} \cap S^*|A_1^* : \hat{A}_1 \cap (A_2^* \cup A_3^*))$. Let $\Theta_3 = (B^*_{\alpha_1\alpha_2} \cup B^*_{\alpha_1\alpha_3}|\hat{A}_2)$ and $\Theta_4 = (B^*_{\alpha_1\alpha_2} \cup B^*_{\alpha_1\alpha_3}|\hat{A}_3)$. After performing the relabel operation $R\langle A_1^*, \alpha_1, \hat{f}\rangle$ the cost of the solution is increased by the cost of edges in Θ_3 and Θ_4. By the local optimality property we get

$C(\Theta_3) + C(\Theta_4) - C(\Theta_1) - C(\Theta_2) \geq 0$. This inequality can be rewritten as follows since $C(\hat{S} \cap S^* | A_1^* : \hat{A}_1 \cap (A_2^* \cup A_3^*)) \geq 0$:

$$C(\hat{S}|A_1^*) \leq C(B^*_{\alpha_1\alpha_2} \cup B^*_{\alpha_1\alpha_3}|\hat{A}_2) + C(B^*_{\alpha_1\alpha_2} \cup B^*_{\alpha_1\alpha_3}|\hat{A}_3). \tag{4}$$

We perform two more relabel operations: $R\langle A_2^*, \alpha_2, \hat{f}\rangle$ and $R\langle A_3^*, \alpha_3, \hat{f}\rangle$. Adding the inequalities obtained from these 3 relabel operations we can show that

$$C(\hat{S}_p) \leq C(S_p^*) + \left[C(B^*_{\alpha_2\alpha_3}|\hat{A}_1) + C(B^*_{\alpha_1\alpha_3}|\hat{A}_2) + C(B^*_{\alpha_1\alpha_2}|\hat{A}_3)\right]. \tag{5}$$

Finally, adding $C(\hat{S} \cap S^*)$ to both sides of (5) we get our second bound,

$$C(\hat{S}) \leq C(S^*) + \left[C(B^*_{\alpha_2\alpha_3}|\hat{A}_1) + C(B^*_{\alpha_1\alpha_3}|\hat{A}_2) + C(B^*_{\alpha_1\alpha_2}|\hat{A}_3)\right]. \tag{6}$$

Theorem 2. *The approximation ratio of algorithm* MULTIWAYCUT *for the 3-way cut problem is* $\frac{4}{3}$.

Proof. We multiply (6) by two and add it to (3) to get: $3C(\hat{S}) \leq 4C(S^*) \Rightarrow C(\hat{S}) \leq \frac{4}{3}C(S^*)$.

5 Experimental Results

We compared our local search algorithm with four other approximation algorithms for the multiway cut problem: the isolation heuristic [8], the algorithm of Calinescu et al. [6], the algorithm of Sharma and Vondrak [13], and the algorithm of Buchbinder et al. [4,5].

We implemented the algorithms in Java. The integer program solver Cplex 12.7, configured using default settings, was used to compute optimal solutions for each test instance. The hardware used to run the experiments were a computer using an Intel Core i5-5200U 220 GHz (4 CPUs) with 16 GB of RAM and SHARCNET's high performance computing clusters Orca, using an AMD Opteron 2.2 GHz (4 CPUs) with 32 GB of RAM; Saw, using an Intel Xeon 2.83 Ghz (4 CPUs) with 16 GB of RAM, and Kraken, using an AMD Opteron 2.2 GHz (4 CPUs) with 8 GB of RAM.

To ensure a polynomial running time for our algorithm we chose the value of ϵ such that $(1 - \frac{\epsilon}{k^2})$ was equal to 99/100. Cplex was used to solve the linear programs needed for the algorithms of Calinescu et al. [6], Sharma and Vondrak [13], and Buchbinder et al. [4,5].

5.1 Input Data

We used inputs from network benchmarks used in DIMACS competitions and randomly generated networks. Even though past DIMACS competitions have not included the multiway cut problem in their challenges, we were able to use network benchmarks for maximum clique, maximum independent set, Steiner tree,

Hamming instances, Keller's conjecture instances, and p-hat generated instances by considering only the largest connected component in each graph and choosing, when needed, random edge capacities and terminals.

The structure of a graph impacts how well a multiway cut algorithm performs. When edges incident on the terminals have much smaller capacities than edges not incident on them, an optimal solution will simply select the edges incident on the terminals. Furthermore, when the number of edges is large, there may be multiple independent isolating cuts of similar cost for each terminal; an optimal solution will select cuts that share edges, while approximation algorithms may choose cuts with a larger number of edges. We used these observations to generate random instances that are difficult for the multiway cut algorithms. Specifically, we generated three types of random graph instances.

Simple Random Graphs: Edges are added between pairs of randomly selected vertices. After a specific number of edges is added, the largest connected component is output.

Linear Decay Random Graphs: These are random graphs where an initial edge density and capacity range is used for edges incident on terminal vertices, and then the edge densities and capacities are linearly decreased as the distance from the terminals increases.

Exponential Decay Random Graphs: These graphs are also created by assigning an initial edge density and capacity range to edges incident on terminal vertices, and then exponentially decreasing the densities and capacities as the distance from the terminals increases.

Linear and exponential decay random graphs could model practical situations such as strength and availability of wireless signals, traffic congestion around popular destinations, or link capacity and topology in client-server networks.

Note that in the linear and exponential decay random graphs edges incident on the terminals have large capacities, ensuring that cuts isolating terminals are not the trivial ones. Edges located far from the terminals are given small capacities, creating a "hot spot" of edges that likely belong to minimum cuts.

5.2 Test Cases

We studied how graph characteristics such as terminal density (k/n), edge capacities, edge density (m/n), and number of vertices impact the solution quality. We also studied the impact of changing the initial labeling scheme and the value of ϵ for MultiwayCut. For all graph instances where terminals were randomly chosen, k was set to be a fraction of the number of vertices. The fractions used were 3/80, 1/16, 1/8, 1/4, 3/8, and 1/2. For all graph instances this ensured that k was at least 3, but not so large that the problem was trivial.

Assigning larger capacities to edges incident on the terminals led to graph instances whose optimal solutions include edges some distance away from the terminals. We generated graph instances using two edge capacity schemes. In the first scheme, edges incident on terminals were assigned rational capacities with values between 30 and 50, while other edges were assigned rational capacities with values between 1 and 25 (in exponential decay random graphs) or between 1 and 45 (in linear decay random graphs). In the second scheme, edges incident on terminals were assigned rational capacities between 1 and 100, while other edges were assigned rational capacities between 1 and 50 (in exponential decay random graphs) or between 1 and 90 (in linear decay random graphs).

We explored the impact of edge density on the solution quality. Both, the graph instances obtained from the DIMACS competitions and the randomly generated graphs with very large edge densities tended to produce simple instances for which all algorithms produced near optimum solutions. Therefore, we concentrated on generating random graphs with a small number of edges. The number of edges chosen for each randomly generated graph were n, $2n$, $3n$, $4n$, $5n$, and $6n$.

We used the following initial labelings for the local search algorithm:

One Each: Each terminal vertex was assigned a different label, and all of the remaining vertices were assigned the label of the first terminal.

Clumps: Each terminal vertex was assigned a different label, and the terminals were added to a queue. While the queue had vertices, the first vertex was removed and the label of that vertex was assigned to all of its unlabeled neighbors, then each of these neighbors was added to the end of the queue.

Random: Each terminal vertex was assigned a different label, and the terminals were added to a queue. While the queue had vertices, the queue was randomly shuffled, the first vertex was removed, and its label was assigned to all of its unlabeled neighbors, which were then added to the queue.

Isolation Heuristic: The vertices were assigned labels corresponding to the partitions selected by the isolation heuristic.

These initial labellings provided a variety of starting points that affected the number of iterations of the algorithm. We also studied the trade-off between solution quality and running time produced by the choice for the value of ϵ.

5.3 Results

For randomly generated graphs, 1,000 experiments were performed for each combination of edge density and terminal density for 80 vertex and 160 vertex graphs. Due to their large running times, only 100 experiments were performed on graphs with 320 vertices. Graph instances obtained from the DIMACS competitions have a large number of edges, hence only 25 to 100 experiments were performed for each terminal density.

Input Networks. Table 2 shows a summary of the results[1] for each of the input networks. The value in each entry of the table is calculated by dividing the value of the solution produced by an approximation algorithm by the value

---

[1] The complete results are available at https://www.csd.uwo.ca/~ablochha/rawdata.pdf.

of the optimum solution produced by Cplex. The column labeled "Avg" lists the mean of all of the ratios in a test case, and the column labeled "Max" lists the largest ratio produced in a test case. The rows are labeled with the name of the input graph and the number of vertices. For each row, solutions from test cases using different values for k, m, and edge capacities have been combined.

All of the approximation algorithms computed near optimum solutions for the instances from the DIMACS competitions. The exponential decay random graphs caused the highest ratios. The isolation heuristic performs the worst, as it computes independent isolating cuts and fails to re-use edges to achieve lower costs. This algorithm does particularly bad on linear and exponential decay random graphs. The algorithms of Calinescu et al., Buchbinder et al., and Sharma and Vondrak, which in the sequel we refer to collectively as the geometric relaxation algorithms, typically compute solutions close to the optimal but occasionally produce solutions near their theoretical worst case performances.

Table 2. Ratios of the solutions computed by approximation algorithms to the optimum, for benchmarks from the DIMACS competitions: maximum independent set (Brock), maximum clique (Gen, C125), Hamming instances, Keller instances, p-hat instances, and Steiner tree instances (ST); and for the randomly generated instances with simple (SR), linear (GL), and exponential (GE) distributions.

Test case		Isolation Heuristic		Local Search		Calinescu		Buchbinder		Sharma and Vondrak	
		Avg	Max	Avg	Max	Avg	Max	Avg	Max	Avg	Max
C125	$n = 125$	1.000	1.002	1.000	1.000	1.000	1.000	1.000	1.000	1.000	1.000
Brock	$n = 200$	1.000	1.005	1.000	1.000	1.000	1.000	1.000	1.000	1.000	1.000
Gen	$n = 200$	1.000	1.005	1.000	1.000	1.000	1.000	1.000	1.000	1.000	1.000
Hamming	$n = 256$	1.000	1.003	1.000	1.000	1.000	1.000	1.000	1.000	1.000	1.000
Keller	$n = 171$	1.000	1.005	1.000	1.000	1.000	1.000	1.000	1.000	1.000	1.000
P-Hat	$n = 320$	1.000	1.000	1.000	1.000	1.000	1.000	1.000	1.000	1.000	1.000
ST	$n = 80$	1.021	1.176	1.001	1.013	1.000	1.000	1.000	1.000	1.000	1.000
	$n = 160$	1.021	1.137	1.000	1.015	1.000	1.000	1.000	1.000	1.000	1.000
	$n = 320$	1.023	1.145	1.000	1.005	1.000	1.000	1.000	1.000	1.000	1.000
SR	$n = 80$	1.031	1.507	1.001	1.133	1.000	1.038	1.000	1.034	1.000	1.024
	$n = 160$	1.039	1.440	1.002	1.118	1.000	1.060	1.000	1.048	1.000	1.039
	$n = 320$	1.032	1.326	1.002	1.128	1.000	1.000	1.000	1.000	1.000	1.000
GL	$n = 80$	1.090	1.500	1.004	1.091	1.000	1.171	1.000	1.154	1.000	1.154
	$n = 160$	1.106	1.600	1.010	1.133	1.000	1.068	1.000	1.089	1.000	1.093
	$n = 320$	1.110	1.643	1.017	1.160	1.000	1.023	1.000	1.044	1.000	1.011
GE	$n = 80$	1.120	1.500	1.006	1.087	1.000	1.292	1.000	1.292	1.000	1.196
	$n = 160$	1.124	1.667	1.015	1.146	1.001	1.125	1.001	1.148	1.001	1.124
	$n = 320$	1.176	1.476	1.031	1.216	1.000	1.074	1.000	1.068	1.000	1.056

Graph Characteristics. Table 3 shows a sample of results from the exponential decay random distributions, which produced the highest ratios across all input networks. While for the exponential decay random graphs the algorithms had the highest ratios, each algorithm only produced these high ratios in a small subset of test cases. The isolation heuristic had its highest ratios when k was in the range of $0.125n$ to $0.25n$, and when the number of edges was $2n$. Our local search algorithm and the geometric relaxation algorithms have similar performance. These algorithms produced their highest ratios when k was in the range of $0.0375n$ to $0.125n$, and when the number of edges was between $2n$ and $5n$.

The local search algorithm has similar performance as the geometric relaxation algorithms. When the number of terminals is small and the number of edges is large, the worst values of the solutions computed by the local search algorithm were close to the average values of its solutions. In contrast, the worst values of solutions computed by the geometric relaxation algorithm were farther from the average values of their solutions.

The local search algorithm performs best when k is small. For networks with a large number of terminals the algorithm does not find global optimal solution due to the large number of label configurations for the vertices. One possible improvement for our algorithm on these network is to consider changing multiple labels at the same time.

Table 3. Average and maximum ratios for several test cases on 80 vertex random graphs with exponential decay distributions.

Test case		Isolation Heuristic		Local Search		Calinescu		Buchbinder		Sharma and Vondrak	
		Avg	Max	Avg	Max	Avg	Max	Avg	Max	Avg	Max
$k = 3$	$m = n$	1.046	1.333	1.000	1.000	1.000	1.000	1.000	1.000	1.000	1.000
	$m = 2n$	1.100	1.261	1.001	1.034	1.000	1.147	1.000	1.151	1.000	1.075
	$m = 3n$	1.082	1.189	1.001	1.046	1.001	1.102	1.001	1.083	1.001	1.124
	$m = 4n$	1.059	1.170	1.001	1.013	1.002	1.115	1.001	1.083	1.001	1.103
	$m = 5n$	1.027	1.130	1.001	1.011	1.001	1.292	1.001	1.292	1.000	1.063
	$m = 6n$	1.006	1.067	1.000	1.013	1.000	1.170	1.000	1.011	1.000	1.014
$k = 5$	$m = n$	1.128	1.500	1.000	1.000	1.000	1.000	1.000	1.000	1.000	1.000
	$m = 2n$	1.230	1.365	1.001	1.038	1.000	1.082	1.000	1.052	1.000	1.153
	$m = 3n$	1.183	1.298	1.001	1.032	1.001	1.124	1.001	1.085	1.000	1.139
	$m = 4n$	1.106	1.216	1.002	1.022	1.004	1.125	1.004	1.106	1.003	1.101
	$m = 5n$	1.065	1.187	1.003	1.024	1.004	1.226	1.003	1.127	1.003	1.091
	$m = 6n$	1.034	1.113	1.002	1.014	1.001	1.061	1.001	1.117	1.001	1.196
$k = 10$	$m = n$	1.076	1.462	1.000	1.000	1.000	1.000	1.000	1.000	1.000	1.000
	$m = 2n$	1.320	1.460	1.000	1.039	1.000	1.000	1.000	1.000	1.000	1.000
	$m = 3n$	1.245	1.331	1.001	1.046	1.000	1.002	1.000	1.003	1.000	1.019
	$m = 4n$	1.168	1.237	1.001	1.033	1.000	1.049	1.000	1.024	1.000	1.065
	$m = 5n$	1.129	1.194	1.002	1.027	1.001	1.075	1.001	1.071	1.001	1.082
	$m = 6n$	1.085	1.149	1.003	1.024	1.002	1.060	1.002	1.053	1.002	1.060

Running Time. A sample of running times for each approximation algorithm is shown in Table 4. The isolation heuristic is the fastest algorithm. The running times for the geometric relaxation algorithms include the time needed for Cplex to compute solutions for the linear programs.

The running times of the algorithms of Calinescu et al., Sharma and Vondrak, and Buchbinder et al. are very similar. The running times of these algorithms scale very well with the size of the input graph. The running time of the local search algorithm depends heavily on how the minimum cut is computed and how many relabel operations are performed. When the number of terminals is small, the local search algorithm is faster than the geometric relaxation algorithms.

Table 4. Running times using 100 experiments for several test cases from the 80 vertex and 160 vertex exponential decay distributions.

Test Case		Isolation Heuristic	Local Search	Calinescu	Buchbinder	Sharma and Vondrak
		Time (ms)				
$n = 80,\ k = 3$	$m = 2n$	1	6	51	51	50
	$m = 3n$	1	14	77	78	77
	$m = 4n$	1	18	91	92	91
$n = 80,\ k = 5$	$m = 2n$	1	15	67	67	67
	$m = 3n$	1	29	81	82	81
	$m = 4n$	1	52	139	141	139
$n = 80,\ k = 10$	$m = 2n$	1	32	93	94	93
	$m = 3n$	1	88	113	114	113
	$m = 4n$	2	126	127	130	127
$n = 160,\ k = 6$	$m = 2n$	2	64	202	203	202
	$m = 3n$	2	160	339	341	339
	$m = 4n$	2	226	534	534	534
$n = 160,\ k = 10$	$m = 2n$	2	137	274	276	274
	$m = 3n$	2	281	336	340	336
	$m = 4n$	3	486	464	470	464
$n = 160,\ k = 20$	$m = 2n$	3	245	352	361	352
	$m = 3n$	4	700	476	483	476
	$m = 4n$	5	970	523	540	523

Initial Labeling and Value of ϵ. Figure 1 shows how the value of epsilon affects the running times and solution quality for the local search algorithm on the exponential decay graph distribution where $m = 4n$ and $k = 10$. When $(1 - \epsilon/k^2) = 0.9975$, the local search algorithm computes solutions very close to the optimum but requires more time, especially in large graphs. As shown in Table 4, we can use $(1 - \epsilon/k^2) = 0.99$ to compute high quality solutions quicker than the other algorithms in small graphs. The local search algorithm can run considerably faster by sacrificing solution quality.

The different initial labellings for the local search algorithm produced similar solutions when $(1 - \epsilon/k^2)$ was between 0.99 and 1.0. When $(1 - \epsilon/k^2)$ was less than 0.99, the initial labellings showed differences in performance. The One Each labeling produced the worst solutions; since our random graph structures have many edges and higher capacities on edges incident on terminals, this initial labeling selected many edges with large capacities for the cuts. In contrast, the Clumps initial labeling was closer to the optimal solution and was not significantly affected when the value of $(1 - \epsilon/k^2)$ decreased below 0.99.

When $(1 - \epsilon/k^2) = 0.99$ the local search algorithm terminates if it cannot improve its previous best solution by at least 1%. Note that when the initial labeling is very close to the optimal solution, the value of ϵ has less of an impact.

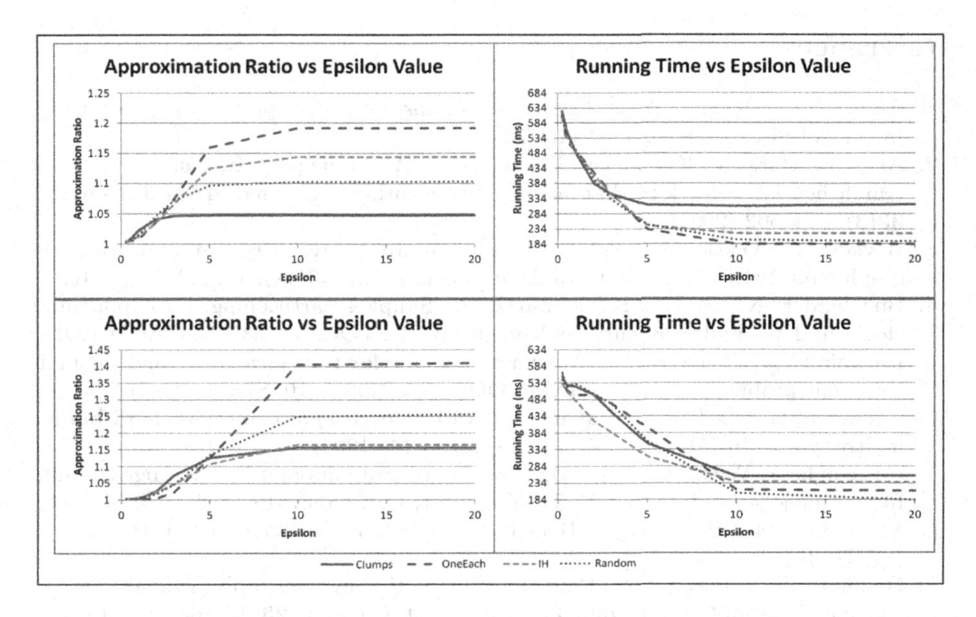

Fig. 1. Results from the 160 vertex exponential decay graph distribution with $m = 4n$ and $k = 10$. Approximation ratios are compared against the epsilon value using the first edge capacity scheme (top-left) and the second edge capacity scheme (bottom-left). Running times are compared against the epsilon value using the first edge capacity scheme (top-right) and the second edge capacity scheme (bottom-right). The x-axis shows the value of ϵ/k^2; the percentage of improvement to the previous best solution required to continue the iterations of the algorithm.

5.4 Conclusion

We compared our local search algorithm with four other approximation algorithms for the multiway cut problem. Even though the local search algorithm has the same worst case approximation ratio as the isolation heuristic, its experimental performance is much better. We observed competitive solution quality of the local search algorithm compared to the algorithms of Calinescu et al., Buchbinder et al., and Sharma and Vondrak. On graphs with exponential decay random distributions with $k <= 0.125$ and $m >= 2n$ the worst solutions produced by our algorithm were much better than the worst solutions produced by the geometric relaxation algorithms.

On networks with 80 vertices, the local search algorithm computed solutions faster than the geometric relaxation algorithms when k was less than $0.125n$, with the smallest test cases being solved significantly faster by the local search algorithm. On networks with 160 vertices, the local search algorithm computed solutions faster than the geometric relaxation algorithms when k was less than $0.0625n$, but did not scale well. Due to the ability to select the value for ϵ, the local search algorithm is more flexible than the other algorithms.

References

1. Alpert, C.J., Kahng, A.B.: Recent directions in netlist partitioning: a survey. Integr. VLSI J. **19**(1–2), 1–81 (1995)
2. Arya, V., Garg, N., Khandekar, R., Meyerson, A., Munagala, K., Pandit, V.: Local search heuristics for k-median and facility location problems. SIAM J. Comput. **33**(3), 544–562 (2004)
3. Boykov, Y., Veksler, O., Zabih, R.: Fast approximate energy minimization via graph cuts. IEEE Trans. Pattern Anal. Mach. Intell. **23**(11), 1222–1239 (2001)
4. Buchbinder, N., Naor, J.S., Schwartz, R.: Simplex partitioning via exponential clocks and the multiway cut problem. In: ACM STOC, pp. 535–544. ACM (2013)
5. Buchbinder, N., Schwartz, R., Weizman, B.: Simplex transformations and the multiway cut problem. In: ACM-SIAM SODA, pp. 2400–2410. SIAM (2017)
6. Călinescu, G., Karloff, H., Rabani, Y.: An improved approximation algorithm for multiway cut. In: ACM STOC, pp. 48–52. ACM (1998)
7. Cunningham, W.H., Tang, L.: Optimal 3-terminal cuts and linear programming. In: Cornuéjols, G., Burkard, R.E., Woeginger, G.J. (eds.) IPCO 1999. LNCS, vol. 1610, pp. 114–125. Springer, Heidelberg (1999). https://doi.org/10.1007/3-540-48777-8_9
8. Dahlhaus, E., Johnson, D.S., Papadimitriou, C.H., Seymour, P.D., Yannakakis, M.: The complexity of multiterminal cuts. SIAM J. Comput. **23**(4), 864–894 (1994)
9. Karger, D.R., Klein, P., Stein, C., Thorup, M., Young, N.E.: Rounding algorithms for a geometric embedding of minimum multiway cut. Math. Oper. Res. **29**(3), 436–461 (2004)
10. Lawler, E.L.: Combinatorial optimization: networks and matroids. Courier Corporation (1976)
11. Lee, C.H., Kim, M., Park, C.I.: An efficient k-way graph partitioning algorithm for task allocation in parallel computing systems. In: First International Conference on Systems Integration, pp. 748–751. IEEE (1990)
12. Lengauer, T.: Combinatorial Algorithms for Integrated Circuit Layout. Springer, Heidelberg (2012). https://doi.org/10.1007/978-3-322-92106-2
13. Sharma, A., Vondrák, J.: Multiway cut, pairwise realizable distributions, and descending thresholds. In: STOC, pp. 724–733. ACM (2014)
14. Stone, H.S.: Multiprocessor scheduling with the aid of network flow algorithms. IEEE Trans. Softw. Eng. **1**, 85–93 (1977)

Algorithmic Analysis of Priority-Based Bin Packing

Piotr Wojciechowski[1], K. Subramani[1(✉)], Alvaro Velasquez[2], and Bugra Caskurlu[3]

[1] LDCSEE, West Virginia University, Morgantown, WV, USA
pwojciec@mix.wvu.edu, k.subramani@mail.wvu.edu
[2] Information Directorate, AFRL, Rome, NY, USA
alvaro.velasquez.1@us.af.mil
[3] TOBB University of Economics and Technology, Ankara, Turkey
caskurlu@gmail.com

Abstract. This paper is concerned with a new variant of Traditional Bin Packing (TBP) called Priority-Based Bin Packing with Subset Constraints (PBBP-SC). In a TBP instance, we are given a collection of items $\{a_1, a_2, \ldots a_n\}$, with $a_i \in (0, 1)$ and a collection of unit-size bins $\{B_1, B_2, \ldots, B_m\}$. One problem associated with TBP is the bin minimization problem. The goal of this problem is to pack the items in as few bins as possible. In a PBBP-SC instance, we are given a collection of unit-size items and a collection of bins of varying capacities. Associated with each item is a positive integer which is called its *priority*. The priority of an item indicates its importance in a (possibly infeasible) packing. As with the traditional case, these items need to be packed in the fewest number of bins. What complicates the problem is the fact that each item can be assigned to only one of a select set of bins, i.e., the bins are not interchangeable. We investigate several problems associated with PBBP-SC. Checking if there is a feasible assignment to a given instance is one problem. Finding a maximum priority assignment in case of the instance being infeasible is another. Finding an assignment with the fewest number of bins to pack a feasible instance is a third. We derive a number of results from both the algorithmic and computational complexity perspectives for these problems.

1 Introduction

In this paper, we study the problem of Priority-Based Bin Packing (PBBP), subject to subset constraints. PBBP shares some aspects with Traditional Bin Packing (TBP) but is different in others. In the traditional bin packing problem, we are given n items $\{a_1, a_2, \ldots, a_n\}$ with each $a_i \in (0, 1)$ and a collection of

This research is supported in part by the Air-Force of Scientific Research through Grant FA9550-19-1-0177 and in part by the Air-Force Research Laboratory, Rome through Contract FA8750-17-S-7007.

A. Mudgal and C. R. Subramanian (Eds.): CALDAM 2021, LNCS 12601, pp. 359–372, 2021.
https://doi.org/10.1007/978-3-030-67899-9_29

unit-sized bins $\{B_1, B_2, \ldots, B_m\}$. The goal is to pack the items into as few bins as possible, while respecting the capacity constraints of the bins. Bin packing is a fundamental problem in combinatorial optimization [12]. Bin packing has been used to model problems in a wide variety of domains such as transportation [20], scheduling [24], VLSI design [16] and supply chain management [22]. Additional application areas include palleting [28,30] and cutting stock problems [7]. The bin packing problem is closely related to problems such as knapsack [18] and subset sum [19].

Several variants and generalizations of the bin packing problem have been studied in the literature. For instance, the generalization in [3] discusses the case where bins have costs associated with them, in addition to capacities. In this paper, we study a variant that finds applications in Security Aware Database Migration (SADM) [26] and palleting [28]. This variant is called Priority-Based Bin Packing with Subset Constraints (PBBP-SC). One of the features of our variant is the uniformity in item sizes, i.e., all items have the same size. An additional feature is that the set of bins into which an item may be placed is constrained, these constraints being termed as *subset constraints*. The third feature of our problem is that associated with each item is a non-negative number called its priority. The priority of an item becomes significant, if the subset constraints cannot be met and we need to choose which items are to be packed. We show that the problems of feasibility and priority maximization (in case of infeasible instances) can be solved in polynomial time, whereas the problem of optimizing the number of bins (in case of feasible instances) is **NP-hard**. We discuss a number of algorithmic strategies for the **NP-hard** variant.

2 Statement of Problems

In this section, we define the problems studied in this paper. These problems are all associated with Priority-based Bin Packing with Subset Constraints (PBBP-SC).

Definition 1 (PBBP-SC). *An instance of* **Priority-based Bin Packing with Subset Constraints (PBBP-SC)** *consists of the following:*

1. *Set of bins B where each bin $b_j \in B$ has capacity c_j.*
2. *Set of items O where each item $o_i \in O$ has priority p_i.*
3. *For each item $o_i \in O$, a set $B_i \subseteq B$ such that item o_i can be packed into any bin in the set B_i, but not into any bin in the set $B \setminus B_i$.*

Throughout this paper, we use n to refer to the number of items and m to refer to the number of bins. In an instance of PBBP-SC, each set B_i is called a subset constraint since it determines which bins item o_i can be packed in.

We study three problems associated with PBBP-SC instances. These are the Feasibility Problem (FP), the Priority Maximization Problem (PMP), and the Bin Minimization Problem (BMP). These problems are defined as follows:

Definition 2 (FP). *The* **Feasibility Problem (FP)***: Given a PBBP-SC instance* **P***, can we pack the items in set O into the bins in set B such that, every item $o_i \in O$ is packed into a bin in set B_i, and every bin b_j contains no more than c_j items?*

Definition 3 (PMP). *The* **Priority Maximization Problem (PMP)***: Given a PBBP-SC instance* **P***, what is the maximum total priority of items in set O that can be packed into the bins in set B such that, every item $o_i \in O$ can only be packed into bins in set B_i, and every bin b_j contains no more than c_j items?*

Note that, if for a given PBBP-SC instance **P** FP is feasible, then PMP is trivial since every item in set O can be packed. Thus, PMP is only interesting on PBBP-SC instances where FP is infeasible.

Definition 4 (BMP). *The* **Bin Minimization Problem (BMP)***: Given a PBBP-SC instance* **P***, what is the smallest subset $B^* \subseteq B$ such that every item $o_i \in O$ is packed into a bin in set $B_i \cap B^*$, and every bin b_j contains no more than c_j items?*

The principal contributions of this paper are as follows:

1. FP and PMP can be solved in polynomial time.
2. BMP is **log-APX-complete**.
3. BMP cannot be solved in time $o(1.41^m)$ or in time $o(1.99^k)$, where k is the smallest number of bins needed to pack all of the items in O, unless the Strong Exponential Time Hypothesis (SETH) fails [10].

3 Motivation

In this section, we briefly discuss two problems that have motivated our analysis in this paper.

1. Security-Aware Database Migration (SADM) - The SADM problem was introduced in [26]. In this problem, we are given a collection of databases (D_i) of various sizes that need to be assigned to migration shifts (S_i). The shifts have varying sizes themselves. Furthermore, each database is constrained by the shifts to which it can be assigned. This models the fact that the expertise for addressing the issues associated with a database can be found only in certain shifts. For instance, it could be the case that database D_1 can be migrated only in shifts S_4 and S_7. We need to assign the databases to the shifts so that these shift assignment constraints for each item are met. At the same time, we wish to minimize the number of shifts used in the assignment, since shifts correspond to man-hours used and are therefore expensive. Database packing albeit with a different objective function is discussed in [27].

2. Palleting - The Pallet Loading (PL) problem is concerned with packing large quantities of cartons onto pallets. This problem finds applications in logistics [13], supply chain optimization [29], and a host of other areas such as container loading in ships [1]. The PL problem is closely related to both the knapsack and the bin packing problems [8]; typically, the cartons in PL are 2-dimensional or 3-dimensional, whereas the items in TBP are uni-dimensional. In its simplest form, the goal is to maximize the number of cartons that are loaded onto a pallet. Alternatively, as argued in [23], the goal is to minimize the "wasted area". In [28], the multi-pallet loading problem is considered. This problem is similar to our bin packing variant; however, they do not consider the subset constraints considered in this paper. In typical Air-Force palleting applications [4], items to be loaded are constrained by which pallets they can be loaded onto. Thus, the work in this paper finds direct applications to Air-Force palleting problems.

4 Feasibility of PBBP-SC

In this section, we show that the feasibility problem for PBBP-SC can be solved in polynomial time. This is done by a reduction to maximum matching.

Let $\mathbf{P}$ be a PBBP-SC instance. From $\mathbf{P}$, we construct the corresponding bipartite graph $\mathbf{G} = \langle U, V, E \rangle$ as follows:

1. For each bin $b_j \in B$, create the vertices $u_{j,1}$ through u_{j,c_j}. We can assume without loss of generality that $c_j \leq n$.
2. For each item $o_i \in O$, create the vertex v_i. Additionally, create the edges $(v_i, u_{j,1})$ through (v_i, u_{j,c_j}) for each $b_j \in B_i$. Observe that for item o_i, we create $\sum_{b_j \in B_i} c_j$ edges.

Example 1. Let $\mathbf{P}$ be the following PBBP-SC instance:

1. $B = \{b_1, b_2\}$ where $c_1 = c_2 = 2$.
2. $O = \{o_1, o_2, o_3\}$ where $B_1 = \{b_1\}$, $B_2 = \{b_1, b_2\}$, and $B_3 = \{b_2\}$.

PBBP-SC instance $\mathbf{P}$ corresponds to the bipartite graph $\mathbf{G}$ in Fig. 1.

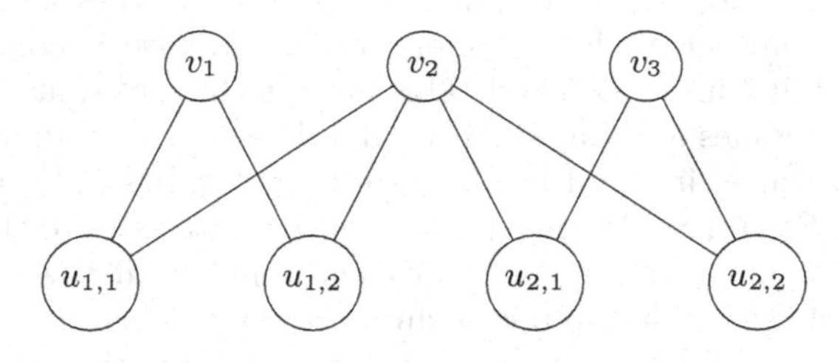

Fig. 1. Bipartite graph $\mathbf{G}$ corresponding to PBBP-SC $\mathbf{P}$.

Theorem 1. **P** *is feasible if and only if the corresponding graph* **G** *has a matching of size* n.

Proof. First, assume that **P** is feasible. Thus, there exists a packing P that assigns every item $o_i \in O$ to a bin $P(o_i) \in B_i$ such that no bin exceeds its capacity. From the packing P, we can construct a matching M in **G** as follows: For each item $o_i \in O$, if o_i is the kth item packed into bin $P(o_i)$, add the edge $(v_i, u_{P(o_i),k})$ to M.

Note that P packs every item in O. Thus, M is a matching in **G** of size n.

Now assume that **G** has a matching M of size n. By construction, **G** is bipartite and $|V| = n$ thus every vertex in V must be matched. From the matching M we can construct a packing P as follows:

1. For every vertex $v_i \in V$ consider the edge $(v_i, u_{j,k}) \in M$.
2. Pack item o_i into bin $P(o_i) = b_j$.

By construction, the edge $(v_i, u_{j,k}) \in E$ if and only if $b_j \in B_i$ and $k \leq c_j$. Thus, P assigns every item $o_i \in O$ to a bin $P(o_i) \in B_i$ and no bin exceeds its capacity. Consequently, P is a valid packing and **P** is feasible. □

From a given PBBP-SC instance **P** the corresponding graph **G** can be constructed in parallel by a CREW-PRAM machine. Note that a CREW-PRAM is a parallel random access machine whose processors can simultaneously read data from the same location but cannot simultaneously write data to the same location [14].

This construction is performed follows:

1. For each item $o_i \in O$, use a single processor to create the vertex v_i. This requires a total of n processors.
2. For each bin $b_j \in B$ and each $l = 1 \ldots c_j$, use a single processor to create the vertex $u_{j,l}$. This requires a total of $\sum_{j=1}^{m} c_j \leq m \cdot n$ processors.
3. For each item o_i, each bin $b_j \in B_i$, and each $l = 1 \ldots c_j$, use a single processor to create the edge $(v_i, u_{j,l})$. This requires a total of $\sum_{i=1}^{n} \sum_{b_j \in B_i} c_j \leq m \cdot n^2$ processors.

Each of these steps takes constant time. Thus, the graph **G** can be constructed in constant time using a CREW-PRAM with $m \cdot n^2$ processors. Consequently, this is an $\mathbf{AC}^0$ reduction [25].

Using the algorithm in [6], a maximum matching in **G** can be found in time $O(\min\{|V| \cdot k, |E|\} + \sqrt{k} \cdot \min\{k^2, |E|\})$ where k is the size of the maximum matching. By construction, in the graph **G**, $|V| = n$, $|U| = \sum_{j=1}^{m} c_j \leq m \cdot n$, and $|E| = \sum_{i=1}^{n} \sum_{b_j \in B_i} c_j \leq m \cdot n^2$. Thus, the maximum matching of **G** can be found in time $O(n^{2.5})$. However constructing **G** takes time $O(|V| + |U| + |E|) \subseteq O(m \cdot n^2)$. Thus, FP can be solved in time $O(n^2 \cdot (m + \sqrt{n}))$.

Additionally, the problem of finding a maximum matching in a bipartite graph can be reduced to FP.

Let $\mathbf{G} = \langle U, V, E \rangle$ be a bipartite graph. From $\mathbf{G}$, we construct the corresponding PBBP-SC instance $\mathbf{P}$ as follows:

1. For each vertex $u_j \in U$, create the bin b_j with capacity $c_j = 1$.
2. For each vertex $v_i \in V$, create the item o_i with subset constraint $B_i = \{b_j : (v_i, u_j) \in E\}$.

Corollary 1. $\mathbf{G}$ *has a matching of size* n *if and only if the corresponding PBBP-SC instance* $\mathbf{P}$ *is feasible.*

Proof. Note that $\mathbf{G}$ is precisely the bipartite graph that would be constructed from $\mathbf{P}$ by the previous reduction, except with each vertex $u_{j,1}$ renamed to u_j. Thus, by Theorem 1, $\mathbf{G}$ has a matching of size n if and only if $\mathbf{P}$ is feasible. □

From a given bipartite graph $\mathbf{G}$ the corresponding PBBP-SC instance $\mathbf{P}$ can be constructed in parallel by a CREW-PRAM machine as follows:

1. For each vertex $u_j \in U$, use a single processor to create the bin b_j. This requires a total of $|U|$ processors.
2. For each vertex $v_i \in V$, use a single processor to create the item v_i. This requires a total of $|V|$ processors.
3. For each vertex v_i and each vertex u_j such that $(v_i, u_j) \in |E|$, use a single processor to add the bin b_j to the set B_i. This requires a total of $|E|$ processors.

Each of these steps takes constant time. Thus, the PBBP-SC instance $\mathbf{P}$ can be constructed in constant time using a CREW-PRAM with $\max\{|V|, |U|, |E|\}$ processors. Consequently, this is an $\mathbf{AC}^0$ reduction [25].

Both of the reductions are $\mathbf{AC}^0$ reductions. Thus, FP is $\mathbf{AC}^0$ equivalent to the maximum matching problem for bipartite graphs. Maximum matching is not currently known to be **P-complete** nor is it known to be in **NC** [2]. If the complexity of maximum matching is established, then $\mathbf{AC}^0$ equivalence means that the same complexity applies to our problem.

4.1 Priority Maximization

The reduction from FP to maximum matching can be adapted to solve PMP by adding weights to the edges in the bipartite graph.

Let $\mathbf{P}$ be a PBBP-SC instance. From $\mathbf{P}$, we construct the corresponding weighted bipartite graph $\mathbf{G_W} = \langle U, V, E, c \rangle$ as follows:

1. For each bin $b_j \in B$, create the vertices $u_{j,1}$ through u_{j,c_j}. We can assume without loss of generality that $c_j \leq n$.
2. For each item $o_i \in O$, create the vertex v_i. Additionally, create the edges $(v_i, u_{j,1})$ through (v_i, u_{j,c_j}) with weight p_i for each $b_j \in B_i$. Observe that for item o_i, we create $\sum_{b_j \in B_i} c_j$ edges.

Example 2. Let $\mathbf{P}$ be the following PBBP-SC instance:

1. $B = \{b_1, b_2\}$ where $c_1 = c_2 = 2$.
2. $O = \{o_1, o_2, o_3\}$ where $B_1 = \{b_1\}$, $B_2 = \{b_1, b_2\}$, and $B_3 = \{b_2\}$. Additionally, $p_1 = 3$, $p_2 = 2$, and $p_3 = 1$.

PBBP-SC instance **P** corresponds to the weighted bipartite graph $\mathbf{G_W}$ in Fig. 2.

Theorem 2. ***P** has a packing that packs items with a total priority of W if and only if the corresponding graph $\mathbf{G_W}$ has a matching of weight W.*

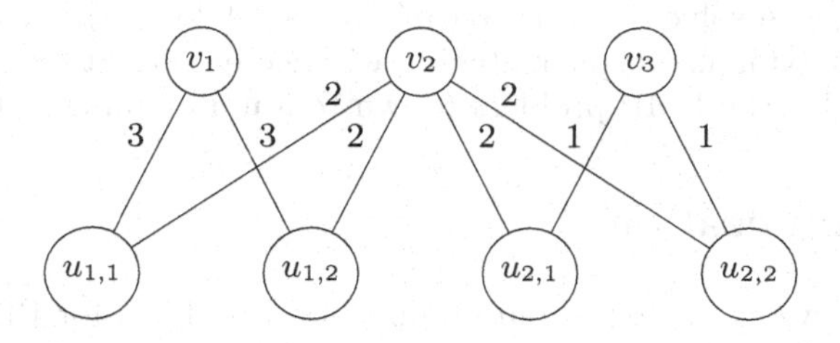

Fig. 2. Weighted bipartite graph $\mathbf{G_W}$ corresponding to PBBP-SC **P**.

Proof. First, assume that there exists a packing P that packs items with a total priority of W. Let $O' \subset O$ be the set of items packed by P. Thus, every item $o_i \in O'$ is packed into bin $P(o_i) \in B_i$ such that no bin exceeds its capacity. Additionally, $\sum_{o_i \in O'} p_i = W$.

From the packing P, we can construct a matching M in $\mathbf{G_W}$ as follows: For each item $o_i \in O'$, if o_i is the kth item packed into bin $P(o_i)$, add the edge $(v_i, u_{P(o_i),k})$ to M.

Consider an item $o_i \in O$. If $o_i \in O'$, then M contains an edge from vertex v_i. By construction, this edge has weight p_i. If $o_i \notin O'$, then M does not contain any edge from vertex v_i. Thus,

$$\sum_{e_l \in M} c(e_l) = \sum_{o_i \in O'} p_i = W.$$

Consequently, M is a matching in $\mathbf{G_W}$ with total weight W.

Now assume that $\mathbf{G_W}$ has a matching M with total weight W. Thus, $\sum_{e_l \in M} c(e_l) = W$. From the matching M, we can construct a packing P as follows:

1. For every vertex $v_i \in V$ matched by M, consider the edge $(v_i, u_{j,k}) \in M$.
2. Pack item o_i into bin $P(o_i) = b_j$. By construction, the edge $(v_i, u_{j,k}) \in E$ if and only if $b_j \in B_i$ and $k \leq c_j$. Thus, no bin exceeds its capacity.

Let O' be the set of items packed by P. Consider the vertex v_i. If v_i is matched by M, then $o_i \in O'$. Additionally, there is an edge $(v_i, u_{j,k}) \in M$ with weight p_i. If v_i is not matched by M, then $o_i \notin O'$. Thus,

$$\sum_{o_i \in O'} p_i = \sum_{e_l \in M} c(e_l) = W.$$

Consequently, P packs items with a total priority of W. □

Using the algorithm in [11], a maximum weight matching in $\mathbf{G_W}$ can be found in time $O(|E| \cdot (|U| + |V|) + (|U| + |V|)^2 \cdot \log(|U| + |V|))$. By construction, in the graph $\mathbf{G_W}$, $|V| = n$, $|U| = \sum_{j=1}^{m} c_j \leq m \cdot n$, and $|E| = \sum_{i=1}^{n} \sum_{b_j \in B_i} c_j \leq m \cdot n^2$. Thus, the maximum weight matching of $\mathbf{G_W}$ can be found in time $O(m^2 \cdot n^2 \cdot (n + \log m))$. However constructing $\mathbf{G}$ takes time $O(|V| + |U| + |E|) \subseteq O(m \cdot n^2)$. Thus, PMP can be solved in time $O(m^2 \cdot n^2 \cdot (n + \log m))$. Note that a faster algorithm for maximum weight matching will decrease the time it takes for this reduction to solve the PMP problem (down to a minimum time of $O(m \cdot n^2)$).

5 Bin Minimization

In this section, we examine the bin minimization problem for PBBP-SC.

First, we look at the complexity of approximating the minimum number of bins. We do this by showing that BMP is complete for the complexity class **log-APX**. Note that **log-APX** is the class of optimization problems for which there is a polynomial time algorithm that approximates the optimal solution to within a factor logarithmic in the size of the input [21]. We do this by relating BMP to the Minimum Set Cover (MSC) problem. This problem is defined as follows:

Given a set $S = \{x_1, \ldots, x_n\}$ and a collection of subsets $S_j \subseteq S$ for $j = 1 \ldots m$, what is the minimum number of subsets whose union is S? This problem is known to be **log-APX-complete** [9].

From an MSC instance SC, we can construct a PBBP-SC instance $\mathbf{P}$ as follows:

1. For each subset S_j, create the bin b_j with capacity $c_j = n$.
2. For item $x_i \in S$, create item o_i with subset constraint $B_i = \{b_j : x_i \in S_j\}$.

Lemma 1. *SC has a cover of size k if and only if $\mathbf{P}$ has a valid packing using k bins.*

Proof. First assume that SC has a cover C of size k. From C, we construct a packing P of $\mathbf{P}$ as follows:

1. For each item $x_i \in S$, let $1 \leq j \leq m$ be the smallest number such that $S_j \in C$ and $x_i \in S_j$.
2. Pack item o_i into bin $P(o_i) = b_j$.

Since $x_i \in S_j$, by construction $b_j \in B_i$. Additionally, each bin in $\mathbf{P}$ has capacity n, thus no bin exceeds its capacity. Since $|C| = k$, only k bins are used by P. Consequently P is a valid packing using only k bins.

Now assume that $\mathbf{P}$ has a valid packing P using only k bins. We construct a cover C of SC as follows: For each bin $b_j \in B$, if b_j is used by P, add S_j to C.

Since P is a valid packing, every item o_i is assigned to a bin $P(o_i)$. By construction, item x_i is covered by the set corresponding to bin $P(o_i)$. Since this set is in C, C is a cover of S. Since P uses only k bins, $|C| = k$ as desired. $\square$

Note that this is a strict reduction since it runs in linear time and preserves the value of the optimization function. Thus, BMP is **log-APX-hard**. Now we need to show that BMP is in **log-APX**. This will be done by providing an approximation algorithm.

Observe that the reduction from set cover only produces PBBP-SC instances where the capacity of the bins is effectively unlimited. Thus, when the capacities matter, the problem may become more difficult to approximate. We show that this is not the case.

We now use the fact that PMP is in **P** to construct an approximation algorithm for BMP. This results in Algorithm 5.1.

For a feasible PBBP-SC instance **P**, let *Opt*(**P**) be the optimal number of bins needed to pack all of the items in O. We now prove that APPROX-MIN-BINS(**P**) $\leq (1 + \log m) \cdot Opt(\mathbf{P})$.

Algorithm 5.1. Approximation Algorithm for BMP

Input: Feasible PBBP-SC instance **P**
Output: Approximate number of bins needed to pack every item in O.

1: **procedure** APPROX-MIN-BINS(**P**)
2: Let $S := \emptyset$ be a set of bins.
3: Let $O(S)$ be a set of objects with maximum cardinality that can be packed into S. ▷ $O(S)$ can be found in polynomial time by solving PMP.
4: **while** $O(S) \neq O$ **do**
5: Let b_j be the bin such that $|O(S \cup \{b_j\})|$ is maximized.
6: $S := S \cup \{b_j\}$.
7: **return** $|S|$.

Theorem 3. APPROX-MIN-BINS(**P**) $\leq (1 + \log m) \cdot Opt(\mathbf{P})$.

Proof. Let P^* be a valid packing of **P** that uses a minimum number of bins and let $B^* \subseteq B$ be the set of bins used by P^*. Note that $|B^*| = Opt(\mathbf{P})$. Additionally, let S be the set of bins constructed by Algorithm 5.1.

Let S_i be the set constructed by the ith iteration of the **while** loop in Algorithm 5.1. Since the objects in $O(S_i)$ can be packed into the bins in S_i and the objects in O can be packed into $S_i \cup B^*$, we have that the objects in $O \setminus O(S_i)$ can be packed into the bins in $B^* \setminus S_i$ by some packing P_i. For each bin $b_j \in B^* \setminus S_i$, let $x_{i,j}$ be the number of objects packed into bin b_j by P_i. We can assume without loss of generality that $x_{i,j} \geq x_{i+1,j}$.

Thus, after the ith iteration of the **while** loop, $\sum_{b_j \in B^* \setminus S_i} x_{i,j} = |O \setminus O(S_i)|$. Consequently, $\sum_{b_j \in B^* \setminus S_i} (x_{i,j} - x_{i+1,j}) = |O(S_{i+1}) \setminus O(S_i)|$.

Let b_j be a bin in B^*. We have that either $b_j \in S$ or every object that could be packed into bin b_j is packed into a bin in S. Thus, there is a smallest k_j such that either:

1. Every object that could be packed into b_j belongs to the set $O(S_{k_j})$.
2. Bin b_j is in S_{k_j+1}.

Thus, for each $i \leq k_j$, bin b_j could have been added to set S_i by Algorithm 5.1. However, in each of these iterations there existed a bin that could add more objects to the current packing. Thus, $|O(S_i)| \geq |O(S_{i-1} \cup \{b_j\})|$. Since at least $x_{i-1,j}$ objects in $O \setminus O(S_{i-1})$ can be packed into bin b_j, $|O(S_{i-1} \cup \{b_j\})| \geq |O(S_{i-1})| + x_{i-1,j}$. Thus, $x_{i-1,j} \leq |O(S_i)| - |O(S_{i-1})|$.

This means that

$$\sum_{i=0}^{k_j-1} \frac{x_{i,j} - x_{i+1,j}}{|O(S_{i+1}) \setminus O(S_i)|} \leq \sum_{i=0}^{k_j-1} \frac{x_{i,j} - x_{i+1,j}}{x_{i,j}} \leq \sum_{i=0}^{k_j-1} \sum_{l=x_{i+1,j}}^{x_{i,j}} \frac{1}{x_{i,j}}$$
$$\leq \sum_{i=0}^{k_j-1} \sum_{l=x_{i+1,j}}^{x_{i,j}} \frac{1}{l} = \sum_{l=x_{k_j,j}}^{x_{0,j}} \frac{1}{l}.$$

We have that $x_{0,j} \leq m$. Additionally, since no objects in $O \setminus O(S_{k_j})$ can be packed into b_j, $x_{k_j,j} = 0$. Thus, $\sum_{i=0}^{k_j-1} \frac{x_{i,j}-x_{i+1,j}}{|O(S_{i+1}) \setminus O(S_i)|} \leq \sum_{l=1}^{m} \frac{1}{l} \leq 1 + \log m$.

This means that

$$|S| = \sum_{i=0}^{|S|-1} 1 = \sum_{i=0}^{|S|-1} \frac{\sum_{b_j \in B^* \setminus S_i} x_{i,j} - x_{i+1,j}}{|O(S_{i+1}) \setminus O(S_i)|}$$
$$= \sum_{b_j \in B^*} \sum_{i=0}^{k_j-1} \frac{x_{i,j} - x_{i+1,j}}{|O(S_{i+1}) \setminus O(S_i)|} \leq |B^*| \cdot (1 + \log m).$$

Thus, Algorithm 5.1 is a $(1 + \log m)$ approximation algorithm for BMP. □

Since, for any feasible PBBP-SC instance **P** the optimal number of bins needed to pack all of the items in O can be approximated to within a factor of $(1 + \log m)$, we have that BMP belongs to the class **log-APX**. We already showed that BMP is **log-APX-hard**. Thus, BMP is **log-APX-complete**.

5.1 Lower Bounds

We now show that BMP is unlikely to be solved in time $o(1.41^m)$ or in time $o(1.99^k)$, where k is the smallest number of bins needed to pack all of the items in O.

These bounds are derived using the Strong Exponential Time Hypothesis (SETH) which is defined as follows:

For each $r \geq 3$, let s_r be the smallest value such that r-SAT can be solved in time $O^*(2^{s_r \cdot n'})$ where n' is the number of variables in the formula. The SETH is the hypothesis that $\lim_{r \to \infty} s_r = 1$ [5]. An immediate consequence of the SETH is that, in general, SAT cannot be solved in time $o((2 - \epsilon)^{n'})$ for any $\epsilon > 0$ [17].

Let Φ be a CNF formula with m' clauses over n' variables. From Φ, we construct a PBBP-SC instance $\mathbf{P}$ as follows:

1. For each variable x_i of Φ, let d_i^+ be the number of clauses which use the literal x_i and let d_i^- be the number of clauses which use the literal $\neg x_i$.
2. For each variable x_i of Φ, create the bin $b_{2\cdot i-1}$ with capacity $c_{2\cdot i-1} = d_i^+ + 1$ and create the bin $b_{2\cdot i}$ with capacity $c_{2\cdot i} = d_i^- + 1$. Additionally, create the item o_i with $B_i = \{b_{2\cdot i-1}, b_{2\cdot i}\}$.
3. For each clause $\phi_j \in \Phi$, create the item $o_{n'+j}$. For each variable x_i:
   (a) If ϕ_j uses the literal x_i, then add bin $b_{2\cdot i-1}$ to $B_{n'+j}$.
   (b) If ϕ_j uses the literal $\neg x_i$, then add bin $b_{2\cdot i}$ to $B_{n'+j}$.

Note that $\mathbf{P}$ has $n = m' + n'$ items and $m = 2 \cdot n'$ bins. We now show that $\mathbf{P}$ has a packing P that uses $k = n'$ bins if and only if Φ is satisfiable.

Theorem 4. $\mathbf{P}$ *has a packing that uses* n' *bins if and only if* Φ *is satisfiable.*

Proof. By construction, the sets B_1 through $B_{n'}$ are mutually disjoint. Thus, the items o_1 through $o_{n'}$ must be packed into separate bins. This means that no packing of the items in O uses fewer than n' bins. Additionally for each $i = 1 \ldots n'$, either bin $b_{2\cdot i-1}$ or bin $b_{2\cdot i}$ must be used.

First assume that $\mathbf{P}$ has a packing P that uses n' bins. Observe that if for any $i = 1 \ldots n'$ both bin $b_{2\cdot i-1}$ and bin $b_{2\cdot i}$ are used, then P must use at least $(n' + 1)$ bins. Thus, P cannot use both of these bins.

From P, we construct an assignment $\mathbf{x}$ to Φ as follows: For each $i = 1 \ldots n'$

1. If P uses bin $b_{2\cdot i-1}$, set variable x_i to **true**.
2. If P uses bin $b_{2\cdot i}$, set variable x_i to **false**.

Now consider a clause $\phi_j \in \Phi$. The item $o_{n'+j}$ is either packed into bin $b_{2\cdot i-1}$ or bin $b_{2\cdot i}$ for some $i = 1 \ldots n'$. If item $o_{n'+j}$ is packed into bin $b_{2\cdot i-1}$, then by construction, clause ϕ_j contains the literal x_i. Additionally, bin $b_{2\cdot i-1}$ is used by packing P. Thus, the assignment $\mathbf{x}$ sets the variable x_i to **true**. Consequently, ϕ_j contains a **true** literal and is satisfied by $\mathbf{x}$.

If item $o_{n'+j}$ is packed into bin $b_{2\cdot i}$, then by construction, clause ϕ_j contains the literal $\neg x_i$. Additionally, bin $b_{2\cdot i}$ is used by packing P. Thus, the assignment $\mathbf{x}$ sets the variable x_i to **false**. Consequently, ϕ_j contains a **true** literal and is satisfied by $\mathbf{x}$.

Since $\mathbf{x}$ satisfies every clause of Φ, Φ is satisfiable.

Now assume that Φ is satisfiable. Thus, there exists an assignment $\mathbf{x}$ that satisfies every clause in Φ. From $\mathbf{x}$, we construct a packing P of $\mathbf{P}$ as follows:

1. For each variable x_i, if x_i is **true**, then pack item o_i into bin $b_{2\cdot i-1}$. Otherwise, pack item o_i into bin $b_{2\cdot i}$.
2. For each clause $\phi_j \in \Phi$, let l_j be a literal in ϕ_j set to **true** by $\mathbf{x}$. If l_j is the literal x_i for some variable x_i, then pack item $o_{n'+j}$ into bin $b_{2\cdot i-1}$. If l_j is the literal $\neg x_i$ for some variable x_i, then pack item $o_{n'+j}$ into bin $b_{2\cdot i}$.

The constructed packing P has the following properties:

1. If $\mathbf{x}$ set the variable x_i to **true**, then the literal $\neg x_i$ is **false**. Thus, no item will be packed into bin $b_{2 \cdot i}$. If $\mathbf{x}$ set the variable x_i to **false**, then the literal x_i is **false**. Thus, no item will be packed into bin $b_{2 \cdot i-1}$. This means that for each variable x_i, P does not use both bin $b_{2 \cdot i-1}$ and bin $b_{2 \cdot i}$. Consequently, P uses at most n' bins.
2. By construction, every item o_j is packed into a bin in B_j.
3. The literal x_i appears in d_i^+ clauses. Thus at most $d_i^+ + 1 = c_{2 \cdot i-1}$ items are packed into bin $b_{2 \cdot i-1}$. Similarly, the literal $\neg x_i$ appears in d_i^- clauses. Thus, at most $d_i^- + 1 = c_{2 \cdot i}$ items are packed into bin $b_{2 \cdot i}$. Consequently, no bin is packed beyond its capacity.

Thus, P is a valid packing of PBBP-SC instance $\mathbf{P}$. Consequently, $\mathbf{P}$ has a packing that uses n' bins. □

Thus, if there were a $o(1.41^m)$ algorithm for BMP, then the feasibility of Φ would be determined in time $o(1.41^{2 \cdot n'}) \subseteq o(1.99^{n'})$. This violates the SETH. Thus, it is unlikely that such an algorithm exists. Similarly, it is unlikely that there is a $o(1.99^k)$ time algorithm for BMP.

6 Conclusion

In this paper, we discussed a new variant of the traditional bin packing problem called priority-based bin packing (PBBP). The PBBP problem is similar to the generalized bin packing problem described in [3], but has several distinct features. The notion of subset constraints introduced in this paper finds applications in several domains other than the ones described in this paper. We showed that the feasibility problem in PBBP is in **P**. This is similar to traditional bin packing but different from generalized bin packing; the feasibility problem in the latter is **NP-hard**. We investigated two optimization problems related to the PBBP problem, viz., the bin minimization problem (for feasible instances) and the priority maximization problem (for infeasible instances). We showed that the bin minimization problem is **log-APX-complete**, while there exists a polynomial time algorithm for the priority maximization problem.

From our perspective, the following open problems are worth pursuing:

1. A new measure of optimization - The focus of the BMP problem is minimizing the total number of bins used. However, there is an alternative metric, viz., minimizing the index of the last bin used. The optimal solutions obtained under the *last* metric are in general different from those obtained under the total number of bins metric. For a detailed discussion of this metric and its application to database migration, see [26].
2. Parameterized algorithms - As mentioned before, in the case of TBP, we cannot hope for a fixed-parameter algorithm, when the parameter is the number of bins, since bin packing is **NP-hard**, even when the number of bins is at

most 2. However, this does not apply in the case of BMP, since in BMP the items have the same size. We believe that the techniques discussed in [15] may be helpful in designing a fixed-parameter tractable algorithm for BMP.
3. Bi-objective optimization - Both the bin minimization problem and the priority maximization problem are special cases of the following bi-objective problem: Given unit-sized items with priorities, a collection of bins, subset constraints relating items and bins, and two numbers K_1 and K_2, is there a packing of items into the bins, such that the subset constraints are satisfied, at most K_1 bins are used and the sum of the priorities of the packed items is at least K_2? Clearly, this problem is **NP-hard**. The question is whether we can develop a dynamic programming algorithm for this problem.

References

1. Alvarez-Valdés, R., Parreño, F., Tamarit, J.M.: A branch-and-cut algorithm for the pallet loading problem. Comput. Oper. Res. **32**, 3007–3029 (2005)
2. Anari, N., Vazirani, V.V.: Matching is as easy as the decision problem, in the NC model. CoRR, abs/1901.10387 (2019)
3. Baldi, M.M., Bruglieri, M.: On the generalized bin packing problem. ITOR **24**(3), 425–438 (2017)
4. Ballew, B.: The distributor's three-dimensional pallet-packing problem: a mathematical formulation and heuristic solution approach, p. 111, March 2000
5. Calabro, C., Impagliazzo, R., Paturi, R.: The complexity of satisfiability of small depth circuits. In: Chen, J., Fomin, F.V. (eds.) IWPEC 2009. LNCS, vol. 5917, pp. 75–85. Springer, Heidelberg (2009). https://doi.org/10.1007/978-3-642-11269-0_6
6. Chandran, B.G., Hochbaum, D.S.: Practical and theoretical improvements for bipartite matching using the pseudoflow algorithm. CoRR, abs/1105.1569 (2011)
7. Csirik, J., Johnson, D.S., Kenyon, C., Orlin, J.B., Shor, P.W., Weber, R.R.: Fast algorithms for bin packing. J. Comput. Syst. Sci. **8**(8), 272–314 (1974)
8. Paul Davies, A., Bischoff, E.E.: Weight distribution considerations in container loading. Eur. J. Oper. Res. **114**(3), 509–527 (1999)
9. Escoffier, B., Paschos, V.T.: Completeness in approximation classes beyond apx. Theor. Comput. Sci. **359**(1), 369–377 (2006)
10. Fomin, F.V., Kratsch, D.: Exact Exponential Algorithm, 1st edn. Springer, Heidelberg (2010)
11. Fredman, M.L., Tarjan, R.E.: Fibonacci heaps and their uses in improved network optimization algorithms. J. ACM **34**(3), 596–615 (1987)
12. Garey, M.R., Johnson, D.S.: Computers and Intractability: A Guide to the Theory of NP-Completeness. W. H. Freeman Company, San Francisco (1979)
13. Hodgson, T.J.: A combined approach to the pallet loading problem. IIE Trans. **14**(3), 175–182 (1982)
14. JaJa, J.: Introduction to Parallel Algorithms, 1st edn. Addison Wesley, Boston (1992)
15. Jansen, K., Kratsch, S., Marx, D., Schlotter, I.: Bin packing with fixed number of bins revisited. J. Comput. Syst. Sci. **79**(1), 39–49 (2013)
16. Jansen, K., Solis-Oba, R.: An asymptotic approximation algorithm for 3D-strip packing. In: Proceedings of the Seventeenth Annual ACM-SIAM Symposium on Discrete Algorithms, SODA 2006, Miami, Florida, USA, 22–26 January 2006, pp. 143–152. ACM Press (2006)

17. Lokshtanov, D., Marx, D., Saurabh, S.: Lower bounds based on the exponential time hypothesis. Bull. EATCS, 41–71 (2011)
18. Martello, S., Toth, P.: Knapsack Problems: Algorithms and Computer Implementations. John Wiley, Hoboken (1990)
19. Nemhauser, G.L., Wolsey, L.A.: Integer and Combinatorial Optimization. John Wiley, New York (1999)
20. Paquay, C., Limbourg, S., Schyns, M.: A tailored two-phase constructive heuristic for the three-dimensional multiple bin size bin packing problem with transportation constraints. Eur. J. Oper. Res. **267**(1), 52–64 (2018)
21. Paschos, V.: An overview on polynomial approximation of NP-hard problems. Yugoslav J. Oper. Res. **19**, 3–40 (2009)
22. Perboli, G., Gobbato, L., Perfetti, F.: Packing problems in transportation and supply chain: new problems and trends. Proc. - Soc. Behav. Sci. **111**(5), 672–681 (2014)
23. Ram, B.: The pallet loading problem: a survey. Int. J. Prod. Econ. **28**, 217–225 (1992)
24. Renault, M.P., Rosén, A., van Stee, R.: Online algorithms with advice for bin packing and scheduling problems. Theor. Comput. Sci. **600**, 155–170 (2015)
25. Stockmeyer, L., Vishkin, U.: Simulation of parallel random access machines by circuits. SIAM J. Comput. **13**, 409–422 (1984)
26. Subramani, K., Caskurlu, B., Acikalin, U.U.: Security-aware database migration planning. In: Brandic, I., Genez, T.A.L., Pietri, I., Sakellariou, R. (eds.) ALGOCLOUD 2019. LNCS, vol. 12041, pp. 103–121. Springer, Cham (2020). https://doi.org/10.1007/978-3-030-58628-7_7
27. Subramani, K., Caskurlu, B., Velasquez, A.: Minimization of testing costs in capacity-constrained database migration. In: Disser, Y., Verykios, V.S. (eds.) ALGOCLOUD 2018. LNCS, vol. 11409, pp. 1–12. Springer, Cham (2019). https://doi.org/10.1007/978-3-030-19759-9_1
28. Terno, J., Scheithauer, G., Sommerweiß, U., Riehme, J.: An efficient approach for the multi-pallet loading problem. Eur. J. Oper. Res. **123**(2), 372–381 (2000)
29. Vargas-Osorio, S., Zuniga, C.: A literature review on the pallet loading problem. Lampsakos **15**, 69–80 (2016)
30. Zhou, K.: The pallet loading method of single category cargo based on railway containerized transport. In: Proceedings of the 2018 10th International Conference on Computer and Automation Engineering, ICCAE 2018, Brisbane, Australia, 24–26 February 2018, pp. 243–249. ACM (2018)

Recursive Methods for Some Problems in Coding and Random Permutations

Ghurumuruhan Ganesan(✉)

Institute of Mathematical Sciences, HBNI, Chennai, India
gganesan82@gmail.com

Abstract. In this paper, we study three applications of recursion to problems in coding and random permutations. First, we consider locally recoverable codes with partial locality and use recursion to estimate the minimum distance of such codes. Next we consider weighted lattice representative codes and use recursive subadditive techniques to obtain convergence of the minimum code size. Finally, we obtain a recursive relation involving cycle moments in random permutations and as an illustration, evaluate recursions for the mean and variance.

Keywords: Locally recoverable codes · Partial locality · Minimum distance · Lattice identification codes · Minimum size · Random permutations · Cycle moments

1 Introduction

Recursive techniques are used quite frequently in coding to obtain bounds on code sizes. As a typical example, the Singleton bound [9] obtains bounds on sizes of $n-$length codes by reducing the problem to that of an $(n-1)-$length code. Similarly, recursive relations are also frequent in terms related to permutations like for example, Stirling numbers of the first kind [8]. In this paper, we study further applications of recursive methods to problems in coding and random permutations.

The paper is organized as follows: In Sect. 2, we consider locally recoverable codes with partial locality and estimate the minimum distance of such codes (Theorem 1) using iterations on subcodes. Next, in Sect. 3, we study lattice representative codes with weights and prove asymptotic convergence of the minimum size, using subadditive techniques (Theorem 2). Finally, in Sect. 4, we establish a recursion for cycle moments of random permutations (Theorem 3) and illustrate our result for the cases of mean and variance (Corollary 1).

2 Locally Recoverable Codes with Partial Locality

Locally recoverable codes for erasures have tremendous applications in distributed storage and retrieval [12] and it is therefore important to understand the

A. Mudgal and C. R. Subramanian (Eds.): CALDAM 2021, LNCS 12601, pp. 373–384, 2021.
https://doi.org/10.1007/978-3-030-67899-9_30

properties of such codes. Typically each erasure correction is performed using a locality set of small size and it is of interest to design codes capable of correcting multiple erasures simultaneously. Such codes are also known as locally repairable codes and storage-bandwidth tradeoff and construction of such codes has been well-studied; for an overview we refer to the recent survey [2]. For distinction, we refer to codes above as *fully* locally recoverable codes since *every* symbol position has a locality set of small size associated with it. In [7], bounds are obtained for the minimum distance of linear fully locally recoverable codes in terms of the size of the locality sets. Later [5] studied bounds on the minimum distance of non-linear systematic fully locally recoverable codes.

In this section, we study minimum distance of locally recoverable codes with partial locality. We assume that only a *subset* of symbol positions have locality set size at most r and obtain bounds on the minimum distance. Let $n \geq k \geq 1$ be integers and let $\mathcal{A}$ be a set of cardinality $\#\mathcal{A} = q$. A set $\mathcal{C} \subseteq \mathcal{A}^n$ of cardinality q^k is defined to be an $(n, k)-$code.

For a set $\mathcal{U} \subseteq \{1, 2, \ldots, n\}$ and an integer $j \in \{1, 2, \ldots, n\} \setminus \mathcal{U}$, we say that position j is determined by $\mathcal{U}$ if there exists a function g_j such that

$$c_j = g_j\left(c_i : i \in \mathcal{U}\right) =: g_j\left(\mathcal{U}\right) \tag{2.1}$$

for all codewords $\mathbf{c} = (c_1, \ldots, c_n) \in \mathcal{C}$. In words, the symbol at the jth position of any codeword can be determined from the symbols with positions in $\mathcal{U}$. The set $\mathcal{F}_{\mathcal{U}}$ of all positions determined by $\mathcal{U}$ is called the *reach* of $\mathcal{U}$. For integer $1 \leq w \leq n$, we define

$$L(w) = L(w, \mathcal{C}) := \max_{\mathcal{U}:\#\mathcal{U}=w} \#\mathcal{F}_{\mathcal{U}}. \tag{2.2}$$

For any $w \geq 1$ we have that $L(w, \mathcal{C}) \leq \Delta(w)$, where $\Delta(w) = q^w$ if $\mathcal{C}$ is a linear code and $\Delta(w) = q^{q^w}$ otherwise. We remark here that q^{q^w} is the *total* number of maps from $\mathcal{A}^w$ to $\mathcal{A}$.

Definition 1. *For integers $\tau, r \geq 2$ and a subset $\Theta \subseteq \{1, 2, \ldots, n\}$, we say that the code $\mathcal{C}$ has $(\Theta, \tau, r)-$local correction capability if for every subset $\mathcal{P} \subseteq \Theta$ of size τ, there exists a set $\mathcal{T}_{\mathcal{P}} \subseteq \{1, 2, \ldots, n\} \setminus \mathcal{P}$ of size at most r such that each position in $\mathcal{P}$ is determined by $\mathcal{T}_{\mathcal{P}}$.*

We define $\mathcal{T}_{\mathcal{P}}$ to be the $r-$locality set corresponding to the set $\mathcal{P}$. Also if $\Theta = \{1, 2, \ldots, n\}$, we say that $\mathcal{C}$ has $(\tau, r)-$local correction capability.

For example, consider the binary linear code $\mathcal{C}$ formed in the following way: For $k \geq 10$ and a word $(c_1, \ldots, c_k) \in \{0, 1\}^k$, let $d_{i_1,i_2,i_3} := c_{i_1} \oplus c_{i_2} \oplus c_{i_3}$ where $\oplus$ denotes addition modulo two. There are $\binom{k}{3} =: n - k - 1$ such terms $\{d_{i_1,i_2,i_3}\}$ which we relabel as $c_{k+1}, \ldots, c_{n-1}$. Finally we let $c_n := \oplus_{i=1}^{k} c_i$. We let $(c_1, \ldots, c_n)$ be the codeword corresponding to the word $(c_1, \ldots, c_k)$. The collection of codewords $\mathcal{C}$ has $(\Theta, \tau, r)-$local correction capability with $\Theta = \{1, 2, \ldots, n-1\}, \tau = 1$ and $r = 3$. For example to recover c_1, we use the relation

$$d_{1,2,3} \oplus d_{1,2,4} \oplus d_{1,3,4} = c_1.$$

In general, each bit $c_j, 1 \leq j \leq k$ can be recovered in a similar manner. Because $k \geq 10$, the bit c_n cannot be recovered by using any three bits of $\{c_1, \ldots, c_{n-1}\}$.

Let $\mathcal{C}$ be any code with $(\Theta, \tau, r)-$local correction capability. For words $\mathbf{x} = (x_1, \ldots, x_n)$ and $\mathbf{y} = (y_1, \ldots, y_n)$ in $\mathcal{C}$ we define the Hamming distance between $\mathbf{x}$ and $\mathbf{y}$ to be $d(\mathbf{x}, \mathbf{y}) := \sum_{i=1}^{n} 1\!\!1(x_i \neq y_i)$, where $1\!\!1(.)$ denotes the indicator function. The minimum distance of $\mathcal{C}$ is then defined as $d(\mathcal{C}) := \min_{\mathbf{x},\mathbf{y} \in \mathcal{C}} d(\mathbf{x}, \mathbf{y})$. We have the following result.

Theorem 1. *Let $\mathcal{C}$ be any $(n, k)-$code with $(\Theta, \tau, r)-$parallel correction capability and let $\theta = \#\Theta$. The minimum distance of $\mathcal{C}$ satisfies*

$$d(\mathcal{C}) \leq n - k + 1 - T \cdot \tau, \tag{2.3}$$

where T is the largest integer t such that

$$t \cdot r \leq k - 1 + \theta - n \text{ and } t \cdot r + \Delta(t \cdot r) \leq \theta - \tau + 1. \tag{2.4}$$

To obtain the bound (2.4), we proceed as in [5] and iteratively construct a sequence of codes with decreasing size, until no further reduction is possible. We use the pigeonhole principle at the end of each step and obtain the sufficient conditions that allow continuation of the iteration procedure. In the proof below, we see that the first estimate in (2.4) determines the maximum number of iterations the procedure can proceed before we run out of codewords to choose from and the second estimate in (2.4) determines the maximum number of iterations for which we are able to choose a "fresh" locality set.

Finally, we recall that $n-k+1$ is the Singleton bound [9] and is the maximum possible minimum distance of an $(n, k)-$code. Therefore the parameter T is in some sense, the "cost" for requiring partial locality.

Proof of Theorem 1. Let $\mathcal{P}_1 \subseteq \Theta$ be any set of size τ and let $\mathcal{I}_1 := \mathcal{T}_{\mathcal{P}_1} = \{l_1, \ldots, l_{m_1}\}, m_1 \leq r$ be the corresponding locality set of cardinality at most r as defined in the paragraph following (2.2) that determines the value of the symbols in positions in $\mathcal{P}_1$. For $\mathbf{x} = (x_1, \ldots, x_{m_1}) \in \mathcal{A}^{m_1}$, let $\mathcal{C}(\mathbf{x}) = \mathcal{C}(\mathbf{x}, \mathcal{I}_1)$ be the set of codewords of $\mathcal{C}$ such that the symbol in position l_j equals x_j for $1 \leq j \leq m_1$.

The number of choices for $\mathbf{x}$ is at most q^{m_1} and there are q^k codewords in $\mathcal{C}$. Therefore by pigeonhole principle, there exists $\mathbf{x}_1$ such that

$$\#\mathcal{C}(\mathbf{x}_1) \geq \frac{\#\mathcal{C}}{q^{m_1}} = q^{k-m_1}. \tag{2.5}$$

We set $\mathcal{C}_1 := \mathcal{C}(\mathbf{x}_1)$ and let $\mathcal{J}_1 := \mathcal{F}_{\mathcal{I}_1}$ be the reach of $\mathcal{I}_1$ (see (2.1)) with cardinality $\tau \leq \#\mathcal{J}_1 \leq \Delta(r)$. The first inequality is true since $\mathcal{P}_1 \subseteq \mathcal{J}_1$ and the second estimate follows from (2.1). By construction, all words in the code $\mathcal{C}_1$ have the same values in the symbol positions determined by $\mathcal{J}_1$; i.e., if $a = (a_1, \ldots, a_n)$ and $b = (b_1, \ldots, b_n)$ both belong to $\mathcal{C}_1$, then $a_j = b_j$ for all $j \in \mathcal{J}_1$.

We now repeat the above procedure with the code $\mathcal{C}_1$ assuming that $r+n-\theta < k$, where $\theta := \#\Theta$. If $\mathcal{R}_1 := \Theta \setminus (\mathcal{I}_1 \cup \mathcal{J}_1)$ is the set of positions not encountered in the first iteration then

$$\#\mathcal{R}_1 \geq \#\Theta - \#\mathcal{J}_1 - \#\mathcal{I}_1 \geq \theta - \Delta(r) - r \tag{2.6}$$

since $\#\mathcal{J}_1 \leq \Delta(r)$. For a set $\mathcal{P} \subseteq \mathcal{R}_1$ of size τ, let $\mathcal{I}(\mathcal{P}) := \mathcal{I}_{\mathcal{P}} \bigcap \mathcal{R}_1$ be the union of positions within the locality sets of the selected τ positions in $\mathcal{P}$, not encountered in the first iteration.

Suppose for every $\mathcal{P} \subseteq \mathcal{R}_1$, we have $\mathcal{I}(\mathcal{P}) = \emptyset$. This means that all symbols with positions in $\mathcal{R}_1$ can simply be determined by the symbol values with positions in $\mathcal{I}_1 \cup \mathcal{J}_1$. This in turn implies that the symbols with positions in $\mathcal{I}_1$ determine *all* the symbols with positions in Θ. Using $r + n - \theta < k$ we then get that the total number of words in the code $\mathcal{C}$ is at most

$$q^{\#\mathcal{I}_1} \cdot q^{n-\#\Theta} = q^{m_1+n-\theta} \leq q^{r+n-\theta} < q^k,$$

a contradiction. Thus there exists $\mathcal{P}_2 \subseteq \mathcal{R}_1$ of size τ whose corresponding set $\mathcal{I}_2 := \mathcal{I}(\mathcal{P}_2)$ is not completely contained in $\mathcal{I}_1 \cup \mathcal{J}_1$.

Letting $1 \leq m_2 \leq r$ denote the cardinality of $\mathcal{I}_2$ and using the pigeonhole principle as before, we get a code $\mathcal{C}_2 \subseteq \mathcal{C}_1$ of size

$$\#\mathcal{C}_2 \geq \frac{\#\mathcal{C}_1}{q^{m_2}} \geq q^{k-m_1-m_2} \geq q^{k-2r}$$

and all of whose words have the same symbol values in the positions determined by $\mathcal{I}_1 \cup \mathcal{I}_2$. In the above, we use the estimate for $\mathcal{C}_1$ obtained in (2.5). As before, let $\mathcal{J}_2 \subseteq \{1, 2, \ldots, n\}$ be the set of positions of the codeword symbols determined by the set $\mathcal{I}_1 \cup \mathcal{I}_2$ so that $\mathcal{P}_1 \cup \mathcal{P}_2 \subseteq \mathcal{J}_2$. The set $\mathcal{I}_1 \cup \mathcal{I}_2$ has cardinality at most $2r$ and so we have from (2.2) that the reach $\mathcal{J}_2$ has cardinality

$$2\tau \leq \#\mathcal{P}_1 + \#\mathcal{P}_2 \leq \#\mathcal{J}_2 \leq \Delta(2r).$$

If $\mathcal{R}_2 := \Theta \setminus (\mathcal{I}_1 \cup \mathcal{I}_2 \cup \mathcal{J}_2)$, then $\#\mathcal{R}_2 \geq \theta - 2r - \Delta(2r)$. Continuing this way, after the end of t iterations, we have a code $\mathcal{C}_t$ of size

$$\#\mathcal{C}_t \geq q^{k-\sum_{j=1}^{t} \#\mathcal{I}_j} \geq q^{k-t\cdot r} \tag{2.7}$$

and a set $\mathcal{R}_t \subseteq \Theta$ of remaining positions not fixed so far, with cardinality

$$\#\mathcal{R}_t \geq \theta - \sum_{j=1}^{t} \#\mathcal{I}_j - \#\mathcal{J}_t \geq \theta - t \cdot r - \Delta(t \cdot r). \tag{2.8}$$

The above procedure can therefore be performed for at least T steps where T is the largest integer t such that

$$t \cdot r \leq k - 1 + \theta - n \text{ and } t \cdot r + \Delta(t \cdot r) \leq \theta - \tau + 1. \tag{2.9}$$

The first condition in (2.9) ensures that $k - T \cdot r \geq 1$ and so the code $\mathcal{C}_T$ has at least two codewords. The second condition in (2.9) ensures that the set $\mathcal{P}_j \subseteq \Theta$ of symbols we pick is at least τ and so

$$\#\mathcal{J}_j \geq \sum_{l=1}^{j} \#\mathcal{P}_l \geq j \cdot \tau \tag{2.10}$$

for each $1 \leq j \leq T$.

Since $\mathcal{C}_T \subseteq \mathcal{C}$, the minimum distance $d(\mathcal{C}_T)$ of $\mathcal{C}_T$ is at least the minimum distance $d(\mathcal{C})$ of $\mathcal{C}$. By definition we recall that the symbol values in positions determined by the set $\bigcup_{1\leq j\leq T}\mathcal{I}_j \bigcup \mathcal{J}_T := \{1,2,\ldots,n\} \setminus \mathcal{Q}_T$ is the same for all the words in $\mathcal{C}_T$. For every word $\mathbf{x} = (x_1,\ldots,x_n) \in \mathcal{C}_T$, we therefore let $\mathbf{x}_T = (x_i)_{i\in\mathcal{Q}_T}$ be the reduced word obtained by just considering the symbols in the remaining positions determined by $\mathcal{Q}_T$. Defining the reduced code $\mathcal{D}_T = \{\mathbf{x}_T : \mathbf{x} \in \mathcal{C}_T\}$ we then have that the minimum distance $d(\mathcal{D}_T) \geq d(\mathcal{C}_T) \geq d(\mathcal{C})$.

The length of the each word in $\mathcal{D}_T$ equals $n - \#\mathcal{Q}_T$ and so using the estimate for $\#\mathcal{D}_T = \#\mathcal{C}_T$ from (2.7) and the Singleton bound we have

$$d(\mathcal{D}_T) \leq \left(n - \sum_{j=1}^{T} \#\mathcal{I}_j - \#\mathcal{J}_T\right) - \left(k - \sum_{j=1}^{T} \#\mathcal{I}_j\right) + 1.$$

Thus $d(\mathcal{D}_T) \leq n - k + 1 - \#\mathcal{J}_T \leq n - k + 1 - T \cdot \tau$, by (2.10). Using the fact that $d(\mathcal{C}) \leq d(\mathcal{D}_T)$, we then get (2.3). □

3 Lattice Representative Codes

Representative codes [10] (also known as hitting sets in some contexts) are important from both theoretical and application perspectives. In [11] the minimum size of hitting sets that intersect all combinatorial rectangles of a given volume are studied. Explicit constructions were described using expander graphs and random walks. Later [14] determined lower bounds for the hitting set size of combinatorial rectangles and also illustrated an application in approximation algorithms. Recently [4] used fractional perfect hash families to study construction of explicit hitting sets for combinatorial shapes.

In this section, we study lattice representative codes for *weighted* rectangles where each vertex is assigned a positive finite weight. We study the minimum size of a representative code that intersects *all* subsets of a given minimum weight. For integers $d, m \geq 1$ let $\mathcal{S} = \mathcal{S}(m) := \{1,2,\ldots,m\}^d$. We refer to elements of $\mathcal{S}$ as points and for a point $v = (v_1,\ldots,v_d) \in \mathcal{S}$, we refer to v_i as the ith entry. For each point $v \in \mathcal{S}$, we assign a finite positive weight $w(v)$. For a set $\mathcal{U} \subseteq \mathcal{S}$ the corresponding weight $w(\mathcal{U}) := \sum_{v\in\mathcal{U}} w(v)$ is the sum of the weights of the points in $\mathcal{U}$. The size of $\mathcal{U}$ is the number of points in $\mathcal{U}$ and is denoted by $\#\mathcal{U}$.

Let

$$1 = \inf_m \min_{v\in\mathcal{S}(m)} w(v) \leq \sup_m \max_{v\in\mathcal{S}(m)} w(v) =: \beta < \infty \tag{3.1}$$

and for $0 < \epsilon < 1$ say that $\mathcal{B} \subseteq \mathcal{S}$ is an $\epsilon-$representative code or simply representative code if $\mathcal{B} \cap \mathcal{U} \neq \emptyset$ for any set $\mathcal{U} \subseteq \mathcal{S}$ of weight $w(\mathcal{U}) \geq \epsilon \cdot m^d$.

The following result obtains an estimate on the minimum size b_m of an $\epsilon-$representative code.

Theorem 2. *For any* $0 < \epsilon < 1$ *and* $\beta \geq 1$ *we have that*

$$m^d \cdot (1-\epsilon) \leq b_m \leq m^d \cdot \left(1 - \frac{\epsilon}{\beta}\right) + 1. \tag{3.2}$$

Suppose the weight function satisfies the following monotonicity relation: If u and v are any two points of $\mathcal{S}$ differing only in the ith entry and $u_i > v_i$, then the weights $w(u) \leq w(v)$. We then have

$$\frac{b_m}{m^d} \longrightarrow \lambda \tag{3.3}$$

as $m \to \infty$ where $1 - \epsilon \leq \lambda \leq 1 - \frac{\epsilon}{\beta}$.

Thus there exists a *fraction* of vertices in a large rectangle that hits *all* sets of a given minimum weight. Moreover, if the weight assignment is monotonic, then the scaled minimum representative code size converges to a positive constant strictly between 0 and 1.

An example of non-trivial weight assignment that satisfies the monotonicity relation is the following: Defining $w(1,1) := 2$ we iteratively assign the weight of each vertex in the set $\{1,2,\ldots,i+1\}^d \setminus \{1,2,\ldots,i\}^d$ as $1+\frac{1}{i}$. The conditions in Theorem 2 are then satisfied with $\beta = 2$.

Proof of Theorem 2. We begin with the proof of (3.2). Throughout we assume that $d = 2$ and an analogous analysis holds for general d. If $\mathcal{F}$ is any representative code of $\mathcal{S}$, then by definition, the weight of the set $\mathcal{S} \setminus \mathcal{F}$ is at most ϵm^2 and since the weight of each vertex is at least one (see (3.1)), we get that the number of points in $\mathcal{S} \setminus \mathcal{F}$ is at most ϵm^2. This implies that the size of $\mathcal{F}$ is at least $(1-\epsilon)m^2$ and so $b_m \geq (1-\epsilon)m^2$.

To find an upper bound on b_m, we let $\mathcal{T} \subseteq \mathcal{S}$ be any "critical" set such that the weight of $\mathcal{T}$ is at most $\epsilon m^2 - 1$ and the weight of $\mathcal{T} \cup \{v\}$ for any point $v \in \mathcal{S} \setminus \mathcal{T}$ is at least ϵm^2. The set $\mathcal{S} \setminus \mathcal{T}$ is then a representative code of $\mathcal{S}$ and since the weight of any point is at most β (see (3.1)), we get that the number of vertices in $\mathcal{T}$ is at least $\frac{\epsilon}{\beta} \cdot m^2 - 1$. This in turn implies that $b_m \leq m^2\left(1 - \frac{\epsilon}{\beta}\right) + 1$. This proves (3.2).

To prove (3.3), we use a subsequence argument analogous to the proof of Fekete's lemma. For integers $m \geq r \geq 1$, we let $m = k \cdot r + s$, where $k \geq 1$ and $0 \leq s \leq r-1$ are integers and split $\{1,2,\ldots,m\}^2$ into four sets

$$\mathcal{S}_1 := \{1,2,\ldots,kr\}^2, \quad \mathcal{S}_2 := \{1,2,\ldots,kr\} \times \{kr+1, kr+2, \ldots, kr+s\},$$

$$\mathcal{S}_3 := \{kr+1,\ldots,kr+s\} \times \{1,2,\ldots,kr\} \text{ and } \mathcal{S}_4 := \{kr+1,\ldots,kr+s\}^2.$$

Thus $\mathcal{S}_2$ is essentially a "rotated" version of $\mathcal{S}_3$. For $1 \leq i \leq 4$, let $\mathcal{G}_i(\epsilon)$ be a representative code of $\mathcal{S}_i$ and let $\mathcal{R}$ be any set in $\{1,2,\ldots,m\}^2$ of weight $w(\mathcal{R}) \geq \epsilon m^2 = \epsilon(kr+s)^2$. We first see that $\bigcup_{i=1}^4 \mathcal{G}_i(\epsilon)$ is a representative code of $\{1,2,\ldots,m\}^2$. Indeed if $\mathcal{R}_i = \mathcal{R} \cap \mathcal{S}_i$, then using the fact that $w(\mathcal{R}) = \sum_{i=1}^4 w(\mathcal{R}_i)$, we get that either $w(\mathcal{R}_1) \geq \epsilon(kr)^2$ or $w(\mathcal{R}_2) \geq \epsilon krs$ or $w(\mathcal{R}_3) \geq \epsilon krs$ or $w(\mathcal{R}_4) \geq \epsilon s^2$. Consequently, we must have that $\mathcal{R} \bigcap \bigcup_{i=1}^4 \mathcal{G}_i(\epsilon) \neq \emptyset$.

If $b^{(i)}$ denotes the minimum size of a representative code of $\mathcal{S}_i$, then from the discussion above we get

$$b_{kr+s} \leq \sum_{i=1}^{4} b^{(i)} \leq b^{(1)} + 2krs + s^2 \leq b^{(1)} + (2k+1)r^2 \tag{3.4}$$

where the second inequality in (3.4) follows from the trivial estimate that the size of any representative code of $\mathcal{R}_i$ is at most the total number of points in $\mathcal{R}_i$ and the final inequality in (3.4) follows from the fact that $s \leq r$.

To estimate $b^{(1)}$, we split $\mathcal{S}_1 := \{1, 2, \ldots, kr\}^2$ into k^2 disjoint rectangles $\mathcal{T}_i$, $1 \leq i \leq k^2$ each containing r^2 points with $\mathcal{T}_1 = \{1, 2, \ldots, r\}^2$. If c_i denotes the minimum size of a representative code of $\mathcal{T}_i$, then using the weight monotonicity relation, we get that $c_i \leq c_1$. To see this is true suppose $\mathcal{T}_2 = \{1, 2, \ldots, r\} \times \{r+1, \ldots, r+2r\}$ so that $\mathcal{T}_2 = \mathcal{T}_1 + (r, 0)$ is obtained by translation of $\mathcal{T}_1$. If $\mathcal{U} \subset \mathcal{T}_2$ is any set of weight at least ϵr^2 then $\mathcal{U} - (r, 0) \subseteq \mathcal{T}_1$ also has weight at least ϵr^2, by the weight monotonicity relation. Consequently if $\mathcal{W}_1$ is a representative code of $\mathcal{T}_1$, then $\mathcal{W}_1 + (r, 0)$ is a representative code of $\mathcal{T}_2$. Thus $c_2 \leq c_1$ and the proof of general c_i is analogous.

From (3.4) and the discussion in the above paragraph, we get that $b_m = b_{kr+s} \leq k^2 b_r + (2k+1)r^2$ and so

$$\frac{b_m}{m^2} \leq \frac{k^2 b_r + (r^2(2k+1))}{m^2} = \left(\frac{kr}{kr+s}\right)^2 \left(\frac{b_r}{r^2} + \frac{2k+1}{k^2}\right).$$

If $m \to \infty$ with r fixed, then $k = k(m, r) = \frac{m-s}{r} \geq \frac{m-r}{r} \to \infty$ as well and so $\frac{kr}{kr+s} \to 1$. This in turn implies that

$$\limsup_m \frac{b_m}{m^2} \leq \limsup_m \left(\frac{b_r}{r^2} + \frac{2k+1}{k^2}\right) = \frac{b_r}{r^2}. \tag{3.5}$$

Since $r \geq 1$ is arbitrary we get from (3.5) that

$$\limsup_m \frac{b_m}{m^2} = \liminf_r \frac{b_r}{r^2} = \inf_r \frac{b_r}{r^2} =: \lambda.$$

Also, the bounds for λ follow from (3.2). □

4 Random Permutations

Random permutations and applications are frequently encountered in computing problems and it is of interest to study the cycle properties of a randomly chosen permutation. The papers [6], [13] studied limiting distributions for the convergence of the number of cycles and cycles lengths of a uniform random permutation, after suitable renormalization. Later [1] used Poisson approximation and estimates on the total variation distance to study the convergence of the overall cycle structure to a process of independent Poisson random variables. Recently [3] have used probability generating functions to study convergence of number of cycles of uniform random permutations conditioned not to have large cycles, scaled and centred, to the Gaussian distribution.

From the combinatorial aspect, Stirling numbers of the first kind and generating functions have been used to study random permutation statistics. Using the Flajolet-Sedgewick theorem it is possible to enumerate permutations with constraints [8]. In this section, we use conditioning to obtain a recursive relation involving cycle moments of random permutations. As an illustration, we compute recursive relation involving the mean and the variance of the number of cycles in a uniformly random permutation.

We begin with a couple of definitions. A permutation π of $\{1,2,\ldots,n\}$ is a bijective map $\pi : \{1,2,\ldots,n\} \to \{1,2,\ldots,n\}$. The total number of possible permutations of $\{1,2,\ldots,n\}$ is therefore

$$n! := n \cdot (n-1) \cdots 2 \cdot 1.$$

A *cycle* of length k in a permutation π is a $k-$tuple $(i_1,\ldots,i_k)$ such that $\pi(i_j) = i_{j+1}$ for $1 \leq j \leq k-1$ and $\pi(i_k) = i_1$. Every number in $\{1,2,\ldots,n\}$ belongs to some cycle of π and this provides an alternate representation of π; for example $(1345)(267)(89)$ is the cycle representation of the permutation π on $\{1,2,\ldots,9\}$ satisfying $\pi(1) = 3, \pi(3) = 4, \pi(4) = 5, \pi(5) = 1, \pi(2) = 6, \pi(6) = 7, \pi(7) = 2, \pi(8) = 9, \pi(9) = 8$.

Let Π denote a uniformly chosen random permutation of $\{1,2,\ldots,n\}$ defined on the probability space $(\Omega_n, \mathcal{F}_n, \mathbb{P}_n)$ so that

$$\mathbb{P}_n(\Pi = \pi) = \frac{1}{n!}$$

for any deterministic permutation π. Let $N_n = N_n(\Pi)$ be the random number of cycles in Π and for integers $n, s \geq 1$, set $\mu_{0,s} := 0$ and $\mu_{n,s} := \mathbb{E}N_n^s$. We have the following result.

Theorem 3. *For integers $n, s \geq 1$ we have*

$$\mu_{n,s} = 1 + \frac{1}{n}\sum_{r=1}^{s}\sum_{j=1}^{n-1}\binom{s}{r}\mu_{j,r}, \tag{4.1}$$

where $\binom{s}{r} = \frac{s!}{r!(s-r)!}$ is the Binomial coefficient.

From the recursive structure of equation (4.1), we then have that $\mu_{n,s}$ could be computed using the previous values $\{\mu_{j,r}\}_{j\leq n-1, r\leq s}$.

As a Corollary of Theorem 3 we have the following recursive relations for the mean and variance of N_n.

Corollary 1. *The mean $\mu_n := \mu_{n,1}$ satisfies $\mu_1 = 1$ and the recursive equation*

$$\mu_n = 1 + \frac{1}{n}\sum_{i=1}^{n-1}\mu_i \tag{4.2}$$

for $n \geq 2$. The sequence $H_n := \sum_{j=1}^{n}\frac{1}{j}$ is the unique sequence satisfying (4.2).

The variance $v_n = var(N_n) := \mu_{n,2} - \mu_{n,1}^2$ satisfies $v_1 = 0$ and the recursive equation

$$v_n = 1 + \frac{1}{n}\sum_{i=1}^{n-1}v_i - \frac{H_n}{n}. \tag{4.3}$$

The sequence $M_n := H_n - \sum_{i=1}^{n}\frac{1}{i^2}$ is the unique sequence satisfying (4.3).

Using (4.2), (4.3) and the recursive relation (4.1), we could similarly compute higher order moments.

We prove Theorem 3 and Corollary 1 in that order.

Proof of Theorem 3. To obtain the desired recursive relation, we condition on the length of the first cycle and study the number of cycles in the remaining set of elements.

Let $\mathcal{S}_1$ denote the cycle of the random permutation Π containing the number 1 and let $L_1 = \#\mathcal{S}_1$ be the length of $\mathcal{S}_1$ so that $\mathcal{S}_1$ is an L_1−tuple. If $L_1 = k \leq n-1$, then Π induces a permutation $\sigma : \{1, 2, \ldots, n-k\} \to \{1, 2, \ldots, n-k\}$ on the remaining $n-k$ numbers $\{1, 2, \ldots, n\} \setminus \mathcal{S}_1$ in the following way. Arrange the numbers in $\{1, 2, \ldots, n\} \setminus \mathcal{S}_1$ in increasing order $j_1 < j_2 < \ldots < j_{n-k}$ and suppose that $\pi(j_l) = m_l$ for $1 \leq l \leq n-k$. The induced permutation σ then satisfies $m_l = j_{\sigma(l)}$ for $1 \leq l \leq n-k$.

Conditional on $L_1 = k$ we now see that σ is uniformly distributed in the sense that for any deterministic permutation $\sigma_0 : \{1, 2, \ldots, n-k\} \to \{1, 2, \ldots, n-k\}$ we have

$$\mathbb{P}_n\left(\sigma = \sigma_0 | L_1 = k\right) = \mathbb{P}_{n-k}(\sigma_0) = \frac{1}{(n-k)!}. \tag{4.4}$$

To see (4.4) is true, we first write

$$\mathbb{P}_n\left(\sigma = \sigma_0 | L_1 = k\right) = \frac{\mathbb{P}_n\left(\{\sigma = \sigma_0\} \cap \{L_1 = k\}\right)}{\mathbb{P}_n(L_1 = k)}. \tag{4.5}$$

If $k = 1$, then the numerator in the right side of (4.5) is $\frac{1}{n!}$. Moreover, if the first cycle simply consists of the single element 1, then the remaining $n-1$ numbers can be arranged in $(n-1)!$ ways and so $\mathbb{P}_n(L_1 = 1) = \frac{(n-1)!}{n!}$. Thus (4.4) is true for $k = 1$.

For $2 \leq k \leq n-1$, we have from (4.5) that $\mathbb{P}_n\left(\sigma = \sigma_0 | L_1 = k\right)$ equals

$$\frac{\sum_{(i_1,\ldots,i_{k-1})} \mathbb{P}_n\left(\{\sigma = \sigma_0\} \cap \{\mathcal{S}_1 = (1, i_1, \ldots, i_{k-1})\}\right)}{\sum_{(i_1,\ldots,i_{k-1})} \mathbb{P}_n(\mathcal{S}_1 = (1, i_1, \ldots, i_{k-1}))}$$

where the summation is over all $k-1$ tuples $(i_1, \ldots, i_{k-1})$ containing distinct elements. For any $(1, i_1, \ldots, i_{k-1})$, the term

$$\mathbb{P}_n\left(\{\sigma = \sigma_0\} \cap \{\mathcal{S}_1 = (1, i_1, \ldots, i_{k-1})\}\right) = \frac{1}{n!} \tag{4.6}$$

and

$$\mathbb{P}_n(\mathcal{S}_1 = (1, i_1, \ldots, i_{k-1})) = \frac{(n-k)!}{n!} \tag{4.7}$$

since there are $(n-k)!$ ways to permute the remaining $n-k$ elements of the set $\{2, \ldots, n\} \setminus \{i_1, \ldots, i_{k-1}\}$. Substituting (4.6) and (4.7) into (4), we get (4.4).

Summing (4.7) over all $k-1$ tuples with distinct entries (for which there are $(n-1)\cdot(n-2)\cdots(n-k+1)$ choices), we also get that $\mathbb{P}_n(L_1 = \#\mathcal{S}_1 = k) = \frac{1}{n}$.

From the discussion in the previous paragraph, we get that the above relation holds for all $1 \leq k \leq n$. Thus

$$\mu_{n,s} = \mathbb{E}_n N_n^s = \sum_{k=1}^{n} \mathbb{E}_n(N_n^s | L_1 = k)\mathbb{P}_n(L_1 = k) = \frac{1}{n}\sum_{k=1}^{n} \mathbb{E}_n(N_n^s | L_1 = k). \quad (4.8)$$

If $k = n$ then $N_n = 1$ and if $1 \leq k \leq n-1$, then $N_n = 1 + M_n$, where M_n is the number of cycles in the induced permutation σ. Therefore we get from (4.8) that

$$\mu_{n,s} = \frac{1}{n} + \frac{1}{n}\sum_{k=1}^{n-1} \mathbb{E}_n\left((1 + M_n)^s | L_1 = k\right). \quad (4.9)$$

Using the conditional distribution equivalence (4.4), we have for $1 \leq k \leq n-1$ that $\mathbb{E}_n\left((1 + M_n)^s | L_1 = k\right)$ equals

$$\mathbb{E}_{n-k}(1 + N_{n-k})^s = 1 + \sum_{r=1}^{s} \binom{s}{r} \mathbb{E}_{n-k} N_{n-k}^r, \quad (4.10)$$

by the Binomial expansion. Substituting (4.10) into (4.9) we get (4.1). □

Proof of Corollary 1. We begin with the proof of (4.2). Setting $s = 1$ in (4.1) and $\mu_n = \mu_{n,1}$, we get that $\mu_1 = 1$ and for $n \geq 2$, we get that μ_n satisfies (4.2). We first see by induction that H_n as defined in Proposition 1 satisfies (4.2). For $n = 2$, this statement is true and suppose H_l satisfies (4.2) for $1 \leq l \leq n-1$. For $l = n$, the right side of (4.2) evaluated with $\mu_i = H_i$ equals

$$1 + \frac{1}{n}\sum_{i=1}^{n-1}\sum_{j=1}^{i}\frac{1}{j} = 1 + \frac{1}{n}\sum_{j=1}^{n-1}\sum_{i=j}^{n-1}\frac{1}{j} = 1 + \frac{1}{n}\sum_{j=1}^{n-1}\frac{n-j}{j}, \quad (4.11)$$

by interchanging the order of summation in the second equality. The final term in (4.11) equals H_n and this proves the induction step.

Suppose now that $\{b_n\}$ is some sequence satisfying (4.2) with $b_1 = 1$ and let $u_n = b_n - H_n$ denote the difference. The sequence $\{u_n\}$ satisfies $u_1 = 0$ and $u_n = \frac{1}{n}\sum_{i=1}^{n-1} u_i$ for all $n \geq 2$. Thus $u_2 = \frac{u_1}{2} = 0$ and iteratively, we get $u_n = u_2 = 0$ for all $n \geq 2$. Thus H_n is the unique sequence satisfying (4.2).

We now obtain the variance estimate as follows. Letting $d_n := \mu_{n,2}$ and $\mu_n := \mu_{n,1} = H_n$, we get from (4.1) that

$$d_n = 1 + \frac{1}{n}\sum_{i=1}^{n-1}(d_i + 2\mu_i) = 2\mu_n - 1 + \frac{1}{n}\sum_{i=1}^{n-1} d_i, \quad (4.12)$$

since $\frac{1}{n}\sum_{i=1}^{n-1}\mu_i = \mu_n - 1$ (see (4.2)). From (4.12) we get that $v_n = d_n - \mu_n^2$ equals

$$\begin{aligned} v_n &= \frac{1}{n}\sum_{i=1}^{n-1}(d_i - \mu_i^2) + \frac{1}{n}\sum_{i=1}^{n-1}\mu_i^2 - (\mu_n - 1)^2 \\ &= \frac{1}{n}\sum_{i=1}^{n-1} v_i + \frac{1}{n}\sum_{i=1}^{n-1}\mu_i^2 - (\mu_n - 1)^2. \end{aligned}$$

It only remains to see that $\frac{1}{n}\sum_{i=1}^{n-1}\mu_i^2 - (\mu_n - 1)^2 = 1 - \frac{H_n}{n}$ and for that we use $\mu_i = H_i = \sum_{j=1}^{i}\frac{1}{j}$ (see (4.2)) to first get that $\frac{1}{n}\sum_{i=1}^{n-1}\mu_i^2$ equals

$$\frac{1}{n}\sum_{i=1}^{n-1}\sum_{j_1=1}^{i}\sum_{j_2=1}^{i}\frac{1}{j_1\cdot j_2} = \frac{1}{n}\sum_{j_1=1}^{n-1}\sum_{j_2=1}^{n-1}\sum_{i=\max(j_1,j_2)}^{n-1}\frac{1}{j_1\cdot j_2}$$

$$\frac{1}{n}\sum_{j_1=1}^{n-1}\sum_{j_2=1}^{n-1}\frac{(n-\max(j_1,j_2))}{j_1\cdot j_2} = \frac{1}{n}\sum_{j_1=1}^{n-1}\Delta(j_1), \tag{4.13}$$

where $\Delta(j_1) = \sum_{j_2=1}^{j_1}\frac{n-j_1}{j_1\cdot j_2} + \sum_{j_2=j_1+1}^{n-1}\frac{n-j_2}{j_1\cdot j_2}$ equals

$$\sum_{j_2=1}^{n-1}\frac{n}{j_1\cdot j_2} - \sum_{j_2=1}^{j_1}\frac{1}{j_2} - \sum_{j_2=j_1+1}^{n-1}\frac{1}{j_1}.$$

Thus $\frac{1}{n}\sum_{i=1}^{n-1}\mu_i^2$ equals

$$\sum_{j_1=1}^{n-1}\sum_{j_2=1}^{n-1}\frac{1}{j_1\cdot j_2} - \frac{1}{n}\sum_{j_1=1}^{n-1}\sum_{j_2=1}^{j_1}\frac{1}{j_2} - \frac{1}{n}\sum_{j_1=1}^{n-1}\sum_{j_2=j_1+1}^{n-1}\frac{1}{j_1}. \tag{4.14}$$

The first term in (4.14) is

$$\sum_{j_1=1}^{n-1}\sum_{j_2=1}^{n-1}\frac{1}{j_1\cdot j_2} = H_{n-1}^2 = \left(H_n - \frac{1}{n}\right)^2$$

and the second term in (4.14) is

$$\frac{1}{n}\sum_{j_1=1}^{n-1}\sum_{j_2=1}^{j_1}\frac{1}{j_2} = \frac{1}{n}\sum_{j_1=1}^{n-1}H_{j_1} = H_n - 1$$

using the fact that $\mu_n = H_n$ satisfies (4.2). The third term in (4.14) equals

$$\frac{1}{n}\sum_{j_1=1}^{n-1}\frac{n-1-j_1}{j_1} = \left(\frac{n-1}{n}\right)\left(H_n - 1 - \frac{1}{n}\right)$$

after rearrangement of terms. Substituting these three expressions into (4.14), we get that $\frac{1}{n}\sum_{i=1}^{n-1}\mu_i^2$ equals $1 + (H_n - 1)^2 - \frac{H_n}{n}$, which is what we wanted to prove. Finally, arguing as before, we also have that M_n is the unique sequence satisfying (4.3). □

Acknowledgements. I thank Professors Rahul Roy, V. Guruswami, C. R. Subramanian and the referees for crucial comments that led to an improvement of the paper. I also thank IMSc for my fellowships.

References

1. Arratia, R., Tavaré, S.: The cycle structure of random permutations. Ann. Probab. **20**, 1567–1591 (1992)
2. Balaji, S.B., Krishnan, M.N., Vajha, M., Ramkumar, V., Sasidharan, B., Kumar, P.V.: Erasure coding for distributed storage: an overview. Sci. China Inf. Sci. **61** (2018)
3. Betz, V., Schäfer, H.: The number of cycles in random permutations without long cycles is asymptotically Gaussian, ALEA. Lat. Am. J. Probab. Stat. **14**, 427–444 (2017)
4. Bhaskara, A., Desai, D., Srinivasan, S.: Optimal hitting sets for combinatorial shapes. Theory Comput. **9**, 441–470 (2013)
5. Forbes, M., Yekhanin, S.: On the locality of codeword symbols in non-linear codes. Discrete Math. **324**, 78–84 (2014)
6. Gončarov, V.: On the field of combinatory analysis. Am. Math. Soc. Trans. **19**, 1–46 (1962)
7. Gopalan, P., Huang, C., Simitci, H., Yekhanin, S.: On the locality of codeword symbols. IEEE Trans. Inf. Theory **58**, 6925–6934 (2012)
8. Graham, R., Knuth D., Patashnik, O.: Concrete Mathematics. Addison-Wesley, Boston (1989)
9. Huffman, W.C., Pless, V.: Fundamentals of Error Correcting Codes. Cambridge University Press, Cambridge (2003)
10. Karpovsky, M.G., Chakrabarty, K., Levitin, L.B.: On a new class of codes for identifying vertices in graphs. IEEE Trans. Inf. Theory **44**, 599–611 (1998)
11. Linial, N., Luby, M., Saks, M., Zuckerman, D.: Efficient construction of a small hitting set for combinatorial rectangles in high dimension. Combinatorica **17**, 215–234 (1997)
12. Rashmi, K.V., Shah, N.B., Kumar, P.V.: Optimal exact-regenerating codes for distributed storage at the MSR and MBR points via a product-matrix construction. IEEE Trans. Inf. Theory **57**, 5227–5239 (2011)
13. Shepp, L.A., Lloyd, S.P.: Ordered cycle lengths in a random permutation. Trans. Am. Math. Soc. **121**, 340–357 (1966)
14. Sunil Chandran, L.: A lower bound for the hitting set size for combinatorial rectangles and an application. Inf. Process. Lett. **86**, 75–78 (2003)

Achieving Positive Rates with Predetermined Dictionaries

Ghurumuruhan Ganesan(✉)

Institute of Mathematical Sciences, HBNI, Chennai, India
gganesan82@gmail.com

Abstract. In the first part of the paper we consider binary input channels that are not necessarily stationary and show how positive rates can be achieved using codes constrained to be within predetermined dictionaries. We use a Gilbert-Varshamov-like argument to obtain the desired rate achieving codes. Next we study the corresponding problem for channels with arbitrary alphabets and use conflict-set decoding to show that if the dictionaries are contained within "nice" sets, then positive rates are achievable.

Keywords: Positive rates · Predetermined dictionaries

AMS 2000 Subject Classification: Primary · 94A15 · 94A24

1 Introduction

Achieving positive rates with low probability of error in communication channels is an important problem in information theory [3]. In general, a rate R is defined to be achievable if there exists codes with rate R and having arbitrarily small error probability as the code length $n \to \infty$. The existence of such codes is determined through the probabilistic method of choosing a random code (from the set of all possible codes) and showing that the chosen code has small error probability.

In many cases of interest, we would like to select codes satisfying certain constraints or equivalently from a predetermined *dictionary* (see [2,7] for examples). For stationary channels, the method of types [4,5] can be used to study positive rate achievability with the restriction that the dictionary falls within the set of words belonging to a particular type. In this paper, we study achievability of positive rates with *arbitrary* deterministic dictionaries for both binary and general input channels using counting techniques.

The paper is organized as follows: In Sect. 2, we study positive rate achievability in binary input channels using predetermined dictionaries. Next in Sect. 3, we describe the rate achievability problem for arbitrary stationary channels and state our result Theorem 2 regarding achieving positive rates using given dictionaries. Finally, in Sect. 4, we prove Theorem 2.

A. Mudgal and C. R. Subramanian (Eds.): CALDAM 2021, LNCS 12601, pp. 385–396, 2021.
https://doi.org/10.1007/978-3-030-67899-9_31

2 Binary Channels

For integer $n \geq 1$, an element of the set $\{0,1\}^n$ is said to be a *codeword* or simply word, of length n. Consider a discrete memoryless symmetric channel with input alphabet $\{0,1\}^n$ that corrupts a transmitted word $\mathbf{x} = (x_1, \ldots, x_n)$ as follows. If $\mathbf{Y} = (Y_1, \ldots, Y_n)$ is the received (random) word, then

$$Y_i := x_i 1\!\!1(W_i = 0) + (1 - x_i) 1\!\!1(W_i = 1) + \varepsilon 1\!\!1(W_i = \varepsilon) \tag{2.1}$$

for all $1 \leq i \leq n$, where $1\!\!1(.)$ denotes the indicator function and ε denotes the erasure symbol. If $W_i = 1$, then the bit x_i is substituted and if $W_i = \varepsilon$, then x_i is erased. The random variables $\{W_i\}_{1 \leq i \leq n}$ are independent with

$$\mathbb{P}(W_i = 1) = p_f(i) \text{ and } \mathbb{P}(W_i = \varepsilon) = p_e(i) \tag{2.2}$$

and so the probability of a bit error (due to either a substituted bit or an erased bit) at "time" index i is $p_f(i) + p_e(i)$. Letting $\mathbf{W} := (W_1, \ldots, W_n)$, we also denote $\mathbf{Y} =: h(\mathbf{x}, \mathbf{W})$ where h is a deterministic function defined via (2.1). We are interested in communicating through the above described channel, with low probability of error, using words from a predetermined (deterministic) dictionary.

Dictionaries

A dictionary of size M is a set $\mathcal{D} \subseteq \{0,1\}^n$ of cardinality $\#\mathcal{D} = M$. A subset $\mathcal{C} = \{\mathbf{x}_1, \ldots, \mathbf{x}_L\} \subseteq \mathcal{D}$ is said to be an $n-$length *code* of size L, contained in the dictionary $\mathcal{D}$. Suppose we transmit a word picked from $\mathcal{C}$, through the channel given by (2.1) and receive the (random) word $\mathbf{Y}$. Given $\mathbf{Y}$ we would like an estimate $\hat{\mathbf{x}}$ of the word from $\mathcal{C}$ that was transmitted. A decoder $g : \{0,1,\varepsilon\}^n \to \mathcal{C}$ is a deterministic map that uses the received word $\mathbf{Y}$ to obtain an estimate of the transmitted word. The probability of error corresponding to the code $\mathcal{C}$ and the decoder g is then defined as

$$q(\mathcal{C}, g) := \max_{1 \leq i \leq L} \mathbb{P}\left(g\left(h\left(\mathbf{x}_i, \mathbf{W}\right)\right) \neq \mathbf{x}_i\right), \tag{2.3}$$

where $\mathbf{W} = (W_1, \ldots, W_n)$ is the additive noise as described in (2.1).

We have the following definition regarding achievable rates using predetermined dictionaries.

Definition 1. *Let $R > 0$ and let $\mathcal{F} := \{\mathcal{D}_n\}_{n \geq 1}$ be any sequence of dictionaries such that each $\mathcal{D}_n$ has size at least 2^{nR}. We say that $R > 0$ is an $\mathcal{F}-$*achievable *rate if the following holds true for every $\epsilon > 0$: For all n large, there exists a code $\mathcal{C}_n \subset \mathcal{D}_n$ of size $\#\mathcal{C}_n = 2^{nR}$ and a decoder g_n such that the probability of error $q(\mathcal{C}_n, g_n) < \epsilon$.*

If $\mathcal{D}_n = \{0,1\}^n$ for each n, then the above reduces to the usual concept of rate achievability as in [3] and we simply say that R is achievable.

For $0 < x < 1$ we define the entropy function

$$H(x) := -x \cdot \log x - (1 - x) \cdot \log(1 - x), \tag{2.4}$$

where all logarithms in this section to the base two and have the following result.

Theorem 1. *For integer $n \geq 1$ let*

$$\mu_f = \mu_f(n) := \sum_{i=1}^{n} p_f(i) \text{ and } \mu_e = \mu_e(n) := \sum_{i=1}^{n} p_e(i)$$

be the expected number of bit substitutions and erasures, respectively in an $n-$length codeword and suppose

$$\min\left(\mu_f(n), \mu_e(n)\right) \longrightarrow \infty \text{ and } p := \limsup_n \frac{1}{n}\left(2\mu_f(n) + \mu_e(n)\right) < \frac{1}{2}. \tag{2.5}$$

Let $H(p) < \alpha \leq 1$ and let $\mathcal{F} := \{\mathcal{D}_n\}_{n\geq 1}$ be any sequence of dictionaries satisfying $\#\mathcal{D}_n \geq 2^{\alpha n}$, for each n. We have that every $R < \alpha - H(p)$ is $\mathcal{F}-$achievable.

For a given α, let $p(\alpha)$ be the largest value of p such that $H(p) < \alpha$. The above result says that every $R < \alpha - H(p)$ is achievable using arbitrary dictionaries. We use Gilbert-Varshamov-like arguments to prove Theorem 1 below.

As a special case, for binary symmetric channels with crossover probability p_f, each bit is independently substituted with probability p_f. No erasures occur and so

$$\mu_f(n) = np_f \text{ and } \mu_e(n) = 0.$$

Thus $p = 2p_f$ and from Theorem 1 we therefore have that if $H(2p_f) < \alpha$, then every $R < \alpha - H(2p_f)$ is achievable.

Proof of Theorem 1

The main idea of the proof is as follows. Using standard deviation estimates, we first obtain an upper bound on the number of possible errors that could occur in a transmitted word. More specifically, if T denotes the number of bit errors in an $n-$length word and $\epsilon > 0$ is given, we use standard deviation estimates to determine $T_0 = T_0(n)$ such that $\mathbb{P}(T > T_0) \leq \epsilon$. We then use a Gilbert-Varshamov argument to obtain a code that can correct up to T_0 bit errors. The details are described below.

We prove the Theorem in two steps. In the first step, we construct the code $\mathcal{C}$ and decoder g and in the second step, we estimate the probability of the decoding error for $\mathcal{C}$ using g. For $\mathbf{x}, \mathbf{y} \in \{0,1\}^n$, we let $d_H(\mathbf{x}, \mathbf{y}) = \sum_{i=1}^{n} 1\!\!1(x_i \neq y_i)$ be the Hamming distance between $\mathbf{x}$ and $\mathbf{y}$, where as before $1\!\!1(.)$ denotes the indicator function. The minimum distance of a code is the minimum distance between any two words in a code.

Step 1: Assume for simplicity that $t := np(1+2\epsilon)$ is an integer and let $d = t+1$. For a word $\mathbf{x}$ let $B_{d-1}(\mathbf{x})$ be the set of words that are at a distance of at most $d-1$ from $\mathbf{x}$. If $\mathcal{C} \subseteq \mathcal{D}$ is a maximum size code with minimum distance at least d, then by the maximality of $\mathcal{C}$ we must have

$$\bigcup_{\mathbf{x}\in\mathcal{C}} B_{d-1}(\mathbf{x}) = \mathcal{D}. \tag{2.6}$$

This is known as the Gilbert-Varshamov argument [6].

The cardinalities of $\mathcal{D}$ and $B_{d-1}(\mathbf{x})$ are $2^{\alpha n}$ and $\sum_{i=0}^{d-1}\binom{n}{i}$ respectively and so from (2.6), we see that the code $\mathcal{C}$ has size

$$\#\mathcal{C} \geq \frac{2^{\alpha n}}{\sum_{i=0}^{d-1}\binom{n}{i}} \tag{2.7}$$

and minimum distance at least d. Also since $p < \frac{1}{2}$, we have for all small $\epsilon > 0$ that $\binom{n}{i} \leq \binom{n}{d-1} = \binom{n}{np(1+2\epsilon)}$ and so $\sum_{i=0}^{d-1}\binom{n}{i} \leq n \cdot \binom{n}{np(1+2\epsilon)}$. Using Stirling approximation we get

$$\binom{n}{np(1+2\epsilon)} \leq 4en \cdot 2^{nH(p+2p\epsilon)}$$

and so from (2.7), we get for $\delta > 0$ that

$$\#\mathcal{C} \geq \frac{1}{4en^2} \cdot 2^{n(\alpha - H(p+2p\epsilon))} \geq 2^{n(\alpha - H(p) - \delta)} \tag{2.8}$$

provided $\epsilon > 0$ is small.

We now use a two stage decoder described as follows: Suppose the received word is $\mathbf{Y}$ and for simplicity suppose that the last e positions in $\mathbf{Y}$ have been erased. For a codeword $\mathbf{x} = (x_1, \ldots, x_n)$, let $\mathbf{x}_{red} := (x_1, \ldots, x_{n-e})$ be the reduced word formed by the first $n - e$ bits. Let $\mathcal{C}_{red} = \{\mathbf{x}_{red} : \mathbf{x} \in \mathcal{C}\}$ be the set of all reduced codewords in the code $\mathcal{C}$ formed by the first $n - e$ bits.

In the first stage of the decoding process, the decoder corrects bit substitutions by collecting all words $\mathcal{S} \subseteq \mathcal{C}_{red}$ whose Hamming distance from $\mathbf{Y}_{red}$ is minimum. If $\mathcal{S}$ contains exactly one word, say $\mathbf{z}_{red}$, the decoder outputs $\mathbf{z}_{red}$ as the estimate obtained in the first step of the iteration. Otherwise, the decoder outputs "decoding error". In the second stage of the decoding process, the decoder uses $\mathbf{z}_{red}$ to correct the erasures. Formally let $\mathcal{S}_e := \{\mathbf{x} \in \mathcal{C} : \mathbf{x}_{red} = \mathbf{z}_{red}\}$ be the set of all codewords whose first $n - e$ bits match $\mathbf{z}_{red}$. If there exists exactly one word $\mathbf{z}$ in $\mathcal{S}_e$, then the decoder outputs $\mathbf{z}$ to be the transmitted word. Else the decoder outputs "decoding error".

Step 2: Suppose a word $\mathbf{x} \in \mathcal{C}$ was transmitted and the received word is $\mathbf{Y}$. Let $\mathbf{W} = (W_1, \ldots, W_n)$ be the random noise vector as in (2.1) and let

$$T_f := \sum_{i=1}^{n} 1\!\!1(W_i = 1)$$

be the number of bits that have been substituted so that

$$\mathbb{E}T_f = \sum_{i=1}^{n} p_f(i) = \mu_f(n),$$

by (2.2). By standard deviation estimates (Corollary $A.1.14$, pp. 312, [1]) we have

$$\mathbb{P}\left(|T_f - \mu_f(n)| \geq \epsilon\mu_f(n)\right) \leq 2e^{-\frac{\epsilon^2}{4}\mu_f(n)} \leq \frac{\epsilon}{2} \tag{2.9}$$

for all n large, by the first condition of (2.5). Similarly if $T_e = \sum_{i=1}^{n} 1\!\!1(W_i = \varepsilon)$ is the number of erased bits, then

$$\mathbb{P}\left(|T_e - \mu_e(n)| \geq \epsilon\mu_e(n)\right) \leq 2e^{-\frac{\epsilon^2}{4}\mu_e(n)} \leq \frac{\epsilon}{2} \tag{2.10}$$

for all n large.

Next, using the second condition of (2.5) we have that

$$(2\mu_f(n) + \mu_e(n))(1+\epsilon) \leq np(1+2\epsilon) = t$$

for all n large and so from (2.9) and (2.10) we get that $\mathbb{P}(2T_f + T_e \geq t) \leq \epsilon$ for all n large. If $2T_f + T_e \leq t$, then by construction the decoder outputs $\mathbf{x}$ as the estimate of the transmitted word. Therefore a decoding error occurs only if $2T_f + T_e \geq t$ which happens with probability at most ϵ. Combining with (2.8) and using the fact that $\delta > 0$ is arbitrary, we get that every $R < \alpha - H(p)$ is $\mathcal{F}-$achievable. □

3 General Channels

Consider a discrete memoryless channel with finite input alphabet $\mathcal{X}$ of size $N := \#\mathcal{X}$, a finite output alphabet $\mathcal{Y}$ and a transition probability $p_{Y|X}(y|x), x \in \mathcal{X}, y \in \mathcal{Y}$. The term $p_{Y|X}(y|x)$ denotes the probability that output y is observed given that input x is transmitted through the channel.

For $n \geq 1$ we define a subset $\mathcal{D}_n \subseteq \mathcal{X}^n$ to be a *dictionary.* A subset $\mathcal{C} = \{x_1, \ldots, x_M\} \subseteq \mathcal{D}_n$ is defined to be an $n-$length code contained within the dictionary $\mathcal{D}_n$. Suppose we transmit the word x_1 and receive the (random) word $\Gamma_{x_1} \in \mathcal{Y}^n$. Given Γ_{x_1} we would like an estimate $\hat{x}$ of the word from $\mathcal{C}$ that was transmitted. A decoder $g : \mathcal{Y}^n \to \mathcal{C}$ is a deterministic map that "guesses" the transmitted word based on the received word Γ_{x_1}. We denote the probability of error corresponding to the code $\mathcal{C}$ and the decoder g as

$$q(\mathcal{C}, g) := \max_{x \in \mathcal{C}} \mathbb{P}\left(g(\Gamma_x) \neq x\right). \tag{3.1}$$

To study positive rate achievability using arbitrary dictionaries, we have a couple of preliminary definitions. Let $p_X(.)$ be any probability distribution on the input alphabet $\mathcal{X}$ and let $H(X) := -\sum_{x \in \mathcal{X}} p_X(x) \log p_X(x)$ be the entropy of a random variable X where the logarithm is to the base N here. Let Y be a random variable having joint distribution $p_{XY}(x,y)$ with the random variable X defined by $p_{XY}(x,y) := p_{Y|X}(y|x) \cdot p_X(x)$. Thus Y is the random output of the channel when the input is X. Letting $p_Y(y) := \sum_x p_{XY}(x,y)$ be the marginal of Y we have that the joint entropy and conditional entropy [3] are respectively given by

$$H(X,Y) = -\sum_{x,y} p_{XY}(x,y) \log p_{XY}(x,y)$$

and

$$H(Y|X) = -\sum_{x,y} p_{XY}(x,y) \log p_{Y|X}(y|x).$$

The following result obtains positive rates achievable with predetermined dictionaries for the channel described above.

Theorem 2. *Let* p_X, p_Y *and* p_{XY} *be as above and let* $0 < \alpha \leq H(X)$*. For every* $\epsilon > 0$ *and for all* n *large, there is a deterministic set* $\mathcal{B}_n$ *with size at least* $N^{n(H(X)-2\epsilon)}$ *and satisfying the following property: If* $\mathcal{D}_n$ *is any subset of* $\mathcal{B}_n$ *with cardinality* $N^{n(\alpha-2\epsilon)}$ *and*

$$R < \alpha - H(Y|X) - H(X|Y) - 7\epsilon \tag{3.2}$$

is positive, then there exists a code $\mathcal{C}_n \subset \mathcal{D}_n$ *containing* N^{nR} *words and a decoder* g_n *with error probability* $q\left(\mathcal{C}_n, g_n\right) < \epsilon$.

Thus if the sequence of dictionaries $\mathcal{F} := \{\mathcal{D}_n\}_{n\geq 1}$ is such that $\mathcal{D}_n \subset \mathcal{B}_n$ for each n, then every $R < \alpha - H(Y|X) - H(X|Y)$ is $\mathcal{F}-$achievable. Also, setting $\alpha = H(X)$ and $\mathcal{D}_n = \mathcal{B}_n$ also gives us that every $R < H(X) - H(X|Y) - H(Y|X)$ is achievable in the usual sense of [3], without any restrictions on the dictionaries. For context, we remark that Theorem 1 holds for *arbitrary* dictionaries.

To prove Theorem 2, we use typical sets [3] together with conflict set decoding described in the next section. Before we do so, we present an example to illustrate Theorem 2.

Example

Consider a binary asymmetric channel with alphabet $\mathcal{X} = \mathcal{Y} = \{0, 1\}$ and transition probability

$$p(1|0) = p_0 = 1 - p(0|0) \text{ and } p(0|1) = p_1 = 1 - p(1|1).$$

To apply Theorem 2, we assume that the input has the symmetric distribution $\mathbb{P}(X_i = 0) = \frac{1}{2} = \mathbb{P}(X_i = 1)$ so that the entropy $H(X)$ equals its maximum value of 1. The entropy of the output $H(Y) = H(q)$ where $q = \frac{1-p_0+p_1}{2}$ and the conditional entropies equal

$$H(Y|X) = \frac{1}{2}\left(H(p_0) + H(p_1)\right) \text{ and } H(X|Y) = \frac{1}{2}\left(H(p_0) + H(p_1)\right) + 1 - H(q).$$

Set $p_0 = p$ and $p_1 = p + \Delta$. If both p and Δ are small, then $H(q)$ is close to one and $H(p_0)$ and $H(p_1)$ are close to zero. We assume that p and Δ are such that

$$\alpha_0 := H(Y|X) + H(X|Y) = H(p) + H(p+\Delta) + 1 - H\left(\frac{1-\Delta}{2}\right)$$

is strictly less than one and choose $\alpha > \alpha_0$. Every $R < \alpha - \alpha_0$ is then $\mathcal{F}-$achievable as in the statement following Theorem 2 and every $R < 1 - \alpha_0$ is achievable without any dictionary restrictions, in the usual sense of [3].

In Fig. 1, we plot $1 - \alpha_0$ as a function of p for various values of the asymmetry factor Δ. For example, for an asymmetry factor of $\Delta = 0.05$ we see that positive rates are achievable for p roughly up to 0.08.

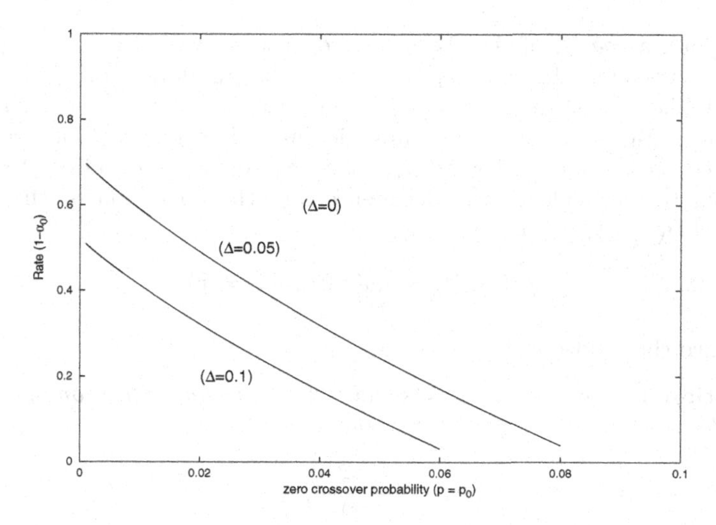

Fig. 1. Plotting the rate $1 - \alpha_0$ as a function of p for various values of the asymmetry factor Δ.

4 Proof of Theorem 2

We use conflict set decoding to prove Theorem 2. Therefore in the first part of this section, we prove an auxiliary result regarding conflict set decoding that is also of independent interest.

4.1 Conflict Set Decoding

Consider a discrete memoryless channel with finite input alphabet $\mathcal{X}_0$ and finite output alphabet $\mathcal{Y}_0$ and transition probability $p_0(y|x), x \in \mathcal{X}_0, y \in \mathcal{Y}_0$. For convenience, we define the channel by a collection of random variables $\theta_x, x \in \mathcal{X}_0$ with the distribution $\mathbb{P}(\theta_x = y) := p_0(y|x)$ for $y \in \mathcal{Y}_0$. All random variables are defined on the probability space $(\Omega, \mathcal{F}, \mathbb{P})$. For $\epsilon > 0, x \in \mathcal{X}_0$ and $y \in \mathcal{Y}_0$ we let $D(x,\epsilon)$ and $C(y,\epsilon)$ be deterministic sets such that

$$\mathbb{P}(\theta_x \in D(x,\epsilon)) \geq 1 - \epsilon \text{ and } C(y,\epsilon) = \{x : y \in D(x,\epsilon)\}. \tag{4.1}$$

We define $D(x,\epsilon)$ to be an $\epsilon-$*probable* set or simply probable output set corresponding to the input x and for $y \in \mathcal{Y}_0$, we denote $C(y,\epsilon)$ to be the $\epsilon-$*conflict set* or simply conflict set corresponding to the output y. There are many possible choices for $D(x,\epsilon)$; for example $D(x,\epsilon) = \mathcal{Y}_0$ is one choice. In Proposition 1 below, we show however that choosing $\epsilon-$probable sets as small as possible allows us to increase the size of the desired code. We also define

$$d_L(\epsilon) := \max_{x \in \mathcal{X}_0} \#D(x,\epsilon) \text{ and } d_R(\epsilon) := \max_{y \in \mathcal{Y}_0} \#C(y,\epsilon) \tag{4.2}$$

where $\#A$ denotes the cardinality of the set A.

As before, a *code* $\mathcal{C}$ of size M is a set of distinct words $\{x_1, \ldots, x_M\} \subseteq \mathcal{X}_0$. Suppose we transmit the word x_1 and receive the (random) word θ_{x_1}. Given θ_{x_1} we would like an estimate $\hat{x}$ of the word from $\mathcal{C}$ that was transmitted. A decoder $g : \mathcal{Y}_0 \rightarrow \mathcal{C}$ is a deterministic map that guesses the transmitted word based on the received word θ_{x_1}. We denote the probability of error corresponding to the code $\mathcal{C}$, the decoder g and the collection of the probable sets $\mathcal{D} := \{D(x,\epsilon)\}_{x\in\mathcal{X}_0}$ as

$$q(\mathcal{C}, g, \mathcal{D}) := \max_{x\in\mathcal{C}} \mathbb{P}\left(g(\theta_x) \neq x\right). \tag{4.3}$$

We have the following Proposition.

Proposition 1. *For $\epsilon > 0$ let $\mathcal{D} = \{D(x,\epsilon)\}_{x\in\mathcal{X}_0}$ be any collection of $\epsilon-$probable sets. If there exists an integer M satisfying*

$$M < \frac{\#\mathcal{X}_0}{d_L(\epsilon)\cdot d_R(\epsilon)}, \tag{4.4}$$

then there exists a code $\mathcal{C} \subseteq \mathcal{X}_0$ of size M and a decoder g whose decoding error probability is $q(\mathcal{C}, g, \mathcal{D}) < \epsilon$.

Thus as long as the number of words is below a certain threshold, we are guaranteed that the error probability is sufficiently small. Also, from (4.4) we see that it would be better to choose probable sets with as small cardinality as possible.

Proof of Proposition 1

Code Construction: We recall that by definition, given input x, the output θ_x belongs to the set $D(x,\epsilon)$ with probability at least $1-\epsilon$. Therefore we first construct a code $\mathcal{C} = \{x_1, \ldots, x_M\}$ containing M distinct words and satisfying

$$D(x_i,\epsilon) \cap D(x_j,\epsilon) = \emptyset \text{ for all } x_i, x_j \in \mathcal{C}. \tag{4.5}$$

Throughout we assume that M satisfies (4.4). To obtain the desired distinct words, we use following the bipartite graph representation. Let $G = G(\epsilon)$ be a bipartite graph with vertex set $\mathcal{X}_0 \cup \mathcal{Y}_0$. We join $x \in \mathcal{X}_0$ and $y \in \mathcal{Y}_0$ by an edge if and only if $y \in D(x,\epsilon)$. The size of $D(x,\epsilon)$ therefore represents the degree of the vertex x and the size of $C(y,\epsilon)$ represents the degree of the vertex y. By definition (see (4.2)) $d_L = \max_{x\in\mathcal{X}} \#D(x,\epsilon)$ and $d_R = \max_{y\in\mathcal{Y}} \#C(y,\epsilon)$ denote the maximum degree of a left vertex and a right vertex, respectively, in G. We say that a set of vertices $\{x_1, \ldots, x_M\}$ is *disjoint* if for all $i \neq j$, the vertices x_i and x_j have no common neighbour (in $\mathcal{Y}_0$). Constructing codes with disjoint $\epsilon-$probable sets satisfying (4.5) is therefore equivalent to finding disjoint sets of vertices in $\mathcal{X}_0$.

We now use direct counting to get a set of M disjoint vertices $\{x_1, \ldots, x_M\}$ in $\mathcal{X}_0$. First we pick any vertex $x_1 \in \mathcal{X}_0$. The degree of x_1 is at most d_L and moreover, each vertex in $D(x_1,\epsilon) \subseteq \mathcal{Y}$ has at most d_R neighbours in $\mathcal{X}_0$. The total number of (bad) vertices of $\mathcal{X}_0$ adjacent to some vertex in $D(x_1,\epsilon)$ is at most

$d_L \cdot d_R$. Removing all these bad vertices, we are left with a bipartite subgraph G_1 of G whose left vertex set has size at least $N_0 - d_L \cdot d_R$ where $N_0 = \#\mathcal{X}_0$. We now pick one vertex in the left vertex set of G_1 and continue the above procedure. After the ith step, the number of left vertices remaining is $N_0 - i \cdot d_L \cdot d_R$ and so from (4.4) we get that this process continues at least for M steps. The words corresponding to vertices $\{x_1, \ldots, x_M\}$ form our code $\mathcal{C}$.

Decoder Definition: Let $\mathcal{C}$ be the code as constructed above. For decoding, we use the *conflict-set decoder* defined as follows: If $y \in D(x_j, \epsilon)$ for some $x_j \in \mathcal{C}$ and the conflict set $C(y, \epsilon)$ does not contain any of word of $\mathcal{C} \setminus \{x_j\}$, then we set $g(y) = x_j$. Otherwise, we set $g(y)$ to be any arbitrary value; for concreteness, we set $g(y) = x_1$.

We claim that the probability of error of the conflict-set decoder is at most ϵ. To see this is true, suppose we transmit the word x_i. With probability at least $1 - \epsilon$, the corresponding output $\theta_{x_i} \in D(x_i, \epsilon)$. Because (4.5) holds, we must necessarily have that $y \notin D(x_k, \epsilon)$ for any $k \neq j$. This implies that the conflict-set decoder outputs the correct word x_i with probability at least $1 - \epsilon$. □

We now prove Theorem 2 using typical sets and conflict set decoding.

4.2 Proof of Theorem 2

For notational simplicity we prove Theorem 2 with $\mathcal{X} = \mathcal{Y} = \{0, 1\}$. An analogous analysis holds for the general case.

The proof consists of three steps. In the first step, we define and estimate the occurrence of certain typical sets. In the next step, we use the typical sets constructed in Step 1 to determine the set $\mathcal{B}_n$ in the statement of the Theorem. Finally, we use Proposition 1 to obtain the bound (3.2) on the rates.

Step 1: Typical Sets: We define the typical set

$$A_n(\epsilon) = (A_{n,1}(\epsilon) \times A_{n,2}(\epsilon)) \bigcap A_{n,3}(\epsilon) \tag{4.6}$$

where

$$A_{n,1}(\epsilon) = \{x \in \mathcal{X}^n : 2^{-n(H(X)+\epsilon)} \leq p(x) \leq 2^{-n(H(X)-\epsilon)}\},$$
$$A_{n,2}(\epsilon) = \{y \in \mathcal{Y}^n : 2^{-n(H(Y)+\epsilon)} \leq p(y) \leq 2^{-n(H(Y)-\epsilon)}\}$$

and

$$A_{n,3}(\epsilon) = \{(x, y) \in \mathcal{X}^n \times \mathcal{Y}^n : 2^{-n(H(X,Y)+\epsilon)} \leq p(x, y) \leq 2^{-n(H(X,Y)-\epsilon)}\}$$

with the notation that if $x = (x_1, \ldots, x_n)$, then $p(x) := \prod_{i=1}^n p(x_i)$.

We estimate $\mathbb{P}(A_{n,1}(\epsilon))$ as follows. If $(X_1, \ldots, X_n)$ is a random element of $\mathcal{X}^n$ with $\{X_i\}$ i.i.d. and each having distribution $p(.)$, then the random

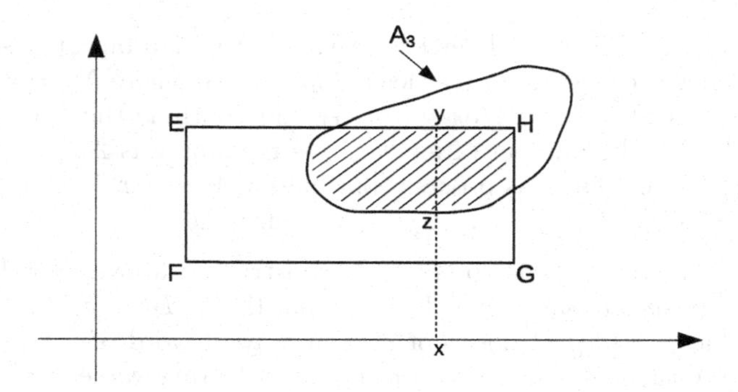

Fig. 2. The set $D_n(x,\epsilon)$ obtained from the sets $A_{n,i}(\epsilon), 1 \leq i \leq 3$.

variable $\log p(X_i)$ has mean $H(X)$ and so by Chebychev's inequality

$$\begin{aligned}\mathbb{P}(A_{n,1}^c(\epsilon)) &= \mathbb{P}\left(\left|\sum_{i=1}^{n} \log p(X_i) - nH(X)\right| \geq nH(X)\epsilon\right)\\ &\leq \frac{1}{n^2H^2(X)\epsilon^2}\mathbb{E}\left(\sum_{i=1}^{n} \log p(X_i) - nH(X)\right)^2\\ &= \frac{1}{nH^2(X)\epsilon^2}\mathbb{E}\left(\log p(X_1) - H(X)\right)^2\end{aligned}$$

which converges to zero as $n \to \infty$. Analogous estimates hold for the sets $A_{n,2}(\epsilon)$ and $A_{n,3}(\epsilon)$ and so

$$\mathbb{P}(A_n(\epsilon)) \geq 1 - \epsilon^2 \tag{4.7}$$

for all n large.

Step 2: Determining the set $\mathcal{B}_n$: We now usethe set $A_n(\epsilon)$ defined above to determine the set $\mathcal{B}_n$ in the statement of the Theorem as follows. For $x \in A_{n,1}(\epsilon)$, let

$$D_n(x,\epsilon) := \{y \in A_{n,2}(\epsilon) : (x,y) \in A_n(\epsilon)\}.$$

In Fig. 2, we illustrate the sets $\{A_{n,i}(\epsilon)\}_{1\leq i\leq 3}$ and the set $A_n(\epsilon)$. The rectangle $EFGH$ denotes $A_{n,1}(\epsilon) \times A_{n,2}(\epsilon)$ and the oval set A_3 represents $A_{n,3}(\epsilon)$. The hatched region represents $A_n(\epsilon)$. The line yz represents the set $D_n(x,\epsilon)$ for $x \in A_{n,1}(\epsilon)$ shown on the $X-$axis.

From Fig. 2 we see that

$$\sum_{x\in A_{n,1}(\epsilon)} \left(\sum_{y\in D_n(x,\epsilon)} p(x,y)\right) = \sum_{(x,y)\in A_n(\epsilon)} p(x,y) \geq 1 - \epsilon^2 \tag{4.8}$$

by (4.7). Letting

$$A_{n,4}(\epsilon) := \left\{ x \in A_{n,1}(\epsilon) : \sum_{y \in D_n(x,\epsilon)} p(y|x) \geq 1 - \epsilon \right\}, \tag{4.9}$$

we split the summation in first term in (4.8) as $L_1 + L_2$ where

$$L_1 = \sum_{x \in A_{n,4}(\epsilon)} \left(\sum_{y \in D_n(x,\epsilon)} p(y|x) \right) p(x) \leq \sum_{x \in A_{n,4}(\epsilon)} p(x) = \mathbb{P}(A_{n,4}(\epsilon)) \tag{4.10}$$

and

$$\begin{aligned} L_2 &= \sum_{x \in A_{n,1}(\epsilon) \setminus A_{n,4}(\epsilon)} \left(\sum_{y \in D_n(x,\epsilon)} p(y|x) \right) p(x) \\ &\leq (1-\epsilon) \sum_{x \in A_{n,1}(\epsilon) \setminus A_{n,4}(\epsilon)} p(x) \\ &\leq (1-\epsilon) \mathbb{P}\left(A^c_{n,4}(\epsilon)\right). \end{aligned} \tag{4.11}$$

Substituting (4.11) and (4.10) into (4.8) we get

$$1 - \epsilon \cdot \mathbb{P}\left(A^c_{n,4}(\epsilon)\right) \geq L_1 + L_2 \geq 1 - \epsilon^2$$

and so $\mathbb{P}\left(A^c_{n,4}(\epsilon)\right) \leq \epsilon$. Because $A_{n,4}(\epsilon) \subseteq A_{n,1}(\epsilon)$, we therefore get that

$$1 - \epsilon \leq \mathbb{P}(A_{n,4}(\epsilon)) = \sum_{x \in A_{n,4}(\epsilon)} p(x) \leq 2^{-n(H(X)-\epsilon)} \# A_{n,4}(\epsilon).$$

Setting $\mathcal{B}_n = A_{n,4}(\epsilon)$ we then get

$$\#\mathcal{B}_n \geq 2^{n(H(X)-\epsilon)} \cdot (1-\epsilon) \geq 2^{n(H(X)-2\epsilon)}$$

for all n large.

Step 3: Using Proposition 1: For $\alpha \leq H(X)$, we let $\mathcal{D}_n$ be any set of size $2^{n(\alpha-2\epsilon)}$ contained within $\mathcal{B}_n$. Let G be the bipartite graph with vertex set $\mathcal{X}_c \cup \mathcal{Y}_c$ where $\mathcal{X}_c := \mathcal{D}_n$, $\mathcal{Y}_c := A_{n,2}(\epsilon)$ and an edge is present between $x \in \mathcal{X}_c$ and $y \in \mathcal{Y}_c$ if and only if $(x,y) \in A_n(\epsilon)$. We now compute the sizes of the probable sets and the conflict sets in that order.

For each $x \in \mathcal{X}_c$ we have by definition (4.9) of $A_{n,4}(\epsilon)$ that

$$\sum_{y \in D_n(x,\epsilon)} p(y|x) \geq 1 - \epsilon \tag{4.12}$$

and so we set $D_n(x,\epsilon)$ to be the $\epsilon-$probable set corresponding to $x \in \mathcal{D}_n$. To estimate the size of $D_n(x,\epsilon)$, we use the fact that $(x,y) \in A_n(\epsilon)$ and so

$$p(y|x) = \frac{p(x,y)}{p(x)} \geq \frac{2^{-n(H(X,Y)+\epsilon)}}{2^{-n(H(X)-\epsilon)}} = 2^{-n(H(Y|X)+2\epsilon)}. \tag{4.13}$$

Thus
$$1 \geq \sum_{y \in D_n(x,\epsilon)} p(y|x) \geq \# D_n(x,\epsilon) \cdot 2^{-n(H(Y|X)+2\epsilon)}$$

and consequently
$$\# D_n(x,\epsilon) \leq 2^{n(H(Y|X)+2\epsilon)}. \tag{4.14}$$

Finally, we estimate the size of the conflict set $C(y,\epsilon)$ for each $y \in \mathcal{Y}_c$. Again we use the fact that if (x,y) is an edge in G then $(x,y) \in A_n(\epsilon)$ and so

$$p(x|y) = \frac{p(x,y)}{p(y)} \geq \frac{2^{-n(H(X,Y)+\epsilon)}}{2^{-n(H(Y)-\epsilon)}} = 2^{-n(H(X|Y)+2\epsilon)}. \tag{4.15}$$

Thus
$$1 \geq \sum_{x \in C(y,\epsilon)} p(x|y) \geq \# C(y,\epsilon) \cdot 2^{-n(H(X|Y)+2\epsilon)}$$

and we get that $\# C(y,\epsilon) \leq 2^{n(H(X|Y)+2\epsilon)}$. Using this and (4.14), we get that the conditions in Proposition 1 hold with

$$N_0 = 2^{n(\alpha-2\epsilon)}, d_L(\epsilon) = 2^{n(H(Y|X)+2\epsilon)} \text{ and } d_R(\epsilon) = 2^{n(H(X|Y)+2\epsilon)}.$$

If $M = 2^{nR}$ with $R < \alpha - H(X|Y) - H(Y|X) - 7\epsilon$, then (4.4) holds and so there exists a code containing $M = 2^{nR}$ words from $\mathcal{D}_n$ giving an error probability of at most ϵ with the conflict set decoder. □

Acknowledgement. I thank Professors Rajesh Sundaresan, C.R. Subramanian and the referees for crucial comments that led to an improvement of the paper. I also thank IMSc for my fellowships.

References

1. Alon, N., Spencer, J.: The Probabilistic Method. Wiley Interscience, Hoboken (2008)
2. Bandemer, B., El Gamal, A., Kim, Y.-H.: Optimal achievable rates for interference networks with random codes. IEEE Trans. Inf. Theory **61**, 6536–6549 (2015)
3. Cover, T., Thomas, J.: Elements of Information Theory. Wiley, Hoboken (2006)
4. Csiszár, I., Körner, J.: Graph decomposition: a new key to coding theorems. IEEE Trans. Inf. Theory **27**, 5–12 (1981)
5. Csiszár, I.: The method of types. IEEE Trans. Inf. Theory **44**, 2505–2523 (1998)
6. Huffman, W.C., Pless, V.: Fundamentals of Error Correcting Codes. Cambridge University Press, Cambridge (2003)
7. Zamir, R.: Lattice Coding for Signals and Networks. Cambridge University Press, Cambridge (2014)

Characterization of Dense Patterns Having Distinct Squares

Maithilee Patawar[1(✉)] and Kalpesh Kapoor[2]

[1] Department of Computer Science and Engineering, Indian Institute of Technology Guwahati, Guwahati, India
maith176101104@iitg.ac.in
[2] Department of Mathematics, Indian Institute of Technology Guwahati, Guwahati, India
kalpesh@iitg.ac.in

Abstract. The square conjecture claims that the number of distinct squares in a word is at most equal to the length of the word. While the conjecture is still open, there have been attempts to define patterns representing a collection of words with a large number of distinct squares. We study the properties of words having the maximum number of distinct squares. These properties are then used to define general criteria to characterize dense patterns. We show that there are infinitely many dense patterns by giving a pattern generator and use the rate of introduction of new squares to compare any two dense patterns. We also give a new dense pattern, P, and prove that it is better than the earlier patterns.

Keywords: Distinct squares · Word patterns · Word combinatorics

1 Introduction

The study of periodic structures reveals many properties of words. Periodicity of words has been an extensively studied topic in word combinatorics, see for example [7]. A square is a concatenation of two identical words. It is the smallest possible periodic structure and most frequently occurring repetitive element in a word. Consequently, the number of squares in a word has been a topic of interest for many researchers and many conjectures about squares have been proposed in the literature [1,3,5]. The earliest conjecture [5] that has remained open is that the number of distinct squares in a word of length n is less than or equal to n.

In general, the proposed conjectures on distinct squares are supported by patterns that attempt to pack as many distinct squares as possible [5,6]. A pattern is a structure that generates a family of words containing similar types of squares. In this work, we investigate the square conjecture using patterns. Our main contributions are listed below:

A. Mudgal and C. R. Subramanian (Eds.): CALDAM 2021, LNCS 12601, pp. 397–409, 2021.
https://doi.org/10.1007/978-3-030-67899-9_32

(a) Identification of properties of words with the maximum number of distinct squares.
(b) We give general criteria to define dense patterns and to compare these patterns.
(c) It is shown that there are infinitely many dense patterns.
(d) We introduce a new pattern that produces words with more number of distinct squares than the existing patterns.

The rest of the paper is organized as follows. Section 2 describes the necessary notations required to understand the concept of dense patterns. We also mention previous conjectures and their supporting patterns. To build new patterns with more squares than the existing patterns, Sect. 3 proves some properties of words containing the maximum number of distinct squares for a given length. Using these properties, we define a dense pattern and give a structure to get many such patterns. Next, Sect. 4 modifies the structures of some existing patterns to make them dense patterns and provide a criterion to compare them. Finally, conclusions and future directions are given in Sect. 5.

2 Preliminaries and Related Work

Let Σ be an alphabet. A word w is a finite sequence of letters drawn from Σ. Any non-empty subsequence of consecutive letters in w is a subword of w. The length of a word w is denoted by $|w|$ and is equal to the number of symbols in w. The word, w, with $|w| = 0$ is referred to as empty word and is denoted by ϵ. We use $\mathbb{N}$ to denote the set of non-negative integers. Let Σ^n be the set of words of length $n \in \mathbb{N}$ defined over an alphabet Σ. Further, $\Sigma^* = \bigcup_{n \in \mathbb{N}} \Sigma^n$ and $\Sigma^+ = \bigcup_{n \in \mathbb{N}-\{0\}} \Sigma^n = \Sigma^* - \{\epsilon\}$.

The concatenation of two words x and y is the word $x.y$ or simply xy. The concatenation is an associative operation. The ith power of a word w, denoted by w^i is defined recursively as $w^1 = w$ and $w^k = w^{k-1}w$ for $k \geq 2$. The concatenation of a collection of words $w_1, \ldots, w_k$ is denoted by $\bigodot_{i=1}^{k} w_i$. Let $w = pvs$ be a word, where p, v and x are words in Σ^*. Then the words p and s are called as prefix and suffix, respectively, of the word w.

A square is a word of the form uu, where $u \in \Sigma^+$. A word, w, is said to be primitive if $w = u^r$ for some u implies $r = 1$. A word, w, is non-primitive if w can be expressed as u^r for some $u \in \Sigma^*$ and $r \in \mathbb{N} - \{0, 1\}$. A square uu is said to have primitive base, if u is a primitive word otherwise it is said to have non-primitive base. A subword u is a border of w if it is both prefix and suffix of w. Suppose $w = xy$ is a word where $x,\ y \in \Sigma^+$. Then, the word yx is said to be a conjugate of the word w.

A word w is said to contain a square uu if w can be written as $xuuy$ for some $x, y \in \Sigma^*$. We will denote the set of all distinct squares in a word w by $DS(w)$. The maximum number of distinct squares in a word of length n is denoted by $MaxNS(n)$. Formally, $MaxNS(n) = \max\{k \mid w \in \Sigma^n \text{ and } |DS(w)| = k\}$.

A number $n \in \mathbb{N}$ is referred to as a no-gain length if $MaxNS(n)$ is equal to $MaxNS(n-1)$, else it is a gain length. Let $MaxDS(n)$ be the set of words with length n having the maximum number of distinct squares. Formally,

$$MaxDS(n) = \{w \mid w \in \Sigma^n \text{ and } \forall u \in \Sigma^n \cdot |DS(w)| \geq |DS(u)|\}.$$

As mentioned earlier, the square conjecture claims that for a word w, $|DS(w)| < |w|$ and a pattern named 'Q' attempts to reach this bound [5]. Let $w = a_1 a_2 \ldots a_n$ be a word. We borrow the notation s_i from [5] to denote the number of squares that starts at a location i for the last time in a word, where $1 \leq i \leq n$. The same study shows that $s_i \leq 2$ for any word over a fixed alphabet. As in [4], we use the term FS double square for a square with $s_1 = 2$. It is proved that the upper bound on the number of distinct squares in a word of length n is $\lfloor \frac{11n}{6} \rfloor$ [4]. Jonoska et al. [6] explored the square conjecture and predicted a different upper bound for the words over a binary alphabet.

Another conjecture claims that the maximum number of primitive base distinct squares in a word w over an alphabet Σ never exceeds $|w| - |\Sigma|$ [3]. The upper bound for the number of distinct squares in a square uu has been shown to be $1.57|u|$ [1]. The claim is supported by a pattern which can produce a square w with at most $0.625|w|$ distinct squares.

In the next section, we identify some characteristics of words for a given length with the maximum number of distinct squares.

3 Square-Maximal Words and Dense Patterns

It is conjectured that the maximum number of distinct squares is achieved for a binary alphabet [8]. In the rest of the paper, we assume that the underlying alphabet is binary containing letters 'a' and 'b'. The exact characterization of the set $MaxDS(n)$ is not known. In other words, to check if a word belongs to the set $MaxDS(n)$, it is required to exhaustively search in the set of all 2^n possible words. A word is called a square-maximal word if it belongs to a set $MaxDS(n)$.

The function $MaxNS$ is non-decreasing, and the difference between two successive values of $MaxNS$ is at most two. Let $n \in \mathbb{N}$ be a no-gain length. Then, for any word $w \in MaxDS(n-1)$, both lw and wl are in $MaxDS(n)$, where $l \in \{a, b\}$. Thus, for a no-gain length n, the cardinality of the set $MaxDS(n)$ is always more than that of the set $MaxDS(n-1)$. Thus, it is important to characterize words for gain lengths because it enables to generate words with no-gain lengths. For this reason, in the following results, we generally focus on the gain lengths.

Consider a word w in $MaxDS(n)$, where n is a gain length. Every letter in the word w must be part of some square in the set $DS(w)$. Otherwise, removing a letter that is not part of any square will give a smaller word with $MaxNS(n)$ squares, which is not possible as n is a gain length. Suppose $w = w_1 w_2$ for some $w_1, w_2 \in \Sigma^+$. We refer to the squares whose last occurrence starts and ends in w_1 and w_2, respectively, by $cross(w_1, w_2)$.

Lemma 1. *Let $n \in \mathbb{N}$ be a gain length and w be a word in $MaxDS(n)$. Further, let $w = w_1w_2$ where $w_1, w_2 \in \Sigma^*$ such that $|DS(w)| = |DS(w_1)| + |DS(w_2)|$. Then, $|DS(w_1) \cap DS(w_2)| = k \geq 0$ if and only if $|cross(w_1, w_2)| = k$.*

Proof. We have, $|DS(w)| = |DS(w_1| + |DS(w_2)| - |DS(w_1) \cap DS(w_2)|$

(if) Suppose the sets $DS(w_1)$ and $DS(w_2)$ have k squares in common. Then, it must be the case that at least k rightmost squares starts in w_1 and ends in w_2 to satisfy the premise $|DS(w)| = |DS(w_1)| + |DS(w_2)|$.

(only if) Suppose there are k rightmost squares that begin and end in words w_1 and w_2, respectively. Define

$$sq_{w_1} = DS(w_1) - DS(w_2), \quad sq_{w_2} = DS(w_2) - DS(w_1),$$
$$sq_{comm} = DS(w_1) \cap DS(w_2), \quad sq_{cross} = DS(w) - (DS(w_1) \cup DS(w_2))$$

Note that sq_{cross} is the set of distinct squares that begin in w_1 and end in w_2 which are not present in w_1 or w_2. Therefore, the number of distinct squares in w is, $|DS(w)| = sq_{w_1} + sq_{w_2} + sq_{comm} + sq_{cross}$. Since $|DS(w_1)| + |DS(w_2)| = sq_{w_1} + sq_{w_2} + 2 * sq_{comm}$, we get

$$|sq_{cross}| = |sq_{comm}|$$

□

If for a gain length n, the relation $MaxNS(n) = MaxNS(n_1) + MaxNS(n - n_1)$ holds for some $n_1 < n$, then n_1 cannot be equal to 1 or $n - 1$. Assume $w = w_1w_2$ where $|w_1| = n_1 > \lceil \frac{n}{2} \rceil$. If $|DS(w_1)| < MaxNS(n_1)$ or $|DS(w_2)| < MaxNS(n - n_1)$, then the set $cross(w_1, w_2)$ is non-empty. In the following lemmas, we find some properties of square-maximal words in which both the subwords w_1 and w_2 are from the respective $MaxDS$ sets.

Lemma 2. *Let a word $w = w_1w_2 \in MaxDS(n)$ for some gain length $n \in \mathbb{N}$ such that $|w_1| \leq |w_2|$. If $|cross(w_1, w_2)| = 0$ then the length of the smallest border of the word w is greater than $|w_1|$.*

Proof. The condition $|cross(w_1, w_2)| = 0$ implies $DS(w_1w_2) = DS(w_2w_1)$. No proper suffix of the word w_1 can be a prefix of the word w_2. Otherwise, $w_1 = w_1'u$ and $w_2 = uw_2'$ will imply $DS(w_1'u.uw_2') = DS(w_1'uw_2')$ for some non-empty words w_1', w_2', u. Similarly, no proper suffix of the word w_2 can be a prefix of the word w_1. Therefore, the length of the smallest border of the words w_1w_2 and w_2w_1 must be greater than $|w_1|$. □

Lemma 3. *Let $n \in \mathbb{N}$ be a gain length and w be a word in $MaxDS(n)$ with $w = w_1w_2$ such that $|DS(w)| = |DS(w_1)| + |DS(w_2)|$ and $DS(w_1) \cap DS(w_2) = \emptyset$. Then, $\{a^2, b^2\} \subseteq DS(w)$.*

Proof (By contradiction). Assume $DS(w)$ contains at most one square from $\{aa, bb\}$. If both a^2 and b^2 are not in w then the word w must be of the form $(ab)^k$ for some positive integer $k > 0$. However, such a word cannot be in $MaxDS(n)$.

Consider the case in which only one among a^2 or b^2 is a subword of w. Without loss of generality, assume $aa \notin DS(w_1)$, $aa \notin DS(w_2)$ and $bb \in w$ implying a square bb is either in set $DS(w_1)$ or $DS(w_2)$. If b^2 is a subword of w_1, then the structure of w_2 depends on the initial and final letters of w_1. As a result, w_2 is either $b(ab)^j$ or $(ab)^j$ for some $j \in \mathbb{N}$. If the subword w_1 starts and ends with the same letter, say 'a' then $w_1 = aua$ where $u \in \Sigma^+$ and $w_2 = b(ab)^j$. However, as the word $w_3 = (ab)^j.a$ contains $|DS(w_2)|$ distinct squares, the word $w' = w_1w_3$ will also be in the set $MaxDS(|w|)$. Here w_3 does not satisfy the constraint on border given in Lemma 2. Therefore, the length $|w|$ must be a no-gain length, which is a contradiction.

Now, suppose w_1 starts and ends with different letters, say $w_1 = aub$. In this case, the only possible structure for w_2 is $(ab)^j$. Similar to the previous case, we have a contradiction as the word $aub.b(ab)^{j-1}$ will also be in the set $MaxDS(|w|)$. □

Lemma 4. *Let $n \in \mathbb{N}$ be a gain length and w be a word in $MaxDS(n)$ with $w = w_1w_2$ such that $|DS(w)| = |DS(w_1)| + |DS(w_2)|$ and $DS(w_1) \cap DS(w_2) = \emptyset$. Then, for some integer $k > 2$ and $u_1, u_2 \in \Sigma^+$, we have $w_1 = a^k u_1 a^k$ and $w_2 = b^k u_2 b^k$. Further, the factorization of a word w as $w_1.w_2$ is unique.*

Proof. We know from Lemma 3 that, for a given gain length, any word containing the maximum number of distinct squares must have the trivial squares, viz. a^2 and b^2. These squares can be in either w_1 or in w_2. Accordingly, we consider two cases for w_1 and w_2 depending on whether they start with the same or the different letters. We show that among all possible structures, only one structure mentioned in Case II satisfies all the given conditions.

Case I Assume w_1 and w_2 start with the same letter. Suppose the words a^2 and b^2 are in the set $DS(w_1)$. Then, these trivial squares cannot be in the set $DS(w_2)$. Thus, the word w_2 must be of the form $(ab)^k$ for some integer $k > 2$. Also, the square $(ba)^2$ cannot be in the set $DS(w_1)$, otherwise the square $(ab)^2$ will also be in the set $DS(w_1)$. So the squares $(ab)^2$ and $(ba)^2$ cannot be in the word w_1. But then we have another word $w' = w_1.(ba)^k$ which has $|DS(w)|$ number of distinct squares implying that w is a no-gain length since the word w' has a border whose length is less than $|w_1|$ which contradicts the assumption.
Consider another alternative wherein the trivial squares aa and bb are subwords of the words w_1 and w_2, respectively. Then the subwords w_1 and w_2 must end and begin, respectively, with 'ab', which again does not satisfy the assumption that n is a gain length.

Case II Suppose w_1 and w_2 start with two different letters. To satisfy the constraint on border mentioned in Lemma 2, suppose w_1 begins and ends with the letter 'a'. Let $a^2 \in DS(w_2)$, then the word w_1 begins with ab and the word w_2 ends with ab, thereby the word w_2w_1 has a border of length smaller than $|w_1|$. Therefore, it must be the case that the structures of the word w_1 and w_2 are $a^k u_1 a^k$ and $b^k u_2 b^k$, respectively, such that $k > 1$ and $DS(u_1) \cap DS(u_2) = \emptyset$. Note that if there is a factorization of the subword w_1 as $w_{11}w_{12}$ such that $cross(w_{11}, w_{12}) = \emptyset$, then a^2 will be a subword of w_{11} and not of w_{12}. However, every subword w_{12} of length more than one will always ends with an a^2. So, the subword $w_1 = w_{11}w_{12}$ cannot have $cross(w_{11}, w_{12}) = \emptyset$. Now, another factorization $w = w_3w_4$ that satisfies the condition $|DS(w)| = |DS(w_3)| + |DS(w_4)|$ is possible if $w_3 = a^k u1a^{k-1}$ and $w_4 = a.b^k u_2 b^k$. However, in such a case, by Lemma 2, n will be a no-gain length. Therefore, the factorization of w as $w_1.w_2$ is unique. □

Let Σ be an alphabet and $w = l_1 \ldots l_n$ be a word, where $l_i \in \Sigma$ for $1 \leq i \leq n$. We refer to the letters l_1 and l_n as terminal letters of w. Any letter that is not a terminal letter is a non-terminal letter of w. In Lemma 4, it is shown that a square-maximal word, say $w = w_1w_2$ of length n can have at most one w_1 that satisfy $cross(w_1, w_2) = \emptyset$. The lemma implies that there are exactly two non-terminal letters of w which cannot be non-terminal letters in any rightmost square of w. We now explore a gain length n for which $MaxNS(n) > MaxNS(n_1) + MaxNS(n - n_1)$ and identify the structure of a square-maximal word.

Lemma 5. *Let $n \in \mathbb{N}$ such that $MaxNS(n) > MaxNS(n_1) + MaxNS(n - n_1)$ for some integer $n_1 \in \{1, \ldots, n-1\}$ and $w \in MaxDS(n)$. The following statements hold:*

(a) $|DS(w)| > |DS(w_1)| + |DS(w_2)|$ for all w_1, w_2 such that $w = w_1w_2$.
(b) $cross(w_1, w_2) \neq \emptyset$ for all w_1, w_2 such that $w = w_1w_2$.

Proof. (a) Suppose $|w_1| = n_1$. Then, the maximum value of the expression $|DS(w_1)| + |DS(w_2)|$ is $|MaxDS(n_1)| + |MaxDS(n - n_1)|$. As $MaxNS(n) > MaxNS(n_1) + MaxNS(n - n_1)$, the relation $|DS(w)| > |DS(w_1)| + |DS(w_2)|$ follows.
(b) We conclude from (a) that the rightmost square starts in w_1 and ends in w_2 for all subwords w_1. So, the set $cross(w_1, w_2)$ is non-empty for every w_1.

□

Following lemma inspects the characteristics of terminal letters in a square-maximal word.

Lemma 6. *Let $n \in \mathbb{N}$ be a gain length and $w \in MaxDS(n)$.*

(a) If $MaxNS(n) = MaxNS(n-1) + 1$, then every terminal letter of w is the terminal letter in exactly one rightmost square of w.
(b) If $MaxNS(n) = MaxNS(n-1) + 2$, then w begins with an FS double square and the last letter of w is a terminal letter of two squares in $DS(w)$.

Proof. As n is a gain length, $MaxNS(n) - MaxNS(n-1) = i \in \{1, 2\}$. The first letter of a word $w \in MaxDS(n)$ must be a part of exactly i distinct squares. Otherwise, removing the first letter will result in a word of $n-1$ length containing more than $MaxNS(n-1)$ distinct squares which is not feasible. Similar argument applies for the last letter of a word w. □

We observe, from Lemma 6, that the square-maximal words for successive gain lengths always begin and end with a square. We have observed in the manual inspection of square-maximal words for lengths up to 40 that if such a word ends with the longest primitive square, then extending it further with the prefix of it's square base results in a longer square-maximal word. In the following lemma, we explain one such way to introduce new square(s) using a prefix of a square base.

Lemma 7. *Let $w = uu$ be a primitive base square such that $|u| > 1$ and v be a proper prefix of u. Then, $|DS(w.v)| \geq |DS(w)| + |v|$.*

Proof. Assume, $u = u_1u_2...u_n$. A square uu has all conjugates of u, and every letter of the first u in w begins with a distinct conjugate. Similarly, for a word $uu.v$ every letter of the word v adds a new square, that is, a conjugate of uu.

The number of new squares added by a prefix v is more than $|v|$ if the word begins with an FS double square. An FS double square is a primitive square that begins with two rightmost squares. The structure of the base of an FS double square is known to be $(xy)^{e_1}(x)(xy)^{e_2}$ [4]. Here, x and $y \in \Sigma^+$ and the integers e_1, e_2 satisfy $e_1 \geq e_2 \geq 0$. A word beginning with an FS double square introduces $|v| + |v'|$ new distinct squares for some non-empty longest common prefix v' of the words x and y, where $|v'| \leq |v|$. □

A pattern is a way to represent a family of words sharing similar characteristics. In the following section, we employ the properties of square-maximal words identified above to define a dense pattern.

3.1 Dense Patterns

We use the notation $T(x)$ to denote a function from $\mathbb{N} \rightarrow \Sigma^+$. For example, $T(x) = a^xb$ generate words $\{b, ab, a^2b, a^3b, \ldots\}$. We refer to a function $T(x)$ as a pattern. For a word, w, the distinct-square density, $\alpha(w)$ is defined as the ratio $\frac{|DS(w)|}{|w|}$ and it is known that no upper bound on $\alpha(w)$ is sharp [8]. We extend the definition of the distinct-square density to a pattern, $T(x)$, and define it as

$$\alpha_T = \lim_{x \to \infty} \frac{|DS(T(x))|}{|T(x)|}$$

The distinct-square density of a pattern depends on the number of no-gain lengths between two successive words generated by the pattern. A high distinct-square density indicates more gain lengths or equivalently less no-gain lengths. The difference between lengths of successive words generated by a pattern need not be a constant. For every positive integer, x, a pattern $T(x)$ introduces $|DS(T(x))| - |DS(T(x-1))|$ new squares. The number of no-gain lengths introduced in $T(x-1)$ to obtain $T(x)$ is defined as $\mathcal{N}_T(x) = (|T(x)| - |T(x-1)|) - (|DS(T(x))| - |DS(T(x-1))|)$. A good pattern should minimize the value of $\mathcal{N}_T(x)$. In Sect. 4, we use $|DS(T(x))|$ and $\mathcal{N}_T(x)$ to compare different patterns.

As mentioned before, we are interested in characterizing the words in a set $MaxDS(n)$, where n is a gain length. To do so, we use the properties of square-maximal words. A word $w = w_1w_2 \in MaxDS(n)$ satisfies either $|DS(w)| > |DS(w_1)| + |DS(w_2)|$ or $|DS(w)| = |DS(w_1)| + |DS(w_2)|$. In the latter case, Lemma 4 shows that such a word has a unique factorization where the subword $w_1 = w_{11}w_{12}$ always satisfy the relation $|DS(w_1)| > |DS(w_{11})| + |DS(w_{12})|$. The only possible structures for this case has $w_1 = a^k u_1 b^k$ and $w_2 = b^k u_2 b^k$. These subwords cannot have any squares in common, so given a subword w_1, it is easy to find w_2. We, therefore, use the relation $|DS(w)| > |DS(w_1)| + |DS(w_2)|$ to obtain a dense pattern. The interpretation of the relation given in Lemma 5 is included in the next definition.

Definition 1 (Dense Pattern). *A pattern, $T(x)$, is said to be a dense pattern if and only if it satisfies the following conditions.*

(a) $\alpha_T \geq 1$, and
(b) For all $x \in \mathbb{N}$, if $T(x) = w_1w_2$ then $cross(w_1, w_2) \neq \emptyset$, where w_1, $w_2 \in \Sigma^+$.

A word produced by a dense pattern is known as a dense word. The following lemma provides an aid to verify the second condition in Definition 1.

Lemma 8. *Let $w = u_1u_2 \dots u_k$ be a word such that for all $i \in \{2, \dots, k-1\}$, the subword $s_{i-1}.u_i.p_{i+1}$ is a rightmost square in $DS(w)$ and u_1p_2 ($s_{k-1}u_k$) is the first (the last) rightmost square of w for some non-empty prefix and suffix, p_i and s_i, respectively, of u_i. Then, $cross(w_1, w_2) \neq \emptyset$ for all w_1, w_2 such that $w = w_1w_2$.*

Proof. The word w begins and ends with a rightmost square. For $1 < i < k$, the structure of a rightmost square $s_{i-1}.u_i.p_{i+1}$ ensures that every non-terminal letter in w is also a non-terminal letter in any rightmost square of w. Thus, for all w_1 and w_2 such that $w = w_1w_2$ implies $cross(w_1, w_2) \neq \emptyset$. □

Now, we define a pattern P as follows:

$$P(x) = a.(a^1ba^2b...a^y).\left\{\bigodot_{i=y-1}^{x-2} (ba^iba^{i+1}ba^{i+2})\right\}.(ba^{x-1}ba^xba^{x-1}bab)a \quad (1)$$

where x and y are positive integers and $y = \lceil \frac{x}{2} \rceil \geq 4$. Similar to the squares in a word obtained from the pattern Q described in [5], the words generated by the pattern P have three types of distinct squares. These are (i) trivial squares having only letter a, squares with exactly two b's, and (ii) squares with exactly four b's. All the squares in the last two types are primitive squares.

A set of primitive square and all its conjugates in the word has the structure described in Lemma 7. Refer Table 1 for the length and the number of distinct squares in a word that can be obtained by the pattern P. We check the pattern against the definition of dense patterns. For this, we first verify that the subwords of this pattern satisfy the criterion (b) of the Definition 1.

Table 1. Properties of the words generated by the pattern P

x mod 2	$\lvert P(x)\rvert$	$\lvert DS(P(x))\rvert$
0	$\frac{1}{8}(10x^2+36x+40)$	$\frac{1}{8}(10x^2+20x+24)$
1	$\frac{1}{8}(10x^2+32x+38)$	$\frac{1}{8}(10x^2+16x+22)$

Lemma 9. *The following subword, w, of a word generated by the pattern P satisfies $cross(w_1, w_2) \neq \emptyset$ for any w_1, w_2, where $w = w_1w_2$, $y \geq 4$ and $x > 6$.*

(a) $w = a^1ba^2b\ldots a^y$

(b) $w = \left\{ \bigodot_{i=y-1}^{x-2} (ba^iba^{i+1}ba^{i+2}) \right\} (ba^{x-1}ba^xba^{x-1}bab)a$

Proof. (a) We can write w according to Lemma 8 by using the rightmost squares in $\{(aba)^2, (abaa)^2, (abaaa)^2 \ldots, (aba^{y-2})^2\}$. Thus, the relation holds true for given w.

(b) Consider the rightmost instances of the squares in a subset R of $DS(w)$, where $R = \{(a^iba^{i+1}ba^2)^2, (a^{i+1}ba^{i+2}ba^2)^2, ..., (a^{x-1}ba^xba^2)^2, (ba^i)^2, (ba)^2\}$. We can use the squares in R to rewrite the word in the structure mentioned in Lemma 8 in which w begins and ends with squares $(ba^i)^2$ and $(ba)^2$, respectively. □

Lemma 10. *The pattern P is a dense pattern.*

Proof. The distinct-square density of P is one (refer Table 1). A word generated by the pattern P is the concatenation of two subwords given in Lemma 9. The lemma shows that these subwords individually qualifies the last condition of the Definition 1. We use the same subset of rightmost squares to write $P(x)$ according to Lemma 8. Thus, P is a dense pattern. □

It is possible to modify a pattern to convert it into a dense pattern. Accordingly, we change the structure of some existing patterns to make them dense patterns. We discuss this later in Sect. 4. We first show that there exist infinitely many dense patterns using a pattern generator.

Theorem 1. *There are infinitely many dense patterns.*

Proof. The following pattern generator generates infinitely many dense patterns.

$$Gen(x,y) = a.(a^1b.a^2...b.a^y).\left\{ \bigodot_{k=(y-1)}^{x-2} (ba^kba^{k+1}ba^{k+2}) \right\} .(ba^{x-1}b.a^x)(ba^{x-1}b.ab).a \quad (2)$$

where $x, y \in \mathbb{N}$ such that $3 \leq y \leq (x - 3)$. Every value of y gives a different pattern and we use $G_3, G_4, ..., G_y$ to denote these patterns.

$$|G_y(x)| = \frac{1}{2}(3x^2 + 9x - 2y^2 + 10) \tag{3}$$

$$|DS(G_y(x))| = \frac{1}{2}(3x^2 + 4x - 2y^2 + 2y + 6 - (x \mod 2)) \tag{4}$$

Equations (3) and (4) shows that the distinct-square density of each G_y is one. Further, the subwords of any of these patterns are as given in Lemma 9. Thus, G_y is a dense pattern. □

Note that every G_y supports the 'stronger' square conjecture [6]. Also, it is possible to get more dense patterns by replacing the letters (a, b) in the generator explained in Theorem 1 with certain words.

We continue to discuss these patterns where we compare the existing best known patterns with the pattern P.

4 Comparison of *P* with the Existing Patterns

The patterns described in Sect. 2 have varying distinct-square densities. We use the Definition 1 to verify existing patterns for a dense pattern. Accordingly, the distinct-square density of a pattern must approach to one. This condition makes us omit the patterns of lower densities. Patterns given in [1,2] have a distinct-square density less than one while the distinct-square density of patterns in [5,6] approaches one. So, we verify only Q and JMS against the definition of a dense pattern. The pattern Q [5] is defined as follows.

$$Q(x) = \bigodot_{i=2}^{x} a^i b a^{i-1} b a^i b$$

We compute the distinct-square density of pattern Q using the Eqs. (5) and (6).

$$|Q(x)| = \frac{1}{2}(3x^2 + 7x - 10) \tag{5}$$

$$|DS(Q(x))| = \frac{1}{2}(3x^2 + 2x - 10 - (x \mod 2)) \tag{6}$$

$$\alpha_Q = \lim_{x \to \infty} \frac{|DS(Q(x))|}{|Q(x)|} = 1 \tag{7}$$

For $|w_1| = 1$, the word $Q(x) = w_1 w_2$ satisfy $cross(w_1, w_2) = \emptyset$. So, Q is not a dense pattern. However, the pattern, Q', obtained by removing the first letter from $Q(x)$ makes it a dense pattern.

$$Q'(x) = (ababa^2b). \bigodot_{i=3}^{x} a^i b a^{i-1} b a^i b$$

We have $|DS(Q'(x))| = |DS(Q(x))|$ and $|Q'(x)| = |Q(x)| - 1$, therefore, the distinct-square density of Q' is one. A set of rightmost squares as mentioned in Lemma 8 exists for Q', that is, $R \subset DS(Q')$ where

$$R = \{(ab)^2, (aba^2baa)^2, (aba^3baa)^2, \ldots, (aba^{i-1}baa)^2, (a^{i-1}ba^ib)^2\}$$

Hence, the pattern Q' is a dense pattern.

A stricter bound for the number of distinct squares is conjectured in [6] and is supported by a pattern called as JMS. It is a simple pattern with the structure:

$$JMS(x) = \bigodot_{i=1}^{x} a^i b$$

Following equations give the length and the number of distinct squares in $JMS(x)$.

$$|JMS(x)| = \frac{1}{2}(x^2 + 3x) \tag{8}$$

$$|DS(JMS(x))| = \frac{1}{2}(x^2 - 2 - (x \mod 2)) \tag{9}$$

The distinct-square density of the pattern obtained with above equations as $\alpha_{JMS} = 1$. Similar to $Q(x)$, a word $JMS(x) = w_1 w_2$ satisfies $cross(w_1, w_2) = \phi$ for $|w_2| = 1$. We remove the last letter of $JMS(x)$ to get a word that satisfies the condition (b) of Definition 1:

$$JMS'(x) = \left\{ \bigodot_{i=3}^{x-1} a^i b \right\} .a^x$$

The distinct-square density of pattern JMS' is one, and we can use Lemma 9 to show that it is a dense pattern. Both the patterns Q' and JMS' construct words using the same principle to increase the number of distinct squares. They maximize the distinct primitive squares to achieve a higher distinct-square density, as mentioned in Lemma 7. Let us see a criterion to compare the dense patterns.

4.1 Comparing Dense Patterns

We obtained two dense patterns Q' and JMS' from the existing patterns. Also, the newly proposed pattern P met all the conditions that are defined for a dense pattern. A pattern that reaches to its distinct-square density quickly is the best pattern. It is evident that if a pattern introduces a lot of no-gain lengths between its successive words, then it will move slowly towards its density. We, therefore, use a notation β_T to determine the rate of a pattern T to arrive its distinct-square density. The notation is valid for a pattern that has at least one no-gain length between its successive words.

Definition 2 (Gain lengths per no-gain length). *Let $x \in \mathbb{N}$. The term $\beta_T(x)$ is the ratio of number of distinct squares in $T(x)$ that are not in $T(x-1)$ to the number of no-gain lengths between $T(x)$ and $T(x-1)$, that is,*

$$\beta_T(x) = \frac{|DS(T(x))| - |DS(T(x-1))|}{\mathcal{N}_T(x)}$$

We get $\beta_{Q'}(x)$ and $\beta_{JMS'}(x)$ from Eqs. (5), (6), (8) and (9) as follows:

$$\beta_{Q'}(x) = \frac{3x}{2} \text{ or } \frac{3x-1}{3} \qquad \text{and} \qquad \beta_{JMS'}(x) = \frac{x}{1} \text{ or } \frac{x-1}{2} \tag{10}$$

Lemma 11. *For all positive integers $x > 4$, there exists $y \in \mathbb{N}$ with $|JMS'(x)| > |Q'(y)|$ and $|DS(JSM'(x))| < |DS(Q'(y)|$.*

Proof. The statement holds since $\beta_{JMS'}(x) < \beta_{Q'}(x)$ (see Eq. (10)). □

Theorem 2. *Pattern P is the lower bound for $MaxNS(n)$.*

Table 2. New Distinct Squares per new No-Gain Length

$\beta_T(x)$	$x \mod 2 = 0$	$x \mod 2 = 1$
$\beta_{Q'}(x)$	$1.5x$	$x - 0.33$
$\beta_P(x)$	$1.5x + 0.5$	$x + 0.5$

Proof. The Lemma 11 shows that the pattern Q' is better than the pattern JMS'. We enlist the β values of patterns Q' and P in Table 2. It shows that the rate of approaching the distinct-square density of pattern P is faster than that of Q'. □

Corollary 1. *For every word, $Q'(x)$, there exists a word, $P(y)$, such that $|Q'(x)| > |P(y)|$ and $|DS(Q'(x))| < |DS(P(y))|$ where x and $y \in \mathbb{N}$ and $x > 5$.*

5 Conclusion and Future Work

We analyzed different patterns and tried to build a structure to pack maximum number of distinct squares. In this context, we found that a square if extended with its square base, adds a conjugate of the square for every newly added letter. We defined a dense pattern, P, using this property and presented a generator to obtain infinite number of dense patterns. The existing patterns are modified to qualify the definition of dense pattern and compared with the proposed dense pattern P. It is concluded that P is the new lower bound for the square conjecture. Since a primitive square has maximum number of conjugates, we conjecture that such a structure has maximum number of distinct primitive squares. The proposed structure of building dense patterns introduces at most one distinct square per letter. Another direction for exploration is to consider extending an FS double square where every added letter adds two new distinct squares.

References

1. Amit, M., Gawrychowski, P.: Distinct squares in circular words. In: Fici, G., Sciortino, M., Venturini, R. (eds.) SPIRE 2017. LNCS, vol. 10508, pp. 27–37. Springer, Cham (2017). https://doi.org/10.1007/978-3-319-67428-5_3
2. Blanchet-Sadri, F., Osborne, S.: Constructing words with high distinct square densities. In: International Conference on Automata and Formal Languages. EPTCS, vol. 252, pp. 71–85 (2017)
3. Deza, A., Franek, F., Jiang, M.: A d-step approach for distinct squares in strings. In: Giancarlo, R., Manzini, G. (eds.) CPM 2011. LNCS, vol. 6661, pp. 77–89. Springer, Heidelberg (2011). https://doi.org/10.1007/978-3-642-21458-5_9
4. Deza, A., Franek, F., Thierry, A.: How many double squares can a string contain? Discret. Appl. Math. **180**, 52–69 (2015)
5. Fraenkel, A.S., Simpson, J.: How many squares can a string contain? J. Comb. Theory Ser. A **82**(1), 112–120 (1998)
6. Jonoska, N., Manea, F., Seki, S.: A stronger square conjecture on binary words. In: Geffert, V., Preneel, B., Rovan, B., Štuller, J., Tjoa, A.M. (eds.) SOFSEM 2014. LNCS, vol. 8327, pp. 339–350. Springer, Cham (2014). https://doi.org/10.1007/978-3-319-04298-5_30
7. Lothaire, M.: Applied Combinatorics on Words, vol. 105. Cambridge University Press, Cambridge (2005)
8. Manea, F., Seki, S.: Square-density increasing mappings. In: Manea, F., Nowotka, D. (eds.) WORDS 2015. LNCS, vol. 9304, pp. 160–169. Springer, Cham (2015). https://doi.org/10.1007/978-3-319-23660-5_14

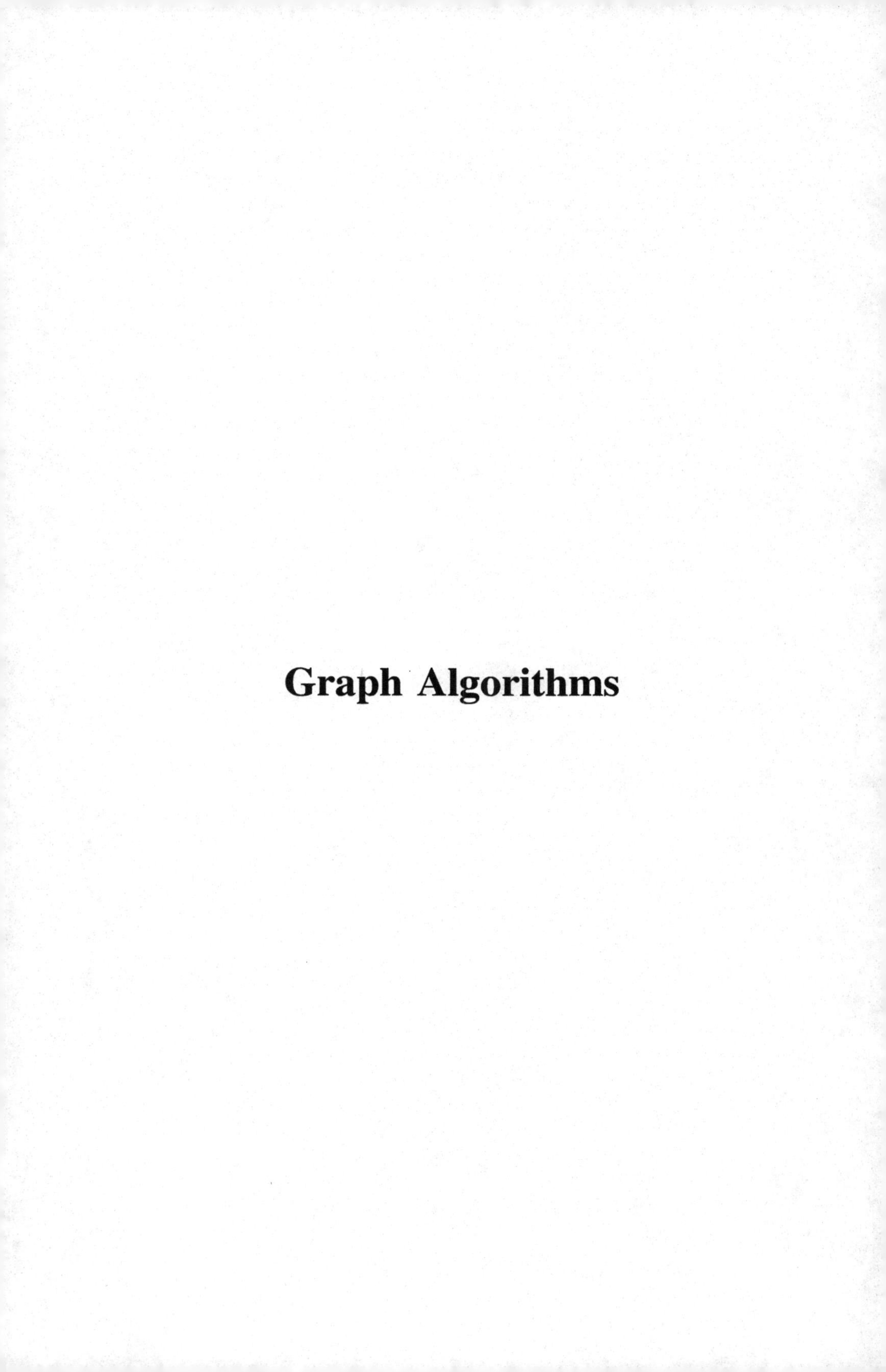

Graph Algorithms

Failure and Communication in a Synchronized Multi-drone System

Sergey Bereg[1(✉)], José Miguel Díaz-Báñez[2], Paul Horn[3], Mario A. Lopez[4], and Jorge Urrutia[5]

[1] Department of Computer Science, University of Texas at Dallas, Richardson, USA
besp@utdallas.edu

[2] Department of Applied Mathematics II, University of Seville, Seville, Spain

[3] Department of Mathematics, University of Denver, Denver, USA

[4] Department of Computer Science, University of Denver, Denver, USA

[5] Instituto de Matemáticas, Universidad Nacional Autónoma de México, Mexico City, Mexico

Abstract. A set of n drones with limited communication capacity is deployed to monitor a terrain partitioned into n pairwise disjoint closed trajectories, one per drone. In our setting, there is a communication link between two trajectories if they are close enough, and drones can communicate provided they visit the link at the same time. Over time, one or more drones may fail and the ability to communicate and stay connected decreases. In this paper we study two properties related to communication: isolation and connectivity. First, we provide efficient algorithms, both centralized and decentralized, for determining the connected components induced by the set of surviving drones. Second, we study isolation and connectivity under a probabilistic failure model and show that, in the case of grids, the system is quite robust in the sense that it can tolerate a large probability of failure before drones become isolated and the system loses full connectivity.

Keywords: Unmanned aerial vehicles · Synchronized communication system · Communication graph · Connectivity · Probabilistic model

S. Bereg—Partially supported by NSF award CCF-1718994.

J. M. Díaz-Báñez—Partially supported by Spanish Ministry of Economy and Competitiveness project MTM2016-76272-R AEI/FEDER, UE and European Union's Horizon 2020 research and innovation programme under the Marie Sk lodowska-Curie grant #734922.

P. Horn—Partially supported by Simons Collaboration Grant #525039.

M. A. Lopez—Partially supported by a University of Denver John Evans Award.

J. Urrutia—Partially supported by PAPIIT Grant IN105221, UNAM.

A. Mudgal and C. R. Subramanian (Eds.): CALDAM 2021, LNCS 12601, pp. 413–425, 2021.
https://doi.org/10.1007/978-3-030-67899-9_33

1 Introduction

Teams of *Unmanned Aerial Vehicles* (UAVs), colloquially known as drones, are becoming a trend in the last few years for their use in a wide variety of applications such as area monitoring, precision agriculture, search and rescue, exploration and mapping, and delivery of products, to name a few; see [10–12] and references therein for a comprehensive survey on the topic. The coordination of a team of autonomous vehicles enables the execution of tasks that no individual autonomous vehicle can accomplish on its own, and thus there has been an increasing interest in studying teams of drones that cooperate with each other. In such multi-drone systems a desired collective outcome arises from the interaction of the drones with each other and with their environment, via a set of installed sensors and communication devices.

The problems raised in this paper assume the framework recently proposed in [7]. A partition of a terrain to be covered is given and every drone is assigned a different section of the partition. Each drone travels on a fixed closed trajectory while performing a prescribed task, such as monitoring its assigned area. In order to allow cooperation, each drone needs to communicate periodically with other drones. Since the UAVs have a limited communication range, two of them need to be in close proximity of each other in order to communicate. In [7] the authors presented a framework to survey a terrain in the scenario described above. As an abstraction, they considered a model in which each drone is modeled by a single point that flies on a unit circle at constant speed, and this speed is the same for all the drones. They assume, w.l.o.g., that one time unit is the time required by a robot to complete a tour of a circle. These circles may intersect at a single point but do not cross. The communication between two robots can take place if their corresponding circles touch, and it is carried out at the point of intersection. They also showed how to generalize the results to a more realistic model. In [7] it is assumed that the unit disk graph defined by the given set of circles (trajectories) is connected, they call it the *communication graph*.

The main problem addressed in [7] is to obtain a synchronization schedule, that is, to assign a starting position and travel direction to each trajectory so that if n drones follow this schedule, every pair of them traveling in two adjacent circles pass through the intersection point of their trajectories at the same time. A set of trajectories with a synchronization schedule conform a *synchronized communication system* (SCS) [7]. In the same paper, the authors also discuss necessary and sufficient conditions for the existence of a synchronization schedule. For an illustration see Fig. 1 and related video[1]. Note that although not every pair of robots can communicate directly, a robot may relay a message to another robot through a sequence of intermediate message exchanges.

If the system is synchronized, as described above, a robot can easily detect the failure of a neighboring robot. If a robot d_i in trajectory C_i arrives at the communication point between C_i and another trajectory C_j, and it fails to meet another robot, it will assume that the robot in C_j is no longer functional. Under

[1] https://www.youtube.com/watch?v=T0V6tO80HOI.

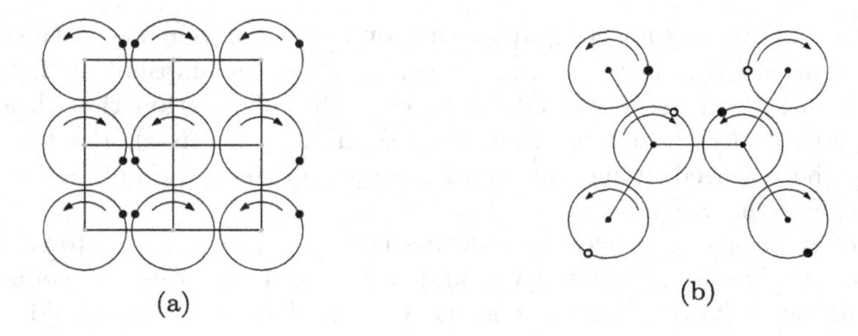

Fig. 1. Examples of synchronized communication systems. The robots in the SCS are represented by solid black points. (a) The communication graph is a grid. (b) The communication graph is a tree. If the white drones leave the system, the black drones become isolated.

such circumstances, a reasonable strategy is for d_i to switch to C_j at this point and take over the task of the missing robot. In [7], this strategy is called the *shifting strategy.* Under the shifting strategy, an undesirable phenomenon, known as *isolation*, may occur. A drone is *isolated* if it fails permanently to meet other drones. The three black drones in Fig. 1(b) never meet, and thus they are isolated. A *ring* is the closed path followed by an isolated drone. Each ring is composed of sections of various trajectories and has a direction of travel determined by the direction of movement in the participating trajectories. Each section of a trajectory between two consecutive link positions participates in exactly one ring, thus the rings in an SCS are pairwise disjoint. The number of rings and its length depends on the communication graph. Figure 2 illustrates some examples. See [2] for a study on rings and the isolation phenomenon.

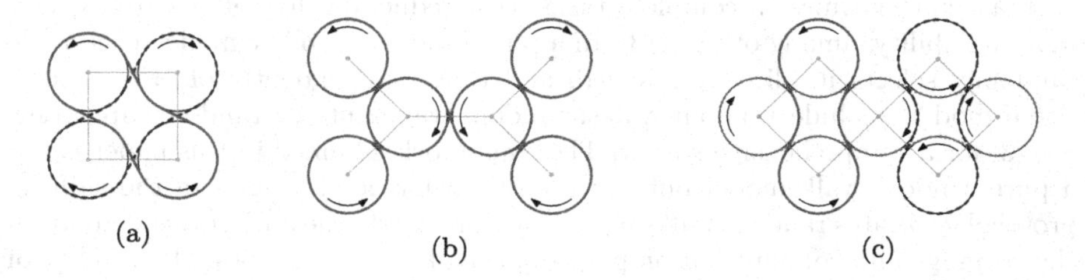

Fig. 2. SCSs with two rings (a); one ring (b) and three rings (c).

In [7], neighboring circles are assigned opposite travel directions (clockwise and counterclockwise) so as to enable the shifting strategy. From now on we work with SCSs where every pair of neighboring circles have opposite travel directions and, consequently, the communication graph is bipartite. Under this model, our main contributions are the following:

1. *Connected components.* Consider a system in which some drones may have failed. Two drones belong to the same connected component if they can exchange messages, possibly through a sequence of intermediaries. We provide

efficient algorithms for computing the connected components of the system, both centralized (where a central server is privy of a snapshot of the system, Theorem 2) and decentralized (using only the information that drones can gather while flying and meeting other drones, Theorem 4). For the case of grids, the required flying time can be proportional to the number of trajectories, and this bound is tight (Theorem 3).
2. *Probabilistic failure model.* We address the robustness of a system in which drones survive with probability p, and study two properties: full connectivity and drone isolation. For $t \times t$ grids, we establish sharp thresholds for the existence of isolated drones (Theorem 5) and connectivity (Theorem 6). These results show that the system is extremely robust to random failure as these thresholds are $o(1)$ as $t \to \infty$. For general grids, we provide less sharp results (Theorems 7 and 8).

2 Related Work

There is a vast literature related to communication strategies for a team of robots monitoring a given area. Our scenario shares similarities with work on patrolling agents [9] where the drones patrol along predefined paths making observations and synchronize with their teammates during a very limited time to share data. Typically, research has focused mostly on construction and validation of working systems, rather than more general and formal analysis of problems. In this paper, we study some algorithmic and probabilistic problems related to communication in the particular framework of a synchronized communication system (SCS) proposed in [7].

Recently, the study of stochastic UAV systems has attracted considerable attention in the field of mobile robots. This approach has several advantages such as shorter times to complete tasks, cost reduction, higher scalability, and more reliability, among others [13]. In a pure random mobility model, each node randomly selects its direction, speed, and time independently of other nodes. Some models include random walks, random waypoints or random directions. See [3] for a comprehensive survey. The framework assumed in this paper is not a pure random walk model but the use of stochastic strategies in the shifting protocol generates random walks [4]. In the same work, the authors evaluate both the coverage and communication performance of a SCS and show the validity of two random strategies compared with the deterministic one.

On the other hand, several algorithmic and combinatorial problems have been studied within a SCS using the deterministic shifting strategy. In [1,2] the authors propose various quality measures for a synchronized system regarding the resilience of a network in the presence of failures. Computing these measures leads to interesting combinatorial and algorithmic problems.

3 Computing the Connected Components

Assume we have a SCS where a subset of drones left the system and the surviving drones apply the shifting protocol. We call the resulting system a *partial* SCS.

While some pairs of drones may communicate directly, communication between other pairs may rely on passing information through other drones; in some cases communication between drones may be impossible. We define the *drone communication graph* G_D as the graph whose vertices are the drones, two of which are adjacent if the corresponding drones communicate directly at some point in time. The connected components of this graph identify which sets of drones can, directly or indirectly, communicate with each other. It is easy to see that communication through other robots can sometimes be faster than direct communication, e.g. it may take a long time for two drones to communicate directly. In this section we show how to compute the connected components in the drone communication graph under two models of computation:

1. **Centralized:** Suppose a central server contains the full information of a SCS, including the set of drone trajectories and the initial locations of the drones. How can the connected components of the SCS be found efficiently?
2. **Decentralized:** Suppose the drones themselves can pass messages when they pass by each other. How can they determine the other drones in their connected component, and how quickly can this be accomplished?

Note that in the second case, the drones do not know how many other drones are active or where they are; they merely learn what drones are active as they meet other drones and exchange information. For that reason, the complexity of both problems is different. The complexity of our algorithm in the first case, is the number of steps the central server needs to be computed, while in the second case it is the *flying time* of the drones before each drone knows its connected component.

Nonetheless, we show that for the $s \times t$ grid both problems can be solved with highly efficient algorithms. The key notion for our results is the use of the *token graph* introduced in [1]. We assume that at time 0 each drone d_i holds a token t_i. This establishes a bijection between the drones and the tokens. When two drones meet, they exchange their tokens. The token graph G_T of a drone system is the graph whose vertices are the tokens, two of which are adjacent if at some time the corresponding tokens are exchanged. Note that each token t_i stays in the same ring of drone d_i. Thus, the token graph can also be defined using drones as vertices where two drones are adjacent if they encounter each other in a system where only two drones exist. We have the following result.

Theorem 1. *Two drones of G_D are in the same component if and only if the corresponding tokens are in the same component in G_T.*

In the case of an $s \times t$ grid with only two drones, the drones encounter each other if and only if they are in the same row or the same column [1]. We call it *RC-property* of the token graph.

Theorem 2. *The connected components in G_D can be computed in polynomial time in the centralized model. Furthermore, they can be computed in $O(st)$ time in the $s \times t$ grid.*

Proof. The token graph can be computed in polynomial time and its connected components can be computed in linear time using breath-first search. For the $s \times t$ grid, we compute a subgraph of the token graph where the vertices are the drones and two drones are adjacent if (i) their circles are in the same row or column, and (ii) there are no other drones in the circles between them. This graph can be computed in $O(st)$ time (by checking every row and column of the grid) and its connected components are the same as in G_D. Again, they can be computed in linear time using breath-first search. □

3.1 Decentralized Computation

The goal in the decentralized model is for each drone d_i to compute $C(d_i)$, its connected component in G_D. We use the following algorithm. Each drone d_i maintains a list $L(d_i)$ of some drones from $C(d_i)$. Initially, we set $L(d_i) := \{d_i\}$. When drones d_i and d_j meet at some time, they replace both $L(d_i)$ and $L(d_j)$ with the union $L(d_i) \cup L(d_j)$. It is clear that if we follow this protocol long enough, all drones will know their connected components (that is $L(d_i) = C(d_i)$ for all i.) Our goal is to give bounds for the running time of this approach.

We emphasize, again, that the time measured here is actually flying time of the drones, i.e. how long do the drones have to fly until they each, individually, know the other drones own components. We ignore, then, the computation of the set unions involved as this is negligible compared with the actual flying time from one communication point to another. We further assume that a unit time is needed to navigate a trajectory. We begin with sharp bounds for the problem on a $t \times t$ grid.

Theorem 3. *On the $t \times t$ grid, at time $t \cdot (t-1)$ we have that $L(d) = C(d)$ for all drones d. Furthermore, there are drone configurations that require $\Omega(t^2)$ time until $L(d) = C(d)$ for all drones d.*

Proof. We use the idea of tokens. At the beginning (time 0), each drone d_i holds token t_i; recall that when two drones encounter each other they exchange tokens (along with taking the union of their respective lists). Let $d(i, m)$ denote the drone holding token t_i at time m. Thus, $d(i, 0) = d_i$. Note that $d(i, m)$ is always in the same component as drone d_i as it holds t_i due to a sequence of interactions with other drones, each passing t_i to the next drone of the sequence. Moreover, $L(d(i, m)) \subseteq L(d(i, m'))$ if $m \leq m'$ as whenever tokens are exchanged, the lists are passed along.

Fix an (arbitrary) drone d_0 and consider any drone d_k in $C(d_0)$. Let $d_0, d_1, \ldots, d_k$ be the shortest path between d_0 and d_k in the token graph. By the construction of the token graph as an RC-graph, it is easy to see that the diameter of the token graph is at most $t-1$, and in particular $k \leq t-1$. Note that t_i and t_{i+1} are in the same row or column, and hence drones holding them meet within time t. This implies that, for instance, $d_1 \in L(d(0, t))$ at time t as when the drones with token t_0 and t_1 meet, the label d_1 is passed to the drone holding token t_0. Inductively, it follows $d_i \in L(d(0, i \cdot t))$ at time $i \cdot t$: the label

d_i is given to the drone carrying token t_{i-1} in the first time t, then to the drone carrying token t_{i-2} in the next time t, until it at last is passed to the drone hopping token t_0. This shows that $L(d(0, t(t-1))$ is complete at time $k \cdot t \leq (t-1)t$ and as d_0 is arbitrary, this completes the proof.

We now provide a set of drones $\{d_1, \ldots, d_k\}$ which show this time can be quadratic. For this set of drones, $d_1 \notin L(d(k, m))$ until time $m = \Omega(t^2)$. The construction involves a set of drones $\{d_1, \ldots, d_k\}$ on the $t \times t$ grid satisfying the following conditions:

1. d_i and d_{i+1}, for $i = 1, 2, \ldots, k$ share the same row or column, and there are no rows or columns with more than two drones.
2. The distances $d_{i,i+1}$ between d_i and d_{i+1} are decreasing, for $i = 1, 2, \ldots, k$.
3. The polygonal chain formed by the union of the segments connecting d_i to d_{i+1} is a *spiral* polygonal chain; see Fig. 3.
4. Drones on the same column move in opposite directions (clockwise and counterclockwise) along their ring, while drones in the same row move in the same direction.

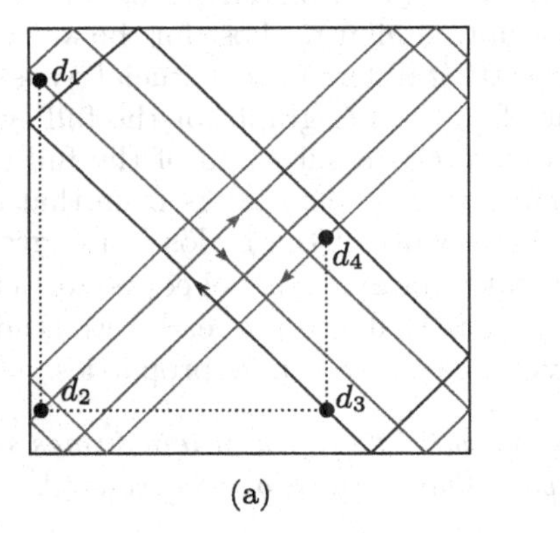

(a)

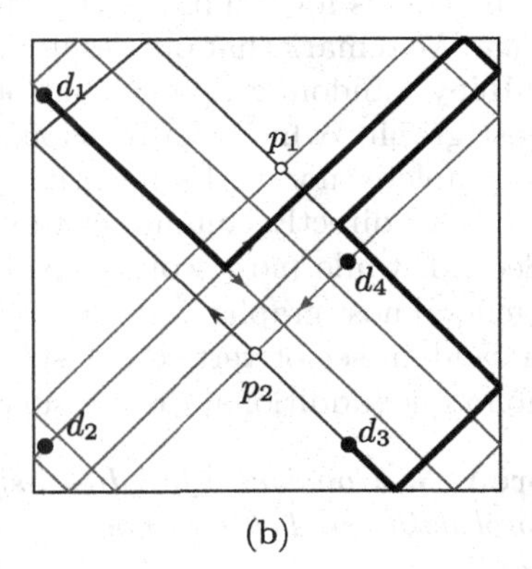

(b)

Fig. 3. (a) Drones are arranged in a spiral polygonal chain. (b) The bold line represents the propagation of the label d_1 through the system for times $\leq t$. Drones holding tokens t_3 and t_4 meet at p_1 and p_2.

The key observation is that the drone holding token t_i will only meet the drones holding token t_{i+1} and t_{i-1} and by placing the drones carefully, the intersections will be set up so that the label of d_1 will only propagate through a small number of consecutive drones in time t.

Consider four consecutive drones in $\{d_1, \ldots, d_k\}$; without loss of generality assume these are d_1, d_2, d_3 and d_4. We claim that at time t, the only elements in $\{d_1, \ldots, d_k\}$ with $d_1 \in L(d(i, t))$ are $i = 1, 2, 3$. To see this, observe that since $d_{1,2} > d_{2,3}$, t_2 meets t_3 the first time before it meets t_1. The label d_1 is thus added to the list of the drone holding token t_3 during the second time the drones holding tokens of t_2 and t_3 meet. At this point t_3 has already swapped with t_4 twice. Hence $d_1 \notin L(d(4,t))$ at time t.

Figures 3a and 3b illustrate first this setup and then the process itself. Figure 3 illustrates how the threshold of knowledge of drone d_1 moves forward through the process Note it never moves from the drone holding token t_1 directly to, say, that holding token t_3 as even though the rings of these drones intersect, drones holding these tokens never directly communicate due to the timing. It follows that to reach the drone holding token t_{2i} label d_1 will take $t(i+1)$ time. □

For a general system, a similar argument can be used to prove the following.

Theorem 4. *Consider a general system of N drones on n trajectories and ring lengths $r_1, r_2, \ldots, r_k$. Then at time $N \cdot \max\{lcm(r_l, r_m) : l \neq m\}$, $L(d_i) = C(d_i)$ for all drones $d_i, i = 1, \cdots, N$.*

4 Communication Within a Probabilistic Failure Model

In this section we study the connectivity of G_D under random failure. We prove sharp thresholds for the properties of containing an isolated vertex and for connectivity. We remark that our results are quite similar to those for the well known Erdős-Rényi random graph [8], but our setting differs in two crucial ways. First the 'host graph' (which can be thought of as the RC-graph for the full system) is not complete; nor is the resulting random G_D a subgraph of the full G_D as which drones directly communicate within a subsystem differs from that of the full. Second, while most work generalizing results of the Erdős-Rényi graph to more general host graphs (see, e.g. [5,6]) takes a random set of edges, we actually take a random set of vertices. A side effect is that the properties we study are not monotone; additional drones surviving may break these properties.

Theorem 5. *Consider a full drone system in the $t \times t$ grid, where drones survive with probability p. Let $\mathcal{I}$ denote the event that some drone is isolated. Fix an arbitrary $\varepsilon > 0$.*

(a) If $p = (1+\varepsilon)\frac{\ln t}{2t}$ then as $t \to \infty$, then $\mathbb{P}(\mathcal{I}) \to 0$.
(b) If $p = (1-\varepsilon)\frac{\ln t}{2t}$ then as $t \to \infty$, then $\mathbb{P}(\mathcal{I}) \to 1$.

Proof. For (a), note that there are t^2 drone locations and in order for a drone to be isolated it must survive and all others in its row and column must fail. Hence the expected number of isolated drones is

$$t^2 p(1-p)^{2t-1} \leq t^2 p e^{-p(2t-1)} = (1+\varepsilon)\frac{t \ln t}{2} \exp\left(-(1+\varepsilon)\frac{(2t-1)}{2t}\ln(t)\right) \to 0,$$

where we note that for t sufficiently large $(1+\varepsilon)\frac{(2t-1)}{2t} > 1$, so that the exponential term is $O(t^{-(1+\varepsilon')})$. (a) then follows by Markov's inequality.

For (b), note that the expected number of isolated drones in this situation is

$$t^2p(1-p)^{2t-1} \leq t^2pe^{-p(2t-1)} = (1-\varepsilon)\frac{t\ln t}{2}\exp\left(-(1-\varepsilon)\frac{(2t-1)}{2t}\ln(t)\right) \geq t^{\varepsilon/2},$$

assuming that t is sufficiently large.

For (b), then it suffices, by Chebyschev's inequality, to show that if X is the number of isolated drones in the system, to show that $\mathrm{Var}(X) = o(\mathbb{E}[X]^2)$. Note that X can be written as $\sum_{(i,j)\in[t]^2} X_{i,j}$, where $X_{i,j}$ is the event that the drone in the $(i,j)th$ position is isolated. Then

$$\mathrm{Var}(X) \leq \mathbb{E}[X] + \sum_{(i,j)\neq(k,l)\in[t]^2} \Big(\mathbb{E}[X_{i,j}X_{k,l}] - \mathbb{E}[X_{i,j}]\mathbb{E}[X_{k,l}]\Big).$$

We bound the sum. If $i = k$ or $j = l$, then $\mathbb{E}[X_{i,j}X_{k,l}] = 0$, and $\mathbb{E}[X_{i,j}] = \mathbb{E}[X_{k,l}] = p(1-p)^{2t-1}$. As the covariance terms being sum are negative these terms can be discarded for upper bounding the variance. For the other terms, where (i,j) and (k,l) are different in both coordinates, $\mathbb{E}[X_{i,j}X_{k,l}] = p^2(1-p)^{4t-4}$ and there are at most t^4 terms of this type and these summands contribute at most

$$t^4\left(p^2(1-p)^{4t-4} - p^2(1-p)^{4t-2}\right) = \mathbb{E}[X]^2((1-p)^{-2}-1) = o(\mathbb{E}[X]^2),$$

where the last equality follows from the form of p. Hence, by Chebyschev's inequality $X \sim \mathbb{E}[X]$ with probability tending to one, and thus are isolated drones. □

Remark 1. Note that for (a), if suffices that $p \geq (1+\varepsilon)\frac{\log t}{t}$ – this follows as the expected number is decreasing in as p increases (assuming that $p \geq \frac{1}{2t-1}$.) Extending the lower bound works so long as the expected number of isolated drones tends to infinity.

Remark 2. Theorem 5(a) implies that, even for fairly small p, the number of isolated drones is 0 with high probability. At the threshold, the number of surviving drones is only $O(t\log t)$, while $t^2 - O(t\log t)$ drones fail in this case. This should be compared with the 1-isolation resilience of the grid, the minimum number of drones whose failure can result in an isolated drone, which is $O(t)$ [1].

Theorem 6. *Consider a full drone system in the $t\times t$ grid, where drones survive with probability p. Let $\mathcal{C}$ denote the event that the system of drones is connected (that is, all drones can communicate with one another). Fix an arbitrary $\varepsilon > 0$.*

(a) If $p = (1+\varepsilon)\frac{\ln t}{2t}$ then as $t \to \infty$, then $\mathbb{P}(\mathcal{C}) \to 1$.
(b) If $p = (1-\varepsilon)\frac{\ln t}{2t}$ then as $t \to \infty$, then $\mathbb{P}(\mathcal{C}) \to 0$.

Proof. Note that (b) follows directly from Theorem 5, as if there is an isolated drone (and more than one drone, as there is at such a p with high probability) then the system is not connected.

We proceed to prove (a). We have already shown that when $p = (1+\varepsilon)\frac{\ln t}{t}$ that there are no components of size 1. We still need to show there is a unique component. To do this, we study a modified breadth first search in the RC-graph, introduced in the previous section. Recall, that performing a breadth-first search in the RC-graph (where vertices are drones and they are joined if they in the same row or column) reveals the connected component of a vertex.

To show that there is precisely one component in this setting, we study a slightly modified tree finding algorithm. An *exposing tree* inside of a component is a rooted tree generated as follows: Choose an initial root vertex (drone) to explore. Add all vertices in its row and column to a queue. Now, each vertex in the queue is iteratively explored. When a vertex is explored, vertices in their row or column are added to the queue *if either their row or column is different from those already added to the queue.* Since every vertex being explored was added to the queue it shares either a row or column with one of the other vertices previously explored, and each vertex is responsible for 'exposing' a new row or new column (with the initial vertex responsible for exposing both.) The set of explored vertices forms the exposing tree.

Generating an exposure tree ends with a subset of a connected component which is both non-empty and possibly proper – but vertices in the component and not in the tree share both a row and a column with vertices in the tree. It also ends with a drone from each of some j columns and k rows (where j and k are determined by the process) and $j+k-1$ vertices. Furthermore, the process ending means that there are no vertices in either of those j columns outside of the k rows and likewise none in the k rows outside of the j columns.

Claim: The probability that the exposing tree process ends with $2 \leq j + k \leq t + 1$ vertices from some starting point tends to zero.

Note that if there are two components, their rows and columns must be disjoint, and hence one of the non-trivial components must have $j+k \leq t$. Thus the claim will complete the proof of the theorem.

Fix $\ell = j + k$. The number of potential exposing trees with $\ell - 1$ vertices in the $t \times t$ grid can be estimated (roughly) as follows. The degrees in the tree can be represented by a sequence of non-negative integers $(a_1, a_2, \ldots, a_\ell)$ with $\sum a_i = \ell - 2$ where a_1, a_2 are the row and column degrees of the first vertex, and a_i is the number of vertices added when the i – 1st vertex from the queue is explored. The number of such solutions is bounded by $\binom{2\ell-2}{\ell} \leq 4^\ell$. There are fewer than $t^2 \cdot t^{(\ell-2)} = t^\ell$ ways of choosing the vertices that are exposed. Note that this is a rather large over-count: it assumes there are t choices each time, when in reality there is a falling factorial type term and also introduces an ordering when exposing the children of a given vertex. None the less, this upper bounds the number of potential processes for a given ℓ is at most $4^\ell t^\ell$.

Now: for a given one of these potential processes, the $j + k - 1 = \ell - 1$ vertices explored must all survive, and the other vertices of their j columns outside of the k rows, and k rows outside of the j columns, must all fail. This has probability $p^{j+k-1}(1-p)^{(t-j)k+(t-k)j} = p^{j+k-1}(1-p)^{t(j+k)-2jk}$. Finally note that $jk \leq \frac{(j+k)^2}{4}$ so that regardless of the individual j, k – for *any* potential process with $\ell = j + k$ fixed the probability of ending is at most

$$p^{j+k-1}(1-p)^{(t-(j+k)/2)\cdot(j+k)} = p^{\ell-1}(1-p)^{(t-\ell/2)\cdot\ell}$$

A union bound over potential exposing trees, shows that the probability that the process ends with a given value of ℓ is at most

$$4^\ell t^\ell p^{\ell-1}(1-p)^{(t-\ell/2)\cdot\ell} = 4^\ell \cdot t((1/2+\varepsilon)\ln(t))^{\ell-1}(1-p)^{t-\ell/2\cdot\ell}$$
$$\leq \exp\left(\ln(t) + \ell\left(\ln(4(1/2+\varepsilon)) + \ln\ln(t) - (1/2+\varepsilon)\frac{\ln t}{t}(t-\ell/2)\right)\right).$$

In the last inequality here, we used the inequality $1 - x \leq e^{-x}$ along withe definition of p. Hence, per a union bound over potential ℓ it suffices to show that

$$\sum_{\ell=2}^{t+1} \exp\left(\ln(t) + \ell\left(\ln(4(1/2+\varepsilon)) + \ln\ln(t) - (1/2+\varepsilon)\frac{\ln t}{t}(t-\ell/2)\right)\right) \to 0$$

as $t \to \infty$. To do this, we note that for $2 \leq \ell \leq 10$, these terms are $o(1)$ individually as for $\ell \leq 10$ $\ell \cdot (1/2+\varepsilon) \cdot \frac{\ln(t)}{t}(t-\ell/2) > (1+\varepsilon/2)\ln(t)$ assuming t is large enough. For $t+1 \geq \ell \geq 8$, the dominant part of the terms comes from

$$\ell \cdot (1/2+\varepsilon)\frac{\ln t}{t}\left(\frac{t-1}{2}\right) > (2+\varepsilon/2)\ln(t).$$

Thus these terms are actually $o(t^{-1})$ and as there are fewer than t such terms in total the sum is $o(1)$ as desired. □

4.1 General Grids

Theorems 5 and 6 above consider the specialized case where the initial setting is a full $t \times t$ grid. The case of general systems, even the case of general $s \times t$ grids is significantly more complicated. Indeed, in $s \times t$ grids, the asymptotic behavior of how s and t are taken to go to infinity in comparison with one another can give rise to a number of different behaviors, depending on the values of s and t.

For instance, when $s > 1$ is fixed, while t goes to infinity the isolation threshold and connectivity threshold differ from each other, and both differ greatly from the above. In this case, we have the following:

Theorem 7. *Consider a full drone system $s \times t$, where drones survive with probability p, where $s > 1$ is fixed as $t \to \infty$.*

1. *If $p = \omega(1/t)$, then $\mathbb{P}(\mathcal{I}) \to 0$.*
2. *If $p = o(1/\sqrt{t})$, then $\mathbb{P}(\mathcal{C}) \to 0$ while if $p = \omega(1/\sqrt{t})$, then $\mathbb{P}(\mathcal{C}) \to 1$*

We only sketch the simple proof.

Proof. For (a), the probability a row contains at most one drone is $tp(1-p)^{t-1} + (1-p)^t$ and if $p = \omega(1/t)$ this tends to zero and the result follows from a union bound. For (b), if $p = o(1/\sqrt{t})$ the expected number of columns containing two drones is $\binom{s}{2} \cdot t \cdot p^2 \to 0$, which implies the resulting communication graph is disconnected as there will be no communication between rows. Once $p = \omega(1/\sqrt{t})$, each of the $\binom{s}{2}$ pairs of rows will have some column where there is a drone in that column in both rows with high probability, and this forces connectivity. □

When both s and t both tend to infinity, the situation becomes more complicated, and we do not pursue a full investigation here. We do note, however, that the following holds:

Theorem 8. *Consider a full drone system in the $s \times t$ grid, where drones survive with probability p. If $s \leq t$ and $s \to \infty$ then if $p = (1+\epsilon)\frac{\ln(s)}{s}$, then $\mathbb{P}(\mathcal{C}) \to 1$.*

This follows as here an $s \times s$ system contained in the grid is both connected and contains a drone in each row and column.

Acknowledgment. This work was initiated at the IX Spanish Workshop on Geometric Optimization, El Rocío, Huelva, Spain, June 18–22, 2018. We thank the other participants of that workshop - L.E. Caraballo, M.A. Heredia and I. Ventura - for helpful discussions and contribution to a creative atmosphere.

References

1. Bereg, S., Brunner, A., Caraballo, L.-E., Díaz-Báñez, J.-M., Lopez, M.A.: On the robustness of a synchronized multi-robot system. J. Comb. Optim. **39**(4), 988–1016 (2020). https://doi.org/10.1007/s10878-020-00533-z
2. Bereg, S., Caraballo, L.E., Díaz-Báñez, J.M., Lopez, M.A.: Computing the k-resilience of a synchronized multi-robot system. J. Comb. Optim. **36**(2), 365–391 (2018)
3. Camp, T., Boleng, J., Davies, V.: A survey of mobility models for ad hoc network research. Wirel. Commun. Mob. Comput. **2**(5), 483–502 (2002)
4. Caraballo, L.E., Díaz-Báñez, J.M., Fabila-Monroy, R., Hidalgo-Toscano, C.: Patrolling a terrain with cooperrative UAVs using random walks. In: 2019 Int'l Conference on Unmanned Aircraft Systems (ICUAS), pp. 828–837. IEEE (2019)
5. Chung, F., Horn, P.: The spectral gap of a random subgraph of a graph. Internet Math. **4**(2–3), 225–244 (2007)
6. Chung, F., Horn, P., Lu, L.: Percolation in general graphs. Internet Math. **6**(3), 331–347 (2009)
7. Díaz-Báñez, J.M., Caraballo, L.E., López, M.A., Bereg, S., Maza, J.I., Ollero, A.: A general framework for synchronizing a team of robots under communication constraints. IEEE Trans. Robot. **33**(3), 748–755 (2017)
8. Erdős, P., Rényi, A.: On random graphs I. Publ. Math. Debrecen **6**, 290–297 (1959)
9. Farinelli, A., Iocchi, L., Nardi, D.: Distributed on-line dynamic task assignment for multi-robot patrolling. Auton. Robots **41**(6), 1321–1345 (2016). https://doi.org/10.1007/s10514-016-9579-8
10. Hayat, S., Yanmaz, E., Muzaffar, R.: Survey on unmanned aerial vehicle networks for civil applications: a communications viewpoint. IEEE Commun. Surv. Tutor. **18**(4), 2624–2661 (2016)

11. Otto, A., Agatz, N., Campbell, J., Golden, B., Pesch, E.: Optimization approaches for civil applications of unmanned aerial vehicles (UAVs) or aerial drones: a survey. Networks **72**(4), 411–458 (2018)
12. Yanmaz, E., Yahyanejad, S., Rinner, B., Hellwagner, H., Bettstetter, C.: Drone networks: communications, coordination, and sensing. Ad Hoc Netw. **68**, 1–15 (2018)
13. Yanmaz, E., Costanzo, C., Bettstetter, C., Elmenreich, W.: A discrete stochastic process for coverage analysis of autonomous UAV networks. In: 2010 IEEE Globecom Workshops, pp. 1777–1782. IEEE (2010)

Memory Optimal Dispersion by Anonymous Mobile Robots

Archak Das, Kaustav Bose(✉), and Buddhadeb Sau

Department of Mathematics, Jadavpur University, Kolkata, India
{archakdas.math.rs,kaustavbose.rs,buddhadeb.sau}@jadavpuruniversity.in

Abstract. Consider a team of $k \leq n$ autonomous mobile robots initially placed at a node of an arbitrary graph G with n nodes. The *dispersion* problem asks for a distributed algorithm that allows the robots to reach a configuration in which each robot is at a distinct node of the graph. If the robots are anonymous, i.e., they do not have any unique identifiers, then the problem is not solvable by any deterministic algorithm. However, the problem can be solved even by anonymous robots if each robot is given access to a fair coin which they can use to generate random bits. In this setting, it is known that the robots require $\Omega(\log \Delta)$ bits of memory to achieve dispersion, where Δ is the maximum degree of G. On the other hand, the best known memory upper bound is $min\{\Delta, max\{\log \Delta, \log D\}\}$ (D = diameter of G), which can be $\omega(\log \Delta)$, depending on the values of Δ and D. In this paper, we close this gap by presenting an optimal algorithm requiring $O(\log \Delta)$ bits of memory.

Keywords: Mobile robots · Dispersion · Depth-first search · Distributed algorithm · Randomized algorithms

1 Introduction

1.1 Background and Motivation

A considerable amount of research has been devoted in recent years to the study of distributed algorithms for autonomous multi-robot system. A multi-robot system consists of a set of autonomous mobile computational entities, called *robots*, that coordinate with each other to achieve some well defined goals, such as forming a given pattern, exploration of unknown environments etc. The robots may be operating on continuous space or graph-like environments. The most fundamental tasks in graphs are GATHERING [3,6,7,11,14–18,23–25,29,30] and EXPLORATION [4,5,9,10,12,13,28]. A relatively new problem which has attracted a lot of interest recently is DISPERSION, introduced by Augustine and Moses Jr. [2]. The problem asks $k \leq n$ robots, initially placed arbitrarily at the nodes of an n-node anonymous graph, to reposition themselves to reach a configuration in which each robot is at a distinct node of the graph. The problem has many practical applications, for example, in relocating self-driven electric cars to recharge

A. Mudgal and C. R. Subramanian (Eds.): CALDAM 2021, LNCS 12601, pp. 426–439, 2021.
https://doi.org/10.1007/978-3-030-67899-9_34

stations where finding new recharge stations is preferable to multiple cars queuing at the same station to recharge. The problem is also interesting because of its relationship to other well-studied problems such as EXPLORATION, SCATTERING and LOAD BALANCING [2].

It is easy to see that the problem cannot be solved deterministically by a set of anonymous robots. Since all robots execute the same deterministic algorithm and initially they are in the same state, the co-located robots will perform the same moves. This is true for each round and hence they will always mirror each other's move and will never do anything different. Hence, throughout the execution of the algorithm, they will stick together and as a result, dispersion cannot be achieved. Using similar arguments, it can be shown that the robots need to have $\Omega(\log k)$ bits of memory each in order to solve the problem by any deterministic algorithm [2]. However, it has been recently shown in [26] that if we consider randomized algorithms, i.e., each robot is given access to a fair coin which can be used to generate random bits, then DISPERSION can be solved by anonymous robots with possibly $o(\log k)$ bits of memory. In [26], two algorithms are presented for DISPERSION from a *rooted configuration*, i.e., a configuration in which all robots are situated at the same node. The first algorithm requires each robot to have $O(max\{\log \Delta, \log D\})$ bits of memory, where Δ and D are respectively the maximum degree and diameter of G. The second algorithm requires each robot to have $O(\Delta)$ bits of memory. In [26], it is also shown that the robots require $\Omega(\log \Delta)$ bits of memory to achieve dispersion in this setting. Notice that while the memory requirement of the second algorithm is clearly $\omega(\log \Delta)$, that of the first algorithm too can be $\omega(\log \Delta)$ depending on the values of Δ and D. In this paper, we close this gap by presenting an asymptotically optimal algorithm that requires $O(\log \Delta)$ bits of memory.

1.2 Related Works

DISPERSION was introduced in [2] where the problem was considered in specific graph structures such as paths, rings, trees as well as arbitrary graphs. In [2], the authors assumed $k = n$, i.e., the number of robots k is equal to the number of nodes n. They proved a memory lower bound of $\Omega(\log k)$ bits at each robot and a time lower bound of $\Omega(\log D)$ rounds for any deterministic algorithm to solve the problem in a graph of diameter D. They then provided deterministic algorithms using $O(\log n)$ bits of memory at each robot to solve DISPERSION on lines, rings and trees in $O(n)$ time. For rooted trees they provided an algorithm requiring $O(\Delta + \log n)$ bits of memory and $O(D^2)$ rounds and for arbitrary graphs, they provided an algorithm, requiring $O(n \log n)$ bits of memory and $O(m)$ rounds (m is the number of edges in the graph). In [19], a $\Omega(k)$ time lower bound was proved for $k \leq n$. In addition, three deterministic algorithms were provided in [19] for arbitrary graphs. The first algorithm requires $O(k\log\Delta)$ bits of memory and $O(m)$ time, ($\Delta =$ the maximum degree of the graph), the second algorithm requires $O(D \log \Delta)$ bits of memory and $O(\Delta^D)$ time, and the third algorithm requires $O(\log(\max(k, \Delta)))$ bits of memory and $O(mk)$ time. Recently, a deterministic algorithm was provided in [20] that runs in $O(\min(m, k\Delta) \log k)$ time and uses $O(\log n)$ bits of memory at each robot. In [22], the problem was studied on grid graphs. The authors presented two deterministic algorithms on anonymous grid graphs that achieve simultaneously optimal bounds with respect to both time and memory complexity. For the first algorithm, the authors considered the local communication model where a robot can only communicate with other robots that are present at the same node. Their second algorithm works in global communication model where a robot can communicate with other robots present anywhere on the graph. In the local communication model, they showed that the problem can be solved in an n-node square grid graph in $O(\min(k, \sqrt{n}))$ time with $O(\log k)$ bits of memory at each robot. In the global communication model, the authors showed that it can be solved in $O(\sqrt{k})$ time with $O(\log k)$ bits of memory at each robot. In [21], the authors extended the work in global communication model to arbitrary graphs. They gave three deterministic algorithms, two for arbitrary graphs and one for trees. For arbitrary graphs, their first algorithm is based on DFS traversal and has time complexity of $O(\min(m, k\Delta))$ and memory complexity of $\Theta(\log(\max(k, \Delta)))$. The second algorithm is based on BFS traversal and has time complexity $O(\max(D, k)\Delta(D + \Delta))$ and memory complexity $O(\max(D, \Delta \log k))$. The third algorithm in arbitrary trees is a BFS based algorithm that has time and memory complexity $O(D \max(D, k))$ and $O(\max(D, \Delta \log k))$ respectively. In [1], the problem was studied on dynamic rings. Fault-tolerant DISPERSION was considered for the first time in [27] where the authors studied the problem on a ring in presence of Byzantine robots. In [26], randomization was used to break the $\Omega(\log k)$ memory lower bound for deterministic algorithms. In particular, the authors considered anonymous robots that can generate random bits and gave two deterministic algorithms that achieve dispersion from rooted configurations on an arbitrary graph. The memory complexity of the algorithms are respectively $O(max\{\log \Delta, \log D\})$ and $O(\Delta)$. For arbitrary initial configurations, they gave a random walk based algorithm that requires $O(\log \Delta)$ bits of memory, but the robots do not terminate.

1.3 Our Results

We study DISPERSION from a rooted configuration on arbitrary graphs by a set of anonymous robots with random bits. In [26], two algorithms with memory complexity $O(max\{\log \Delta, \log D\})$ and $O(\Delta)$ were reported. The question of whether the problem can be solved with $O(\log \Delta)$ bits of memory at each robot was left as an open problem. In this paper, we answer this question affirmatively by presenting an algorithm with memory complexity $O(\log \Delta)$. The lower bound result presented in [26] implies that the algorithm is asymptotically optimal with respect to memory complexity.

Organization. In Sect. 2, we describe the model and introduce notations that will be used in the paper. In Sect. 3, we describe the main algorithm. In Sect. 4, we prove the correctness of our algorithm and establish the time and memory complexity. In Sect. 5, we discuss some future research directions.

2 Technical Preliminaries

Graph. We consider a connected undirected graph G of n nodes, m edges, diameter D and maximum degree Δ. For any node v, its degree is denoted by $\delta(v)$ or simply δ when there is no ambiguity. The nodes are anonymous, i.e., they do not have any labels. For every edge connected to a node, the node has a corresponding port number for the edge. For every node, the edges incident to the node are uniquely identified by port numbers in the range $[0, \delta - 1]$. There is no relation between the two port numbers of an edge. If u, v are two adjacent nodes then $\text{port}(u, v)$ denotes the port at u that corresponds to the edge between u and v.

Robots. Robots are anonymous, i.e., they do not have unique identifiers. Each robot has $O(\log \Delta)$ bits of space or memory for computation and to store information. Each robot has a fair coin which they can use to generate random bits. Each robot can communicate with other robots present at the same node by message passing: a robot can broadcast some message which is received by all robots present at the same node. The size of a message is no more than its memory size because it can not generate a message whose size is greater than its memory size. Therefore, the size of a message must be $O(\log \Delta)$. Also, when there are many robots (co-located at a node) broadcasting their messages, it is not possible for a robot to receive all of these messages due to limited memory. When there is not enough memory to receive all the messages, it receives only a subset of the messages. The view of a robot is local: the only things that a robot can 'see' when it is at some node, are the edges incident to it. The robots have access to the port numbers of these edges. It cannot 'see' the other robots that may be present at the same node. The only way it can detect the presence of other robots is by receiving messages that those robots may broadcast. The robots can move from one node to an adjacent node. Any number of robots are allowed to move via an edge. When a robot moves from a node u to node v, it is aware of the port through which it enter v.

Time Cycle. We assume a fully synchronous system. The time progresses in rounds. Each robot knows when a current round ends and when the next round starts. Each round consists of the following.

- The robots first performs a series of synchronous computations and communications. These are called *subrounds*. In each subround, a robot performs some local computations and then broadcasts some messages. The messages received in the ith subround are read in the $(i + 1)$th subround. The local computations are based on its memory contents (which contains the messages that it might have received in the last subround and other information that it had stored) and a random bit generated by the fair coin.
- Then robots move through some port or remains at the current node.

Problem Definition. A team of k $(\leq n)$ robots are initially at the same node of the graph G. The DISPERSION problem requires the robots to re-position themselves so that i) there is at most one robot at each node, and ii) all robots have terminated within a finite number of rounds.

3 The Algorithm

3.1 Local Leader Election

Before presenting our main algorithm, we give a brief description of the LEADERELECTION() subroutine. We adopt this subroutine from [26]. When $k \geq 1$ robots are co-located together at a node, LEADERELECTION() subroutine allows exactly one robot to be selected as the leader within one round. Formally, 1) if $k = 1$, the robot finds out that it is the only robot at the node, 2) if $k > 1$, after finitely many rounds (with high probability), i) exactly one robot is elected as leader, ii) all robots can detect when the process is completed. Each robot starts off as a candidate for leader. In the first subround, every robot broadcasts 'start'. If a robot finds that it has received no message, it then concludes that it is the only robot at the node. Otherwise, it concludes that there are multiple robots at the node and does the following. In each subsequent subround, each candidate flips a fair coin. If heads, it broadcasts 'heads', otherwise it does not broadcast anything. If a robot gets tails, and receives at least one ('heads') message, it stops being a candidate. This process is repeated until exactly one robot, say r, broadcasts in a given sub-round. In this subround, r broadcasts 'heads', but receives no message, while all other non-candidate robots have not broadcasted, but received exactly one message. So r elects itself as the leader, and all robots detect that the process is completed. The process requires $O(1)$ bits of memory at each robot and terminates in $O(\log k)$ subrounds with high probability.

3.2 Overview of the Algorithm

In this subsection, we present a brief overview of the algorithm. The execution of our algorithm can be divided into three stages. In the first stage, the robots, together as a group, perform a DFS traversal in the search of empty nodes, starting from the node where they are placed together initially. We shall call this node the *root* and denote it by v_R. Whenever the group reaches an empty node, they perform the LEADERELECTION() subroutine to elect a leader. The leader settles at that node, while the rest of the group continues the DFS traversal. Note that the settled robot does not terminate. This is because when the robots that are performing the DFS return to that node, they need to detect that the node is occupied by a settled robot. Recall that a robot cannot distinguish between an empty node and a node with a terminated robot. Therefore, the active settled robot helps the travelling robots to distinguish between an occupied node and an empty node, and also provides them with other information that are required to correctly execute the DFS. The size of the travelling group decreases by one, each time the DFS traversal reaches an empty node. The first stage completes when each robot has found an empty node for itself. Let r_L denote the last robot that finds an empty node, v_L, for itself. Although dispersion is achieved, this robot will not terminate. The other settled robots do not know that dispersion is achieved and will remain active. Therefore r_L needs to revisit those nodes and ask the settled robots to terminate. First r_L will return to the root v_R via the *rootpath* which is the unique path in the DFS tree from v_L to v_R. This is the second stage of the algorithm. In the third stage, r_L performs a second DFS traversal and asks the active settled robots to terminate. Since the active settled robots play a crucial role in the DFS traversal, r_L needs to be careful about the order in which it should ask the settled robots to terminate. Finally, r_L terminates after it returns to v_L.

In Table 1, we give details of the variables used by the robots. If *variable_name* is some variable, then we shall denote the value of the variable stored by r as *r.variable_name*.

Table 1. Description of the variables used by the robots

Variable	Description
role	It indicates the role that the robot is playing in the algorithm. It takes values from {`explore`, `settled`, `return`, `acknowledge`, `done`}. Initially, *role* ← `explore`
entered	It indicates the port through which the robot has entered the current node. Initially, *entered* ← ∅. For simplicity, assume that it is automatically updated when the robot entered a node
received	It indicates the message(s) received by the robot in the current subround. After the end of each subround, the messages are erased, i.e., it is reset to ∅. Initially, *received* ← ∅
direction	It indicates the direction of movement of a robot during a DFS traversal. It takes values from {`forward`, `backward`}. Initially, *role* ← `forward`
parent	For a settled robot on some node, it indicates the port number towards the parent of that node in the DFS tree. Initially, *parent* ← ∅
child	For a settled robot on some node that is on the rootpath, it indicates the port number towards the child of that node in the DFS tree that is on the rootpath. Initially, *child* ← ∅
visited	For a settled robot on some node, it indicates whether the node where the robot is settled has been visited by r_L in the third stage. Initially, *visited* ← 0

3.3 Detailed Description of the Algorithm

A pseudocode description of the algorithm is presented in the full version [8] of the paper. In the starting configuration, all robots are present at the root node

v_R. Initially, *role* of each robot is `explore`. In the first stage, the robots have to perform a DFS traversal together as a group. This group of robots is called the *exploring group*. Whenever the exploring group reaches an empty node (a node with no settled robot), one of the robots will settle at that node, i.e., it will change its *role* to `settled` and remain at that node. For the rest of the algorithm, it does not move. However, it stays active and checks for any received messages. A settled robot can receive three types of messages:

- it may receive a query about the contents of its internal memory
- it may receive an instruction to change the value of some variable
- it may be asked to terminate.

When queried about its memory, it broadcasts a message containing its *role*, *parent*, *child* and *visited*. If it is asked to change the value of some variable or terminate, then it does so accordingly. Any robot with role `explore`, `return` or `acknowledge`, in the first sub-round of any round, broadcasts a message querying about internal memory of any settled robot at the node. If it receives no message in the second sub-round, then it concludes that there is no settled robot at that node. Whenever the robots find that there is no settled robot at the node, during the first stage, they start the LeaderElection() subroutine to elect a leader. For any robot r, LeaderElection() results in one of the following outcomes:

- it is elected as the leader
- it is not elected as the leader
- it finds that it is the only robot at that node.

In the first case, it changes *r.role* to `settled` and sets *r.parent* equal to *r.entered*. Recall that *r.entered* is the port through which it entered the current node and in the beginning, *r.entered* is set to $\emptyset$. We shall call *r.parent* the *parent port* of the node where r resides. We shall refer to a robot that has set its *role* to `settled` as a *settled robot*. In the second case, it will continue the DFS: if $r.entered = \emptyset$, it leaves via port 0 and if $r.entered \neq \emptyset$, it leaves via port $(r.entered + 1) \bmod \delta$. If $(r.entered + 1) = r.entered \bmod \delta$, it changes its variable *direction* to `backward` before exiting the node. Recall that the variable *direction* is used to indicate the direction of the movement during a DFS traversal. In the third case, it changes *r.role* to `return`.

Now consider the case where the robots find that there is a settled robot at the node. If the *direction* is set to `forward` when they encounter the settled robot, it indicates the onset of a cycle. So the robots change the *direction* to `backward` and leave the node via the port through which they entered it. Now suppose that the *direction* is set to `backward` when they encounter the settled robot. Recall that the robots have received from the settled robot, say a, a message which contains *a.parent*. The robots check if *a.parent* is equal to the port number through which it entered, say z, plus 1 (modulo the degree of the node). If yes, it implies that the robots have moved through all edges adjacent to the node, and hence they leave the node via *a.parent* which is the port through which they entered for the first time. If no, then it means that they have not moved through the port $(z + 1)\text{mod}\delta$ before. So they change the direction to `forward` and leave via $(z + 1)\text{mod}\delta$.

The DFS traversal in the first stage ends when a robot, say r_L, with *role* set to `explore`, finds that it is the only robot at a node, say v_L. Recall that when this happens, r_L changes its *role* to `return`. At this point, the first stage ends, and the second stage starts. It then leaves v_L via the port through which it entered. In each of the following rounds where the *role* of r_L is `return`, it does the following. In the first subround, it broadcasts a query. In the next subround, it receives a message from the settled robot at that node which contains its *parent*. If the obtained value of *parent*, say x, is not $\emptyset$, it means that r_L is yet to reach the root v_R. Then r_L broadcasts an instruction for the settled robot to change the value of its *child* to the port via which r_L entered the node. This value of *child* will be called the *child port* of the node. After broadcasting the instruction, r_L leaves through the port x. If $x = \emptyset$, then it means that r_L has reached the root v_R. In this case, r_L broadcasts the same instruction and then changes the values of $r_L.role$, $r_L.direction$ and $r_L.entered$ to respectively `acknowledge`, `forward` and $\emptyset$. At this point the second stage ends, and the third stage starts.

In the following rounds, r_L with *role* `acknowledge` does the following. In the first sub-round, it broadcasts a query. It either receives a reply or does not. If it receives a message, then it contains the values of *parent*, say x, and *child*, say y, and *visited* of the settled robot at that node. Now, the value of variable *visited* can be 0 or 1. If the value of *visited* is 0, it denotes that the settled robot is visited for the first time in the third stage. The robot r_L then broadcasts a message instructing the settled robot to change the value of its variable *visited* to 1. Now $(r_L.entered + 1)\text{mod}\delta$ can be equal to $r_L.entered$ (the case of one degree node) or y or neither of them. In the former case, it changes its variable *direction* to `backward`. In the first two cases, it broadcasts a message instructing the settled robot to terminate and leaves through port $(r_L.entered + 1)\text{mod}\delta$. If $(r_L.entered + 1)\text{mod}\delta$ is neither equal to $r_L.entered$, nor equal to y, r_L just exits through $(r_L.entered+1)\text{mod}\delta$ without broadcasting any message for termination. If the value of *visited* is 1, it denotes that the settled robot has been visited before in the third stage. If the value of variable *direction* of r_L is `forward`, it changes the value of *direction* to `backward` and exits through the port through which it entered the node at the previous round. Otherwise the value of variable *direction* of r_L is `backward`. In this case, three sub-cases arise. If $(r_L.entered + 1)\text{mod}\delta$ is equal to x, then r_L broadcasts a message instructing the settled robot to terminate, and then r_L exits through the port $(r_L.entered + 1)\text{mod}\delta$. Otherwise if, $(r_L.entered + 1)\text{mod}\delta$ is equal to y, then also r_L broadcasts a message instructing the settled robot to terminate, changes the variable *direction* to `forward` and then r_L exits through the port $(r_L.entered+1)\text{mod}\delta$. If $(r_L.entered+1)\text{mod}\delta$ is neither equal to x, nor y, then r_L changes *direction* to `forward` and exits through $(r_L.entered+1)\text{mod}\delta$. Now, we consider the case where r_L in third stage does not receive any answer to its query. If its *direction* is set to `forward`, it changes its *direction* to `backward` and then exits through the same port by which it entered the node in the previous round. If its direction is set to `backward`, then it means that r_L was at v_L in the previous round. So r_L changes its *role* to `done` and leaves the node through the port via which it entered. Then it will reach v_L in the next round and it will find that its *role* is `done` and terminate.

4 Correctness Proof and Complexity Analysis

In this section, we give a brief outline of the proof. Readers are referred to the full version [8] of the paper for the detailed proof.

The first stage of our algorithm is the same as that of [26]. The robots simply perform a DFS traversal. Whenever a new node is visited, one of the robots settle there. The DFS continues until k distinct nodes are visited. To see that the DFS traversal can be correctly executed in our setting, it suffices to verify that the robots can correctly ascertain 1) if a node is previously visited and 2) if all neighbors of a node have been visited. For 1), observe that the presence of settled robot at a node indicates that the node has already been visited. So, when the robots with *direction* `forward` go to a node which has a settled robot, it backtracks, i.e., it changes its *direction* to `backward` and leaves the node via the port through which it had entered. For 2), observe that the port p through which robots first enters a node v is set as its parent port, i.e., the robot settled at v sets its variable *parent* to p. Then the robots will move through all other ports with *direction* `forward` in the order $p+1, p+2, \ldots, \delta-1, 0, 1, \ldots, p-1$ (unless the DFS is stopped midway for k distinct nodes have been visited). This is because if the robots leaves via a port q (with *direction* `forward`), it re-enters v via the same port q after some rounds (with *direction* `backward`) and then leaves via $(q+1)\text{mod}\delta(v)$ (with *direction* `forward`) in the next round if $(q+1)\text{mod}\delta(v) \neq p$. Clearly, when $(q+1)\text{mod}\delta(v) = p$, it indicates that the robots have moved through all ports other than p with *direction* `forward`, i.e., all neighbors of v have been visited. The robots can check if $(q+1)\text{mod}\delta(v) = p$ because their variable *entered* is equal to q and the variable *parent* of the robot settled at v is equal to p. Therefore, have the following result.

Theorem 1. *There is a round t_1, at the beginning of which*

1. *each node of G has at most one robot*
2. role *of exactly one robot r_L is* `explore` *and the role of the remaining $k-1$ robots is* `settled`
3. *if $V' \subseteq V$ is the set of nodes occupied by robots, then $G[V']$ (the subgraph of G induced by V') is connected*
4. *if $E' \subseteq E$ is the set of edges corresponding to the variable* parent *of robots in $\mathcal{R}\backslash\{r_L\}$ and variable* entered *of r_L, then the graph $T = T(V', E')$ is a DFS spanning tree of $G[V']$,*
5. *r_L is at a leaf node v_L of T.*

There is a unique path, i.e., the rootpath $v_R = v_1, v_2, \ldots, v_s = v_L$ in $T(V', E')$ from v_R to v_L. Furthermore, for any consecutive vertices v_i, v_{i+1} on the path 1) if $i+1 < s$, the variable *parent* of the settled robot at v_{i+1} is set to port(v_{i+1}, v_i) and 2) if $i+1 = s$, the variable *entered* of robot r_L at v_{i+1} is set to port(v_{i+1}, v_i). So, according to our algorithm, r_L will move along this path to reach v_R. For each node $v_i, i < s$, on the rootpath, when r_L reaches v_i along its way to v_R, it instructs the settled robot at v_i to set its variable *child* to port(v_i, v_{i+1}). Therefore, we have the following result.

Theorem 2. *There is a round t_2, at the beginning of which*

1. *r_L is at v_R with r_L.role =* **return**
2. *each node of $T(V', E') \backslash \{v_L\}$ has a settled robot*
3. *if $v_R = v_1, v_2, \ldots, v_s = v_L$ is the rootpath and r_i is the settled robot at $v_i, i < s$, then r_i.child = port(v_i, v_{i+1})*
4. *if r is a settled robot on a non-rootpath node, then r.child $= \emptyset$.*

From round $t_2 + 1$, r_L will start a second DFS traversal. This DFS traversal is trickier than the earlier one because the settled robots will one by one terminate during the process. Recall that the settled robots played important role in the first DFS. We shall prove that r_L will correctly execute the second DFS traversal. In fact, we shall prove in Lemma 1 (see the full version [8]) that the DFS traversal in the first stage is exactly same as the DFS traversal in the third stage in the sense that if the exploring group is at node v at round $i < t_1$ (in the first stage), then r_L is at node v at round $t_2 + i$ (in the third stage).

Let us first introduce a definition. In the following definition, whenever we say 'at round', it is to be understood as 'at the beginning of round'. Round i in the first stage is said to be *identical* to round j in the third stage if the exploring group at round i and r_L at round j are at the same node, say u and one of the following holds:

I1 At round i, there is no settled robot at u, the exploring group contains more than one robot and the variable *direction* for each robot in the exploring group is set to `forward`. At round j there is a settled robot at u with its variable *visited* set to 0. The variables *direction* and *entered* of r_L at round j are equal to those of each robot in the exploring group at round i.

I2 At round i, there is a settled robot at u, the variable *direction* for each robot in the exploring group is set to `forward` and the variable *entered* for each robot in the exploring group is $\neq \emptyset$. At round j, either there is a terminated robot at u, or there is a settled robot at u with its variable *visited* set to 1. The variables *direction* and *entered* of r_L at round j are equal to those of each robot in the exploring group at round i.

I3 At round i, there is a settled robot at u, the variable *direction* for each robot in the exploring group is set to `backward` and the variable *entered* for each robot in the exploring group is $\neq \emptyset$. At round j, there is an active settled robot at u with its variable *visited* set to 1. The variables *direction* and *entered* of r_L at round j are equal to those of each robot in the exploring group at round i.

Lemma 1. *Round i is identical to $t_2 + i$ for all $1 \leq i < t_1$.*

Theorem 3. *By round $t_2 + t_1$ all the settled robots have terminated.*

Proof. A settled robot at a non-rootpath node will terminate if r_L moves from that node to its parent and a settled robot at a rootpath node will terminate if r_L moves from that node to its child. Recall that if v is a non-rootpath node, then in Phase 1 the exploring group leaves it via its parent port once. Also, if v is a rootpath node other than v_L, then in Phase 1 the exploring group leaves it via its child port once. So the result follows Lemma 1.

Theorem 4. *At round $t_2 + t_1 + 2$, r_L terminates at v_L.*

Proof. It follows from Lemma 1 that r_L will be at v_{s-1} at round $t_2 + t_1 - 1$. It is easy to see that it will then move to $v_s = v_L$ with *direction* `forward`. So at round $t_2 + t_1$, r_L is at v_L with *direction* set to `forward`. Since v_L has no settled robot, r_L will change its *direction* to `backward` and exit through the port through which it entered v_L. But moving through this port leads to v_{s-1}. The settled robot at this node is already terminated by Theorem 3. Hence at round $t_2 + t_1 + 1$, r_L enters with *direction* `backward` a node where it does not receive any message. So it will then change its *role* to `done` and exit the port through which it entered the node. Therefore, at round $t_2 + t_1 + 2$, r_l again returns to v_L, this time with *role* `done`, and hence will terminate.

Theorem 5. *The algorithm is correct and requires $O(\log \Delta)$ bits of memory at each robot, which is optimal in terms of memory complexity. The worst case round complexity is $\Theta(k^2)$ rounds (assuming that* LeaderElection() *terminates each time).*

Proof. The correctness follows from the above results. The LeaderElection() subroutine costs $O(1)$ bits of memory for each robot. Among the variables, *role*, *visited* and *direction* costs $O(1)$ bits of memory, and the variables *entered*, *parent*, *child*, *received* costs $O(\log \Delta)$ bits of memory for each robot. Hence the algorithm requires $O(\log \Delta)$ bits of memory at each robot. The optimality follows from the lower bound result proved in [26].

Exactly k distinct nodes are visited in our algorithm. In the first stage, movement of the exploring group takes $O(m')$ rounds where m' is the number of edges in the subgraph of G induced by these k vertices. Clearly $m' = O(k^2)$. So the first stage requires $O(k^2)$ rounds. The third stage requires (k^2) rounds as well since apart from the last two rounds, it is exactly identical to the first stage. Clearly the second round takes $O(k)$ rounds. So the overall round complexity is $O(k^2)$. To see that our analysis is tight, we show an instance where $\Omega(k^2)$ rounds will be required. Consider the graph of size k in Fig. 1. Here port$(v_R, v_1) = 0$, port$(v_1, v_R) = 0$, port$(v_1, v_2) = 1$, port$(v_1, v_L) = 2$. Here, after reaching v_1, the exploring group will go v_2. It is easy to see that $\Theta(k^2)$ rounds will be spent inside the $(k-3)-$clique. Finally the last robot will return to v_1 and then move to v_L.

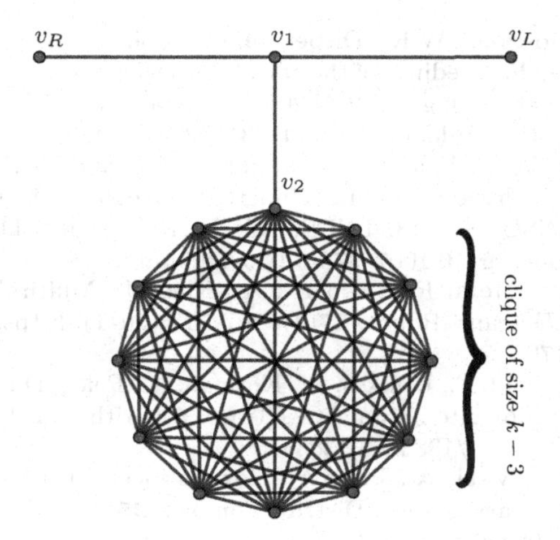

Fig. 1. An example where the algorithm requires $\Theta(k^2)$ rounds to complete.

5 Concluding Remarks

We have presented a memory optimal randomized algorithm for DISPERSION from rooted configuration by anonymous robots. This resolves an open problem posed in [26]. Time complexity of our algorithm is $\Theta(k^2)$ rounds in the worst case (assuming that LEADERELECTION() terminates each time). Any algorithm that solves the problem requires $\Omega(k)$ rounds in the worst case. To see this, consider a path with $n \geq k$ nodes with all robots initially at one of its one degree nodes. An interesting open problem is to close this gap.

For arbitrary configuration, the random walk based algorithm presented in [26] requires the robots to stay active indefinitely. Therefore an interesting open question is whether it is possible to solve the problem by anonymous robots from non-rooted configurations without requiring robots to stay active indefinitely.

Acknowledgement. We would like to thank Pritam Goswami for valuable discussions. The first two authors are supported by UGC, Govt. of India, and NBHM DAE, Govt. of India respectively. We would like to thank the anonymous reviewers for their valuable comments which helped us to improve the quality and presentation of the paper.

References

1. Agarwalla, A., Augustine, J., Moses Jr, W.K., Madhav, S.K., Sridhar, A.K.: Deterministic dispersion of mobile robots in dynamic rings. In: Bellavista, P., Garg, V.K. (eds.) Proceedings of the 19th International Conference on Distributed Computing and Networking, ICDCN 2018, Varanasi, India, January 4–7, 2018, pp. 19:1–19:4. ACM (2018). https://doi.org/10.1145/3154273.3154294

2. Augustine, J., Moses Jr, W.K.: Dispersion of mobile robots: a study of memory-time trade-offs. In: Proceedings of the 19th International Conference on Distributed Computing and Networking, ICDCN 2018, Varanasi, India, January 4–7, 2018, pp. 1:1–1:10 (2018). https://doi.org/10.1145/3154273.3154293
3. Bose, K., Kundu, M.K., Adhikary, R., Sau, B.: Optimal gathering by asynchronous oblivious robots in hypercubes. In: Gilbert, S., Hughes, D., Krishnamachari, B. (eds.) ALGOSENSORS 2018. LNCS, vol. 11410, pp. 102–117. Springer, Cham (2019). https://doi.org/10.1007/978-3-030-14094-6_7
4. Brass, P., Cabrera-Mora, F., Gasparri, A., Xiao, J.: Multirobot tree and graph exploration. IEEE Trans. Robot. **27**(4), 707–717 (2011). https://doi.org/10.1109/TRO.2011.2121170
5. Cohen, R., Fraigniaud, P., Ilcinkas, D., Korman, A., Peleg, D.: Label-guided graph exploration by a finite automaton. ACM Trans. Algorithms **4**(4), 42:1–42:18 (2008). https://doi.org/10.1145/1383369.1383373
6. Czyzowicz, J., Kosowski, A., Pelc, A.: How to meet when you forget: log-space rendezvous in arbitrary graphs. Distrib. Comput. **25**(2), 165–178 (2012). https://doi.org/10.1007/s00446-011-0141-9
7. D'Angelo, G., Stefano, G.D., Navarra, A.: Gathering on rings under the look-compute-move model. Distrib. Comput. **27**(4), 255–285 (2014). https://doi.org/10.1007/s00446-014-0212-9
8. Das, A., Bose, K., Sau, B.: Memory optimal dispersion by anonymous mobile robots. CoRR abs/2008.00701 (2020). https://arxiv.org/abs/2008.00701
9. Das, S., Dereniowski, D., Karousatou, C.: Collaborative exploration of trees by energy-constrained mobile robots. Theory Comput. Syst. **62**(5), 1223–1240 (2018). https://doi.org/10.1007/s00224-017-9816-3
10. Dereniowski, D., Disser, Y., Kosowski, A., Pajak, D., Uznanski, P.: Fast collaborative graph exploration. Inf. Comput. **243**, 37–49 (2015). https://doi.org/10.1016/j.ic.2014.12.005
11. Dieudonné, Y., Pelc, A.: Anonymous meeting in networks. Algorithmica **74**(2), 908–946 (2016). https://doi.org/10.1007/s00453-015-9982-0
12. Diks, K., Fraigniaud, P., Kranakis, E., Pelc, A.: Tree exploration with little memory. J. Algorithms **51**(1), 38–63 (2004). https://doi.org/10.1016/j.jalgor.2003.10.002
13. Duncan, C.A., Kobourov, S.G., Kumar, V.S.A.: Optimal constrained graph exploration. ACM Trans. Algorithms **2**(3), 380–402 (2006). https://doi.org/10.1145/1159892.1159897
14. Izumi, T., Izumi, T., Kamei, S., Ooshita, F.: Mobile robots gathering algorithm with local weak multiplicity in rings. In: Patt-Shamir, B., Ekim, T. (eds.) SIROCCO 2010. LNCS, vol. 6058, pp. 101–113. Springer, Heidelberg (2010). https://doi.org/10.1007/978-3-642-13284-1_9
15. Kamei, S., Lamani, A., Ooshita, F., Tixeuil, S., Wada, K.: Gathering on rings for myopic asynchronous robots with lights. In: 23rd International Conference on Principles of Distributed Systems, OPODIS 2019, December 17–19, 2019, Neuchâtel, Switzerland. LIPIcs, vol. 153, pp. 27:1–27:17. Schloss Dagstuhl - Leibniz-Zentrum für Informatik (2019). https://doi.org/10.4230/LIPIcs.OPODIS.2019.27
16. Klasing, R., Kosowski, A., Navarra, A.: Taking advantage of symmetries: gathering of many asynchronous oblivious robots on a ring. Theor. Comput. Sci. **411**(34–36), 3235–3246 (2010). https://doi.org/10.1016/j.tcs.2010.05.020
17. Klasing, R., Markou, E., Pelc, A.: Gathering asynchronous oblivious mobile robots in a ring. Theor. Comput. Sci. **390**(1), 27–39 (2008). https://doi.org/10.1016/j.tcs.2007.09.032

18. Kowalski, D.R., Malinowski, A.: How to meet in anonymous network. Theor. Comput. Sci. **399**(1–2), 141–156 (2008). https://doi.org/10.1016/j.tcs.2008.02.010
19. Kshemkalyani, A.D., Ali, F.: Efficient dispersion of mobile robots on graphs. In: Proceedings of the 20th International Conference on Distributed Computing and Networking, ICDCN 2019, Bangalore, India, January 04–07, 2019, pp. 218–227 (2019). https://doi.org/10.1145/3288599.3288610
20. Kshemkalyani, A.D., Molla, A.R., Sharma, G.: Fast dispersion of mobile robots on arbitrary graphs. In: Algorithms for Sensor Systems - 15th International Symposium on Algorithms and Experiments for Wireless Sensor Networks, ALGOSENSORS 2019, Munich, Germany, September 12–13, 2019, Revised Selected Papers, pp. 23–40 (2019). https://doi.org/10.1007/978-3-030-34405-4_2
21. Kshemkalyani, A.D., Molla, A.R., Sharma, G.: Dispersion of mobile robots in the global communication model. In: Mukherjee, N., Pemmaraju, S.V. (eds.) ICDCN 2020: 21st International Conference on Distributed Computing and Networking, Kolkata, India, January 4–7, 2020, pp. 12:1–12:10. ACM (2020). https://doi.org/10.1145/3369740.3369775
22. Kshemkalyani, A.D., Molla, A.R., Sharma, G.: Dispersion of mobile robots on grids. In: Rahman, M.S., Sadakane, K., Sung, W.-K. (eds.) WALCOM 2020. LNCS, vol. 12049, pp. 183–197. Springer, Cham (2020). https://doi.org/10.1007/978-3-030-39881-1_16
23. Luna, G.A.D., Flocchini, P., Pagli, L., Prencipe, G., Santoro, N., Viglietta, G.: Gathering in dynamic rings. Theor. Comput. Sci. **811**, 79–98 (2020). https://doi.org/10.1016/j.tcs.2018.10.018
24. Miller, A., Pelc, A.: Fast rendezvous with advice. Theor. Comput. Sci. **608**, 190–198 (2015). https://doi.org/10.1016/j.tcs.2015.09.025
25. Miller, A., Pelc, A.: Time versus cost tradeoffs for deterministic rendezvous in networks. Distrib. Comput. **29**(1), 51–64 (2016). https://doi.org/10.1007/s00446-015-0253-8
26. Molla, A.R., Moses, W.K.: Dispersion of mobile robots: the power of randomness. In: Gopal, T.V., Watada, J. (eds.) TAMC 2019. LNCS, vol. 11436, pp. 481–500. Springer, Cham (2019). https://doi.org/10.1007/978-3-030-14812-6_30
27. Molla, A.R., Mondal, K., Moses, W.K.: Efficient dispersion on an anonymous ring in the presence of weak Byzantine robots. In: Pinotti, C.M., Navarra, A., Bagchi, A. (eds.) ALGOSENSORS 2020. LNCS, vol. 12503, pp. 154–169. Springer, Cham (2020). https://doi.org/10.1007/978-3-030-62401-9_11
28. Panaite, P., Pelc, A.: Exploring unknown undirected graphs. J. Algorithms **33**(2), 281–295 (1999). https://doi.org/10.1006/jagm.1999.1043
29. Stefano, G.D., Navarra, A.: Optimal gathering of oblivious robots in anonymous graphs and its application on trees and rings. Distrib. Comput. **30**(2), 75–86 (2017). https://doi.org/10.1007/s00446-016-0278-7
30. Ta-Shma, A., Zwick, U.: Deterministic rendezvous, treasure hunts, and strongly universal exploration sequences. ACM Trans. Algorithms **10**(3), 12:1–12:15 (2014). https://doi.org/10.1145/2601068

Quantum and Approximation Algorithms for Maximum Witnesses of Boolean Matrix Products

Mirosław Kowaluk[1(✉)] and Andrzej Lingas[2]

[1] Institute of Informatics, University of Warsaw, Warsaw, Poland
kowaluk@mimuw.edu.pl

[2] Department of Computer Science, Lund University, 22100 Lund, Sweden
Andrzej.Lingas@cs.lth.se

Abstract. The problem of finding maximum (or minimum) witnesses of the Boolean product of two Boolean matrices (MW for short) has a number of important applications, in particular the all-pairs lowest common ancestor (LCA) problem in directed acyclic graphs (dags). The best known upper time-bound on the MW problem for $n \times n$ Boolean matrices of the form $O(n^{2.575})$ has not been substantially improved since 2006. In order to obtain faster algorithms for this problem, we study quantum algorithms for MW and approximation algorithms for MW (in the standard computational model). Some of our quantum algorithms are input or output sensitive. Our fastest quantum algorithm for the MW problem, and consequently for the related problems, runs in time $\tilde{O}(n^{2+\lambda/2}) = \tilde{O}(n^{2.434})$, where λ satisfies the equation $\omega(1, \lambda, 1) = 1 + 1.5\,\lambda$ and $\omega(1, \lambda, 1)$ is the exponent of the multiplication of an $n \times n^{\lambda}$ matrix by an $n^{\lambda} \times n$ matrix. Next, we consider a relaxed version of the MW problem (in the standard model) asking for reporting a witness of bounded rank (the maximum witness has rank 1) for each non-zero entry of the matrix product. First, by adapting the fastest known algorithm for maximum witnesses, we obtain an algorithm for the relaxed problem that reports for each non-zero entry of the product matrix a witness of rank at most ℓ in time $\tilde{O}((n/\ell)n^{\omega(1,\log_n \ell,1)})$. Then, by reducing the relaxed problem to the so called k-witness problem, we provide an algorithm that reports for each non-zero entry $C[i,j]$ of the product matrix C a witness of rank $O(\lceil W_C(i,j)/k \rceil)$, where $W_C(i,j)$ is the number of witnesses for $C[i,j]$, with high probability. The algorithm runs in $\tilde{O}(n^{\omega}k^{0.4653} + n^{2+o(1)}k)$ time, where $\omega = \omega(1,1,1)$.

1 Introduction

If A and B are two $n \times n$ Boolean matrices and C is their Boolean matrix product then for any entry $C[i,j] = 1$ of C, a *witness* is an index k such that $A[i,k] \wedge B[k,j] = 1$. The largest (or, smallest) possible witness for an entry is called the *maximum witness* (or *minimum witness*, respectively) for the entry.

The problem of finding "witnesses" of Boolean matrix product has been studied for decades mostly because of its applications to shortest-path problems [1, 2].

A. Mudgal and C. R. Subramanian (Eds.): CALDAM 2021, LNCS 12601, pp. 440–451, 2021.
https://doi.org/10.1007/978-3-030-67899-9_35

The problem of finding maximum witnesses of Boolean matrix product (MW for short) has been studied first in [6] in order to obtain faster algorithms for all-pairs lowest common ancestor (LCA) problem in directed acyclic graphs (dags) [9]. It has found many other applications since then including the all-pairs bottleneck weight path problem [18] and finding for a set of edges in a vertex-weighted graph heaviest triangles including an edge from the set [19]. The fastest known algorithm for the MW problem and the aforementioned problems runs in $O(n^{2+\lambda})$ time [6], where λ satisfies the equation $\omega(1, \lambda, 1) = 1 + 2\lambda$. The currently best bounds on $\omega(1, \lambda, 1)$ follow from a fact in [4,11] (see Fact 4 in Preliminaries) combined with the recent improved estimations on the parameters $\omega = \omega(1, 1, 1)$ and α (see Preliminaries) [13,15]. They yield an $O(n^{2.569})$ upper bound on the running time of the algorithm (originally, $O(n^{2.575})$ [6]). For faster algorithms in sparse cases, see [5].

In this paper, we study two different approaches to deriving faster algorithms for the problem of computing maximum (or minimum) witnesses of the Boolean product of two $n \times n$ Boolean matrices (MW for short). The first approach is to consider the MW problem in the more powerful model of quantum computation. The other approach is to relax the MW problem (in the standard model) by allowing its approximation.

In the first part of our paper, we present quantum algorithms for the MW problem assuming a Quantum Random Access Machine (QRAM) model [17]. First, we consider a straightforward algorithm for MW that uses the quantum minimum search due to Dürr and Høyer [7] for each entry of the product matrix separately in order to find its maximum witness[1]. It runs in $\tilde{O}(n^{2.5})$ time. By adding as a preprocessing a known output-sensitive quantum algorithm for Boolean matrix product, we obtain an output-sensitive quantum algorithm for MW running in $\tilde{O}(n\sqrt{s} + s\sqrt{n})$ time, where s is the number of non-zero entries in the product matrix. By refining the straightforward algorithm in a different way, we obtain also an input-sensitive quantum algorithm for MW running in $\tilde{O}(n^2 + n^{1.5}m^{0.5})$ time, where m is the number of non-zero entries in the sparsest among the two input matrices. Then, we combine the idea of multiplication of rectangular submatrices of the input Boolean matrices with that of using the quantum minimum search of Dürr and Høyer in order to obtain our fastest quantum algorithm for MW running in $\tilde{O}(n^{2+\lambda/2})$ time, where λ satisfies the equation $\omega(1, \lambda, 1) = 1 + 1.5\lambda$. By the currently best bounds on $\omega(1, \lambda, 1)$, the running time of our algorithm is $O(n^{2.434})$. We obtain the same asymptotic upper time-bounds for the aforementioned problems related to MW.

In the second part of our paper, we consider a relaxed version of the MW problem (in the standard model) asking for reporting a witness of bounded rank (the maximum witness has rank 1) for each non-zero entry of the matrix product. First, by adapting the fastest known algorithm for maximum witnesses, we obtain an algorithm for the relaxed problem that reports for each non-zero entry of the product matrix a witness of rank at most ℓ in time $\tilde{O}((n/\ell)n^{\omega(1,\log_n \ell,1)})$.

[1] For somewhat related applications of the quantum minimum search of Dürr and Høyer to shortest path problems see [16].

Then, by reducing the relaxed problem to the so called k-witness problem, we provide an algorithm that reports for each non-zero entry $C[i,j]$ of the product matrix C a witness of rank $O(\lceil W_C(i,j)/k \rceil)$ with high probability, where $W_C(i,j)$ is the number of witnesses for $C[i,j]$. The algorithm runs in $\tilde{O}(n^{\omega}k^{0.4653} + n^{2+o(1)}k)$ time.

Organization. In Preliminaries, we provide some basic notions and/or facts on matrix multiplication and quantum computation. In Sect. 3, we present our basic procedure for searching an interval of indices for the maximum witness, the straightforward quantum algorithm for MW implied by the procedure, and the output-sensitive and input-sensitive refinements of the algorithm. In Sect. 4, we present and analyze our fastest in the general case quantum algorithm for MW. In Sect. 5, we present our approximation algorithms for MW in the standard computational model. For applications of our quantum algorithms to the problems related to MW as well as additional remarks the reader is referred to the full version [12].

2 Preliminaries

For a positive integer r, we shall denote the set of positive integers not greater than r by $[r]$.

For a matrix D, D^t denotes its transpose.

A *witness* for a non-zero entry $C[i,j]$ of the Boolean matrix product C of a Boolean $p \times q$ matrix A and a Boolean $q \times r$ matrix B is any index $k \in [q]$ such that $A[i,k]$ and $B[k,j]$ are equal to 1. The number of witnesses for $C[i,j]$ is denoted by $W_C(i,j)$. A witness k for $C[i,j]$ is of rank h if there are exactly $h-1$ witnesses for this entry greater than k. The witness of rank 1 is the *maximum witness* for $C[i,j]$. The *witness problem* is to report a witness for each non-zero entry of the Boolean matrix product of the two input matrices. The *maximum witness problem* (MW) is to report the maximum witness for each non-zero entry of the Boolean matrix product of the two input matrices.

Recall that for natural numbers p, q, r, $\omega(p,q,r)$ denotes the exponent of fast matrix multiplication for rectangular matrices $n^p \times n^q$ and $n^q \times n^r$, respectively. The following recent upper bound on $\omega = \omega(1,1,1)$ is due to Le Gall [13].

Fact 1. *The fast matrix multiplication algorithm for $n \times n$ matrices runs in $O(n^{\omega})$ time, where ω is not greater than 2.3728639 [13] (cf. [20]).*

Alon and Naor provided almost equally fast solution to the witness problem for square Boolean matrices in [2]. It can be easily generalized to include the Boolean product of two rectangular Boolean matrices of sizes $n \times n^q$ and $n^q \times n$, respectively. The asymptotic matrix multiplication time n^{ω} is replaced by $n^{\omega(1,q,1)}$ in the generalization.

Fact 2. *For $q \in (0,1]$, the witness problem for the Boolean matrix product of an $n \times n^q$ Boolean matrix with an $n^q \times n$ Boolean matrix can be solved (deterministically) in $\tilde{O}(n^{\omega(1,q,1)})$ time.*

Let α stand for $sup\{0 \leq r \leq 1 : \omega(1,r,1) = 2 + o(1)\}$. The following recent lower bound on α is due to Le Gall and Urrutia [15].

Fact 3. *The inequality* $\alpha > 0.31389$ *holds [15].*

Coppersmith [4] and Huang and Pan [11] proved the following fact.

Fact 4. *The inequality* $\omega(1,r,1) \leq \beta(r)$ *holds, where* $\beta(r) = 2 + o(1)$ *for* $r \in [0,\alpha]$ *and* $\beta(r) = 2 + \frac{\omega-2}{1-\alpha}(r-\alpha) + o(1)$ *for* $r \in [\alpha, 1]$ *[4,11].*

It will be the most convenient to formulate our quantum algorithms in the model of Quantum Random Access Machine (QRAM) saving the reader a lot of technical details of alternative formulations in the quantum circuit model [17]. Thus, our quantum algorithm can access any entry of any input matrix A in an access random manner (cf. [14,16]). More precisely, following [16], we assume that there is an oracle O_A which for $i,\ j \in [n]$ and $z \in \{0,\ 1\}^*$, maps the state $|i\rangle|j\rangle|0\rangle|z\rangle$ into the state $|i\rangle|j\rangle|A[i,j]\rangle|z\rangle$. When a whole table T is stored in the random access memory of QRAM such an oracle o_T corresponding to T is implicit. We shall estimate the time complexity of our quantum algorithms in the unit cost model, in particular we shall assign unit cost to each call to an oracle. In case the time complexity of our quantum algorithm exceeds the size of the input matrices, we may assume w.l.o.g. that the input matrices are just read into the QRAM memory.

Following Le Gall [14], we can generalize the definition of a quantum algorithm for Boolean matrix product to include the MW problem.

Definition 1. *A quantum algorithm for witnesses of Boolean matrix product (or the MW problem) is a quantum algorithm that when given access to oracles* O_A *and* O_B *corresponding to Boolean matrices* A *and* B*, computes with probability at least* $2/3$ *all the non-zero entries of the product* $A \times B$ *along with one witness (the maximum witness, respectively) for each non-zero entry.*

Note the probability of at least $\frac{2}{3}$ can be enhanced to at least $1 - n^{-\gamma}$, for $\gamma \geq 1$, by iterating the algorithm $O(\log n)$ times. When the size of the input is bounded by $poly(n)$, one uses the term *almost certainly* for the latter probability.

In fact, all our quantum algorithms for MW but the output sensitive one report also "No" for each zero entry of the product matrix.

3 Quantum Search for the Maximum Witness

One can find the maximum witness for a given entry of the Boolean product of two Boolean $n \times n$ matrices in $\tilde{O}(\sqrt{n})$ time with high probability by recursively using Grover's quantum search [10] interleaved with a binary search. However, the most convenient is to use a specialized variant of Grover's search due to Dürr and Høyer [3,7] for finding an entry of the minimum value in a table.

Fact 5 *(Dürr and Høyer [7]). Let $T[k]$, $1 \leq k \leq n$ be an unsorted table where all values are distinct. Given an oracle for T, the index k for which $T[k]$ is minimum can be found by a quantum algorithm with probability at least $\frac{1}{2}$ in $O(\sqrt{n})$ time.*

Using this fact, we can design the following procedure $MaxWit(A, B, i, j)$ returning the maximum witness of the entry $C[i, j]$ (if any) of the product C of two Boolean $n \times n$ matrices A and B.

Procedure. $MaxWit(A, B, i, j)$

Input: oracles corresponding to a Boolean $p \times q$ matrix A and a Boolean $q \times r$ matrix B, and indices $i \in [p]$, $j \in [r]$.
Output: if the $C[i, j]$ entry of the Boolean product C of A and B has a witness then its maximum witness in $[q]$ otherwise "No".

1. $n \leftarrow \max\{p, q, r\}$
2. Define an oracle for a virtual, one-dimensional integer table $T[k]$, $k \in [q]$ by $T[k] = 2n - A[i, k]B[k, j]n - k$.
3. Iterate $O(\log n)$ times the algorithm of Dürr and Høyer for T and set k' to the index minimizing T.
4. If $T[k'] < n$ then return k' as the maximum witness otherwise return "No".

Lemma 1. *Let β be a positive integer. By repetitively using the algorithm of Dürr and Høyer, $MaxWit(A, B, i, j)$ can be implemented in $\tilde{O}(\beta\sqrt{n})$ time such that it returns a correct answer with probability at least $1 - n^{-\beta}$.*

Proof. To begin with observe that for k, $k' \in [q]$, if $k \neq k'$ then $T[k] \neq T[k']$. This obviously holds for k, $k' \in [q]$ if $A[i, k]B[k, j] = A[i, k']B[k', j]$ as well when $A[i, k]B[k, j] \neq A[i, k']B[k', j]$. Furthermore, the value of $T[k]$ can be computed with the help of the oracles for A, B in constant time in the QRAM model. Next, suppose that the minimum value of T is achieved for the index k'. It is easily seen that if $T[k'] < n$ then k' is the maximum witness of $C[i, j]$ and otherwise $C[i, j]$ does not have any witness. By running the minimum search algorithm of Dürr and Høyer $O(\beta \log n)$ times, we can identify the maximum witness of $C[i, j]$ with probability at least $1 - n^{-\beta}$ in $\tilde{O}(\beta\sqrt{n})$ time. □

By Lemma 1, a straightforward $\tilde{O}(n^{2.5})$-time method for MW is just to run the procedure $MaxWit(A, B, i, j)$ with appropriately large constant β for each entry $C[i, j]$ of the product matrix C. We shall refer to this method as Algorithm 1. (see the full version [12] for a pseudo-code). Note that this returns also "No" for zero entries of C.

An Output-Sensitive Algorithm for MW. By adding as a preprocessing a known output-sensitive quantum algorithm for the Boolean product of the matrices A and B, Algorithm 1 can be transformed into an output-sensitive one too.

Algorithm 2

Input: oracles corresponding to Boolean $n \times n$ matrices A, B.

Output: maximum witnesses for all non-zero entries of the Boolean product of A and B.

1. Run an output-sensitive quantum algorithm for the Boolean product C of A and B;
2. **for** all non-zero entries $C[i, j]$ **do**
   $MaxWit(A, B, i, j)$

Theorem 1. *The MW problem can be solved by a quantum algorithm in* $\tilde{O}(n\sqrt{s} + s\sqrt{n})$ *time, where* s *is the number of non-zero entries in the product.*

Proof. Consider Algorithm 2. Due to Step 1, the procedure $MaxWit$ is called only for non-zero entries of C. Hence, the total time taken by Step 2 is $\tilde{O}(s\sqrt{n})$ by Lemma 1 with any fixed β. It is sufficient now to plug in the output-sensitive quantum algorithm for Boolean matrix product due to Le Gall [14] running in $\tilde{O}(n\sqrt{s} + s\sqrt{n})$ time to implement Step 1. In order to obtain enough large probability of the correctness of the whole output, we can iterate the plug in algorithm a logarithmic number of times and pick enough large β in Lemma 1. We obtain the output-sensitive upper bound claimed in the theorem. □

An Input-Sensitive Algorithm for MW. We can also refine the straightforward quantum algorithm for MW in order to obtain an input-sensitive quantum algorithm for MW.

Algorithm 3

Input: Boolean $n \times n$ matrices A, B,
Output: maximum witnesses for all non-zero entries of the Boolean product of A and B and "No" for all zero entries of the product.

1. For each column j of the matrix B compute the sequence K_j of indices $k \in [n]$ in decreasing order such that $B[k, j] = 1$ by using the oracle for the matrix B. Construct a one dimensional integer table S_j of length $|K_j|$ such that for $s \in [|K_j|]$, $S_j[s]$ is the s-th largest element in K_j.
2. **for** all i, $j \in [n]$ **do**
   (a) Define an oracle for a virtual, one-dimensional integer table $T_{i,j}$ of length $|K_j|$ such that for $s \in [|K_j|]$, $T_{i,j}[s] = 2n - A[i, S_j[s]]n - S_j[s]$. (The value of $T_{i,j}[s]$ can be retrieved in constant time by using the oracle for the matrix A and the table S_j.)
   (b) Iterate $O(\log n)$ times the algorithm of Dürr and Høyer for $T_{i,j}$ and set s' to the index minimizing $T_{i,j}$.
   (c) If $T_{i,j}[s'] < n$ then return $S_j[s']$ (i.e., $n - T_{i,j}[s']$) as the maximum witness for $C[i, j]$ otherwise return "No" for $C[i, j]$.

Theorem 2. *The MW problem for the Boolean product of two Boolean* $n \times n$ *matrices, with* m_1 *and* m_2 *non-zero entries respectively, admits a quantum algorithm running in* $\tilde{O}(n^2 + n^{1.5}\sqrt{\min\{m_1, m_2\}})$ *time.*

Proof. Consider Algorithm 3. Its correctness follows from the definition of the tables $T_{i,j}$, in particular the fact that each of them has distinct values. Let us estimate the time complexity of Algorithm 3. We may assume w.l.o.g. that the number of non-zero entries in the matrix B is m_2. Steps 1, 2(a) and 2(c) can be easily done in $\tilde{O}(n^2)$ total time. In Step 2(b), computing the maximum witnesses for the entries in the i-th row of the product matrix takes $\tilde{O}(\sum_{j=1}^{n}\sqrt{|K_j|})$ time by Fact 5. Since $\sum_{j=1}^{n}|K_j| \le m_2$ and the arithmetic mean does not exceed the quadratic one, we obtain $\sum_{j=1}^{n}\sqrt{|K_j|} \le n\sqrt{\frac{m_2}{n}}$. Consequently, Algorithm 2 runs in $\tilde{O}(n^2 + n^2\sqrt{\frac{m_2}{n}})$ time.

As in case of Algorithms 1 and 2, we can pick enough large constant at $\log n$ in the upper bound on the number of iterations of the algorithm of Dürr and Høyer in order to guarantee that the whole output of Algorithm 3 is correct with probability at least $\frac{2}{3}$. Hence, by the time analysis of Algorithm 3 and $A \times B = (B^t \times A^t)^t$, we obtain the theorem. □

4 The Fastest Method: Combining Rectangular Boolean Matrix Multiplication with Quantum Search

The best known algorithm for MW from [6] relies on the multiplication of rectangular submatrices of the input matrices. We can combine this idea with that of our procedure $MaxWit$ based on the quantum search for the minimum in order to obtain our fastest quantum algorithm for MW.

Algorithm 4

Input : oracles corresponding to Boolean $n \times n$ matrices A, B, and a parameter $\ell \in [n]$.
Output: maximum witnesses for all non-zero entries of the Boolean product of A and B, and "No" for all zero entries of the product.

1. Divide A into $\lceil n/\ell \rceil$ vertical strip submatrices $A_1, ..., A_{\lceil n/\ell \rceil}$ of width ℓ with the exception of the last one that can have width $\le \ell$.
2. Divide B into $\lceil n/\ell \rceil$ horizontal strip submatrices $B_1, ..., B_{\lceil n/\ell \rceil}$ of width ℓ with the exception of the last one that can have width $\le \ell$.
3. **for** $p \in [\lceil n/\ell \rceil]$ compute the Boolean product C_p of A_p and B_p
4. **for** all $i,\ j \in [n]$ **do**
   (a) Find the largest p such that $C_p[i,j] = 1$ or set $p = 0$ if it does not exist
   (b) **if** $p > 0$ **then** return $\ell(p-1) + MaxWit(A_p, B_p, i, j)$ **else** return "No".

Lemma 2. *Algorithm 4 runs in time* $\tilde{O}((n/\ell)n^{\omega(1,\log_n \ell,1)} + n^3/\ell + n^2\sqrt{\ell})$.

Proof. Steps 1, 2, take $O(n^2)$ time. Step 3 requires $O((n/\ell)n^{\omega(1,\log_n \ell,1)})$ time. Step 4(a) takes $O(n^2 \times n/\ell)$ time totally. Finally, Step 4(b) requires $\tilde{O}(n^2\sqrt{\ell})$ time totally by Lemma 1. □

By Lemma 1 with sufficiently large β and the time analysis in Lemma 2, we can obtain trade-offs between preprocessing time and answering a maximum witness query time depending on ℓ. See Theorem 3 in the full version [12].

Finding ℓ Minimizing the Total Time. By Lemma 2, the total time taken by Algorithm 4 for maximum witnesses is

$$\tilde{O}((n/\ell) \cdot n^{\omega(1,\log_n \ell,1)} + n^3/\ell + n^2\sqrt{\ell}).$$

By setting r to $\log_n \ell$ our upper bound transforms to $\tilde{O}(n^{1-r+\omega(1,r,1)} + n^{3-r} + n^{2+r/2})$. Note that by assuming $r \geq \frac{2}{3}$, we can get rid of the additive n^{3-r} term. Hence, by solving the equation $1 - \lambda + \omega(1, \lambda, 1) = 2 + \lambda/2$ implying $\lambda \geq \frac{2}{3}$ by $\omega(1, \lambda, 1) \geq 2$ and setting sufficiently large β in Lemma 1, we obtain our main result.

Theorem 3. *Let λ be such that $\omega(1, \lambda, 1) = 1 + 1.5\,\lambda$. The maximum witnesses for all non-zero entries of the Boolean product of two $n \times n$ Boolean matrices can be computed almost certainly by a quantum algorithm in $\tilde{O}(n^{2+\lambda/2})$ time.*

Note that by Fact 4, the solution λ of the equation $\omega(1, \lambda, 1) = 1 + 1.5\,\lambda$ is satisfied by $\lambda = \frac{1-\alpha\,(\omega-1)}{1.5\,(1-\alpha)-(\omega-2)} + o(1)$. Note also that λ is increasing in ω and decreasing in α. Hence, the inequality $\lambda < 0.8671$ holds by Fact 1 and Fact 3. We obtain the following concrete corollary.

Corollary 1. *The maximum witnesses for all non-zero entries of the Boolean product of two $n \times n$ Boolean matrices can be computed almost certainly by a quantum algorithm in $\tilde{O}(n^{2.4335})$ time.*

For the applications of our quantum algorithms to the related important problems in graph algorithms listed in Introduction see the full version [12].

5 Approximation Algorithms

In this section, we present two approximation approaches to MW in a standard computational model. The first approach follows the idea of the fastest known algorithm for MW [6] but instead of searching the final index intervals where the respective maximum witnesses are localized some witnesses from the intervals are reported. The second approach relies on the repetitive applying the deterministic algorithm for multiple witnesses from [8] and the goodness of its approximation for a matrix product entry depends on the number of witnesses for the entry.

The Method Based on Rectangular Matrix Multiplication. By slightly modifying the algorithm for MW [6] (or, the quantum Algorithm 4) based on fast rectangular multiplication, we can obtain a faster approximation algorithm. For a given ℓ, it reports for each non-zero entry of the Boolean matrix product a witness of rank not exceeding ℓ instead of the maximum witness. In the time analysis of the approximation algorithm, we rely on the fact that witnesses for non-zero entries of the Boolean product of two Boolean matrices can be reported in time proportional to the time taken by fast Boolean matrix multiplication up to polylogarithmic factors (see Fact 2).

Algorithm 5

Input : Boolean $n \times n$ matrices A, B, and a parameter $\ell \in [n]$.
Output: witnesses for all non-zero entries of the Boolean product of A and B having rank not exceeding ℓ and "No" for all zero entries of the product.

1. Divide A into $\lceil n/\ell \rceil$ vertical strip submatrices $A_1, ..., A_{\lceil n/\ell \rceil}$ of width ℓ with the exception of the last one that can have width $\leq \ell$.
2. Divide B into $\lceil n/\ell \rceil$ horizontal strip submatrices $B_1, ..., B_{\lceil n/\ell \rceil}$ of width ℓ with the exception of the last one that can have width $\leq \ell$.
3. **for** $p \in [\lceil n/\ell \rceil]$ **do**
4. Compute the Boolean product C_p of A_p and B_p along with single witnesses for all positive entries of the product
5. **for** all $i,\ j \in [n]$ **do**
   (a) Find the largest p such that $C_p[i,j] = 1$ or set $p = 0$ if it does not exist
   (b) **if** $p > 0$ **then** return the found witness of $C_p[i,j]$ **else** return "No"

Lemma 3. *Algorithm 5 runs in time* $\tilde{O}((n/\ell)n^{\omega(1,\log_n \ell,1)})$.

Proof. Steps 1, 2, take $O(n^2)$ time. Step 3 requires $\tilde{O}((n/\ell)n^{\omega(1,\log_n \ell,1)})$ time by a straightforward generalization of the $\tilde{O}(n^\omega)$-time algorithmic solution to the witness problem for square Boolean matrices given in Fact 2 to include rectangular Boolean matrices. Step 4(a) takes $O(n^2 \times n/\ell)$ time totally. Finally, Step 4(b) requires $O(n^2)$ time totally. It remains to observe that the term $\tilde{O}((n/\ell)n^{\omega(1,\log_n \ell,1)})$ dominates the asymptotic time complexity of the algorithm by $\omega(1, \log_n \ell, 1) \geq 2$. □

Theorem 4. *For all non-zero entries of the Boolean matrix product of two Boolean* $n \times n$ *matrices, witnesses of rank not exceeding* ℓ *can be reported in time* $\tilde{O}((n/\ell)n^{\omega(1,\log_n \ell,1)})$.

The Method Based on Multi-witnesses. A straightforward method to obtain single witnesses of rank $O(\lceil W_C(i,j)/k \rceil)$ for the nonzero entries $C[i,j]$ of the Boolean product C of two Boolean $n \times n$ matrices is to iterate a randomized algorithm for single witnesses for the entries of C [2]. After $O(k \log n)$ iterations such witnesses can be reported with high probability. This straightforward method takes $\tilde{O}(n^\omega k)$ time. We provide a more efficient algorithm for this problem based on the algorithm for the so called k-witness problem from [8].

The *k-witness problem* for the Boolean matrix product of two $n \times n$ Boolean matrices is to produce a list of r witnesses for each positive entry of the product, where r is the minimum of k and the total number of witnesses for this entry.

In the following fact from [8], the upper bounds have been updated by incorporating the more recent results on the parameters ω (Fact 1) and α [15].

Fact 6 [8]. *There is a randomized algorithm solving the k-witness problem almost certainly in time* $\tilde{O}(n^{2+o(1)}k + n^\omega k^{(3-\omega-\alpha)/(1-\alpha)})$, *where* $\alpha \approx 0.31389$ *(see Fact 3). One can rewrite the upper time bound as* $\tilde{O}(n^\omega k^\mu + n^{2+o(1)}k)$, *where* $\mu \approx 0.46530$.

Algorithm 6

Input: Boolean $n \times n$ matrices A, B, and a parameter $k \in [n]$ not less than 4.
Output: single witnesses $Wit[i,j]$ for all non-zero entries $C[i,j]$ of the Boolean product C of A and B such that $rank(Wit[i,j]) \le 4\lceil W_C(i,j)/k\rceil$ with probability at least $\frac{1}{2} - e^{-1}$.

1. $D \leftarrow B$;
2. initialize $n \times n$ integer matrix Wit by setting all its entries to 0;
3. **for** $q = 1, ..., O(\log n)$ **do**
   (a) run an algorithm for the k-witness problem for the product F of the matrices A and D;
   (b) for all $1 \le i,\ j \le n$, set $Wit[i,j]$ to the maximum of $Wit[i,j]$ and the maximum among the reported witnesses for $F[i,j]$;
   (c) uniformly at rand om set each 1 entry of D to zero with probability $\frac{1}{2}$.

$TW(n,k)$ will stand for the running time of the k-witness algorithm for the Boolean product of the two input Boolean matrices of size $n \times n$ used in Algorithm 6.

Lemma 4. *Algorithm 6 runs in* $\tilde{O}(TW(n,k) + n^2k)$ *time.*

Proof. The block of the while loop can be implemented in $O(TW(n,k) + n^2k)$ time. It is sufficient to observe that the block is iterated $O(\log n)$ times. □

Lemma 5. *For* $1 \le i,\ j \le n$ *and* $k \ge 4$, *the final value of* $Wit[i,j]$ *in Algorithm 5 is a witness of* $C[i.j]$ *with rank at most* $4\lceil W_C(i,j)/k\rceil$ *with probability not less than* $\frac{1}{2} - e^{-1}$.

Proof. We may assume with out loss of generality that $W_C(i,j)/k > 1$ since otherwise the maximum witness for $C[i,j]$ is found already in the first iteration of the block of the while loop. Let $\ell = \lceil \log_2 W_C(i,j)/k\rceil$. A witness of the entry $C[i,j]$ survives $\ell + 1$ iterations of the block of the while loop with probability $2^{-\ell-1}$. Hence, after $\ell + 1$ iterations of the block of the while loop the expected number of witnesses of the entry $C[i,j]$ that survive does not exceed $k/2$. Consequently, the number of witnesses of $C[i,j]$ that survive does not exceed k with probability at least $\frac{1}{2}$. They are reported as witnesses of $F[i,j]$ in the $\ell + 2$ iteration. On the other hand, the probability that none of witnesses not greater than $4W_C(i,j)/k$ survives the $\ell + 1$ iterations is at most $(1 - \frac{1}{2^{\ell+1}})^{4W_C(i,j)/k} \le e^{-1}$ by $k \ge 4$. Observe that for events A and B, $Prob(A \cap B) \ge 1 - Prob(\bar{A} \cup \bar{B}) \ge 1 - Prob(\bar{A}) - Prob(\bar{B})$. Hence, at least one witness of rank at most $4W_C(i,j)/k$ survives $\ell + 1$ iterations and it is reported in the $\ell + 2$ iteration with probability at least $1 - \frac{1}{2} - e^{-1} \ge \frac{1}{2} - e^{-1}$. □

Theorem 5. *Let C be the Boolean product of two Boolean $n \times n$ matrices and let k be an integer not less than 4. One can compute for all non-zero entries $C[i,j]$ single witnesses of rank* $O(\lceil W_C(i,j)/k\rceil)$ *in* $\tilde{O}(n^{\omega}k^{0.4653} + n^{2+o(1)}k)$ *time almost certainly.*

Proof. By Lemma 5, it is sufficient to iterate Algorithm 5 $O(\log n)$ times to achieve the probability of at least $1 - n^{-\beta}$, $\beta \geq 1$. The time complexity bound follows from Lemma 4 by the upper bound on $TW(n, k)$ from Fact 6. □

Assuming the notation from the theorem, we obtain the following corollary.

Corollary 2. *There is a randomized algorithm that for $4 \leq k \leq n^{0.4212}$ computes for all non-zero entries $C[i, j]$ single witnesses of rank $O(\lceil W_C(i, j)/k \rceil)$ almost certainly in time substantially subsuming the best known upper time bound for computing maximum witnesses for all non-zero entries of C. In particular, if the number of witnesses for each entry of C is upper bounded by $w \leq n^{0.4212}$ then by setting $k = w$, we obtain for all non-zero entries of C a witness of rank $O(1)$ almost certainly, substantially faster then maximum witnesses for these entries.*

Acknowledgments. The authors thank Francois Le Gall for a useful clarification of the current status of quantum algorithms for Boolean matrix product. The research has been supported in part by Swedish Research Council grant 621-2017-03750.

References

1. Alon, N., Galil, Z., Margalit, O., Naor, M.: Witnesses for Boolean matrix multiplication and for shortest paths. In: Proceedings of 33rd Symposium on Foundations of Computer Science (FOCS), pp. 417–426 (1992)
2. Alon, N., Naor, M.: Derandomization, witnesses for Boolean matrix multiplication and construction of perfect hash functions. Algorithmica **16**, 434–449 (1996)
3. Ambainis, A.: Quantum search algorithms. SIGACT News **35**(2), 22–35 (2004)
4. Coppersmith, D.: Rectangular matrix multiplication revisited. J. Symb. Comput. **1**, 42–49 (1997)
5. Cohen, K., Yuster, R.: On minimum witnesses for Boolean matrix multiplication. Algorithmica **69**(2), 431–442 (2014)
6. Czumaj, A., Kowaluk, M., Lingas, A.: Faster algorithms for finding lowest common ancestors in directed acyclic graphs. Theor. Comput. Sci. **380**(1–2), 37–46 (2007)
7. Dürr, C., Høyer, P.: A quantum algorithm for finding the minimum. arXiv: 9607.014 (1996/1999)
8. Gąsieniec, L., Kowaluk, M., Lingas, A.: Faster multi-witnesses for Boolean matrix product. Inf. Process. Lett. **109**, 242–247 (2009)
9. Grandoni, F., Italiano, G.F., Lukasiewicz, A. Parotsidis, N., Uznanski, P.: All-Pairs LCA in DAGs: breaking through the $O(n^{2.5})$ barrier. To Appear in Proc. SODA 2021. CoRR abs/2007.08914 (2020)
10. Grover. L.K.: A fast quantum mechanical algorithm for database search. In: Proceedings of Annual ACM Symposium on Theory of Computing (STOC), pp. 212–219 (1996)
11. Huang, X., Pan, V.Y.: Fast rectangular matrix multiplications and applications. J. Complex. **14**, 257–299 (1998)
12. Kowaluk, M., Lingas, A.: Quantum and approximation algorithms for maximum witnesses of Boolean matrix products. CoRR abs/2004.14064 (2020)
13. Le Gall, F.: Powers of tensors and fast matrix multiplication. In: Proceedings of 39th International Symposium on Symbolic and Algebraic Computation, pp. 296–303 (2014)

14. Gall, F.: A time-efficient output-sensitive quantum algorithm for Boolean matrix multiplication. In: Chao, K.-M., Hsu, T., Lee, D.-T. (eds.) ISAAC 2012. LNCS, vol. 7676, pp. 639–648. Springer, Heidelberg (2012). https://doi.org/10.1007/978-3-642-35261-4_66
15. Le Gall, F., Urrutia, F.: Improved rectangular matrix multiplication using powers of the Coppersmith-Winograd tensor. In: Proceedings of SODA 2018, pp. 1029–1046 (2018)
16. Navebi, A., Vassilevska Williams, V.: Quantum algorithms for shortest path problems in structured instances. arXiv:1410.6220 (2014)
17. Nielsen, M., Chuang, I.: Quantum Computation and Quantum Information. Cambridge University Press, Cambridge (2000)
18. Shapira, A., Yuster, R., Zwick, U.: All-pairs bottleneck paths in vertex weighted graphs. Algorithmica **59**, 621–633 (2011)
19. Vassilevska, V., Williams, R., Yuster, R.: Finding heaviest H-subgraphs in real weighted graphs, with applications. ACM Trans. Algorithms **6**(3), 441–4423 (2010)
20. Vassilevska Williams, V.: Multiplying matrices faster than Coppersmith-Winograd. In: Proceedings of 44th Annual ACM Symposium on Theory of Computing (STOC), pp. 887–898 (2012)

Template-Driven Rainbow Coloring of Proper Interval Graphs

L. Sunil Chandran[1], Sajal K. Das[2], Pavol Hell[3], Sajith Padinhatteeri[4], and Raji R. Pillai[1(✉)]

[1] Indian Institute of Science, Bengaluru, India
{sunil,rajipillai}@iisc.ac.in
[2] Missouri University of Science and Technology, Rolla, USA
sdas@mst.edu
[3] Simon Fraser University, Burnaby, Canada
pavol.hell@gmail.com
[4] Birla Institute of Technology and Science Pilani, Pilani, India
sajith@hyderabad.bits-pilani.ac.in

Abstract. For efficient design of parallel algorithms on multiprocessor architectures with memory banks, simultaneous access to a specified subgraph of a graph data structure by multiple processors requires that the data items belonging to the subgraph reside in distinct memory banks. Such "conflict-free" access to parallel memory systems and other applied problems motivate the study of *rainbow coloring* of a graph, in which there is a fixed template $\mathcal{T}$ (or a family of templates), and one seeks to color the vertices of an input graph G with as few colors as possible, so that each copy of $\mathcal{T}$ in G is rainbow colored, i.e., has no two vertices the same color. In the above example, the data structure is modeled as the *host* graph G, and the specified subgraph as the template $\mathcal{T}$. We call such coloring a *template-driven rainbow coloring* (or TR-coloring). For large data sets, it is also important to ensure that no memory bank (color) is overloaded, i.e., the coloring is as balanced as possible. Additionally, for fast access to data, it is desirable to quickly determine the address of a memory bank storing a data item. For arbitrary topology of G and $\mathcal{T}$, finding an optimal and balanced TR-coloring is a challenging problem. This paper focuses on rainbow coloring of proper interval graphs (as hosts) for cycle templates. In particular, we present an $O(k \cdot |V| + |E|)$ time algorithm to find a TR-coloring of a proper interval graph G with respect to k-length cycle template, C_k. Our algorithm produces a coloring that is (i) *optimal*, i.e., it uses minimum possible number of colors in any TR-coloring; (ii) *balanced*, i.e, the vertices are evenly distributed among the different color classes; and (iii) *explicit*, i.e., the color assigned to a vertex can be computed by a closed form formula in constant time.

Keywords: Rainbow coloring · Template · TRB-coloring · Proper interval graph

A. Mudgal and C. R. Subramanian (Eds.): CALDAM 2021, LNCS 12601, pp. 452–470, 2021.
https://doi.org/10.1007/978-3-030-67899-9_36

1 Introduction

Efficient and scalable implementation of parallel algorithms on multiprocessor architectures with multiple memory banks depends on how fast the items in the underlying data structure can be accessed in parallel. To simultaneously access the data items required for a computation by multiple processors, the pertinent data must reside in different memory banks. This problem of "conflict-free" access to parallel memory systems can be formulated as *rainbow coloring* of templates corresponding to the items to be accessed in a host graph representing the data structure [2,6].

For example, following an insertion or deletion in a binary heap constructed as a complete binary tree, the re-heapification (i.e., readjusting the heap) always takes place along a single path from the root to some of the leaves. Thus, parallel (re-)heapification, which may appear very sequential in nature due to level-by-level adjustment of the heap property, will be efficient if all data items lying on any path from the root to any of the leaves can be accessed simultaneously [8]. Here the host graph G is the heap data structure represented as a complete binary tree while the template is any root-to-leaf path.

As another example, the problem is to color the nodes of the k-dimensional binary hypercube, Q_k (host graph), having numerous applications in multiprocessor interconnection networks, databases, and coding theory, such that every d-dimensional subcube Q_d (template), for $1 \leq d \leq k$, is rainbow colored [7].

In the following let us provide yet another example suggesting the problem is a natural one, even for proper interval graphs and cycle templates. Time-dependent applications like human contact networks or social networks can be modeled as proper interval graphs when there is a need to satisfy some constraints like "First Come First Served". Certain sub-structures (e.g., paths or cycles) in such graphs may provide useful information, for example, circular spreading of a disease along a long cycle of a contact network.

For illustration, let each member x of a community of people (say, a club or organization), denoted as S, visit a common meeting place for a certain interval of time, say $f(x)$, every day. Two members $x, y \in S$ interact with each other only if their corresponding intervals intersect, i.e., $f(x) \cap f(y) \neq \emptyset$. Therefore, the contact graph corresponding to the members of this community represents an interval graph. Furthermore, this is a proper interval graph if the meeting place follows certain rules like the first member to arrive always leaves first (e.g., if they come to consult a doctor). Consider the spread of a disease based on contacts at such a meeting place. In particular, some epidemiology studies examine how the disease spreads along contact paths or cycles. The goal is to compute the probability of the disease infecting a person when multiple contact paths exist between an individual and an infected person. When the disease is a global pandemic like COVID-19, the contact network can presumably be very large due to the fact that the community/organization may consist of a large number of members. Therefore, it is required to process a huge amount of data belonging to long cycles or multiple paths in the contact network in a short period of time, requiring high performance computing in a multiprocessor environment. Specifically, we need to retrieve cycles of a specified length from the memory banks as efficiently as possible so as to process various data connected to the spread of the pandemic along with the details of patients belonging to various cycles.

The above examples motivate the following definition. Let G be a *host graph* and $\mathcal{T}$ be a *template*, which is a fixed graph (or a family of graphs). A *template-driven rainbow coloring* (in short, *TR-coloring*) of G with respect to $\mathcal{T}$ is a vertex coloring of G such that all subgraphs of G isomorphic to $\mathcal{T}$ (or any member of the family $\mathcal{T}$ of graphs) are *rainbow colored*, i.e., no two vertices of $\mathcal{T}$ are assigned the same color. The minimum number of colors in a *TR*-coloring of a host graph G for a given template $\mathcal{T}$ will be denoted by $\chi^R_{\mathcal{T}}(G)$.

We observe that *TR*-coloring generalizes the usual notion of proper coloring since a proper vertex coloring of the host graph G is actually a *TR*-coloring of G for the template $\mathcal{T} = P_2$, the path on two vertices (i.e., an edge). If $\mathcal{T}$ is the set of all paths on at most $k+1$ vertices, the *TR*-coloring becomes the well-known k-distance coloring of graphs, where the vertices have to obtain different colors if they are at a distance at most k. (This can also be seen as simply a proper coloring of the power graph G^k, obtained from G by adding all edges between vertices of distance at most k.) Hence the notion of *TR*-coloring with path templates is reasonably well-explored [13]. In this paper, we consider what may be considered the next natural template family, namely the *cycles*.

Load Balancing. In the parallel memory system example, if the graph is relatively large, it becomes important to ensure that no memory bank (hence a color) is overloaded. In other words, the vertices of the graph have to be almost equally shared among different memory banks (i.e., color classes), recalling that the vertices having the same color will be stored in the same memory bank. We may capture the requirement of balancing the color classes by demanding that the difference between the cardinality of the largest and smallest color classes is at most τ, a constant or a slow growing function of the number of vertices $|V|$ in G. Such a load balanced coloring of G is called a τ-balanced *TR*-coloring, or simply *TRB*-coloring when $\tau = 1$. The minimum number of colors required for τ-balanced *TR*-coloring of G with respect to a template $\mathcal{T}$ will be denoted by $\chi^{\tau\text{-}RB}_{\mathcal{T}}(G)$, or simply $\chi^{RB}_{\mathcal{T}}(G)$ when $\tau = 1$.

Explicit Computation of Colors. Another requirement from practical considerations is the following. It is very advantageous to know an explicit formula for the memory bank address to which the data items corresponding to a vertex i are stored. In our model, this means that the color of any vertex i is computed easily by a closed form expression. Such a coloring where the color assigned to a vertex i has an explicit formula in terms of i will be called an *explicit coloring*.

In summary, for *TR*-coloring of a graph, our goal is to design *efficient* algorithms that guarantee that the resulting coloring is (i) *optimal*, i.e., it uses the minimum possible number of colors; (ii) *balanced*, i.e., the vertices are evenly distributed among the different color classes; and (iii) *explicit*, i.e., the color assigned to a vertex can be computed in constant time (by a closed form formula).

1.1 Related Work

The graph theory community has traditionally studied different variants of rainbow coloring problems, both for vertex and edge coloring. Several of these problems analyze the existence of certain types of rainbow colored template graphs (e.g., rainbow paths, cycles, subtrees, or matchings) with maximum possible number of vertices (edges) in a host graph that has been colored by a proper (or some other) coloring [1,4,12,14]. Moreover, some of the rainbow coloring problems for induced subgraphs as templates, are closely related to certain fundamental open problems on the chromatic number of graphs [11,18].

In the literature, there also exist problems seeking vertex or edge coloring (not necessarily proper) of the host graph in which copies of certain templates are required to be rainbow colored. A related problem, the so-called rainbow connection number studied in the last ten years, seeks an optimal coloring of the vertices (or edges) of the host graph such that between any pair of vertices there is a rainbow path [15]. A variant of this concept is known as the strong rainbow coloring, where a rainbow shortest path is sought between any pair of vertices. Finally, very strong rainbow coloring implies all shortest paths of the host graph are required to be rainbow colored [3].

Concerning the conflict-free data access in a multiprocessor architecture, there are algorithms for *TRB*-coloring when the host graph G and the template $\mathcal{T}$ have special topological structure. For example, in [8] the authors proposed optimal and explicit *TRB*-coloring algorithms where the host graph G is a q-ary tree or binomial tree, and the template $\mathcal{T}$ is a path or a subtree. In [2], an optimal *TR*-coloring of path templates is proposed in hosts like two-dimensional arrays, circular lists, and complete trees. An optimal and explicit *TRB*-coloring of tori and hypercube graphs (as hosts) for star templates is presented in [6]. Most of the earlier work focuses on *TRB*-coloring of special graphs that possess certain characteristics useful for specified templates. In general, the problem is highly challenging for arbitrary topology of G and $\mathcal{T}$, due to the overlapping of different instances of $\mathcal{T}$ in G.

1.2 Our Contributions

In this paper, we consider *proper interval* graphs as hosts, and *cycles* C_k of length k as templates.

We emphasize that finding a balanced template-driven rainbow coloring is challenging for arbitrary topology of G and $\mathcal{T}$. It also appears hard in the full class of interval graphs, even for simple templates such as cycles. We view this work on proper interval graphs as the first step towards treating the more general case of interval graphs. Our major contributions are as follows.

- We present an algorithm that computes an optimal and explicit *TRB*-coloring of a proper interval graph G with respect to the cycle template C_k. Our most technical contribution is proving the correctness of the proposed algorithm. While the algorithm is greedy-like and simple to state, the proof of correctness requires a detailed structural analysis.

- The time complexity of the algorithm is $O(k \cdot |V| + |E|)$, which is $O(|V| + |E|)$ when k is a constant. Furthermore, in many of the applied contexts, the average degree is a bounded by a constant. For such sparse graphs, $|E| = O(|V|)$, and thus our algorithm runs in linear time.

The paper is organized as follows. Section 2 introduces preliminary concepts and terminology. Section 3 describes our algorithm for *TRB*-coloring of a proper interval graph G for template C_k. Section 4 proves the correctness and Sect. 5 summarizes the algorithm and time complexity. Finally we conclude the paper.

2 Preliminaries

An *interval graph* is an undirected graph $G(V, E)$ such that the vertices can be represented by intervals on the real line, where two vertices are adjacent if and only if the corresponding intervals intersect. The corresponding set of intervals is called the *interval representation* of G. We denote by $left(v_i)$ and $right(v_i)$ the left and right endpoints, respectively, of an interval $v_i \in V(G)$. The left end ordering of an interval graph G is an ordering of the vertices of G according to the left end point of vertices in the corresponding interval representation. That is, if $v_1 v_2 \ldots v_n$ is a left end ordering of $V(G)$, then $left(v_i) < left(v_{i+1})$ for $1 \leq i \leq n-1$. (We may assume that the endpoints of the representing intervals are distinct.) For each vertex $v_i \in V(G)$, a higher (respectively, lower) indexed adjacent vertex of v_i in the ordering is called a *right* (respectively, *left*) *neighbor* of v_i. For v_i, v_j with $i < j$, the symbol $[v_i, v_j]$ denotes the ordered set of vertices from v_i to v_j in the ordering including both v_i and v_j. Similarly, $(v_i, v_j]$ is the set $[v_i, v_j]$ excluding v_i and including v_j.

A *proper interval graph* G is an interval graph which admits an interval representation such that no interval is properly contained in another interval. It follows from the definition of a proper interval graph that if we use the right end ordering, we would obtain the same ordering of vertices, since $left(v_i) < left(v_j)$ if and only if $right(v_i) < right(v_j)$. Thus we shall call $v_1 v_2 \ldots v_n$ *the ordering* of G and use the symbol $<$ in this sense. The following well-known fact provides much useful information about proper interval graphs.

Proposition 1 [16]. *For any connected proper interval graph G, if u, v and w are any three vertices of $V(G)$ such that $u < v < w$ with respect to the left end ordering of the graph, then $uw \in E(G)$ implies $uv \in E(G)$ and $vw \in E(G)$.*

In particular, the set of left neighbors of a vertex u in a proper interval graph G form a clique; the same is true for its right neighbors. Moreover, all right (respectively, left) neighbors of u occur consecutively in the ordering immediately after (respectively, before) vertex u. In other words, if v is the last right neighbor of u, then the set of right neighbors of u is exactly $(u, v]$. The closed right neighborhood of a vertex $u \in V(G)$ is denoted by $N_{right}(u)$. That is $N_{right}(u) = \{u\} \cup \{\text{right neighbors of } u\}$. We also cite the following useful fact.

Proposition 2 [17]. *For any connected proper interval graph G with the ordering $v_1 \ldots v_n$, there exists a Hamiltonian path $P_n = v_1, v_2, \ldots, v_n$.*

Table 1 in Appendix A.1 lists the summary of notations used in this paper.

3 *TRB*-Coloring of Proper Interval Graphs with Cycle Templates

In this section we describe an efficient algorithm for optimal *TRB*-coloring of proper interval graphs when the template is a cycle C_k.

When the template $\mathcal{T} = C_3 = K_3$, the *TR*-coloring is easy for all perfect graphs. A graph G is called *perfect* if the chromatic number of each induced subgraph of G is equal to the size of the largest clique in that subgraph. Observe that interval graphs are perfect [10], so this takes care of our problem for the template C_3. Note that $\chi^{RB}_{K_3}(G) = \chi^{R}_{K_3}(G) = 1$ if G has no triangles, so we may focus on perfect graphs G with at least one triangle.

Theorem 1. *If G is any perfect graph with at least one triangle, then for the template $\mathcal{T} = K_3$, we have $\chi^{R}_{K_3}(G) = \chi(G)$. In particular, if G is a proper interval graph then $\chi^{RB}_{K_3}(G) = \chi^{R}_{K_3}(G) = \chi(G)$.*

Proof. If G does not have a triangle, then obviously $\chi^{RB}_{K_3}(G) = \chi^{R}_{K_3}(G) = 1$. If it contains a triangle then let ω be the size of the largest clique. Since G is a perfect graph, it admits a proper ω coloring and $\chi(G) = \omega$. Observe that any proper coloring of G is a *TR*-coloring for template $\mathcal{T} = K_3$. Therefore, we have $\chi^{R}_{K_3} \leq \omega$. Moreover, every proper interval graph G has a balanced proper coloring with $\chi(G)$ colors, according to [19]. Thus $\chi^{RB}_{K_3} \leq \omega$ also holds. On the other hand, $\chi^{R}_{K_3}$ (and hence also $\chi^{RB}_{K_3}$) is also at least ω. This is because the vertices of any maximum clique K_ω has to be colored with distinct colors since otherwise some triangles will not be rainbow colored. Therefore $\chi^{R}_{K_3} = \chi(G) = \omega$. □

Thus we obtain an optimal *TR*-coloring of perfect graphs and *TRB*-coloring for proper interval graphs with template C_3. (The coloring is also explicit, see [10,19].) The case of template C_k with $k > 3$ is significantly more involved, and does not appear to be easy to solve for perfect graphs, or even for interval graphs. However, we are able to propose an efficient algorithm for optimal *TRB*-coloring with template $C_k, k \geq 4$, when the host graph is a proper interval graph.

Let G be a proper interval graph and let $\mathcal{T} = C_k$ be the template, for $k \geq 4$. We first make some simplifying assumptions. Since every cycle of G is contained in a unique biconnected component, we may assume that G is biconnected. A minimum *TR*-coloring of G can easily be constructed from minimum *TR*-coloring of the biconnected components, since these components form a tree structure, in fact a path structure in the case of proper interval graphs [17]. The biconnected components of a graph are easily computed by a depth-first search. (Edges of G that do not lie in any cycles are not in any biconnected component, but they can be ignored, as can all biconnected components with fewer than k vertices.)

We state a property of biconnected proper interval graphs G from [17].

Proposition 3 *(Lemma 4.1, [17]). Let G be a biconnected proper interval graph. Any vertex except the last two vertices in the left end ordering of G has at least* two *right neighbors.*

The preceding property implies the existence of a k-cycle C_k containing any k consecutive vertices in a left end ordering of G. To see this, consider a biconnected proper interval graph G with a left end ordering $\sigma = v_1 v_2 \ldots v_n$. Let $v_i, v_{i+1}, \ldots, v_{i+k-1}$ be k consecutive vertices in σ. By Proposition 2 the vertices $v_i, v_{i+1}, \ldots, v_{i+k-1}$ constitute a path, say P in G. Observe that any vertex $v_j \in V(P)$, for $i \leq j \leq i+k-3$, is adjacent to vertex v_{j+2} by Proposition 3. Hence we have one of the following k-cycles in G, depending on the parity of k.

$$C_k = \begin{cases} v_i, v_{i+2}, v_{i+4} \ldots, v_{i+k-3}, v_{i+k-1}, v_{i+k-2}, v_{i+k-4} \ldots, v_{i+1}, v_i, & \text{if } k \text{ is odd,} \\ v_i, v_{i+2}, v_{i+4} \ldots, v_{i+k-2}, v_{i+k-1}, v_{i+k-3} \ldots, v_{i+1}, v_i, & \text{otherwise.} \end{cases}$$

If there exists a k-cycle containing a pair of vertices u, v in G, then the following lemma (see Appendix A.2 for a proof) shows the existence of a k-cycle with any pair of vertices in the interval $[u, v]$ in the left end ordering of G.

Lemma 1. *In any biconnected proper interval graph G, if u precedes v in a left end ordering of G and there is a k-cycle C_k of G containing both u and v, then there exists a k-cycle C_k of G containing both u' and v' for any u' and v' such that $u \leq u' \leq v' \leq v$.*

Let G be a biconnected proper interval graph with ordering $\sigma = v_1 v_2 \ldots v_n$. We denote by $p_k(G)$ the maximum of $j - i + 1$ such that there exists a k-cycle C_k of G containing v_i, v_j with $i < j$. We call the parameter $p_k(G)$ the *k-span* of G. It turns out to be a lower bound on the number of colors required for both *TR*-coloring and *TRB*-coloring of G.

Corollary 1. *When $T = C_k$ and G is a biconnected proper interval graph, we have the lower bounds $\chi_T^{RB}(G) \geq \chi_T^R(G) \geq p_k(G)$.*

Proof. The first inequality is trivial. To prove the second inequality, note that by definition of $p_k(G)$ there exist two vertices u and v in G such that they are $p_k(G)$ vertices apart with respect to the ordering, and some k-cycle of G contains both u and v. It now follows from Lemma 1 that all vertices in the interval $[u, v]$ in the ordering must obtain different colors. Therefore, $\chi_T^R(G) \geq p_k(G)$. □

In fact, this lower bound is achievable, and we have $\chi_T^{RB}(G) = p_k(G)$.

Theorem 2. *Let G be a biconnected proper interval graph with ordering $\sigma = v_1 v_2 \ldots v_n$. The function color$(v_i) = i \bmod p_k(G)$ is an explicit balanced* TR-*coloring of G with respect to template C_k. Therefore $\chi_T^{RB}(G) = \chi_T^R(G) = p_k(G)$.*

Indeed, this is a proper coloring, since no copy of C_k involves vertices further apart than $p_k(G)$ and therefore all copies of C_k are rainbow colored. It is optimal by Corollary 1, and it is balanced because of the way the colors are calculated.

To compute the k-span $p_k(G)$, we calculate for each vertex v_i its *pivotal* vertex $pivot(v_i)$, defined as the vertex v_j with the highest index j for which there exists a k-cycle of G containing both v_i and v_j. Clearly, the k-span $p_k(G)$ is the maximum distance from any v_i to its pivotal vertex $pivot(v_i)$. Note that the value $pivot(v)$ is computed only for vertices $v \in V(G)$ such that $||[v, v_n]|| \geq k$.

It remains to explain how to find the pivotal vertices. We find $pivot(v_j)$, $v_j \in V(G)$ by locating a vertex w to the right of v_j in the ordering, such that there exist two internally disjoint paths of length $\lfloor \frac{k}{2} \rfloor$ and $\lceil \frac{k}{2} \rceil$, each between v_j and w. For this we compute $2 \cdot \lfloor \frac{k}{2} \rfloor$ distinct special vertices S_a^i and S_b^i, $1 \leq i \leq \lfloor \frac{k}{2} \rfloor$. This requires an iterative process explained in the next subsection.

3.1 Special Vertices and Pivots

To find $pivot(v)$, $v \in V(G)$, we compute special vertices S_a^i and S_b^i, $1 \leq i \leq \lfloor \frac{k}{2} \rfloor$. Let $N_{right}(v)$ be the set consisting of v and all its right neighbors ordered by $<$. Let S_a^1 and S_b^1 be the last two vertices in $N_{right}(v)$, where $S_a^1 < S_b^1$. (By Proposition 3, there always exist two such neighbors other than v.) For $i \geq 1$, let us define three operations for computing S_a^i and S_b^i.

- *rightmost*(x) = the last vertex of $N_{right}(x)$, where x is either S_a^i or S_b^i;
- *a-shift*(i) = the operation to reset S_a^i to its immediate left neighbor; and
- *b-shift*(i) = the operation to reset S_b^i to its immediate left neighbor.

The special vertices S_a^i and $S_b^i, 1 \leq i \leq \lfloor \frac{k}{2} \rfloor$ with respect to a vertex v are iteratively computed by Procedure 1, which is illustrated in Appendix A.3.

We note that Procedure 1 may fail to produce $k-1$ distinct special vertices, as it is possible that $S_a^i = v_n$ or $S_b^i = v_n$ for some $i < \lfloor k/2 \rfloor$, or $S_a^{\lfloor k/2 \rfloor} = v_n$ when k is odd. In that case we set $pivot(v) = v_n$. Otherwise, if there exist $k-1$ distinct vertices S_a^i and S_b^i in G, we define $pivot(v)$ as follows. (Note that even though the exponent in the expression is always $\lfloor \frac{k}{2} \rfloor$, the subscripts differ for odd and even values of k.)

$$pivot(v) = \begin{cases} S_b^{\frac{k-1}{2}}, & \text{if } k \text{ is odd,} \\ S_a^{\frac{k}{2}}, & \text{if } k \text{ is even.} \end{cases} \tag{3.1}$$

The main technical difficulty is in proving the correctness of these calculations. They are the backbone of the algorithm, as once the value $p_k(G)$ is known an optimal coloring of G is described in Theorem 2.

We first show that the execution of *a-shift* and *b-shift* operations cannot occur simultaneously. Consider a biconnected proper interval graph G and let $v \in V(G)$ be the vertex for which S_a^i and S_b^i, $1 \leq i \leq \lfloor \frac{k}{2} \rfloor$, are computed.

Procedure 1: *Computing_special_vertices(v)*

```
1 Initialize S_a^0 := S_b^0 := v
2 Let v_n be the last vertex in the ordering of G
3 for 1 ≤ i ≤ ⌊k/2⌋ do
4     Set S_a^i = rightmost(S_a^{i-1})
5         If S_a^i = v_n then exit
6         If S_a^i = S_b^{i-1} then do b-shift(i − 1)
7     Set S_b^i = rightmost(S_b^{i-1})
8         If S_b^i = v_n then exit
9         If S_b^i = S_a^i then do a-shift(i)
```

Proposition 4. *For* $i \geq 2$*,* a-shift(i) *is mutually exclusive with* b-shift$(i-1)$ *and* b-shift(i) *operations.*

The proof is given in Appendix A.2.

We now show that Proposition 4 together with the property of proper interval graphs (Proposition 1) guarantee the adjacency and a strict ordering among the vertices S_a^i and S_b^i, for all i. Let $S_a^0 = S_b^0 = v$. For simplicity of presentation, let $A_0 = S_b^0, A_1 = S_a^1, A_2 = S_b^1, A_3 = S_a^2, A_4 = S_b^2, \ldots$. That is $S_a^i = A_{2i-1}, 0 < i \leq \lfloor \frac{k}{2} \rfloor$ and $S_b^i = A_{2i}, 0 \leq i \leq \lfloor \frac{k}{2} \rfloor$.

Proposition 5. $A_0 < A_1 < A_2 < \ldots < A_t$*, where* $t = 2 \cdot \lfloor \frac{k}{2} \rfloor$*, and* A_{i-1} *is adjacent to* A_i*, for* $0 < i \leq t$.

The proof is given in Appendix A.2.

4 Proof of Correctness

In Subsect. 4.1 we show that the vertex v along with these $k-1$ special vertices forms a k-cycle which we call the 'canonical cycle'. (The shifting in Procedure 1 is necessary to ensure the existence of a k-cycle using these special vertices.) In Subsect. 4.2 we show that no k-cycle of G contains both v and any vertex $w > pivot(v)$ in the ordering. These two facts then imply that the computed value $pivot(v)$ is indeed the pivotal vertex for v.

4.1 Constructing the Canonical Cycle with Special Vertices

Suppose first that vertex $v \in V(G)$ has $k-1$ distinct special vertices defined by Procedure 1. We construct the following canonical k-cycles C_k^o (for odd k) and C_k^e (for even k).

$$C_k^o = v, S_a^1, \ldots, S_a^{\frac{k-1}{2}}, S_b^{\frac{k-1}{2}}, S_b^{\frac{k-1}{2}-1}, \ldots, S_b^1, v, \text{if } k \text{ is odd,} \tag{4.1}$$

$$C_k^e = v, S_a^1, \ldots, S_a^{\frac{k}{2}-1}, S_a^{\frac{k}{2}}, S_b^{\frac{k}{2}-1}, \ldots, S_b^1, v, \text{if } k \text{ is even.} \tag{4.2}$$

Proposition 5 guarantees the adjacency of consecutive vertices in C_k^o and C_k^e. Since Procedure 1 defines *pivot*(v) as $S_b^{\frac{k-1}{2}}$ for odd k and $S_a^{\frac{k}{2}}$ for even k, Eqs. (4.1) and (4.2) give k-cycles containing v and *pivot*(v) in the respective cases.

Note that sometimes $k-1$ special vertices are not defined. In this case *pivot*(v) is assigned to be the last vertex v_n in the ordering of G. Construction of C_k containing v and *pivot*(v) in this special case is explained in Appendix A.3 with some illustrative examples. The following Corollary summarizes the above discussion.

Corollary 2. *Let G be a biconnected proper interval graph and $v \in V(G)$. The value pivot(v) computed by Procedure 1 has the property that there exists a k-cycle containing both v and pivot(v).*

4.2 Absence of k-Cycles Beyond the Pivots

To prove the correctness of our algorithm, it only remains to show that the algorithm correctly determines the pivotal vertices. Recall that Corollary 2 shows that there exists a k-cycle containing v and *pivot*(v), for each $v \in G$. Hence it remains to prove the following fact.

Proposition 6. *Let G be a biconnected proper interval graph and $v \in V(G)$. Then G has no k-cycle containing v and any vertex $w >$ pivot(v) in the ordering.*

Since this is trivial if *pivot*(v) $= v_n$, we may assume that v has $k-1$ distinct special vertices S_b^i and S_a^i. To prove the proposition, we use the lemma below, which matches any possible cycle C involving both v and w, with the existing special vertices $S_a^0 = v, S_a^1, \ldots, S_a^{\lfloor \frac{k}{2} \rfloor}, S_b^0 = v, S_b^1, \ldots, S_b^{\lfloor \frac{k}{2} \rfloor}$ for the vertex v.

Lemma 2. *Suppose that C is a cycle including both v and w, where $w >$ pivot(v). There exist two disjoint sets $X, Y \subseteq C$ with exactly $\lfloor \frac{k}{2} \rfloor$ vertices each, such that X contains exactly one vertex of each interval $(S_a^{i-1}, S_a^i]$, and Y contains exactly one vertex of each interval $(S_b^{i-1}, S_b^i]$, for $i = 1, 2, \ldots, \lfloor \frac{k}{2} \rfloor$.*

The proof is given in Appendix A.2.

By Lemma 2 there exists a unique representative of C in each of the ranges $(S_a^{i-1}, S_a^i], (S_b^{i-1}, S_b^i]$, for $1 \leq i \leq \lfloor \frac{k}{2} \rfloor$. Thus there are $k-1$ distinct vertices in the cycle C lying in $(v, \textit{pivot}(v)]$. Since $w >$ *pivot*(v), and C is assumed to be a cycle through v and w, there are at least $(k-1)+2 = k+1$ vertices in C. Therefore the cycle C contains more than k vertices, proving Proposition 6.

5 The Overall Algorithm and Its Complexity

We summarize our algorithm as follows. The input is a proper interval graph H.

1. Compute biconnected components G of H, and process each G separately.
2. Find the ordering of each biconnected proper interval graph G.
3. Compute for each vertex v of G its special vertices and its *pivot*(v).

4. Compute the k-span $p_k(G)$.
5. Set $p_k := \max_G p_k(G)$, the maximum of all $p_k(G)$ on biconnected components G in H.
6. Color the vertices of H, from left to right in the ordering, with p_k colors iteratively. That is a vertex v_i in the ordering gets $color(v_i) := i \bmod p_k$.

We have shown that the algorithm is correct on each biconnected component G of H. Since each cycle in H is completely contained inside a unique biconnected component, a *TR*-coloring of each G ensures a *TR*-coloring of the graph H. Since $p_k(G)$ colors are enough to get a *TR*-coloring of G, any $p_k \geq p_k(G)$-coloring as described in Theorem 2 also ensures a *TR*-coloring of the component G. Therefore a p_k coloring of H as described in Step 6 above is a *TR*-coloring with respect to the template C_k. Since $p_k = \max_G p_k(G)$, there exists a biconnected component G_i such that $p_k = p_k(G_i)$ and hence the *TR*-coloring of the graph H is optimal. At last, the iterative distribution of p_k colors among the vertices of H ensures the balancing part.

Now we consider the complexity. Step 1 is computed by a simple depth first search in time $O(|V| + |E|)$. Step 2 is performed by a recognition algorithm for proper interval graphs, in time $O(|V|+|E|)$ [5,9]. The special vertices and pivots are computed by Procedure 1 and the remarks following it, as is the span of G.

Theorem 3. *The proposed algorithm produces an optimal* TRB-*coloring of a proper interval graph G, for the template C_k, in $O(k \cdot |V| + |E|)$ time.*

Proof. By the preceding remarks, it only remains to discuss Steps 3–6. The right neighborhood $N_{right}(v)$ of each vertex $v \in G$ can be obtained by a linear traversal of the ordered list of vertices in $O(|V|)$ time, and the computation of the special vertices requires at most $\lfloor \frac{k}{2} \rfloor$ iterations, each of which takes constant time. Hence, the computation of special vertices requires $O(k)$ time. The span is computed from the pivots in $O(|V|)$ time. The bound given incorporates all these.

6 Conclusions

Motivated by several applied problems, we formulated a template-driven rainbow coloring problem, and proposed an efficient *TR*-coloring algorithm of proper interval graphs for k-cycle templates. For fixed k, the algorithm runs in $O(|V|)$ time for sparse graphs. Thus it is a practical algorithm and can be experimentally evaluated at scale. It remains for future investigation to explore *TR*-coloring and *TRB*-coloring of general interval graphs, for cycle templates and for other templates that may occur in applications, such as stars.

Acknowledgements. This collaboration started while S. K. Das and P. Hell were visiting the Indian Institute of Science, Bangalore in fall 2019. S. K. Das was partially supported by the Satish Dhawan Visiting Chair Professorship at IISc (September-December, 2019) and the US National Science Foundation grant CCF-1725755. P. Hell was supported by the Smt Rukmini Gopalakrishnachar Chair Professorship at IISc (November-December, 2019). S. Padinhatteeri was supported by the grant PDF/2017/002518 from Science and Engineering Research Board, India.

A Appendix

A.1 Notations

The notation used in this paper is summarized in Table 1.

Table 1. Notations

G or $G(V,E)$	: A simple undirected graph G with vertex set V and edge set E
$\mathcal{T}$	: A template; finite graph or a family of finite graphs
C_k	: Cycle (template) of length k
TR-coloring	: Template rainbow coloring of a graph
TRB-coloring	: 1-balanced template rainbow coloring of a graph
$\chi(G)$	: Chromatic number of G
$\chi^{R}_{\mathcal{T}}(G)$	: Minimum number of colors in a TR-coloring with template $\mathcal{T}$ in G
$\chi^{RB}_{\mathcal{T}}(G)$	: Minimum number of colors in a TRB-coloring with template $\mathcal{T}$ in G
$[v_i, v_j]$	: Ordered set of all vertices from v_i to v_j in G, including both v_i, v_j
(v_i, v_j)	: Ordered set of all vertices from v_i to v_j in G, excluding both v_i, v_j
$v+1$	: The immediate right neighbor of a vertex v in an ordering of $V(G)$
$v-1$	: The immediate left neighbor of v in an ordering of $V(G)$
$color(v)$	: Color of a vertex v

A.2 Proofs of Lemmas and Propositions

Proof **(Proof of Lemma 1).**

Let $\sigma = (v_1 v_2 \ldots v_n)$ be a left end ordering of G. Suppose there exists a k-cycle C^1_k containing both the vertices u and v such that $v_1 \leq u < v \leq v_n$. Let $u', v' \in [u, v]$ such that $u' < v'$. We show how to construct a k-cycle C^2_k containing the vertices u' and v' by modifying C^1_k. We may assume that not both u' and v' already are in C^1_k.

Suppose $v' \notin C^1_k$. Let y be the first vertex among the right neighbors of v' in the ordering such that $y \in V(C^1_k)$. Consider the neighbors of y in $V(C^1_k)$, say x and x'. Let us assume that $x < y < x'$. Since y is the first right neighbor of v' in C^1_k we have $x < v' < y$ and the adjacency of x to y implies x and v' are adjacent (See Proposition 1). Let w_1 be the leftmost and w_2 be the rightmost vertices of $V(C^1_k)$ with respect to the ordering. (Note that in some cases $w_1 = u$ and $w_2 = v$.) Hence both the neighbors of w_2 in $V(C^1_k)$, say z and z', are left neighbors of w_2 and assume $z < z'$. Therefore $z < z' < w_2$ and since w_2 is adjacent to z, the vertices z and z' are adjacent by Proposition 1. Thus we construct the k-cycle, C^2_k from C^1_k by the following steps when $y \neq w_2$.

(i) replace the edge xy with the path $xv'y$, (ii) delete the vertex w_2 and (iii) add the edge zz'.

If $y = w_2$ then replacing y by v' in C^1_k gives a k-cycle. This is because in this case the neighbors of w_2 in C^1_k are also adjacent to v' by Proposition 1.

Thus we have successfully constructed a k-cycle containing the vertices v' and u. If $u' \notin C_k^1$ then we follow similar arguments as above to add the vertex u' in C_k^2. In this case y becomes the last vertex among the left neighbors of u' in the ordering and replace w_1 with w_2 in the above arguments. Thus the new cycle C_k^2 is a k-cycle containing both u' and v'. If $u' \in C_k^1$ or $v' \in C_k^1$ then addition of v' or u' is sufficient to get C_k^2. □

Proof **(Proof of Proposition** 4**).** Suppose while determining S_a^i the operation *a-shift*(i) is invoked. This implies $rightmost(S_b^{i-1}) = rightmost(S_a^{i-1})$ and the procedure *Compute_special_vertices(v)* assigns S_b^i to the vertex $rightmost(S_b^{i-1})$ and reassigns S_a^i to the immediate left vertex of $rightmost(S_b^{i-1})$ in the ordering of $V(G)$. Moreover, *a-shift*(i) makes the vertices corresponding to S_a^i and S_b^i consecutive in the ordering. Now suppose *b-shift*(i) is invoked. This implies $rightmost(S_a^i) = S_b^i$. Since S_b^i and S_a^i are consecutive, $rightmost(S_a^i) = S_b^i$ implies S_a^i has only one right neighbor and it is S_b^i. since S_a^i has only one right neighbor, S_b^i, it means that $S_a^i = v_{n-1}$ and $S_b^i = v_n$ by Propositions 2 and 3. But then, Procedure 1 is terminated at the ith iteration of the loop, so *b-shift*(i) cannot have been invoked, a contradiction. Therefore, *a-shift*(i) and *b-shift*(i) are mutually exclusive.

Now assume *b-shift*$(i-1)$ is invoked. By similar arguments as above, this implies $rightmost(S_a^{i-1}) = S_b^{i-1}$ and the procedure *Compute_special_vertices(v)* assigns S_a^i as $rightmost(S_a^{i-1})$ itself and reassigns S_b^{i-1} as immediate left neighbor of S_a^i. This makes S_b^{i-1} and S_a^i as consecutive vertices in the ordering. Therefore, $rightmost(S_b^{i-1})$ can never be S_a^i; otherwise it must be by Propositions 2 and 3 that $(S_b^{i-1}) = v_{n-1}$ and $S_a^i = v_n$. But then, Procedure 1 is terminated before the calculation of S_b^i, so *a-shift*(i) cannot have been invoked, a contradiction. Thus *a-shift*(i) and *b-shift*$(i-1)$ are mutually exclusive. □

Proof **(Proof of Proposition** 5**).** The proof is by induction on i, where $i \leq t$. As base case take $i = 2$. It is easy to verify that $A_0 < A_1 < A_2$ since $A_2 = rightmost(A_0)$ and $A_1 = rightmost(A_0) - 1$. Moreover by Proposition 1, A_1 and A_2 are adjacent. Thus for $i = 2$, the statement is true. Now let $i > 2$. As induction hypothesis, assume that $A_0 < A_1 < \ldots < A_{i-1}$ and A_{i-2} is adjacent to A_{i-1}. Since A_{i-1} is adjacent to A_{i-2}, we can infer that $A_{i-2} < A_{i-1} \leq rightmost(A_{i-2})$. According to the the way the special vertices S_a^j and S_b^j are calculated, (see Procedure 1), we have $A_i = rightmost(A_{i-2})$ or $rightmost(A_{i-2}) - 1$; the second case occurs if and only if a corresponding *a-shift*(i) or *b-shift*(i) occurs. We consider both the cases below.

Case 1. $A_i = rightmost(A_{i-2})$: From $A_{i-1} \leq rightmost(A_{i-2})$, we infer $A_{i-1} \leq A_i$. Note that $A_{i-1} = rightmost(A_{i-3})$ or $rightmost(A_{i-3}) - 1$. Since $rightmost(A_{i-3}) \leq rightmost(A_{i-2}) = A_i$, we have $A_{i-1} = A_i$ only if $rightmost(A_{i-3}) = rightmost(A_{i-2})$. However, in this case, there is either *a-shift*(j) or *b-shift*(j) depending on whether $A_{i-1} = S_a^j$ or S_b^j, for $j = \lfloor \frac{i-1}{2} \rfloor$; and A_{i-1} will be fixed to $rightmost(A_{i-3}) - 1$ which is same as $rightmost(A_{i-2}) - 1 = A_i - 1$ in this special case. It follows that $A_{i-1} \neq A_i$ and therefore $A_{i-1} < A_i$, as required.

Case 2. $A_i = rightmost(A_{i-2}) - 1$. This occurs because $rightmost(A_{i-1}) = rightmost(A_{i-2})$, and either *a-shift* or *b-shift* occurs. $A_i = rightmost(A_{i-2}) - 1$ which is the same as $rightmost(A_{i-1}) - 1$ in this special case. Given that $rightmost(A_{i-1}) \geq (A_{i-1} + 1) + 1$ (by Proposition 3), we have $A_i = rightmost(A_{i-1}) - 1 \geq A_{i-1} + 1 > A_{i-1}$, as required.

Since $A_{i-2} < A_{i-1} < A_i$ and A_i is a neighbor of A_{i-2}, by Proposition 1, we see that A_{i-1} is adjacent to A_i as claimed. □

Proof **(Proof of Lemma 2).** Partition C into two internally disjoint paths P_1, P_2 from v to w, and let, for any special vertex z of v, the symbol $\mathrm{NEXT}_t(z), t = 1, 2$, denotes the first vertex of P_t that strictly follows z in the ordering of G. We describe below a useful property of the functions $\mathrm{NEXT}_t(x)$, $t = 1, 2$.

- Suppose first that $z = S_a^i$ for some i.
  - If *a-shift*(i+1) occurs and $\mathrm{NEXT}_t(z) = rightmost(z)$, then $\mathrm{NEXT}_t(z) = S_b^{i+1}$.
  - Otherwise, $\mathrm{NEXT}_t(z) \in (S_a^i, S_a^{i+1}]$.
- Now suppose that $z = S_b^i$.
  - If *b-shift*(i+1) occurs and $\mathrm{NEXT}_t(z) = rightmost(z)$, then $\mathrm{NEXT}_t(z) = S_a^{i+2}$.
  - Otherwise, $\mathrm{NEXT}_t(z) \in (S_b^i, S_b^{i+1}]$.

In the first case $\mathrm{NEXT}_t(z) \leq rightmost(z)$ and unless *a-shift*(i+1) occurs we have $S_a^{i+1} = rightmost(z)$. Even in the case *a-shift*(i+1) occurs, we will have $S_a^{i+1} = rightmost(z) - 1$ and therefore if $\mathrm{NEXT}_t(z) < rightmost(z)$, we still have $\mathrm{NEXT}_t(z) \leq S_a^{i+1}$.

In the second case, we have $rightmost(z) = S_b^{i+1}$, unless *b-shift*(i+1) occurs. Since $\mathrm{NEXT}_t(z) \leq rightmost(z)$, in this case $\mathrm{NEXT}_t(z) \leq S_b^{i+1}$. Even if *b-shift*(i+1) occurs, we will have $S_b^{i+1} = rightmost(z) - 1$.
Therefore if $\mathrm{NEXT}_t(z) < rightmost(z)$ we have $\mathrm{NEXT}_t(z) \leq S_b^{i+1}$ as claimed.

We now proceed to construct the sets X, Y by induction on i. For $i = 1$, we take $X = \{w_1\}$ and $Y = \{u_1\}$, where w_1, u_1 are the neighbors of v on the paths P_1 and P_2. Since both $u_1, w_1 \leq rightmost(v) = S_b^1$ and $S_a^1 = S_b^1 - 1$, it is easy to see that one of them belongs to $(S_a^0, S_a^1]$, and so it is placed in X as x_1, and the other is in $(S_b^0, S_b^1]$, so it can be placed in Y as y_1. Now assume that for $i = p$ the statement of the lemma is true, and we already have distinct vertices $x_1, \ldots, x_p$ in X and distinct vertices $y_1, \ldots, y_p$ in Y. We proceed to define x_{p+1} and y_{p+1}.

Note that for y_{p+1} to be different from all previous x_i, y_i, it is sufficient to ensure that $y_{p+1} \in (S_b^p, S_b^{p+1}]$ since this interval does not intersect any of the intervals $(S_\alpha^{j-1}, S_\alpha^j]$ for $j \leq p$ where $\alpha = a, b$. Moreover $(S_a^p, S_a^{p+1}]$ intersects only with $(S_b^{p-1}, S_b^p]$ from the previously considered intervals, and therefore the only element x_{p+1} has to avoid in the previously constructed x_i, y_i is y_p. It follows that once we establish that x_{p+1}, y_{p+1} are in the appropriate intervals, we just have to show that $x_{p+1} \neq y_p$ and $x_{p+1} \neq y_{p+1}$ in order to complete the proof. In the following let y_p belong to path P_t where $t \in \{1,2\}$, and let P_r be the other path, i.e. $y_p \notin P_r$.

Case 1. *a-shift*$(p+1)$ occurs and $\text{NEXT}_r(S_a^p) = rightmost(S_a^p)$.

Let $x_{p+1} = \text{NEXT}_t(S_b^p)$ and $y_{p+1} = \text{NEXT}_r(S_a^p) = S_b^{p+1} \in (S_b^p, S_b^{p+1}]$. Now clearly $x_{p+1} = \text{NEXT}_t(S_b^p) \neq y_{p+1} = S_b^{p+1}$ since they belong to different paths. Since *a-shift*$(p+1)$ occurs, *b-shift*$(p+1)$ cannot occur. Thus $rightmost(S_b^p) = S_b^{p+1} = y_{p+1} \neq x_{p+1}$. It follows that $x_{p+1} = \text{NEXT}_t(S_b^p) \leq rightmost(S_b^p) - 1 \leq S_b^{p+1} - 1 = S_a^{p+1}$, due to *a-shift*(p+1). It follows that $x_{p+1} \in (S_b^p, S_a^{p+1}] \subseteq (S_a^p, S_a^{p+1}]$. Moreover, since $x_{p+1} > S_b^p$ and $y_p \leq S_b^p$, we have $x_{p+1} \neq y_p$.

Case 2. *b-shift*$(p+1)$ occurs and $\text{NEXT}_t(S_b^p) = rightmost(S_b^p)$.

Let $x_{p+1} = \text{NEXT}_r(S_a^p)$ and $y_{p+1} = \text{NEXT}_r(S_a^{p+1})$. We first note that $x_{p+1} \in (S_a^p, S_a^{p+1}]$: This is because since *b-shift*$(p+1)$ occurs, *a-shift*$(p+1)$ cannot occur. Clearly $x_{p+1} \neq y_p$ since both belong to different paths. Moreover $x_{p+1} \neq y_{p+1}$ since $y_{p+1} > S_a^{p+1}$. Note that $y_{p+1} \leq rightmost(S_a^{p+1}) = S_a^{p+2} = rightmost(S_b^p)$ (by definition of *b-shift*(p+1)), but in this situation, *rightmost*$(S_b^p) = \text{NEXT}_t(S_b^p)$, which is a vertex of path P_t and thus cannot be y_{p+1} which belongs to path P_r. So, $y_{p+1} \leq rightmost(S_b^p) - 1 = S_b^{p+1}$, by the definition of *b-shift*$(p+1)$. Therefore $y_{p+1} \in (S_a^{p+1}, S_b^{p+1}] \subset (S_b^p, S_b^{p+1}]$, as required.

We note that Cases 1 and 2 cannot occur simultaneously by Proposition 4.

Case 3. Neither Case 1 nor Case 2 occurs.

We set $x_{p+1} = \text{NEXT}_r(S_a^p)$ and $y_{p+1} = \text{NEXT}_t(S_b^p)$. Clearly $x_{p+1} \neq y_{p+1}$ since they do not belong to the same path. Also $y_p \neq x_{p+1}$ since $y_p \in P_t$ and $x_{p+1} \in P_r$. Moreover, $x_{p+1} \in (S_a^p, S_a^{p+1}]$, since the special case where this is not applicable is handled in Case 1; and $y_{p+1} \in (S_b^p, S_b^{p+1}]$ since the special case where this is not applicable is handled in Case 2. □

A.3 Examples: Constructing a *k*-cycle with Special Vertices

First we explain the construction of a k-cycle containing v and *pivot*(v) when $k-1$ special vertices are not defined. This happens when the last vertex v_n in the ordering of G is assigned to some special vertex. Let $S_x^j, x \in \{a,b\}$ be the first special vertex computed such that $S_x^j = v_n$. Observe that the value *pivot*(v) is computed only for vertices $v \in V(G)$ such that $|[v, v_n]| \geq k$. Here *pivot*$(v) = v_n = S_x^j$. If $x = a$, then $C' = v, S_a^1, \ldots, S_a^j, S_b^{j-1}, \ldots, S_b^1, v$ or if $x = b$, then $C' = v, S_a^1, \ldots, S_a^j, S_b^j, S_b^{j-1}, \ldots, S_b^1, v$ is a cycle containing the vertices v and *pivot*(v). But this cycle C' may not be a k-cycle since $j \leq \lfloor \frac{k}{2} \rfloor$. (Note that $j < \lfloor \frac{k}{2} \rfloor$ except when $S_x^j = S_a^{\lfloor \frac{k}{2} \rfloor} = v_n$ in the case of odd cycles.) To have a k-cycle C_k containing v and *pivot*(v) we add vertices to C' in the following way.

Let $|V(C')| = k'$ and $k' < k$. Then we add $m = k - k'$ vertices from (v, v_n) to the cycle C' to form the k-cycle C_k. Since $|[v, v_n]| \geq k$ there are m vertices from (v, v_n) that are not in $V(C')$. Let $y_1 y_2 \ldots y_m$ be such m vertices such that

$v < y_1 < y_2 < \cdots < y_m < v_n$ with respect to the ordering. Since the special vertices S_x^i partition the set $[v, v_n]$ into $(S_x^{i-1}, S_x^i], 1 \leq i \leq j$, the m vertices $y_1 y_2 \ldots y_m$ belong to some of these parts. If $\{y_i, y_{i+1}, \ldots, y_{i+t}\} \subseteq [S_x^{p-1}, S_x^p]$, for some $1 \leq p \leq j$, $1 \leq i \leq m$ and $0 \leq t \leq m - i$, then we replace the edge $S_x^{p-1} S_x^p$ in C' by the path $S_x^{p-1}, y_i, y_{i+1}, y_{i+2}, \ldots, y_{i+t}, S_x^p$. The adjacency of these vertices is guaranteed by Proposition 5 and Proposition 1. This process is continued until all the m vertices are added to the cycle C' transforming it to a k-cycle containing both v and $pivot(v) = v_n$.

The following examples illustrate the construction of a k-cycle containing v and $pivot(v)$. Example 1 illustrates the computation of special vertices S_a^i and S_b^i for finding the pivotal vertex with respect to each vertex $v \in V(G)$. It also shows the construction of the canonical even cycle and odd cycle. Example 2 shows the existence of a k-cycle when the computation of S_a^i and S_b^i ends before $i = \lfloor \frac{k}{2} \rfloor$.

Example 1: Consider the proper interval graph G in Fig. 1. The vertices of G are labeled according to the left end ordering. Vertex v_{11} is a cut vertex and hence, G has two biconnected components H_1 and H_2 with ordered vertex sets $\{v_1, \ldots, v_{11}\}$ and $\{v_{11}, \ldots, v_{14}\}$ respectively. Suppose the template is $\mathcal{T} = C_8$ and consider the vertex v_1 in H_1. Then for $1 \leq i \leq 4$, the special vertices S_a^i and S_b^i with respect to v_1 are computed iteratively as follows. (See Fig. 2).

$I_0 : S_a^1 = v_2, S_b^1 = v_3$
$I_1 : S_a^2 = \textit{rightmost}(v_2) = v_5,\ S_b^2 = \textit{rightmost}(v_3) = v_5$
$: \text{Execute } \textit{a-shift}(2) \text{ as } S_a^2 = S_b^2 \text{ and reset } S_a^2 = v_4$
$I_2 : S_a^3 = \textit{rightmost}(v_4) = v_6,\ S_b^3 = \textit{rightmost}(v_5) = v_8$
$I_3 : S_a^4 = \textit{rightmost}(v_6) = v_8$
$: \text{Execute } \textit{b-shift}(3) \text{ as } S_a^4 = S_b^3 \text{ and reset } S_b^3 = v_7$
$: S_b^4 = \textit{rightmost}(v_7) = v_{11}$

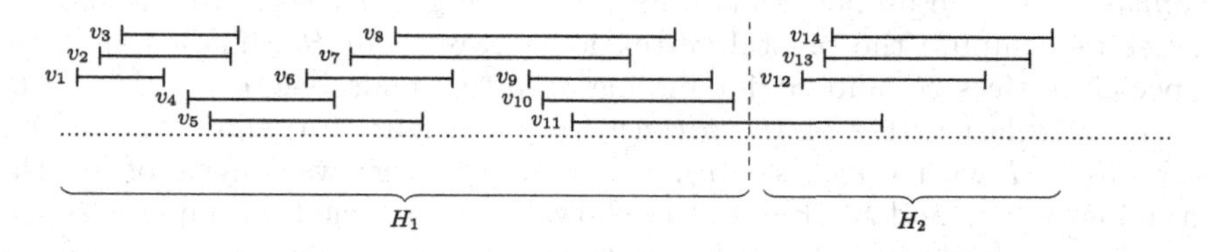

Fig. 1. A proper interval graph G with *two* biconnected components, H_1 and H_2

The canonical cycles for template $\mathcal{T} = C_8$ and $\mathcal{T} = C_9$ are shown in Fig. 3, as (a) and (b) respectively. The solid lines are the edges of the cycles.

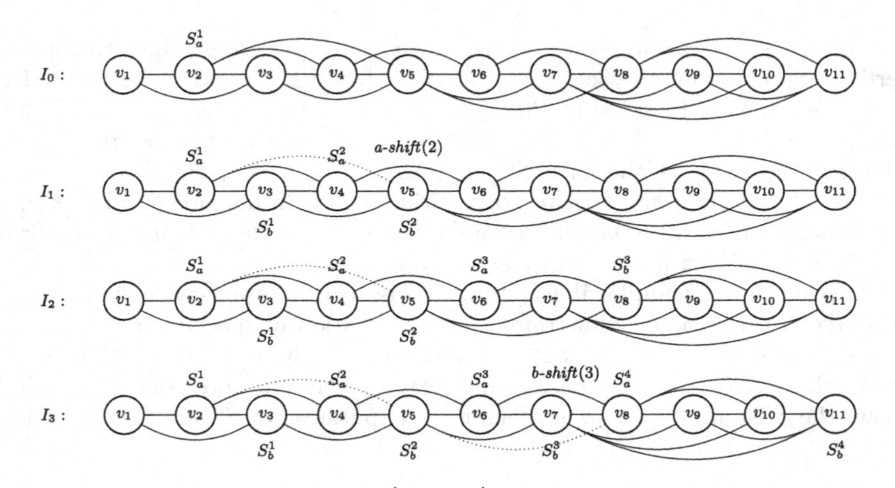

Fig. 2. Computational stages of S_a^i and S_b^i, $1 \leq i \leq \lfloor \frac{k}{2} \rfloor$ for the vertex $v_1 \in V(H_1)$

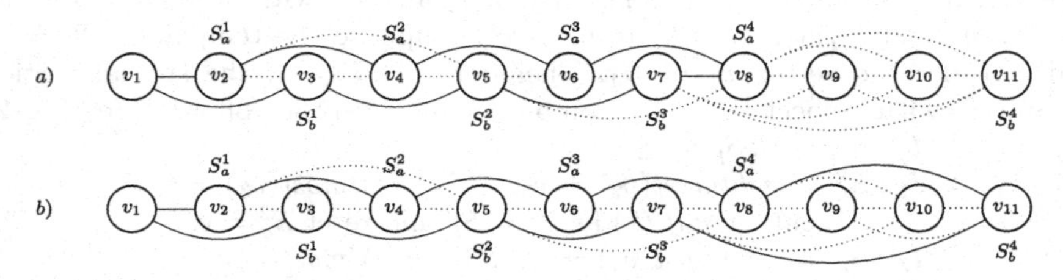

Fig. 3. a) The canonical cycle for template $\mathcal{T} = C_8$ containing v_1 and *pivot*(v_1) b) the canonical cycle for template $\mathcal{T} = C_9$ containing v_1 and *pivot*(v_1)

Example 2: For the proper interval graph G in Fig. 1, let the template be $\mathcal{T} = C_4$. Let us compute the pivotal vertex for vertex v_7 in H_1. The computation of special vertices S_a^i and S_b^i for finding *pivot*(v_7) ends before $i = \lfloor \frac{k}{2} \rfloor$. Since $S_b^1 = v_{11}$, the last vertex in H_1, *pivot*(v_7) $= v_{11}$. Figure 4(a) shows the induced subgraph of H with vertex set $[v_7, v_{11}]$. Figure 4(b) shows a cycle of length 3 containing v_7, S_a^1, and S_b^1. Figure 4(c) shows an even cycle for template $\mathcal{T} = C_4$ containing v_7 constructed by replacing the edge $v_7 S_a^1$ by the path v_7, v_8, S_a^1.

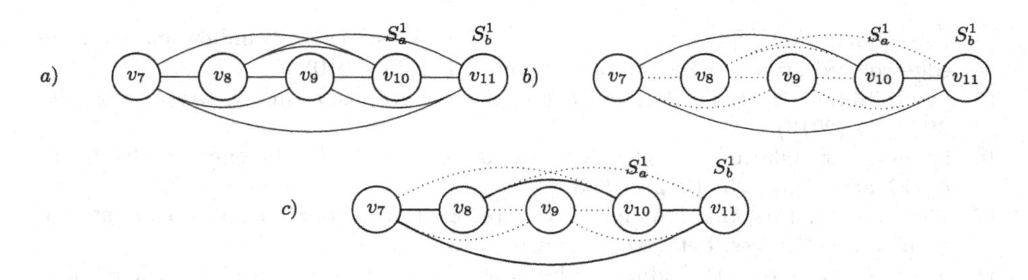

Fig. 4. a) S_a^i and S_b^i vertices defined for the vertex $v_7 \in V(H_1)$ b) A cycle of length 3 (solid lines) containing v_7, S_a^1 and S_b^1 c) An even cycle for template $\mathcal{T} = C_4$ containing v_7 and $pivot(v_7)$

References

1. Balogh, J., Molla, T.: Long rainbow cycles and hamiltonian cycles using many colors in properly edge-colored complete graphs. Eur. J. Combin. **79**, 140–151 (2017)
2. Bertossi, A.A., Pinotti, M.C.: Mappings for conflict-free access of paths in bidimensional arrays, circular lists, and complete trees. J. Parallel Distrib. Comput. **62**, 1314–1333 (2002)
3. Chandran, L.S., Das, A., Issac, D., van Leeuwen, E.J.: Algorithms and bounds for very strong rainbow coloring. In: Bender, M.A., Farach-Colton, M., Mosteiro, M.A. (eds.) LATIN 2018. LNCS, vol. 10807, pp. 625–639. Springer, Cham (2018). https://doi.org/10.1007/978-3-319-77404-6_46
4. Chen, H.: Long rainbow paths and rainbow cycles in edge colored graphs - a survey. Appl. Math. Comput. **317**, 187–192 (2018)
5. Corneil, D.G., Kim, H., Natarajan, S., Olariu, S., Sprague, A.P.: A simple linear time recognition of unit interval graphs. Inf. Process. Lett. **55**, 99–104 (1995)
6. Das, S.K., Finocchii, I., Petreschi, R.: Conflict-free star-access in parallel memory systems. J. Parallel Distrib. Comput. **66**, 1431–1441 (2006)
7. Das, S.K., Pinotti, M.C.: Conflict-free access to templates of trees and hypercubes in parallel memory systems. In: Jiang, T., Lee, D.T. (eds.) COCOON 1997. LNCS, vol. 1276, pp. 1–10. Springer, Heidelberg (1997). https://doi.org/10.1007/BFb0045066
8. Das, S.K., Pinotti, M.C.: Optimal mappings of q-ary and binomial trees into parallel memory modules for fast and conflict-free access to path and subtree templates. J. Parallel Distrib. Comput. **60**, 998–1027 (2000)
9. Deng, X.T., Hell, P., Huang, J.: Linear-time representation algorithms for proper circular arc graphs and proper interval graphs. SIAM. J. Comput. **25**, 390–403 (1996)
10. Golumbic, M.C.: Algorithmic Graph Theory and Perfect Graphs. Annals of Discrete Mathematics, p. 57. Elsevier, Amsterdam (2004)
11. Gyárfás, A., Sárkäzy, G.N.: Induced colorful trees and paths in large chromatic graphs. Electron. J. Combin. **23**, P4.46 (2016)
12. Kostochka, A.V., Yancey, M.: Large rainbow matchings in edge-colored graphs. Combin. Probab. Comput. **21**, 255–263 (2012)
13. Kramer, F., Kramer, H.: A survey on the distance-coloring of graphs. Discrete Math. **308**, 422–426 (2008)

14. LeSaulnier, T.D., Stocker, C.J., Wenger, P.S., West, D.B.: Rainbow matching in edge-colored graphs. Electron. J. Combin. **17**, N26 (2010)
15. Li, X., Shi, Y., Sun, Y.: Rainbow connections of graphs: a survey. Graph. Combin. **29**, 1–38 (2013)
16. Looges, P.J., Olariu, S.: Optimal greedy algorithms for indifference graphs. Comput. Math. Appl. **25**, 15–25 (1993)
17. Panda, B.S., Das, S.K.: A linear time recognition algorithm for proper interval graphs. Inf. Process. Lett. **87**, 153–161 (2003)
18. Scott, A., Seymour, P.: Induced subgraphs of graphs with large chromatic number IX: rainbow paths. Electron. J. Combin. **24**, 2.53 (2017)
19. de Werra, D.: Some uses of hypergraph in timetabling. Asia-Pac. J. Oper. Res. **2**, 2–12 (1985)

Minimum Consistent Subset of Simple Graph Classes

Sanjana Dey[1(✉)], Anil Maheshwari[2], and Subhas C. Nandy[1]

[1] ACM Unit, Indian Statistical Institute, Kolkata, India
info4.sanjana@gmail.com
[2] School of Computer Science, Carleton University, Ottawa, Canada

Abstract. In the minimum consistent subset (MCS) problem, a connected simple undirected graph $G = (V, E)$ is given whose each node is colored by one of the colors $\{c_1, c_2, \ldots, c_k\}$, and the objective is to compute a subset $\mathcal{C} \subseteq V$ such that for each node $v \in V$, its set of nearest neighbors in $\mathcal{C}$ (with respect to the hop-distance) contains at least one vertex of the same color as v. The decision version of the MCS problem is NP-complete for general graphs. Even for planar graphs, the problem remains NP-complete. We will consider some simple graph classes like path, caterpillar, bi-chromatic spider, bi-chromatic comb, etc., and propose polynomial-time algorithms for solving the problem on those graphs.

1 Introduction

The geometric variation of the consistent subset problem was first introduced by Hart [4]. Let P be a set of multi-colored points in the plane. A consistent subset of P is a set $S \subseteq P$ such that for every point $p \in P \setminus S$, the closest point among the points in S has the same color as that of p. In the minimum consistent subset problem, the objective is to find a consistent subset of P with minimum cardinality. In [6], it is shown that the decision version of this problem is NP-complete even for three-colored point sets in $\mathbb{R}^2$. In the same paper, the author proposed an $O(n^2)$ time algorithm for the minimum consistent subset problem with two colored point sets where one set is a singleton. The general version of the minimum consistent subset problem for bicolored points is NP-hard [5]. In [1], the consistent subset problem for collinear points is solved in $O(n^2)$ time. Recently, a sub-exponential time algorithm for the consistent subset problem in $\mathbb{R}^2$ is proposed in [2]. It is also shown that in $O(n \lg n)$ time one can test whether the size of the minimum consistent subset of a bicolored point set in $\mathbb{R}^2$ is 2 or not. In the same paper, an $O(n)$ time algorithm is presented for the collinear points, thereby improving the previous running time by a factor of $\Theta(n)$. They also propose an $O(n^6)$ time dynamic programming algorithm for points arranged on two parallel lines.

The general definition of the *minimum consistent subset* (MCS) problem was given in [3] where a ground set X and a constraint t is given. The objective is

A. Mudgal and C. R. Subramanian (Eds.): CALDAM 2021, LNCS 12601, pp. 471–484, 2021.
https://doi.org/10.1007/978-3-030-67899-9_37

to compute subsets $X' \subseteq X$ that satisfy the constraint t. They proposed the following application for reducing the data communication overheads: *Transmit the pair (t, X') to a user. The user can classify (its color) each element of the ground set X using X' and the constraint t.* In the geometric variation of MCS an appropriate *distance* measure serves as the constraint. In this paper, we study the following graph-theoretic version of the consistent subset problem:

> Let $G = (V, E)$ be a graph whose nodes are classified into k classes, namely $V_1, \ldots, V_k$. The objective is to choose subsets $V_i' \subseteq V_i$, $i = 1, \ldots, k$ such that for each member $v \in V$, if $v \in V_i$ then among its nearest neighbors in $\cup_{i=1}^{k} V_i'$ there is a node of V_i', and $\sum_{i=1}^{k} |V_i'|$ is minimum. The distance between a pair of nodes u and v is the number of nodes in the shortest path from u to v, and will be referred to as hop-distance(u, v).

In [3], an application of the geometric version of consistent subset problem is mentioned in the context of reverse clustering. Here, a ground (point) set $P \in \mathbb{R}^2$ and a training (colored point) set P' are given. The points in P are classified (colored) according to their respective nearest neighbor in P'. In the context of graphs, this application is not directly applicable since a node $v \in V$ may have several nearest neighbors of different colors in V'. Thus, the applications suggested may not be strictly applicable here. For each such element, one needs to compare its properties with a limited number of elements, to decide its correct group. As an example, we can mention an application of our problem in the citation networks. Each author is a node and there is an edge between two nodes if an author cites the other. Also, each author is assigned with a category stating his/her research interests. One can use the consistent subset of this graph to prepare the editorial board of a journal so that each paper is assigned to an editor of the approximately similar research interest.

Our Contribution: To the best of our knowledge, there is not enough study about the minimum consistent subset problem in graphs. The status of the problem is even unknown for trees. In [1], it was shown that the decision version of the minimum consistent subset problem is NP-complete for general graphs $G = (V, E)$. The hardness reduction uses the minimum dominating set problem on an undirected graph. Using the same reasoning, one can claim that the MCS problem for a planar graph is NP-hard. In this paper, we present polynomial-time algorithms for the MCS problem for some simple graph classes, namely, (i) path, (ii) caterpillar, (iii) spider and (iv) comb. We first consider the bichromatic version of these problems. We introduce the concept of *run*, *gate* and *block* to partition the graph into subgraphs so that each subgraph can be handled independently with limited interactions with its 'neighboring' subgraphs. The non-trivial part of this approach using these structures is to design techniques to carefully handle the limited interaction. The basic idea is to use these structures from the given graph G to create a new graph H, called *overlay graph.* In doing so, we reduce the minimum consistent subset problem on G to finding a shortest s-t path in this new graph H. Minor tailoring of the algorithm for path works for the cycle graph. The multi-chromatic version for paths and cycles can also be solved using the same algorithm in linear time. Algorithms for the bichromatic version of the MCS problem are proposed for caterpillar graph, spider graph and comb graph with time complexity $O(n)$, $O(n^2)$, and $O(n^2)$ respectively, where n is the number of nodes in the graph. In the rest of the paper, we will use $\mathcal{C}$ to denote a minimum consistent subset of the input graph G.

2 Bichromatic Paths

In a path (or line) graph $G = (V, E)$, the nodes in V are listed in the order $p_1, p_2, \ldots, p_n$. Each pair of consecutive nodes define an edge of the graph, i.e., $E = \{(p_i, p_{i+1}), i = 1, 2, \ldots, n-1\}$. Each node has degree two except the two terminal nodes of the path that have degree one. In a bichromatic path graph, each node is assigned with one color in {*red*, *blue*}. A *run* is a consecutive set of nodes of the same color on the path (see Fig. 1). In a run of length 1, the node itself is always selected.

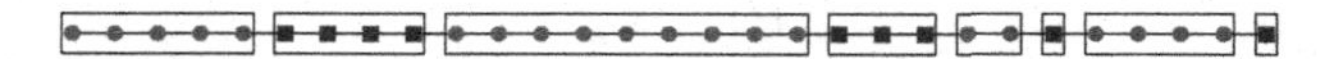

Fig. 1. Runs in a path graph: each black rectangle denotes a run. (Color figure online)

Lemma 1. *In the minimum consistent subset of a path graph, each run will have at least one and at most two nodes in $\mathcal{C}$. Moreover, exactly one node will be sufficient*[1] *from the first and the last run.*

Algorithm: Consider a pair of adjacent runs R_j and R_{j+1}. Assume, without loss of generality, $|R_j| \leq |R_{j+1}|$. For each member $p_i \in R_j$, there exists at most three members, say $p_k, p_{k+1}, and\, p_{k+2} \in R_{j+1}$, such that if p_i is included in $\mathcal{C}$ then any one of those three members of R_{j+1} must be included in $\mathcal{C}$ to satisfy the consistency property of the boundary nodes of R_j and R_{j+1} that are adjacent to each other. Thus, $(p_i, p_{k+\theta})$ forms a *valid-pair* for $\theta = 0, 1, 2$ (Fig. 2).

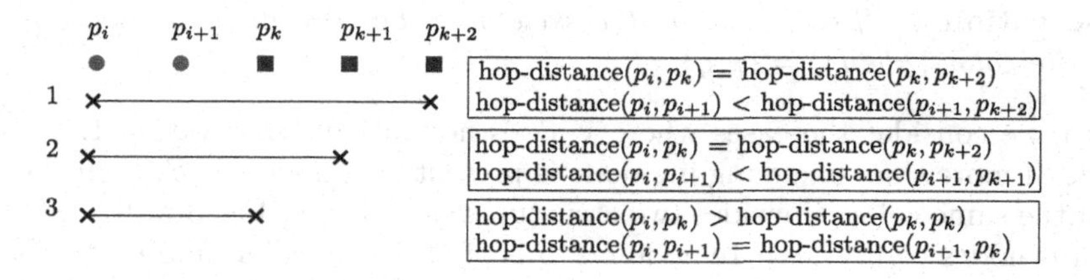

Fig. 2. Valid pairs: (p_i, p_k), (p_i, p_{k+1}) and (p_i, p_{k+2}).

[1] As the first run will have just one other run to its right, we need to select a node depending on its closest node of opposite color in the second run. Similarly for the last run.

We define the *overlay graph* $H = (V \cup \mathcal{D}, \mathcal{F})$ as follows. The nodes of H are the nodes of G, and k dummy nodes $\mathcal{D} = \{d_1, \ldots, d_k\}$, where k is the number of runs. The edges in the set $\mathcal{F}$ are of two types. For each valid-pair, we add a directed *type-1* edge in $\mathcal{F}$. For each node p_i in a run R_j we add two directed *type-2* edges (p_i, d_j) and (d_j, p_i) in $\mathcal{F}$. The weight of each *type-1* edge is 0. The *type-2* edges incident to $\mathcal{D} \backslash \{d_1, d_k\}$ have weight 1. Each *type-2* edge incident to $\{d_1, d_k\}$ has weight 0. For the complete demonstration of the graph H. A *forward s-t* path is a path from s to t where the indices of the p_i nodes appear in increasing order. Now we find the shortest forward s-t path with $s = d_1$ and $t = d_k$ in the graph H, and remove the d_j's to obtain MCS of the original path graph G.

Theorem 1. *The shortest s-t path of the* overlay graph H *gives the minimum consistent subset of the path G, and it executes in* $O(n)$ *time.*

Note: A minor tailoring of the same algorithm works for a bichromatic cycle graph $G = (V, E)$, where the nodes in V are connected in a closed chain.

3 Bichromatic Caterpillar Graph

A caterpillar $G = (V, E)$ is a tree in which every node is within distance 1 from a path in G, called *skeleton*. The nodes in V that are not on the skeleton are termed as *dangling nodes*. Thus, $V = S \cup D$, where the nodes in S are on the skeleton, and D contains the dangling nodes from all nodes in S. I.e., $D = \bigcup_{v \in S} D_v$, where D_v is the set of nodes dangling at the node $v \in S$. Each dangling node in D_v is at distance 1 from a node $v \in S$. In this section, we will consider the MCS problem for a bichromatic caterpillar where each node of V is colored by *red* or *blue*. The cases $|V| = 1$ or 2 can be solved trivially. If $|V| \geq 3$, then we assume that the first and the last node of the skeleton consist of at least one dangling node. If the first (resp. last) node v of the skeleton does not have any dangling node and the node adjacent to v is u, we can consider node v as the dangling node of u.

Observation 1. *If any node on the skeleton S has two dangling nodes p, q of opposite colors, then* $\mathcal{C} = \{p, q\}$.

So, we consider the cases where V does not satisfy Observation 1. In other words, if more than one dangling node is present at a node $v \in S$, then they are all of the same color. Consider two structures as shown in Fig. 3(a, b) consisting of three nodes (p, q, r), where node $r \in S$ is of arbitrary color, and is attached to two nodes p and q of opposite colors. As we are considering instances that do not satisfy Observation 1, both p and q cannot be dangling at node r. Without loss of generality, let us assume that $p \in S$ and $q \in D_r$. We now define the concept of *gate* where node r is called the *base* of the gate. The two cases shown in Fig. 3(a) and 3(b) are referred to as *left-gate* and *right-gate* respectively. The existence of a left-gate (resp. right-gate) (p, q, r) implies that by choosing $\{p, q\}$ in $\mathcal{C}$ all the

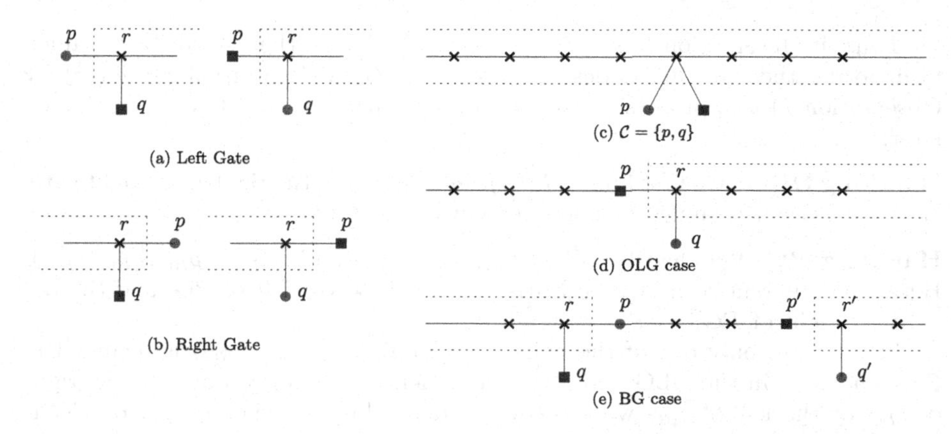

Fig. 3. (a) Left gate. (b) Right gate. (c) Visulalizing Observation 1. (d) A caterpillar with only left gate (OLG). (e) A caterpillar with both gates (BG). Dotted regions signify the part of the caterpillar covered by the gates.

nodes to the right (resp. left) of that gate, including the base node r, are covered[2], and we need to compute the minimum consistent subset of the subgraph of G that is attached with $p \in S$ at its left (resp. right) side. Here, the following four situations need to be considered depending on the occurrence/non-occurrence of left and/or right-gate.

NG: There is *no* gate in G.
OLG: Only *left-gate*(s) is/are present in G. The leftmost left-gate is called LG.
ORG: Only *right-gate*(s) is/are present in G. The rightmost right-gate is called RG.
BG: Both *left-gate*(s) and *right-gate*(s) are present in G. Here LG and RG are defined as above.

Handling OLG: We find $LG = (p_{LG}, q_{LG}, r_{LG})$. Let S_R and S_L be two sets of nodes in S that are to the right of r_{LG} and to the left of p_{LG}, respectively.

Observation 2. *(a) If $\mathcal{C}$ contains $\{p_{LG}, q_{LG}\}$ and no node from S_R then all the nodes in $S_R \bigcup (\cup_{u \in S_R} D_u)$, irrespective of their colors, are covered by $\mathcal{C}$.*

(b) If $D_{p_{LG}} \neq \emptyset$ or the color of all the dangling nodes ($D_{p_{LG}}$) of p_{LG} are of same color as that of p_{LG}, then no members of $D_{p_{LG}}$ need to be included in $\mathcal{C}$; otherwise all the members of $D_{p_{LG}}$ are included in $\mathcal{C}$.

As LG is the left-most left gate, there does not exist any gate in S_L. Thus, we solve the MCS problem for S_L as the NG case.

Observation 3. *If all the nodes in S_R are of color(q_{LG}), then instead of including $\{p_{LG}, q_{LG}\}$ in $\mathcal{C}$ an appropriate pair $\{p', q'\}$ may be included in $\mathcal{C}$, where $p' \in S_L$ and $q' \in S_R$, to reduce the size of the MCS $\mathcal{C}$.*

[2] By the term "a node $v \in V$ is covered by $\mathcal{C}$" we mean that the nearest (or one of the nearest) node of node v in $\mathcal{C}$ is of color(v).

Thus, if Observation 3 is satisfied in the OLG case then $\mathcal{C}$ can be obtained by ignoring the dangling nodes at node r of LG (all are of color color(q) by Observation 1) and processing the problem instance as the NG case, explained later.

Handling ORG: We first find $RG = (p_{RG}, q_{RG}, r_{RG})$, the right-most right gate. Next, this case is handled analogously as the case OLG.

Handling BG: We identify $LG = (p_{LG}, q_{LG}, r_{LG})$ and $RG = (p_{RG}, q_{RG}, r_{RG})$. Here, two different scenarios may arise: (i) LG is to the left of RG, and (ii) LG is to the right of RG.

In Case (i), only one of the pairs (p_{LG}, q_{LG}) and (p_{RG}, q_{RG}) is included in $\mathcal{C}$ as was done in the OLG and ORG case. Hence, the portion of S to the right of r_{LG} or the left of r_{RG} will be covered depending on whether LG or RG is considered for inclusion in $\mathcal{C}$. If (p_{LG}, q_{LG}) (resp. (p_{RG}, q_{RG})) be considered for inclusion in $\mathcal{C}$ then S_L (resp. S_R) satisfies an NG case, and needs to be solved separately. In our algorithm, we will obtain the solutions by considering LG and RG for inclusion in $\mathcal{C}$ separately, and then choose the solution of smaller size.

In Case (ii), The nodes $\{p_{LG}, q_{LG}, p_{RG}, q_{RG}\}$ need to be included in $\mathcal{C}$. As mentioned earlier, if the color of all the nodes in $D_{p_{LG}}$ (if any) are different from color(p_{LG}), then $D_{p_{LG}}$ needs to be included in $\mathcal{C}$. Similarly, if the color of all the nodes in $D_{p_{RG}}$ is different from color(p_{RG}), then $D_{p_{RG}}$ needs to be included in $\mathcal{C}$. Thus, the portion to the left of r_{RG} and to the right of r_{LG} are covered. Here also Observation 3 may apply for LG or RG or both. Accordingly, the unsolved part S_{mid} (the portion between p_{RG} and p_{LG}) will be defined. We process S_{mid} as the NG case.

Handling NG: We now describe the method of handling a part S' of S that is not covered by any gates. It may happen that $S = S'$. Similar to the concept of *run* in Sect. 2, here we define *block* as a connected component in S' of the same color (see Fig. 4, where each block is highlighted by a box). Let us recall from Observation 1 that the dangling nodes (if any) attached to any node of S' are of the same color. As S' does not contain any gates, we have the following observation.

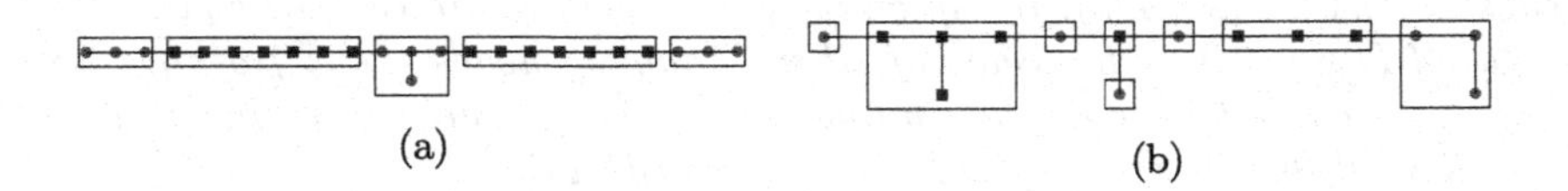

Fig. 4. Caterpillars with the blocks highlighted.

Observation 4. *(a) If a pair of adjacent nodes (u and v say) on the skeleton are of opposite color then each of them (u and v) can not have any dangling node of its own color.*

(b) *If a block on the skeleton has exactly one node (say w), then the dangling nodes (if any) of w are all of color different from that of w (as mentioned in part (a)). We will name such a node w as split-node.*

Lemma 2. *Each block in the NG scenario will have at least one and at most two representatives in $\mathcal{C}$.*

By Observation 4 and Lemma 2, each split node (along with its dangling nodes) is included in $\mathcal{C}$. Thus, S' is further divided using split-nodes. Let us now consider each of these unsolved parts separately and solve as in Sect. 2.

Theorem 2. *The proposed algorithm for the caterpillar is correct, and produces optimum result in $O(n)$ time.*

4 Bichromatic Spider Graph

In a bichromatic *spider graph* $G(V, E)$, $V = \{v\} \cup_{i=1}^{k} V_i$; v is called the *center*, which may be of color *red* or *blue*, and each V_i is a path of bi-colored nodes of length n_i whose one end is connected with v. The set E are the edges that form the k paths. $|V| = \sum_{i=1}^{k} |V_i| + 1$, and $|E| = \sum_{i=1}^{k} |V_i|$. We will refer each path V_i as the *leg* of the spider. In each V_i, the run attached to v is referred to as *first run* of that leg, and will be denoted as ρ_i. The subsequent runs in V_i are numbered accordingly. We will use $\mathcal{C}$ to denote the minimum cardinality consistent subset for the spider graph G. We will also use $\mathcal{C}(u)$ to denote a minimum size consistent subset of G among all possible consistent subsets that contain u.

Observation 5. *$\mathcal{C}$ must contain (i) at least one member of the set $(\cup_{i=1}^{k} \rho_i) \bigcup \{v\}$, and (ii) at least one member from each run, excepting its* first run, *in every leg of the spider.*

In order to compute the minimum consistent subset, we need to consider the following three situations depending on the color of the first run ρ_i of each leg V_i in the spider: (i) ρ_i, $i = 1, 2, \ldots, k$ are of same color, but it is different from that of v, (ii) ρ_i, $i = 1, 2, \ldots, k$ are of same color as that of v, and (iii) ρ_i, $i = 1, 2, \ldots, k$ are of different colors.

If the input instance satisfies case (i), then v must be included in $\mathcal{C}$. We compute a minimum consistent subset $\mathcal{C}'$ as $\mathcal{C}' = (\cup_{i=1}^{k} \mathcal{C}_i) \bigcup \{v\}$, where $\mathcal{C}_i$ is the minimum consistent subset of the path (leg) $V_i \cup \{v\}$ assuming that v is chosen in $\mathcal{C}_i$. This can be computed as in Sect. 2 with $t = v$. We set $\chi = \sum_{i=1}^{k} |\mathcal{C}_i| + 1$.

Now, we will consider case (ii), where v and ρ_i, $i = 1, 2, \ldots, k$ are of same color. We initialize $\mathcal{C}$ and χ as in the earlier case, i.e., $\mathcal{C} = (\cup_{i=1}^{k} \mathcal{C}_i) \bigcup \{v\}$. Next, we test whether the size of the minimum consistent subset can be improved if v is not chosen in $\mathcal{C}$.

Let $u \in \cup_{j=1}^{k} \rho_j$, and U be the path segment from v to u. We use (i) $\mathcal{C}_i(u)$ to denote the minimum sized consistent subset of $V_i \cup U$ where u is included, and it is the node in $\mathcal{C}_i(u)$ closest to v, and (ii) $\hat{\mathcal{C}}_i(u)$ is the minimum sized consistent subset of the path $(V_i \setminus U) \cup \{u\}$ that includes u (see Fig. 5).

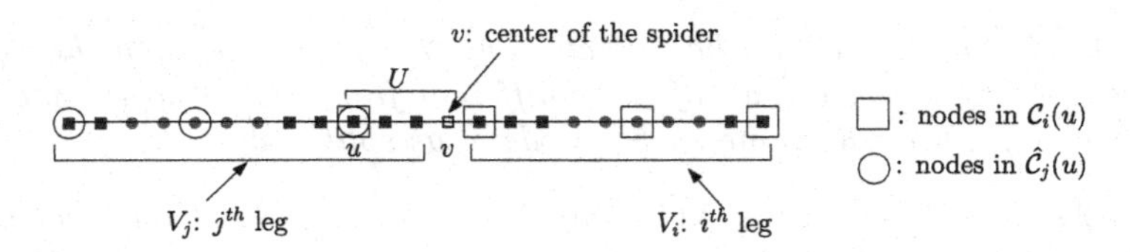

Fig. 5. $\mathcal{C}_i(u)$: optimum solution for $V_i \cup U$, and $\hat{\mathcal{C}}_j(u)$: optimum solution for $V_j \setminus U \cup \{u\}$:

Lemma 3. *If $v \notin \mathcal{C}$ then there exists at least one node $u \in \cup_{i=1}^{k} \rho_i$ in $\mathcal{C}$. If the node $u \in \rho_j$ belongs to $\mathcal{C}$, and is closest to v with respect to hop-distance, then $\mathcal{C} = \hat{\mathcal{C}}_j(u) \bigcup (\cup_{i=1,\ldots,k, i \neq j} \mathcal{C}_i(u))$.*

$\mathcal{C}_i(u)$ is also computed using the algorithm of Sect. 2 for the graph G constructed for the path $V_i \cup U$. Note that, $\mathcal{C}_i(u)$ may contain another node $w \in \rho_i$ (in addition to u). But, as u is closest to v among the nodes in $\mathcal{C}_i(u)$, w must satisfy hop-distance$(v, u) \leq$ hop-distance(w, u). Thus, while constructing the overlay graph, we add the edges (w, u) for $w \in \rho_i$ satisfying hop-distance$(v, u) \leq$ hop-distance(w, u). If there exists any edge (w, u) with color$(w) \neq$ color(u) (from the run of V_i adjacent to ρ_i to the node u), we must have hop-distance$(v, u) \leq$ hop-distance(w, u) (for the consistency of node v in $\mathcal{C}_i(u)$). Needless to say no edge (w, u) will be present in the graph for $w \in U$. Now, the shortest path in this overlay graph will produce a minimum size consistent subset $\mathcal{C}_i(u)$ for $V_i \cup U$.

If case (iii) appears, one needs to consider the gates around the node v. Let us name the run adjacent to the node v along V_i (resp. V_j) as $R \equiv \rho_i$ of color *red* (resp $B \equiv \rho_j$ of color *blue*). If two nodes $p \in R$ and $q \in B$ in $\mathcal{C}$ with hop-distance(p, v) = hop-distance(q, v), then all the nodes in $V \setminus \{V_i \cup V_j\}$ are covered with respect to their consistency. We need to add the minimum consistent subsets $\mathcal{C}_i(p)$ for V_i that includes p, and $\mathcal{C}_j(q)$ for V_j that includes q. The size of the consistent subset χ is updated if the existing $\chi > |\mathcal{C}_i(p)| + |\mathcal{C}_j(q)|$. Note that, the hop-distance of u and w from v may be anything in $\{1, 2, \ldots, \min(|R|, |B|)\}$. A single execution of the algorithm for the path (Sect. 2) returns the size of all the consistent subsets of V_i (resp. V_j) with every node u_α, $\alpha \in \{1, \ldots, |R|\}$ (resp. w_β, $\beta \in \{1, \ldots, |B|\}$) as its element that is closest to v. We run the algorithm of Sect. 2 for every leg V_i of the spider, and the total time needed is $O(|V|)$. Now, we consider each pair of legs, say V_i and V_j whose runs $R = \rho_i$ and $B = \rho_j$ adjacent to v are of different colors, say *red* and *blue*. Assume that color(v) = red. For each $\alpha \in \{1, 2, \ldots, \min(|R|, |B|)\}$, compute $\chi_\alpha^{ij} = \min(\mathcal{C}(v_{\alpha-1}^i), \mathcal{C}(v_\alpha^i), \mathcal{C}(v_\alpha^j))$, and set $\mathcal{C}^{ij} = \min_{\alpha=1}^{\min(|R|,|B|)} \chi_\alpha^{ij}$. If color$(v)$ = blue then similar steps are needed to compute $\mathcal{C}^{ij}$. Finally, report the optimum solution $\mathcal{C}$ satisfying $|\mathcal{C}| = \min\{|\mathcal{C}_{ij}| \,\big|\, \text{for all } i, j \text{ such that color}(\rho_i) \neq \text{color}(\rho_j)\}$. The entire execution needs time $O(n + (\max_{i=1}^{k} \rho_i))^2 = O(n^2)$.

Theorem 3. *The proposed algorithm correctly computes $\mathcal{C}$ of a spider graph G in $O(n^2)$ time, where n is the number of nodes in the graph G.*

5 Bichromatic Comb Graph

A *comb graph* $G = (V, E)$ consists of a path S of m nodes, called *skeleton*, and each node $p_i \in S$ is attached with a path $D(p_i)$ (called *leg*) of size n_i (≥ 1) that includes p_i also. Thus, $|D(p_i) \cap S| = 1$. Here, $V = \cup_{i=1}^{m} D(p_i)$ with $|V| = \sum_{i=1}^{m} n_i = n$. The edges in E are defined by the edges in the path S and the dangling path $D(p_i)$ at each skeleton vertex p_i, $i = 1, \ldots, m$. As in other problems, we assume that the nodes in G are bicolored. The objective is to choose a minimum consistent subset $\mathcal{C} \subseteq V$ for the graph G.

We will use the following notations to describe our algorithm:

- A *run* in the skeleton is a maximal set of consecutive nodes having the same color. Assume that $S = S_1 \cup \ldots \cup S_k$, where S_j is the j-th run; k is the number of runs in S.
- For a node $p_i \in S$, $\Psi(p_i)$ denotes the run in $D(p_i)$ attached to the node p_i.
- As in Sect. 3, here also we define a *block* for a connected set of nodes of the same color. For an element $p_i \in S_j$, $B(p_i)$ is the block of nodes $\cup_{q \in S_j} \Psi(q)$, each of color(p_i). Thus $B(p_i)$ is same for each $p_i \in S_j$.
- $L(p_i) \subset S$ is the subset of nodes in S that are to the left of $p_i \in S$. Similarly, $R(p_i) \subset S$ is the subset of nodes in S that are to the right of $p_i \in S$.

We will use the idea of Sect. 2 to formulate the problem as the shortest s-t path problem of an overlay graph $H = (U, F)$ whose nodes $U = \cup_{i=0}^{m+1} \Psi(p_i)$, where $\cup_{i=1}^{m} \Psi(p_i)$ is the union of the run of all the legs attached to the skeleton, $\Psi(p_0) = s$, and $\Psi(p_{m+1}) = t$. An edge $(q, r) \in F$, $q \in \Psi(p_\ell)$, $r \in \Psi(p_{\ell'})$ if $\ell \neq \ell'$, i.e., q, r are not in the same leg. The cost of the edge $w(q, r)$ is computed as follows. For any two nodes p_ℓ and $p_{\ell'}$, we have the following two scenarios:

- If $B(p_\ell) = B(p_{\ell'})$, i.e., p_ℓ and $p_{\ell'}$ belong to the same block. We identify a pair of consecutive vertices $\theta, \theta' \in S$ such that θ (resp. θ') is nearer to p_ℓ (resp. $p_{\ell'}$).
- If p_ℓ and $p_{\ell'}$ belong to two adjacent runs, and there exists a pair of consecutive vertices $\theta, \theta' \in S$ with color(θ) $\neq$ color(θ') such that θ (resp. θ') is nearer to p_ℓ (resp. $p_{\ell'}$)

Now, $w(q, r)$ is the sum of the sizes of the consistent subsets of (i) $D(p_\alpha)$ for all nodes p_α on the skeleton from p_ℓ to θ with q as the only node in the last run of the path $D(p_\alpha) \oplus \{p_\alpha, \ldots, q\}$, and (ii) $D(p_\beta)$ for all nodes p_β on the skeleton from $p_{\ell'}$ to θ' with r as the only node in the last run of the path $D(p_\beta) \oplus \{p_\beta, \ldots, r\}$, where $\oplus$ is the concatenation operator.

Before describing the algorithm, we first state the following preprocessing phase.

Preprocessing: For each node $q \in \cup_{i=1}^{m} \Psi(p_i)$, we create an array σ_q of size m as stated below. Let p_i and p_j belong to the same run, say S_α, of the skeleton, and $q \in \Psi(p_j)$. The i-th node of σ_q contains the size of the *constrained* minimum consistent subset of $D(p_i)$, denoted by $\mathcal{C}(D(p_i), q)$, with the constraint that only q ($\in B(p_i)$) from the last run is in that consistent subset (see Fig. 7(b)).

Step 1: Generation of $\mathcal{C}(D(p_i), q)$ for all the nodes $q \in \cup_{i=1}^{m} \Psi(p_i)$.

- Let $p_i \in S_\alpha$. Compute $\mu = \max_{q \in B(p_i)}$ hop-distance(p_i, q), and let X be a chain of μ nodes of *color*(p_i). Create a path $\Pi = D(p_i) \oplus X$. The first run of Π starts from the leaf node of $D(p_i)$ and its last run is $\Psi(p_i) \oplus X$.
- Create the overlay graph H for the path Π as in Sect. 2; its s node is connected to all the members in the first run of Π, and t node is connected to all the nodes in the last run of Π. Each node $q \in \Pi$ is attached with a weight field $w(q)$, initialized with ∞.
- We execute the algorithm of Sect. 2 on the path Π. After execution of the algorithm, it will contain the size of the minimum consistent subset of $D(p_i)$ that contains the node q of the last run.
- Note that, as in Sect. 2, here also in the minimum consistent subset of Π, the members present from both the first run and the last run are exactly one. Thus, for each node $q \in \Pi$, if the weight $w(q) = \infty$ then it implies that there does not exist any consistent subset of $D(p_i)$ with only the node q in the last run of Π.

Step 2: Assign the value of $\sigma_q(i)$ $(= \mathcal{C}(D(p_i), q))$ for each node $q \in \cup_{p_j \in B(p_i)} \Psi(p_j)$ with the w value of the θ-th elements of Π where $\theta = |\Psi(p_i)| +$ hop-distance(q, p_i).

Algorithm: As in Sect. 3, here also we define three types of *gates*. Each gate is a tuple (a, b, c) of nodes in V, where $b \in S$ (colored black in Fig. 6), may be of any color; nodes a and c are of opposite colors (say red and blue). Unlike the case of caterpillar, here a, c may not always be adjacent to b. However, a and c are equidistant from b and all the nodes on the path from b to a (resp. b to c), excluding b, are of same color.

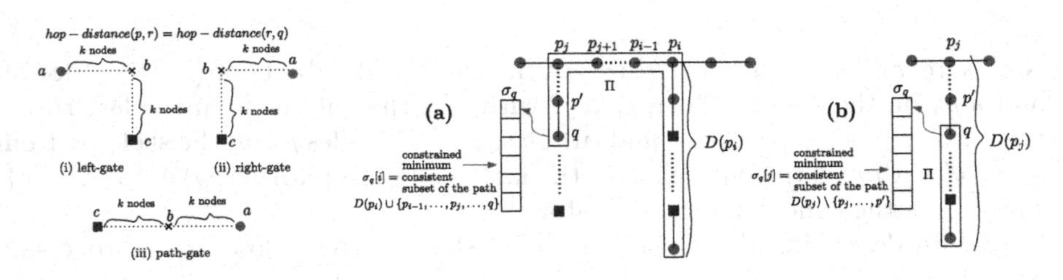

Fig. 6. Gates in *comb graph*. **Fig. 7.** Data structure σ_q: (a) $\sigma_q(i)$ for $j \neq i$, (b) $\sigma_q(i)$ for $j = i$

left-gate $\mathcal{L}_{q,p_i,r}$: It is a tuple (q, p_i, r), where $p_i \in S$, $r \in D(p_i)$, $q \in D(p_j)$, $p_j \in L(p_i)$ such that hop-distance(q, p_i) = hop-distance(p_i, r), and all the nodes on the path from q to p_{i-1} are of color(q), and all the nodes on the path from r to $D(p_i) \setminus \{p_i\}$ are of color(r) (see Fig. 7(a (i)). Here p_i is referred to be the *base node* for this left-gate. If $\{q, r\}$ of this left-gate is included in $\mathcal{C}$, then all the nodes in $R(p_i)$ along with their dangling legs are covered.

right-gate $\mathcal{R}_{q,p_i,r}$: It is a tuple (q, p_i, r), where $p_i \in S$ is the *base node*, $r \in D(p_i)$, $q \in D(p_j)$ where $p_j \in R(p_i)$, hop-distance(q, p_i) = hop-distance(p_i, r), and all the nodes on the path from q to p_{i-1} are of color(q) (see Fig. 7(a (ii)). Here p_i is referred to be the *base node* for this right-gate. If $\{q, r\}$ of this right-gate is included in $\mathcal{C}$, then all the nodes in $L(p_i)$ along with their dangling legs are covered.

path-gate $\mathcal{P}_{q,p_i,r}$: It is a tuple (q, p_i, r), where $p_i \in S$ is the *base node*, $q \in D(p_j)$, $r \in D(p_{j'})$; $p_j, p_{j'} \in S$ are respectively in the left and right sides of p_i (Fig. 7(a (iii)). Inclusion of this gate in $\mathcal{C}$ implies inclusion of the following nodes in $\mathcal{C}$.

$$\{q,r\} \bigcup \Big(\bigcup_{p_\theta \in L(p_j)} \mathcal{C}(D(p_\theta), q) \Big) \bigcup \Big(\bigcup_{\theta=j}^{i-1} \mathcal{C}(D(p_\theta), q) \Big) \bigcup \Big(\bigcup_{\theta=i+1}^{j'} \mathcal{C}(D(p_\theta), r) \Big) \bigcup \Big(\bigcup_{p_\theta \in R(p_{j'})} \mathcal{C}(D(p_\theta), r) \Big)$$

The elements in $D(p_i)$ are already covered for the choice of q, r in $\mathcal{C}$.

Arguing as in Sect. 3, here also if left-gates (resp. right-gates) are present in the problem instance, then the *base node* of the left-most (resp. right-most) left-gates (resp. right gates) needs to be considered. However, in both the cases, the q and r node need to be appropriately chosen to minimize the size of $\mathcal{C}$.

In a sequential scan from left to right, we can identify the *base node* p_ℓ of the left-most left-gate, and the *base node* p_r of the right-most right-gate. If left gate (resp. right gate) is not present, then we set $\ell = -\infty$ (resp. $r = \infty$). As in Sect. 3, here also, we need to consider the four situations, namely **NG**, **OLG**, **ORG** and **BG**.

5.1 Handling OLG and ORG Case

If $\mathcal{L}_{q,p_i,r}$ is the only gate present in V then we include $\{q, r\}$ in $\mathcal{C}$. Thus, the nodes in $R(p_i)$ are covered, and we have $\mathcal{C} = \{q, r\} \bigcup \mathcal{C}(D(r), r) \bigcup_{p_\theta \in L(p_i)} (D(p_\theta), q)$. Note that, if $q \in \Psi(p_j)$ and $p_i, p_j \in S_\alpha$ then for all $\theta \in \{i-1, \ldots, j\}$, $\mathcal{C}(D(p_\theta), q)$ is already computed in the preprocessing step, and its size is available in $\sigma_q(\theta)$. Thus, we need to compute only $\mathcal{C}(L(p_j), q)$, and

$$\mathcal{C} = \{q, r\} \bigcup \mathcal{C}(D(p_r), r) \bigcup_{\theta=i-1}^{j} \mathcal{C}(D(p_\theta), q)) \bigcup \mathcal{C}(L(p_j), q),$$

It needs to be mentioned that $L(p_j)$ (the portion of S to the left of p_j) does not contain any gate, and is to be processed as the **NG** case.

The **ORG** case is similarly handled for a right gate $\mathcal{R}_{q,p_i,r}$. Here,

$$\mathcal{C} = \{q, r\} \bigcup \mathcal{C}(D(p_r), r) \bigcup_{\theta=i+1}^{j} \mathcal{C}(D(p_\theta), q) \bigcup \mathcal{C}(R(p_j), q),$$

where (i) the size of $\mathcal{C}(D(p_\theta), q))$ are available in $\sigma_q(\theta)$ for $\theta \in \{i+1, \ldots, j\}$ and (ii) $R(p_j)$ is to be processed as an NG case to compute $\mathcal{C}(R(p_j), q)$.

5.2 Handling BG Case

- if $\ell \leq r$, then we need to compute (i) the size of $\mathcal{C}$ considering only the left-most left-gate using the method for **OLG** case, and also (ii) the size of $\mathcal{C}$ considering only the right-most right-gate as in **ORG** case. The minimum of them is reported as the optimum $\mathcal{C}$.
- if $\ell > r$, then the $R(p_\ell)$ and the $L(p_r)$ part are already covered. We need to compute the consistent subset of the portion $L(p_\ell) \bigcap R(p_r)$ as the **NG** case.

5.3 Handling NG Case

Now, we will explain the processing of the portion of S as the *NG* instance. From now onwards, we will use S for this portion of the given instance of the problem. We create a multi-partite overlay graph $H = (U, F)$ with the nodes $U = U_0 \cup U_1 \cup U_2 \cup \ldots \cup U_m \cup U_{m+1}$ where U_i corresponds to $p_i \in S$, and its nodes correspond to the elements of $\Psi(p_i)$, $i = 1, 2, \ldots, m$; $U_0 = \{s\}$ and $U_{m+1} = \{t\}$. Let us remind that, if the last node of the minimum consistent subset $(\mathcal{C}(D(p_i), q))$ of the leg $D(p_i)$ is $q \in \Psi(p_j)$ (p_i, p_j belongs to the same run of S) then $|\mathcal{C}(D(p_i), q)|$ is available in $\sigma_q(i)$.

Needless to say, there is no edge between any pair of nodes in U_i, $i = 1, \ldots, m$. For a pair of sets U_ℓ and $U_{\ell'}$, we add an edge between every pair of nodes (q, r), where $q \in U_\ell$ and $r \in U_{\ell'}$. The edge weights are computed as follows:

Type-0 edge: The node $s \in U_0$ is connected with every node of $U_1 \cup \ldots U_{r_1}$, where $p_1, \ldots, p_{r_1} \in S_1$ (where r_1 is the length of S_1, the first run of S). If $q \in U_\ell$, where $\ell \leq r_1$, then the weight of the directed edge (s, q) is $\omega(s, q) = \sum_{i=1}^{\ell-1} \sigma_q(i) + \frac{1}{2}\sigma_q(\ell)$.

Type-1 edge: For a pair of nodes (q, r) where $q \in U_\ell$ and $r \in U_{\ell'}$ and the corresponding elements $p_\ell \in S_\alpha$ and $p_{\ell'} \in S_{\alpha+1}$ then the cost $\omega(q, r)$ of the *type-1* edge (q, r) is computed as follows (see Fig. 8(a)):

- Let p_θ and $p_{\theta+1}$ be two consecutive elements in S that belong to S_α and $S_{\alpha+1}$ respectively. If the consistency condition for p_θ $(\theta - \ell \leq \ell' - \theta)$ and for $p_{\theta+1}$ $((\theta+1) - \ell \leq \ell' - (\theta+1))$ are satisfied then we set $\omega(q, r) = \frac{1}{2}\sigma_q(\ell) + \sum_{j=\ell+1}^{\alpha} \sigma_q(j) + \sum_{j=\alpha+1}^{\ell'-1} \sigma_r(j) + \frac{1}{2}\sigma_r(\ell')$;
- otherwise we set $\omega(q, r) = \infty$.

Type-2 edge: For a pair of nodes (q, r) where $q \in U_\ell$ and $r \in U_{\ell'}$ and both the corresponding elements $p_\ell, p_{\ell'} \in S_\alpha$ then the cost $\omega(q, r)$ of the *type-2* edge (q, r) is computed as follows (see Fig. 8(b)):

- compute $\lambda = \lceil \frac{\text{hop-distance}(q,r)}{2} \rceil$.
- If $\lambda > \max(\text{hop-distance}(q, p_\ell), \text{hop-distance}(r, p_{\ell'}))$ then we can get $p_\theta \in S$ such that hop-distance$(q, p_\theta) = \lambda$. In such a case, for all elements in $\Psi(p_\ell), \ldots \Psi(p_\theta)$ the nearest element will be q, and for all elements $\Psi(p_{\theta+1}), \ldots \Psi(p_{\ell'})$ the nearest element will be r, and we set $\omega(q,r) = \frac{1}{2}\sigma_q(\ell) + \sum_{j=\ell+1}^{\lambda} \sigma_q(j) + \sum_{j=\lambda+1}^{\ell'-1} \sigma_r(j) + \frac{1}{2}\sigma_r(\ell')$;
- otherwise we set $\omega(q,r) = \infty$.

Type-0′ edge: $U_\beta \cup \ldots \cup U_m$ is connected with $t \in U_{m+1}$, where $p_\beta, \ldots, p_m \in S_k$ (last run of S). If $q \in U_\ell$, where $\beta \le \ell \le m$, then the weight of the directed edge (q,t) is $\omega(q,t) = \frac{1}{2}\sigma_r(\ell) + \sum_{i=\ell+1}^{m} \sigma_q(i)$.

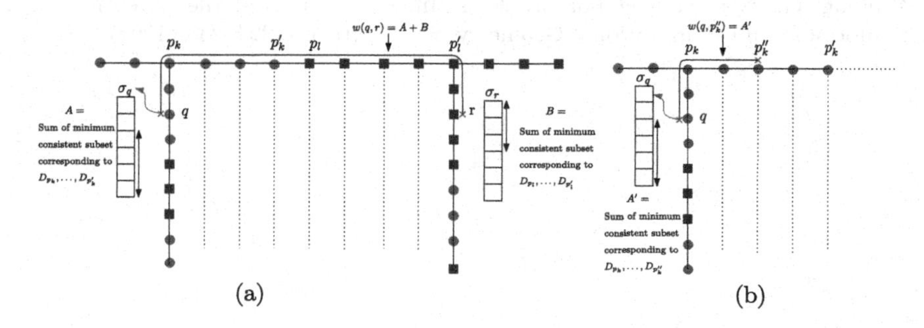

Fig. 8. (a) Type 1 edge. (b) Type 2 edge.

Now, a shortest path from s to t in the overlay graph H will give the size of the minimum consistent subset. The following two notes are important for the correctness of the algorithm:

Note-1: We need to specifically mention that the weight of an edge is equal to the sum of the size of the minimum consistent subset of the legs whose corresponding elements in S are covered by that edge. Thus, for a valid edge (q,r) (i.e., $\lambda > \max(\text{hop-distance}(q, p_\ell), \text{hop-distance}(r, p_{\ell'}))$), if any one of $\sigma_q(j)$ in the first sum or any one of $\sigma_r(j)$ in the second sum is ∞, then $\omega(q,r)$ is set to ∞.

Note-2: In order to avoid duplicate counting, we have added 1/2 of the cost of each (one or two) terminal leg covered by an edge to the weight of that edge.

Theorem 4. *The aforesaid algorithm correctly computes the minimum consistent subset of a comb graph in $O(m(m+n))$ time, where $m = |S|$ (the size of the skeleton) and $n = |V|$ (the total number of nodes in the input graph G).*

References

1. Banerjee, S., Bhore, S., Chitnis, R.: Algorithms and hardness results for nearest neighbor problems in bicolored point sets. In: Bender, M.A., Farach-Colton, M., Mosteiro, M.A. (eds.) LATIN 2018. LNCS, vol. 10807, pp. 80–93. Springer, Cham (2018). https://doi.org/10.1007/978-3-319-77404-6_7

2. Biniaz, A., et al.: On the minimum consistent subset problem. In: Friggstad, Z., Sack, J.-R., Salavatipour, M.R. (eds.) WADS 2019. LNCS, vol. 11646, pp. 155–167. Springer, Cham (2019). https://doi.org/10.1007/978-3-030-24766-9_12
3. Gao, B.J., Ester, M., Cai, J.-Y., Schulte, O., Xiong, H.: The minimum consistent subset cover problem and its applications in data mining. In: Proceedings of the 13th ACM SIGKDD International Conference on Knowledge Discovery and Data Mining, KDD 2007, pp. 310–319 (2007)
4. Hart, P.: The condensed nearest neighbor rule (corresp.). IEEE Trans. Inf. Theory **14**(3), 515–516 (1968)
5. Khodamoradi, K., Krishnamurti, R., Roy, B.: Consistent subset problem with two labels. In: Panda, B.S., Goswami, P.P. (eds.) CALDAM 2018. LNCS, vol. 10743, pp. 131–142. Springer, Cham (2018). https://doi.org/10.1007/978-3-319-74180-2_11
6. Wilfong, G.: Nearest neighbor problems. In: Proceedings of the Seventh Annual Symposium on Computational Geometry, SCG 1991, pp. 224–233 (1991)

Computational Complexity

Balanced Connected Graph Partition

Satyabrata Jana[1(✉)], Supantha Pandit[2], and Sasanka Roy[1]

[1] Indian Statistical Institute, Kolkata, India
satyamtma@gmail.com, sasanka.ro@gmail.com
[2] Dhirubhai Ambani Institute of Information and Communication Technology, Gandhinagar, Gujarat, India
pantha.pandit@gmail.com

Abstract. We study a variation of the graph partition problem on colored graphs called the *k-Balanced Connected Graph Partition (k-BCGP)* problem. We are given a connected non-unicolor graph G where the vertices in G have colored either *red* or *blue* and a positive integer k. The *k-BCGP* problem seeks a partition of G into k non-empty connected subgraphs such that each subgraph is as balanced (contains the same number of red and blue vertices) as possible i.e., minimizing the maximum *unbalancedness* across all k connected subgraphs where the unbalancedness of a connected subgraph of G is defined as the absolute difference between the number of red and blue vertices in that subgraph.

We target some special classes of graphs namely, paths, trees, bipartite graphs, planar bipartite graphs, and chordal graphs with different values of k. For each of these classes either we prove NP-hardness or design a polynomial-time algorithm. More specifically, on the positive side, we design a polynomial-time algorithm for the *k-BCGP* problem on paths. For trees, we present a polynomial-time algorithm only for any fixed k. On the negative side, we prove that the *k-BCGP* problem is NP-hard for the bipartite graphs when the value of k is 2. For planar bipartite graphs, we prove the NP-hardness of the *k-BCGP* problem where k is a part of the input. We further prove that the *k-BCGP* problem is also NP-hard for the chordal graphs when $k = 2$. We also show that these NP-hard problems do not admit any constant factor approximation algorithms.

Keywords: NP-hardness · Balanced partition · Trees · Paths · Bipartite graphs · Planar graphs · Chordal graphs

1 Introduction

Partition graphs into non-empty subgraphs and optimize certain parameters on those subgraphs are well studied in the literature [2,6–8,12,14]. We study a new variation, the *k-BCGP* problem, of the graph partition problem on a connected *non-unicolor* graph G where the vertices of G have colored either red or blue (we call it as a red-blue graph). By saying non-unicolor, we mean all vertices in G should not have the same color. We say that a connected red-blue graph

A. Mudgal and C. R. Subramanian (Eds.): CALDAM 2021, LNCS 12601, pp. 487–499, 2021.
https://doi.org/10.1007/978-3-030-67899-9_38

is *balanced* if it contains the same number of red and blue vertices. Otherwise, we say it is *unbalanced*. For a graph, we define the *unbalancedness*, in short *ubn* as the absolute difference between the number of red and blue vertices in that graph. Note that for a balanced graph, *ubn* is zero.

We define a *partition of a graph* $G(V, E)$ with vertex set V and edge set E as follows: For a graph $G(V, E)$, a k-partition of G is a set of k subgraphs G_i of G such that each G_i is a non-empty connected subgraph satisfying $\bigcup_{i=1}^{k} V(G_i) = V$ and $V(G_i) \cap V(G_j) = \emptyset$ for all $i, j\ (\neq i)$ where $1 \leq i,\ j \leq k$. Further, if we assume that G is a red-blue graph, a subgraph of G is called *balanced connected subgraph* if it is connected and contains the same number of red and blue vertices. We define the *unbalancedness (UBN)* of a partition $\{G_1, G_2, \ldots, G_k\}$ on G as the maximum *ubn* over all the k parts in that partition i.e., $\max\{ubn(G_i): 1 \leq i \leq k\}$, where $ubn(G_i)$ is the absolute difference between the number of red and blue vertices in G_i. *k-BCGP* problem seeks a partition that is having minimum unbalancedness. We use some standard notations throughout the paper. For any graph G, $V(G)$ denotes the set of all vertices in G. Also for $U \subseteq V(G)$, we denote $G[U]$ and $N[U]$ to define the subgraph induced by U in G and neighbors of U in G including U itself, respectively. We now formally define the *k-BCGP* problem as follows.

k-Balanced Connected Graph partition (k-*BCGP*): Given a connected non-unicolor graph G, where the vertices of G are colored either red or blue, and a positive integer k, the objective is to partition G into k non-empty connected subgraphs $G_1, G_2, \ldots, G_k$ such that it minimizes $\max_{1 \leqslant i \leqslant k} ubn(G_i)$, where $ubn(G_i)$ is the absolute difference between the number of red and blue vertices in G_i. We call this optimal value as *UBN* of *k-BCGP* on G.

The *k-BCGP* problem is a variation of the graph partition problem. Chen et al. [6] mention that the graph partitioning problem and its closely related problems have applications in various fields including image processing, clustering, computational topology, information, library processing, etc. Some of these applications may be well-suited for the *k-BCGP* problem. More specifically, partitioning of a connected graph into a certain number of balanced connected subgraphs can be used for modeling several problems in databases, operating systems and cluster analysis [17].

We can motivate the *k-BCGP* problem with an application as follows. Let there are n districts (assuming together all form a connected region) and each district is governing by a political party from $\{X, Y\}$. Now we want to partition these districts into k parts such that each part is connected and can be as balanced (contains the same number of districts run by X and Y) as possible. This implies that each party has the same dominant power in each connected part.

1.1 Previous Work

Graph partitioning is a well-studied problem in the literature. Many variants of this problem has been studied in the last four decades. Chen et al. [6] introduced the following problem called the cardinality k-*BGP*: given a connected graph $G = (V, E)$, cardinality k-*BGP* problem seeks a partition of V into k non-empty subsets $V_1, V_2, \ldots, V_k$ such that $G[V_i]$ is connected for every $i = 1, 2, \ldots, k$, and $\max_{1 \leq i \leq k} |V_i|$ is minimized. This problem is NP-hard on bipartite graphs,for any fixed $k \geq 2$ [8]. k-*BCGP* problem is closely related to the cardinality k-*BGP* problem. In our problem definition, we restrict our input to be a non-unicolor graph. The reason behind this restriction is that k-*BCGP* problem on a unicolor graph is exactly the same as the cardinality k-*BGP* problem which is a well-studied problem. Chen et al. [6] also studied the weighted version of this problem and provided approximation algorithms for various values of k.

Chlebíková [7] showed that for any $\epsilon > 0$, it is NP-hard (even for bipartite graphs) to approximate the maximum balance $(\min(|V_1|, |V_2|), V_1 \cup V_2 = V, V_1 \cap V_2 = \emptyset)$ of the partition for $G = (V, E)$ into two connected subgraphs with an absolute error guarantee of $|V|^{1-\epsilon}$. Chlebíková [7] also give a PTAS within ratio $4/3$ for vertex-weighted 2-*BGP* problem. Chataigneret al. [17] proved that for any fixed $k \geq 2$, cardinality max-min k-*BGP* problem is NP-hard on k-connected graphs unless P = NP. Also, this problem cannot be approximated within $6/5$ unless P = NP, when k is part of the input. In the literature, the weighted version of the k-*BGP* problem is also studied as the minimum spanning k-forestproblem. This problem on trees is well studied in the literature [3,14], which admits a linear time exact algorithm. Dyer et al. [8] introduced another type of partition problem, called the $\pi_1 Vk(\pi_2)$: that checkswhether the vertices of a graph with property the π_1 can be partitionedinto equal sized subsets (having size k) so that each subset induces a graph satisfying property the π_2. Previously it was known that when π_1 has no restriction then for $k \geq 3$, Vk(connected) is NP-complete [10]. Dyer et al. [8] showed that planar Vk(connected) is NP-complete. He also proved that each of the following problems is NP-complete for $k \geq 3$: planar bipartite Vk (connected), planar bipartite Vk(tree), planar bipartite Vk(path). Madkour et al. [12] introduced the edge-weightedvariant of the k-*BGP* problem. Here the input is an edge -weighted graph and an integer k, and the goal is to partition the vertex set into k sets such that each subset induces a connected graph that minimizes the weight of the heaviest subgraph, where the weight of each subgraph isthe weight of a minimum spanning tree of the graph induced by the vertices in that subgraph. Madkour et al. [12] proved that this problem is NP-hardon general graphs for any fixed $k \geq 2$, and proposed two k-approximation algorithms. Later Vaishali et al. [16] showed that for $k = 2$ this problem remains NP-hard on graphs having all edges of equal weight, and presented a lineartime algorithm for the problem when the input graph is a tree. There is another variant of the partitioning problem in literature [11] called as *k-balanced partitioning problem.* Here the goal is to partition the vertices of a graph into k parts of equal size such that each part induce a connected graph that it minimizes the total weight of the edges connecting different subgraphs. The k-balanced partitioning problem can be considered in the exact form (all parts need to be of equal size) and in a relaxed (or bi-criteria) variant, where their sizes are at most $c * n/k$. The bound of $\mathcal{O}(\sqrt{\log n \log k})$ presented by Krauthgamer et al. [11] is for a relaxed variant. For an exact variant, the special case $k = 2$ is the famous *minimum bisectionproblem*, which is known to be NP-hard [15]. Andreev and Raecke [1] also worked on a relaxed part and presented a bicriteria polynomial time approximation algorithm with an $\mathcal{O}(\log^2 n)$-approximation algorithm. Matić et al. [13] considered another problem, called the *maximally balanced connected partition* (MBCP) problem in graphs. Given a vertex-weighted graph with vertex set V, the goal of MBCP is to partition V into two nonempty sets V_1 and V_2, such that subgraphs of G induced by V_1 and V_2 are connected and that minimize the difference between the sums of the weights of vertices from V_1 and V_2. MBCP is proved to be NP hard [13]. Matić et al. also describes two heuristic algorithms: greedy algorithm and genetic algorithm, for the same problem. Recently, Bhore et al. [4,5] studied a problem of finding a largest size (cardinality of the vertex set) balanced connected subgraph in a simple connected bicolored graph. They showed polynomial-time algorithms and NP-hardness results for various classes of graphs.

1.2 Our Contributions

- We design polynomial-time algorithms for the *k-BCGP* problem on paths ($\mathcal{O}(kn^2)$ time) and trees ($\mathcal{O}(n^k)$ time) with n vertices (Sect. 2).
- We prove that 2-*BCGP* problem is NP-hard in bipartite graphs (Sect. 3).
- We prove that the *k-BCGP* problem is NP-hard for the planar bipartite graphs where k is not fixed (Sect. 4).
- We prove that 2-*BCGP* problem is NP-hard in chordal graphs (Sect. 5).

In this paper, we prove that in each of the NP-hard problems (items 2–4 above), it is NP-hard to find a partition that has a solution of cost zero i.e., $UBN = 0$. This implies that, these NP-hard problems do not admit any constant factor approximation algorithms.

2 *k-BCGP* Problem on Trees and Paths

We give polynomial-time algorithm for the *k-BCGP* problem on trees. We further provide an improved algorithm for the *k-BCGP* problem on paths.

2.1 Trees

Let T be a tree that contains n vertices. Observe that removing $k - 1$ edges (arbitrarily choosen) from T produces k many connected subgraphs. So we try removing each possible $k-1$ edges from the $n-1$ edges of T. For each choice, we calculate the maximum unbalancedness over all connected subgraphs. Finally, we return the subgraphs for which the maximum unbalancedness is minimum.

We now calculate the time taken by the algorithm. There are $\mathcal{O}(n^{k-1})$ ways of choosing $k-1$ edges from $n-1$ edges of T. Computing the unbalancedness in a connected subgraph of size m takes $\mathcal{O}(m)$ time. So for a partition, the time needed for calculating maximum unbalancedness is $\mathcal{O}(n)$. Hence, the total time requirement is $\mathcal{O}(n^k)$.

2.2 Paths

Here we present a dynamic programming based algorithm that solve the *k-BCGP* problem for paths in $\mathcal{O}(kn^2)$ time. Let $P_{1,n}$ be a path with n vertices $v_1, v_2, \ldots, v_n$, where $E(P_{1,n}) = \{(v_i, v_{i+1})\colon 1 \leq i < n\}$. For a pair of integers i, j where $1 \leq i \leq j$, we denote $P_{i,j}$ to define the subgraph of $P_{1,n}$ induced by the vertices $v_i, v_{i+1}, \ldots, v_j$. For each pair (i, t), where $1 \leq i \leq n$ and $t \leq k$, we define $\mathcal{D}[i, t]$ to be the *UBN* of *t-BCGP* problem on $P_{1,i}$. Our main goal is to compute $\mathcal{D}[n, k]$. We now describe how to compute $\mathcal{D}[i, t]$. For the base cases $\mathcal{D}[1, t] = 1$ when $t = 1$ and $\mathcal{D}[1, t] = 0$ when $t > 1$. For a pair (i, t), to compute $\mathcal{D}[i, t]$, we make use of (i) $\mathcal{D}[i', t-1]$ and (ii) $ubn(P_{i'+1,i})$ for all i' where $1 \leq i' < i$. Then we use the following recurrence for computing $\mathcal{D}[i, t]$.

$$\mathcal{D}[i, t] = \min_{\forall i', 1 \leq i' < i} \{\max(\mathcal{D}[i', t-1],\ ubn(P_{i'+1,i}))\}$$

Correctness: We have to show that $\mathcal{D}[n, t]$ produce minimum unbalancedness over all k partitions. i.e., *UBN* of *k-BCGP* on $P_{1,n}$. Let us assume that in an optimal solution, say S of *k-BCGP* problem on $P_{1,n}$, H be the subgraph in S that contains the vertex v_n and j be the smallest index such that $v_j \in H$. Clearly, *UBN* of $(k-1)$-*BCGP* problem on $P_{1,j-1} \leq$ *UBN* of *k-BCGP* on $P_{1,n}$, and $ubn(P_{j,n}) \leq$ *UBN* of *k-BCGP* on $P_{1,n}$. So, we can correctly compute the value of $\mathcal{D}[n, t]$, by comparing the value of $\mathcal{D}[i-1, k-1]$ and $ubn(P_{i,n})$ for all i where $1 \leq i < n$.

Time Complexity: We need to fill two tables. Table 1 is for computing $ubn(P_{i,j})$ for all i, j where $1 \leq i \leq j \leq n$. Table 2 is for computing $\mathcal{D}[i, t]$, where $1 \leq i \leq n$ and $t \leq k$. First we fill all the entries in table 1. The total number of entries in this table is $\mathcal{O}(n^2)$. We can use the following process to compute $ubn(P_{i,j})$ efficiently. Let $r_{i,j}$ and $b_{i,j}$ denote the number of red and blue vertices in $P_{i,j}$. Clearly $ubn(P_{i,j}) = |r_{i,j} - b_{i,j}|$. Now $r_{i,j} = r_{1,j} - r_{1,i-1}$ and $b_{i,j} = b_{1,j} - b_{1,i-1}$. Initially computing $r_{1,i}$ and $b_{1,i}$ for all $i, 1 \leq i \leq n$ takes linear time. As for each pair (i, j), computing $ubn(P_{i,j})$ takes constant time, hence we can fill all the entries in table 1 in $\mathcal{O}(n^2)$ time. Next, the number of entries in table 2 is nk. By the recurrence for computing $\mathcal{D}[i, t]$, to fill one entry it needs $\mathcal{O}(n)$ comparisons. Hence we can fill all the entries in table 2 in $\mathcal{O}(kn^2)$ time. Hence we conclude with the following theorem.

Theorem 1. *The k-BCGP problem on a path with n vertices can be solved in*$\mathcal{O}(kn^2)$ *time.*

3 2-*BCGP* Problem on Bipartite Graphs

We show that the 2-*BCGP* problem is NP-hard for the bipartite graphs. We give a reduction from the *Dominating Set (DOM)* problem, which is known to be NP-complete on general graphs [10]. Given a graph G with vertex set V, edge set E, and an integer k. The dominating set problem asks whether there exists a set $U \subseteq V$ such that $|U| \leq k$ and $N[U] = V$, where $N[U]$ denotes neighbors of U in G including U itself.

During the reduction, we first generate a graph $H = (R \cup B, E')$ from an instance $X(G, k)$ of the dominating set problem. Next, we show that H is bipartite.

Reduction: Let $G = (V, E)$ be graph with n vertices $V = \{v_1, v_2, \ldots, v_n\}$. We construct a connected bicolored graph $H = (R \cup B, E')$, where R and B are the sets of red and blue vertices, respectively in H, with the following way:

- For each vertex v_i, $1 \leq i \leq n$, we add a blue vertex $v_i \in B$, a red vertex $v_i' \in R$, and add an edge $(v_i, v_i') \in E$. For each edge $(v_i, v_j) \in E$, we add two edges $(v_i, v_j'), (v_i', v_j)$ in E'.
- Add a path of k red vertices starting at r_1 and ending at r_k. Then we add n edges (r_1, v_i) in E' for all $i, 1 \leq i \leq n$.
- Add a path of $(n - k)$ blue vertices starting at u_1 and ending at u_{n-k}. Then we add n edges (u_1, v_i) in E' for all $i, 1 \leq i \leq n$.
- Add a path of $(n^2 - kn + 2k)$ blue vertices starting at b_1 and ending at $b_{(n^2-kn+2k)}$. Then we add the edge (r_k, b_1) in E'.
- For each red vertex $v_i' \in R$, where $1 \leq i \leq n$, take a path of $(n - k - 1)$ red vertices starting at r_1^i and ending at r_{n-k-1}^i. Then we add the edges $\{(r_1^i, v_i'): 1 \leq i \leq n\}$ into E'.

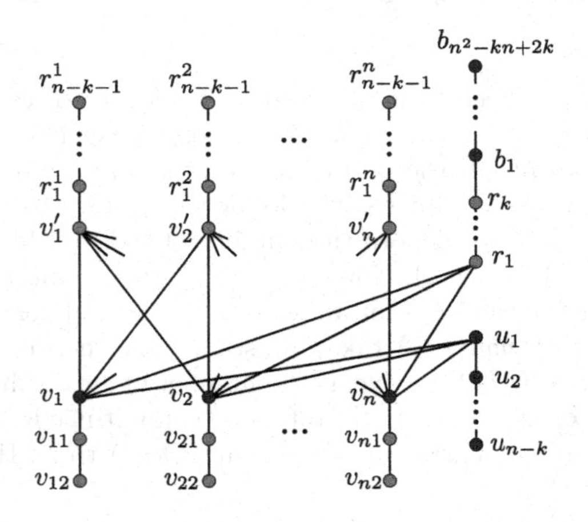

Fig. 1. Construction of the instance $H = (R \cup B,\ E')$ of the 2-*BCGP* problem, where R and B are the sets of red and blue vertices, respectively in H.

- For each blue vertex $v_i \in B$, where $1 \leq i \leq n$, we add two red vertices v_{i1}, v_{i2} in R and add two edges (v_{i1}, v_{i2}) and (v_i, v_{i1}) into E'.

This completes the construction. See Fig. 1 for an illustration of this construction. The number of vertices and edges are $2(n^2+2n-nk+k)$ and $(2|E|+2n^2-2kn+5n+2k-2)$. Clearly, the construction of H can be done in polynomial time. Also, it is easy to verify that the graph H is balanced with $(n^2+2n-nk+k)$ vertices of each color. Now we prove the following lemmas.

Lemma 1. *If H' is a balanced connected subgraph of H such that $u_1 \in V(H')$ and $(V\backslash V(H'))$ induces a non-empty balanced connected subgraph of H then $r_1 \notin V(H')$.*

Proof. We prove it by contradiction. Let us assume that H' is a balanced connected subgraph of H such that H' contains both u_1 and r_1 where $(V\backslash V(H'))$ induces a non-empty balanced connected subgraph of H. Also we assume that $|V(H') \cap \{v'_1, \ldots, v'_n\}| = t_1$ and $|V(H') \cap \{v_1, \ldots, v_n\}| = t_2$. Clearly, $t_2 \neq 0$ (because of connectivity between u_1 and r_1 in H'). As $u_1 \in V(H')$ and $G[V\backslash V(H')]$ is connected so $V(H')$ contains all the blue vertices in $\{u_1, \ldots, u_{n-k}\}$. Also if $v_i \in V(H')$ then $V(H')$ must contain both the red vertices v_{i1} and v_{i2}. Similarly, if $v'_j \in V(H')$ then $V(H')$ should include all the red vertices $\{r^j_1, r^j_2, \ldots, r^j_{n-k-1}\}$. As $r_1 \in V(H')$ so $\{r_1, \ldots, r_k, b_1, \ldots, b_{n^2-kn+2k}\} \subset V(H')$.Now the number of red and blue vertices in H' is $2t_2+t_1(n-k)+k$ and $t_2+(n-k)+(n^2-kn+2k)$, respectively. As H' is balanced, so $2t_2+t_1(n-k)+k = t_2+(n-k)+n^2-kn+2k$. This implies $t_2+t_1(n-k) = n^2-kn+n$. Now this will be satisfied only when $t_2 = t_1 = n$holds. This implies $V = V(H')$, that is a contradiction. □

By a similar argument, we can prove the following lemma.

Lemma 2. *If H' is a balanced connected subgraph of H such that $u_1 \in V(H')$ and $(V\backslash V(H'))$ induces a non-empty balanced connected subgraph of H then $V(H') \cap \{v'_1, v'_2, \ldots, v'_n\} = \emptyset$.*

Lemma 3. *G has a dominating set of size k if and only if H has a zero solution i.e., $UBN = 0$ in 2-BCGP. More specifically, exactly one way we can partition H into two balanced connected subgraphs, one of them contains $4(n-k)$ vertices and the other contains $2(n^2-nk+3k)$ vertices.*

Proof. Assume that G has a dominating set U of size k. Let $X = \{v_i, v_{i1}, v_{i2} \colon v_i \notin U\} \cup \{u_j \colon 1 \leq j \leq (n-k)\}$ and $Y = V(H)\backslash X$. Now $|X| = 4(n-k)$ and $|Y| = 2(n^2-nk+3k)$. It is easy to verify that both $G[X]$ and $G[Y]$ are balanced connected. Hence, H can be partitioned into two balanced connected subgraphs, one of cardinality $4(n-k)$ and other $2(n^2-nk+3k)$. Hence H has a zero solution in 2-*BCGP*.

On the other hand, assume that H has a zero solution in 2-*BCGP*, i.e., there exists a partition of H into two balanced connected subgraphs, say H_1 and H_2. Without loss of generality, assume that $u_1 \in V(H_1)$. By Lemma 1 and 2, $r_1 \notin V(H_1)$ and $V(H_1) \cap \{v'_1, v'_2, \ldots, v'_n\} = \emptyset$. Also as $u_1 \in V(H_1)$ so $\{u_1, u_2, \ldots, u_{n-k}\} \subset V(H_1)$. Let $|V(H_1) \cap \{v_1, v_2, \ldots, v_n\}| = t$. If $v_i \in V(H_1)$ then $V(H_1)$ must contain v_{i1} and v_{i2}. As H_1 is balanced, so $2t = t + (n-k)$, that implies $t = (n-k)$. Therefore H_1 is a balanced connected subgraph of H with size $4(n-k)$. Let $U = \{v_j : v_j \notin V(H_1)\}$ with $|U| = k$. As H has a zero solution in 2-*BCGP*, so $G[V(H) \backslash V(H_1)]$ is balanced connected. As $G[V(H) \backslash V(H_1)]$ is connected so the subgraph of H induced by $U \cup \{v'_i : 1 \leq i \leq n\}$ is connected. Hence the corresponding vertices of U in G are a dominating set of size k. □

It is easy to see that the graph we constructed from the *DOM* problem in Fig. 1 is indeed a bipartite graph. Hence we conclude the following theorem.

Theorem 2. *The2-BCGP problem is NP-hard for bipartite graphs.*

4 *k-BCGP* Problem on Planar Bipartite Graphs

In this section, we prove that the *k-BCGP* problem is NP-hard for planar bipartite graphs where k is not fixed.

We give a reduction from the *Planar 3-Dimensional Matching (P3DM)* problem, that is known to be NP-complete [9]. In a *3-Dimensional Matching (3DM)* problem, we are given three sets $\mathcal{A}$, $\mathcal{B}$, and $\mathcal{C}$ of cardinality n each and a set T of triples from $\mathcal{A} \times \mathcal{B} \times \mathcal{C}$. The objective is to decide whether there exists a subset $T' \subseteq T$ of triples such that $T' = n$ and T' contains all the elements in $\mathcal{A}$, $\mathcal{B}$, and $\mathcal{C}$. We now construct a bipartite graph on the instance $X(\mathcal{A} \cup \mathcal{B} \cup \mathcal{C},\ T)$ of the *3DM* problem as follows. For each element of $\mathcal{A} \cup \mathcal{B} \cup \mathcal{C}$ and each triple in T we take a vertex. There is an edge between a vertex corresponding to a triple in T and a vertex corresponding to an element in $\mathcal{A} \cup \mathcal{B} \cup \mathcal{C}$ if and only if the element in $\mathcal{A} \cup \mathcal{B} \cup \mathcal{C}$ belongs to that triple in T. A *P3DM* problem is a *3DM* problem whose underline bipartite graph is planar.

During the reduction, we generate an instance $G = (R \cup B,\ E)$ of the *m-BCGP* problem from an instance $X(\mathcal{A} \cup \mathcal{B} \cup \mathcal{C},\ T)$ of the *P3DM* problem. Next, we show that G is a planar bipartite graph.

Reduction: Let $X(\mathcal{A} \cup \mathcal{B} \cup \mathcal{C},\ T)$ be an instance of the *P3DM* problem where $\mathcal{A} = \{a_1, \ldots, a_n\}, \mathcal{B} = \{b_1, \ldots, b_n\}, \mathcal{C} = \{c_1, \ldots, c_n\}$, and $T = \{t_1, t_2, \ldots, t_m\}$. In each triple $t_i \in T$, we denote the elements from $\mathcal{A}, \mathcal{B}$, and $\mathcal{C}$ as t_{i1}, t_{i2}, and t_{i3}, respectively. We construct a connected bicolored graph $G = (R \cup B,\ E)$, where R and B are the sets of red and blue vertices, respectively in G, with the following way:

- For each element $\alpha \in \mathcal{A} \cup \mathcal{B} \cup \mathcal{C}$, add a red vertex $\alpha \in R$.
- For each $t_i \in T$, add a blue vertex $t_i \in B$.
- For each $t_i = (t_{i1}, t_{i2}, t_{i3})$, add three edges $(t_i, t_{i1}), (t_i, t_{i2})$, and (t_i, t_{i3}) to the edge set E.
- For each vertex t_j, $1 \leq j \leq m$, add a red vertex r_j and join the edge (t_j, r_j).
- For each vertex $a_i \in R$, $1 \leq i \leq n$, take a path of three blue vertices starting with b^i_1 and ending with b^i_3. Then we join the edge (a_i, b^i_1).

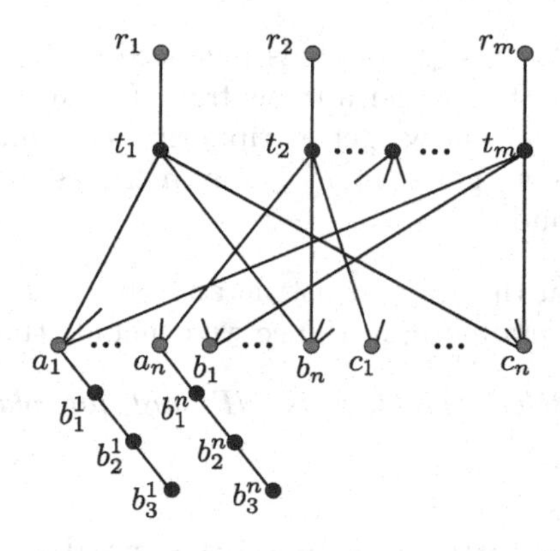

Fig. 2. Construction of the instance $G = (R \cup B,\ E)$ of the *m-BCGP* problem, where R and B are the sets of red and blue vertices, respectively in G. (Color figure online)

This completes the construction. See Fig. 2 for the complete construction. Clearly, the numbers of vertices and edges in G are polynomial in terms of the numbers of elements and triples in X; hence, the construction can be done in polynomial time. We now have the following observation.

Observation 1. *There exists no balanced connected subgraph H in G such that $V(H) \cap \{t_1, \ldots, t_m\} = \emptyset$ and each connected subgraphs in $G[V(G) \backslash V(H)]$ is balanced.*

Lemma 4. *$X(\mathcal{A} \cup \mathcal{B} \cup \mathcal{C},\ T)$ is a yes instance of the P3DM problem if and only if G has a zero solution i.e., UBN$= 0$ in the m-BCGP problem. More specifically, exactly one way we can partition G into m balanced connected subgraphs such that n of them contain exactly 8 vertices and the remaining $(m - n)$ connected subgraphs contain exactly 2 vertices.*

Proof. Assume that $X(\mathcal{A} \cup \mathcal{B} \cup \mathcal{C},\ T)$ is an yes instance in the *P3DM* problem, i.e., there is a $T' \subseteq T$ with $|T'| = n$ such that T' contains all the elements of $\mathcal{A}, \mathcal{B}$ and $\mathcal{C}$. Let $T' = \{t_{x_1}, t_{x_2}, \ldots, t_{x_n}\} \subseteq \{t_1, \ldots, t_m\}$. Recall that $t_{x_i} = (t_{x_i 1}, t_{x_i 2}, t_{x_i 3})$, for all $i, 1 \leq i \leq n$. Let X be the set $\{G_{x_1}, \ldots, G_{x_n}\}$ of n connected subgraphs, where G_{x_i} is the connected subgraph induced by the vertices $\{t_{x_i}, t_{x_i 1}(= a_\beta), t_{x_i 2}, t_{x_i 3}, r_{x_i}, b_1^\beta, b_2^\beta, b_3^\beta\}$ in G. Also let Y be the set $\{G_\alpha \colon t_\alpha \notin T'\}$ of $(m - n)$ connected subgraphs, where G_α is a path of two vertices $t_\alpha \in B$ and $r_\alpha \in R$. Now it is easy to verify that $X \cup Y = G_1 \cup G_2 \cup \ldots \cup G_m$ where each $G_i, 1 \leq i \leq m$ is balanced connected and $\bigcup_{i=1}^m V(G_i) = R \cup B$. Hence G has a zero solution in *m-BCGP*.

On the other hand, assume that G has a zero solution in *m-BCGP*, i.e., there exists a partition of G into m balanced connected subgraphs, say $G_1, \ldots, G_m$. Using Observation 1 we can say that, each G_i contains exactly one vertex from $\{t_1, \ldots, t_m\}$. Without loss of generality assume that G_1 contains b_3^1.Clearly $a_1 \in G_1$. Also G_1 contain exactly one vertex, say t_i from $\{t_1, \ldots, t_m\}$. Now $t_i \in G_1$

implies that r_i must be in G_1. As G_1 is balanced so G_1 must contain two red vertices t_{i2} and t_{i3}. By this we get a triple from T that contain a_1. By a similar argument with $b_3^j, 2 \leq j \leq n$, we get a triple that contain a_j. As $G_1, \ldots, G_m$ is a partition of G, hence we get a set $T' \subseteq T$ of n triples such that T' contains all the elements of $\mathcal{A}, \mathcal{B}$ and $\mathcal{C}$. □

It is easy to see that the graph constructed from the *P3DM* problem in Fig. 2 is indeed a planar bipartite graph. Hence we conclude the following theorem.

Theorem 3. *Thek-BCGP problem is NP-hard for planar bipartite graphs, where k is not fixed.*

5 2-*BCGP* Problem on Chordal Graphs

In this section, we prove that the 2-*BCGP* problem is NP-hard for chordal graphs. We give a reduction from the *Exact-Cover-by-3-Sets (EC3Set)* problem, which is known to be NP-complete [10]. In this *EC3Set* problem, we are given a set U with $3k$ elements and a collection S of m subsets of U such that each $s_i \in S$ contains exactly 3 elements. The objective is to find an exact cover for U (if one exists), i.e., a sub-collection $S' \subseteq S$ such that every element of U occurs in exactly one member of S'. During the reduction, we generate an instance $G = (R \cup B,\ E)$ of the 2-*BCGP* problem from an instance $X(S, U)$ of the *EC3Set* problem. Next, we show that G is a chordal graph.

Reduction: Let $X(S, U)$ be an instance of the *EC3Set* problem where $S = \{s_1, s_2, \ldots, s_m\}$ and $U = \{u_1, u_2, \ldots, u_{3k}\}$. We construct a connected bicolored graph $G = (R \cup B,\ E)$, where R and B are the sets of red and blue vertices, respectively in G, with the following way:

- For each set $s_i \in S$, $1 \leq i \leq m$, we add a blue vertex $s_i \in B$.
- For each element $u_j \in U$, $1 \leq j \leq 3k$, we add a red vertex $u_j \in R$.
- For each set $s_i = \{u_\alpha, u_\beta, u_\gamma\} \in S$, we add three edges (s_i, u_α), (s_i, u_β), and (s_i, u_γ) to the edge set E.
- Add a path of $k(2 + 3k + 3m)$ blue vertices starting at b_1 and ending at $b_{k(2+3k+3m)}$. Then we add m edges (b_1, s_i) in E for all $i, 1 \leq i \leq m$.
- Add a path of $(m - k)$ red vertices starting at r_1 and ending at r_{m-k}. Then we add m edges (r_1, s_i) in E for all $i, 1 \leq i \leq m$.
- For each red vertex u_i where $1 \leq i \leq 3k$, take a path of $(k + m)$ red vertices starting at r_1^i and ending at r_{k+m}^i. Then we add the edges $\{(r_1^i, u_i) \colon 1 \leq i \leq 3k\}$ into E.
- We add the edges, between each pair of vertices s_i and s_j, for all $i, j (\neq i)$ where $1 \leq i, j \leq m$, into E.

This completes the construction. See Fig. 3 for the complete construction. Clearly, the numbers of vertices and edges in G are polynomial in terms of the numbers of elements and sets in X; hence, the construction can be done in polynomial time. We now prove the following lemma.

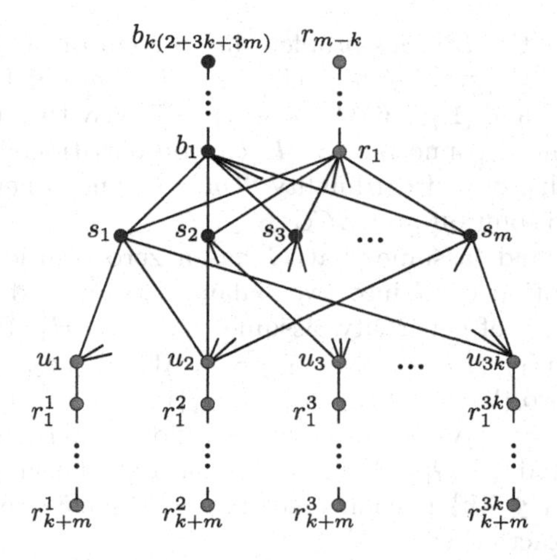

Fig. 3. Construction of the instance $G = (R \cup B,\ E)$ of the 2-*BCGP* problem. For the clarity of the figure we omit the edges between each pair of vertices s_i and s_j, for all $i, j (\neq i)$ where $1 \leq i, j \leq m$. (Color figure online)

Lemma 5. *If H is a balanced connected subgraph of G such that $r_1 \in V(H)$ and $(V \backslash V(H))$ induces a non-empty balanced connected subgraph of G then $b_1 \notin V(H)$.*

Proof. We prove it by contradiction. Let us assume that H is a balanced connected subgraph of G such that H contains both r_1 and b_1, where $(V \backslash V(H))$ induces a non-empty balanced connected subgraph of G. Also we assume that $|V(H) \cap \{s_1, \ldots, s_m\}| = t_1$ and $|V(H) \cap \{u_1, \ldots, u_{3k}\}| = t_2$. Clearly, $t_1 \neq 0$ (because of connectivity between b_1 and r_1 in H). As $b_1 \in V(H)$ and $G[V \backslash V(H)]$ is a connected subgraph of G so $V(H)$ contains all the blue vertices in $\{b_1, \ldots, b_{k(2+3k+3m)}\}$. Also if $u_i \in V(H)$ then $V(H)$ must contain all the red vertices from $\{r_j^i : 1 \leq j \leq (k+m)\}$. Now the number of red and blue vertices in $V(H)$ is $(m-k) + t_2(k+m+1)$ and $t_1 + 2k + 3k(k+m)$, respectively. As H is balanced, so $(m-k) + t_2(k+m+1) = t_1 + 2k + 3k(k+m)$. This implies $m + t_2(k+m+1) = 3k + t_1 + 3k(k+m)$. Now this will be satisfied only when $t_1 = m$ and $t_2 = 3k$ holds. This implies that $V = V(H)$, a contradiction. □

By a similar argument, we can prove the following lemma.

Lemma 6. *If H is a balanced connected subgraph of G such that $r_1 \in V(H)$ and $(V \backslash V(H))$ induces a non-empty balanced connected subgraph of G then $V(H) \cap \{u_1, u_2, \ldots, u_{3k}\} = \emptyset$.*

Now, we prove the following lemma.

Lemma 7. *The instance X of the EC3Set problem has a solution if and only if H has a zero solution i.e., $UBN = 0$ in the 2-BCGP.*

Proof. Assume that the *EC3Set* problem has a solution. Let S^* be an optimal solution in it. Let $X = \{s_i : s_i \notin S^*\} \cup \{r_j : 1 \leq j \leq (m-k)\}$ and $Y = V(G) \setminus X$. Now $|X| = 2(m-k)$ and $|Y| = 6k(1+k+m)$. It is easy to verify that both $G[X]$ and $G[Y]$ are balanced connected. So H can be partitioned into two balanced connected subgraphs, one of cardinality $2(m-k)$ and other is $6k(1+k+m)$. Hence G has a zero solution in 2-*BCGP*.

On the other hand, assume that G has a zero solution in 2-*BCGP*, i.e., there exists a partition of G into two balanced connected subgraphs, say G_1 and G_2. Without loss of generality, assume that $r_1 \in H_1$. By Lemma 5 and 6, $b_1 \notin V(G_1)$ and $V(G_1) \cap \{u_1, u_2, \ldots, u_{3k}\} = \emptyset$. Also $\{r_1, r_2, \ldots, r_{m-k}\} \subset V(G_1)$. As G_1 is balanced so $|V(G_1) \cap \{s_1, s_2, \ldots, s_m\}| = m - k$. Let $S^{**} = \{s_j : s_j \notin V(G_1)\}$ with $|S^{**}| = k$. As G has a zero solution in 2-*BCGP*, so $G[V \setminus V(G_1)]$ is balanced connected. As $G[V \setminus V(G_1)]$ is connected, the subgraph of G induced by $S^{**} \cup \{u_i : 1 \leq i \leq 3k\}$ is connected. Hence, the corresponding elements of S^{**} in S give an exact cover for U. □

Notice that, each cycle in G must contain a pair of vertices from $\{s_i : 1 \leq i \leq m\}$. As E contains the edges between each pair of vertices s_i and s_j, for all $i, j (\neq i)$ where $1 \leq i, j \leq m$, so the graph we constructed in Fig. 3 is indeed a chordal graph. Hence we conclude the following theorem.

Theorem 4. *The* 2-*BCGP problem is NP-hard for chordal graphs.*

References

1. Andreev, K., Racke, H.: Balanced graph partitioning. Theory Comput. Syst. **39**(6), 929–939 (2006)
2. Apollonio, N., Becker, R., Lari, I., Ricca, F., Simeone, B.: Bicolored graph partitioning, or: gerrymandering at its worst. Discrete Appl. Math. **157**(17), 3601–3614 (2009)
3. Becker, R.I., Schach, S.R., Perl, Y.: A shifting algorithm for min-max tree partitioning. J. ACM **29**(1), 58–67 (1982)
4. Bhore, S., Chakraborty, S., Jana, S., Mitchell, J.S.B., Pandit, S., Roy, S.: The balanced connected subgraph problem. In: Pal, S.P., Vijayakumar, A. (eds.) CALDAM 2019. LNCS, vol. 11394, pp. 201–215. Springer, Cham (2019). https://doi.org/10.1007/978-3-030-11509-8_17
5. Bhore, S., Jana, S., Pandit, S., Roy, S.: Balanced connected subgraph problem in geometric intersection graphs. In: Li, Y., Cardei, M., Huang, Y. (eds.) COCOA 2019. LNCS, vol. 11949, pp. 56–68. Springer, Cham (2019). https://doi.org/10.1007/978-3-030-36412-0_5
6. Chen, Y., Chen, Z.-Z., Lin, G., Xu, Y., Zhang, A.: Approximation algorithms for maximally balanced connected graph partition. In: Li, Y., Cardei, M., Huang, Y. (eds.) COCOA 2019. LNCS, vol. 11949, pp. 130–141. Springer, Cham (2019). https://doi.org/10.1007/978-3-030-36412-0_11
7. Chlebíková, J.: Approximating the maximally balanced connected partition problem in graphs. Inf. Process. Lett. **60**(5), 223–230 (1996)
8. Dyer, M.E., Frieze, A.M.: On the complexity of partitioning graphs into connected subgraphs. Discret. Appl. Math. **10**(2), 139–153 (1985)

9. Dyer, M.E., Frieze, A.M.: Planar 3DM is NP-complete. J. Algorithms **7**(2), 174–184 (1986)
10. Garey, M.R., Johnson, D.S.: Computers and Intractability: A Guide to the Theory of NP-Completeness. W. H. Freeman, New York (1979)
11. Krauthgamer, R., Naor, J., Schwartz, R.: Partitioning graphs into balanced components. In: SODA 2009, pp. 942–949. SIAM (2009)
12. Madkour, A.R., Nadolny, P., Wright, M.: Finding minimal spanning forests in a graph. arXiv preprint arXiv:1705.00774 (2017)
13. Matić, D., Božić, M.: Maximally balanced connected partition problem in graphs: application in education. Teach. Math. **29**, 121–132 (2012)
14. Perl, Y., Schach, S.R.: Max-min tree partitioning. J. ACM **28**(1), 5–15 (1981)
15. Stockmeyer, L.: Some simplified NP-complete graph problems. Theoret. Comput. Sci. **1**, 237–267 (1976)
16. Vaishali, S., Atulya, M.S., Purohit, N.: Efficient algorithms for a graph partitioning problem. In: Chen, J., Lu, P. (eds.) FAW 2018. LNCS, vol. 10823, pp. 29–42. Springer, Cham (2018). https://doi.org/10.1007/978-3-319-78455-7_3
17. Wakabayashi, Y., Chataigner, F., Salgado, L.B.: Approximation and inapproximability results on balanced connected partitions of graphs. Discret. Math. Theor. Comput. Sci. **9**(1), 177–192 (2007)

Hardness Results of Global Roman Domination in Graphs

B. S. Panda(✉) and Pooja Goyal

Computer Science and Application Group, Department of Mathematics,
Indian Institute of Technology Delhi, Hauz Khas, New Delhi 110016, India
bspanda@maths.iitd.ac.in, poojaagoyal92@gmail.com

Abstract. A Roman dominating function (RDF) of a graph $G = (V, E)$ is a function $f : V \rightarrow \{0, 1, 2\}$ such that every vertex assigned the value 0 is adjacent to a vertex assigned the value 2. A global Roman dominating function (GRDF) of a graph $G = (V, E)$ is a function $f : V \rightarrow \{0, 1, 2\}$ such that f is a Roman dominating function of both G and its complement $\overline{G}$. The weight of f is $f(V) = \Sigma_{u \in V} f(u)$. The minimum weight of a GRDF in a graph G is known as *global Roman domination number* of G and is denoted by $\gamma_{gR}(G)$. Minimum Global Roman Domination is to find a global Roman dominating function of minimum weight and Decide Global Roman Domination is the decision version of Minimum Global Roman Domination. In this paper, we show that Decide Global Roman Domination is NP-complete for bipartite graphs and chordal graphs. We also show that Minimum Global Roman Domination cannot be approximated within a factor of $(\frac{1}{2} - \epsilon) \ln |V|$ for any $\epsilon > 0$ unless P = NP. On the positive side, we propose an $O(\ln |V|)$-approximation algorithm for Minimum Global Roman Domination for any graph $G = (V, E)$.

Keywords: Global Roman domination · NP-complete · Approximation algorithm

1 Introduction

Let $G = (V, E)$ be a finite, simple and undirected graph with vertex set V and edge set E. A set $D \subseteq V$ is called a dominating set of G if every vertex $v \in V \backslash D$ is adjacent to at least one vertex in D. The *domination number* of G is the minimum cardinality among all dominating sets of G and it is denoted by $\gamma(G)$. Minimum Domination is to find a dominating set of minimum cardinality and Decide Domination is the decision version of Minimum Domination. Domination in graphs has been studied extensively and has several applications (see [6,7]). A set $D \subseteq V$ is called a global dominating set (GD-set) of G if D is dominating set for both G and its complement $\overline{G}$. The *global domination number* of G is the minimum cardinality among all global dominating sets of G and it is denoted by $\gamma_g(G)$. Minimum Global Domination is to find a global dominating set of minimum cardinality.

A. Mudgal and C. R. Subramanian (Eds.): CALDAM 2021, LNCS 12601, pp. 500–511, 2021.
https://doi.org/10.1007/978-3-030-67899-9_39

A Roman dominating function (RDF) of a graph G is a function $f : V \rightarrow \{0, 1, 2\}$ such that any vertex u with $f(u) = 0$ has at least one neighbor v with $f(v) = 2$. The weight of f is $f(V) = \Sigma_{u \in V} f(u)$. The minimum weight of a RDF in a graph G is known as *Roman domination number* of G and is denoted by $\gamma_R(G)$. MINIMUM ROMAN DOMINATION is to find a Roman dominating function of minimum weight and DECIDE ROMAN DOMINATION is the decision version of MINIMUM ROMAN DOMINATION. The concept of Roman domination was defined by Ian Stewart in an article entitled "Defend the Roman Empire!" [11]. Further, Cockayne et al. [3] initiated the study of Roman domination. Roman domination has been studied from algorithmic point of view in [2]. It has been proved that DECIDE ROMAN DOMINATION is NP-complete in bipartite graphs and chordal graphs (see [9]).

A global Roman dominating function (GRDF) of a graph G is a function $f : V \rightarrow \{0, 1, 2\}$ such that f is a Roman dominating function of both G and its complement $\overline{G}$. The weight of f is $f(V) = \Sigma_{u \in V} f(u)$. The minimum weight of a GRDF in a graph G is known as *global Roman domination number* of G and is denoted by $\gamma_{gR}(G)$. The concept of global Roman domination has been introduced in [1,10].

Minimum global Roman domination and its decision version are defined as follows:

MINIMUM GLOBAL ROMAN DOMINATION (Minimum GRD)
Instance: A graph $G = (V, E)$.
Solution: A minimum global Roman dominating function f of G.

DECIDE GLOBAL ROMAN DOMINATION (Decide GRD)
Instance: A graph $G = (V, E)$ and a positive integer r.
Question: Deciding $\gamma_{gR}(G) \leq r$?

In this paper, we study the complexity of MINIMUM GLOBAL ROMAN DOMINATION. The rest of the paper is organized as follows. In Sect. 2, we present some pertinent definitions and some preliminary results. In Sect 3, we show the NP-completeness result of DECIDE GLOBAL ROMAN DOMINATION in bipartite graphs and chordal graphs. In Sect 4, we propose an approximation algorithm with approximation ratio $4(1+\ln|V|)$, to approximate MINIMUM GLOBAL ROMAN DOMINATION in a graph. In Sect 5, we show that MINIMUM GLOBAL ROMAN DOMINATION cannot be approximated within $(\frac{1}{2} - \epsilon) \ln |V|$ for any $\epsilon > 0$ unless P = NP.

2 Preliminaries

Let $G = (V, E)$ be a finite, simple and undirected graph with no isolated vertex. The open neighborhood of a vertex v in G is $N_G(v) = \{u \in V \mid uv \in E\}$ and the closed neighborhood is $N_G[v] = \{v\} \cup N_G(v)$. The degree of a vertex v is $|N_G(v)|$ and is denoted by $d_G(v)$. If $d_G(v) = 1$, then v is called a *pendant* vertex and the neighbor of a pendant vertex is called a *support* vertex. The minimum and the maximum degree of G will be denoted by $\delta(G)$ and $\Delta(G)$, respectively. For $D \subseteq V$, $G[D]$ denote the subgraph induced by D. We use the standard notation $[k] = \{1, 2, \ldots, k\}$.

A *bipartite graph* is an undirected graph $G = (X, Y, E)$ whose vertices can be partitioned into two disjoint sets X and Y such that every edge has one end vertex in X and the other in Y. A bipartite graph $G = (X, Y, E)$ is *complete bipartite* if for every $x \in X$ and $y \in Y$, there is an edge $xy \in E$. A complete bipartite graph with partitions of size $|X| = m$ and $|Y| = n$, is denoted $K_{m,n}$. In particular for $m = 1$, $K_{1,n}$ is known as *star* graph. A graph $G = (V, E)$ is said to be a *chordal graph* if every cycle of length at least four has a chord, i.e., an edge joining two non-consecutive vertices of the cycle.

For a graph $G = (V, E)$, a Roman dominating function $f : V \to \{0, 1, 2\}$ can be denoted by (V_0, V_1, V_2), where $V_i = \{v \in V \mid f(v) = i\}$ for $i \in \{0, 1, 2\}$. Note that there exists a one to one correspondence between the function $f : V \to \{0, 1, 2\}$ and the ordered partition (V_0, V_1, V_2) of V. Thus, we will write $f = (V_0, V_1, V_2)$.

Observation 1 *(see [10]). For any graph G, $\gamma_g(G) \leq \gamma_{gR}(G) \leq 2\gamma_g(G)$.*

Observation 2 *(see [10]). Let G be any graph. Then $\gamma_{gR}(G) = \gamma_R(G)$ if and only if there exists a $\gamma_R(G)$-function $f = (V_0, V_1, V_2)$ such that for every vertex in V_0 there is a vertex in V_2 such that they are not adjacent.*

Observation 3 *(see [3]). For any graph $G = (V, E)$, $\gamma(G) \leq \gamma_R(G) \leq 2\gamma(G)$.*

3 NP-Completeness Results

In this section, we show that decide global Roman domination is NP-complete for bipartite graphs and chordal graphs. For this we recall the definition of Exact-3-Cover.

Exact-3-Cover (X3C)
Instance: A finite set X with $|X| = 3q$ and a collection $\mathcal{C}$ of 3-element subsets of X.
Question: Does $\mathcal{C}$ contain an exact cover for X, that is, a subcollection $C' \subseteq \mathcal{C}$ such that every element in X occurs in exactly one member of C'?

Theorem 4. Decide Global Roman Domination *is NP-complete for bipartite graphs.*

Proof. Given a function $f : V \to \{0, 1, 2\}$ of weight at most r for a bipartite graph $G = (V, E)$, it can be checked in polynomial time whether f is a global Roman dominating function of G. Hence, Decide Global Roman Domination is in NP for bipartite graphs. To show the hardness, we give a polynomial reduction from Exact-3-Cover, which is known to be NP-complete (see [8]). Given an arbitrary instance $(X, \mathcal{C})$ of $X3C$, $X = \{x_1, x_2, \ldots, x_{3q}\}$ and $\mathcal{C} = \{C_1, C_2, \ldots, C_t\}$. We construct a bipartite graph $G = (V, E)$ from the instance $(X, \mathcal{C})$ as follows:

- For each vertex $x_j \in X$, we construct a subgraph H_j obtained from a vertex y_j and three stars $K_{1,3}$ centered at u_j, v_j and w_j, by adding edges $y_ju_j, u_jv_j, v_jw_j, w_jy_j$. Further, join vertex x_j with y_j.
- For each $C_i \in \mathcal{C}$, we add a vertex c_i and a star $K_{1,3}$ centered at d_i. Add edge c_id_i.
- Finally add edges x_jc_i if and only if $x_j \in C_i$.

Clearly, the graph G is a bipartite graph. We show an example in Fig. 1, bipartite graph G is obtained from the system $(X, \mathcal{C})$, where $X = \{x_1, x_2, x_3, x_4, x_5, x_6\}$ and $\mathcal{C} = \{\{x_1, x_2, x_3\}, \{x_2, x_4, x_5\}, \{x_3, x_5, x_6\}, \{x_4, x_5, x_6\}\}$.

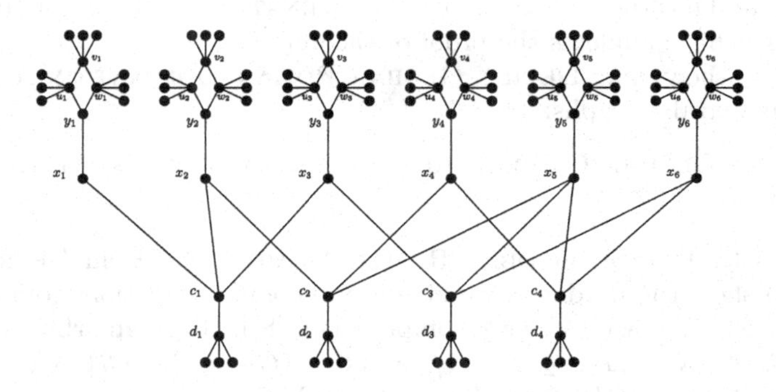

Fig. 1. An illustration of the construction of G from system $(X, \mathcal{C})$ in the proof of Theorem 4.

Now to complete the proof, it suffices to prove the following claim:

Claim. The system $(X, \mathcal{C})$ has an exact cover if and only if the graph G has a global Roman dominating function with weight at most $20q + 2t$.

Proof. Suppose that C' is a solution of $(X, \mathcal{C})$. We construct a GRDF h on G with $h(V) = k$, where $k = 20q + 2t$. We label all pendant vertices by 0. All u_j's, v_j's, w_j's and d_i's are labelled by 2. Also, every x_j and y_j are labelled by 0. For any i, we label c_i by 2 if $C_i \in C'$ and label c_i by 0 if $C_i \notin C'$. Observe that since C' exists, its cardinality is exactly q, and so the number of c_i's with weight 2 is q. Since C' is a solution for $X3C$, any vertex of X has a neighbor labelled by 2. Every vertex c_i has a neighbor d_i labelled by 2. Hence, h is a RDF of G with $h(V) = 2q + 2t + 3 \cdot 6q = k$. It can be easily observed that every vertex v of the graph G of label 0 is adjacent to a vertex of label 2 as well as non-adjacent to a vertex of label 2. Thus, h is a GRDF of the graph G of weight $20q + 2t$.

Conversely assume that G has a GRDF $h_1 = (V_0, V_1, V_2)$ of weight $k = 20q + 2t$. Clearly, $h_1(V(H_j)) \geq 6$ for every $j \in [3q]$; in particular $h_1(u_j) = h_1(v_j) = h_1(w_j) = 2$. Thus, $h_1(y_j) = 0$. Also, $h_1(d_i) = 2$ for every $i \in [t]$ and for every pendant vertex p of G, $h_1(p) = 0$. Next we show that no x_i needs to be assigned a positive value.

Claim. If $h_1(V) = k$ then for each $x_j \in X$, $h_1(x_j) = 0$.

Proof. Assume $h_1(V) = k$ and there exist $l(\geq 1)$ x_j's such that $h_1(x_j) \neq 0$. The number of x_j's with $h_1(x_j) = 0$ is $3q - l$. Since h_1 is a GRDF, each x_j with $h_1(x_j) = 0$ should have a neighbor c_i with $h_1(c_i) = 2$. So the number of c_i's required with $h_1(c_i) = 2$ is $\lceil \frac{3q-l}{3} \rceil$. Hence $h_1(V) = 2t + 18q + l + 2\lceil \frac{3q-l}{3} \rceil$, which is greater than k, a contradiction. Therefore for each $x_j \in X$, $h_1(x_j) = 0$. □

Since each c_i has exactly three neighbors in X, clearly, there exist q number of c_i's with weight 2 such that each x_j is adjacent to c_i of weight 2. Consequently, $C' = \{c_i \mid h_1(c_i) = 2\}$ is an exact cover for $\mathcal{C}$. This completes the proof of claim. □

Therefore, Decide Global Roman Domination is NP-complete for bipartite graphs. This completes the proof of theorem. □

Next, we show that Decide Global Roman Domination remains NP-complete in chordal graphs.

Theorem 5. Decide Global Roman Domination *is NP-complete for chordal graphs.*

Proof. Clearly, Decide Global Roman Domination is in NP for chordal graphs. To show the hardness, we give a polynomial reduction from Exact-3-Cover, which is known to be NP-complete (see [8]). Given an arbitrary instance $(X, \mathcal{C})$ of $X3C$, $X = \{x_1, x_2, \ldots, x_{3q}\}$ and $\mathcal{C} = \{C_1, C_2, \ldots, C_t\}$. We construct a chordal graph $G = (V, E)$ from the instance $(X, \mathcal{C})$ as follows:

- For each vertex $x_j \in X$, we construct a subgraph H_j obtained from a vertex y_j and three stars $K_{1,3}$ centered at u_j, v_j and w_j, by adding edges $y_j u_j, u_j v_j, v_j w_j, w_j y_j, y_j v_j$. Further, join vertex x_j with y_j.
- For each $C_i \in \mathcal{C}$, we add two vertices b_i, c_i and a star $K_{1,3}$ centered at d_i. Add edge $c_i b_i$, $b_i d_i$ and $c_i c_k$ for every $i, k \in [t]$.
- Finally add edges $x_j c_i$ if and only if $x_j \in C_i$.

Illustration of the construction of G from the set system $(X, \mathcal{C})$ is given in Fig. 2. Clearly, the graph G is a chordal graph.

Now to complete the proof, it suffices to prove the following claim:

Claim. The system $(X, \mathcal{C})$ has an exact cover if and only if the graph G has a global Roman dominating function with weight at most $20q + 2t$.

Proof. Due to space constraint, we omit the proof of claim.

Therefore, Decide Global Roman Domination is NP-complete for chordal graphs. This completes the proof of theorem. □

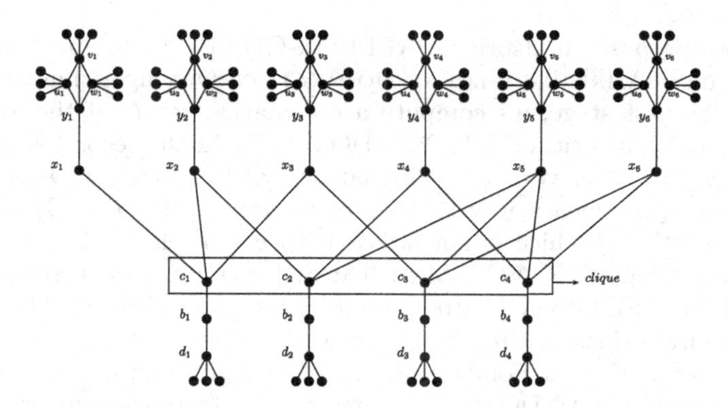

Fig. 2. An illustration of the construction of G from system $(X, \mathcal{C})$ in the proof of Theorem 5.

4 Approximation Algorithm

In this subsection, we propose an approximation algorithm for MINIMUM GLOBAL ROMAN DOMINATION whose approximation ratio is a logarithmic factor in the size of the input. To obtain the approximation ratio of MINIMUM GLOBAL ROMAN DOMINATION, we require approximation ratio of MINIMUM GLOBAL DOMINATION. Further, to obtain approximation ratio of MINIMUM GLOBAL DOMINATION, we use known approximation algorithms for the MINIMUM DOMINATION and MINIMUM SET COVER.

The following approximation result is known for MINIMUM DOMINATION.

Theorem 6 [4]. MINIMUM DOMINATION *in graph G with maximum degree $\Delta(G)$ can be approximated with an approximation ratio of* $1 + \ln(\Delta(G) + 1)$.

By Theorem 6, there exists a polynomial time algorithm, APPROX-DOM-SET algorithm, that outputs a dominating set D of a graph G of order n and achieves the approximation ratio of $1 + \ln(\Delta(G) + 1)$; that is, $|D| \leq (1 + \ln(\Delta(G) + 1))\gamma(G) \leq (1 + \ln|V|)\gamma(G)$.

Next we need to recall MINIMUM SET COVER. Let X be any non-empty set. Let $\mathcal{F}$ be a collection of subsets of X. A set $\mathcal{C} \subseteq \mathcal{F}$ is called a set cover of X, if every element of X belongs to at least one element of $\mathcal{C}$. MINIMUM SET COVER is to find a minimum cardinality set cover of X. The following approximation result is known for MINIMUM SET COVER.

Theorem 7 [4]. MINIMUM SET COVER *for the instance $(X, \mathcal{F})$ can be approximated with an approximation ratio of* $1 + \ln q$, *where* $q = \max\{|S| : S \in \mathcal{F}\}$.

By Theorem 7, there exists an APPROX-SET-COVER algorithm with input $(X, \mathcal{F})$, that outputs a set cover $\mathcal{C}$ in polynomial time such that cardinality of the set cover $\mathcal{C}$ is at most $1 + \ln q$ times the cardinality of the optimal set cover of $(X, \mathcal{F})$, where q is the cardinality of the maximum cardinality set of $\mathcal{F}$.

Next, we propose an algorithm APPROX-GD-SET to compute an approximate solution of Minimum Global Domination. Our algorithm works in two stages: in the first stage, we compute a dominating set D of the given graph $G = (V, E)$ using algorithm APPROX-DOM-SET. In the second stage, we find an additional set Q of vertices such that $D \cup Q$ becomes a GD-set of G. We select the set Q in such a way that every vertex $v \in V \setminus (D \cup Q)$ has at least one vertex in $D \cup Q$ which is not adjacent to v. To select this set Q, we first construct an instance $(X, \mathcal{F})$ of Minimum Set Cover, and then use the algorithm APPROX-SET-COVER to compute a set cover $\mathcal{C}$ of X. Thereafter, we construct Q from $\mathcal{C}$ such that $D \cup Q$ becomes a GD-set of G.

Assume that D is a dominating set of the given graph G obtained by the algorithm APPROX-DOM-SET. Next, we illustrate the construction of an instance $(X, \mathcal{F})$ of Minimum Set Cover. Let X be the subset of all vertices v in $V \setminus D$ such that v does not have any non-adjacent vertex in D, that is, $|D \cap (V \setminus N_G(v))| = 0$. For a vertex v in the graph G, the set of all vertices which are non-adjacent to v is denoted by $V \setminus N_G(v)$. Let $V \setminus D = \{u_1, \ldots, u_p\}$. Let $S_i = X \cap (V \setminus N_G(u_i))$ for $i \in [p]$. Thus, S_i is the set of all vertices in X which are non-adjacent to u_i in G. We now define $\mathcal{F} = \{S_1, \ldots, S_p\}$. Thus, $(X, \mathcal{F})$ forms an instance of Minimum Set Cover. Now, let $\mathcal{C}$ be the set cover obtained by the algorithm APPROX-SET-COVER. Let $Q = \{u_i \in V \setminus D \mid S_i \in \mathcal{C}\}$. Note that for each vertex $x \in X$, there must exist at least one vertex $w \in Q$ such that $N_G(x) \cap \{w\} = \emptyset$, because $\mathcal{C}$ is a set cover of X. Thus, $D \cup Q$ is a GD-set of G. Next, we summarize the approximation algorithm APPROX-GD-SET.

Algorithm 1. APPROX-GD-SET

Input: A graph $G = (V, E)$.
Output: A GD-set D_g of graph G.

```
begin
    Compute a dominating set D of G using algorithm APPROX-DOM-SET;
    Construct an instance (X, F) of MINIMUM SET COVER as defined above;
    if X == ∅ then
        D_g = D;
    else
        Compute a set cover C of (X, F) using algorithm APPROX-SET-COVER;
        Q = {u_i ∈ V\D | S_i ∈ C};
        D_g = D ∪ Q;
    return D_g;
```

We note that the algorithm APPROX-GD-SET returns a GD-set of a given graph $G = (V, E)$ in polynomial time. We are now in a position to prove the following theorem.

Theorem 8. Minimum Global Domination *in a graph G can be approximated with an approximation ratio of* $2(1 + \ln |V|)$.

Proof. In order to prove the theorem, we show that the GD-set D_g returned by our algorithm APPROX-GD-SET is an approximate solution of MINIMUM GLOBAL DOMINATION with an approximation ratio of $2(1 + \ln|V|)$; that is, $|D_g| \leq 2(1 + \ln|V|)\gamma_g(G)$.

The algorithm APPROX-DOM-SET returns a dominating set D of G with an approximation ratio $1 + \ln|V|$; that is, $|D| \leq (1 + \ln|V|)\gamma(G)$.

Similarly, for the instance $(X, \mathcal{F})$ of MINIMUM SET COVER, the algorithm APPROX-SET-COVER returns a set cover $\mathcal{C}$ of X with an approximation ratio $(1 + \ln q)$, where q is the cardinality of a largest set in $\mathcal{F}$. Since the graph G has minimum degree $\delta(G)$, $|V \backslash N_G(u_i)| \leq |V| - \delta(G)$ for each $i \in [p]$. Thus, $|S_i| \leq |V| - \delta(G) \leq |V|$ for all $i \in [p]$. Hence, the cardinality of each set of $\mathcal{F}$ is at most $|V|$, implying that $q \leq |V|$. Thus, if $\mathcal{C}^*$ is an optimal set cover of the instance $(X, \mathcal{F})$, then $|\mathcal{C}| \leq (1 + \ln|V|)|\mathcal{C}^*|$.

Recall that $Q = \{u_i \in V \backslash D \mid S_i \in \mathcal{C}\}$, and so $|Q| = |\mathcal{C}|$. Let $Q^* = \{u_i \in V \backslash D \mid S_i \in \mathcal{C}^*\}$. Then $|Q^*| = |\mathcal{C}^*|$ and $|Q^*|$ denotes the minimum number of vertices needed to extend a dominating set D of G to a GD-set of G. We note that the minimum number of vertices needed to extend a set of vertices to a GD-set of G is no more than the global domination number of G. Hence, $|\mathcal{C}^*| \leq \gamma_g(G)$.

$$\begin{aligned}|D_g| = |D \cup Q| = |D| + |Q| = |D| + |\mathcal{C}| &\leq (1 + \ln|V|)\gamma(G) + (1 + \ln|V|)|\mathcal{C}^*| \\ &\leq 2(1 + \ln|V|)\gamma_g(G).\end{aligned}$$

The GD-set D_g, returned by the algorithm APPROX-GD-SET, is therefore an approximate solution of MINIMUM GLOBAL DOMINATION with an approximation ratio of $2(1 + \ln|V|)$. This completes the proof of the Theorem 8. □

Next, we propose an algorithm APPROX-GRDF to compute an approximate solution of MINIMUM GLOBAL ROMAN DOMINATION. In our algorithm, first we compute a global dominating set D_g of the input graph G using the approximation algorithm APPROX-GD-SET. Next, we construct an ordered partition $f = (V_0, V_1, V_2)$ of V in which every vertex in D_g will be labelled by 2 and the remaining vertices will be labelled by 0.

Now, let $f = (V \backslash D_g, \emptyset, D_g)$ be the ordered partition returned by APPROX-GRDF algorithm. It can be easily seen that every vertex $v \in V$ is assigned with weight either 0 or 2. Since D_g is a global dominating set of G, every vertex $v \in V \backslash D_g$ labelled by 0 is adjacent to a vertex $u \in D_g$ labelled by 2 as well as non-adjacent to a vertex $w \in D_g$ labelled by 2. Thus, f gives a global Roman dominating function of G. We note that the algorithm APPROX-GRDF computes a global Roman dominating function of a given graph G in polynomial time. Hence, we have the following result.

Theorem 9. MINIMUM GLOBAL ROMAN DOMINATION *in a graph* $G = (V, E)$ *can be approximated with an approximation ratio of* $4(1 + \ln|V|)$.

Proof. Let D_g be the global dominating set returned by the algorithm APPROX-GD-SET and f be the global Roman dominating function returned by the algorithm APPROX-GRDF. It can be observed that $f(V) = 2|D_g|$. It is known

Algorithm 2. APPROX-GRDF

Input: A graph $G = (V, E)$.
Output: A global Roman dominating function $f = (V_0, V_1, V_2)$ of graph G.
begin
  Compute a global dominating set D_g of G using algorithm APPROX-GD-SET;
  $f = (V \backslash D_g, \emptyset, D_g)$;
  return f;

that $|D_g| \leq 2(1 + \ln(|V|))\gamma_g(G)$. Therefore, $f(V) \leq 4(1 + \ln(|V|))\gamma_g(G)$. Thus Observation 1 leads to $f(V) \leq 4(1 + \ln(|V|))\gamma_{gR}(G)$. Thus, Minimum Global Roman Domination in a graph G can be approximated with an approximation ratio of $4(1 + \ln|V|)$. □

5 Lower Bound on Approximation Ratio

In this subsection, we obtain a lower bound on the approximation ratio of Minimum Global Roman Domination for any graph. To obtain our lower bound we give an approximation preserving reduction from Minimum Set Cover. We need the following theorem proved in [5].

Theorem 10 [5]. Minimum Set cover *for set system* $(U, \mathcal{C})$ *cannot be approximated within* $(1 - \epsilon) \ln |U|$ *for any* $\epsilon > 0$ *unless* $P = NP$.

We are ready to prove the inapproximability of Minimum Global Roman Domination.

Theorem 11. Minimum Global Roman Domination *for any graph* $G = (V, E)$ *cannot be approximated within* $(\frac{1}{2} - \varepsilon) \ln |V|$ *for any* $\varepsilon > 0$ *unless* $P = NP$.

Proof. Given an instance $(U, \mathcal{C})$ of Minimum Set Cover, where $U = \{u_1, u_2, \ldots, u_q\}$. $\mathcal{C} = \{C_1, C_2, \ldots, C_t\}$. Now we construct a graph $G = (V, E)$ in polynomial time as follows.

- For each element u_j in the set U, add three vertices x_j, y_j and z_j in G.
- For each set C_i in the collection $\mathcal{C}$, add a vertex c_i in the vertex set of G. Also, add edges $c_i c_k$ for every $i, k \in [t]$ and $i \neq k$.
- Next we add two stars $K_{1,3}$ centered at r and s. Add edges $\{rs, rc_i \mid i \in [t]\}$.
- If an element u_j belongs to set C_i, then add edges $x_j c_i$, $y_j c_i$ and $z_j c_i$ in G.

We show an example in Fig. 3, where graph G is obtained from the set system $(U, \mathcal{C})$ with $U = \{u_1, u_2, u_3\}$ and $\mathcal{C} = \{\{u_1, u_2\}, \{u_2, u_3\}, \{u_3, u_1\}, \{u_3\}\}$.

Claim. $\gamma_{gR}(G) = 2|S^*| + 4$, where S^* is the minimum cardinality set cover of the system $(U, \mathcal{C})$.

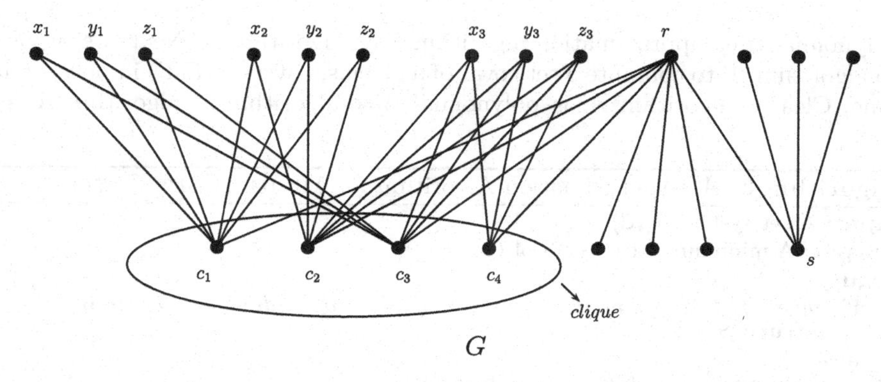

Fig. 3. An illustration of the construction of G from the system $(U, \mathcal{C})$ in the proof of Theorem 11.

Proof. Let S^* be a minimum set cover of G. We define a function $f : V \rightarrow \{0, 1, 2\}$ as follows.

$$f(v) = \begin{cases} 2, & \text{if } v \in \{u \mid u \in S^*\} \cup \{r, s\} \\ 0, & \text{otherwise.} \end{cases}$$

Since S^* is a set cover for $(U, \mathcal{C})$, every element of U belongs to at least one element of S^*. So, we label all vertices of the set $\{x_i, y_i, z_i \mid i \in [q]\}$ by 0. Next, we label r and s by 2 and every pendant vertex adjacent to r and s by 0. Since $\{c_i \mid C_i \in \mathcal{C}\}$ forms a clique. Hence, every $c_i \notin S^*$ of label 0 is adjacent to a $c_k \in S^*$ of label 2. Hence, f is a RDF of H with $f(V) = 4 + 2|S^*|$. It can be easily observed that every vertex of the graph G of weight 0 is adjacent to a vertex of weight 2 as well as non-adjacent to a vertex of weight 2. Thus, f is a GRDF of the graph G of weight $4 + 2|S^*|$. Thus, for a minimum weight GRDF f^* of G, $f^*(V) \leq 4 + 2|S^*|$.

Conversely, let $f^* = (V_0, V_1, V_2)$ be a minimum weight GRDF of G. Clearly, weight of a subgraph $K_{1,3}$ must be at least 2. Hence, $f^*(r) = f^*(s) = 2$ and every pendant vertex of r and s will be labelled by 0. Since each element $u_k \in U$ corresponds to three vertices x_k, y_k and z_k in G, any optimal solution of G will never assign non-zero weights to these three vertices. Let $u_k \in U$. Then there exists a set $C_l \in \mathcal{C}$ such that $u_k \in C_l$ and $f^*(c_l) = 2, f^*(x_k) = f^*(y_k) = f^*(z_k) = 0$. Hence, for every $u_k \in U$, there exists a $C_l \in \mathcal{C}$ such that weight of c_l is 2 under f^*, i.e. $f^*(c_l) = 2$. Hence, all weight 2 vertices of $V \cap \{c_j \mid j \in [t]\}$ will form a set cover of $(U, \mathcal{C})$. So, $S = \{C_i \mid c_i \in V_2\}$. Thus, for a minimum cardinality set cover S^*, $2|S^*| \leq 2|S| \leq f^*(V) - 4$.

Hence, $f^*(V) = 2|S^*| + 4$. This completes the proof of claim. □

Now, for the resulting graph G one can now confine to GRDF h consisting of $\{r, s\}$ and a subset S of $\mathcal{C}$ corresponding to a set cover, hence we have $h(V) = 2|S| + 4$.

Suppose that Minimum Global Roman Domination can be approximated within a ratio of α, where $\alpha = (\frac{1}{2} - \epsilon)\ln(|V|)$ for some fixed $\epsilon > 0$, by some polynomial time approximation algorithm, say Algorithm A. Next, we propose an algorithm A' to compute a set cover of a given set system $(U, \mathcal{C})$ in polynomial time. Clearly, algorithm A' is polynomial time algorithm as algorithm A is a

Algorithm 3. A': Approximation Algorithm for Minimum Set Cover

Input: A set system $(U, \mathcal{C})$
Output: A minimum set cover S of $(U, \mathcal{C})$.
begin
  if *(there exists a minimum set cover S of $(U, \mathcal{C})$ of cardinality $< l$)* **then**
    return S;
  else
    Construct a graph G as described above;
    Compute a GRDF f of G using algorithm A;
    $S = \{C_i \in \mathcal{C} \mid f(c_i) \neq 0\}$;
    return S;

polynomial time algorithm. Since l is a constant, step 1 of the algorithm can be executed in polynomial time. Note that if S is computed in Step 1, then S is optimal. So we analyze the case where $|S| \geq l$.

Let S^* be an optimal set cover in $(U, \mathcal{C})$. It is clear that $|S^*| \geq l$. Let S be the set cover computed by algorithm A'. Then

$$|S| = \frac{h(V) - 4}{2} \leq h(V) \leq \alpha \cdot \gamma_{gR}(G) \leq \alpha(2|S^*| + 4) \leq 2\alpha\Big(1 + \frac{2}{|S^*|}\Big)|S^*|$$
$$\leq 2\alpha\Big(1 + \frac{2}{l}\Big)|S^*|$$

Hence, algorithm A' approximates Minimum Set Cover for given set system $(U, \mathcal{C})$ within the ratio $2\alpha(1 + \frac{2}{l})$.

Let l be a positive integer such that $\frac{2}{l} < \frac{\epsilon}{2}$. Then algorithm A' approximates Minimum Set Cover for given set system $(U, \mathcal{C})$ within the ratio $2\alpha(1 + \frac{2}{l}) \leq 2(\frac{1}{2} - \epsilon)(1 + \frac{\epsilon}{2}) \ln |V| = (1 - \epsilon') \ln |U|$ for $\epsilon' = \frac{\epsilon^2}{2} + \frac{3\epsilon}{2}$ as $\ln |V| = \ln(3|U| + |\mathcal{C}| + 8) \approx \ln |U|$ for sufficiently large value of $|U|$.

Therefore, the Algorithm A' approximates Minimum Set Cover within ratio $(1 - \epsilon) \ln(|U|)$ for some $\epsilon > 0$. By Theorem 10, if Minimum Set Cover can be approximated within ratio $(1 - \epsilon) \ln(|U|)$ for some $\epsilon > 0$, then $\mathsf{P} = \mathsf{NP}$. Hence, if Minimum Global Roman Domination can be approximated within ratio $(\frac{1}{2} - \epsilon) \ln(|V|)$ for some $\epsilon > 0$, then $\mathsf{P} = \mathsf{NP}$. This proves that Minimum Global Roman Domination cannot be approximated within $(\frac{1}{2} - \epsilon) \ln(|V|)$ for any $\epsilon > 0$ unless $\mathsf{P} = \mathsf{NP}$. □

6 Conclusion

In this paper, we have shown that Decide Global Roman Domination is NP-complete for bipartite graphs and chordal graphs. We have proposed a $4(1+$

$\ln |V|$)-approximation algorithm for finding minimum weight GRDF in any graph G. Further, we have presented inapproximability result of MINIMUM GLOBAL ROMAN DOMINATION.

References

1. Atapour, M., Sheikholeslami, S.M., Volkmann, L.: Global roman domination in trees. Graph. Combin. **31**(4), 813–825 (2015)
2. Cockayne, E.J., Dreyer Jr., P.A., Hedetniemi, S.M., Hedetniemi, S.T., McRae, A.A.: The algorithmic complexity of roman domination
3. Cockayne, E.J., Dreyer Jr., P.A., Hedetniemi, S.M., Hedetniemi, S.T.: Roman domination in graphs. Discrete Math. **278**(1–3), 11–22 (2004)
4. Cormen, T.H., Leiserson, C.E., Rivest, R.L., Stein, C.: Introduction to Algorithms. MIT Press, Cambridge (2009)
5. Dinur, I., Steurer, D.: Analytical approach to parallel repetition. In: Proceedings of the Forty-sixth Annual ACM Symposium on Theory of Computing, pp. 624–633. ACM (2014)
6. Haynes, T., Hedetniemi, S., Slater, P.: Domination in Graphs: Advanced Topics. Marcel Dekker Inc., New York (1998)
7. Haynes, T., Hedetniemi, S., Slater, P.: Fundamentals of Domination in Graphs. Marcel Dekker Inc., New York (1998)
8. Johnson, D.S., Garey, M.R.: Computers and Intractability: A Guide to the Theory of NP-Completeness. W.H. Freeman, New York (1979)
9. Liu, C.-H., Chang, G.J.: Roman domination on strongly chordal graphs. J. Combin. Optim. **26**(3), 608–619 (2013)
10. Pushpam, P.R.L., Padmapriea, S.: Global roman domination in graphs. Discrete Appl. Math. **200**, 176–185 (2016)
11. Stewart, I.: Defend the Roman empire!. Sci. Am. **281**(6), 136–138 (1999)

References

Author Index

Printed in the United States
By Bookmasters